CLIMATE CHANGE – GEOPHYSICAL FOUNDATIONS AND ECOLOGICAL EFFECTS

Edited by **Juan Blanco** and **Houshang Kheradmand**

Climate Change – Geophysical Foundations and Ecological Effects
Edited by Juan Blanco and Houshang Kheradmand

Published by InTech

Janeza Trdine 9, 51000 Rijeka, Croatia

Publishing Process Manager Iva Lipovic

Technical Editor Teodora Smiljanic

Cover Designer InTech Design Team

Additional hard copies can be obtained from orders@intechopen.com

Climate Change – Geophysical Foundations and Ecological Effects,
Edited by Juan Blanco and Houshang Kheradmand
p. cm.
ISBN 978-953-307-419-1

For EU product safety concerns:
IN TECH d.o.o., Prolaz Marije Krucifikse Kozulić 3, 51000 Rijeka, Croatia,
info@intechopen.com or visit our website at intechopen.com.

Meet the editors

Dr. Blanco is a Research Associate at the University of British Columbia. His work is focused on the development and evaluation of ecological models to simulate the influences of management, climate and other ecological factors on tree growth. He is currently collaborating with research teams from Canada, USA, Spain, Cuba, and China in using ecological models to explore the effects of climate change, atmospheric pollution and alternative forest practices in natural and planted forest in boreal, temperate and tropical forests. His research has been applied in mining to optimize reclamation plans, in forestry to assess the potential for carbon sequestration and by government agencies to define local guidelines for long-term sustainable forest management. Among other topics related to forest ecology, Dr. Blanco has studied the influence of climate variations on tree growth and estimated the possible ecological consequences of climate change in forest ecosystems. He has also co-authored the first book dedicated exclusively to the use of hybrid ecological models in forest management, entitled "Forecasting Forest Futures" (Earthscan, London).

Dr. Kheradmand was born on March 1953 and received his masters in chemistry in 1980 and a doctorate in petro-chemistry in 1983 from ULP "Louis Pasteur University -Strasbourg", followed by a 2nd doctorate in polymer sciences in 1987 from Strasbourg Polymer Research Center. Since 1987 Dr. Kheradmand has worked in the European chemical industry with different responsibilities, such as Quality lab manager, R&D group leader, European product stewardship manager, European sustainable development manager, European Technology Awareness and Innovation Manager. He is Life Cycle thinking, Life Cycle Assessment and Sustainable Development expert. He has an active participation in different European organizations such as CEFIC, CEPE, EPDLA - EPDLA Chairman, ERMA - Member of Executive Committee and Chairman of Technical Committee, French Ecolabel - Member of Paint Ecolabel technical committee, FFC (Fédération Française pour les sciences de la Chimie) - Member of FFC steering Committee and Member of ChDD ''Chimie pour un Développement Durable®' working groupe committee , IUPAC - Member of IYC 2011 launch ceremony organization team. Dr. Kheradmand has presented different topics in WW seminars and exhibitions.

Contents

Preface

Climate is a fundamental part of the world as we know it. The landscape and everything on it are determined by climate acting over long periods of time (Pittock 2005). Therefore, any change on climate will have effects sooner or later on the world around us. These changes have happened before in the past, and they will likely happen again in the future. Climate variability can be both natural or anthropogenic (Simard and Austin 2010). In either case, the change in the current climate will have impacts on the biogeophysical system of the Earth. As all human activities are built on this system, our society will be impacted as well. As a consequence, climate change is increasingly becoming one of the most important issues, generating discussions in economy, science, politics, etc. There is no discrepancy among scientists that climate change is real and it has the potential to change our environment (Oreskes and Conway 2010), but uncertainty exists about the magnitude and speed at which it will unfold (Moss et al. 2010). The most discussed effect of global warming is the increase of temperatures, although this increase will not be homogeneous through the seasons, with the winters expected to warm up significantly more than the summers. In addition, changes in precipitation are also expected, that could lead to increase or decrease of rainfall, snowfall and other water-related events. Finally, a change in the frequency and intensity of storm events could be possible, although this is probably the most uncertain of the effects of global warming. These uncertainties highlight the need for more research on how global events have effects at regional and local scales, but they also indicated the need for the society at large to assume a risk-free approach to avoid the worse effects of climate change in our socio-economical and ecological systems (IPCC 2007).

Humans have been dealing with risk-related activities for a long time. For example, when buying a car or home insurance, the discussion is not about whether the adverse effects will happen or not, but on how to reduce its effects and recover and if they happen. In many countries, having car insurance is compulsory to drive a car, even if only a small percentage of drivers suffer car accidents compared to the total number of cars. In addition, the most risky manoeuvres (i.e. excessive speed, not stopping on red light, etc.) are banned to reduce the risks of accidents. Similarly, developing policies and practices that reduce and minimize the risks and effects of climate change are needed, even if the worse situations will never happen. If not, we will be in the equivalent of driving without insurance and without respecting the signals. All policies and practices for economic, industrial and natural resource management need to be founded on sound scientific foundations. This volume offers an interdisciplinary view of the biophysical issues related to climate change, and provides glimpse of the state-of-the-art research carried out around the world to inform scientists, policymakers and other stakeholders.

Any scientific discipline learns from experience, and the science of climate change is not different. Climate change is defined as a phenomenon by which the long-term averages of weather events (i.e. temperature, precipitation, wind speed, etc.) that define the climate of a region are not constant but change over time. Climate is also the result of very complex interactions between physical, chemical and biological variables. As a result, at geological scales of time, climate is constantly going through periods of relatively stable conditions followed by periods of change. There have been a series of past periods of climatic change, registered in historical or paleoecological records that can be studied from different geophysical variables. In the first section of this book, a series of state-of-the-art research projects explore the biophysical causes for climate change and the techniques currently being used and developed for its detection in several regions of the world. In this section, Dameris and Loyola describe the interactions between physical, dynamical, and chemical processes in Earth atmosphere. Manfredi et al. provide a new statistical methodology to study changes in historical climatic data. Winguth discusses the feedbacks between climate change and biogeochemical cycles during the Paleocene-Eocene Thermal Maximum. Historical and current changes in climatic and geophysical variables can be found around the globe. In North America, Assani et al. study the temporal variability of rain-induced floods in southern Quebec (Canada), whereas Ramírez-Beltrán et al. showcase a study to detect the change in climatic conditions at global scale and in the Caribbean basin. In South America, Hamza et al. studied the changes in soil temperature to detect variability in climate for the last decades, while Penalba and Bettolli analyze the changes in atmospheric circulation and daily precipitation caused by climate change in the Argentinean Pampas. In Asia, Yi studies the pollen records in Korea to establish the changes in climate during the Holocene, whereas Kim and Nam analyzing the records of beryllium deposits in the marine sediment, and Kyoung et al. discuss the effects of climate change in droughts in the Korean peninsula.

The knowledge of past changes in the environment will be of great value to try to understand what will happen under future climatic conditions different from the current ones. However, the effects of climate change on ecosystems around the world are not something of the future. They are happening now all around the globe. Ecological changes in the phenology and distribution of plants and animals are occurring in all aquatic and terrestrial ecosystems. Predator-prey and plant-insect interactions have been disrupted when interacting species have responded differently to warming (Parmesan 2006). The second section of the book explores the effects that have already been reported on the flora and fauna. These changes affect all types of ecosystems and creatures. In the Pacific Ocean, Salvadeo et al. report changes in the movement of marine mammals along the North American coast, and Povilitis examines the worrying situation of birds in Hawaii, threatened by the changing climate. In Asia, the changes in Monsoon patterns could change the number of generations of mosquitoes, as Ohta and Kaga report. In Europe, Itämies et al. describe how the populations of moths have shifted in the last decades in Finland, while Rakonczai explores the ecological changes already observed in the Carpathians. In Africa, Guo et al. study the consequence of changes in climatic patters on quiver trees growth and distribution. To finish this section, Sternberg et al. discuss the use of climatic gradients as similes of climate change and Drégelyi-Kiss and Hufnagel provide a theoretical study on climate-induced changes in freshwater ecosystems.

Being the ecosystems most potentially affected by climate change, the arctic and alpine regions are already experiencing some of the most noticeable and fastest changes. Range-restricted species, particularly polar and mountaintop species, show severe range contractions and have been the first groups in which entire species have gone extinct due to recent climate change (Parmesan 2006). The last section of the book provides some of the latest research in these ecosystems. Mazepa et al. describe some of the changes detected in the stand structure of forests in arctic Russia. Bender et al. describe the research being done in the Alps related to climate change. Also in the Alps, Jomelli et al. study if avalanches are already being affected by the change in climate, whereas Bajracharya et al. report on the shrinking of Himalayan glaciers in Nepal. Yde et al. review the different environmental effects that climate change can cause in glacial ecosystems. Among these changes, the increasing melt of glaciers could have important effects. Braithwaite explains the reasons for this trend, and Fettweis et al. estimate the increase in sea level rise by the melting of Greenland ice sheet.

All things considered, these 25 chapters provide a good overview of the different changes that have already been detected in all the regions of the world. They are an introduction to the research being done around the globe in connection to this topic. However, climate change is not just a theoretical issue only important for scientists or environmentalists. It also has direct implications in our socio-economical systems. The other two books of this series "Climate change – Socioeconomic effects" and "Climate Change – Research and Technology for Adaptation and Mitigation" explore these topics in detail, and we encourage the reader to consult them as well.

The Editors want to finish this preface acknowledging the collaboration and hard work of all the authors. We are also thankful to the Publishing Team of InTech for their continuous support and assistance during the creation of this book. Special thanks are due to Ms Ana Pantar for inviting us to lead this exciting project, and to Ms Iva Lipovic for coordinating the different editorial tasks.

Dr. Juan Blanco
Dep. Forest Sciences,
Faculty of Forestry University of British Columbia,
Canada

Dr. Houshang Kheradmand
LCT/LCA and Sustainable Development Expert
Scientific and Steering Committee member
Fédération Française pour les sciences de la Chimie
France

References

IPCC, 2007: Summary for Policymakers. In: Climate Change 2007: The Physical Science Basis. Contribution of Working Group I to the Fourth Assessment Report of the Intergovernmental Panel on Climate Change [Solomon, S., D. Qin, M. Manning, Z. Chen, M. Marquis, K.B. Averyt, M.Tignor and H.L. Miller (eds.)]. Cambridge University Press, Cambridge, United Kingdom and New York, NY, USA.

Moss, R.H., Edmonds, J.A., Hibbard, K.A., Manning, M.R., Rose, S.K., van Vuuren, D.P., Carter, T.R., Emori, S., Kainuma, M., Kram, T., Meechl, G.A., Mitchell, J.F.B., Nakicenovic, N., Riahi, K., Smith, S.J., Stouffer, R.J., Thomson, A.M., Weyant, J.P., Wilbanks, T.J. (2010). The next generation of scenarios for climate change research and assessment. *Nature*, Vol 463, p747-756.

Oreskes, N., Conway, E.M. (2010). Merchants of Doubt: How a Handful of Scientists Obscured the Truth on Issues from Tobacco Smoke to Global Warming. Bloomsbury Press, New York. ISBN 9781596916104.

Parmesan, C. (2006). Ecological and evolutionary responses to recent climate change. *Annual Reviews of Ecology and Evolutionary systematic,* Vol 37, p637-669.

Pittock, A.B. (2005). Climate change. Turning up the heat. Earthscan, London. ISBN 0643069343.

Simard, S.W., Austin, M.E. (2010). Climate change and variability. InTech, Rijeka. ISBN 978-953-307-144-2.

Part 1

Climate Variability

1

Chemistry-Climate Connections – Interaction of Physical, Dynamical, and Chemical Processes in Earth Atmosphere

Martin Dameris[1] and Diego Loyola[2]
[1]*Deutsches Zentrum für Luft- und Raumfahrt, Institut für Physik der Atmosphäre*
[2]*Deutsches Zentrum für Luft- und Raumfahrt, Institut für Methodik der Fernerkundung*
Oberpfaffenhofen,
Germany

1. Introduction

The climate system of the Earth atmosphere is affected by a complex interplay of dynamical, physical and chemical processes acting in the troposphere (atmospheric layer reaching from the Earth surface up to about 12 km height) and the Middle Atmosphere, i.e. the stratosphere (from about 12 to 50 km) and the mesosphere (from 50 to 100 km). Moreover, mutual influences between these atmospheric layers must be taken into account to get a complete picture of the Earth climate system. An outstanding example which can be used to describe some of the complex connections of atmospheric processes is the evolution of the ozone layer in the stratosphere and its interrelation with climate change.

The stratospheric ozone layer (located around 15 to 35 km) protects life on Earth because it filters out a large part of the ultraviolet (UV) radiation (wavelength range between 100 nm and 380 nm) which is emitted by the sun. The almost complete absorption of the energy-intensive solar UV-B radiation (280-320 nm) is especially important. UV-B radiation particularly affects plants, animals and people. Increased UV-B radiation can, for example, adversely impact photosynthesis, cause skin cancer and weaken the immune system. In addition, absorption of solar UV radiation by the stratospheric ozone layer causes the temperature of the stratosphere to increase with height, creating a stable layer that limits strong vertical air movement. This plays a key role for the Earth's climate system. Approximately 90% of the total ozone amount is found in the stratosphere. Only 10% is in the troposphere; ozone concentrations in the troposphere are much lower than in the stratosphere.

Data derived from observations (measurements from satellites and ground-based instruments) and respective results from numerical simulations with atmospheric models are used to describe and explain recent alterations of the dynamics and chemistry of the atmosphere.

Since the beginning of the 1980ies in each year the ozone hole develops over Antarctica during spring season (i.e. September to November), showing a decrease in the total amount of ozone of up to 70% (see Figure 4). Especially in the lower stratosphere (about 15-25 km altitude), ozone is almost completely destroyed during this season. Relatively shortly after the discovery of the ozone hole, the extreme thinning of the ozone layer in the south-polar stratosphere was explained as a combination of special meteorological conditions and changed chemical composition induced by industrially manufactured (anthropogenic) chlorofluorocarbons (CFCs) and halons.

1.1 Ozone chemistry

In the atmosphere, ozone (O_3) is produced exclusively by photochemical processes. Ozone formation in the stratosphere is initiated by the photolysis of molecular oxygen (O_2). This produces two oxygen atoms (O) which recombine with molecular oxygen to form ozone. Since ozone is created by photochemical means, it is mainly produced in the tropical and subtropical stratosphere, where sunshine is most intensive throughout the year. At the same time, the ozone molecules formed in this way are destroyed again by the photolysis of ozone and by reaction with an oxygen atom. These reactions form the basis of stratospheric ozone chemistry, the so-called Chapman mechanism (Chapman, 1930). But if stratospheric ozone amounts are determined via this simple reaction system and the known rate constants and photolysis rates, the results obtained are about twice as high as the measured values. Since the early 1950ies, it has been known that fast so-called catalytic cycles reduce the determined ozone amounts to the observed values. By the early 1970ies, the catalysts had been identified as the radical pairs OH/HO_2 and NO/NO_2, which are formed from water vapour (H_2O) and nitrous oxide (N_2O) respectively (Bates and Nicolet, 1950; Crutzen, 1971; Johnston, 1971). In the mid-1970ies, the radical pairs Cl/ClO (from CFCs) and Br/BrO (from halons) were identified as further significant contributors (Molina and Rowland, 1974; Wofsy et al., 1975). The important point is that a catalyst can take part in the reaction cycle several thousand times and therefore is very effective in destroying ozone molecules. The increased occurrence of CFCs and halons due to anthropogenic emissions has significantly accelerated stratospheric ozone depletion cycle over recent decades, triggering a negative stratospheric ozone trend which is most obvious in the Southern polar stratosphere during spring time where the ozone hole is found. In the troposphere, CFCs and halons are mostly inert. Over time (several years), they are transported into the stratosphere. Only there they are photolysed and converted into active chlorine or bromine compounds.

In particular, ozone is depleted via the catalytic Cl/ClO-cycle in polar spring. However, the kinetics of these processes are very slow, because the amount of UV radiation is limited due to the prevailing twilight conditions. In the polar stratosphere, it is mainly chemical reactions on the surface of stratospheric ice particles that are responsible for activating chlorine (and also bromine) and then driving ozone depletion immediately after the end of polar night (Solomon et al., 1986). In the very cold lower polar stratosphere, polar stratospheric clouds (PSCs) form during polar night (Figure 1). PSCs develop at temperatures below about 195 K (= -78 °C) where nitric acid trihydride crystals form (NAT, $HNO_3 \cdot 3H_2O$). Under the given conditions in the lower stratosphere ice particles develop at temperatures below approx. 188 K (= -85 °C). Due to different land-sea distributions on the Northern and Southern Hemisphere, the lower stratosphere over the south pole cools significantly more in winter (June – August) than the north polar stratosphere (December – February) (see Section 1.2). The climatological mean of polar winter temperatures of the lower Arctic stratosphere is around 10 K higher than that of the lower Antarctic stratosphere. While the Antarctic stratosphere reaches temperatures below PSC-forming temperatures for several weeks every year, there is a pronounced year-on-year variability in the north polar stratosphere: relatively warm winters, where hardly any PSCs develop are observed, as well as very cold winters, with conditions similar to that of Antarctica. This means that expansive PSC fields develop in the Antarctic stratosphere every year, but are seldom seen over the Arctic (see Section 2.1). A detailed description of chemical processes affecting ozone is given by Dameris (2010).

Fig. 1. Polar stratospheric clouds over Finland. The picture was taken on January 26, 2000 from the DLR research aircraft *Falcon*.

1.2 Importance of stratospheric dynamics

Since the 1990ies it became obvious that the ozone layer was not just getting thinner over Antarctica, but over many other regions, too, although to a lesser extent (see Figure 5). Many observations from satellite instruments and ground based techniques (incl. radiosondes) have shown a clear reduction of the amount of stratospheric ozone, e.g. in middle geographical latitudes (about 30°-60°) of both hemispheres. From that time on observational evidences and the actual state of understanding have been reviewed in WMO/UNEP Scientific Assessments of Ozone Depletion (WMO, 1992; 1995; 1999; 2003; 2007; 2011). It turned out that the thickness of the stratospheric ozone layer is not solely controlled by chemical processes in the stratosphere. Physical and dynamic processes play an equally important role.

The polar stratosphere during winter is dominated by strong west wind jets, the polar vortices. Due to the different sea-land distribution in the Northern and Southern Hemisphere these wind vortices develop differently in the two hemispheres. Large-scale waves with several hundreds kilometres wavelength are generated in the troposphere, for example during the overflow of air masses over mountain ridges. These waves propagate upward into the stratosphere and affect the dynamics there including the strength of the polar wind jets. The polar vortex in the Southern Hemisphere is less disturbed and therefore the mean zonal wind speed is stronger than in the Northern Hemisphere. In the Southern Hemisphere this leads to a stronger isolation of stratospheric polar air masses in winter and a more pronounced cooling of the polar stratosphere during polar night (see Section 1.1).

Additionally, atmospheric trace gas concentrations are affected by air mass transports, which are determined by wind fields (wind force and direction). The extent to which such a transport of trace gases takes place depends on the lifetime of the chemical species in question. Only if the chemical lifetime of a molecule is longer than respective dynamical timescales, the transport contributes significantly to the distribution of the chemical substance. For example, in the lower stratosphere the chemical lifetime of ozone is long enough that transport processes play a key role in geographical ozone distribution there. At these heights, ozone can be transported to latitudes where, photochemically, it is only produced to an insignificant extent. In this way, ozone generated at tropical (up to about 15°), sub-tropical (about 15°-30°) and middle latitudes is transported particularly effectively in the direction of the winter pole (i.e. towards the north polar region from December to February and towards the south polar region from June to August), due to large-scale meridional (i.e. north-south) circulation. There, it is mixed in with the local air. This leads to an asymmetric global ozone distribution with peaks at higher latitudes during the corresponding spring months and not over the equator (see Figures 7 and 8). At higher stratospheric latitudes, it is thus particularly difficult to separate chemical influences on ozone distribution (ozone depletion rates) from the changes caused by dynamic processes.

With respect to climate change due to enhanced greenhouse gas concentrations (i.e. in particular carbon dioxide, CO_2, methane, CH_4, and nitrous oxide, N_2O) caused by human activities, it is expected that temperatures in the troposphere will further increase (IPCC, 2007) and that they will further decrease in the stratosphere due to radiation effects (Chapter 4 in WMO, 2011). Since the reaction rates of many chemical reactions are directly depending on atmospheric temperature, climate change will directly influence chemical processes and therefore the amount and distribution of chemical substances in Earth atmosphere. Moreover, changes in atmospheric temperature and temperature gradients are modifying dynamic processes that drive the circulation system of the atmosphere. This would result in changing both, the intensity of air mass transports and the transportation routes, with possible long-term consequences for the atmospheric distribution of radiatively active gases, including ozone. Changes in distribution of the climate-influencing trace gases in turn affect the Earth's climate.

2. Measuring ozone with satellite instruments and numerical modelling

Since many years ozone distribution in the stratosphere is observed by ground-based and satellite instruments (see Section 2.1). In particular measurements from space help to get a global view of the state of the stratospheric ozone layer and its temporal evolution including short-term fluctuations and long-term changes (i.e. trends). An outstanding task is to combine multi-year observations derived from different sensors flown on different satellites in a way that at the end one gets consistent and homogeneous data products which enable solid scientific investigations of processes causing the basic state of the atmosphere and its variability. In addition to a detailed analysis of existing measurements, numerical models of the atmosphere are used to reproduce as best as possible recent atmospheric conditions and the modulation in space and with time. Sensitivity studies help to identify those processes most relevant to describe climatological mean atmospheric conditions as well as spatial and temporal changes. For example, changes in climate, the temporal evolution of the ozone layer and the connections between them are simulated by atmospheric models which consider all known and relevant dynamical, physical as well as chemical processes (see

Section 2.2). In such numerical studies, it is important to consider natural processes and their variations, as well as human activities relevant to atmospheric processes. A comprehensive evaluation of data derived from numerical model simulations with respective observations helps to identify the strength and weaknesses of the applied model systems which to a great part reflect the current state of the knowledge about processes acting in Earth atmosphere (see Section 3). A good understanding of all crucial processes is necessary, for example, for reliably estimating the future development of the ozone layer (Section 4). In this context, alterations in atmospheric processes due to climate change must be considered.

2.1 Observations from satellite

Satellite remote sensing of ozone started in 1970 with the Backscatter Ultraviolet Spectrometer (BUV) onboard the NASA satellite Nimbus-4. The first Total Ozone Mapping Spectrometer (TOMS) was launched in 1978 onboard the Nimbus-7 satellite and was followed by a series of Solar Backscatter UV Instrument (SBUV). TOMS measured the total column of atmospheric ozone content whereas the SBUV measured height resolved stratospheric ozone profiles. The last TOMS instrument operated until 2007, the Ozone Mapper Profiler Suite (OMPS) to be launched in 2011 will continue this data record.

The European contribution to satellite base measurements of atmospheric composition started with the Global Ozone Monitoring Experiment (GOME) sensor onboard the ESA satellite ERS-2 launched in 1995. GOME measured not only ozone (total column, profiles and tropospheric column) but also a number of atmospheric composition gases like nitrogen dioxide, sulphur dioxide, bromine monoxide, water vapour, formaldehyde, chlorine dioxide, glyoxalin as well as clouds and aerosols (see Burrows et al., 1999). The GOME data record is continued with the SCIAMACHY sensor onboard the ESA satellite ENVISAT launched in 2002, with the Dutch sensor OMI onboard the NASA satellite AURA launched in 2004, and with the GOME-2 sensor onboard the EUMETSAT satellite MetOp-A launched in 2006. This 16 years data record will be continued with the GOME-2 sensors on the EUMETSAT satellites MetOp-B (to be launched in 2012) and MetOp-C (to be launched in 2017). The ESA's Sentinel 5 precursor mission (to be launched in 2015), Sentinel 4 and Sentinel 5 with further extend this data record with similar sensor systems in the next decades.

Remote Sensing in the UV/VIS spectral range between 280 nm and 450 nm is based on measurements of backscattered radiation from the Earth-atmosphere system. The Differential Optical Absorption Spectroscopy (DOAS) fitting technique is used to derive trace gas slant column amounts along the viewing path of the GOME-type instruments. The spectral structure of ozone in the Huggings bands (Figure 2) measured by a satellite sensor is compared to laboratory measurements to quantify the ozone content on the atmosphere.

The slant columns determined with DOAS are finally converted to geometry-independent vertical column amounts through division by appropriate air mass factors (Van Roozendael et al., 2006) which result from radiative transfer calculations (see Figure 3). Air mass factors describe the enhanced absorption of a given trace gas due to slant paths of incident light in the atmosphere. The ozone retrieval must also take into account the influence of clouds and other atmospheric effects (Loyola et al., 2011).

Satellite total ozone measurements are systematically compared with ground-based measurements and the differences are typically lower than 1%. Nevertheless satellite ozone data from different instruments may show spatial and temporal differences due to sensor

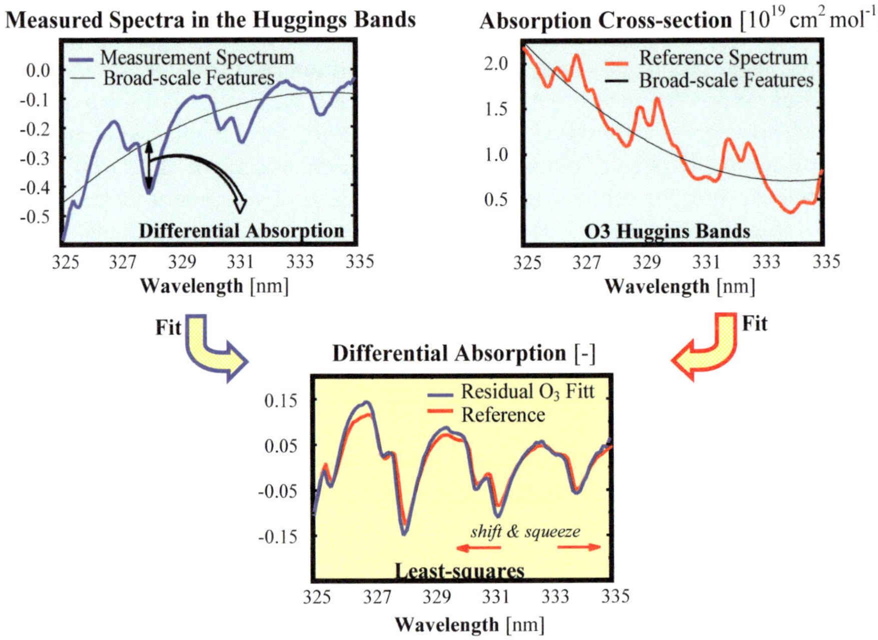

Fig. 2. Schematically representation of the DOAS principle used for the retrieval of ozone content from the Huggings bands between 325 nm and 335 nm. The differential structure of satellite measurements (top left) and laboratory measurements (top right) are fitted together (low panel) to determine the current ozone amount.

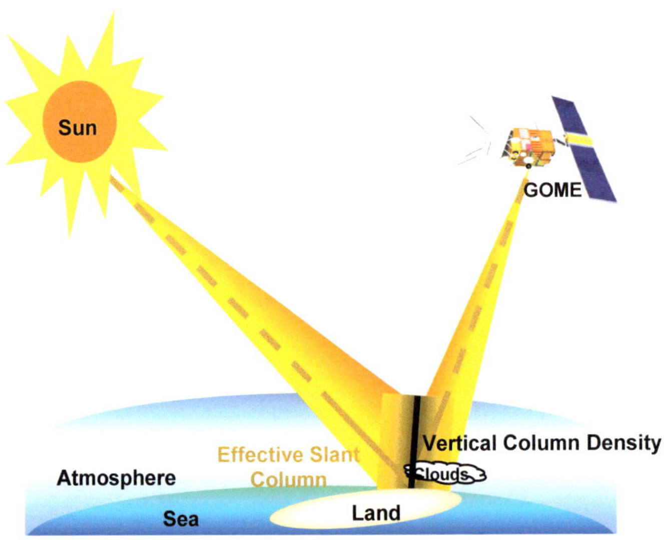

Fig. 3. The satellite measured ozone slant column (brown path) is converted to viewing geometry independent vertical column of ozone (black path).

specific characteristics and drifts. Therefore some corrections are needed before merging data from different satellites to create long-term homogenous climate data records that can be used for ozone trend studies. In this chapter we use the merged satellite TOMS/OMI data record (Stolarski et al., 2006) starting in 1979 and the merged GOME/SCIAMACHY/GOME-2 data record (Loyola et al., 2009) starting in 1995.

An ozone hole is said to exist when the total ozone column sinks to values below 220 DU, which is around 30% under the norm. Dobson Units are column densities – a measure of the total amount of ozone in a column over a specific place. At standard temperature and pressure (1000 hPa, 0 °C), a 0.01-mm thick ozone layer corresponds to 1 DU. A 300-DU thick ozone layer at the Earth's surface would thus correspond to a pure ozone column of 3 mm. Figure 4 shows the evolution of ozone hole as measured by the TOMS sensor onboard the Nimbus 7 satellite between 1979 to 1992, TOMS data from the Meteor satellite between 1993 to 1994, GOME data from the ERS-2 satellite between 1995 and 2002, SCIAMACHY data from the ENVISAT satellite between 2003 and 2006, and GOME-2 data from the MetOp-A satellite between 2007 and 2010. The average ozone from October 1st to 3rd is plotted for all the years with the exception of 1993 and 2002 where data from September 23rd to 25th are used. In 1993 no TOMS data were available at the beginning of October and in 2002 the data from September are plotted to show the atypical split of the ozone hole due to the unusual meteorological conditions in the stratosphere occurring only in 2002.

Corresponding results for Northern Hemisphere spring time conditions are presented in Figure 5. There, average total ozone column from March 25th to 27th is plotted for all years between 1979 and 2011 except 1995 where no satellite data is available. Obviously the ozone depletion is not as strong as in the Southern Hemisphere and the trend towards lower ozone amount is much less visible. The interannual variability is high which can be explained by the variability of stratospheric dynamics (see Sections 1.1 and 1.2). Nevertheless, most clearly seen in years like 1997 and 2011, the dynamic situation of the Arctic stratosphere can be very similar to the Antarctic, i.e. showing a well-pronounced and undisturbed polar vortex in winter with temperatures low enough to form PSCs in large extent. Other years which also show a significant reduction of total ozone in northern spring are 1990, 1993, 1996, and 2007. On the other hand, in years like 1998 and 2010 when stratospheric temperatures are enhanced due to disturbed stratospheric dynamic conditions, total ozone values are much higher. It is also obvious that total ozone values at low latitudes (i.e. tropical and sub-tropical regions) are naturally low.

2.2 Simulations with chemistry-climate models

Chemistry-climate models (CCMs) are numerical tools which are used to study connections between atmospheric chemistry and climate (Figure 6). They are composed of two basic modules: An Atmospheric General Circulation Model (AGCM) and a Chemistry Model.

An AGCM is a three-dimensional model describing large-scale (i.e. spatial resolution of a few hundred km) physical, radiative, and dynamical processes in the atmosphere over years and decades. It is used to study changes in natural variability of the atmosphere and for investigations of climate effects of radiatively active trace gases (greenhouse gases) and aerosols (i.e. natural and anthropogenic particles in the atmosphere), along with their interactions and feedbacks. Usually, AGCM calculations employ prescribed concentrations of radiatively active gases, e.g. CO_2, CH_4, N_2O, CFCs, and O_3. Changes of water vapour (H_2O) concentrations due to the hydrological cycle are directly simulated by an AGCM. The

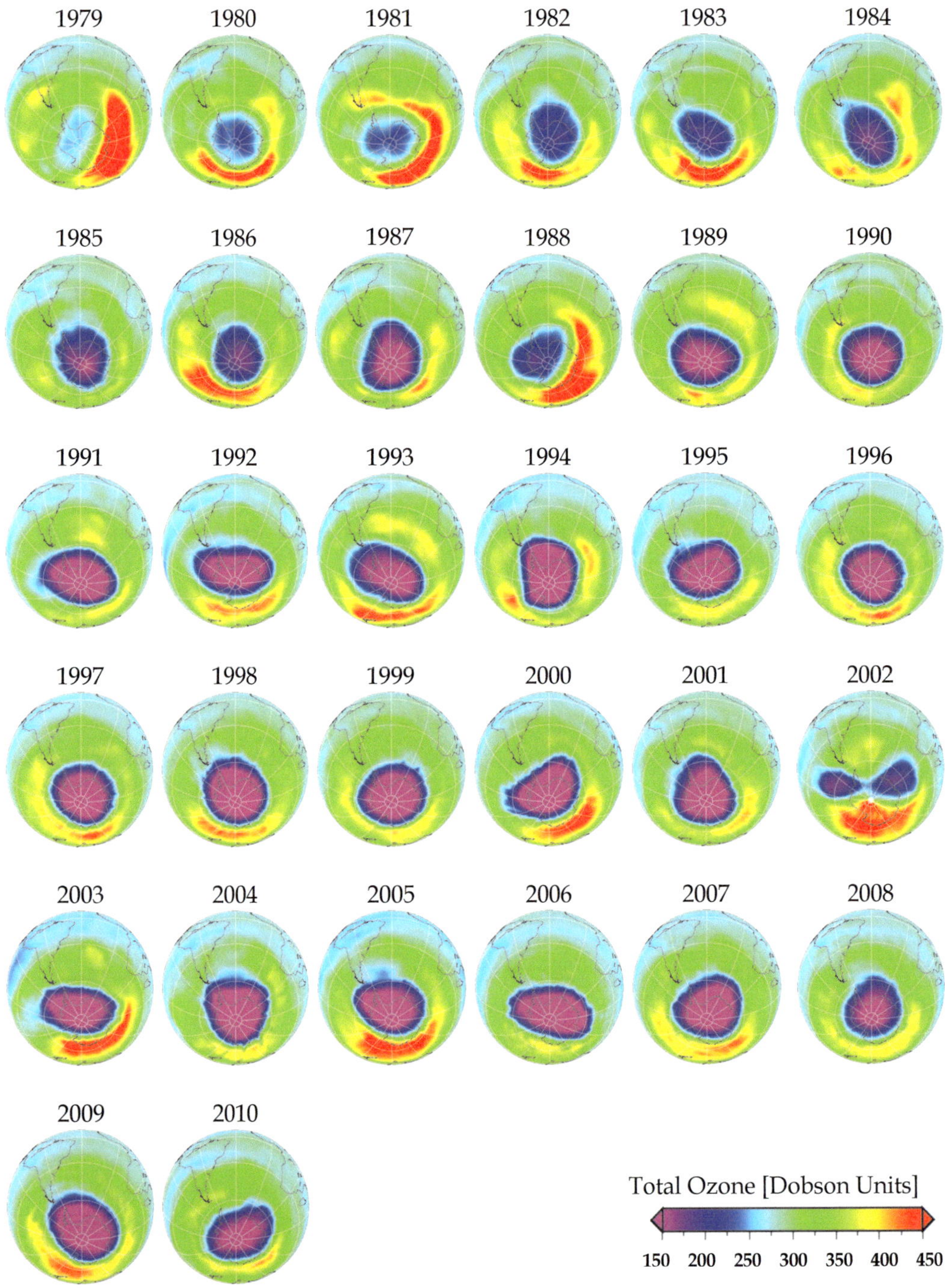

Fig. 4. Evolution of the ozone hole derived from satellite measurements in early October from 1979 until 2010. The purple area over the South Polar Region indicates the area of the ozone hole (see text).

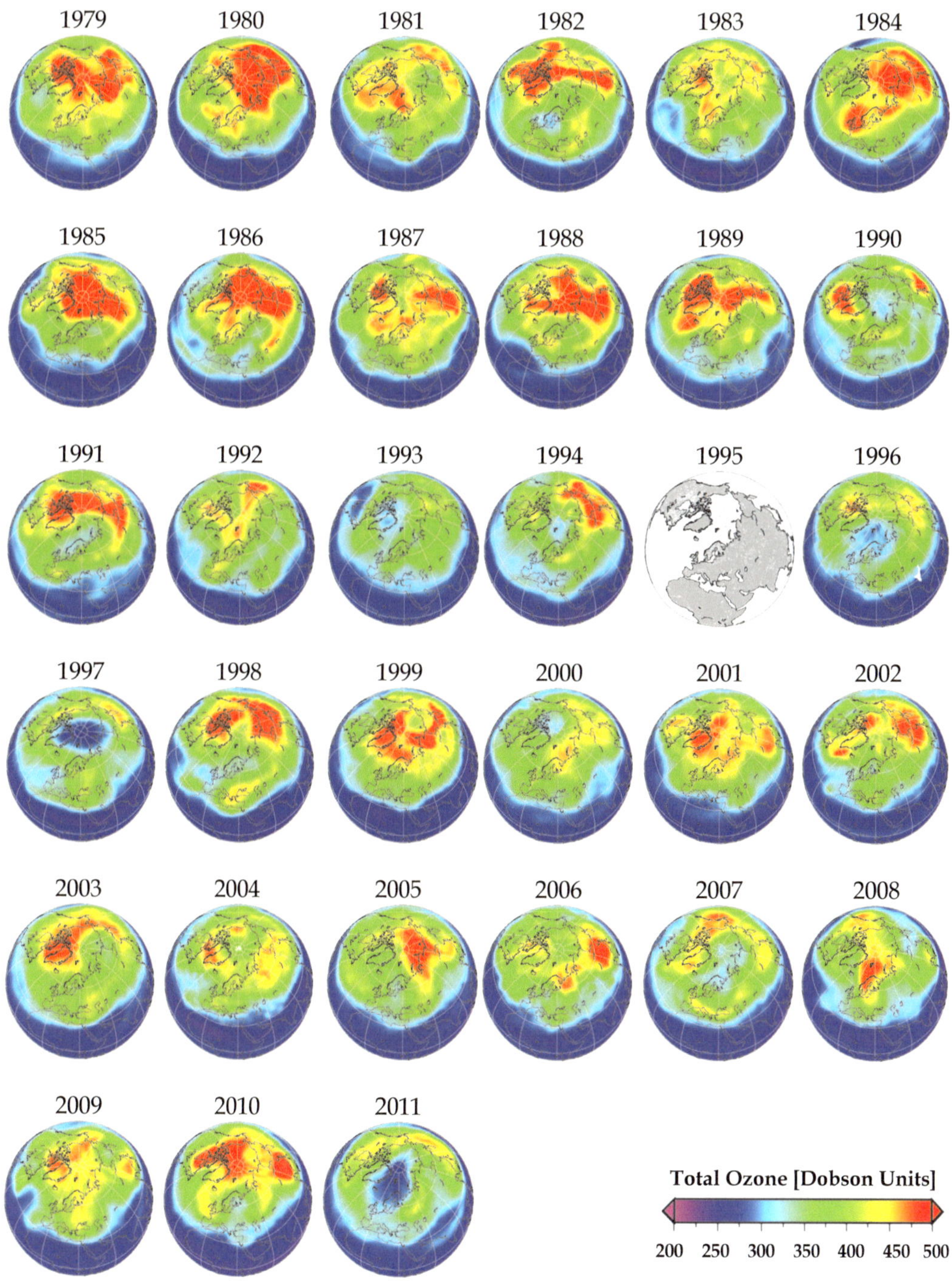

Fig. 5. As Figure 4, but for the Northern Hemisphere and using a different colour scale. Evolution of the ozone derived from satellite measurements in late March from 1979 until 2011 (no data available for 1995, see text).

temporal development of sea surface temperatures (SSTs) and sea ice coverage are prescribed in these models. The chosen boundary conditions for concentrations of radiatively active gases and SSTs (incl. sea ice) represent a specific period of time, e.g. some years or decades.

In a CCM, i.e. an AGCM interactively coupled to a model describing in detail atmospheric chemistry, the simulated concentrations of the radiatively active gases are used in the calculations of heating rates (e.g. due to the absorption of short-wave solar radiation) and cooling rates (e.g. due to the emission of long-wave infrared radiation). Changes in the abundance of these gases due to chemistry and advection influence heating and cooling rates and, consequently, variables describing atmospheric dynamics such as temperature and wind. This gives rise to a dynamical-chemical coupling in which the chemistry influences the dynamics (via radiative heating and cooling) and vice versa (via temperature and advection). As an example, ozone represents the dominant radiative-chemical feedback in the stratosphere. Simulations with CCMs also consider gradual changes in concentrations of radiatively active gases and other boundary conditions (e.g., emissions). The temporal development of source gas emissions are prescribed for a specific episode (years to decades).

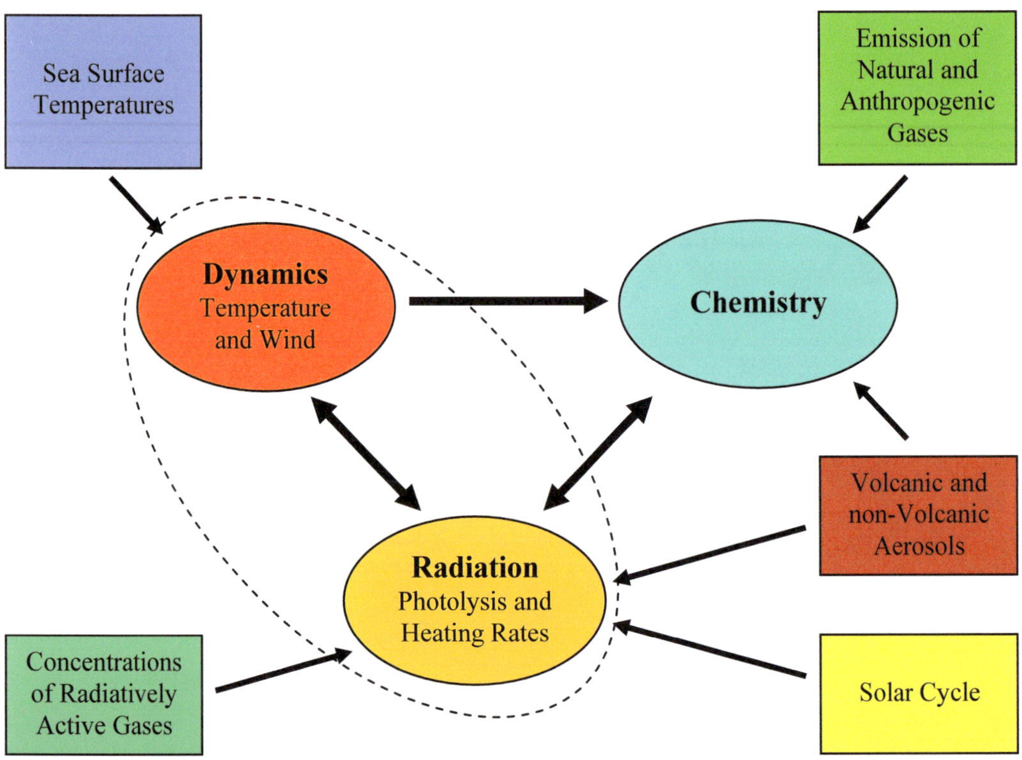

Fig. 6. Schematic of a Chemistry-Climate Model (CCM). The core of a CCM (oval symbols) consists of an atmospheric general circulation model (AGCM; i.e. dynamics and radiation part) that includes calculation of the heating and cooling rates and a detailed chemistry module. They are interactively coupled. Arrows indicate the direction of effect. Rectangular boxes denote external forcing.

As an example, in the following a brief description of the CCM E39CA is presented providing some useful details to better understand how such a model system works and respective simulations are performed. E39CA consists of the dynamic part E39 and the chemistry module CHEM. "E39" is an AGCM, based on the circulation model ECHAM4 (Roeckner et al., 1996). It has a vertical resolution of 39 levels from the Earth surface up to the top layer centred at 10 hPa (Land et al., 2002). The horizontal model grid on which the tracer transport, model physics and chemistry are calculated, has a mesh size of approximately 3.75° · 3.75° (latitude, longitude). The chosen time step for model integration is 24 minutes. The chemistry module "C" (Steil et al., 1998) is based on a family concept which combines related chemical constituents with short lifetimes (shorter than that of the dynamics or the model time-step used) into one family with a life-time larger than the time-step. "C" includes stratospheric homogeneous and heterogeneous ozone chemistry and the most relevant chemical processes for describing the tropospheric background chemistry with 107 photochemical reactions, 37 chemical species and four heterogeneous reactions on PSCs and on sulphate aerosols. Heating and cooling rates and photolysis rates are calculated on-line from the modelled distributions of the radiatively active gases O_3, CH_4, N_2O, H_2O and CFCs, and the actual cloud distribution.

In the following some boundary conditions of an E39CA model simulation are described covering the years from 1960 to 2050 (simulation "R2"). The model simulation considers various natural and anthropogenic forcings like the 11-year activity cycle of the sun with varying intensity of solar radiation (particularly in the ultraviolet which affects ozone chemistry), the quasi-biennial oscillation (QBO) of tropical zonal winds in the lower stratosphere (i.e. the direction of the zonal wind is alternating between west and east with a mean period of 28 months; see Baldwin et al., 2001), chemical and direct radiative effects of gases and aerosols (i.e. particles) emitted during major volcanic eruptions, and the increase in greenhouse gas concentrations. R2 is a simulation performed to assess the past and future evolution assuming a consistent set of boundary conditions which are partly based on observations (for the past) and on particular assumptions for future developments. For example, the future concentrations of long-lived greenhouse gases (CO_2, CH_4, and N_2O) are based on a 'business as usual' scenario (i.e. the A1B future scenario described in IPCC, 2001); future concentrations of ozone depleting substances follow the A1 scenario prescribed in WMO (2003), e.g. assuming a phase out of the CFCs in the troposphere and stratosphere over the coming decades leading to a continuous decrease of stratospheric chlorine content in future. Moreover, the wind phases of the QBO which were observed in the past are consistently continued. The solar activity signal observed between 1977 and 2007 is continually repeated until 2050. Furthermore, no major volcanic eruptions have been adopted in years up to 2050. Sea surface temperatures (SSTs) and sea ice coverage are prescribed using a continuous dataset derived from the coupled atmosphere-ocean model HadGEM1, provided by the UK Met Office Hadley Centre (Johns et al., 2006). The results from HadGEM1 are taken from a transient simulation with prescribed anthropogenic forcing as observed in the past, and following the SRES-A1B scenario in the future.

More details about the CCM E39CA and the respective assumptions made in the numerical simulation are provided in Stenke et al. (2009) and Garny et al. (2009).

3. Confronting model results with observations – a basement for predictions

In this section data derived from multi-year space-borne measurements are compared with respective data derived from simulation R2 with the CCM E39CA. It should provide some insight into current capabilities of numerical modelling of atmospheric processes and how model results are evaluated on the basement of observations. The evaluation of results derived from numerical modelling with observations gives indications about the quality of the applied model which partly reflects our current understanding of atmospheric processes and the cause and effect relationships leading to changes in atmospheric behaviour. The following examples of E39CA are only exemplary; more detailed comparisons including evaluations with other CCMs are provided in the SPARC CCMVal report (2010).

The evolution of the total ozone column in the atmosphere and respective standard deviation (both in Dobson Units, DU) as a function of latitude and time derived from GOME/SCIAMACHY/GOME-2 satellite measurements and the E39CA R2 simulation are presented in Figure 7. Note that the colour bars of the total ozone (upper two figures) are different for satellite and model data to better compare the latitude-time patterns (see discussion below regarding Figure 8). It is obvious that the overall variations of total ozone with latitude and time are well reproduced by E39CA.

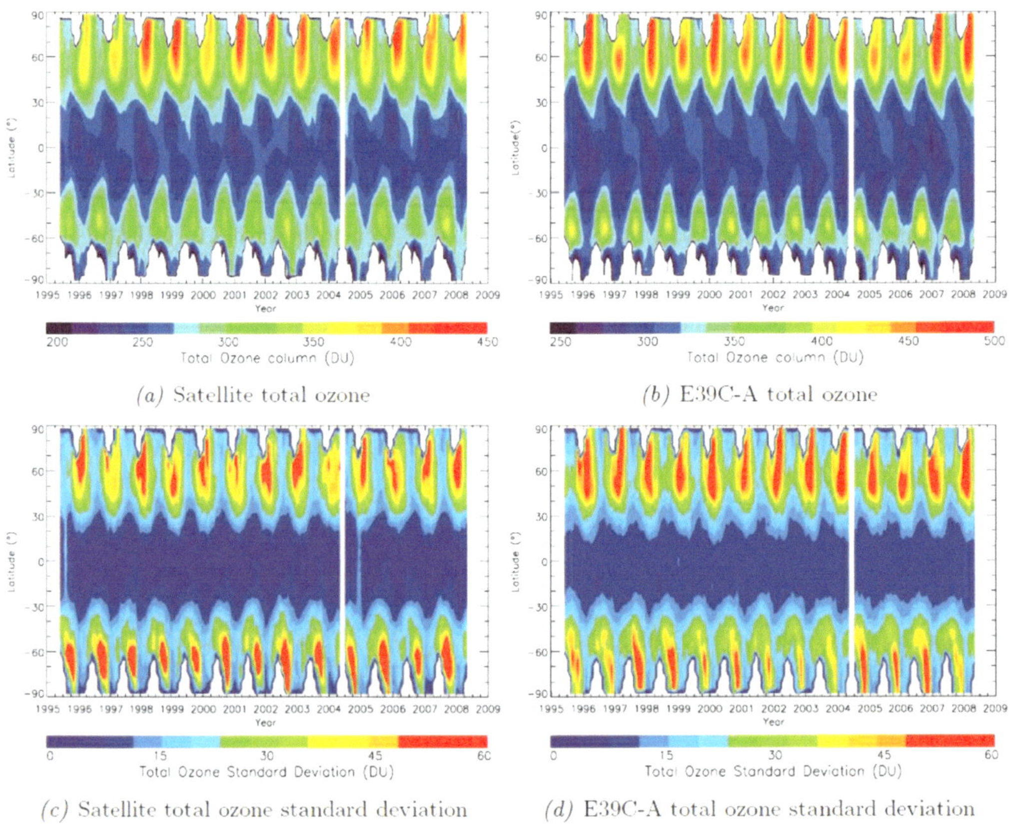

(a) Satellite total ozone
(b) E39C-A total ozone
(c) Satellite total ozone standard deviation
(d) E39C-A total ozone standard deviation

Fig. 7. Latitudinal evolution of total ozone (top) and standard deviation (bottom) from June 1995 to May 2008. GOME/SCIAMACHY/GOME-2 satellite data are presented on the left side (a, c) and E39CA model on the right side (b, d). Satellite measurements from April 2004 are not available; the corresponding model data are therefore also neglected (Figure 8 in Loyola et al., 2009).

The standard deviation of a given quantity (here total ozone in the lower two figures) is a measure for variability of the respective system, describing the range of variability in a specific region and period of time. Again, the agreement between model results and observations is satisfactory, i.e. the spatial and temporal structures are well reproduced. The latter result is important because it indicates that E39CA is able to reproduce adequately the internal variability of stratospheric dynamics and chemistry which is different in the Northern and Southern Hemisphere and the tropics (see Sections 1.1 and 1.2).

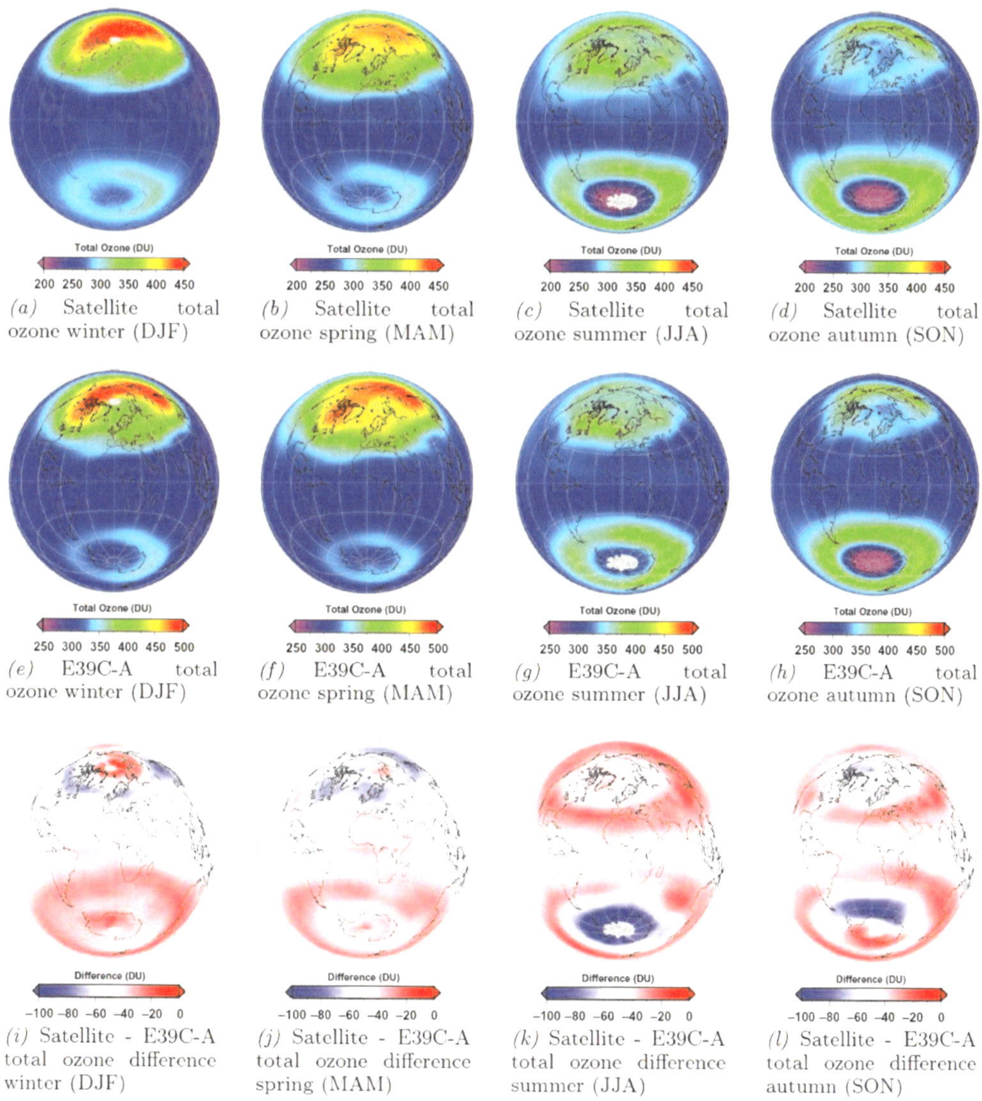

(a) Satellite total ozone winter (DJF)
(b) Satellite total ozone spring (MAM)
(c) Satellite total ozone summer (JJA)
(d) Satellite total ozone autumn (SON)
(e) E39C-A total ozone winter (DJF)
(f) E39C-A total ozone spring (MAM)
(g) E39C-A total ozone summer (JJA)
(h) E39C-A total ozone autumn (SON)
(i) Satellite - E39C-A total ozone difference winter (DJF)
(j) Satellite - E39C-A total ozone difference spring (MAM)
(k) Satellite - E39C-A total ozone difference summer (JJA)
(l) Satellite - E39C-A total ozone difference autumn (SON)

Fig. 8. Seasonal mean values of total ozone (June 1995 to May 2008) from GOME/SCIAMACHY/GOME-2 satellite instruments (top), the E39CA simulation (middle), and the difference between satellite measurements and model results (bottom) (Figure 6 in Loyola et al., 2009).

Figure 8 provides a more detailed evaluation of the absolute accuracy of total column ozone values as derived from E39CA simulations. Here, seasonal mean values of total ozone derived from satellite instrument measurements and E39CA are once again presented for the time period from June 1995 to May 2008. Please note that the colour bars here are also different for satellite and model data since E39CA total ozone values have a positive bias: A general shift to higher total ozone values is found ranging from about 5 DU in high northern latitudes during winter (DJF) to about 100 DU in high southern latitudes during winter (JJA). This finding indicates that there are still some weaknesses in the applied model system leading to an overall overestimation of total column ozone. Nevertheless, it is obvious that the meridional structure is well represented by E39CA in all seasons. The seasonal changes are well reproduced by the model. Particularly in the Northern Hemisphere, the latitudinal structure compares in a reasonable way. For example, the position of the polar vortex during winter and spring, which is indicated by lower ozone values over Eurasia, is correctly simulated by E39CA. While the Northern Hemisphere is dominated by a clear zonal wave number 1 pattern (i.e. one maximum and one minimum along a latitudinal circle), the distribution of ozone in the Southern Hemisphere has a much more zonally symmetric structure during all seasons which is captured by the model.

In addition to Figure 7, Figure 9 shows seasonal means of the standard deviation of total ozone, again for satellite data and model results. The overall seasonal change and the hemispheric patterns of the standard deviation in the model follow quite well the respective values from observations, but there are some differences in details. For example, in the distribution of the standard deviation in northern winter (DJF) high latitudes show some obvious differences: While in E39CA, the variability is low in the centre of the polar vortex (approximately between northern Europe and the North Pole) and higher in the surroundings, the satellite data show high variability in the vortex centre and a lower standard deviation over North America and eastern Asia. This finding can be explained by the fact that the polar vortex is too stable in E39CA, i.e. the number of minor and major warmings is lower than observed (e.g. Stenke et al., 2009). In the summer hemisphere (DJF in the Southern Hemisphere) the standard deviation is much higher in the model, but the region of maximum variability agrees again well with those derived from observed values. Another clear difference is found in the Southern Hemisphere spring months (SON) indicating a weaker variability in the South Polar Region (see also lower part of Figure 11). This model behaviour is explained by a too cold polar lower stratosphere in E39CA ('cold pole problem') reducing the dynamical variability in this region strongly (Stenke et al., 2009).

The comparisons shown so far were based on climatological and seasonal mean values. They are mainly used to rate the basic state of a numerical model system. Also important is the evaluation of the temporal evolution of an atmospheric quantity. The adequate reproduction of short-term variability and long-term changes, i.e. trends, is another prerequisite for robust assessment of future developments. Some examples are presented in the following section.

4. Prognostic studies

As a result of international agreements on protecting the stratospheric ozone layer (Montreal Protocol in 1987 and its amendments), the rapid increase in concentrations of the main CFCs in the troposphere has been stopped. Since the mid-1990ies, a decline in

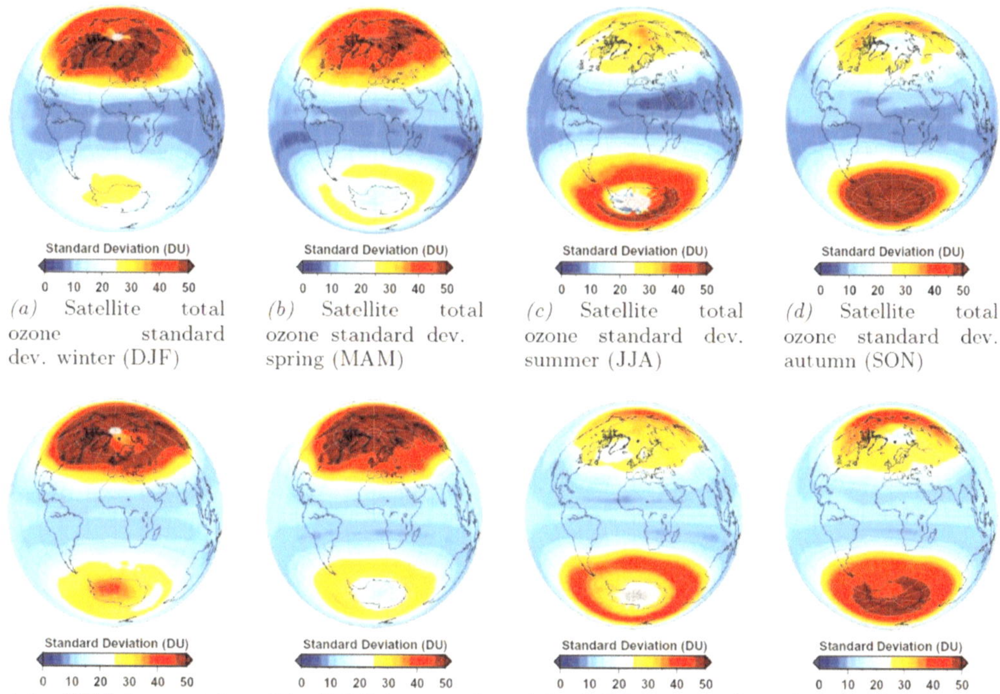

(a) Satellite total ozone standard dev. winter (DJF)
(b) Satellite total ozone standard dev. spring (MAM)
(c) Satellite total ozone standard dev. summer (JJA)
(d) Satellite total ozone standard dev. autumn (SON)
(e) E39C-A total ozone standard dev. winter (DJF)
(f) E39C-A total ozone standard dev. spring (MAM)
(g) E39C-A total ozone standard dev. summer (JJA)
(h) E39C-A total ozone standard dev. autumn (SON)

Fig. 9. Seasonal mean values of total ozone standard deviations (June 1995 to May 2008) from GOME/SCIAMACHY/GOME-2 satellite instruments (top) and the E39CA simulation (bottom) (Figure 7 in Loyola et al., 2009).

tropospheric CFC content has been observed (WMO, 2011). Consequently, a slight decrease in stratospheric chlorine concentrations has also been detected for several years now. However, due to the long lifetimes of CFCs in the atmosphere, it will take until the middle of this century for the stratosphere's chlorine content to go back to values resembling those observed in the 1960ies. Therefore, it is expected that the strong chemical ozone depletion observed over the past three decades will decrease in the near future. In this context, a solid assessment of the timing of the ozone layer recovery and particular the closure of the ozone hole is not a trivial task since the future evolution of the ozone layer is affected by several processes. In particular ongoing climate change will have an influence on atmospheric dynamics (including the transport of ozone) and ozone chemistry (via temperature changes). CCMs are suitable tools to perform prognostic studies regarding the evolution of stratospheric ozone content, because they are considering the complex interactions of dynamical, physical and chemical processes.

Based on prognostic studies with CCMs it is expected that the ozone layer will build up again in the next decades and that the ozone hole over Antarctica will be closed (see Chapter 3 in WMO, 2011; Chapter 9 in SPARC CCMVal, 2010). Figure 10 shows an example of the temporal evolution of total ozone deviations regarding a mean ozone value (1995-2009) for the near global mean (i.e. global mean values neglecting polar regions). Looking into the past it is obvious that E39CA is able to reproduce seasonal and interannual fluctuations in a sufficient manner, although the amplitudes of ozone anomalies are slightly underestimated. Model data and data derived from satellite observations clearly show the signal of the 11-year solar cycle. The absolute minimum ozone values observed in years 1993-95, which are caused by the eruption of the volcano Pinatubo, are not adequately reproduced by E39CA. The simulated increase in stratospheric ozone amount after year 2010 is a direct consequence of the prescribed decrease of stratospheric chlorine content.

The speed at which the ozone layer will rebuild in future depends on a range of other factors, however. Rising atmospheric concentrations of radiatively active gases (such as CO_2, CH_4 and N_2O) do not just cause the conditions in the troposphere to change (i.e. the greenhouse effect warms the troposphere), but also in the stratosphere which cools down with increasing CO_2 concentrations. The regeneration of the ozone layer thus takes place under atmospheric conditions different to those prevailing during the ozone depletion processes of recent decades. Due to climate change, it is highly unlikely that the ozone layer will return to exactly the way it was before the time of increased concentrations of ozone-depleting substances.

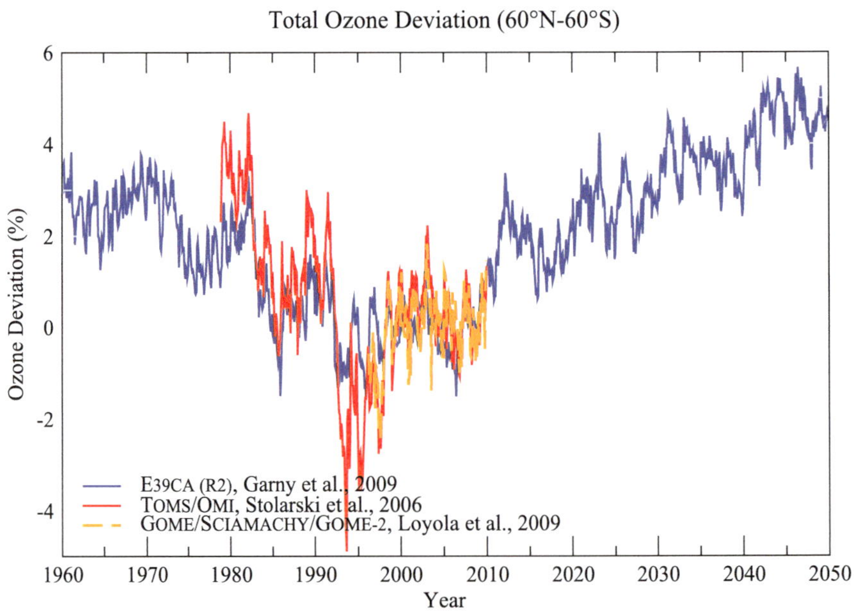

Fig. 10. Average deviations of the total ozone column (in %) for the region between 60°N and 60°S. The mean annual cycle for the period 1995-2009 was subtracted in each case. The orange and red curves represent data obtained from satellite instruments (TOMS, GOME, SCIAMACHY and GOME-2). The blue curve shows results from a numerical simulation (R2) using a chemistry-climate model (E39CA).

Due to further increasing greenhouse gas concentrations, global atmospheric temperatures will continue to change over the coming decades, i.e. it is expected that the troposphere will continue to warm up and that the stratosphere will cool down further due to radiation effects. In addition, it must also be taken into account that, due to the expected build-up of the ozone layer, stratospheric ozone heating rates (absorption of solar ultraviolet radiation by ozone) will increase again, to some extent counteracting the increased cooling due to rising greenhouse gas concentrations. However, as the ozone concentration depends largely on the background temperature, there will be some feedback. Since climate change also involves a change in the stratosphere's dynamics, "dynamic" heating of the stratosphere can also occur, depending on the time of year and place, which leads to local stratospheric heating, rather than cooling. That's why it is important to take the interactions between chemical, physical and dynamic processes into account, both for the interpretation of observed changes in the ozone layer and for prognostic studies. It must be always considered that ozone and climate connections are influencing each other in both directions. Therefore, estimates of future stratospheric ozone concentration developments and climate change are not trivial and bring uncertainties with them.

Future prognostics with CCMs also clearly indicate that ozone regeneration will be faster in some areas than in others, where it's quite possible that the recovery of the ozone layer will be delayed (Chapter 3 in WMO, 2011). The results of E39CA also show that the regeneration of the ozone layer will vary from region to region and does not represent a simple reversal of the depletion observed over recent years. Examples are presented in Figure 11, showing the evolution of the stratospheric ozone layer in the Northern and Southern polar regions.

In contrast to Figure 10, only the data for respective spring months are shown when ozone depletion maximises. First of all, E39CA reproduces nicely the different evolution of the ozone layer in the Northern and Southern Hemisphere in the past showing a more pronounced thinning of the ozone layer in the Southern Hemisphere due to the formation of the ozone hole. Interannual fluctuations are well captured by E39CA in the Northern Hemisphere while they are underestimated in the Southern Hemisphere. Here, for example, the model does not create dynamical situations leading to weak polar vortices in late winter and early spring and therefore higher ozone values as particularly observed in 1988 and 2002 (see also Figure 4). Obviously, the recreation speed of the ozone layer is different in the Northern and in the Southern Hemisphere: In the Northern Hemisphere ozone values found in the 1960ies and 1970ies are reached again around 2030 and further increase afterwards. In the Southern Hemisphere the 2050-values are still below the values found in the 1960ies (see also Figure 12). Looking again to Figure 10, here the ozone values after 2030 are also higher than before 1980.

What is the reason for this different behaviour in different stratospheric regions? Due to the continued increase in greenhouse gas concentrations in the atmosphere, the stratosphere will further cool down, resulting in faster ozone-layer regeneration especially in the middle and upper stratosphere. Here, lower temperatures slow down ozone destroying (temperature depending) chemical reactions. In the lower polar stratosphere, in particular in the Southern Hemisphere, the rebuilding of the ozone layer may slow down during spring. There, lower temperatures lead to an increased formation of polar stratospheric clouds (PSCs), which are the necessary prerequisite to start ozone depletion. On the other hand, climate change will affect ozone transport to higher latitudes. E39CA predict enhanced transport of stratospheric air masses from tropical to extra-tropical regions. Caused by the isolation of the Southern Hemisphere polar stratosphere (originated by the high zonal wind speed there; see Section 1.2) this change does not lead to a faster closure of the ozone hole (see also Figure 12); contrastingly, in the Northern Hemisphere the transport of ozone rich air towards polar latitudes is intensified, leading to a quicker back formation of the ozone layer. Overall, model simulations indicate that the ozone layer is expected to recreate faster in the Arctic than in the Antarctic stratosphere.

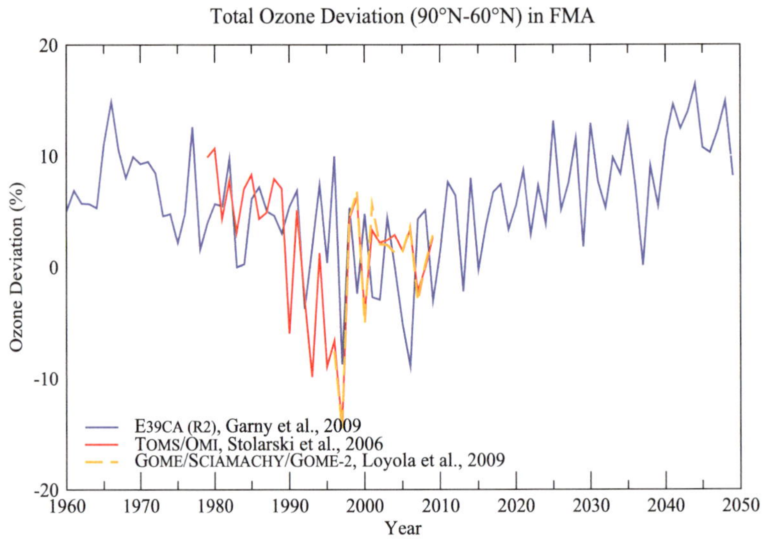

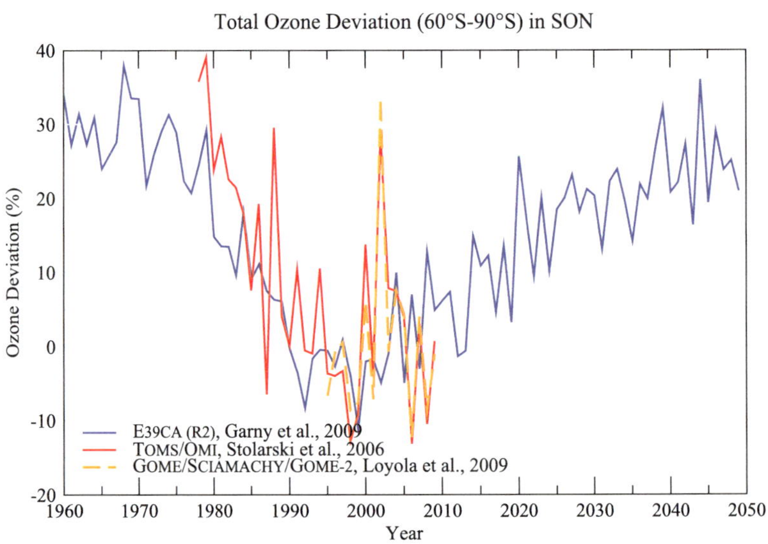

Fig. 11. As Figure 10, but now for the polar regions (top: Northern Hemisphere for months February, March, and April; bottom: Southern Hemisphere for months September, October, and November). Deviations are given with regard to the mean value of the period 1995-2009 (in %) for the region between 60° and 90°. Notice the different scales on the y-axis.

Figure 12 provides another view of the temporal evolution of the ozone hole as it is simulated by E39CA from the 1960ies to the 2040ies. Here, decadal mean total ozone values for October are presented. Respective mean values derived from satellite instruments are shown for comparison (notice again that the satellite and the CCM plots use different colour bars representing Dobson Units.) These images of simulated total ozone columns also indicate that the closure of the ozone hole will be delayed regarding the prescribed temporal decrease of the stratospheric chlorine content, i.e. will be not completed before 2050 (see also lower part of Figure 11).

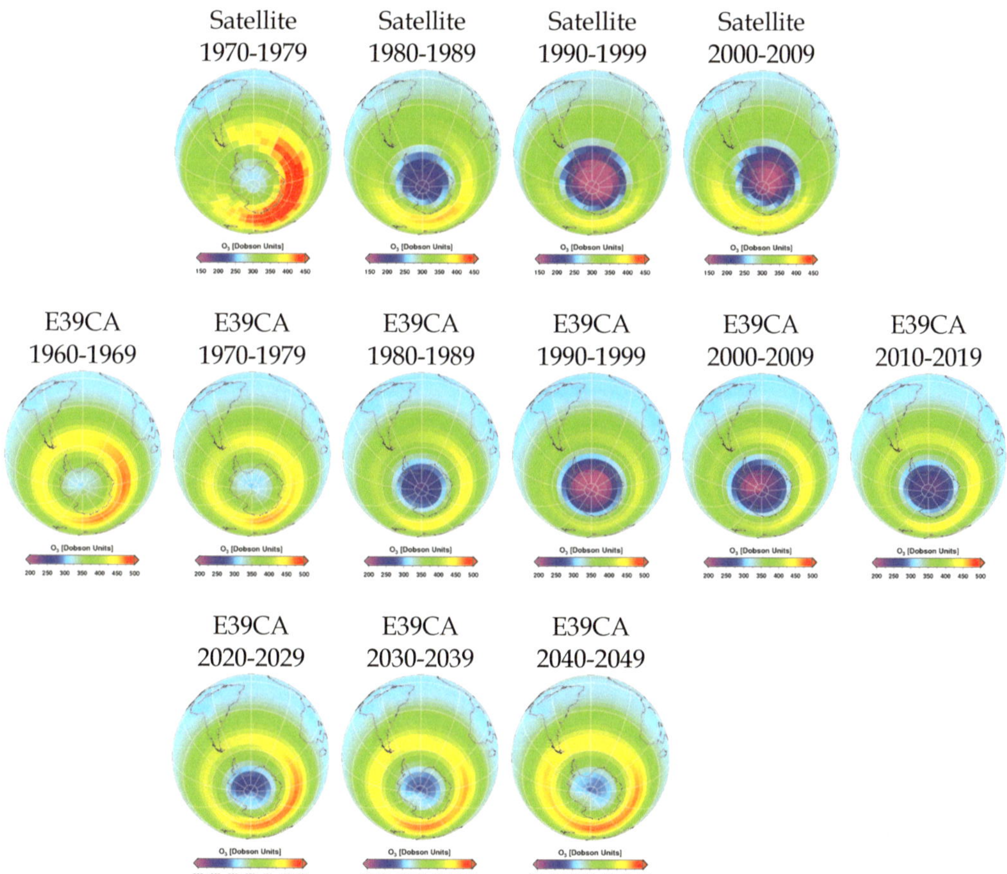

Fig. 12. Decadal evolution of the ozone hole in October (monthly mean value) derived from satellite measurements (TOMS, GOME, SCIAMACHY, GOME-2) from 1970 to 2009 and the simulation with the CCM E39CA from 1960 until 2049.

5. Discussion of uncertainties and conclusion

Obviously, reliable estimates of the future evolution of atmospheric behaviour are difficult because of strong interactions between physical, dynamical and chemical processes which, moreover, are expected to be modified in a future climate with enhanced greenhouse gas concentrations.

Further uncertainties must be taken into account regarding the future development of both the stratospheric ozone layer and the climate. The future development of stratospheric water vapour concentrations is currently highly uncertain. Prognostic studies with numerical models indicate that tropospheric water vapour concentrations would increase with increasing temperatures in the troposphere, which would also enhance the amount of water vapour being transported from the troposphere into the lower tropical stratosphere (e.g. WMO, 2007). Higher stratospheric water vapour concentrations could lead to an increased PSC-forming potential in the polar regions during the winter months and thus enhance the ozone depletion potential. As water vapour is also an important greenhouse gas, changes of atmospheric concentrations would affect atmospheric radiation balance. Stratospheric water vapour concentrations would also increase with rising methane (CH_4) concentration (e.g. caused by rice cultivation, intensive livestock farming, thawing of permafrost soils) due to oxidation of CH_4, which would itself increase ozone production in the lower stratosphere. On the other hand, rising CH_4 concentrations would bind reactive chlorine in the atmosphere. However, CH_4 is also an important greenhouse gas. Therefore, higher atmospheric CH_4 concentrations would influence both the climate and atmospheric chemistry. A further rise in atmospheric nitrous oxide (N_2O) concentrations would increase the amount of atmospheric nitrogen oxides, thus decreasing the ozone content of the middle and upper stratosphere. N_2O (although known as laughing gas) stems from both natural (for example oceans and tropical forests) and anthropogenic sources (for example emissions from cultivated soil, industrial processes, the combustion of fossil fuels, biomass, and biofuels). N_2O emissions near the Earth's surface are the most important source of nitrogen oxides in the stratosphere. N_2O is also a key greenhouse gas. This is another example which nicely illustrates the close relationship between problems associated with climate change and those relating to changes in the stratospheric ozone layer. Regulating the atmospheric N_2O content in the atmosphere is not just important for protecting the Earth's climate, but also for the future evolution of the stratospheric ozone layer. A reduction in N_2O emissions would both lower the anthropogenic greenhouse effect and positively influence the recovery of the ozone layer.

To summarise: Clearly, natural effects such as the variability of solar radiation and particle emissions that are due to strong volcanic eruptions influence the stratospheric ozone concentration. Internal fluctuations of stratospheric circulation affect the thermal structure of the stratosphere and air mass transport. Chemical production and depletion of ozone are determined by photochemical processes, homogeneous gas-phase reactions, and heterogeneous chemistry on the surface of particles (aerosols and PSCs). Understanding atmospheric processes and the interconnections between the various processes is made even more difficult by the fact that atmospheric conditions change over the long term owing to increased greenhouse gas concentrations. Climate change influences the overall stratospheric ozone production (i.e. the sum of ozone depletion and production) both directly and indirectly, and thus determines the rate of ozone regeneration, which will vary with altitude and latitude. Furthermore, changes in stratospheric circulation can potentially modify the development of the ozone layer in the 21st century. For example, a stronger meridional circulation in an atmosphere with increased greenhouse gas concentrations could cause the stratospheric wind systems to change during the winter months, thus for example resulting in decreased zonal wind speeds. This could lead to higher mean stratospheric temperatures in the polar regions and thus less PSCs.

The climate-chemistry connections presented in this chapter clearly demonstrate that in assessing atmospheric changes it is not enough to look at the processes independent of each other. Changes in climate and in the chemical composition of the atmosphere are closely interrelated, sometimes in very complex ways. Therefore, surprising developments cannot be ruled out in the future, either.

6. Acknowledgment

Thanks to NASA for the provision of the TOMS satellite data, ESA/DLR for the provision of the GOME and SCIAMACHY data, O3M-SAF/EUMETSAT/DLR for the provision of the GOME-2 data. We would like to thank colleagues from BIRA-IASB (Belgium), DLR (Germany), RT Solutions Inc. (USA) and AUTH (Greece) for their work on ozone retrieval

algorithms from the European satellites and corresponding geophysical validation. Special thanks to the DLR colleagues, M. Coldewey-Egbers for merging the GOME/SCIAMACHY/GOME-2 measurements and providing the ozone deviation data, and A. Stenke and H. Garny for the development of the E39CA model and providing the simulation data.
This work was partially supported by the ESA Climate Change Initiative project on Ozone (Ozone CCI).

7. References

Baldwin, M.P.; Gray, L.J.; Dunkerton, T.J.; Hamilton, K.; Haynes, P.H.; Randel, W.J.; Holton, J.R.; Alexander, M.J.; Hirota, I.; Horinouchi, T.; Jones, D.B.A.; Kinnersley, J.S.; Marquardt, C.; Sato, K.; and Takahashi, M. (2001). The Quasi-Biennial Oscillation, *Rev. Geophys.*, Vol. 39, 179–229

Bates, D.R.; and Nicolet, M. (1950). The photochemistry of atmospheric water vapor, *J. Geophys Res.*, Vol. *55*, 301–327

Burrows, J.; Weber, M; Buchwitz, M; Rozanov, V.; Ladstätter-Weißenmayer, A.; Richter, A.; DeBeek, R.; Hoogen, R.; Bramstedt, K.; Eichmann, K.; Eisinger, M.; Perner, D. (1999). The Global Ozone Monitoring Experiment (GOME): Mission Concept and First Scientific Results, *J. Atmos. Sci.*, Vol. 56, No. 2, pp. 151–175

Chapman, S. (1930). A theory of upper-atmosphere ozone, *Mem. Roy. Meteor. Soc.*, Vol. 3, pp. 103–125

Crutzen, P.J. (1971). Ozone production rates in an oxygen-hydrogen-nitrogen atmosphere, *J. Geophys. Res.*, Vol. *76*, 7311–7327

Dameris, M. (2010). Climate change and atmospheric chemistry: How will the stratospheric ozone layer develop?, *Angew. Chem. Int. Ed.*, Vol. 49, pp. 8092–8102, doi: 10.1002/anie.201001643

Garny, H.; Dameris, M.; Stenke, A. (2009). Impact of prescribed SSTs on climatologies and long-term trends in CCM simulations, *Atmos. Chem. Phys.*, Vol. 9, pp. 6017–6031

IPCC, Intergovernmental Panel on Climate Change (2001). Climate Change 2001: The Scientific Basis, Cambridge University Press, New York, USA

IPCC, Intergovernmental Panel on Climate Change (2007). Climate Change 2007: The Physical Basis 2007, ISBN 978-0-521-70596-7, pp. 996

Johnston, H.S. (1971). Reduction of stratospheric ozone by nitrogen oxide catalysts from SST exhaust, *Science*, Vol. 173, 517–522

Johns, T.C.; Durman, C.F.; Banks, H.T.; Roberts, M.J.; McLaren, A.J.; Ridley, J.K.; Senior, C.A.; Williams, K.D.; Jones, A.; Rickard, G.J.; Cusack, S.; Ingram, W.J.; Crucifix, M.; Sexton, D.M.H.; Joshi, M.M.; Dong, B.-W.; Spencer, H.; Hill, R.S.R.; Gregory, J.M.; Keen, A.B.; Pardaens, A.K.; Lowe, J.A.; Bodas-Salcedo, A.; Stark, S.; and Searl, Y. (2006). The New Hadley Centre Climate Model (HadGEM1): Evaluation of Coupled Simulations, *J. Climate*, Vol. 19, 1327–1353

Land C.; Feichter, J.; and Sausen, R. (2002). Impact of vertical resolution on the transport of passive tracers in the ECHAM4 model, *Tellus B*, 54, 344–360

Loyola, D.; Coldewey-Egbers, M.; Dameris, M.; Garny, H.; Stenke, A.; Van Roozendael, M.; Lerot, C.; Balis, D. & Koukouli M. (2009). Global long-term monitoring of the ozone layer - a prerequisite for predictions, *Int. J. of Remote Sensing*, Vol. 30, No. 15, pp. 4295–4318

Loyola, D.; Koukouli, M.; Valks, P.; Balis, D.; Hao, N.; Van Roozendael, M.; Spurr, R.; Zimmer, W.; Kiemle, S.; Lerot, C.; Lambert, J.-C. (2011). The GOME-2 total column ozone product: Retrieval algorithm and ground-based validation, *J. Geophys. Res.*, Vol. 116, D07302, doi: 10.1029/2010JD014675

Molina, M.J.; and Rowland, F.S. (1974). Atomic-catalysed destruction of ozone, *Nature*, Vol. 249, 810-812

Rayner, N.A.; Parker, D.E.; Horton, E.B.; Folland, C.K.; Alexander, L.V.; Rowell, D.P.; Kent, E.C.; and Kaplan, A. (2003). Global Analyses of sea surface temperatures, sea ice, and night marine air temperature since the late nineteenth century, *J. Geophys. Res.*, Vol. 108, 4407, doi: 10.1029/2002JD002670

Roeckner, E.; Arpe, K.; Bengtsson, L.; Christoph, M.; Claussen, M.; Dümenil, L.; Esch, M.; Giorgetta, M.; Schlese, U.; and Schulzweida, U. (1996). The atmospheric general circulation model ECHAM-4: Model description and simulation of present-day climate, Report No. 218, Max-Planck-Institut für Meteorologie, Hamburg

Solomon, S.; Garcia, R.R.; Rowland, F. S.; and Wuebbles, D. J. (1986). On the Depletion of Antarctic Ozone, *Nature*, Vol. *321*, 755-758

SPARC CCMVal (2010). SPARC Report on the Evaluation of Chemistry-Climate Models, V. Eyring, T.G. Shepherd, D.W. Waugh (Eds.), SPARC Report No. 5, WCRP-132, WMO/TD-No. 1526, http://www.atmosp.physics.utoronto.ca/SPARC

Steil, B., Dameris, M.; Brühl, C.; Crutzen, P.J.; Grewe, V.; Ponater, M.; and Sausen, R. (1998). Development of a chemistry module for GCMs: first results of a multiannual integration, *Ann. Geophysicae*, Vol. 16, 205–228

Stenke, A.; Dameris, M.; Grewe, V.; and Garny, H. (2009). Implications of Lagrangian transport for simulations with a coupled chemistry-climate model, *Atmos. Chem. Phys.*, Vol. 9, 5489–5504.

Stolarski, R.; Frith, S. (2006). Search for evidence of trend slow-down in the long-term TOMS/SBUV total ozone data record: the importance of instrument drift uncertainty, *Atmos. Chem. Phys.*, Vol. 6, 4057-4065

Van Roozendael, M.; Loyola, D.; Spurr, R.; Balis, D.; Lambert, J-C.; Livschitz, Y.; Valks, P.; Ruppert, T.; Kenter, P.; Fayt, C.; Zehner, C. (2006). Ten years of GOME/ERS-2 total ozone data: the new GOME Data Processor (GDP) Version 4: I. Algorithm Description, *J. Geophys.Res.*, Vol. 111, D14311

WMO/UNEP, World Meteorological Organization, Scientific Assessments of Ozone Depletion: 1991 (1992). Global Ozone Research and Monitoring Project, Report No. 25, Geneva

WMO/UNEP, World Meteorological Organization, Scientific Assessments of Ozone Depletion: 1994 (1995). Global Ozone Research and Monitoring Project, Report No. 37, Geneva

WMO/UNEP, World Meteorological Organization, Scientific Assessments of Ozone Depletion: 1998 (1999). Global Ozone Research and Monitoring Project, Report No. 44, Geneva

WMO/UNEP, World Meteorological Organization, Scientific Assessments of Ozone Depletion: 2002 (2003). Global Ozone Research and Monitoring Project, Report No. 47, Geneva, pp. 498

WMO/UNEP, World Meteorological Organization, Scientific Assessments of Ozone Depletion: 2006 (2007). Global Ozone Research and Monitoring Project, Report No. 50, Geneva, pp. 572

WMO/UNEP, World Meteorological Organization, Scientific Assessments of Ozone Depletion: 2010 (2011). Global Ozone Research and Monitoring Project, Report No. 52, Geneva, pp. 516

Wolfsy, S.C.; Mc Elroy, M.B.; Yung, Y.L. (1975). The chemistry of atmospheric bromine, *Geophys. Res. Lett.*, Vol. 2, 215-218

2

Time Correlation Laws Inferred from Climatic Records: Long-Range Persistence and Alternative Paradigms

Maria Lanfredi, Tiziana Simoniello,
Vincenzo Cuomo and Maria Macchiato
*Institute of Methodologies for the Environmental Analysis-IMAA-CNR,
Department of Physics- University "Federico II",
Italy*

1. Introduction

Observational time series of climatic variables exhibit substantial changeability on spatial and temporal scales over many orders of magnitude. In statistical terms, this implies a continuous variance distribution involving all resolvable time scales (frequencies), starting from those comparable with the age of the Earth.

A correct causal interpretation of such a variability is very difficult even in the context of a cognitive approach (e.g., von Storch, 2001) to the problem.

Cognitive models are minimum complexity models aiming at the scientific understanding of the most relevant processes occurring at any given temporal and spatial scale. Although generally they cannot be useful for management decisions straightforwardly, their role is fundamental especially for understanding the internal climatic variability that cannot be passively related to external forcing factors. The concept of stochastic process is essential in this framework, since it synthesizes collective behaviours which contribute as a whole to the overall dynamics. As stochastic processes are the macroscopic result of many degrees of freedom, the characterization of their correlation properties across different scales through the analysis of observational data is a problem of statistical inference and their modelling is usually a mechanical-statistical problem.

Maybe, the most famous early effort aiming to summarize the climate variance distribution among different frequencies, which is commonly referred as climate spectrum, is the ideal sketch proposed by Mitchell (1976) (see Fig. 1).

All the features of this spectrum that deviate from the flat behaviour typical of white noise (pure random process) deserve dynamical interpretation in order to understand climate. Within the traditional picture of the climate dynamics, the variance distribution among different temporal scales is seen as the superposition of oscillations generated by astronomical cycles (spectral spikes), quasi-periodic or aperiodic fluctuations with a preferred scale (broad spectral peaks), and internal stochastic processes whose temporal correlation decays according to characteristic time scales. These last are responsible for all the continuous broad-band deviations of the spectrum from flatness. Within this picture, the variance accumulations that do not appear in the form of peaks and spikes, such as that we can observe in-between the red vertical lines of Fig. 1 by scanning the figure from the short to the long periods, are due to the superposition of stochastic processes with different scales. This "redness" (optical analogy: dominance of low frequencies) would reflect the thermal inertia of slow climatic subsystems, such as ocean and cryosphere, and would be the result of a progressive addition of variance "shelves" (Mitchell, 1976) generated by ever slower scale-dependent processes. Hasselmann (1976) proposed an interesting interpretation of this redness by assuming that the heat-storage capacity of "slow" Earth's sub-systems act to integrate random "fast" disturbances in a dynamical context that is therefore characterized by separation between short and long scales. As an example, ocean would act as a long term integrator of the meteorological atmospheric forcing (white noise on climatic scales) thus providing "memory" to the atmosphere-ocean system in the form of non zero correlation among different scales, that is redness. The resulting simplest paradigm for this integration is Brownian motion (random walk) that is a non-stationary scale free process whose variance increases linearly with scale (Mandelbrot & van Ness, 1968). Such an ideal motion is able to produce random trends of any length but within climate dynamics the presence of dissipative phenomena is expected to dump such integration. Then, dissipation introduces a characteristic time scale that marks the temporal horizon for the decay of the fluctuations toward the mean value and also oceanic processes approach white noise asymptotically (e.g., Von Storch et al, 2001).

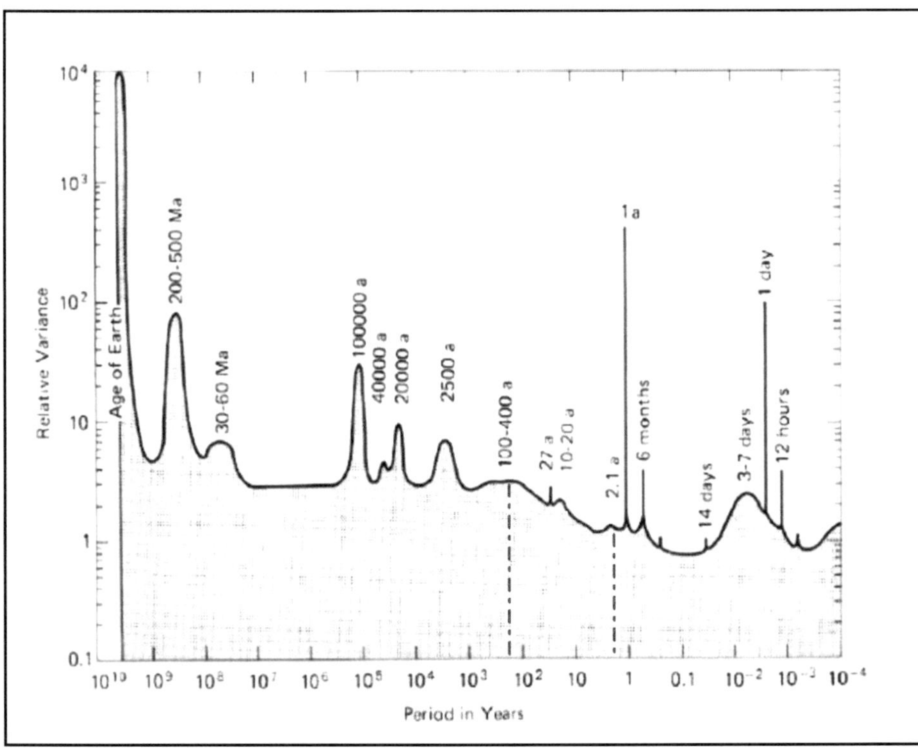

Fig. 1. Idealized sketch of the planetary climate variance spectrum (after Mitchell 1976).

To this day, the necessity of understanding the actual origin of the climate variability in the entire spectral range is still recognised as a "primary goal" of the climate research, especially if we aim to address the impact of human activities on climate. Currently, the availability of historical records of atmospheric temperature, which is the key variable of any terrestrial process, and the possibility of enlarging the observational time window back to about 400,000 years ago (proxy paleoclimatic data) give us the unique opportunity to get realistic insights into the correlation structures that characterise climate regimes from the meteorological to the glacial-interglacial domain.

Contextually, the development of new mathematical-statistical tools, devised for enhancing specific correlation features (e.g. fractal persistence), make it possible to better discriminate such correlations from structures ascribable to more traditional superpositions of fluctuations and cycles. For many years, the scientific community has worked to rightly interpret the collection of observational data in order to improve the current understanding of the climate dynamics, evaluate the performances of models, and detect signatures of climate change blurred within regime variability. In particular, many works have focused on red spectral patterns in order to explore the possibility that the scale free dynamics typical of fractals, either non stationary (fractal Brownian motion) or stationary (fractional Brownian noise) (Mandelbrot & van Ness, 1968), could provide a description of climate better than the traditional one. A wide literature, based on both classical and new mathematical-statistical tools, is now available which reports analysis results and possible dynamical scenarios able to explain the sample time scale laws (e.g.,Konscielny-Bunde et al., 1996, 1998; Govindan et al., 2001;Eichner et al., 2003; Kurnaz, 2004; Varotsos et al., 2006; Vecchio & Carbone, 2010). These works suggest long range persistence (power law correlation) rather than scale dependence (exponential correlation) as a good statistical paradigm for explaining the climate spectrum redness on scales up to about 10^2 years. Also some analyses of pre-historical records (Pelletier, 1998; Huybers & Curry, 2006) support scale-invariance, since a random walk spectrum appears in the time scale range from 10^2 to 10^4 years. In both historical and pre-historical climate, scale separation seems to fail giving place to a continuum of self-organized scales. In this case, weather would be the only dynamical framework where it works well.

In spite of the wide consensus around these studies, there are contradictory results about the universality of the scaling and the dependence of the exponent on the distance to sea (e.g., Vyushin et al., 2004a, 2004b; Blender and Fraedrich, 2004). More in general the interpretation of such a scaling is rather controversial because of the many drawbacks of the methodologies adopted (e.g., Hu, 2001; Kantelhardt, 2001; Metzler, 2003; Mauran et al., 2004; Gao et al., 2006; Rust, 2006; Lanfredi et al, 2009, Simoniello et al, 2009).

This chapter discusses the state of the art of the studies of historical time series of atmospheric temperature, particularly focused on the interpretation of redness, and provides new analysis results for enhancing the debate on paleoclimatic observations. The core of the chapter is the discussion of the correlation structures estimated from observational data and their reliability. This is a typical problem of statistical inference that is crucial for identifying the right class of dynamical models to be used in the climate modelling. It is shown that the most popular recent interpretations, supporting power-law correlation, are not the only possible. The traditional simpler explanations are also acceptable and may work better than the complex ones. The discussion is inserted into the framework of the stochastic approach to the climate approximation, although our arguments are useful for climate modelling also within a non-stochastic approach to the problem.

The chapter is organised according to the following principal points:

Section 2 summarises the main physical and statistical concepts and tools used in the chapter. A short overview of the basic models and operational implications concerning scale separation and scale invariance is provided; analysis tools and their potential weak points are discussed. Section 3 discusses the analyses of historical and pre-historical data. Detailed statistical estimates and literature results are provided in order to support the discussion. Then, the debate on the dynamical nature of redness is extended to millennial time scales. Finally, section 4 concerns the conclusive part of the chapter.

2. Basic concepts and statistical tools

In this Section we summarise the main physical and statistical concepts and tools used in the chapter. These substantially concern the main general forms of correlation, scale dependence (short-range correlation) and scale invariance (long-range correlation), which are useful for the selection of the right class of stochastic models for climate. Of course, the discussion is not exhaustive but merely aims to provide the basic background that is necessary for the understanding of the chapter's content.

2.1 Autoregressive processes, scale dependence, and their role in the traditional stochastic climate

Stationary stochastic processes are often fruitfully modelled by means of autoregressive processes, which are filters whose input is a Gaussian independent process (white noise) ε_t (e.g., Jenkins & Watts, 1968). The output of an autoregressive process AR(p) of order p is:

$$X_t = \sum_{i=1}^{p} a_i X_{t-i} + \varepsilon_t \tag{1}$$

where (a_i) are the autoregressive coefficients and ε_t is a Gaussian random process with zero mean and variance σ^2. In particular, the paradigmatic model of the meteorological fluctuations is the first order autoregressive process AR(1):

$$X_t = aX_{t-i} + \varepsilon_t ; \qquad |a| < 1 \tag{2}$$

where the index t indicates the daily step. The autocovariance function is:

$$< X_t X_{t+n} > = \frac{|a|^n}{1-a^2}\sigma^2 \tag{3}$$

This last decays with the characteristic length τ=-1/ln(a). For continuous processes, the correlation function is:

$$\rho(\tau) = e^{-t/\tau} \tag{4}$$

For very long time scales AR(1) is completely stationary with variance:

$$< X_t^2 > = \frac{1}{1-a^2}\sigma^2 \qquad t >> \tau \tag{5}$$

AR(1) is the most simple example of scale-dependent process: for $t \ll \tau$ it is strongly correlated whereas it becomes a white noise for $t \gg \tau$. Within the traditional approach to climate approximation, this white noise describes the variability of meteorological variables in a scale range satisfying the condition:

$$\tau_m \ll \Delta t \ll \tau_c \qquad (6)$$

where τ_m is the meteorological characteristic scale (a few days) and τ_c is the closest characteristic scale of climate (e.g. that of the oceans). More in general, it describes elementary stochastic processes whose superposition can generate redness through the progressive addition of variance. In fact, according to Eqs. 3 and 5, the fluctuations of an AR(1) produce low variance (high covariance) on scales shortest than its characteristic one; such a variance increases with scale up to the value in Eq. 5 on asymptotic scales. Roughly speaking, if we consider the superposition of different first order autoregressive processes AR_i (1) ($i=1,\ldots,n$) and separated time scales $\tau_1 \ll \ldots \ll \tau_n$, the total process behaves as AR_1 (1) for $t \ll \tau_2$ and its variance increases of a step $\sigma_i^2/(1-a_i^2)$ any time we exceed the scale τ_i thus producing a red accumulation.

2.2 Fractional Brownian motion, fractional Gaussian Noise and the concept of scale invariance

Fractional Brownian motion and fractional Gaussian noise, which were defined in (Mandelbrot & Van Ness, 1968), generalize Brownian motion and white noise, respectively. The time trace $B(t)$ of a Brownian motion (random walk) is characterized by independent increments $B(t+\tau)-B(t)$ having a Gaussian distribution. Such increments have mean zero and variance $|\tau|$; the mean separation between two points is proportional to the square root of the time separation:

$$|B(t+\tau)-B(t)| \propto |\tau|^{1/2} \qquad (7)$$

Mandelbrot & Van Ness (1968) introduced the family of fractional Brownian motions (fBm's) by generalizing the Eq. 7.
FBm's are random variables with Gaussian increments satisfying the condition:

$$|B^H(t+\tau)-B^H(t)| \propto |\tau|^H \qquad (0<H<1) \qquad (8)$$

where the exponent H is the Hurst's coefficient. Thus, ordinary random walk coincides with an *fBm* with $H=0.5$. For discrete times it can be approximated by summing up a white noise $w=\{w_k : k=0,1,\ldots\}$:

$$B_n = \sum_0^n w_k \qquad (n=0,1,\ldots) \qquad (9)$$

Equivalently, we can define the incremental process $z^H=\{z_k^H : k=0,1,\ldots\}$ of an *fBm* such that:

$$B_n^H = \sum_0^n z_k^H \qquad (n=0,1,\ldots) \qquad (10)$$

thus obtaining a fractal generalization of white noise which is called fractional Gaussian noise (*fGn*). FGn has a standard normal distribution for every *k*; the corresponding autocorrelation function $\rho(.)$ is:

$$\rho(\Delta k) = \tfrac{1}{2}\left[\left|\Delta k - 1\right|^{2H} - 2\Delta k^{2H} + \left|\Delta k + 1\right|^{2H}\right] \tag{11}$$

If $H = 0.5$, $\rho(\Delta k) = 0$ for $\Delta k \neq 0$. This condition brings back to ordinary Brownian motion, which has independent increments. In all the other cases *fGn* is a stationary process whose covariance is non zero for any finite Δk.
The Hurst coefficient *H* provides a measure of the persistence properties of the process according to the following scheme:

- $0<H<0.5$ $\rho(\Delta k) < 0$: all points of an *fGn* separated by a lag time Δk are negatively correlated ; both *fGn* and the corresponding integral *fBm* are **anti-persistent**.
- $H=0.5$ $\rho(\Delta k) = 0$: *fGn* is **independent** and the corresponding motion is the classical random walk.
- $0.5<H<1$ $\rho(\Delta k) > 0$: all points of *fGn* separated by a lag time Δk are positively correlated; both *fGn* and the corresponding *fBm* are **persistent**.

The correlation of *fGn* expresses scale free interdependence and decays as a power law. For continuous times $\rho(\tau) \propto \tau^{-\gamma}$ ($\gamma=2-2H$). The case $\gamma>0.5$ characterizing persistent processes is particularly interesting since the theoretical correlation implies non zero probability that disturbances survive on times as long as infinity (long range memory).
Such ideal processes may be useful within empirical studies aiming to describe observational stationary time series which show interdependence between very distant samples without approaching white noise. In these cases, the most classical models that are characterized by exponential decorrelation $\rho(\tau) = e^{-\frac{t}{\tau}}$ (e.g., autoregressive processes) could fail to account for such a long-range dependence.

2.3 Drawbacks of time series analysis for the detection of scale invariance; detrended fluctuation analysis

Generally, the investigation of time series aims to identify a class of theoretical processes able to synthesize some given correlation features of observational data: the class of the processes is assumed to be unknown. As a consequence, in order to propose a given model as a realistic descriptor of the investigated dynamics, we have to demonstrate both the compatibility of the tested theoretical correlation structure with that estimated from data (necessary condition) and to exclude any other alternative forms of correlation (sufficient condition).
Actually, the procedures that are used to identify the existence of power-law correlation do not allow us to satisfy both these conditions. It is well known that the variance spectrum is very sensitive to any form of non stationary behaviour. It is suitable for investigating stationary or cyclo-stationary signals or, more in general, signals with weak local features. As far as climatic time series, this condition cannot be guaranteed. Any external forcing such as volcanic eruptions and externally induced temporary warming/cooling trends can produce misleading results.

In order to avoid these drawbacks, some authors developed alternative tools, such as Detrended Fluctuation Analysis (DFA) (Peng et al, 1995), aiming to minimize externally-induced non-stationary effects describable in the form of low-order polynomials. We shortly recall how this methodology works. The time series to be analysed is integrated and divided into N boxes of length n. In each box, a least square polynomial $y_n(k)$, representing the trend in that particular box, is fitted to the integrated data $y(k)$. Then, the root-mean-square fluctuation:

$$F(n) = \sqrt{\sum_{k=1}^{N} \left[y(k) - y_n(k) \right]^2 / N} \tag{12}$$

is calculated. This computation is repeated on many time-scales (box sizes) in order to characterize $F(n)$ as a function of n. Power-law (fractal) scaling implies a linear relationship in a log-log plot. Under such conditions fluctuations can be characterized by a scaling exponent α ($\alpha=H$ for *fGn*). In this chapter the 2nd-order Detrending (DFA2) is adopted in order to minimize the effects of discontinuities and linear trends.

This methodology, that is generally considered the most powerful for identifying *fGn*, may produce many false positive results. This point is well stressed in Mauran et al., (2004). This is a method developed to discover fractals blurred in noise. In practice, it intrinsically postulates that a fractal is present and try to estimate the scaling coefficient minimizing external disturbances. It satisfies the necessary condition above (if a fractal is present it is generally able to find it) but is not able to satisfy the sufficient condition, since if there is not any fractal the estimation of a linear best fit in a log-log plot of sample statistics is not sufficient for supporting the actual existence of a power law. In particular, log-log collinearity should be carefully verified.

3. Results from time series analysis of atmospheric temperature

In this Section we discuss some examples of analyses of temperature time series aiming to detect long range persistence. We refer to bibliography for in-depth information.

3.1 Historical data

The rationale behind most of the investigations on historical data is the more or less explicit use of white noise as null hypothesis.

Within the classical stochastic approach to climate approximation the fastest processes we deal with are the meteorological processes, whose time scale is considered well-separated from all the slower climatic time scales. Such a meteorological variability has been traditionally explained by low-order autoregressive processes such as the paradigmatic first-order autoregressive process (AR1):

$$X_t = aX_{t-i} + \varepsilon_t \; ; \tag{13}$$

where X_i is the meteorological variable, a is the first-order autocorrelation coefficient, and ε_i represents white noise. According to this model, the parameter a accounts for rapid inter-day correlation decay so that the asymptotic behaviour, starting from scales of a few weeks, is uncorrelated and unpredictable: $X_i \sim \varepsilon_i$

More recently, in the wake of the great success of empirical fractal tools devised for enhancing power-law correlation in noised and biased observational data (e.g., Peng et al., 1995; Konscielny-Bunde et al., 1998; Freeman et al., 2000; Matsoukas et al., 2000; Haggerty et al., 2002; Bunde et al., 2002; Kandelhardt et al., 2003, 2006; Blender and Fraedrich, 2003), many researches have focused on historical atmospheric temperature time series for exploring the possibility that long range persistence characterizes climate after the meteorological correlation is decayed (e.g., Konscielny-Bunde et al., 1996, 1998; Govindan et al., 2001; Eichner et al., 2003; Kurnaz, 2004; Varotsos et al., 2006).
Their analyses, based on the estimation of the Hurst coefficient prevalently by means of DFA, seem to put into evidence slightly long range persistent features and their conclusion is that the asymptotic noise ε_l is not white but is a power law correlated noise (see Kiraly & Janosi, 2002 for a fractal version of Eq. 13).
According to these works Fractional Gaussian noise has been suggested as a realistic model for explaining the statistical dependence of atmospheric temperature anomalies (deviations from the mean annual trend) on climatic time scales.

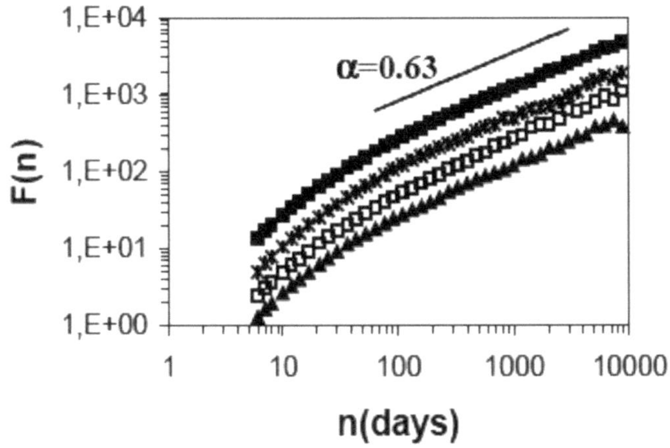

Fig. 2. Plot of the detrended fluctuation function for daily atmospheric temperature time series (Klein Tang, 2002) recorded in Prague (filled squares), Wien (stars), St. Petersburg (empty squares), Potsdam (triangles) (after Lanfredi et al, 2009).

Fig. 2 shows the results of DFA applied to four atmospheric temperature time series widely analysed literature (Lanfredi et al, 2009 and references therein).
The apparent linear behaviour of the fluctuation function on decadal scales is rather evident and the value of the Hurst coefficient greater than 0.5 indicates a long range persistent behaviour. Nevertheless, just the well known redness of the climatic spectrum suggests that white noise is not the right null hypothesis against long range persistence. The actual problem is to establish whether the power law is the best representation for the atmospheric temperature correlation or instead alternative time scale laws are acceptable. In practice there is a problem of functional form goodness for the linear fit. Fig.3 (Lanfredi et al., 2009) shows the residuals from the power law best fit of Fig. 2 which should be a stationary noise in the time range where the time series is fractal. On the contrary, the residuals are arranged in a non-linear way in all the cases.

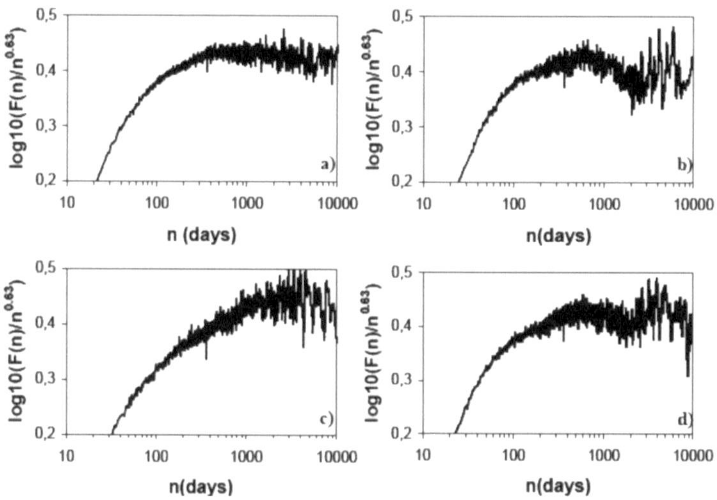

Fig. 3. Plots of the ratio $F(n)/n^{\alpha}$ in logarithmic scale for the four time series of Fig. 2: (a) Prague; (b)Wien; (c) St. Petersburg; (d) Potsdam (after Lanfredi et al., 2009).

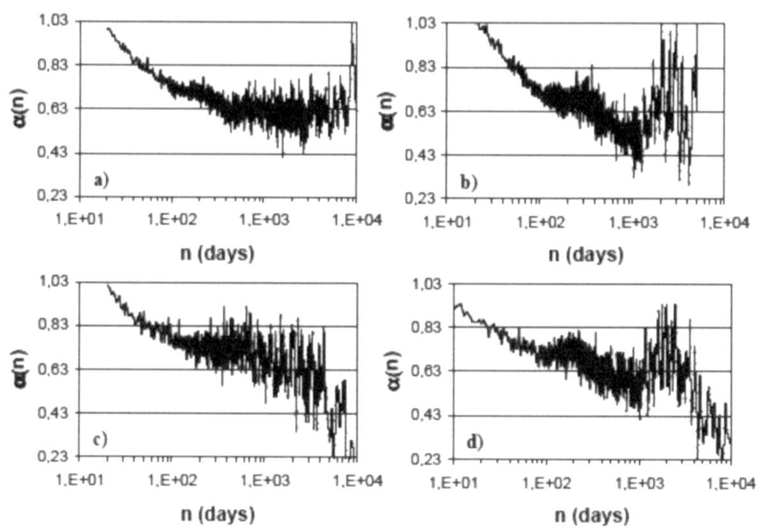

Fig. 4. Estimates of $\alpha(n)$ for (a) Prague; (b)Wien; (c) St. Petersburg; (d) Potsdam (after Lanfredi et al., 2009).

In addition, within the scaling regime, the scale invariant law $F(kn)=kF(n)$ should hold for any k. Thus, the function $\alpha(n)=log_k[F(kn)/F(n)]$ should provide an estimation of the local scaling coefficient. Again, $\alpha(n)$ should be a stationary noise where a scaling regimes occurs. Fig. 4 shows the estimates of $\alpha(n)$ for the four time series of Fig. 2. On short time scales the high value of $\alpha(n)$ accounts for a strong correlation that progressively decays approaching a noised and irregular behaviour that does not allow us to detect scaling regimes unquestionably. Most likely, the apparent scaling is due to the emergence of slower fluctuations that add "shelves" (Mitchell,1976) to the time series variance.

In order to assess how short range dependent processes appear when examined by means of fractal tools, we can investigate time series simulated on the basis of observational data and modelled according to scale separation (Lanfredi et al, 2009).
Fig. 5 and 6 show the analysis results of a simple two-scales (weather-climate) process, modelled on the basis of the autocorrelation function of the Prague's data. The analogies with the real data (Figs 2,3 and 4) are very impressive. The two-scale model is able to account for the whole results obtained from the fractal investigation. The mechanism that produces scaling is clear. Correctly, the total fluctuation function *F(n)* ends as a white noise (Fig. 5b) only in the latest part of the plot. Nevertheless, since the variance produced by the slow climatic variable emerges only on the long time scales, if we try to fit the function globally from the short to the long time scales (Fig. 5a), a spurious scaling occurs for the presence of the variance shelf.

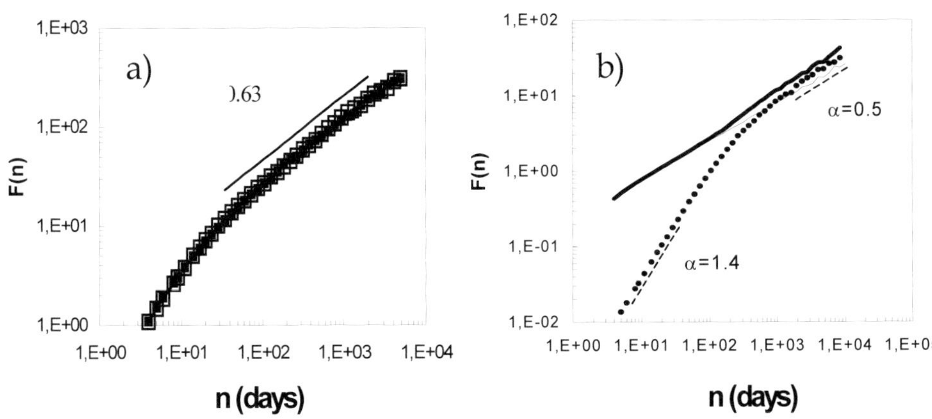

Fig. 5. a) Results of DFA for real (filled squares) and simulated (empty squares) anomalies. The continuous line shows the empirical power law reported in literature; b) effect of an hidden long scale within an asymptotic noise, the high scaling coefficient of the hidden process α=1.4 on short scales is an indication of strong correlation and is compatible with values estimated for the ocean (after Lanfredi et al, 2009).

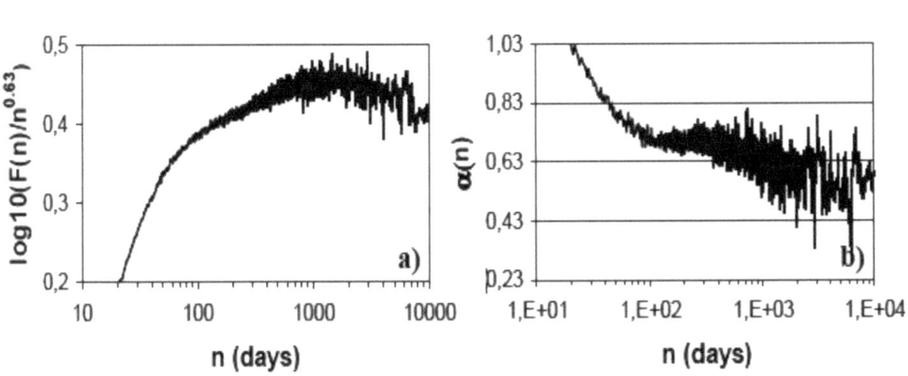

Fig. 6. Residuals from the linear best fit, and estimation of the local scaling exponent of a two-scales (weather-climate) autoregressive process (after Lanfredi et al, 2009).

3.2 Paleoclimatic data

The temperature time series obtained from the Vostok ice core dataset (Petit et al, 1999) provides a unique source of information about climate changes over glaciological scales.
Although unevenly sampled in time and affected by reconstruction errors, such as non-temperature effects, observational uncertainty, age-model uncertainty, etc., it includes structures generated by those time scale laws we are searching for. Above all, they can inform us about possible common correlation structures unifying climate dynamics on historical and paloclimatic eras.
Fig. 7 shows this paleorecord that describes temperature variability for the past 420,000 years. The time series appears to be rather noised even if some near systematic behaviours are detectable. Among them, the longest oscillations (Milankovitch cycles) account for the alternation between glacial and interglacial eras. Although the astronomical variability that drives them are known to be a combination of cyclical changes of the Earth-Sun geometry (eccentricity, obliquity, precession), there is not yet a shared interpretation of the underlying dynamics (e.g., Meyers et al 2008). These data include information on the effects of the so-called "Pacemaker of the Ice Ages " (Hayes, et al., 1976) on the terrestrial internal climatic variability. Just this variability under the action of the astronomical forcing could provide useful insights on the mechanisms that govern the mutual interactions between the different climatic subsystems. Also in this case, we do not discuss this specific dynamical problem but illustrate the difficulty related to the inference of time scale laws from this dataset.

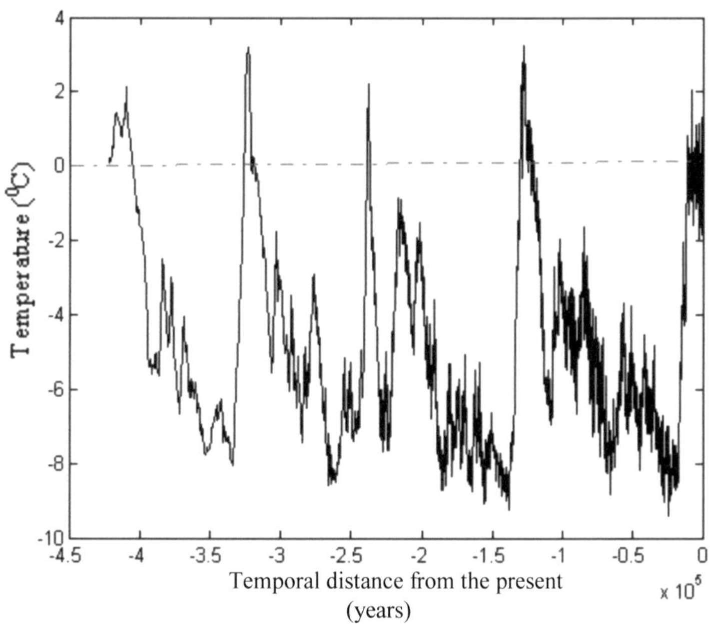

Fig. 7. Reconstructed temperature data from the Vostok Ice Core dataset (Petit et al., 1999). Temperature is the difference from the mean recent time value (red line).

Maybe, the most famous work proposing scale invariance as the main tool for explaining climate variability over millennia is that by Huybers & Curry, (2006). It gathers both historical and paleoclimatic data and discusses their power spectrum within an unified theory based on a fractal continuum of time scales. The estimated variance spectrum is reported in Fig. 8.
The low frequency scaling coefficient for the paleorecord corresponds to a value of the Hurst's coefficient H=0.32, which is signature of anti-persistent fractional Brownian motion. Quite similar results were also found by Pelletier, (1998) who estimated a coefficient compatible with random walk.

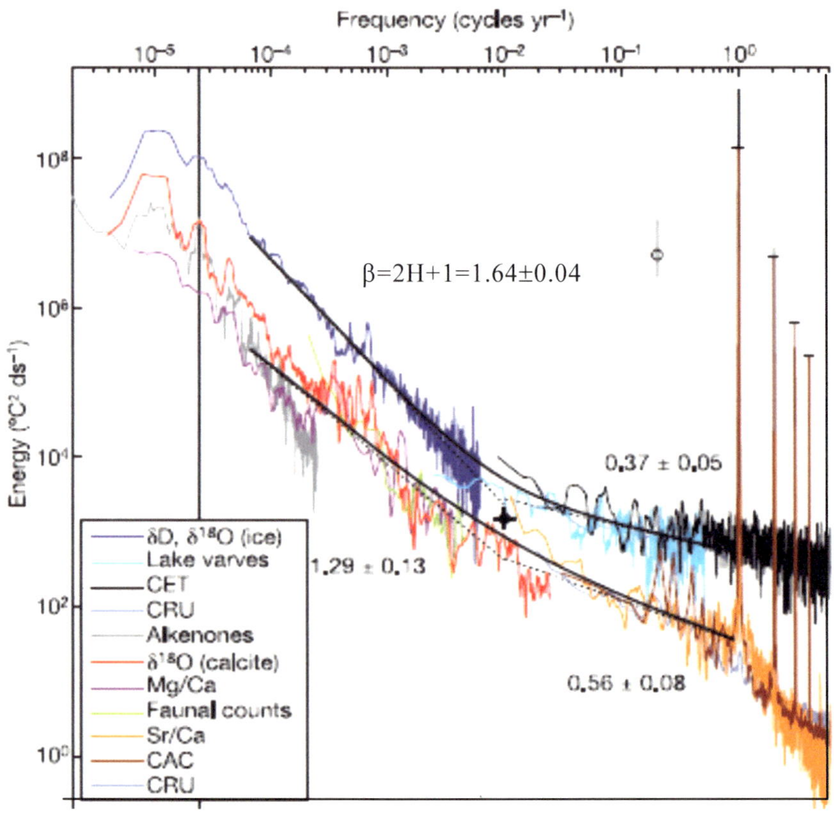

Fig. 8. Sample estimation of the planetary climate spectrum (after Huybers & Curry,2006). The parameter β is the estimated scaling coefficient.

The visual inspection of this spectrum in the low frequency range , so as it is, raises some questions. Differently from the high frequency cycles (annual frequency and sub-harmonics), which appear as spikes well separated from the continuous spectrum of the stochastic component, the millennial cycles are difficult to be separated from noise: it is necessary to know them a priori for interpreting the spectrum correctly. As already specified above, the variance spectrum is not the best tool for investigating complex signals where trends and oscillations could introduce spurious scaling (Gao et al, 2006). Huybers & Curry (2006) estimated the paleorecord scaling in the frequency range between 1/100 and 1/15,000 years to minimize the influence from the Milankovitch bands. Nevertheless, cyclic trends occur in the analysed band too (e. g., Kerr, 1996).

In order to delve into this problem we can investigate time scale laws in the time domain by estimating the second order structure function (Kolmogorov, 1941):

$$\gamma(\tau) = <[X(t+\tau) - X(t)]^2> \tag{14}$$

which is the best statistical tool for studying fractional Brownian motion, since it can be applied to non-stationary data and $\gamma(\tau) \propto \tau^{2H}$ when the time series *X(t)* is an *fBm*. Second order structure function coincides with the variogram used in Geostatistics (Cressie,1993), which is a well known tool for investigating time scale dependence also when data are unevenly sampled.

Fig. 9 illustrates the structure function of the Vostok time series normalised to $2\sigma^2_X$.

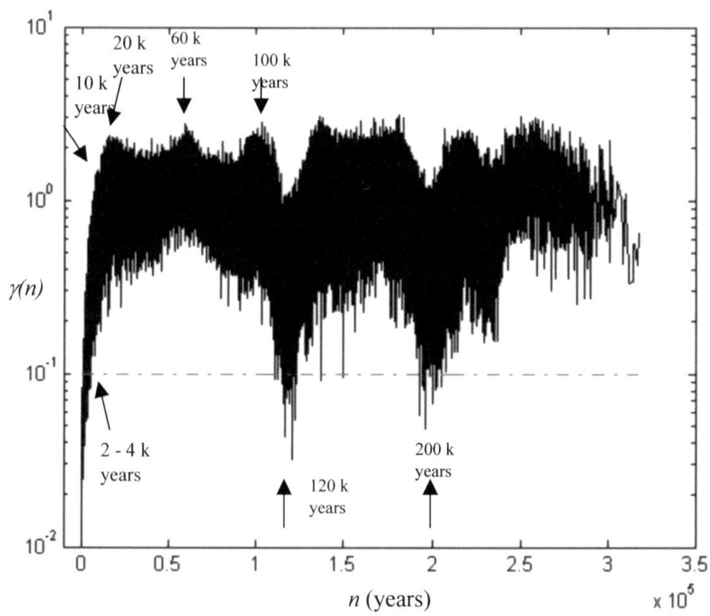

Fig. 9. Second order structure function of the Vostok Ice Core dataset. Arrows indicates approximately the time scales where cycles exhibit minimum or maximum values. The level $\gamma(\tau) = 1$ corresponds to the variance of the total time series.

Differently from the sample spectrum, the structure function reveals long time oscillations explicitly. They are clear in spite of the strong noised character of the estimations due to the uneven and limited sampling etc.. In the time domain, maximum values are associated to odd multiple of semiperiods whereas minimum values correspond to multiple periods. In a composition of cycles and noises, the minimum values reached in the periodic part of $\gamma(n)$ (red dashed line in Fig. 8) mark the percentage contribution due to pure noise. The scale where this plateau is intercepted for the first time (a few thousand of years) marks the crossover between the scales where the truly stochastic noise is observable and that where the contribution of the cycles starts to appear. Then the scaling would occur in a scale range where the non stationary character of the oscillations contaminates the variance of the noise. By looking at Fig. 10 in the temporal range where scaling should appear (red line from 10^2 to 1.5×10^4 years), a direct estimation provides the value H=0.53, which is in a rather good agreement with the estimations of Pelletier (1998). Nevertheless, we can observe that the scales shorter than 10^3 years are evidently not collinear with the subsequent ones. The same is true above 10^4 years, where $\gamma(n)$ appears flatter. If we estimate H by progressively shortening the Huybers & Curry range from the short scale side, its value increases. The same is true if we shorten it by starting from the long time scales. The maximum value $H=0.6$ is obtained about in the middle of the initial range but this does cover not even one decade, which is the minimum requirement for keeping confidence in scale invariance. In addition, we can note that the apparent linearity ends with a maximum value that corresponds to one half period of the ~20 k years oscillation. The central about linear behaviour seems to be an inflection transient between the short time scales (concavity up) and those belonging to cycles where the function $\gamma(n)$ exhibits a different curvature (concavity down).

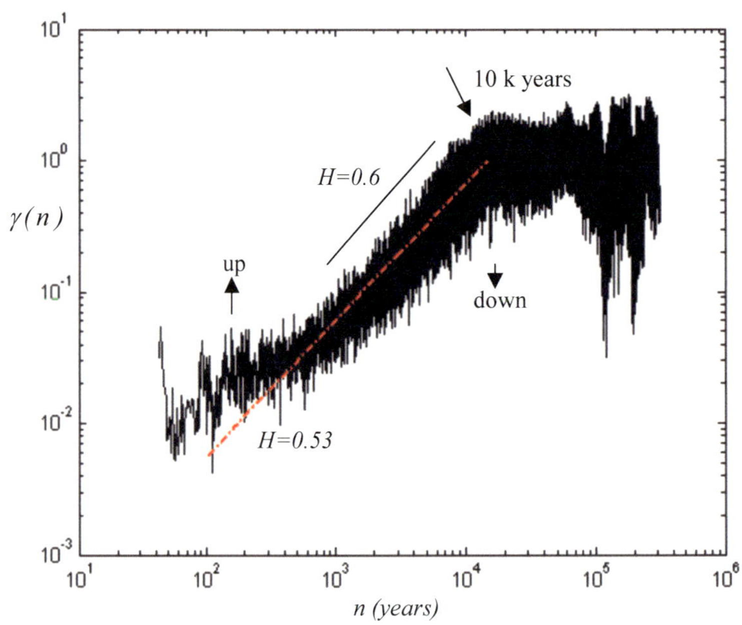

Fig. 10. As Fig. 9 but in double logarithmic scale. The peak at about 10 k years is the maximum anticorrelation associated to a cycle of about 20 k years. The red line is the best fit computed on the time scales indicated by Huybers & Curry (2006): 10^2 years- 1.5×10^3 years. Arrows indicate the concavity semi-planes.

4. Conclusion

In all the studies on the dynamics of the natural world, observational time series play a fundamental role since they are the main source of information for inferring the underlying causal mechanisms. Especially in a stochastic context, when the number of degree of freedom is high, observational time series can provide those time scale laws that rule temporal correlation thus helping us to identify the right reference class of theoretical models. Nevertheless, the interpretation of time laws estimated from real data can be rather difficult because the analysis results can be rather ambiguous in many cases. Dynamical inferences from climate observations fall just in this class. The analysis results illustrated in this chapter put into evidence that no conclusive interpretation of the sample variance spectrum is available yet. It is clear that the analyses performed have to be carefully supervised, preferring those mathematical and statistical tools that are less sensitive to local (in time) disturbances, trends, and cycles that can trick analysts. Spurious scaling can easily appear thus suggesting an erroneous modelling of the deep dynamical characteristics of the climate. Future work should address the problem of temporal persistence not only by demonstrating that climate redness is compatible with scale invariance (necessary condition) but also by demonstrating that it "is not" compatible with a progressive coming out of a few ever slower scale-dependent processes (sufficient condition).

5. Acknowledgment

This work was developed in the framework of "TeRN" project (Tecnologie per le Osservazioni della Terra e i Rischi Naturali) (Rif. Miur DM28424) supported by the Italian Ministry of University and Research (MiUR)

6. References

Blender, R. & Fraedrich, K. (2003). Long time memory in global warming simulations, *Geophys. Res. Lett.*, Vol. 30, No 14, (July 2003),pp. 1769, ISSN 0094-8276.

Blender, R., & K. Fraedrich (2004). Comment on "Volcanic forcing improves atmosphere-ocean coupled general circulation model scaling performance" by D. Vyushin, I. Zhidkov, S. Havlin, A. Bunde, and S. Brenner, *Geophys. Res. Lett.*, Vol. 31, (November 2004), pp.L22213, ISSN 0094-8276.

Bunde, A., Schellnhuber, H.-J., a & J. Kropp (Eds.).(2002). *The Science of Disasters*, SpringerScience, ISBN 3540413243, Potsdam-Germany, 2002

Cressie N., *Statistics for spatial data* (1993), Wiley, NY, pp.900, ISBN 978047100255.

Eichner, J. F., Koscielny-Bunde, E., Bunde, A., Havlin, S., & Schellnhuber, H. J.(2003). Power-law persistence and trends in the atmosphere: A detailed study of long temperature records *Phys. Rev. E*, Vol. 68, (October 2003), pp. 046133, ISSN 1550-2376.

Freeman, M. P., Watkins, N. W., & Riley, D. J. (2000), Evidence for a solar wind origin of the power law burst lifetime distribution of the AE indices, *Geophys. Res. Lett.*, Vol. 27, No 8, (April 2000) pp.1087–1090, ISSN 0094-8276.

Gao, J., Tung, W.-W., Hu, J., Cao, Y., Sarshar, N., & Roychowdhury, V. P. (2006) Assessment of long-range correlation in time series: How to avoid pitfalls, *Phys. Rev. E*, Vol. 73, No 1 , (January 2006), pp. 016117, ISSN 1550-2376.

Govindan, R. B., Vjushin, D., Brenner, S., Bunde, A., Havlin, S., & Schellnhuber, H.-J (2001) Long-range correlations and trends in global climate models: Comparison with real data, *Physica A*, Vol.294. No 1-2, pp. 239–248, ISSN 0378-4371.

Haggerty, R., Wondzell, S. M., & Johnson, M. A. (2002) Powerlaw, residence time distribution in the hyporheic zone of a 2nd-order mountain stream, *Geophys. Res. Lett.*, Vol. 29, (July 2002), No 13, pp. 1640, ISSN 0094-8276.

Hasselmann, K.(1976), Stochastic climate models, Part I, Theory, *Tellus*, Vol. 28, pp. 473–485.

Hayes, D. J., Imbrie, J., Shackleton, N. J., (1976) Variations in the Earth's Orbit: Pacemaker of the Ice Age, *Science, New Series*, Vol. 194, No. 4270, (December 1976), pp. 1121-1132

Hu, K., Ivanov, P. Ch., Chen, Z., Carpena, P., & Stanley, H. E. (2001). Effect of trends on detrended fluctuation analysis, *Phys. Rev. E*, Vol. 64, (June 2001), pp. 011114, ISSN 1550-2376.

Huybers P. & Curry H. (2006). Links between annual, Milankovitch and continuum temperature variability. *Nature*, Vol 44, No 18, (May 2006), pp 329-332, ISSN 0028-0836.

Jenkins, G. M. & Watts D. (1968) *Spectral Analysis and its Applications*, ISBN 0816244642, Holden-Day, pp. 332.

Kantelhardt, J.W., Koscielny-Bunde, E., Rego, H. H. A., Havlin, S., & Bunde, A.(2001) Detecting long-range correlations with detrended fluctuation analysis, *Physica A*, Vol. 295, pp.441–454, ISSN 0378-4371.

Kantelhardt, J. W., Havlin, S., & Ivanov, P. Ch. (2003) Modeling transient correlations in heartbeat dynamics during sleep, *Europhys. Lett.*, Vol. 62, No 2, (April 2003), pp.147–153, ISSN 1286-4854.

Kantelhardt, J. W., Koscielny-Bunde, E., Rybski, D., Braun, P., Bunde, A. & Havlin S. (2006) Long-term persistence and multifractality of precipitation and river runoff records, *J. Geophys. Res.*, Vol. 111, No D1, (January 2006), pp. 1-13, ISSN 01480227.

Kerr R. A., (1996), Climate: Ice Rhythms—Core Reveals a Plethora of Climate Cycles, *Science* Vol. 274 No. 5287 (October 1996) , pp. 499-500, ISSN 1095-9203

Király, A. & Jánosi, I. M. (2002) Stochastic modelling of daily temperature fluctuations, *Phys. Rev. E*, Vol. 65, (April 2002), pp.051102, ISSN 1550-2376.

Klein Tank, A. M. G., Wijngaard, J. B., K¨onnen, G. P., B¨ohm, R., et al.(2002) Daily dataset of 20th-century surface air temperature andprecipitation series for the European Climate Assessment, *Int. J. Climatol.*,Vol 22, pp 1441–1453, ISSN 1097-0088.

Kolmogorov, A. N. *Dokl. Akad. Nauk SSSR Vol.* 32, No. 16 (1941); [an accessible English translation can be found in (1991) *Proc. R. Soc. Lond*on Ser. A Vol 434, pp. 15-17, ISSN 0950–1207]

Koscielny-Bunde, E., Bunde, A., Havlin, S. & Goldreich, Y. (1996) Analysis of daily temperature fluctuations, *Physica A*, Vol. 231, (October 1996), pp. 393–396, ISSN 0378-4371.

Koscielny-Bunde, E., Bunde, A., Havlin, S., Roman, H. E., Goldreich,Y. and Schellnhuber, H-J. (1998) Indication of a Universal Persistence Law Governing Atmospheric Variability, *Phys. Rev. Lett.*, Vol. 81, No3, (January 1998), pp. 729–732, ISSN 1079-7114.

Kurnaz, M. L. Application of Detrended Flucuation Analysis to monthly average of the maximum daily temperature to resolve different Climates (2004), *Fractals*, Vol. 12, No 4, pp. 365–373, ISSN 1793-6543.

Lanfredi M., Simoniello T., Cuomo V. & Macchiato M. (2009), Discriminating low frequency components from long range persistent fluctuations in daily atmospheric temperature variability *Atmos. Chem. Phys.*, Vol. 9, (July 2009), pp. 4537-4544,ISSN 1680-7316.

Mandelbrot, B. B. & van Ness, J. W.: (1968) Fractional Brownian motions, fractional noises and applications, *SIAM Rev.*, Vol. 10, pp 422–437.

Matsoukas, C., Islam, S., and Rodriguez-Iturbe, I.(2000) Detrended fluctuation analysis of rainfall and streamflow time series, *J. Geophys. Res.*, Vol. 105, No D23-29, pp 165–29 172, ISSN 01480227.

Mauran, D., Rust, H.W., & Timmer, J. (2004) Tempting long-memory– on the interpretation of DFA results, *Nonlin. Processes Geophys.*, Vol. 11, (November 2004), pp. 495–503, ISBN 1607-7946.

Metzler, R, (2003) Comment on "Power-law correlations in the southernoscillation-index fluctuations characterizing El Niño", *Phys. Rev. E*, Vol. 67, (January 2003), pp. 018201, ISSN 1550-2376.

Meyers, S. R., Sageman, Bradley B. & Pagani, M. (2008) Resolving Milankovitch: Consideration of signal and noise. *Am J Sci* Vol 308, (June 2008), pp. 770-786, ISSN 1945-452X

Mitchell, J.M. (1976). An Overview of Climatic Variability and Its Causal Mechanism. *Quaternary Research,* Vol.6, No.4, (September 1976) npp. 481-493, ISSN 0033-5894.

Pelletier, J. (1998) The power-spectral density of atmospheric temperature from time scales of 10^{-2} to 10^{6} yr. *Earth Planet. Sci. Lett.* Vol.158, (May 1998), pp. 157--164, ISSN 0012-821X

Peng, C. K., Havlin, S., Stanley, H. E. and Goldberger, A. L. (1995) Quantification of scaling exponents and crossover phenomena in heartbeat time series, *Chaos,* 5(1), 82-87,.

Petit J.R, Jouzel J., Raynaud D. et al. M. (1999), Climate and Atmospheric History of the Past 420,000 years from the Vostok Ice Core, Antarctica, *Nature,* (June 1999)Vol. 399, pp.429-436, ISSN ISSN 0028-0836..

Rust, H. (2006) Interactive comment on "Long-memory processes in global ozone and temperature variations" by C. Varotsos and D. Kirk-Davidoff, *Atmos. Chem. Phys.* Discuss., Vol. 6, (May 2006), pp. S1182–S1185, ISSN 1680-7367.

Simoniello T., Coppola R., Cuomo V., D'Emilio M., Lanfredi M.,Liberti M. & Macchiato M. (2009), Searching for persistence in atmospheric temperature time series: a re-visitation of results from detrend fluctuation analysis, *Int. J. Mod. Phys.* B, Vol 23, (November 2009) No 28-29, pp. 5417-5423, ISSN 0217-9792.

Varotsos, C. & Kirk-Davidoff, D. (2006) Long-memory processes in ozone and temperature variations at the region 60_ S–60_ N, *Atmos. Chem. Phys.*, Vol. 66, (September 2006), pp. 4093–4100, ISSN 1680-7316.

Vecchio A., Carbone V. (2010) Amplitude-frequency fluctuations of the seasonal cycle, temperature anomalies, and long-range persistence of climate records. *Phys. Rev. E* Vol. 82, (December 2010), pp. 066101, ISSN 1550-2376.

Von Storch J. S., Muller P. & Bauer E. (2001) Climate variability in millennium integrations with coupled atmosphere-ocean GCMs: a spectral view. *Climate Dynamics,* Vol. 17, No. 5-6, pp. 375-389, ISSN 1432-0894.

Vyushin, D., Zhidkov, I., Havlin, S., Bunde, A., & Brenner, S. (2004a) Volcanic forcing improves atmosphere-ocean coupled general circulation model scaling performance, *Geophys. Res. Lett.*, Vol. 31,(January 2004), pp. L10206, ISSN 0094-8276.

Vyushin, D., Zhidkov, I., Havlin, S., Bunde, &, and Brenner, S. (2004b) Reply to the Comment on "Volcanic forcing improves atmosphere-ocean coupled general circulation model scaling performance", *Geophys. Res. Lett.*, Vol. 31, (January 2004),pp. L22210, ISSN 0094-8276.

3

The Paleocene-Eocene Thermal Maximum: Feedbacks Between Climate Change and Biogeochemical Cycles

Arne Max Erich Winguth
University of Texas at Arlington
USA

1. Introduction

It is predicted that by the year 2300, the atmospheric CO_2 concentration will exceed ~2000 ppmv (Caldeira & Wickett, 2003; Mikolajewicz et al., 2007), corresponding to a release of 4000 x 10^{15} g carbon (PgC) by fossil fuel emissions and land use changes since the beginning of the industrial revolution. The anthropogenic carbon will eventually sequester on time scales of 100,000 yrs as organic carbon into the ocean and land biosphere and as $CaCO_3$ into the geosphere (Archer et al., 1998). This carbon transfer in the atmosphere-ocean system is comparable to that at the Paleocene-Eocene boundary (55 Ma), when a massive release of carbon into the climate system led to a prominent global warming event referred to as the Paleocene-Eocene Thermal Maximum (PETM). The PETM is characterized by a major (>3.0‰) negative carbon isotope excursion, documented in marine and terrestrial fossils (e.g. Koch et al., 1992; Kelly et al., 1998; Handley et al., 2008), and a worldwide seafloor carbonate dissolution horizon (e.g. Bralower et al., 1997; Lu et al., 1998; Schmitz et al., 1996; E. Thomas et al., 2000) as well as shoaling of the lysocline and carbonate compensation depth (Zachos et al., 2005). These changes are consistent with the release of more than 2000 PgC of isotopically depleted carbon into the ocean-atmosphere system within less than 10,000 years (Panchuk et al., 2008; Zachos et al., 2007, 2008), pointing to a greenhouse gas-driven warming (see Fig. 1). Recent estimates from Cui et al. (2011) indicate a slow emission rate of 0.3-1.7 PgC yr^{-1} as compared to the present-day emission of carbon dioxide of ~9.9 PgC yr^{-1} from fossil fuel emissions (Boden et al., 2010) and land-use changes (Houghton, 2008). Surface temperatures increased by 5°C in the tropics (Tripati & Elderfield, 2005; Zachos et al., 2005) and mid-latitudes (Wing et al., 2005), and by 6-8°C in the ice-free Arctic and sub-Antarctic (Hollis et al., 2009; Kennett & Stott, 1991; Moran et al., 2006; Sluijs et al., 2006, 2007, 2008a, 2011; E. Thomas et al., 2000; Weijers et al., 2007), and deep-sea temperatures increased by 4-6°C (Tripati and Elderfield, 2005; Zachos et al., 2008), relative to Paleocene temperatures (see Fig. 1). At the same time, large-scale changes in the climate system occurred, for example in the patterns of atmospheric circulation, vapor transport, precipitation (Robert & Kennett, 1994; Pagani et al., 2006a; Brinkhuis et al., 2006; Sluijs et al., 2008a, 2011; Wing et al., 2005), intermediate and deep-sea circulation (Nunes & Norris 2006; D.J. Thomas, 2004; D.J. Thomas et al., 2008) and a rise in global sea level (Sluijs et al., 2008b; Handley et al., 2011). The sea level rise is caused by various factors, including thermal expansion, decrease in ocean basin volume, decrease in mountain glaciers, as well as local tectonic changes. Topography and bathymetry during the PETM differed significantly from today with respect to the distribution of landmasses, sizes of ocean basins and width and depth of seaways.

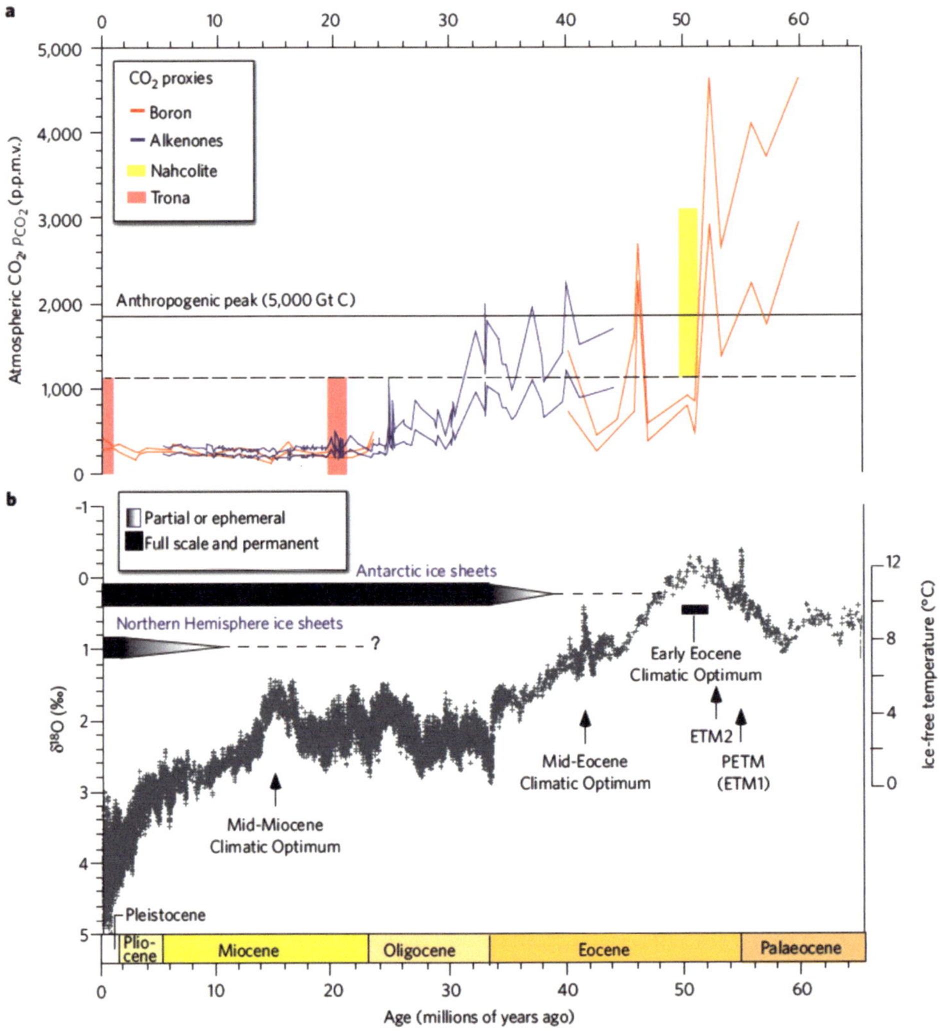

Fig. 1. Evolution of atmospheric pCO_2 concentration and deep-sea temperature reconstruction over the past 65 million years (from Zachos et al., 2008). a) Atmospheric pCO_2 for the period 0 to 65 million years ago. The dashed horizontal line shows the minimum pCO_2 for the early Eocene (1,125 ppmv), as given by calculations of equilibrium with Na–CO_3 mineral phases (vertical bars, where the length of the bars indicates the range of pCO_2 over which the mineral phases are stable) that are found in Neogene and early Eocene lacustrine deposits. The vertical distance between the upper and lower colored lines shows the range of uncertainty for the alkenone and boron proxies. b) Deep-sea benthic foraminiferal oxygen-isotope curve based on records from Deep Sea Drilling Project and Ocean Drilling Program sites. [Reproduced by permission of AAAS; copyright 2008 AAAS.]

2. Climate change and variability at the beginning of the PETM

The causes leading to the warming event at the Paleocene-Eocene boundary are still controversial (Fig. 2).

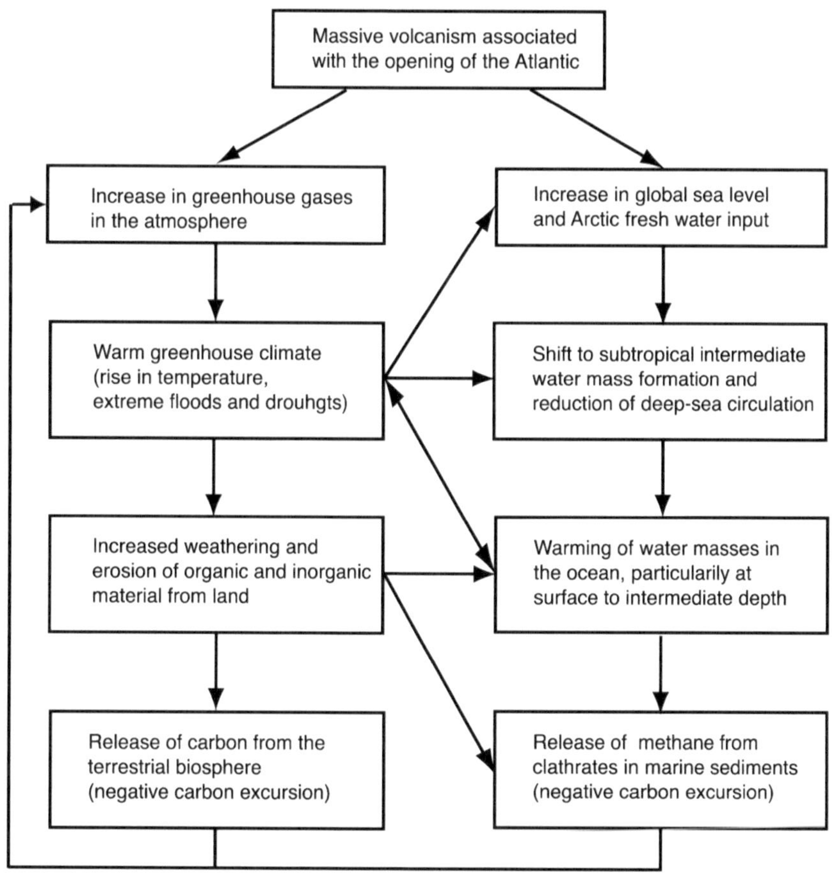

Fig. 2. Major feedbacks for the initial warming at the Paleocene-Eocene boundary. Note that the feedbacks and their magnitudes are still controversial (see e.g. Bowen and Zachos, 2010).

One possible sequence of events inferred from paleoproxies begins with a volcanically induced greenhouse gas (water vapor, CO_2, CH_4, and other constituents) increase that would have produced a global increase in surface temperature (Bralower et al., 1997; Kennett & Stott, 1999; Sluijs et al., 2007, 2011; E. Thomas et al., 2000). Various climate-modeling studies have investigated the warming event at the Paleocene-Eocene boundary in response of the elevated greenhouse gas concentrations. These studies utilized atmospheric general circulation models (Sloan & Barron, 1992; Sloan & Rea, 1995; Huber & Sloan, 1999; Shellito et al., 2003; Shellito & Sloan, 2006), ocean general circulation models (Bice et al., 2000; Bice & Marotzke, 2002), or more recently coupled comprehensive climate models (Heinemann et al., 2009; Huber & Sloan, 2001; Huber & Caballero, 2003; Huber & Caballero, 2011; Lunt et al., 2010; Shellito et al., 2009; Winguth et al., 2010) to simulate the mean climate and its variability during the Eocene, but they have not been able to reproduce the high temperatures of the PETM in the high latitudes, and were controversial regarding the cause of this warming (Pagani et al., 2006b; Zeebe et al., 2009).

Some of the more recent studies have investigated the climate feedbacks with a sequence of different greenhouse gas concentrations (e.g. Heinemann et al., 2009; Lunt et al., 2010; Winguth et al., 2010). In the following, we summarize key findings of the paper of Winguth et al. (2010), using a complex earth system model, the comprehensive Community Climate System Model version 3 (CCSM-3; Collins et al., 2006), in order to investigate PETM climate feedbacks in response to rises in the greenhouse gas concentrations. Huber & Caballero (2011) used the same model, but with a different dust concentration in the atmosphere. The simulated increase by 2.5°C from $4xCO_2$ to $8xCO_2$ in CCSM-3 could be explained by CO_2 emissions due to enhanced volcanic activity at the beginning of the PETM (Fig. 3).

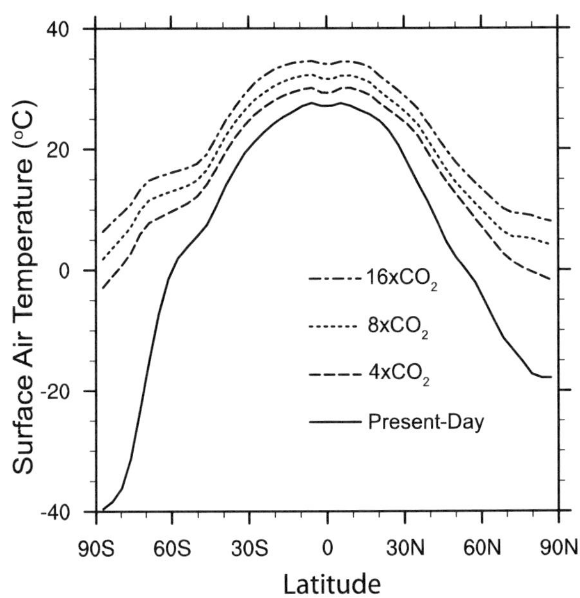

Fig. 3. Zonally averaged (50-yr mean) surface air temperature (in °C) for present-day (solid), $4xCO_2$ PETM (long-dashed), $8xCO_2$ PETM (short-dashed), and $16xCO_2$ PETM (dashed-dotted)(from Winguth et al., 2010).

Surface temperatures in the tropics rise by only ~2°C from $4xCO_2$ to $8xCO_2$, in agreement with temperature reconstructions (Pearson et al., 2007) and with future climate predictions (IPCC, 2007) of a more extreme warming at high latitudes vs. low latitudes in a warmer world. Temperature increase over land exceeds that over the ocean (Fig. 4) due to reduced latent heat fluxes and lower heat capacity. Over the continents, the 30°C isotherm in the $8xCO_2$ simulation reaches up to 30° latitude, about 5° more poleward than for the present-day simulation. Maximum simulated temperatures, comparable to extreme temperatures in the present-day Sahara, are simulated over subtropical Africa and South America (~50°C for $8xCO_2$), resulting in warm sea-surface temperatures in the adjacent oceans through advection. Simulated minimum surface temperatures (for $8xCO_2$) are between 3°C and 7°C over the Arctic and about -10°C over northeast Asia.

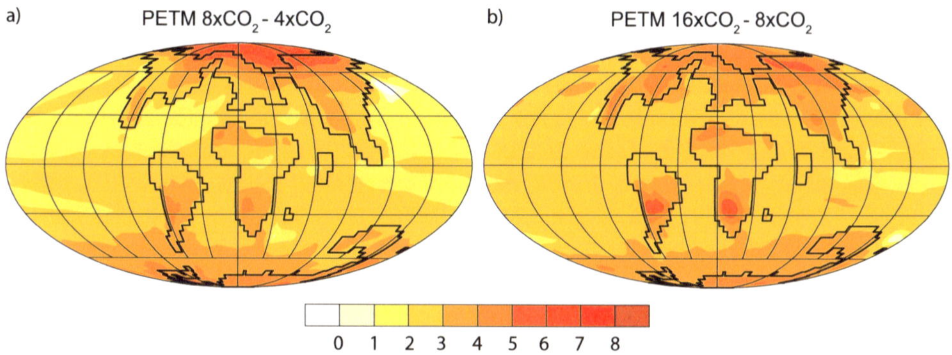

Fig. 4. Surface air temperature (SAT) in °C for a) the difference between the 8x and the $4xCO_2$ PETM experiment corresponding to opening of the Atlantic by massive volcanism and b) differences between the 16x and the $8xCO_2$ PETM experiment by the release of carbon from the marine and terrestrial carbon stocks (100-year mean).

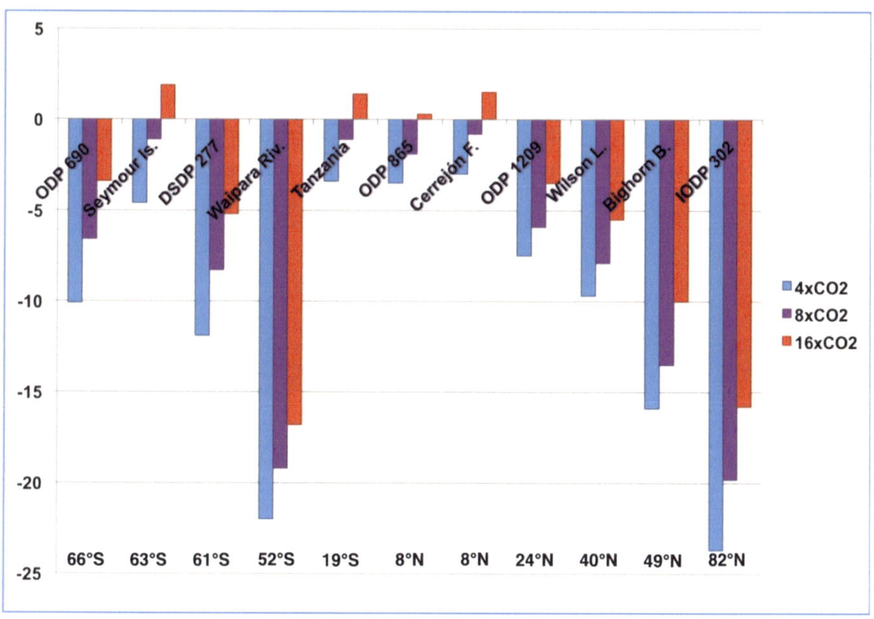

Fig. 5. Differences between reconstructed surface temperatures (in °C) and 50-year annual mean temperature from the CCSM-3 climate simulation with an atmospheric CO_2 concentration of 4x (blue), 8x (purple), and 16x (red) the preindustrial level (see Winguth et al., 2010). For reference, paleolatitude is listed for each location.

While data-inferred paleotemperatures are relatively well represented in the tropical regions, a significant bias between model results and data remains for the Arctic Ocean (Sluijs et al., 2006) and for the area around New Zealand (Fig. 5, Waipara River; Hollis et al., 2009). The bias in the northern polar region (IODP core 302 A; Sluijs et al., 2006) is of complex nature and could for example be associated with the concentration of cloud condensation nuclei used in CCSM-3 (Huber & Caballero, 2011; Kump & Pollard, 2008), the uncertainties in paleolocations (N-S position, or distance form shore), or with skewing of data towards summer temperatures (Sluijs et al., 2006). The causes for model-data discrepancies at high southern latitudes remain controversial.

A positive climate-carbon cycle feedback loop leading to further PETM warming due to destabilization of methane hydrates is shown in Fig. 2. There is sufficient evidence from various sites around the globe, including the New Jersey shelf (Sluijs et al., 2007), the North Sea (Bujak & Brinkhuis, 1998; Sluijs et al., 2007), the Southern Ocean (Kennett and Stott, 1991), and New Zealand (Hollis et al., 2009) that ^{13}C-depleted carbon in the form of isotopically light CO_2 and/or CH_4 was released from the sea floor (Dickens et al., 1995, 1997; Higgins & Schrag, 2006; Pagani et al., 2006a) or from wetlands (Pancost et al., 2007) into the atmosphere-ocean-biosphere system (Sluijs et al., 2007; Bowen and Zachos, 2010). As shown in Figs. 3 and 4, such a change in the radiative forcing from 8xCO_2 to 16xCO_2 leads to a simulated additional warming of ~2°C globally, with 4°C at the poles, 5°C over South America and South Africa, and 2°C at the equator. For the Southern Ocean, cool water masses moderate the climate over the polar southern hemisphere, so that south of 60°, the temperature increase in response to the increase of CO_2-radiative forcing is smaller than in the northern hemisphere (Fig. 4b). In mid-latitudes, the bias between the 16xCO_2 simulation and reconstructed PETM surface temperatures is reduced compared to simulations with a lower atmospheric CO_2 level with high dust concentration; the values are comparable to the 8xCO_2 scenario with lower dust concentration in Huber & Caballero (2011). For the tropics, evidence from fossil remains of a giant boid snake in northeastern Colombia (Head et al., 2009) and modeling studies (Winguth et al., 2010; Huber & Caballero, 2011) support warm average temperatures of 30-34°C.

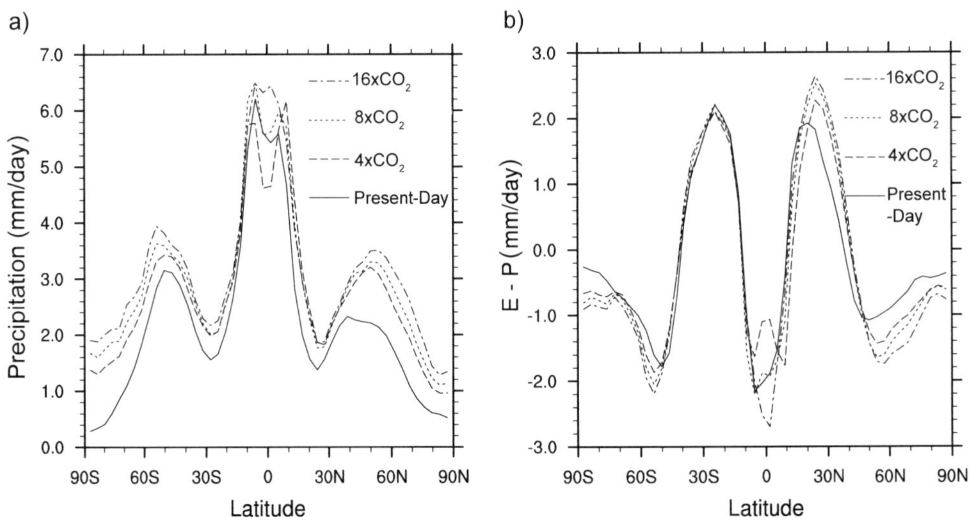

Fig. 6. Zonally averaged (50-yr mean) precipitation (in mm/day) (a) and evaporation minus precipitation (in mm/day) (b) for present-day (solid), 4xCO_2 PETM (long-dashed), 8xCO_2 PETM (short-dashed), and 16xCO_2 PETM (dashed-dotted) (from Winguth et al., 2010). The increase in global warming during the PETM leads to extremes in the hydrological cycle (droughts in the subtropics and higher precipitation and flooding in the tropics and high latitudes).

Rapid warming at the beginning of the Eocene has been inferred from the widespread distribution of dinoflagellate cysts (or dynocysts). The abundance of one dynocyst species, Apectodinium, dramatically increased at different locations worldwide (Bujak & Brinkhuis, 1998; Crouch et al., 2001; Heilmann-Clausen & Egger, 2000; Sluijs et al., 2007), implying a change in environmental conditions such as warmer sea surface temperatures and increased food availability in form of phytoplankton (Burkholder et al., 1992) due to increased nutrient delivery by weathering (Ravizza et al., 2001; Zachos & Dickens, 2000) and erosion (Fig. 2).
The climate-carbon cycle feedback associated with an increase in greenhouse gases (Fig. 2) might also have been enhanced by an increase in the atmospheric water vapor fluxes (Figs. 6 and 7); for instance, latent heat flux by evaporation and precipitation rises with the warming of the surface (Fig. 6). Higher precipitation and lower sea surface salinity values are derived for the Arctic from isotopic measurements as well as from dinocyst assemblages (Pagani et al., 2006a; Sluijs et al., 2008a). The enhanced precipitation at high latitudes is consistent to patterns simulated for future climate scenarios (e.g. Cubasch et al., 2001; Meehl et al., 2006; Mikolajewicz et al., 2007). For the southern high latitudes, a simulated increase in precipitation is confirmed by clay-mineral indicators from the Antarctic continent, pointing towards humid conditions at the PETM (Robert & Kennett, 1994). Compared to present-day, differences in the geography and mountain height cause remarkable changes. For instance, a higher than present-day ratio of tropical land-to-ocean area at the PETM reduces the tropical ocean surface and hence the oceanic source of atmospheric moisture (Barron et al., 1989). This change in tropical surface area not only reduces significantly tropical precipitation, but also poleward moisture transport from the tropics. However, increase in precipitation by higher than present-day greenhouse gases counteracted this effect during the PETM.

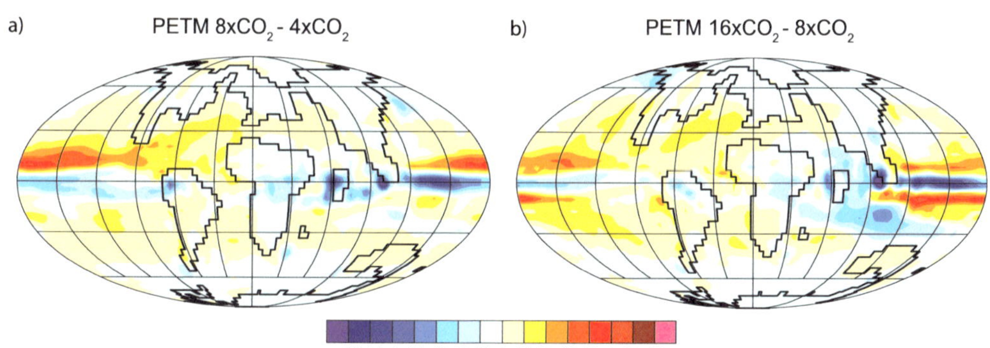

Fig. 7. Evaporation minus precipitation (in mm/day) for a) the difference between the 8x and the 4xCO_2 PETM experiment, and b) the difference between the 16x and the 8xCO_2 PETM simulation. The increase in global warming during the PETM leads to extremes in the hydrological cycle (increases in aridity in the subtropics are shaded red, and in humidity in the tropics are shaded blue).

An initial increase in CO_2 in the atmosphere by volcanic outgassing would have increased the strength of the hydrological cycle. Model simulations suggest that the subtropics at ~30° became drier and that precipitation at 60° increased significantly (Fig. 6), which is consistent to future climate projections (IPCC, 2007). Over North America during summer, the simulated total amount of rainfall decreases from lower to mid-latitudes in response to a northward-directed monsoonal moisture transport over the Mississippi watershed from the Gulf (Sewall & Sloan, 2006; Winguth et al., 2010). Sedimentary records from the mid-latitudes of the North American continent have produced conflicting evidence for hydrological changes in this region. For example, a ~25% increase in relative humidity for the northern continental mid-latitudes (Bighorn Basin, Wyoming, paleolatitude ~49 °N) has been inferred from an amplification in the carbon isotope excursion in soil organic matter (Bowen et al., 2004), but vegetation analysis inferred a decrease of ~40% in precipitation at the beginning of the PETM (Wing et al., 2005). Drier PETM conditions occurred probably in Utah (USA, paleolatitude ~45 °N; Bowen & Bowen, 2008); these findings are, however, controversial, since other studies (Retallack, 2005) suggest enhanced rainfall for this region (Bowen & Bowen, 2009; Retallack, 2009). Droughts by reduced soil moisture and biomass burning through wildfires in the subtropics during the PETM could have provided a significant carbon release to the atmosphere (Fig. 7).

In western Europe, sedimentary records from the Spanish Pyrenees (Schmitz & Pujalte, 2007) indicate seasonally increased precipitation during the PETM, leading to enhanced runoff into the Tethys Ocean and thus enhanced productivity by a rise in nutrient availability in the near-shore areas (Schmitz et al., 1996; Speijer & Wagner, 2002; Gavrilov et al., 2003). Increased precipitation over England (~+1 mm/day change from $4xCO_2$ to $16xCO_2$) would also have generated a feedback on the carbon cycle, for example enhanced carbon emission from wetlands (Pancost et al., 2007).

3. Feedbacks associated with the PETM ocean circulation

In this section, two feedback loops involving the carbon cycle and climate are discussed. The first is associated with the rise of greenhouse gas concentrations (water vapor, CO_2, CH_4, and other gases) in the atmosphere due to tectonic changes such as volcanism (Bralower et al., 1997; Kennett & Stott, 1991; Lyle et al., 2008; Sluijs et al., 2007; Storey et al., 2007; Svensen et al., 2004) and the second with the response of the climate system to regional or global sea level change by tectonic uplift and climatic changes (Fig. 2). Evidence of marine transgression during the PETM (Handley et al., 2011; Maclennan & Jones, 2006; Schmitz & Pujalte, 2003; Sluijs et al., 2008b) related to changes in spreading rate, volcanism, and regional perturbations as well as climatic changes (melting of glaciers, thermal expansion, and changes in ocean circulation) suggests that sea levels rose by approximately 20-30 m. The increase in surface temperatures and freshening of the sea surface by enhanced poleward moisture transport in response to a rise of greenhouse gases changes the regional buoyancy and momentum fluxes, leading to changes in vertical density gradients and stratification of the deep sea. Warmer and more saline subtropical water masses are modeled associated with the initial PETM warming (~0.2 psu higher in salinity for the $8xCO_2$ than for the $4xCO_2$ experiment), originating near the Gulf of Mexico and mixed via the eastern North Atlantic into intermediate layers. For intermediate water masses in the North Atlantic Ocean, simulated temperature rises by ~4°C (from ~11°C to ~15°C) (Fig. 8a).

In the Pacific, the increase of the atmospheric CO_2 to $8xCO_2$ results in an increase in the vertical density gradient, since surface waters become significantly lighter with the warming. Deep-sea temperatures increase by ~2.5°C due to the global warming (Fig. 8b). The simulated Pacific circulation in the $4xCO_2$ scenario is nearly symmetric about the equator, with deep-sea ventilation occurring in the polar regions of the northern and

southern hemisphere (Fig. 9a), in agreement with analyses of Nd isotope data that indicated a bimodal ventilation (D.J. Thomas et al., 2008). The northward-directed Atlantic deep-sea circulation of ~4 Sv (1 Sv = 10^6 m^3 s^{-1}) in the $4xCO_2$ scenario with a source of deep-water formation in the South Atlantic is comparable in strength with the modern but reversed.
With an increase of the CO_2-radiative forcing to $8xCO_2$, the ventilation of the deep sea is reduced and the age of water masses in intermediate depth is increased (Fig. 9b), particularly in the southern high latitudes. The Atlantic deep-sea circulation in the $8xCO_2$ scenario remains reversed, in agreement with Zeebe & Zachos (2007), who used inferred $[CO_3^{2-}]$ gradients in the deep sea, but in contrast with the reconstruction of an abrupt shift in the deep-sea circulation during the PETM to a North Atlantic deep-water source, based on benthic carbon isotope records (Nunes & Norris, 2006).

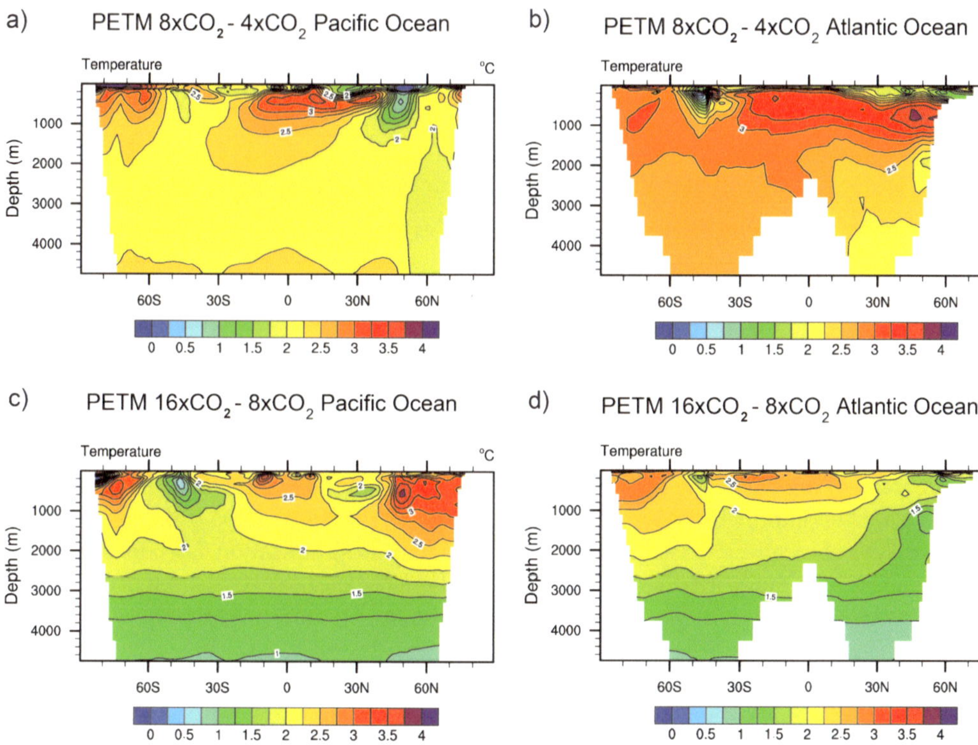

Fig. 8. Vertical sections of the potential temperature (50-year mean) differences of the $8xCO_2$ and $4xCO_2$ PETM experiments for the Pacific Ocean (a) and the Atlantic Ocean (b) for the beginning of the warming, and differences of the $16xCO_2$ and the $8xCO_2$ PETM experiments for the Pacific Ocean (c) and the Atlantic Ocean (d) (from Winguth et al., 2010).

The warming of intermediate and deep-water masses could have had a positive feedback on the ocean circulation (Fig. 2), as proposed in Bice & Marotzke (2002). The warming of the ocean by changes in the buoyancy forcing (heat and freshwater fluxes) and circulation lowers the depth of methane hydrate stability, which depends on pressure, temperature, salinity, and gas composition, from ~900 m to ~1500 m (Fig. 10; Dickens et al., 1995). This change might have triggered a massive methane hydrate release into the atmosphere-ocean system, which in turn accelerated the global warming (Archer & Buffett, 2005). The potential consequences of such an amplification are displayed in Figs. 8c and d, for the assumption that the carbon release corresponded to ~4400 PgC (16xCO_2 experiment) relative to the 8xCO_2 experiment. A temperature increase of >3.5°C is simulated for high-latitude intermediate water masses in the Pacific due to an increase in vertical density gradients. The increase in the ideal age of water masses is shown in Fig. 9c.

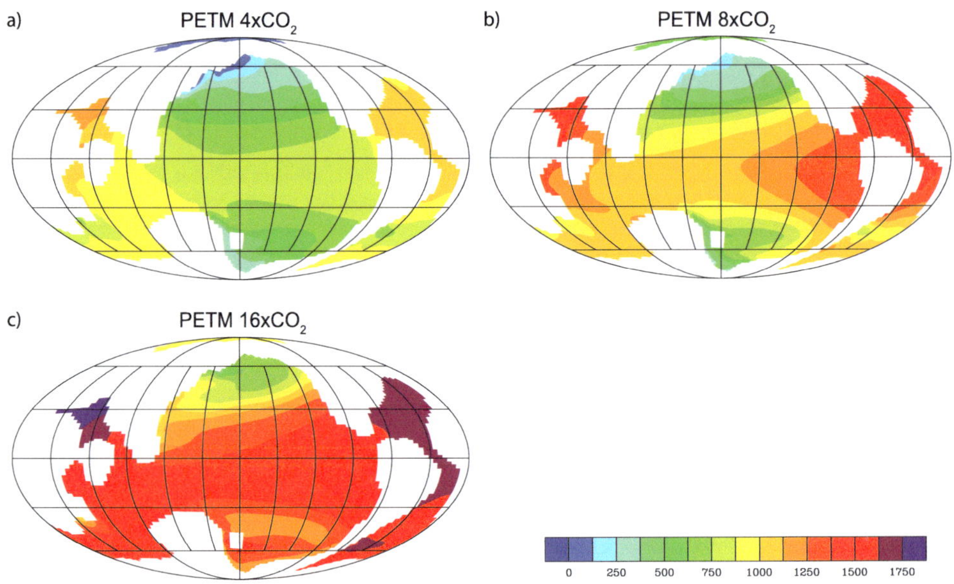

Fig. 9. Change of residence time of water masses in ~1800 m depth (idealized age in yrs) for a) 4xCO_2 PETM simulation, b) 8xCO_2 PETM simulation, and c) 16xCO_2 PETM simulation (100-yr mean). Increase of idealized age corresponds to an increase in stratification. Locations of deep-sea ventilation are in the Northern and Southern Pacific. The deep-sea ventilation decreases particularly in the southern ocean with higher atmospheric CO_2 radiative forcing.

The feedback loop associated with the sea level changes at the PETM would have affected the oceans by an enhanced freshing from Arctic Ocean. An increased flow via the Turgay Strait, the passage between the Arctic and Tethys Ocean, has been inferred from the abundance of dinoflagellate cysts (e.g. Iakokleva et al., 2001). Higher sea levels might also have allowed a throughflow via the Fram and Bering Straits, as supported by Nd-Sr isotopes in fish fossils (Gleason et al., 2009; Roberts et al., 2009), paleogeographic reconstructions (Scotese, 2011) and climate simulations (Cope & Winguth, 2011; Heinemann et al., 2009).

A freshwater input from the Arctic Ocean into the North Pacific Ocean (Marincovich & Gladenkov, 1999) would have produced an increase in the vertical density gradients and led to a weakening of the North Pacific intermediate water masses by 2.5 Sv at 30°N, and a comparable increase in the Pacific deep-sea circulation. The opening of the Bering Strait would have shifted formation of intermediate water masses in the North Pacific more equatorward towards the arid subtropics, by that increasing temperature and salinity of intermediate water masses (Fig. 11). Such a temperature change in intermediate waters at the beginning of the PETM warming could have contributed to the release of methane hydrates (e.g. Kennett & Stott, 1991; Sluijs et al., 2007). Most of the methane released from the hydrates would have ultimately reached the atmosphere or oxidized as CO_2 and thus increased the greenhouse gas radiative forcing during the PETM (Fig. 2).

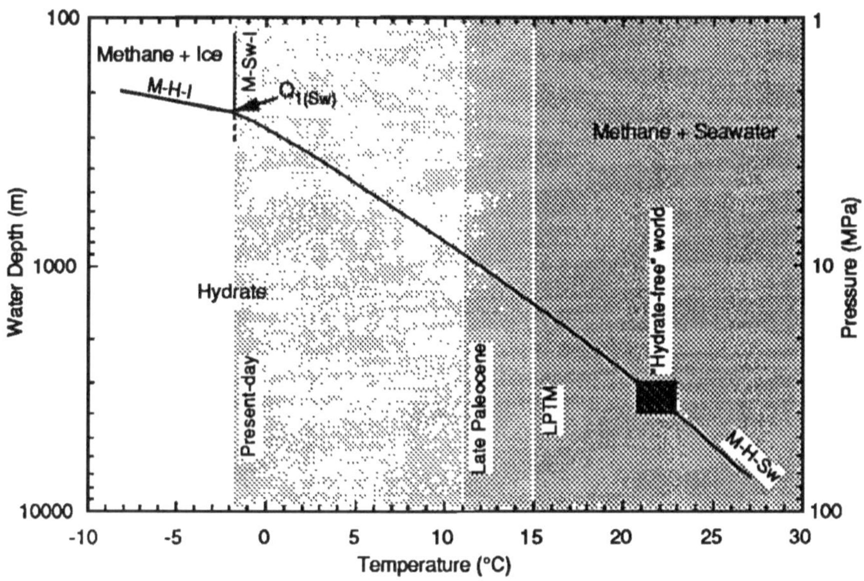

Fig. 10. Methane-hydrate temperature-depth (pressure) diagram from Dickens et al. (1995), adapted after Dickens & Quinby-Hunt (1994). The triple point (Q_1) is the point where all three phases (methane and sea ice, methane and sea water, methane hydrates) meet. Above the triple point to the left are conditions where methane and sea ice and to the right where methane and seawater exist. The area below the curve denotes the conditions under which methane hydrates are stable (at present-day with bottom water temperatures of -1.5 °C and depths below 250 m). For the pre-PETM, the critical depth below which methane hydrates were stable was around 900 m. A 4°C water temperature increase at the PETM would have lowered the critical depth by ~600 m to ~1500 m. [Reproduced by permission of American Geophysical Union; copyright 1995 American Geophysical Union.]

4. Feedbacks associated with the atmospheric chemistry during the PETM

Many potentially important feedback processes are associated with atmospheric chemistry (Beerling et al., 2007).

Possible changes associated with clouds at the beginning of the PETM are for example cloud albedo, cloud optical depth, or heat transport by tropical cyclones. Clouds interfere with the transfer of radiation because they reflect a certain amount of radiation back to space and they act as a blanket for thermal radiation. The reflectivity of clouds is influenced by cloud condensation nuclei (CCN). While today's major source for CCN over land is due to pollution, CCN concentrations over remote ocean areas are linked to marine productivity via dimethyl sulfide (DMS) emission from the ocean. DMS emitted from certain phytoplankton groups is mixed into the troposphere and is oxidized to sulfate particles, which then act as CCN for marine clouds. The CCN concentration affects cloud droplet size and distribution, which influences cloud reflectivity and hence the climate. Climate change on the large scale, in turn, affects the ocean circulation, nutrient cycles and consequently the phytoplankton concentration in the oceans and thereby closes via DMS emission the feedback loop, as first hypothesized by Charlson et al. (1987). If global productivity had declined during the PETM by ocean stagnation and reduced equatorial upwelling, the concentration of CCN would also have been reduced.

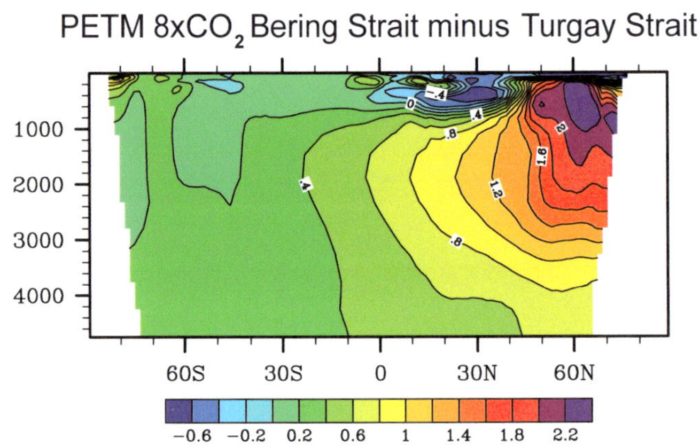

Fig. 11. Vertical section of the potential temperature (50-year mean) difference of the 8xCO_2 PETM simulation with a passage between the Arctic Ocean and the Pacific (Bering Strait) minus the simulation with a passage between the Arctic Ocean and the Indian Ocean. The changes in throughflow might have been caused by sea level rise due to tectonic and climatic changes (Fig. 2). A change of freshwater input from the Arctic might ultimately have caused warming of water masses and thus triggered a positive feedback loop between the climate and the carbon cycle.

This would have affected the cloud optical depth (Kump and Pollard, 2008), leading to high-latitude warming and a further increase in ocean stratification and stagnation of the deep-sea circulation, probably similar to the one modeled in the 16xCO_2 experiment by Winguth et al. (2010). Polar stratospheric clouds (Sloan & Pollard, 1998; Kirk-Davidoff et al., 2002) or intensified tropical cyclone activity (Korty et al., 2008) could have further exaggerated warming at the PETM.

Another feedback between the carbon cycle and the climate that may have played an important role during the PETM are volatile organic compounds (VOCs; Beerling et al., 2007). VOCs are emitted by plants, for example isoprene with present-day emission rates comparable to that of methane (Guenther et al., 2006; Prather & Erhalt, 2001). Isoprene is a major player in the oxidative chemistry of the troposphere and influences the formation of tropospheric ozone (Fehsenfeld et al., 1992), decreases the hydroxyl radical concentration, increases the residence time of CH_4, and is involved in forming organic aerosols influencing the climate by acting as CCN (Beerling et al., 2007).

High CH_4 emissions during the PETM could have increased the atmospheric methane concentration and enhanced radiative forcing (with a ~21 times higher global warming potential than CO_2 over a time span of 100 years; IPCC, 1990). Methane in the atmosphere is typically either reduced by oxidation to CO_2 or interacts with other chemical components.

Emission scenarios for the PETM considering atmospheric chemistry involving NOx and ozone reactions indicate that the life-time of methane in the atmosphere increases with increasing emission of methane, thus leading to an increased radiative forcing influencing the climate and methane hydrate destabilization within a positive feedback loop (Schmidt & Shindell, 2003).

5. Feedbacks associated with weathering during the PETM

While the feedbacks listed in the previous sections illustrate the complexity of the PETM warming, the rapid recovery phase after the CIE remains controversial as well. Rapidly regrowing organic carbon stocks on land and in the ocean on climatic time scales <10^4 years may have contributed to a draw-down of the atmospheric CO_2 concentration (Bowen & Zachos, 2010), thus creating a positive feedback between a cooler climate and a more vigorous ocean circulation with reduced vertical density gradients and enhanced ventilation from high latitudes (Fig. 12). Intensification of wind-driven upwelling and enhanced high-latitude mixing stimulate global productivity through higher nutrient availability in the euphotic zone. Such an increase in the productivity (Bains et al., 1999; Stoll et al., 2007; Sluijs et al., 2006) could eventually have accelerated the draw-down of the atmospheric CO_2.

On longer geologic timescales (>10^4 yrs), carbon sequestration by weathering of continental rocks becomes an important process. Atmospheric CO_2 and H_2O reacts with rocks and is converted into aqueous bicarbonate that is transferred to the oceans via riverine discharge and eventually deposited on the seafloor as biogenic carbonates (Walker et al., 1981; Berner, 2004).

The hothouse climate during the PETM with an increase in precipitation and plant growth likely accelerated weathering. The associated large input of dissolved bicarbonates into the ocean would have neutralized the oceans' acidity and led to post-CIE deepening of the lysocline (Zachos et al., 2005), and preservation of calcareous marine sediments (Fig. 12; Kelly et al., 2010). This negative weathering feedback would ultimately have led to a draw-down of atmospheric CO_2, climatic cooling and reduced weathering.

6. Conclusive remarks

The PETM, represented by the largest perturbation in climate and carbon cyle during the last 60 million years (Fig. 1; Pearson & Palmer, 2000; Royer et al., 2007) can be considered as an analog for future climate change. The analysis of ice bubbles trapped in the Antarctic suggests a variability of the atmospheric CO_2 concentration over the last 800,000 yrs ranging from 172 ppmv to 300 ppmv for the preindustrial period. As a result of human activities, CO_2 in the atmosphere rose over the last couple of hundred years with a pace not seen in recent geological history. In the year 2011, the atmospheric CO_2 concentration exceeded 390 ppmv (Tans & Keeling, 2011), and a doubling of the pre-industrial atmospheric CO_2 level is expected by the end of this century. The climate sensitivity for this doubling in CO_2 is estimated to be 1.9–6.2 K due to the positive forcings, i.e. the rise in greenhouse gases, and including the negative forcing arising from the cooling effects of aerosols (IPCC, 2007; Andreae, 2007). A release of ~2000 PgC into the atmosphere in the next couple of hundred years could eventually trigger the release of an additional 2000-4000 PgC from marine sediments (Archer & Buffett, 2005), a flux comparable to that observed at the PETM (Zachos et al., 2008) and more than 10 times higher than observed during the last million years. The additional carbon release would act as a positive feedback, accelerating the warming.

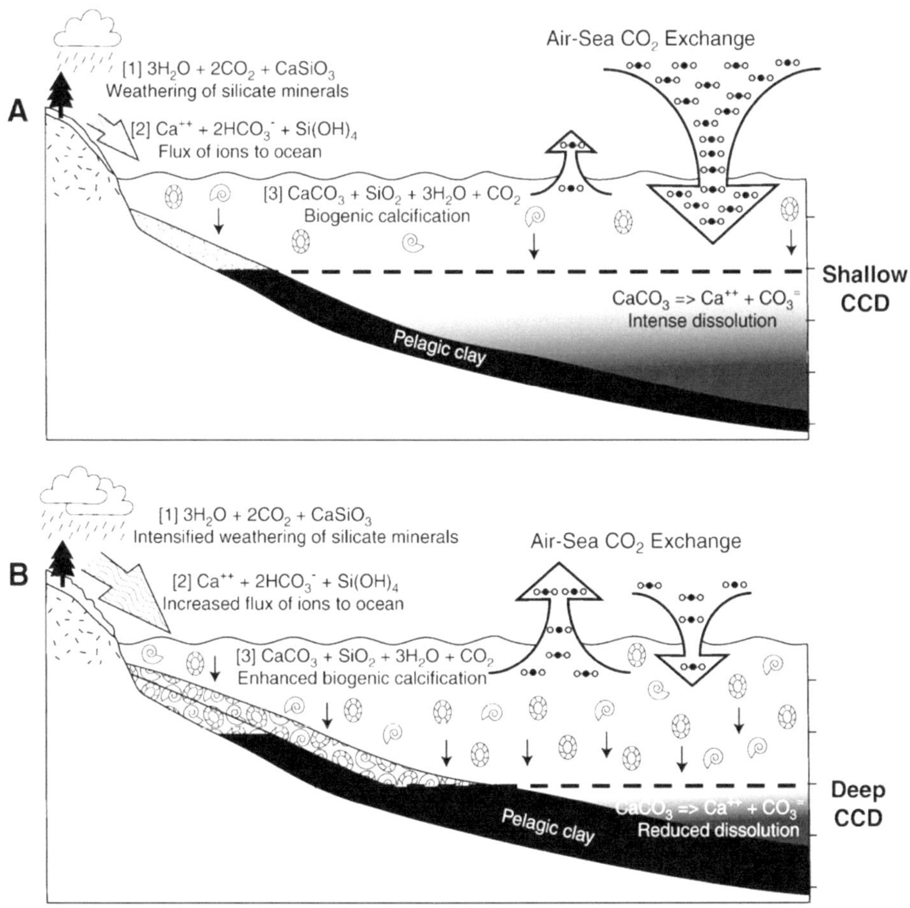

Fig. 12. Schematic changes in the carbonate-silicate geochemical cycle associated with the PETM (from Kelly et al., 2010). a) A rapid release of massive amounts of carbon into the ocean–atmosphere-biosphere system raises atmospheric pCO_2 levels, increases the carbon flux into the oceanic reservoir, thus raises the calcium carbonate compensation depth (CCD), and reduces the biogenic calcification. Preservation of carbonates is restricted to shallow areas on the seafloor. b) In the recovery phase of the atmosphere-ocean-biosphere system, silicate weathering is accelerated, reducing the atmospheric partial pressure of CO_2 and increasing the flux of dissolved bicarbonate ions and silicic acid to the ocean, thus neutralizing ocean acidification, and leading to a deepening of the CCD and preservation of carbonates in deeper areas on the seafloor. [Reproduced by permission of Elsevier; copyright 2010 Elsevier.]

The climatic and biogeochemical response to remarkable carbon emissions would likely be severe, for example a more frequent occurrence of climate extremes (heat waves, droughts and floods), particularly over the continents and at high latitudes, as well as ocean warming and stagnation. Another likely effect is ocean acidification and a rise of the calcite dissolution depth (Zachos et al., 2005), affecting marine organisms with calcareous shells (E. Thomas, 1998, 2003, 2007). The increased vertical gradients in the ocean together with warmer temperatures might produce near-anoxic conditions in the oxygen minimum zone (comparable with dead zones in the Black Sea or the Gulf of Mexico). Geochemical evidence for the PETM supports a downward expansion of the oxygen-minimum zone below 1500 m (Chun et al., 2010; Nicolo et al., 2011) in agreement with foraminiferal evidence.

7. Acknowledgments

All model simulations were done on NCAR computers, supported by NSF. The work is supported by NSF Grant EAR-0628336.

8. References

Andreae, M.O., 2007: Atmospheric aerosols versus greenhouse gases in the twenty-first century. *Phil. Trans. R. Soc. London, A*, 365, 1915–1923. doi:10.1098/rsta.2007.2051.

Archer, D., & Buffett, B., 2005: Time-dependent response of the global ocean clathrate reservoir to climatic and anthropogenic forcing. *Geochemistry, Geophysics, Gesosystems*, 6, Q03002, doi:10.1029/2004GC000854.

Archer, D., Kheshgi, H., & Maier-Reimer, E., 1998: Dynamics of fossil fuel CO_2 neutralization by marine CaCO3. *Global Biogeochem. Cycles*, 12, 259–276.

Bains, S.R., Corfield, R., & Norris, R.D., 1999: Mechanisms of climate warming at the end of the Paleocene. *Science*, 285, 724–727, doi: 10.1126/science.285.5428.724.

Barron, E.J., Hay, W.W. & Thompson, S.,1989: The hydrologic cycle: a major variable during Earth history. *Palaeogeogr., Palaeoclimatol., Palaeoecol.*, 75, 157–174.

Beerling, D.J., Hewitt, C.N., Pyle, J.A., & Raven, J.A., 2007: Critical issues in trace gas biogeochemistry and global change. *Phil. Trans. R. Soc. London, A*, 365, 1629–1642.

Berner, R.A. (Ed.), 2004: The Phanerozoic Carbon Cycle: CO_2 and O_2, Oxford University Press, 150 pp.

Bice, K.L., & Marotzke, J., 2002: Could changing ocean circulation have destabilized methane hydrate at the Paleocene/Eocene boundary? *Paleoceanography*, 17, doi:10.1029/2001PA000678.

Bice, K.L., Scotese, C.R., Seidov, D., & Barron, E.J., 2000: Quantifying the role of geographic change in Cenozoic ocean heat transport using uncoupled atmosphere and ocean models. *Palaeogeogr,. Palaeoclimatol., Palaeoecol.*, 161, 295-310.

Boden, T.A., Marland, G., & Andres, R.J., 2010: Global, Regional, and National Fossil-Fuel CO_2 Emissions. Carbon Dioxide Information Analysis Center, Oak Ridge National Laboratory, U.S. Department of Energy, Oak Ridge, Tenn., U.S.A. doi: 10.3334/CDIAC/00001_V2010.

Bowen, G.J., & Bowen, B.B., 2008: Mechanisms of PETM global change constrained by a new record from central Utah. *Geology*, 36, 379-382.

Bowen, G.J., & Bowen, B.B., 2009: Mechanisms of PETM global change constrained by a new record from central Utah: Reply. *Geology*, 37, e185.

Bowen, G.J., & Zachos, J.C., 2010: Rapid carbon sequestration at termination of the Palaeocene-Eocene Thermal Maximum. *Nature Geoscience*, 3, 866-869, doi:10.1038/NGEO1014.

Bowen, G.J., Beerling, D.J., Koch, P.L., Zachos, J.C., & Quattlebaum, T., 2004: A humid climate state during the Paleocene/Eocene thermal maximum. *Nature*, 432, 495-499.

Bralower, T.J., Thomas, D.J. Zachos, J.C., Hirschmann, M.M., Röhl, U., Sigurdsson, H., Thomas, E., & Whitney, D.L., 1997: High-resolution records of the late Paleocene thermal maximum and circum-Caribbean volcanism: Is there a causal link? *Geology,* 25, 963-967.

Brinkhuis, H., et al., 2006: Episodic fresh surface waters in the Eocene Arctic Ocean. *Nature,* 441, 606-609.

Bujak, J.P., & Brinkhuis, H., 1998: Global warming and dinocyst changes across the Paleocene/Eocene epoch boundary. In: *Late Paleocene-early Eocene biotic and climatic events in the marine and terrestrial records*, Aubry, M.-P., Lucas, S., & Berggren, W.A. (Eds.), p. 277-295, Columbia University Press.

Burkholder, J.M., Noga, E.J., Hobbs, C.H., & Glascow, H.B., 1992: New phantom dinofagellate is the causative agent of major estuarine fish kills. *Nature,* 358, 407-410.

Caldeira, K., & Wickett, M.E., 2003: Anthropogenic CO_2 and ocean pH. *Nature,* 425, 365.

Charlson, R.J., Lovelock, J.E., Andreae, M.O. & Warren, S.G. 1987: Oceanic phytoplankton, atmospheric sulphur, cloud albedo and climate. *Nature,* 326, 661–665, doi:10.1038/326655a0.

Chun, C.O.J., Delaney, M.L., & Zachos, J.C., 2010: Paleoredox changes across the Paleocene-Eocene thermal maximum, Walvis Ridge (ODP Sites 1262, 1263, and 1266): Evidence from Mn and U enrichment factors. *Paleoceanography,* 25, doi:10.1029/2009PA001861.

Collins, W.D., et al., 2006: The Community Climate System Model Version 3 (CCSM3). *J. Climate,* 19, 2122-2143.

Cope, J.T., and Winguth, A., 2011: On the sensitivity of the Eocene ocean circulation to Arctic freshwater pulses. *Palaeogeogr., Palaeoclimatol., Palaeoecol.,* 306, 82–94.

Crouch, E.M., Heilmann-Clausen, C., Brinkhuis, H., Hugh E.G., Morgans, H.E.G., Rogers, K.M., Hans Egger, H., & Schmitz, B., 2001: Global dinoflagellate event associated with the late Paleocene thermal maximum. *Geology,* 29, 315-318, doi: 10.1130/0091-7613.

Cubasch, U., et al., 2001: Projections of future climate change. In: *Climate Change 2001: The Scientific Basis. Contribution of Working Group I to the Third Assessment report of the Intergovernmental Panel on Climate Change,* Houghton, J.T., Ding, Y., Griggs, D.J., Noguer, M., van der Linden, P.J., Dai, X., Maskell, K., & Voss, C.A. (Eds.), p. 525-582. Cambridge University Press, Cambridge, United Kingdom.

Cui, Y., Kump, L.R., Ridgwell, A.R., Charles, A.J., Junium, C.K., Diefendorf, A.F., Freeman, K.H., Urban, N.M. & Harding, I.C., 2011: Slow release of fossil carbon during the Palaeocene–Eocene Thermal Maximum, *Nature Geoscience,* 4, 481–485, doi: 10.1038/ngeo1179.

Dickens, G.R., & Quinby-Hunt, M.S., 1994: Methane hydrate stability in seawater. *Geophys. Res. Lett.,* 21, 2115-2118.

Dickens, G.R., O'Neil, J.R., Rea, D.K., & Owen, R.M., 1995: Dissociation of Oceanic Methane Hydrate as a Cause of the Carbon Isotope Excursion at the End of the Paleocene. *Paleoceanography,* 10, 965–971.

Dickens, G.R., Castillo, M.M., & Walker, J.C.G., 1997: A blast of gas in the latest Paleocene: Simulating first-order effects of massive dissociation of methane hydrate. *Geology,* 25, 259-262.

Fehsenfeld, F., et al., 1992: Emissions of volatile organic compounds from vegetation and their implications for atmospheric chemistry. *Global Biogeochem. Cycles*, 6, 389–430.

Gavrilov, Y., Shcherbinina, E.A., & Oberhänsli, H. 2003: Paleocene/Eocene boundary events in the northeastern Peri-Tethys. In: *Causes and Consequences of Globally Warm Climates in the Early Paleogene*, Wing, S.L., Gingerich, P.D., Schmitz, B., & Thomas, E. (Eds.), Geological Society of America, Special Paper, v. 369, p. 147-168.

Gleason, J.D., Thomas, D.J., Moore Jr., T.C., Blum, J.D., Owen, R.M., & Haley, B.A., 2009: Early to Middle Eocene History of the Arctic Ocean from Nd-Sr Isotopes in Fossil Fish Debris, Lomonosov Ridge. *Paleoceanography*, 24, PA2215, doi:10.1029/2008PA001685.

Guenther, A., Karl, T., Harley, P., Wiedinmyer, C., Palmer, P.I., & Geron, C. 2006: Estimates of global terrestrial isoprene emissions using MEGAN (model of emissions of gases and aerosols from nature). *Atmos. Chem. Phys. Discuss.*, 6, 107–173.

Handley, L., Pearson, P.N., McMillan, I.K., and Pancost, R.D., 2008: Large terrestrial and marine carbon and hydrogen isotope excursions in a new Paleocene/Eocene boundary section from Tanzania. *Earth Planet. Sci. Lett.*, 275, 17-25.

Handley, L., Crouch E.M., and Pancost R.D., 2011: A New Zealand record of sea level rise and environmental change during the Paleocene–Eocene Thermal Maximum. *Palaeogeogr., Palaeoclimatol., Palaeoecol.*, 305, 185–200.

Head, J.J., Bloch, A.J., Hastings, I.K., Bourque, J.R., Cadena, E.A., Herrera, F.A., Polly, P.D. & Jaramillo, C.A., 2009: Giant boid snake from the Palaeocene neotropics reveals hotter past equatorial temperatures. *Nature*, 457, 715-717.

Heilmann-Clausen, C., & Egger, H., 2000: The Anthering outcrop (Austria): a key-section for correlation between Tethys and northwestern Europe near the Paleocene/Eocene boundary. *GFF*, 122, 69.

Heinemann, M., Jungclaus, J.H., & Marotzke, J., 2009: Warm Paleocene/Eocene Climate as simulated in ECHAM5/MPI-OM. *Climate of the Past Discussions*, 5, 1297-1336.

Higgins, J.A., & Schrag, D.P., 2006: Beyond methane: Towards a theory for Paleocene-Eocene Thermal Maximum. *Earth Planet. Sci. Lett.*, 245, 523-537.

Hollis, J.H., Handley, L., Crouch, E.M., Morgans, H.E.G., Baker, J.A., Creech, J., Collins, K.S., Gibbs, S.J., Huber, M., Schouten, S., Zachos, J.C., & Pancost, R.D., 2009: Tropical sea temperatures in the high-latitude South Pacific during the Eocene. *Geology*, 37, 99-102.

Houghton, R.A., 2008: Carbon Flux to the Atmosphere from Land-Use Changes: 1850-2005. In: *TRENDS: A Compendium of Data on Global Change*. Carbon Dioxide Information Analysis Center, Oak Ridge National Laboratory, U.S. Department of Energy, Oak Ridge, Tenn., U.S.A.

Huber, M., & Sloan, L.C., 1999: Warm climate transitions: A general circulation modeling study of the Late Paleocene Thermal Maximum. *J. Geophys. Res.*, 104, 16633-16655.

Huber, M., & Sloan, L.C., 2001: Heat transport, deep waters, and thermal gradients: Coupled simulation of an Eocene greenhouse climate. *Geophys. Res. Lett.*, 28, 3481-3484.

Huber, M., & Caballero, R., 2003: Eocene El Niño: Evidence for robust tropical dynamics in the "hothouse". *Science*, 299, 877-881.

Huber, M., & Caballero, R., 2011: The early Eocene equable climate problem revisited. *Clim. Past*, 7, 603-633doi:10.5194/cp-7-603-2011.

Iakokleva, A.I., Brinkhuis, H., & Cavagnetto, C., 2001: Late Paleocene-Early Eocene dinoflagellae cysts from the Turgay Strait, Kazakhstan; correlations across ancient seaways. *Paleogeogr., Paleoclimatol., Paleoecol.*, 172, 243-268.

IPCC, 1990: *Report prepared for Intergovernmental Panel on Climate Change by Working Group I.* Houghton, J.T., Jenkins G.J. & Ephraums J.J. (Eds.), Cambridge University Press, Cambridge, Great Britain, New York, NY, USA and Melbourne, Australia, 410 pp.

IPCC, 2007: *Climate Change 2007: The Physical Science Basis. Contribution of Working Group I to the Fourth Assessment Report of the Intergovernmental Panel on Climate Change.* Solomon, S., Qin, D., Manning, M., Chen, Z., Marquis, M., Averyt, K.B., Tignor, M., & H.L. Miller (Eds.), Cambridge University Press, Cambridge, United Kingdom and New York, NY, USA, 996 pp.

Kelly, D.C., Bralower, T.J. & Zachos, J.C. 1998: Evolutionary consequences of the latest Paleocene Thermal Maximum for tropical planktonic foraminifera. *Paleogeogr., Paleoclimatol., Paleoecol.*, 141, 139-161.

Kelly, D.C., Nielsen, T.M.J., McCarren, H.K., Zachos, J.C., & Röhl, U., 2010: Spatiotemporal patterns of carbonate sedimentation in the South Atlantic: Implications for carbon cycling during the Paleocene–Eocene thermal maximum. *Paleogeogr., Paleoclimatol., Paleoecol.*, 293, 30-40.

Kennett, J.P., & Stott, L.D., 1991: Abrupt deep-sea warming, paleoceanographic changes and benthic extinctions at the end of the Paleocene. *Nature,* 353, 225-229.

Kirk-Davidoff, B.D., Schrag, D.P., & Anderson, J.G., 2002: On the feedback of stratospheric clouds on polar climate. *Geophys. Res. Lett.*, 29, 1556, doi:10.1029/2002GL014659.

Koch, P.L., Zachos, J.C., & Gingerich, P.D., 1992: Correlation between isotope records in marine and continental carbon reservoirs near the Palaeocene/Eocene boundary. *Nature,* 358, 319-322.

Korty, R.L., Emanuel, K.A., & Scott, J.R., 2008: Tropical cyclone-induced upper ocean mixing and climate: application to equable climates. *J. Climate,* 21, 638-654.

Kump, L.R., & Pollard, D., 2008: Amplification of Cretaceous warmth by biological cloud feedbacks. *Science,* 320, 195.

Lu, G., Keller, G. & Pardo, A., 1998: Stability and change in Tethyan planktic foraminifera across the Paleocene-Eocene transition. *Mar. Micropaleontol.*, 35, 203-233.

Lunt, D.J., Valdes, P.J., Jones, T.D., Ridgwell, A., Haywood, A.M., Schmidt, D.N., Marsh, R., & Maslin, M., 2010: CO_2-driven ocean circulation changes as an amplifier of Paleocene-Eocene thermal maximum hydrate destabilization. *Geology,* 38, 875-878, doi: 10.1130/G31184.1.

Lyle, M., Barron, J., Bralower, T.J., Huber, M., Olivarez Lyle, A., Ravelo, A.C., Rea, D.K., & Wilson, P.A., 2008: Pacific Ocean and Cenozoic evolution of climate. *Rev. Geophys.*, 46, 1-47.

Maclennan, J., & Jones, S.M., 2006: Regional uplift, gas hydrate dissociation and the origins of the Paleocene-Eocene Thermal Maximum. *Earth Planet. Sci. Lett.*, 245, 65-80.

Marincovich, L., Jr., & Gladenkov, A.Y., 1999: Evidence for an early opening of the Bering Strait. *Nature,* 397, 149–151.

Meehl, G.A., Washington, W.M., Santer, B.D., Collins, W.D., Arblaster, J.M., Hu, A., Lawrence, D.M., Teng, H. Buja, L.E. & Strand, W.G., 2006: Climate change projections for the twenty-first century and climate change commitment in the CCSM3. *J. Climate,* 19, 2597-2616.

Mikolajewicz, U., Gröger, M., Maier-Reimer, E., Schurgers, G., Vizcaíno, M., & Winguth, A., 2007: Long-term effects of anthropogenic CO_2 emissions simulated with a complex earth system model. *Clim. Dyn.*, 28, 599-633.

Moran, K., et al., 2006: The Cenozoic palaeoenvironment of the Arctic Ocean. *Nature,* 441, 601-605.

Nicolo, M.J., Dickens, G.R., & Hollis, C.J., 2011: South Pacific intermediate water oxygen depletion at the onset of the Paleocene-Eocene Thermal Maximum as depicted in New Zealand margin sections. *Paleoceanography,* in press.

Nunes, F., & Norris, R.D. 2006: Abrupt reversal in ocean overturning during the Paleocene/Eocene warm period. *Nature,* 439, 60-63.

Pagani, M., Pedentchouk, N. Huber, M., Sluijs, A., Schouten, S., Brinkhuis, H., Sinninghe, J.S., Damsté, Dickens, G.R., & the Expedition 302 Scientists, 2006a: Arctic hydrology during global warming at the Paleocene/Eocene thermal maximum. *Nature,* 442, 671-675.

Pagani, M., Caldeira, K., Archer, D., & Zachos, J.C., 2006b: An ancient carbon mystery. *Science,* 314, 1556-1557.

Panchuk, K., Ridgwell, A. & Kump, L.R., 2008: The sedimentary response to Paleocene-Eocene Thermal Maximum carbon release: A model-data comparison. *Geology,* 36, 315-318.

Pancost, R.D., Steart, D.S., Handley, L., Collinson, M.E., Hooker, J.J., Scott, A.C., Grassineau, N.V., & Glasspool, I.J., 2007: Increased terrestrial methane cycling at the Palaeocene-Eocene thermal maximum. *Nature,* 449, 332-335.

Pearson, P.N., & Palmer, M.R. 2000: Atmospheric carbon dioxide concentrations over the past 60 million years. *Nature,* 406, 695-699.

Pearson, P.N., van Dongen, B.E., Nicholas, C.J., Pancost, R.D., Schouten, S., Singano, J.M., & Wade, B.S., 2007: Stable warm tropical climate through the Eocene Epoch. *Geology,* 35, 211-214.

Prather, M., & Ehhalt, D., 2001: Atmospheric chemistry and greenhouse gases. In: *Climate change 2001. The scientific basis,* Houghton, J.T., Ding, Y., Griggs, D.J., Noguer, M., van der Linden, P.J., Dai, X., Maskell, K., & Johnson, C.A. (Eds.), p. 239–287. Cambridge, UK: Cambridge University Press.

Ravizza, G., Norris, R.N., Blusztajn, J., & Aubry, M.-P., 2001: An osmium isotope excursion associated with the late Paleocene thermal maximum: Evidence of intensified chemical weathering. *Paleoceanography,* 16, 155-163.

Retallack, G.J., 2005: Pedogenic carbonate proxies for amount and seasonality of precipitation in paleosols. *Geology,* 33, 333-336.

Retallack, G.J., 2009: Mechanisms of PETM global change constrained by a new record from central Utah: Comment. *Geology,* 37, e184-e185.

Robert, C., & Kennett, J.P., 1994: Antarctic subtropical humid episode at the Paleocene-Eocene boundary: Clay-mineral evidence. *Geology,* 22, 211-214.

Roberts, C.D., LeGrande, A.N., & Tripati, A.K., 2009. Climate sensitivity to Arctic seaway restriction during the Early Paleogene. *Earth and Planetary Science Letters,* 286, 576-585.

Royer, D.L., Berner, R.A., & Park, J., 2007: Climate sensitivity constrained by CO_2 concentrations over the past 420 million years. *Nature,* 446, 530-532.

Schmidt, G.A., & Shindell, D.T., 2003: Atmospheric composition, radiative forcing, and climate change as a consequence of a massive methane release from gas hydrates. *Paleoceanography*, 18, 1004, doi:10.1029/2002PA000757.

Schmitz, B., & Pujalte, V., 2007: Abrupt increase in seasonal extreme precipitation at the Paleocene-Eocene boundary. *Geology*, 35, 215-218.

Schmitz, B., Speijer, R.P., & Aubry, M.P., 1996: Latest Paleocene benthic extinction event on the southern Tethyan shelf (Egypt): Foraminiferal stable isotopic ($\delta^{13}C$, $\delta^{18}O$) records. *Geology*, 24, 347-350.

Scotese, C.R., 2011: PALEOMAP, date of access: 07/05/2011, available from: <http://www.scotese.com>.

Sewall, J. O., & Sloan, L.C. 2006: Come a little bit closer: A high-resolution climate study of the early Paleogene Laramide foreland. *Geology*, 34, 81-84.

Shellito C.J., & Sloan, L.C., 2006: Reconstructing a lost Eocene paradise: Part I. Simulating the change in global floral distribution at the initial Eocene thermal maximum. *Global Planet. Ch.*, 50, 1-17.

Shellito, C.J., Sloan, L.C., & Huber, M., 2003: Climate model sensitivity to atmospheric CO_2 levels in the Early-Middle Paleogene. *Paleogeogr., Paleoclimatol., Paleoecol.*, 193, 113-123.

Shellito, C.J., Lamarque, J.-F. & Sloan, L.C., 2009: Early Eocene Arctic climate sensitivity to pCO_2 and basin geography. *Geophys. Res. Lett.*, 36, L09707, doi:10.1029/ 2009GL037248.

Sloan, L.C. & Barron, E.J., 1992: Eocene climate model results: Quantitative comparison to paleo-climatic evidence. *Palaeogeogr., Palaeoclim., Palaeoecol.*, 93, 183-202.

Sloan, L.C., & Pollard, D., 1998: Polar stratospheric clouds: A high latitude warming mechanism in an ancient greenhouse world. *Geophys. Res. Lett.*, 25, 3517–3520.

Sloan, L.C., & Rea, D.K., 1995: Atmospheric carbon dioxide and early Eocene climate: A general circulation modeling sensitivity study. *Palaeogeogr., Palaeoclim., Palaeoecol.*, 119, 275-292.

Sluijs, A., 14 others, & the Expedition 302 Scientists, 2006: Subtropical Arctic Ocean temperatures during the Paleocene/Eocene thermal maximum. *Nature*, 441, 610-613.

Sluijs, A., Brinkhuis, H., Schouten, S., Bohaty, S.M., John, C.M., Zachos, J.C., Reichart, G.-J., Sinninghe Damsté, J.S., Crouch, E.M., & Dickens, G.R., 2007: Environmental precursors to rapid light carbon injection at the Palaeocene/Eocene boundary. *Nature*, 450, 1218-1221.

Sluijs, A., Röhl, U., Schouten, S., Brumsack, H.-J., Sangiorgi, F., Sinninghe Damsté, J.S., & Brinkhuis, H., 2008a: Arctic Late Paleocene - Early Eocene paleoenvironments with special emphasis on the Paleocene - Eocene thermal maximum (Lomonosov Ridge, IODP Expedition 302). *Paleoceanography*, 23, PA1S11, doi:10.1029/2007PA001495.

Sluijs, A., et al., 2008b: Eustatic variations during the Paleocene-Eocene greenhouse world. *Paleoceanography*, 23, PA4216, doi:10.1029/2008PA001615.

Sluijs, A. Bijl, P.K., Schouten, S., Roehl, U., Reichart, G.-J., and Brinkhuis, H., 2011: Southern ocean warming, sea level and hydrological change during the Paleocene-Eocene thermal maximum. *Climate of the Past*, 7, 47–61.

Speijer, R.P., & Wagner, T., 2002: Sea-level changes and black shales associated with the late Paleocene Thermal Maximum (LPTM): Organic geochemical and micropaleontologic evidence from the southern Tethyan margin (Egypt-Israel). In: *Catastrophic events and mass extinctions*: Impacts and beyond, Koeberl, C., and MacLeod, K.G. (Eds.), Geological Society of America Special Paper, 356, p. 533–549.

Stoll, H.M., Shimizu, N., Ziveri, P., & Archer, D., 2007: Coccolithophore productivity response to greenhouse event of the Paleocene-Eocene Thermal Maximum. *Earth and Planetary Science Letters*, 258, 192-206.

Storey, M., Duncan, R.A., & Swisher III, C.C., 2007: Paleocene-Eocene Thermal Maximum and the opening of the Northeast Atlantic. *Science*, 316, 587-589.

Svensen, H., Planke, S., Malthe-Sørenssen, A., Jamtveit, B., Myklebust, R., Eidem, T.R., & Rey, S.S., 2004: Release of methane from a volcanic basin as a mechanism for initial Eocene global warming. *Nature*, 429, 542-545.

Tans, P., & Keeling, R., 2011: Trends in atmospheric carbon dioxide., *NOAA/ESRL and Scripps Institution of Oceanography*, date of access: 07/05/2011, available from: <http://www.esrl.noaa.gov/gmd/ccgg/trends/>

Thomas, D.J., 2004: Evidence for deep-water production in the North Pacific Ocean during the early Cenozoic warm interval. *Nature*, 430, 65-68.

Thomas, D.J., Lyle, M., Moore, T.C., Jr., & Rea, D.K., 2008: Paleogene deep-water mass composition of the tropical Pacific and implications for thermohaline circulation in a Greenhouse World. *Geochemistry, Geophysics, Geosystems*, 9, 1-13, doi:10.1029/2007GC001748.

Thomas, E., 1998: The biogeography of the late Paleocene benthic foraminiferal extinction, In: *Late Paleocene-early Eocene biotic and climatic events in the marine and terrestrial records*, Aubry, M.-P., Lucas, S., & Berggren, W.A. (Eds.), , p. 214-243, Columbia University Press.

Thomas, E., 2003: Extinction and food at the seafloor: A high-resolution benthic foraminiferal record across the initial Eocene Thermal Maximum, Southern Ocean Site 690. *GSA Spec. Paper*, 369, 319-332.

Thomas, E., 2007: Cenozoic mass extinctions in the deep sea; what disturbs the largest habitat on Earth? In: *Large Ecosystem Perturbations: Causes and Consequences:* Monechi, S., Coccioni, R., & Rampino, M. (Eds.), *GSA Special Paper*, 424, 1-24.

Thomas, E., Zachos, J.C., & Bralower, T. J., 2000: Deep-Sea Environments on a Warm Earth: latest Paleocene - early Eocene. In: *Warm Climates in Earth History*, Huber, B., MacLeod, K., and Wing, S. (Eds.), pp. 132-160, Cambridge University Press, Cambridge, United Kingdom.

Tripati, A., & Elderfield, H. 2005: Deep-sea temperature and circulation changes at the Paleocene-Eocene Thermal Maximum. *Science*, 308, 1894-1898.

Walker, J.C.G., Hays, P.B., & Kasting, J.F., 1981: A negative feedback mechanism for thelong-term stabilization of earth's surface temperature. *Journal of Geophysical Research*, 86 (C10), 9776–9782.

Weijers, J.W.H., Schouten, S., Sluijs, A., Brinkhuis, H., & Sinninghe Damsté, J.S., 2007: Warm arctic continents during the Paleocene-Eocene thermal maximum. *Earth Planet. Sci. Lett.*, 261, 230-238.

Wing, S.L., Harrington, G.J., Smith, F.A., Bloch, J.I., Boyer, D.M., & Freeman, K.H., 2005: Transient floral change and rapid global warming at the Paleocene-Eocene boundary. *Science*, 310, 993-996.

Winguth, A.M.E., Shellito, C., Shields, C., & Winguth, C., 2010: Climate response at the Paleocene-Eocene Thermal Maximum to greenhouse gas forcing - A model study with CCSM3. *J. Climate,* 23, 2562-2584, doi:10.1175/2009JCLI3113.1.

Zachos, J.C., & Dickens, G.R., 2000: An assessment of the biogeochemical feedback response to the climatic and chemical perturbations of the LPTM. *Gff,* 122, 188-189.

Zachos, J.C., et al., 2005: Rapid acidification of the ocean during the Paeocene-Eocene thermal maximum. *Science,* 308, 1611-1615.

Zachos, J.C., Bohaty, S.M., John, C.M., McCarren, H., Kelly, D.C., & Nielsen, T., 2007: The Paleocene-Eocene carbon isotope excursion: constraints from individual shell planktonic foraminifer records. *Phil. Trans. R. Soc. London, A,* 365, 1829-1842.

Zachos, J.C., Dickens, G.R., & Zeebe, R.E., 2008: An early Cenozoic perspective on greenhouse warming and carbon cycle dynamics. *Nature,* 451, 279-283.

Zeebe, R.E., & Zachos, J.C., 2007: Reversed Deep-Sea Carbonate Ion Basin Gradient During Paleocene-Eocene Thermal Maximum. *Paleoceanography,* 22, PA3201, doi:1029/2006PA001395.

Zeebe, R.E., Zachos, J.C., & Dickens, G.R., 2009: Carbon dioxide forcing alone insufficient to explain Palaeocene-Eocene Thermal Maximum warming. *Nature Geoscience,* 2, 576-580.

4

Temporal Variability of Rain-Induced Floods in Southern Quebec

Assani Ali Arkamose, Landry Raphaëlle,
Quessy Jean-François and Clément Francis
Université du Québec à Trois-Rivières, Québec
Canada

1. Introduction

Although the impact of climate warming on streamflow in general and on floods in particular is a much-debated topic (e.g. Koutsoyiannis et al., 2008; Kundzewicz et al., 2008; Räisanen, 2007; Sun et al., 2007), there is general agreement about the geographical variability of these changes. In Quebec, a region characterized by a temperate, continental- and maritime-type climate, a consensus is forming in light of results of climate and hydrological modeling regarding the effects of climate warming on flood magnitude. These effects will depend on the season and the underlying cause of floods. Thus, whereas climate and hydrological models predict a decrease in the magnitude of spring floods resulting from snowmelt (freshets) due to a decrease in the amount of snow falling in winter, they also predict an increase in the magnitude of rain-induced floods as a result of increasing rainfall intensity during summer and fall. Thus, Roy et al. (2001) predicted a significant increase in the intensity of heavy precipitation (20 and 100-year recurrence intervals) which will result in a much greater increase in the magnitude of summer floods. In the Châteauguay River, for instance, peak flow for a flood induced by a 20-year rainfall event will double or triple, depending on initial soil moisture conditions, by the end of the century. However, according to Zhang et al. (2000), no significant increase in rainfall intensity has been observed in Quebec or Canada over the past century, which would explain the absence of any significant change in the interannual variability of the magnitude of rain-induced floods observed in many regions of Canada (Cunderlik & Ouarda, 2009).

While Assani et al. (2011) have shown that the amount of rainfall from August to November has significantly increased in southeastern Quebec, south of parallel 46°N on the South Shore of the St. Lawrence River, no study has looked at the impact of this increase in rainfall on the interannual variability of rain-induced floods. Analysis of the interannual variability of snowmelt-induced spring floods (spring freshets) has revealed no generalized significant decrease in their magnitude (Assani et al., 2010), despite a recorded increase in temperature since the 1970's in Quebec. On the contrary, a significant increase in the magnitude of spring floods on the North Shore of the St. Lawrence is recorded from 1934 to 2000, which is thought to result from the continental nature of climate in this region rather than from increasing temperature. Assani et al. (2010) have also shown that the interannual variability of snowmelt-induced spring floods is significantly correlated with the AMO climate index on the North Shore, and with the SOI and AO climate indices on the South Shore, north and south of parallel 47°N, respectively.

In Quebec, rain-induced floods can sometimes be more devastating than snowmelt-related floods. For instance, in July 1996, the Saguenay-Lac-Saint-Jean area of central Quebec, north of the St. Lawrence River, was the scene of a series of floods which caused more than 1 billion dollars in damage and 10 fatalities. More recently, in the Rivière-au-Renard area of eastern Quebec, south of the St. Lawrence, floods caused several million dollars in damage and two deaths in August 2007. These events led to speculation about the likely increase in the magnitude of rain-induced flood due to climate warming, as predicted by climate and hydrological models. To see whether or not these speculations are supported by observational data, the following two main goals were set for this study:

1. Analyze the interannual variability of the magnitude of rain-induced floods in different regions of Quebec, to see whether this magnitude increased significantly over time, particularly in the southeastern part of the province, south of parallel 46°N on the South Shore of the St. Lawrence River, where the amount of rainfall has increased over time.
2. Determine which climate factors (climate indices) affect the interannual variability of the magnitude of rain-induced floods, in order to see if snowmelt-induced (spring) floods and rain-induced floods are affected by the same climate indices.

2. Methods

2.1 Streamflow data sources and selection of climatic indices

The streamflow data come from Environment Canada's Hydat CD-ROM (2004). Eighteen rivers for which the daily streamflows are measured continuously over a relatively long period wherever possible were selected (table 1 and fig.1). To be able to analyze a greater number of stations, we delimited the study period between 1934 and 2004. Streamflow in these rivers is not affected by the presence of dams. To constitute the seasonal maximum flow series for each year, we selected the highest daily average flow measured in the period from July to October, as this is the period with the highest frequency of rain-induced floods in Quebec.

Several authors have already analyzed the relationship between the climatic indices and streamflows in Québec, Canada and North America. For instance, in Québec, Anctil and Coulibaly (2004) observed a positive correlation between the annual average flows and the PNA (Pacific-North America) and AO indices in the St. Lawrence watershed during the 1938-2000 period. Apart from these two indices, Déry and Wood (2004, 2005) also observed a correlation between streamflows and the PDO (Pacific Decadal Oscillation), the NINO3.4 and SOI (Southern Oscillation index) indices in the other two major Québec watersheds (Hudson Bay and Ungava Bay) during the 1964-2000 period. On the scale of Eastern Canada, a correlation was observed between NAO (North Atlantic Oscillation) and the river flows (see Kingston et al., 2006). On the scale of the North American continent, AMO and PDO are correlated to the annual average flows in many regions of the United States (Enfield et al., 2001; McCabe et al., 2004). Curtis (2008) observed a correlation between AMO and heavy summer rains in the United States and Mexico. At the daily scale, Assani et al. (2010) noted a significant correlation between maximum spring flows and AMO on the North Shore, and AO and ENSO on the South Shore. In this study, we correlated the annual maximum flows to all the climatic indices already correlated to streamflows: AMO, AO, NAO, NINO3.4, PDO and SOI. The data for the AMO, SOI, NINO3.4 and PDO indices are taken from the following websites: http://www.cdc.noaa.gov/ClimateIndices/List. (2007-10-08), and NAO is taken from http://www.cgd.ucar.edu/cas/jhurrell/indices.data.html (2006-10-08), and

AO is taken from http://jisao.washington.edu/data/ao/ (2006-10-08). For each climatic index and for each year, we first derived the seasonal mean of the monthly indices over four months (April-July, May-August, June-September, July-October, and August-November), then over three months (April-June, May-July, June-August, July-September, August-October, and September-November).

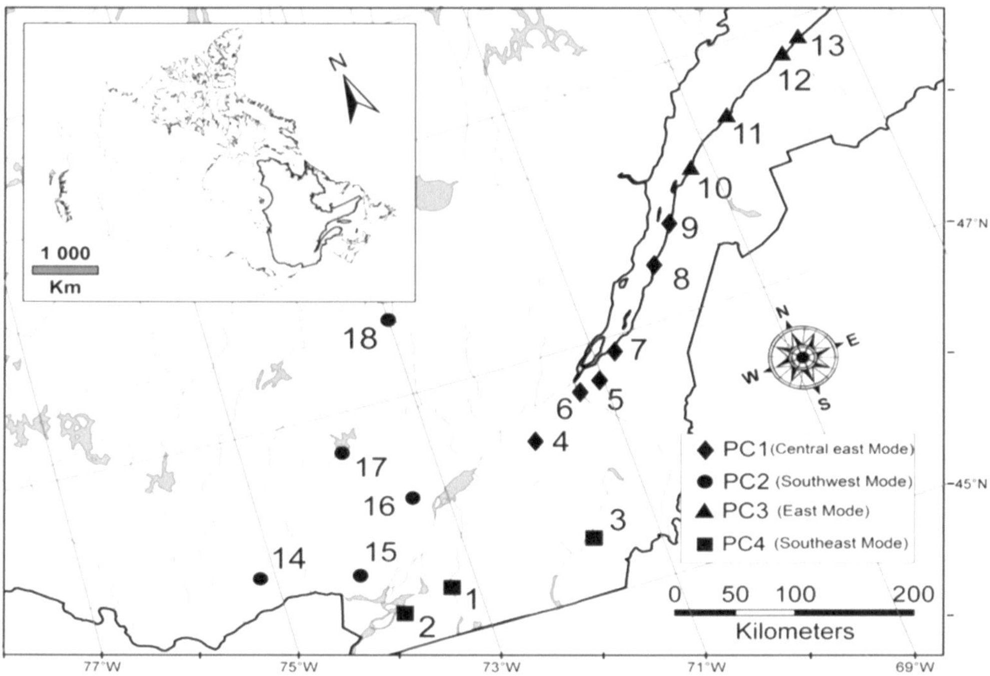

Fig. 1. Location of stations grouped in four modes derived from principal component analysis.

2.2 Statistical data analysis

2.2.1 Interannual variability modes of seasonal daily maximum flows

The first step was to apply principal components analysis (PCA) to the seasonally maximum flow (August to November) data measured at different stations (e.g. Hannachi et al., 2007). We chose this method for comparison with the results of previous work (e.g.Anctil & Coulibaly, 2004; Assani et al., 2010). Moreover, it is widely used in hydroclimatology to analyze the influence of climate factors on the temporal variability of precipitations and streamflows (e.g., McCabe et al., 2004; Vicente-Serrano, 2005). To determine the number of mode patterns (temporal variability modes of annual maximum flow), principal component analysis (mode S) was applied to the correlation matrix (and not to the covariance matrix), based on the correlations calculated between the seasonal flows measured at the different (individual) stations in order to eliminate the influence of extreme values (Bigot et al. 1997), and the effect provoked by a site's local variability (Siew-Yan-Yu et al. 1998). Thus, we analyzed a matrix consisting of 71 lines (number of years of streamflow measurement, from 1934 to 2004) and 18 columns (number of rivers analyzed). We applied the Varimax rotation method to maximize the saturation values of the stations on the principal components and obtain more stable and physically more robust mode patterns.

No	River	Area (km²)	Latitude (N)	Longitude (W)	MAF (m^3/s)
1	Richelieu	22000	45°18′	73°15′	327
2	Chateaugay	2500	45°17′	73°48′	26.0
3	Eaton	642	45°28′	71°39′	8.6
4	Etchemin	1130	46°38′	71°02′	28.5
5	Nicolet de sud-ouest	544	45°47′	71°58′	8.4
6	Beaurivage	709	46°39′	71°17′	10.5
7	Du Sud	826	46°49′	70°45′	14.7
8	Ouelle	802	47°25′	69°56′	15.6
9	Du Loup	1050	47°49′	69°31′	15.4
10	Trois-Pistoles	932	48°05′	69°11′	13.4
11	Rimouski	1610	48°24′	68°33′	26.5
12	Matane	826	48°46′	67°32′	33.4
13	Blanche	208	48°46′	67°39′	4.8
14	De La Petite Nation	1330	45°47′	75°05′	19.1
15	Du Nord	1170	45°47′	74°00′	21.5
16	L'Assomption	1340	46°00′	73°25′	23.5
17	Matawin	1390	46°41′	73°54′	22.4
18	Vermillon	2670	47°39′	72°57′	37.2

MAF = Mean annual flow

Table 1. Rivers analyzed

The rationale for this maximization is the fact that the criterion used for grouping rivers into modes or homogeneous hydrological regions is based on the values of loadings of rivers on the significant principal components using the "maximum loading" rule. According to this rule, a station is associated with a significant principal component if its loading value on this component is larger than on other components (Vicente-Serrano, 2005). Thus, all stations for which the loadings values are largest on a given principal component define a variability mode or homogeneous hydrological region. However, this loading value must be statistically significant, and since it is not possible rigorously to test a loading value using a statistical test, the correlation between streamflow in each river and the scores of each significant principal component was calculated. This correlation, whose value corresponds to the loading value of the river on a significant component, was thus tested using Student's t test (Assani et al., 2010). A river is therefore correlated to a principal component if its loading value, whose significance has been indirectly tested using the corresponding correlation coefficient, is statistically significant at the 5% level. The Kaiser (1960) criterion based on the eigenvalues of principal components was used to determine the number of significant components, since any principal component with eigenvalue equal to or larger than 1 is considered significant. The grouping of the stations in temporal variability modes was based on "the maximum loading rule". According to this rule, a station is associated with a principal component when the value of its loading is higher on this component than on the others (Vicente-Serrano, 2005).

2.2.2 Analysis of the interannual variability of the magnitude of rain-induced seasonal daily maximum flows

To test the stationarity of the temporal modes, we first applied the regression method (a parametric test) to the scores of the significant principal components (Kundzewicz et al., 2005). However, because it is not possible, using this method, to determine the exact date of a shift in the mean of a hydrologic series, nor whether this shift is abrupt or progressive, the Lombard (1987) method was used to derive these two parameters (date and type of shift).

Suppose we have a series of observations, noted $X_1,...,X_n$, where X_i is the observation taken at time $T = i$. these observations are supposed to be independent. One question of interest is to see whether the mean of this series has changed. If μ_i refers to the theoretical mean of X_i, then a possible pattern for the mean is given by Lombard's smooth-change model, where

$$\mu_i = \begin{cases} \theta_1 & if\ 1 \le i \le T_1; \\ \theta_1 + \dfrac{(i-T_1)\ (\theta_2-\theta_1)}{T_2-T_1}, & if\, T_1 < i \le T_2; \\ \theta_2 & if\, T_2 < i \le n. \end{cases} \tag{1}$$

In other words, the mean changes gradually from θ_1 to θ_2 between times T_1 and T_2. As a special case, one has the usual abrupt-change model when $T_2 = T_1 + 1$.

In order to test formally whether the mean in a series is stable, or rather follows model (1), one can use the statistical procedure introduced by Lombard (1987). To this end, define R_i as the rank of X_i among $X_1,...X_n$. Introduce the Wilcoxon score function $\phi(u) = 2u - 1$ and define the rank score of X_i by

$$Z_i = \frac{1}{\sigma_\phi}\left\{\phi\left(\frac{R_i}{n+1}\right) - \bar{\phi}\right\}, \qquad i \in \{1,...,n\}, \tag{2}$$

where

$$\phi = \frac{1}{n}\sum_{i=1}^{n} \phi\left(\frac{i}{n+1}\right) \quad \text{and} \quad \sigma_\phi^2 = \frac{1}{n}\sum_{i=1}^{n}\left\{\phi\frac{i}{n+1} - \bar{\phi}\right\}^2 \tag{3}$$

Lombard's test statistic is

$$S_n = \frac{1}{n^5}\sum_{T_1=1}^{n-1}\ \sum_{T_2=T_1+1}^{n} L_{T_1 T_2}^2, \tag{4}$$

where

$$L_{T_1,T_2} = \sum_{j=T_1+1}^{T_2}\ \sum_{i=1}^{j} Z_i \tag{5}$$

At the 5% level of significance, one concludes that the mean of the series changes significantly according to a pattern of type (1) whenever $S_n > 0.0403$. Note that the test is suitable for the detection of all kinds of patterns in equation (1), including abrupt changes. A complete investigation of the power and robustness of S_n and of five other test statistics proposed by Lombard is given in Quessy *et al.* (2011).

2.2.3 Comparison of the magnitude of rain-induced floods with the magnitude of two- and five-year floods

To determine the extent of a potential change in the magnitude of rain-induced floods, this magnitude was compared with that of annual floods with a 2-year recurrence interval in each watershed. Two-year flood flows were estimated from annual series of measured daily maximum flows for each watershed using the regional approach developed by Anctil et al. (1998). This method is based on the regionalized law of general extreme values (GEV). This estimate was produced follows:

- First, we calculated the quantiles (Q_R) corresponding to the two-year recurrence by means of the formulas developed by Anctil et al. (1998) in the natural homogeneous hydrologic regions. These have been defined in Québec by means of the Hosking and Wallis method. The following equations were used:

$$Q_R = \xi + \frac{(\alpha B)}{\kappa} \tag{6}$$

$$B = 1 - \left\{ -\ln\left[\frac{(T-1)}{T} \right] \right\}^{\kappa} \tag{7}$$

where T is the return period; κ, α and ξ respectively are the shape, location and scale parameters of the standardized parameters of the regional GEV distribution. These parameters are estimated by means of the L-moments method, for which the values were calculated by Anctil et al.(1998) in the natural homogeneous hydrologic regions defined in Québec.

- Finally, we estimated the two-year recurrence quantile (Q_2) downstream from the dams by means of the following equation:

$$Q_2 = Q_R Q_m \tag{8}$$

Q_m is the mean of daily maximum flows for a given river, derived from an annual series compiled from daily flow data measured from October (yr-1) to September of each hydrological year. Using equation 9, it is possible to compare the intensity of the magnitude of rain-induced floods with that of snowmelt-induced floods.

2.2.4 Analysis of the relationship between the annual maximum flow and the climatic indices

The relationship between the seasonal maximum daily flows and the climatic indices was calculated by means of canonical correlation analysis (CANCOR). Compared to other methods of multivariate analysis, CANCOR takes into account both intra-group relationships and the cross correlations between variables of two groups. Indeed, it creates factors (linear transformations of variables, commonly called canonical variables) in the first group (dependent variables) simultaneously to factors in the second group (independent variables). It requires those factors to be orthogonal to each other within the same group, so they are interpreted as independent dimensions of the phenomenon expressed by a group of variables. Thus, the canonical analysis helps to maximize the correlation between the first factor of the first group and the first factor of the second group than between the second factors of the two groups, each one considered orthogonal to the two factors of the first pair, than between the third factors and so on. Each pair of factors expresses a type of relationship between variables in both groups. The intensity of the relationship of a determined type is measured by a canonical correlation coefficient, which is the correlation coefficient between the factors of the same pair. This method allows simultaneous correlation of several dependent variables (streamflows) and several independent variables (climatic indices). It is widely used in climatology (e.g., Chen & Chen, 2003; Dukenloh & Jacobett, 2003; Repelli & Nobre, 2004). The two correlation methods were calculated between the annual climatic indices and the principal components scores (McCabe et al., 2004).

3. Results

3.1 Modes and long-term trend of the variability of rain-induced maximum seasonal flows in Quebec

Using principal component analysis, it was possible to group the 18 rivers into four variability modes each defined by a statistically significant principal component (Fig. 1 and Table 3). The first principal component (East-Central Mode) is correlated with rivers located between 45°30′N and 48°N on the South Shore. The second principal component (Southwest Mode) is correlated with all rivers located on the North Shore. And the last two principal components (East and Southeast Modes) are correlated with rivers on the South Shore located respectively north of 48°N and south of 45°30′N. The total variance explained by the four components exceeds 70%.

The interannual variability of the principal component (PC) scores is shown in Figure 2, and linear regression and Lombard method results are summarized in Tables 3 and 4. Recall that the interannual variability of PC scores reflects the interannual variability of streamflow in rivers with which the principal components are significantly correlated. A statistically significant increase in streamflow is only observed in the Southeast Mode, south of 45°30′N on the South Shore (Figs. 2-3 and Tables 3 and 4). However, analysis of regression results reveals a significant decrease in the interannual variability of the scores of the first principal component, which is correlated with rivers located between 45°30′N and 48°N (East-Central Mode), and a significant increase in the interannual variability of the scores of the second principal component (Southwest Mode). However, the Lombard method could not confirm these changes in principal components I and II (PC I and II). Hence, the above results do not point to any generalized, province-wide increase in the interannual variability of the magnitude of rain-induced daily maximum flows. To determine the extent of the increase observed in the Southeast Mode, the number of times the two-year flood flow calculated from an annual daily maximum flow series was reached or exceeded was determined for the four modes (Table 5). The two-year flood flow was reached or exceeded in 50% of the watersheds, and analysis of its geographical distribution reveals that the two-year flood magnitude is attained more frequently on the South Shore south of 48°N (reached or exceeded in 7 out of 8 watersheds) than elsewhere.

N°	Rivers	PCI	PCII	PCIII	PCIV
		Southeast Mode			
1	Richelieu	0.040	0.145	0.055	**0.848**
2	Chateaugay	0.074	0.202	0.116	**0.848**
3	Eaton	0.420	0.202	0.070	**0.533**
		East-Central Mode			
4	Etchemin	**0.876**	0.170	0.108	-0.089
5	Nicolet de sud-ouest	**0.766**	0.162	-0.156	0.386
6	Beaurivage	**0.721**	0.175	0.210	0.067
7	Du Sud	**0.652**	0.081	0.235	0.127
8	Ouelle	**0.689**	0.098	0.389	-0.001
9	Du Loup	**0.710**	0.112	0.509	0.015
		East Mode			
10	Trois-Pistoles	0.377	0.068	**0.818**	-0.073
11	Rimouski	0.310	0.240	**0.814**	0.025
12	Matane	0.132	0.218	**0.727**	0.157
13	Blanche	0.031	0.092	**0.753**	0.108
		Southwest Mode			
14	De La Petite Nation	0.023	**0.829**	0.140	0.317
15	Du Nord	0.320	**0.816**	0.053	0.133
16	L'Assomption	0.262	**0.804**	0.172	0.152
17	Matawin	0.082	**0.890**	0.166	0.180
18	Vermillon	0.092	**0.772**	0.145	-0.041
Explained variance (%)		22.2	20.7	17.2	11.8

The higher values of Rivers loadings on PCs show in the bold.

Table 2. Principal components Loadings of Rivers

PC (Mode)	a	b	R^2	Fc
PC I (East-Central Mode)	-0.013	25.28	0.070	**5.194**
PC II (Southwest Mode)	0.011	22.19	0.0541	**3.946**
PC III (East Mode)	0.007	13.29	0.0194	1.365
PCIV (Southeast Mode)	0.019	36.63	0.148	**11.986**

a = slope of the curve; b = y-intercept; R^2 = coefficient of determination; Fc = value of the Fisher-Snedecor test statistic. Fc values which are statistically significant at the 95% level are shown in bold.

Table 3. Regression parameters for curves fitted to the factorial scores of the principal components

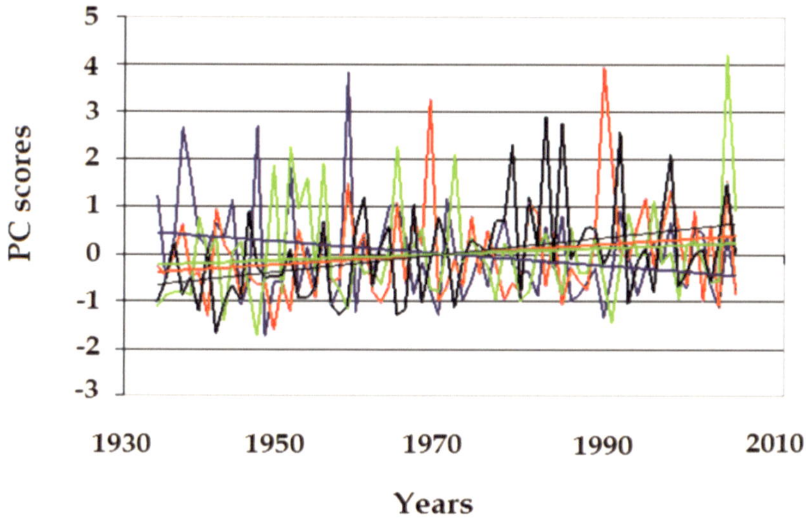

Fig. 2. Interannual variability of the Principal components scores (1934-2004). PCI = East-Central Mode (blue curve); PCII = Southwest Mode (red curve); PCIII = East Mode (red curve); PCIV = Southeast Mode (black curve);

Principal Components (Mode)	Sn	Year of change
PC I (East-Central)	0.0257	-
PC II (Southwest)	0.0281	-
PC III (East)	0.0132	-
PC IV (Southeast)	0.1152	1958

The value of Sn > 0.043, shown in bold, is statistically significant at the 95% level.

Table 4. Analysis of the interannual variability of PC scores (1934-2004). Lombard test results.

3.2 Relationship between climate indices and streamflow (PC scores)

Results of the canonical analysis of correlations are shown in Table 6. Each of the principal components, which represent the variability of streamflow in the four modes, is correlated with a canonical variable. This reflects the fact that the principal components are independent from one another, the first one being correlated with V3, the second, with V2, the third, with V4 and the last, with V1. As for climate indices, only quarterly indices derived from the means of the September to November indices show a significant correlation with principal components. The AMO index is correlated with the canonical variable W1. Since the V1 and W1 canonical variables are correlated, AMO is correlated with the last principal component, which represents streamflow variability (PC scores) in the Southeast Mode. This correlation is negative. AO is positively correlated with the second principal component (PC II), which encompasses rivers on the North Shore (Southwest Mode). The SOI climate index is correlated with the third principal component (PC III) linked to rivers located north of 48°N. Finally, the first principal component is not significantly correlated with any climate index. As for explained variance, it is larger for canonical variables correlated with principal components than for those correlated with climate indices.

No	River	Qmax	Q2-year (m^3/s)	Fr
Southeast Mode				
1	Richelieu	896	865	0
2	Châteauguay	418	403	3
3	Eaton	183	177	2
Center-east Mode				
4	Etchemin	253	244	2
5	Nicolet	142	137	2
6	Beaurivage	179	173	0
7	Du Sud	246	237	1
8	Ouelle	119	112	3
9	Du Loup	170	160	0
East Mode				
10	Trois-Pistoles	212	200	0
11	Rimouski	268	252	0
12	Matane	377	355	1
13	Blanche	41	38	4
Southwest Mode				
14	De La Petite Nation	82	79	0
15	Du Nord	191	184	0
16	L'Assomption	155	150	2
17	Matawin	142	137	0
18	Vermillon	224	216	0

Qmax = mean of annual daily maximum flows (October (yr-1) to September) calculated for the 1934-2004 interval; Q2 = two-year flood flow estimated from the annual series using the regional method; Fr= number of times Q2 was reached or exceeded from August to October during the 1934-2004 interval.

Table 5. Number of times (Fr) the two-year annual flood flow (estimated using the regional method) was reached or exceeded in the various watersheds, from 1934 to 2004.

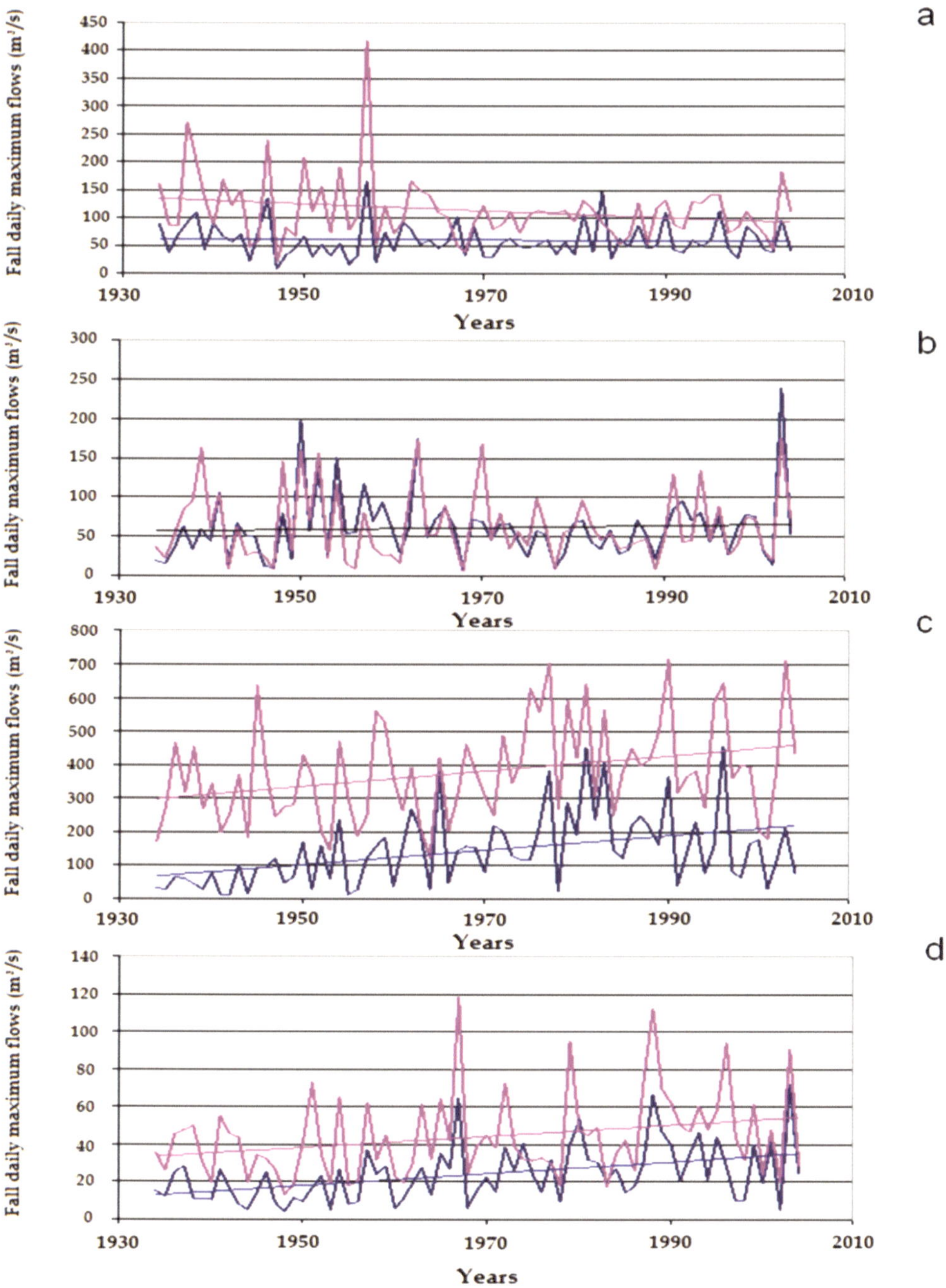

Fig. 3. Interannual variability of the fall daily maximum flow of a few river. a = Nicolet Sw (blue curve) and Etchemin (red curve) rivers in Center-East Mode; Rimouski (blue curve) and Trois-Pistoles (red curve) Rivers in East Mode; Châteaugay (blue curve) and Richelieu (red curve) Rivers in Southeast Mode; De la Petite Nation (blue curve) and Matawin (red curve) Rivers in Southwest Mode

Variables	V1	V2	V3	V4	W1	W2	W3	W4
PC I	-0.127	0.440	**0.862**	0.286				
PC II	0.528	**0.840**	-0.314	-0.054				
PC III	-0.371	0.142	-0.410	**0.821**				
PC IV	**0.873**	-0.408	0.160	0.291				
AMOfall					**-0.864**	0.388	0.435	-0.236
AOfall					0.316	**0.832**	-0.363	0.369
NAOfall					-0.249	0.197	0.032	0.187
PDOfall					0.477	0.035	0.429	-0.599
SOIfall					-0.028	-0.195	0.328	**0.884**
EV (%)	29.9	27.0	25.9	20.4	22.7	18.4	12.3	26.0

EV = explained variance. The higher values of coefficient of correlation show in bold.

Table 6. Correlation between the principal components and canonical variables (V), and correlation between climatic indices and canonical variables (W).

4. Discussion and conclusion

In light of the interannual variability of snowmelt-induced floods (Assani et al., 2010), analysis of the interannual variability of rain-induced floods in Quebec during the period from 1934 to 2004 led to four significant results:

(i) The 18 rivers analyzed were grouped into four modes: one on the North Shore and three on the South Shore, the latter being located south of 45°30′N, between 45°30′N and 48°N, and north of 48°N, respectively. For spring snowmelt-induced floods, the same rivers defined three modes: one on the North Shore and the other two on the South Shore, on either side of parallel 47°N. The effect of local factors on the origin of floods could account for the presence of an extra mode for rain-induced floods. Thus, rain-induced floods may be caused by three factors: summer storms resulting from convective motion (convective rainfall), polar front-induced rainfall (frontal rainfall), and rainfall caused by other tropical cyclones in the Atlantic basin. Local factors have a stronger effect on convective rainfall than on the other two types of rainfall. However, because the necessary data were not available, the effect of each of the above three factors on flood genesis in Quebec could not be quantified. In springtime, floods are almost exclusively caused by snowmelt. As such, the effect of local factors on snowmelt-induced flood is limited.

(ii) Analysis of the interannual variability of streamflow only revealed a significant increase in the southeast, south of 45°30′N on the South Shore (Southeast Mode). Two factors may account for this increase:

- An increase in agricultural surface area, as this is a region of Quebec in which agricultural lands make up more than 20% of all watersheds. This high proportion of farmland could lead to significant runoff which, in turn, would result in an increase in flood magnitude over time. However, since Muma et al. (2011) showed that in increase in agricultural surface area in a watershed has no impact on the magnitude of rain-induced flood flows in Quebec, this factor cannot account for the increase in magnitude of flows observed over time in the region.

- An increase in precipitation. Analysis of the interannual variability of seasonal precipitation from August to November revealed a significant increase in the amount of rainfall south of parallel 46°N from 1950 to 2000 on the South Shore, as shown in Figure 3 for a number of stations in the Richelieu and Châteauguay watersheds. This increase is thought to be the main cause of the increase in rain-induced flows observed in that part of the province, and could also account for the higher frequency of attainment and/or exceedance of the two-year annual flood magnitude in the Southeast Mode than in the other three modes.

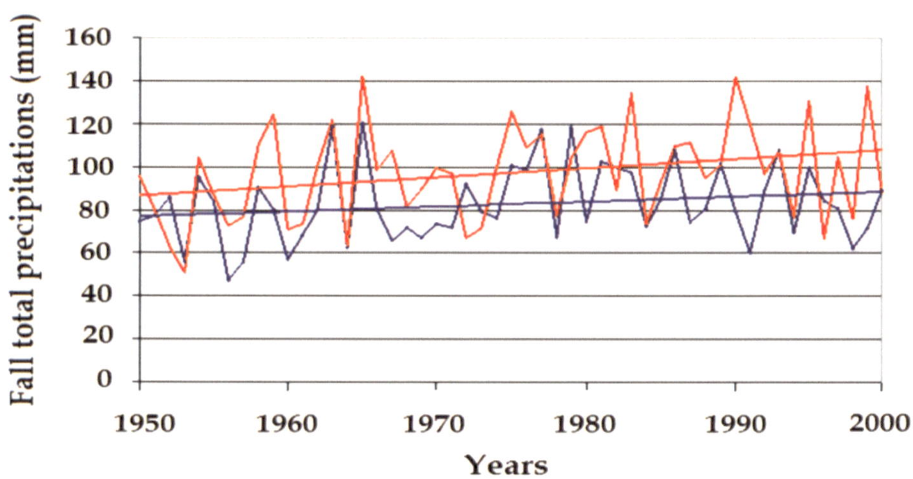

Fig. 5. Interannual variability of August to November precipitation over the period from 1950 to 2000 at two stations in the Châteauguay and Richelieu rivers watersheds, south of 45°30′N. Les Cèdres station: blue curve; Magog station : red curve

(iii) Analysis of the relationship between rain-induced flood flows and climate indices revealed a correlation between the interannual variability of flood flows and climate indices in three modes. In the Southeast Mode, located south of 45°30′N and characterized by a significant increase in streamflow and precipitation, flood flows show a negative correlation with the AMO index; in the Southwest Mode, flows are positively correlated with the AO index; and in the East-Central Mode, flows are negatively correlated with AO. For snowmelt-induced spring flows, AMO is negatively correlated with streamflow in North Shore rivers (Southwest Mode), this mode being characterized by a significant increase in spring flood flows over the 1930-2000 interval. AO is negatively correlated with streamflow in South Shore rivers located south of 47°N (Southeast and East-Central Modes) and SOI is positively correlated with streamflow in rivers located north of 47ºN (Assani et al., 2010). This comparison leads to the conclusion that rain- and snowmelt-induced floods are not correlated with the same climate indices in the Southeast and Southwest Modes. However, it also shows that AMO is correlated with rain-induced (Southeast Mode) and snowmelt-induced (Southeast Mode) floods characterized by a significant increase in flow over time in Quebec.
The effect of these three indices (AMO, AO, and SOI), which show a significant correlation with streamflow in the three modes, on the interannual variability of streamflow in Quebec

has been described by Assani et al. (2010). AMO is correlated negatively to precipitation and streamflow in many regions of North America (e.g. Curtis, 2008; Enfield et al., 2001; McCabe et al., 2004). In fact, positive values of the index (positive phase) coincide with a decrease in precipitation and streamflow in many regions of North America in general, and in Quebec in particular, whereas negative values (negative phase) of the index are associated with an increase in precipitation and streamflow. During a positive AMO phase, more frequent changes in the circulation and shear of the westerly and a weakening of cyclonic activities and transfer of water vapour from the Atlantic Ocean to the continent are observed. These factors trigger a decrease in precipitation and streamflow in Quebec. As for AO, it is positively correlated with rain-induced flood flows in Quebec. According to the scheme proposed by Kingston et al. (2006), when AO is in positive phase (high values), an increase in SSTs (surface ocean temperatures) is observed (more northerly Gulf Stream position), along with a reduced influence of the East Coast trough. As a result, the frequency of southerly airflow increases, and storm tracks coincide with the coast more often. Thus, streamflow increase in Québec. Finally, the influence of ENSO would lead to an increase in atmospheric humidity and cyclonic activities in the region during El Niño episodes. These two factors are responsible for an increase in summer and winter precipitation in Québec.
This study shows that, in the region characterized by a significant increase in rainfall, the magnitude of flood flows has significant increased. Moreover, the frequency of flows larger than the two-year annual flood flow has also increased in the watersheds. These findings confirm climate model predictions about the impact of climate warming on the intensity of rain-induced floods in Quebec. However, this increase is not a generalized, province-wide phenomenon.

5. References

Anctil, F., Martel, F., Hoang, V.D. (1998). Analyse régionale des crues journalières de la province du Québec. *Canadian Journal of Civil Engineering,* Vol. 25; pp.125–146, ISNN 0315-1468

Anctil, F. & Coulibaly, P. (2004). Wavelet analysis of the interannual variability in Southern Québec Streamflow. *Journal of climate, Vol.* 17, pp. 163-73, ISNN 0894-8755

Assani, A.A., Charron, S., Matteau, M. & Mesfioui, M. (2010b) Temporal variability modes of floods for catchements in the St.lawrence Watershed (Quebec, Canada). *Journal of Hydrology,* Vol. 385, pp.292-299, ISNN 0022-1694

Assani, A.A., Landry, R. & Laurencelle, M. (2011) Comparison of interannual variability modes and trends of spring streamflow, fall precipitation and streamflow, and winter streamflow in southern Québec (Canada). *Rivers Research and Application* (accepted), ISNN 1534-1459

Bigot, S., Camberlin, P., Moron, V. & Richard, Y. 1997. Structures spatiales de la variabilité des précipitations en Afrique : une transition climatique à la fin des années 1960 ? *Comptes Rendus de l'Académie des Sciences,* Paris, série II a, Vol. 324, pp.181-188.

Chen, D. & Chen, Y. (2003). Association between winter temperature in China and upper air circulation over East Asia revealed by canonical correlation analysis. *Global and Planetary change,* Vol.37, pp.315-325, ISNN: 0921-8181

Cunderlik, J.M. & Ouarda, T.B.J.M. (2009). Trends in the timing and magnitude of floods in Canada. *Journal of Hydrology, Vol.* 375, pp.471-480, ISNN: 0022-1694

Curtis, S. (2008). The Atlantic multidecadal oscillation and extreme daily precipitation over the US and Mexico during the Hurricane season. *Climatic Dynamics*, Vol.30, pp.343-351.

Déry, S.J. & Wood, E.F. (2004). Teleconnection between the Arctic Oscillation and Hudson Bay river discharge. *Geophysical Research Letters*, Vol. 31: LI18205, doi: 1029/2004GL020729, ISNN 0094-8276

Muma, M., Assani, A.A., Landry, R., Quéssy, J-F. & Mesfioui, M. (2011). Effects of the change from forest to agriculture land use on the spatial variability summer extreme daily flow characteristics in Southern Quebec (Canada). *Journal of Hydrology*, ISNN: 0022-1694 (Accepted)

Déry, S.J. & Wood, E.F. (2005). Decreasing river discharge in northern Canada. *Geophysical Research Letters*, Vol. 32: L10401, doi: 10.1029/2005GL022845, ISNN 0094-8276

Dunkeloh, A. & Jacobett, J. (2003). Circulation dynamics of Mediterranean precipitation variability 1948-98. *International Journal of Climatology*, Vol. 23, pp.1843-1866, ISNN 0899-8418.

Enfield, D.B., Mestas-Nuñez, A.M. & Trimble, P.J. (2001). The Atlantic multidecadal oscillation and its relation to rainfall and river flows in the continental US. *Geophysical Research Letters*, Vol. 28, pp.2077-2080, ISNN 0094-8276

Environment Canada. (2004). Données sur les débits de rivières. Province du Québec, CD-Rom, Ottawa (Canada).

Hannachi, A., Jolliffe, I.T. & Stephenson, D.B. 2007. Empirical orthogonal function and related techniques in atmospheric science: a review. *International Journal of Climatology*, Vol.27, pp.1119-1152, ISNN 0899-8418

Kaiser, H.F. (1960) The application of electronic computers to factor analysis. *Educational and Psychological Measurement*, Vol. 20, pp. 141-151, ISNN 0013-1644

Kingston, D.G., Lawler, D.M. & McGregor GR (2006) Linkages between atmospheric circulation, climate and streamflow in the northern North Atlantic: research prospects. *Progress in Physical Geography*, Vol. 30, pp.143-174, ISNN 0309-1333

Koutsoyiannis, D., Montanari, A., Lins, H.F. & Cohn, T.A. (2008). Discussion of "The implications of projetcted climate change for freshwater resources and their management". *Hydrological Sciences-Journal*, Vol. 54, pp.394-405, ISNN 0262-6667

Kundzewicz, Z.W., Graczyk, D., Maureer, T., Pinskwar, I., Radziejewski, M., Svensson, C. &Szwed, M. 2005. Trend detection in river flow series: 1. Annual maximum flow. *Hydrological Sciences Journal*, Vol.50, pp.797-810, ISNN 0262-6667

Kundzewicz, Z.W., Mata L.J., Arnell, N.W., Döll, P., Jimenez, B., Miller, K., Oki, T., Sen, Z. & Shiklomanov, I. (2008). The implications of projected climate change for freshwater resources and their management. *Hydrological Sciences Journal*, Vol.53, pp.3-10, ISNN 0262-6667

McCabe, G.J., Palecki, M.A. & Betancourt JL. (2004) Pacific and Atlantic Ocean influences on multidecadal drought frequency in the United States. *Proceedings of National Academy of Sciences*, USA, Vol. 101, pp. 4136-4141.

Lombard, F. (1987). Rank tests for changepoint problems. *Biometrika*, Vol. 74, pp.615-624, ISNN 0006-3444

Quessy, J.-F., Favre, A.-C., Saïd M. & Champagne, M. (2011) Statistical inference in Lombard's smooth-change model. *Environmetrics*, doi: 10.1002/env.1108 (in press), ISNN 1180-4009

Räisänen, J. (2007). How reliable are climate models? *Tellus, Series A: Dynamic Meteorology and Oceanography*, Vol. 59, pp.2-29, ISNN 0280-6509

Ripelli, C.A. & Nobre, P. (2004). Statistical prediction of sea-surface temperature over the Tropical Atlantic. *International Journal of Climatology*, Vol. 24, pp.45-55, ISNN 0899-8418

Roy, L., Leconte, R., Brissette, F.P. & Marche, C. (2001). The impact of climate change on seasonal floods of a southern Quebec River Basin. *Hydrological Processes*, Vol.15, pp.3167-3179, ISNN 0885-6087

Siew-Yan-Yu, T.O., Rousselle, J., Jacques, D. & Nguyen, V.-T.-V. (1998). Régionalisation du régime des précipitations dans la région des Bois-francs et de l'Estrie par l'analyse en composantes principales. *Canadian Journal of Civil Engineers*, Vol.25, pp.105-1058 ISNN 0315-1468

Vicente-Serrano, S.M. (2005) El Niño and La Niña influence on droughts at different timescales in the Iberian Peninsula. *Water Resources Research*, Vol. 41: W12415, doi: 10.1029/2004WR003908, ISNN 0043-1397

Zhang, X., Vincent, L.A., Hogg, W.D. & Niitsoo, A. (2000). Temperature and precipitation trends in Canada during the 20th century. *Atmosphere-Ocean*, Vol. 38, pp.395-429, ISNN 0705-5900

5

Detecting of a Global and Caribbean Climate Change

Nazario D. Ramirez-Beltran, Joan Manuel Castro and Oswaldo Julca
University of Puerto Rico at Mayaguez,
Puerto Rico

1. Introduction

Weather is defined as what is happening to the atmosphere at any given time while climate is what would be expected to occur at any given time of the year based on many years of meteorological observations. Change in climate constitutes shifts in meteorological conditions lasting a few years or longer. The climate change can occur in a single meteorological variable or in a group of variables affecting a region or the entire Earth (Burroughs, 2001). It is expected that a climate change can be expressed by the behaviour of time series of meteorological variables and in this study, a meteorological variable that expresses a climate change is called a climate indicator. Over the years, the climate of the Earth has changed due to natural or anthropogenic factors, and the research community is concentrating on the identification of the evidences of these changes. However, there are some uncertainties about the occurrence of a significant climate change and especially the time when the changes have become evident. The main purpose of this chapter is to introduce a statistical test to determine when a significant climate change has occurred assuming that a climate indicator is available. A climate indicator is a meteorological variable that reveals the climate of a region or a given part of the Earth. The suggested statistical test will be applied to detect climate changes at the global and Caribbean scale using several climatic indicators.
During the last 140 years the Earth has been experimented several climate changes, which have been documented by several researchers (Huntingford et al., 2006; Hansen 2005; Easterling et al., 2000; Battisti el al., 1997; She and Krueger 2004; and Barnett et al., 1999). For example, Easterling et al., (1997) reported that the global mean surface air temperature has risen about 0.5° C during the 20th century. A large part of the world ocean has shown coherent changes of heat content during the last 50 years (Leuliette et al., 2004). Frich et al., (2002) claim that during the second half of the 20th century the world has become both warmer and wetter for global land areas and currently wet periods produce significantly larger rainfall than a few decades ago. These observed extreme events are in line with the expected changes due to the new greenhouse conditions. Global warming is affecting human lives, and in particular is severely impacting the agriculture forestry and in general the economy (Salinger 2005). For instance, Easterling et al., (2000) pointed out that in the United States since 1987 more than 360 weather events have produced losses in excess of $5 million each event with several catastrophic consequences. The temperature in globe has increased during the last 140 years, because that the number of heat waves has increased (Schar et al., 2004; Changnon et al., 2000). Global warming is a real process that is leading to

catastrophic consequences. It has been documented that the global warming is mainly due to anthropogenic factors (Huntingford et al., 2006; Hansen 2005; Easterling et al., 2000). The International Project of Climate Change (IPCC) has established that most of the observed increment in global temperatures since the mid-20th century is very likely due to the observed increase in anthropogenic greenhouse gas concentrations. The anthropogenic activities that affect the global warming are the emission of greenhouse gases and changes in land use, such as urbanization and agriculture. Recent simulation results have shown that the global warming during the last 20th century cannot only be explained by external forces, but also by natural variability that play an important role (Huntingford 2006, IPCC 2001, IPCC 2007). Battist et al. (1997) pointed out that the global warming may be attributed to natural variability, which can be observed in the energy transported by the atmosphere and ocean circulation. Kruger (2004) and Barnett et al., (1999) have established that the natural variability is due to volcanic eruption (Robock A., 2000), and the solar flux variability. Atmospheric dynamics simulations at global and regional scales have been conducted over different scenarios to predict the most likely future climate impacts (IPCC 2001; Angeles 2005; Huntingford et al., 2006; Hansen 2005). Stott and Ketteborough (2002) claim that predicting Earth surface temperature is almost impossible since the anthropogenic and the natural variability include a large amount of uncertainties that may be difficult to predict. However, Angeles et al., (2007) uses global outputs and a regional atmospheric model to project that during the next five decades there will be large concentrations of rainfall episodes in smaller areas across the Caribbean basin.

Climate change detection and attribution techniques usually apply global or regional circulation models and/or statistical techniques to detect climate changes (Easterling et al., 1997, 2000; Barnett et al., 1999; Schar et al., 2004; Smith et al., 2002; Santer et al., 2005; Meehl et al., 2004; Menne et al., 2005; Tomé and Miranda 2004; Feldstein 2002; IPCC 2001; Smith et al., 2002). A climate change may be expressed as a change in the mean or in the autocorrelation function of the underlying climate indicator. A statistical algorithm for climate change detection is introduced here with the intention of providing a tool to determine when a significant climate change has occurred. The algorithm is based on determining when the mean of the underlying climate indicator exhibits a significant deviation from a selected reference. The algorithm will divide the climate indicator in two parts the reference data and the testing data. The reference data will be used to identify the deterministic and stochastic components of the reference data, and the testing part is used to measure the deviation from the reference. Thus if a significant deviation from the reference data is found a climate change is detected at the identified point in time. The algorithm can detect climate changes that occur in the trend, in the seasonal or in the stochastic component. A simulation technique was used to design a climate indicator with three components and a postulated change was used to validate the performance of the proposed test. Real climate indicators were also used to detect climate change. Regression techniques were used to model the trend and seasonal components, and a time series model was used to represent the stochastic component of real time series data. Climate changes at global and Caribbean scale were studied.

The second section of this chapter will present the basis of the proposed algorithm for detecting changes in a climate indicator. The third section presents a simulation exercise to illustrate the performance of the detection test. The fourth section will describe the data and sources of information, as well as some applications of the test at the Global and Caribbean climate scale, and the last section will present some conclusions.

2. Methodology

The deterministic components of a climate indicator, such as trend, and periodicity will be identified and removed from the data to estimate the stochastic component, which will be modelled by using a time series model. The identified time series model is also removed from the data to obtain a white noise process. Finally, a sequential statistical test will be implemented to detect whether or not a significant deviation from the white noise process has occurred. The algorithm is based on the fact that if a climate indicator does not contain any climate change then the entire time series will behave as a white noise process. On the other hand, if a climate indicator involves a climate change the stochastic behaviour of the testing part will show that the underlying climate change caused a significant deviation from the white noise process. The algorithm includes six major steps: (1) collect the largest time series of a climate indicator; (2) divide the data sets in two parts: the first par will be called the reference data and the second part as the testing data; (3) identify periodicity and trend components based on the reference data, and remove periodicity and trend components from the entire time series and call the resulting time series the estimated of the stochastic component; (4) identify an autoregressive moving average (ARMA) model to the first part of the estimated stochastic component; (5) compute the ARMA fingerprint; and (6) use a sequential hypothesis testing procedure to determine whether or not a significant change has occurred on the mean or in the autocorrelation function of the process. This study will focus on detecting changes on the mean of the process, where the mean of the process may be a constant or a time dependent function.

2.1 Step one: select a climate indicator

It is assumed that climate properties of a given part of the world are expressed by a sequence of a meteorological variable, which we referred as a climate indicator. Thus, a climate indicator can be a time series of air temperature, sea level, rainfall, etc. It is required that the selected time series has no missing values and observations have been obtained at equal time intervals. The climate change can occur at different time scales and to be able to detect a climate change it is required to select a climate indicator that contains observations before and after the climate change. It is desirable that the time series will be large enough to identify the deterministic and stochastic components of the underlying meteorological process and leaving a significant part of data for testing. The length of the data will be established in step two.

2.2 Step two: dividing the time series

The time series will be divided into two parts. The first part will be called the reference data and the second part will be the testing data. The reference data will be used as a reference level to measure the deviation (if any) of the testing data with respect to the reference data. If the underlying time series is a periodic series, the length of the reference data must contain at least three times the length of the period. On the other hand, if the time series is not a periodic series, it is recommended that reference data would contain at least 50 observations. The reference data will be located on the left (older values of the series) and the testing part on the right hand side (more recent values) of the series. Typically, the reference data is located at the beginning of the time series; however, it could be placed in almost any part of the series as long as enough observations are available. The testing part will be at least 50 observations and will be used to measure whether or not there exists a

significant change with respect to the reference data. It should be noted that the change detection test will depend on the meteorological properties of the selected reference and testing data, which can be expressed as follows:
Reference data:

$$\{x_t,\ t = 1, 2,\ \dots, m\} \tag{1}$$

Testing data:

$$\{x_t,\ t = m+1, m+2, \dots, n, \qquad m \geq 50,\ \text{and}\ n \geq 2m\} \tag{2}$$

where x_t represents the underlying climate indicator at time t; m is sample size of the reference data, and n is the total number of observations considered for climate change detection, and $n \geq 2m$.

2.3 Step three: identifying the deterministic components

A climate indicator may be stationary or a nonstationary process and consequently it may have deterministic and stochastic components. The deterministic component may be a trend and/or a periodic component. Thus, the reference data of the climate indicator may be expressed as follows:

$$x_t = T_t + P_t + s_t, \qquad t = 1, \dots, m \tag{3}$$

where T_t and P_t are the deterministic trend and periodic components, respectively at time t, and s_t is the stochastic components at time t. The trend component can be modeled by a polynomial in time and the periodic component by a sinusoidal function of time. The autocorrelation function and the periodogram of x_t can be used to identify the trend and the periodic components; i. e., if the autocorrelation function dies out very fast, the underlying process is stationary and the process do not exhibit trend; otherwise, a trend should be identified. Regression techniques can be used to estimate the parameters of the trend and the sinusoidal function. The Fourier-based method or a more accurate method based on wavelet techniques (Nicolay, et al., 2010) can be used to estimate the size of the required periods. The trend and periodic components are removed from the original time series to estimate the stochastic component, $\hat{s}_t$.

$$\hat{s}_t = x_t - \widehat{T}_t - \widehat{P}_t, \qquad t = 1, \dots, n \tag{4}$$

where $\widehat{T}_t$ and $\widehat{P}_t$ are the estimates of the trend and periodic components, respectively.
In time series analysis is customary to remove the trend and periodic components by using the appropriate difference operators. For example:

$$\nabla^d x_t = (1 - B)^d x_t \tag{5}$$

where ∇^d is the difference-of-order d operator and B^k is the backshift operator, which is defined as follows: $B^k x_t = x_{t-k}$. The operator to remove periodicity can be written as follows:

$$\nabla_p^D x_t = (1 - B^p)^D x_t \tag{6}$$

where D is the order of seasonal difference and p is the period of the time series. Unfortunately, these operators cannot be applied in this case since the application of these operators will remove not only the trend and periodic components but also will remove the climate change signal; and consequently, these operators are not suitable for the purposes of detecting a climate change.

2.4 Step four: identifying the stochastic component

The first m values of the stochastic component, $\hat{s}_t$, are used to identify the autocorrelation structure of the time series. Most of the stochastic components of climate indicators are a sequence that can be represented by autoregressive moving average (ARMA) models (Box and Jenkins 1976). The reference data will be tested first to determine whether or not the reference data is an autocorrelated or a white noise process. If the underlying process is a white noise there is no need of removing the autocorrelation structure. On the other hand, if the autocorrelated structure of the reference data is significant, this dependent structure can be identified by using an ARMA model, which can be expressed as follows:

$$\hat{s}_t = \frac{1-\theta_1 B-\theta_2 B^2-\cdots-\theta_q B^q}{1-\phi_1 B-\phi_2 B^2-\cdots-\phi_r B^r} a_t, \quad t = 1, 2, \dots, m \tag{8}$$

where $\hat{s}_t$, B and m were defined previously; θ's and ϕ's are the parameters of the moving average and the autoregressive components of the model, respectively; a_t is a sequence of independent random variables with mean equal to zero and a constant variance. It should be noted that the transformed reference data should be a stationary process since the trend and periodicity components have been removed from the original time series. Stationary in the sense that the mean and the autocorrelation function will not change over time.

The identification of an ARMA model consists of determining the values of r and q. The identification is accomplished by using the autocorrelation and partial autocorrelation functions of the stochastic component and a numerical parameters estimation algorithm, as described in several time series textbooks (Box and Jenkins 1976; Brockwell and Davis 2002; Wei, 1990; Pandit and Wu 1983). Nonlinear regression techniques are used to estimate the parameters of the ARMA model and several statistical programs are available to perform this estimation task, for instance: Statgraphics, Minitab, Matlab, etc. It should be noted that the main purpose of identifying an ARMA model is to remove the autocorrelation structure, as shown in the next step.

2.5 Step five: computing the ARMA fingerprint

The ARMA fingerprint is the sequence created by the difference at each point in time between the estimated of stochastic component, $\hat{s}_t$, and the estimated stochastic component from the ARMA model, $\tilde{s}_t$. The ARMA fingerprint can be computed as follows:

$$f_t = \hat{s}_t - \tilde{s}_t, \qquad t = 1,2, \dots, n \tag{8}$$

$$\tilde{s}_t = \frac{1-\tilde{\theta}_1 B-\tilde{\theta}_2 B^2-\cdots-\tilde{\theta}_q B^q}{1-\tilde{\phi}_1 B-\tilde{\phi}_2 B^2-\cdots-\tilde{\phi}_r B^r} \tilde{a}_t, \quad t = 1, 2, \dots, m \tag{9}$$

where $\hat{s}_t$ is an estimated of the stochastic component and is computed by using eq. (4); whereas, $\tilde{s}_t$ is an estimated of $\hat{s}_t$ and is computed by evaluating eq. (9), and f_t is the ARMA fingerprint; $\tilde{a}_t$ are the residuals for the stochastic component; $\tilde{\theta}$'s and $\tilde{\phi}$'s are parameter estimated that must be computed with the m values of the stochastic component, $\hat{s}_t$. It should be noted that the model fitting is computed with t=1,…,m; however, the finger print is computed for the entire time series.

Thus, if no change has occurred in the underlying process then the fingerprint will behave as a white noise sequence; where a white noise process is a sequence formed by independent random variables with zero mean and constant variance. However, if the process exhibits a significant deviation from the white noise, the ARMA model will show a unique characteristic which will be exhibited either in the mean or in the autocorrelation function of the given sequence and this special sequence will be called here the ARMA fingerprint. Thus, if a significant change occurs in the mean of the process, the ARMA fingerprint will also exhibit a significant deviation from the mean. On the other hand, if a change occurs in the second moment of the process, the fingerprint may also exhibits a significant deviation in the autocorrelation function.

2.6 Step six: sequential hypothesis testing

If external forces affected the climate indicator, its ARMA fingerprint will present an autocorrelation function with a significant deviation from the autocorrelation of the white noise process. The suggested procedure will detect changes on the mean, and changes in the autocorrelation function. The exponentially weighted moving average (EWMA) test is adopted to detect the change on the mean of the process at every point in time. EWMA test was proposed by Roberts (1959) and adopted here because it is an efficient test to detect a small shift in the mean and also because it is a robust test in the senses that it is not affected by moderate deviations from the Gaussian process as well as because it is not affected by weak autocorrelated time series. Thus, if a climate change induced a strong autocorrelation function it will be detected by EWMA test. The exponentially weighted average, z_t, of the fingerprint is defined as:

$$z_t = \lambda f_t + (1 - \lambda) z_{t-1}, \qquad t = 1, \dots, n \tag{10}$$

A significant increment in the mean occurs at time t if $z_t > U_t$ and a significant decrement occurs in the mean at time t if $z_t < L_t$, where

$$U_t = \mu + M\sigma\sqrt{\frac{\lambda}{2-\lambda}[1 - (1 - \lambda)^{2t}]} \qquad t = 1, \dots, n \tag{11}$$

$$L_t = \mu - M\sigma\sqrt{\frac{\lambda}{2-\lambda}[1 - (1 - \lambda)^{2t}]} \qquad t = 1, \dots, n \tag{12}$$

where f_t is the ARMA fingerprint at time t, μ and σ are the mean and the standard deviation of the ARMA fingerprint sequence during the reference data period (t=1,2,…,m); the initial value of z_t can be estimated by the average of the fingerprint during the reference data period; λ is a weighted constant and varies between zero to one. M is a constant that depend on λ; however, to have better results it is recommended to take the following values of $\lambda = 0.2$ and $M = 3$ Montgomery, (2001).

3. Simulation

A simulation approach is used to generate a climate indicator to measure the capabilities of the suggested detection test. Mote Carlo simulation technique was used to generate a synthetic time series to represent a climate indicator. It is expected that the climate indicator may be the result of ocean and atmospheric interactions and it may include several patterns such as trend, periodicity, and stochastic components. The climate indicator can be represented by the following expression:

$$x_t = \begin{cases} T_t + P_t + s_t, & t = 1, \dots, m \\ T_t + P_t + s_t + \psi_t & t = m+1, m+2, \dots, n \end{cases} \quad (13)$$

where

$$T_t = 5 + 0.05t + 0.0004t^2 \quad (14)$$

$$P_t = 8\cos\left(\frac{2\pi t}{12}\right) + \left(4 + \frac{\ln(t+1)}{16}\right)\cos\left(\frac{2\pi}{30}t\left(1 + \frac{\ln(t+1)}{100}\right)\right) \quad (15)$$

$$s_t = 0.2s_{t-1} + 0.6s_{t-2} - 0.5a_{t-1} + a_t \quad (16)$$

$$\psi_t = \begin{cases} 3\sigma, & t > m \\ 0, & \text{otherwise} \end{cases} \quad (17a)$$

$$\psi_t = \begin{cases} 0.1t & t > m \\ 0, & \text{otherwise} \end{cases} \quad (17b)$$

The variable x_t is the climate indicator at time t; the variable T_t is a quadratic trend; P_t is a periodic component and is composed by two functions, the first function includes a large amplitude with period equal to twelve, and the second function has a moderated amplitude with a function of time and a fixe period equal to 30; the stochastic component is an ARMA(2,1) process, a_t is a Gaussian noise sequence with zero mean and variance 9. A similar function was also used by Nicolay (2010) to represents a climate indicator. The variable ψ_t represents the induced climate change, which may have a step (17a) or a ramp function (17b) that starts when $t = m + 1$. The step function represents a sudden increment of a meteorological variable whereas the ramp function is a slow and persistent increment of climate change. It is assumed that the step change occurred at time m+1 and will persist up to the end of the series with a magnitude of 3 times the standard deviation of the white noise, σ. When the ramp function is used the persistent climate change will be very small at the beginning of the change and the simulated climate change will become evident after the time approaches to n. It is expected that the climate change will be detected right after the change has occurred for the case of the step function; however, when the ramp function is simulated the change is detected after some delay (about 35 units of time).

Figure 1 shows the different components of the simulated time series. The top panel on the right of Figure 1 shows the simulated trend and the top panel on the left shows the periodic function. The middle panel on the left of Figure 1 shows the stochastic component whereas the middle panel on the right shows the step function. The bottom panel of Figure 1 shows the simulated climate indicator, which is the sum of the previously plotted variables.

As described in the previous section, the procedure to implement the detection test includes six steps. The first step consists on selecting a climate indicator to be studied and in this case, the simulated climate indicator x_t, is the selected time series.

The second step consists on dividing the time series in two sets, the first set includes one hundred values, which were assigned to the reference data; i.e., m=100, and the remaining observations correspond the testing data.

The third step consists on identifying the deterministic components, which includes the trend and the periodic component. Figure 2 shows the sample autocorrelation function of the simulated reference data, x_t. This autocorrelation function indicates that the underlying time series is not stationary and includes a strong periodic component. Thus, it is necessary to remove first the trend component to properly identify the periodic functions, and finally, remove the periodic component to develop a stationary time series.

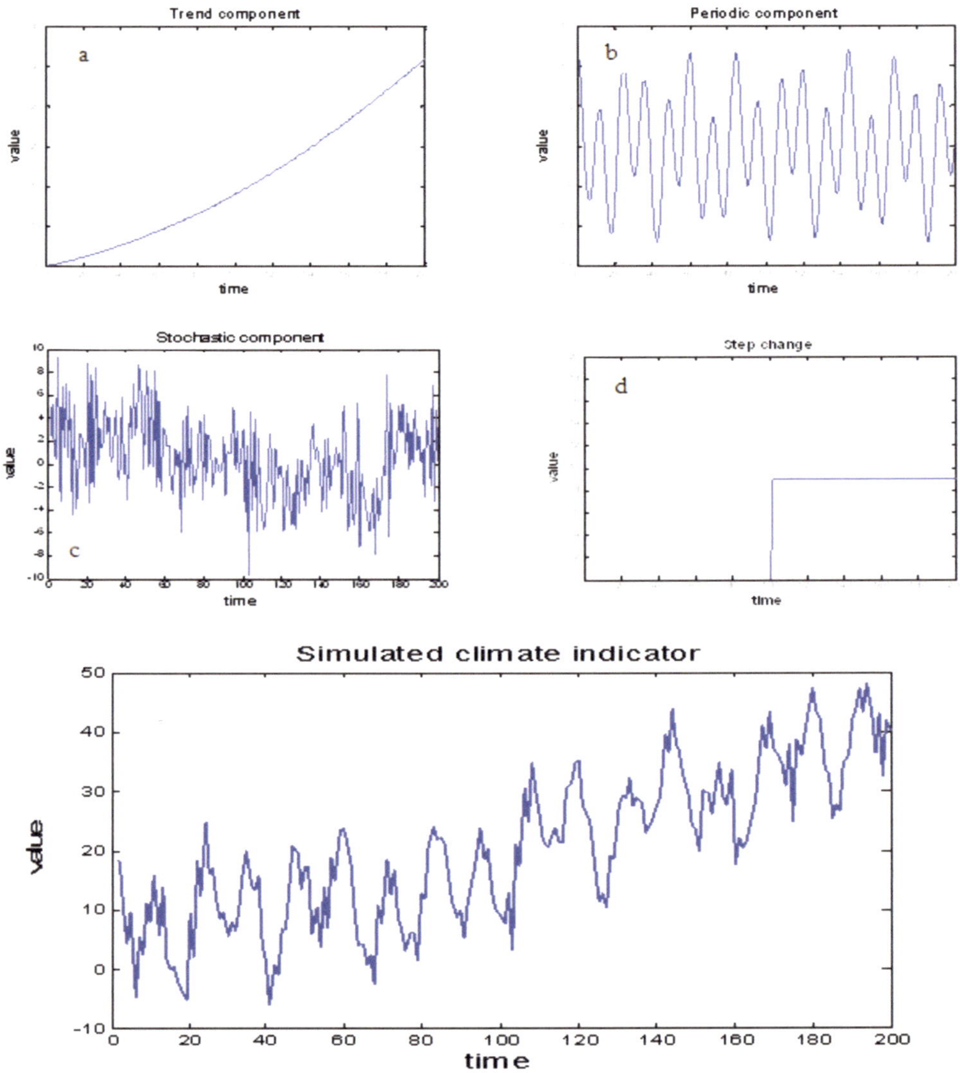

Fig. 1. Panels a, b, c and d show the trend component, the periodic component, the stochastic component, and the step change, respectively. Panel at the bottom shows the climate indicator, which is the result of the sum of the described components.

A quadratic model was used to estimate the trend of the climate indicator. Thus, the following regression model was identified:

$$\hat{T}_t = 7.5 + 0.041t + 0.0003t^2, \qquad t = 1, \ldots, 100 \tag{18}$$

Statistical analysis shows that the included parameters are significant at 5% level. The simulated reference data and the estimated trend are given on Figure 3. In this figure, the continuous and smooth line shows the estimated trend and the broken and continues line is the simulated reference data.

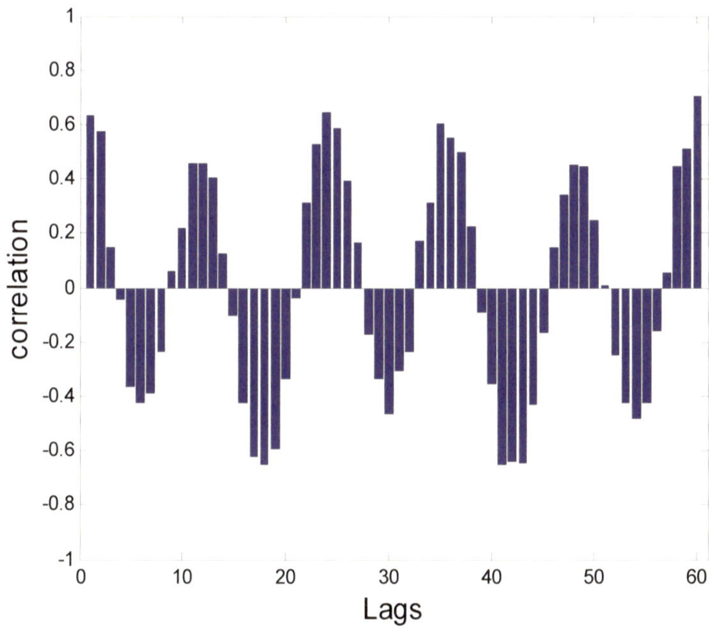

Fig. 2. Autocorrelation function of the simulated reference data.

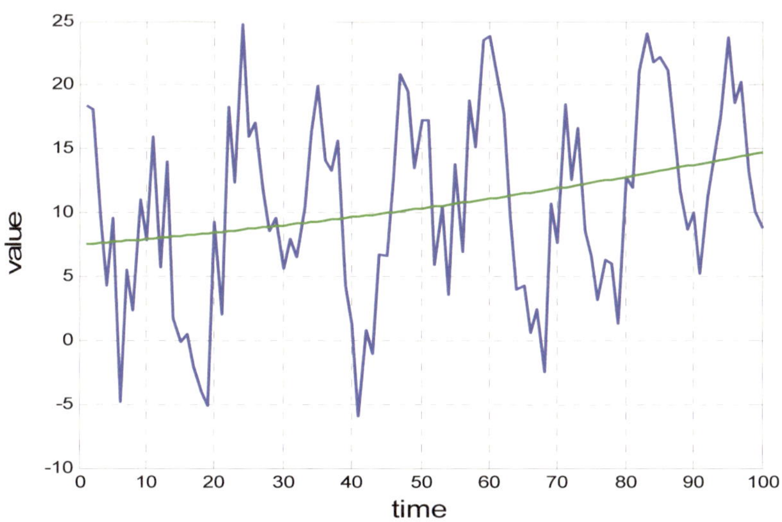

Fig. 3. Simulated reference data and estimated trend.

The trend was removed from the entire time series to create the detrended sequence, which is given below:

$$y_t = x_t - \widehat{T}_t,\ t = 1, \dots, 200 \tag{19}$$

The Fast Fourier Transform was used to develop the periodorgram and two significant ordinates were found indicating that the periodic component includes two harmonic functions, as exhibits in Figure 4. The estimated periods of these functions are 11.9 and 28.4.

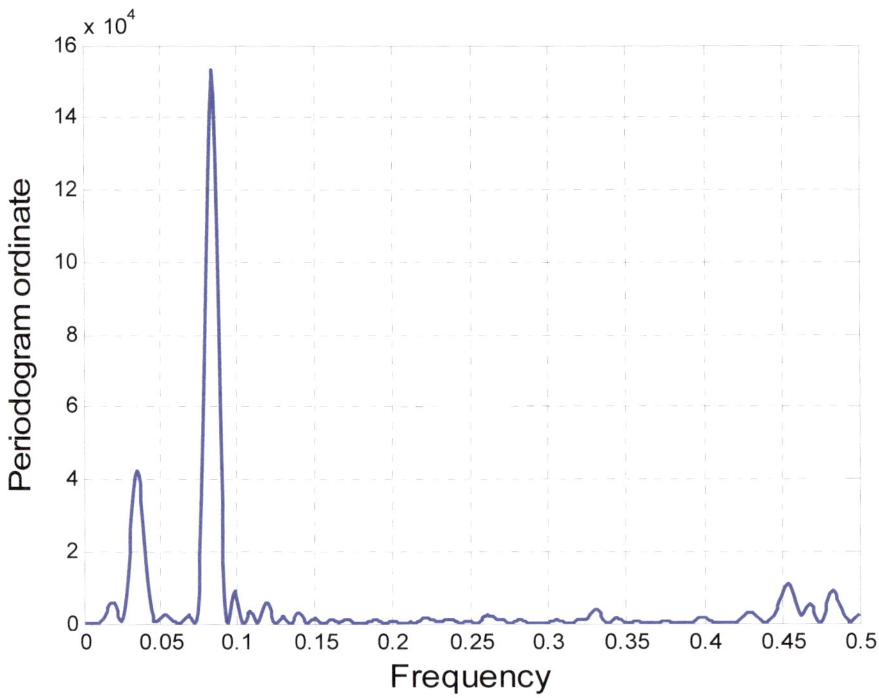

Fig. 4. The periodogram of the deterended time series shows that there are two significant harmonics with the largest ordinates at the frequencies of 0.084 and 0.035 that correspond to the period of 11.9 and 28.4, respectively.

Periodogram and autocorrelation function show that the detrended reference data has a periodic component that can be modeled by using the following sinusoidal function:

$$\widehat{P}_t = 0.0098 - 0.5705\sin\left(\frac{2\pi t}{11.9}\right) + 8.1084\cos\left(\frac{2\pi t}{11.9}\right) - 0.9133\sin\left(\frac{2\pi t}{28.4}\right) + 4.3805\cos\left(\frac{2\pi t}{28.4}\right) \tag{20}$$

Figure 5 shows the reference data with the estimated periodic component. The sinusoidal continuous line shows the estimated periodic component while the broken and continuous line shows the stochastic data.

The trend and the periodic components are removed from the simulated reference data to compute the stochastic component as follows:

$$\widehat{s}_t = x_t - \widehat{T}_t - \widehat{P}_t \tag{21}$$

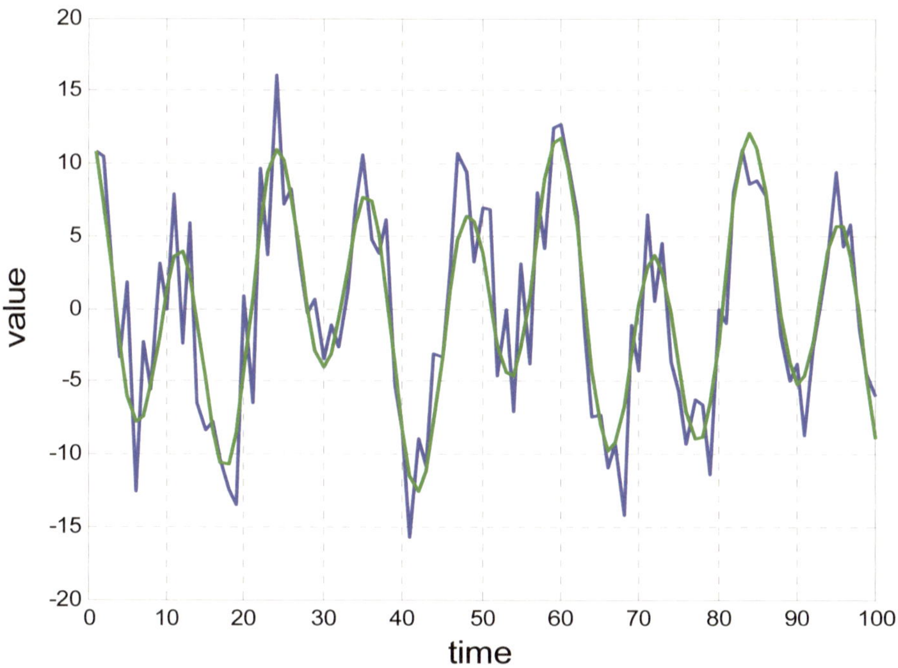

Fig. 5. This figure shows the reference data after removing the trend component and the estimated periodic component. The smooth and continuous line shows the estimated periodic component.

The fourth step consists on identifying a time series model for the stochastic component. The sample autocorrelation function and a parameter estimation algorithm are used simultaneously to identify the corresponding ARMA model. It can be shown that the theoretical autocorrelation function for and ARMA(2,1) can be written as follows (Ramirez-Beltran and Sastri 1997):

$$\rho_k = \frac{\gamma_k}{\gamma_0} \tag{22}$$

where

$$\begin{bmatrix}\gamma_0\\ \gamma_1\\ \gamma_2\end{bmatrix} = \begin{bmatrix}1 & -\phi_1 & -\phi_2\\ -\phi_1 & 1-\phi_2 & 0\\ -\phi_2 & -\phi_1 & 1\end{bmatrix}^{-1} \begin{bmatrix}1+\theta(\phi_1+\theta)\\ \theta\\ 0\end{bmatrix}\sigma^2 \tag{23}$$

$$\gamma_k = \phi_1\gamma_{k-1} + \phi_2\gamma_{k-2}, \qquad k \geq 3 \tag{24}$$

where ρ_k and γ_k are the autocorrelation and autocovariance functions at lag k of the stochastic component, respectively; ϕ's are the autoregressive parameters, and θ is the moving average parameter, and σ^2 is the variance of the white noise. The simulation was based on the following values: $\phi_1 = 0.2$, $\phi_2 = 0.6$, $\theta = -0.5$ and $\sigma = 3$.

The left panel of Figure 6 shows the theoretical autocorrelation function of and ARMA(2,1) model, and the right panel of Figure 6 shows the sample autocorrelation function of $\hat{s}_t$.

The Matlab computer software (MathWorks, 2000) was used to estimate the parameters of the ARMA(2,1), and the estimated model can be written as follows:

$$\tilde{s}_t = 0.19\tilde{s}_{t-1} + 0.56\tilde{s}_{t-2} - 0.68\tilde{a}_{t-1}, \qquad t = 1, \dots, 100 \tag{26}$$

Data also provide information to estimate the standard deviation of the noise, which is $\tilde{\sigma} = 2.74$.

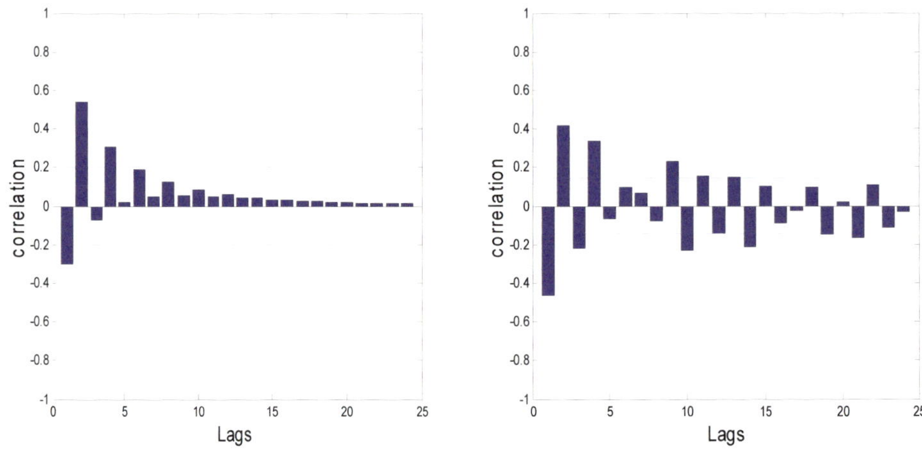

Fig. 6. Theoretical (left) and sample (right) autocorrelation functions for the stochastic component.

The fifth step consists of deriving the ARMA fingerprint. The estimated parameters in the previous step are used to compute the stochastic time series and removed the autocorrelation structure from the simulated stochastic component. Thus, the ARMA fingerprint was computed as follows:

$$f_t = \hat{s}_t - \tilde{s}_t, \qquad t = 1, \dots, 200 \tag{27}$$

Figure 7 shows the ARMA fingerprint, and the first one hundred values resembles to the pattern of a white noise process. However, the last part of the fingerprint shows a significant deviation from the white noise.

The sixth step consists on applying the sequential hypothesis test for detecting the induced climate change. Essentially, the EWMA test includes a 95% confidence interval for the mean of a white noise process. Thus, the values that are outside of the interval indicate that a significant deviation has occurred on the mean; and consequently, that particular observation shows the time when the climate change becomes evident. The left panel of Figure 8 shows that the induced climate change was detected at time t=102, and the red stats that fall beyond the 95 confidence interval indicate that the climate change is evident during this period of time. In this particular exercise, the size of the step change was three time the standard deviation of the noise ($3\sigma = 9$). A climate change was also simulated by using the ramp function. Thus, the equation (17b) was used to induce a slowly increasing climate change. The right panel of Figure 8 shows the results for the ramp function, which indicates that after 35 time units the change becomes evident. The right panel of Figure 8 also shows the sequential hypothesis test, which indicates the climate change is large enough to be detected, and the detection occurs at time t=135.

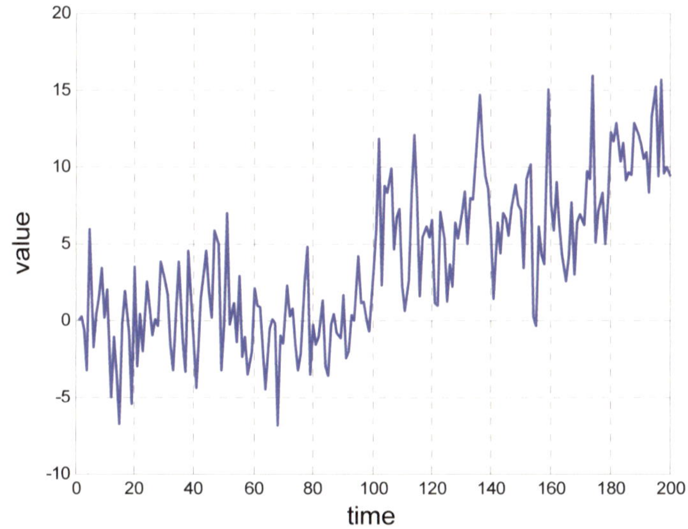

Fig. 7. The ARMA fingerprint of the simulated process.

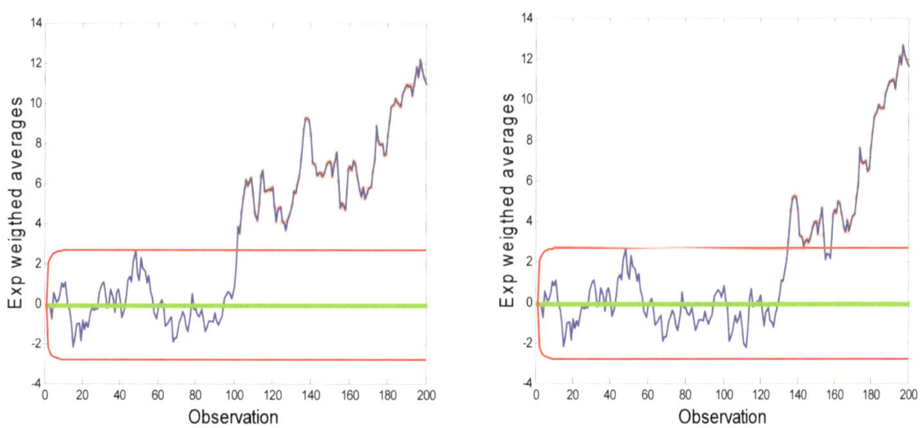

Fig. 8. The left panel shows the performance of the detection test when a step change has been induced and this change has been detected at t=102. On the other hand, the right panel shows results when the climate change was induced by a ramp function. In this case the change was detected at t=135; i.e., after 35 units of delay.

4. Climate change detection

Four data sets were used to implement the climate change detection test. The first two sets are associated with the factors that induced a climate change, and other two are climate indicators related to meteorological variables that exhibited the vestiges of a climate change embedded along the time series.

4.1 Attribution variables

The factors that induced a climate change are known as attribution variables and in this study, we selected two attribution variables (IPCC, 2001). The sunspots are considered as a natural attribution variable; i.e., the induced changes are the result of sun energy variations that directly impact the Earth climate conditions, whereas the carbon dioxide (CO_2) emissions are considered as an anthropogenic attribution variable.

4.1.1 Sunspots

Sun exhibits signs of varying activity in the form sunspots. These are dark areas, which are seen at lower latitudes and crossing the phase of the sun as it rotates, and are cooler than the surrounding chromospheres. A sunspot consist of two regions a dark central umbra at temperature of around 4,000°K and surrounding lighter penumbra at around 5,000°K. Thus, the darkness is purely a matter of contrast that appears dark compared to the general brightness of the sun. A spot may be from $1x10^3$ to $2x10^5$ km in diameter with a life cycle from hours to months (Burroughs, 2001). If the amount of energy emitted by the sun varies over the time and the Earth is receiving a radiation from the sun; consequently, some changes on the Earth surface temperature may be attributed to variation of solar radiation. In 1843 Heinrich Schwabe (Burroughs, 2001) discovered that the number of sun spots exhibits a periodically behaviour; however, it was until 1980 when satellite data confirmed such discovery. She et al. (2004) pointed out that there is a relationship between temperatures observed in mesopause and the effect of solar cycle. Satellite data has been used as a medium to support that an indirect measurement of solar radiation can be obtained by studying the behaviour of sunspots. For instance, Julca (2007) shows that sunspots and observed solar irradiances exhibit 0.77 of correlation coefficient and this result confirms that the solar activity may be studied by analysing the behaviour of sunspots. Figure 9 exhibits the comparison of observed sunspot pattern and solar irradiance during the period 1979-2005. The analysed time series of sunspots was obtained from the Royal Observatory of Belgium (http://sidc.oma.be/sunspot-data/), and the studied period was from January 1750 until February 2011. Figure 10 shows the patterns of the sunspots over the studied period.

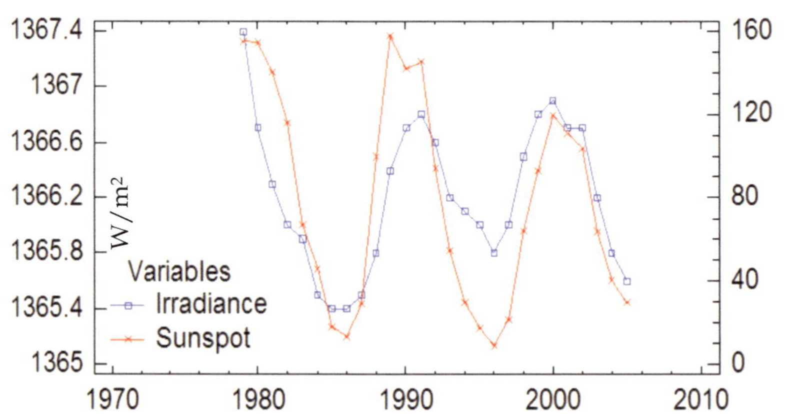

Fig. 9. Sunspots and solar radiation. The left vertical axis shows the scale of solar radiation and is given in W/m²; whereas, the right scale shows the number of sunspots.

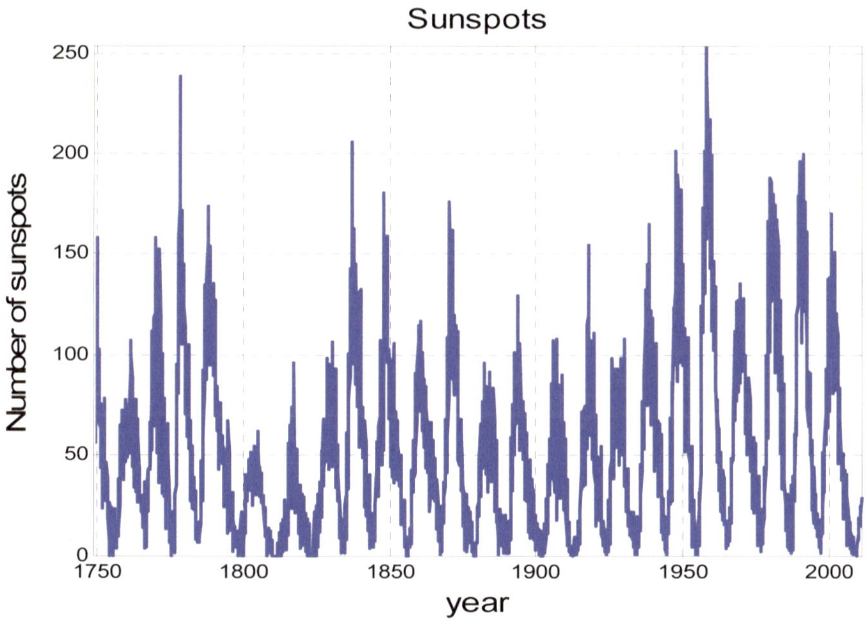

Fig. 10. The monthly sunspots form January 1749 to February 2011.

The algorithm for detecting climate change was applied to identify whether or not the sunspots attribution variable exhibits a significant change on the mean. The half of the available observations was used as the reference data (from January 1749 to December 1880) and the remaining part of the series was used as the testing data (from January 1881 to February 2011). The selected reference data do not exhibit a significant trend; and consequently, no trend was removed. However, sunspots show an unstable variance; i.e., a data transformation should be explored to stabilize the variance. Logarithmic transformation was discarded since it produces extreme lower values and a significant bias is introduced to the data. The squared root transformation was applied to data and better results were found. Six harmonic functions were needed to model the periodic component. Table 1 shows the parameter estimation of the sinusoidal functions. After removing the periodic component, an ARMA(2,1) model was identified into the stochastic component. Thus, the fingerprint of sunspots was computed as follows:

$$f_t = \hat{s}_t - \tilde{s}_t \tag{28}$$

where

$$\hat{s}_t = \sqrt{x_t} - \widehat{P}_t \tag{28}$$

$$\widehat{P}_t = -0.201 + \sum_{i=1}^{6} \hat{b}_i \sin\left(\frac{2\pi t}{p_i}\right) + \hat{c}_i \cos\left(\frac{2\pi t}{p_i}\right) \tag{29}$$

$$\tilde{s}_t = \left(\frac{1-0.7057B}{1-1.164B-0.1903B^2}\right)\tilde{a}_t \tag{30}$$

where f_t is the ARMA fingerprint and x_t is the number of sunspots at the month t ($t = 1, \ldots, n$), $n = 3146$, and $m = 1573$; the standard deviation of the estimated white noise process is 1.099. After removing the seasonal component the ARMA fingerprint technique was implemented and the monthly stochastic component shows a significant increment during about three decades from 1955 to 1989 and especially during the first decade (1955-1965) it was detected a significant increment of sunspots as shown in Figure 11.

i	$\hat{b}_i$	$\hat{c}_i$	p_i (months)
1	-0.6842	1.8636	135
2	1.3737	-1.1725	899
3	-0.1579	1.2634	121
4	0.19	1.2516	166
5	-0.4951	-0.9182	101
6	0.4223	-0.3006	111

Table 1. Parameter estimation of the sinusoidal function

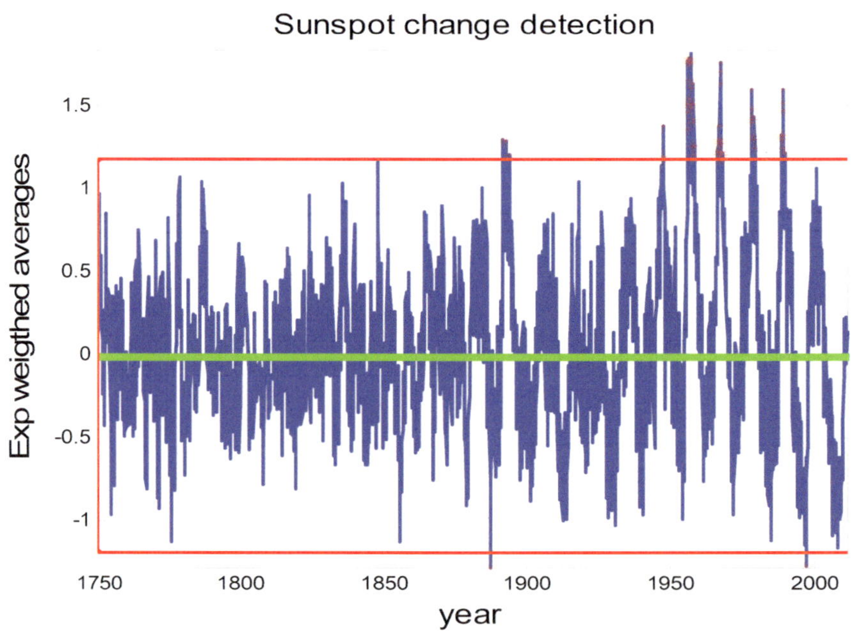

Fig. 11. Sunspots show an evident increasing amount of sunspots from 1955 to 1989, with a significant increment during the decade (1955 to 1965). The reference data were from January 1749 to December 1880, and the testing data from January 1881 to February 2011.

The direct link between sunspot number and solar output fits with the hypothesis that cold period known as the Little Ice Age and the colder weather of the late seventeen century was the result of an almost complete absence of sunspots known as the Maunder minimum (Burroughs, 2001). The climate impact of changes in solar radiance in the ultraviolet (UV) region has been suggested. Because wavelengths between 200 to 300 nm are absorbed high in the stratosphere by oxygen and ozone, they initiate photochemical reactions which influence the weather at lower levels. It has been shown that the amount of solar radiation entering the lower atmosphere varies with solar activity, as a result of alterations in stratospheric ozone concentrations caused by the changing UV flux. This reduces the amount of solar energy reaching the lower atmosphere in middle and high latitudes in winter when solar activity is high. These changes could have a significant impact on global circulation; the increase solar UV radiance in the lower tropical stratosphere will expand the Hadley circulation leading to a pole ward shift of the sub-tropical westerly jet (Burroughs, 2001). Another consequence of changing UV fluxes reaching the lower atmosphere is to affect the formation of free-radicals in the lower atmosphere (hydroxyl radical). This alters the production of condensation nuclei and hence the formation of clouds. In effect, more UV radiation reaching the troposphere will increase the concentration of condensation nuclei and hence make it cloudier environment. Thus, varying of solar activity it will modify the climate over the Earth.

4.1.2 Carbon dioxide

The carbon dioxide time series was obtained from Mauna Loa station Hawaii. Data are reported as a dry air mole fraction defined as the number of molecules of carbon dioxide divided by the number of all molecules in air, including carbon dioxide (CO_2) itself, after water vapor has been removed. Thus, a mole fraction of CO_2 is expressed as parts per million (ppm) and is the number of molecules of CO_2 in every one million molecules of dried air. The CO_2 monthly data is the monthly mean CO_2 mole fraction determined from daily averages. This data set includes about 53 years of data from March 1958 to February 2011. The underlying data includes a few missing values which were estimated by the data source (http://www.esrl.noaa.gov/gmd/ccgg/trends/). Information from this station has the longest continuous record of CO_2 concentrations in the world. This climate attribution variable has been considered as one of the most favourable locations for measuring undisturbed air because of the Hawaii environmental conditions. It should be noted that the volcanic events were excluded from the records (Keeling and Whorf 2005). Figure 12 shows the behaviour of the observed CO_2.

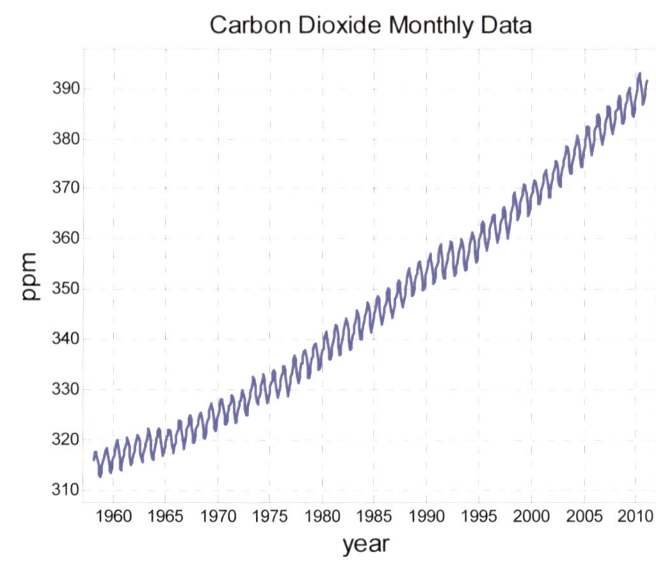

Fig. 12. Monthly carbon dioxide at Mauna Loa Hawaii, from March 1958 to February 2011.

The carbon dioxide shows a strong trend and seasonality components. The reference data includes 312 monthly values and correspond to the period of 26 years from March 1958 to February 1984 and the testing data includes 27 years of data from March 1984 to February 2011. To perform a climate change detection analysis there may be two possibilities, depending of the identified trend. It can be fitted either a linear or a quadratic trend to the reference data, and the parameters of both the linear and the quadratic trend are significant and also the proportions of explained variance are about the same; therefore, it is justifiable the application of a linear or a quadratic trend. However, the results are quite different and the interested reader should be aware about the interpretation of results.
The analyses of linear and quadratic trend are given in Table 2. The first column indicates the component to be analysed and the second and the third columns refer to results from the linear and the quadratic analysis, respectively.

Component	Linear Trend				Quadratic Trend			
Fingerprint	$f_t = \hat{s}_t - \tilde{s}_t$				$f_t = \hat{s}_t - \tilde{s}_t$			
Stochastic	$\hat{s}_t = x_t - \hat{T}_t - \hat{P}_t$				$\hat{s}_t = x_t - \hat{T}_t - \hat{P}_t$			
Trend	$\hat{T}_t = 312.95 + 0.0918t$				$\hat{T}_t = 315.34 + 0.0461t + 0.000145t^2$			
Periodic component	$\hat{P}_t = \sum_{i=1}^{2} \hat{b}_i \sin\left(\frac{2\pi t}{p_i}\right) + \hat{c}_i \cos\left(\frac{2\pi t}{p_i}\right)$				$\hat{P}_t = \sum_{i=1}^{2} \hat{b}_i \sin\left(\frac{2\pi t}{p_i}\right) + \hat{c}_i \cos\left(\frac{2\pi t}{p_i}\right)$			
	i	$\hat{b}_i$	$\hat{c}_i$	p_i (months)	i	$\hat{b}_i$	$\hat{c}_i$	p_i (months)
	1	2.5017	0.971	12	1	2.5011	0.969	12
	2	-0.6331	-0.3541	6	2	-0.6334	-0.3545	6
Stochastic	$\tilde{s}_t = \left(\frac{1 - 0.5882B}{1 - 1.235B - 0.2431B^2}\right)\tilde{a}_t$				$\tilde{s}_t = \left(\frac{1 - 0.6327B}{1 - 1.217B + 0.2723B^2}\right)\tilde{a}_t$			
Standard deviation of $\tilde{a}_t$	0.3146				0.3036			

Table 2. Analysis for a linear and a quadratic trend.

The left panel of Figure 13 shows the linear trend that was fitted to the reference data and this function was evaluated for t during the testing period ($t = 313, 314, \ldots, 636$). This figure indicates that the expected mean (straight line) is smaller than the actual mean during the testing period. Therefore, the detection test should indicate that there is a significant increment on the mean during the testing data. This figure indicates that during the reference data the mean of CO_2 has an increasing rate of 0.0918 ppm per month; however, during the testing period the increasing rate growth larger than the linear trend, since the sequential test detects an increment with respect to the mean (linear trend). On the other hand, the right panel of Figure 13 shows a quadratic trend that was fitted to the reference data and this function was evaluated for t during the testing period. This figure indicates that the expected mean (parabola) is larger than the actual mean during the testing period. Therefore, the detection test should indicate that there is a reduction with respect to the mean (quadratic trend). Apparently, the emissions of CO_2 during the last decade have been reduced with respect to the quadratic emission rate. However, the reduction on emission is still larger than the linear trend.

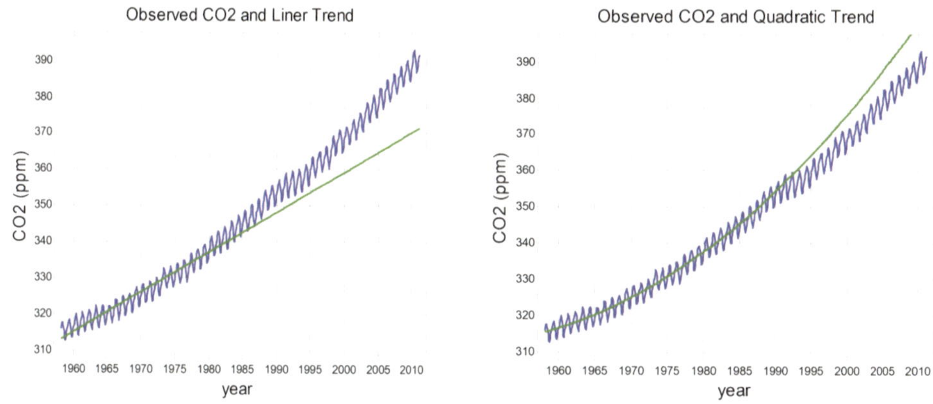

Fig. 13. The left panel shows the observed CO_2 and a linear trend, which was fitted to the reference data and evaluated during the testing data. The right panel shows the observed CO_2 and the quadratic trend, which was fitted to the reference data and evaluated during the testing data. The reference data is from March 1958 to February 1984 and the testing data is from March 1984 to February 2011.

The left panel of Figure 14 shows that in 1988, the emission rate of CO_2 is larger than the mean rate 0.0918 ppm per month and this event becomes evident during the period of 1995-2011. The right panel of Figure 14 shows that the reduction with respect to the mean (parabola) is evident during the period 1990-2011. Thus, the CO_2 during the testing period exhibited an increasing emission rate that is larger than the linear but smaller than the quadratic rate and this phenomenon is clearly exhibited during the period 1990-2011.

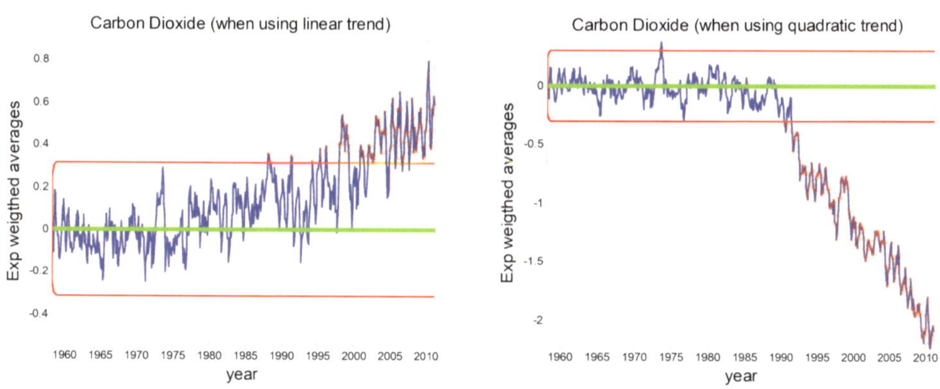

Fig. 14. The left panel shows that in 1988, the emissions of CO_2 are larger than 0.0918 ppm per month and this event becomes evident during 1995-2008. The right panel shows the relative reduction of concentration of CO_2 when using a quadratic trend. The relative reduction occurred during 1992-2008.

4.2 Climate indicators

A climate indicator is a time series of a meteorological variable that contains the behaviour of a climate of the Earth and may exhibit a climate change. The selected climate indicators were the surface temperature, and cloud cover. The information contained in the climate indicators come from ground meteorological stations, satellite observations, and numerical weather prediction models.

4.2.1 Global surface temperatures

The global surface temperature is a climate indicator that has been used to show evidence that the Earth is warming and has become much stronger during recently years as reported in the Third Assessment Report (IPCC 2001). The IPCC (2007) also reported that from 1995-2006 were the warmest years since 1850. The global surface temperatures were provided by the Goddard Institute for Space Studies (GISS), and include the period from January 1880 to December 2010. The surface data set was developed based on the Global Historical Climatology Network (GHCN). This analysis included observations from 6,300 ground stations located in different parts of the world (Hansen et al. 2010). Data from stations were confirmed with satellite data, and global climate model. Global temperature analyses are routinely either omitting urban stations or adjusting their long-term trends to try to eliminate or minimize the urban effect. The GISS analysis used 1951–1980 as the base period to develop the global temperature anomalies and the detailed description of data analysis is presented by Hansen et al. (2010). The monthly anomalies of the global land-ocean surface (AGLO) temperatures are shown in Figure 15, this product was selected because involves both the land and ocean measurements. These data were acquired at the web site http://data.giss. nasa.gov/gistemp/.

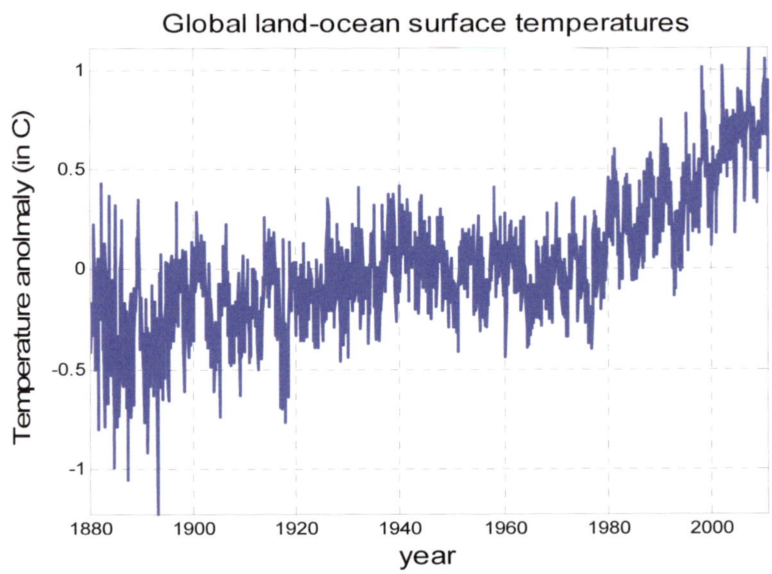

Fig. 15. Monthly anomalies of global land-ocean surface temperatures (1880-2010).

The AGLO temperatures were divided in two parts. The reference data includes the first one hundred years from January 1880 to December 1979 (m = 1200) and thirty-one years for the testing data from January 1980 to December 2010. The reference data show a significant linear trend with and increment of 0.00029°C per month. After removing the linear trend, the autocorrelation function and the periodogram were computed and it was found that this data do not include a significant periodic component, this is because that climatology was

removed by GISS working group. The Autocorrelation function shows that the stochastic component can be represented by and ARMA(1,1).
Thus, the ARMA fingerprint can be computed as follows:

$$f_t = \hat{s}_t - \tilde{s}_t \tag{31}$$

where

$$\hat{s}_t = x_t - \widehat{T}_t = x_t + 0.28 - 0.00029t \tag{32}$$

$$\tilde{s}_t = \left(\frac{1-0.4023B}{1-0.827B}\right)\tilde{a}_t \tag{33}$$

The variable x_t represents the AGLO temperatures; the standard deviation of $\tilde{a}_t$ for $t = 1, \ldots, m$ was 0.169°C. Figure 16 shows that the EWMA sequential test indicating a significant increment of AGLO temperature, which becomes evident in 1998. During the reference part (1880-1890), there are a few observations that exhibit some cooling behaviour or may be the presence of outliers.

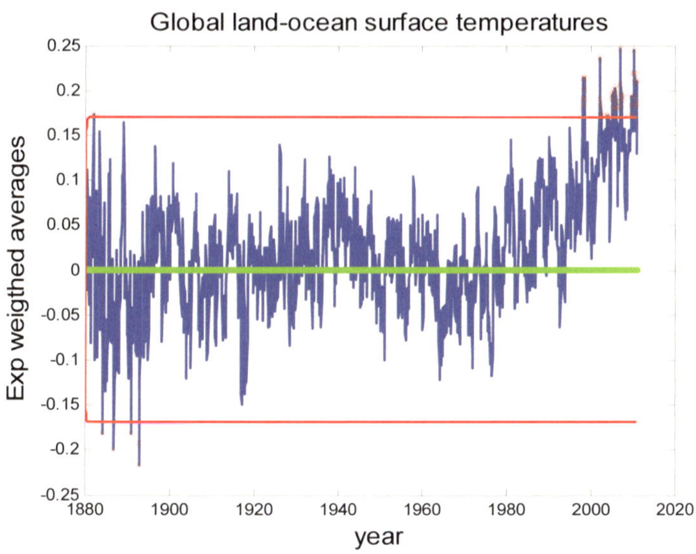

Fig. 16. The AGLO temperatures show a significant increment that becomes evident en 1998.

4.2.2 Caribbean surface air temperatures

Air temperatures for the major Caribbean islands were obtained from GHCN version 3. (ftp://ftp.ncdc.noaa.gov/pub/data/ghcn/v3/). The studied Caribbean islands are Cuba (CU), Jamaica (JA), Puerto Rico (PR), and La Espanola, which includes Dominican Republic (DR) and Haiti (HA). Table 3 shows the summary of stations that were used in this work. The monthly air temperature used in this analysis includes the following period: from January 1948 to February 2011. The quality of the data set was improved by removing inhomogeneities from the data record associated with non-climatic influences such as changes in instrumentation, and station environment (Peterson and Easterling, 1994). The monthly surface air temperatures from the National Center for Environmental Prediction

(NCEP) reanalysis data were also used as a proxy variable to estimate some of the missing values that were encounter in some stations. The nearest grid point to each island was used to derive a regression equation between the temperature at a given station and the temperature of the nearest NCEP-grid-point. Regression equations exhibit an average of 0.9 of correlation coefficient between station temperature and NCEP data.

Country	Cuba	Dominican Republic	Haiti	Jamaica	Puerto Rico
Number of stations	14	28	1	5	15

Table 3. Number of stations over the Caribbean area.

Thirty years of data (from January 1948 to December 1977) were used to estimate climatology and anomalies; and after performing this calculations some of the data do not exhibit seasonal component. Figure 17 shows the monthly temperature anomalies for the major Caribbean islands from January 1948 to February 2011. Since Haiti provides only one station, the analysis for this country was omitted.

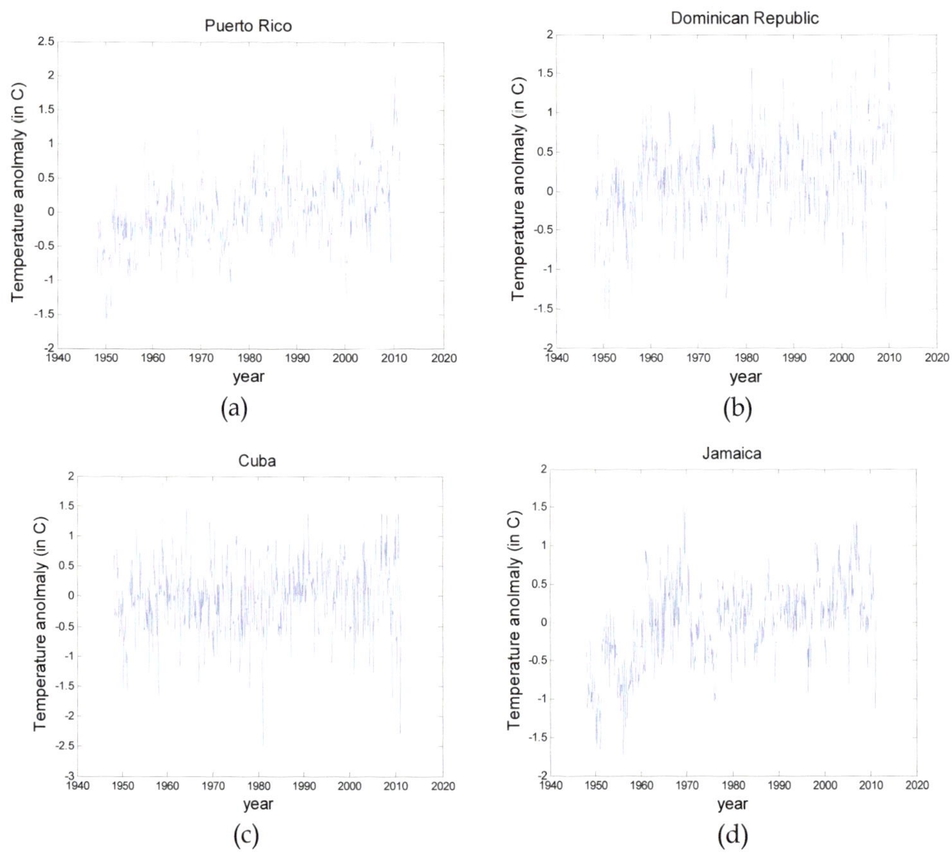

Fig. 17. Monthly temperature anomalies for the major Caribbean island. Data set includes observations from January 1948 to February 2011.

The reference data of monthly anomalies for Puerto Rico exhibit a linear trend with an increasing rate of temperature of about 0.00095°C per month. The trend was removed and the periodicity component was very weak since climatology was subtracted from data; and consequently, this component was therefore deleted. The Autocorrelation function shows that the stochastic component can be represented by and autoregressive process (AR) of order one, AR(1). The AR process is a particular case of the ARMA model in which the moving average component is missing. The procedure to calculate the fingerprint is outlined in Table 4. The sequential statistical test was implemented to detect if there is any deviation from the mean. Figure 18a shows that Puerto Rico air temperatures indicate an additional increment in temperature that become evident in year 2010. This figure also shows some false alarms that occurred on 1988 and 2000. This pattern shows a weak climate-change signal over the island of Puerto Rico.

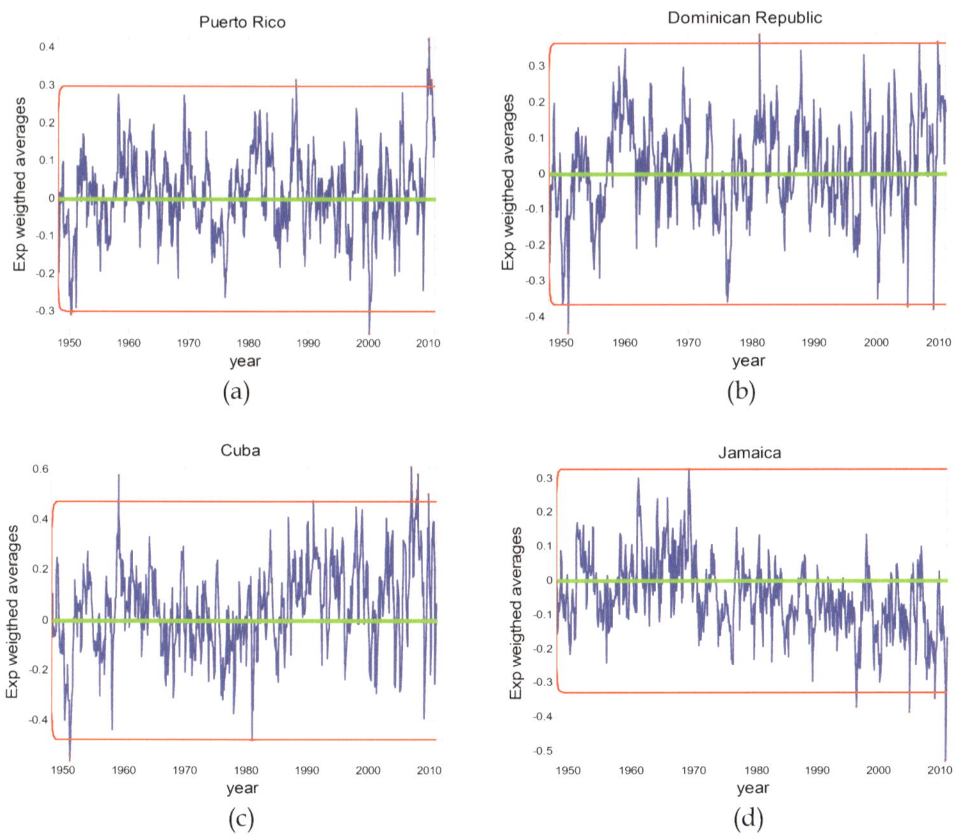

Fig. 18. a) This figure shows that in 2010 a significant climate change occurred in Puerto Rico. This figure also shows two false alarms occurred on 1988 and 2000; b) For Dominican Republic there was a significant trend during the reference data and no climate change during testing data, only a couple of false alarms occurred on 1981, 2005, 2009, and 2010; c) Cuba shows no trend during the reference data; however, there was a significant climate change that occurred during 2007-2010; d) Jamaica shows the largest trend during the reference data and a reduction on surface air temperature during the testing data, which indicates that the there were an over estimation of the linear trend during the reference data.

The first thirty years of Dominican Republic were used to estimate climatology and also were used as the reference data (from January 1948 to December 1977). The reference data of temperature anomalies for Dominican Republic exhibit a linear trend with an increasing rate of about 0.00073°C per month. The harmonic analysis and the autocorrelation function show that there is a significant periodic component with period equal to 12 months. The stochastic component was computing after removing the trend and periodic component, as shown in Table 4. The autocorrelation function of the stochastic component shows that this process can be represented by an autoregressive (AR) process of orders one. Figure 18b shows that there was no significant climate change for Dominican Republic. This figure also exhibits a couple of false alarms that occurred on the following years: 1981, 2005, 2009, and 2010.

Component	Puerto Rico	Dominican Republic
Fingerprint	$f_t = \hat{s}_t - \tilde{s}_t$	$f_t = \hat{s}_t - \tilde{s}_t$
Stochastic	$\hat{s}_t = x_t - \widehat{T}_t$	$\hat{s}_t = x_t - \widehat{T}_t - \widehat{P}_t$
Trend	$\widehat{T}_t = -0.40 + 0.00095t$	$\widehat{T}_t = -0.1372 + 0.00073t$
Periodic component	none	$\widehat{P}_t = -0.0057 \sin\left(\frac{2\pi t}{12}\right) - 0.0054 \cos\left(\frac{2\pi t}{12}\right)$
Stochastic	$\tilde{s}_t = \left(\frac{1}{1 - 0.703B}\right) \tilde{a}_t$	$\tilde{s}_t = \left(\frac{1}{1 - 0.626B}\right) \tilde{a}_t$
standard deviation of $\tilde{a}_t$	0.30 C	0.36 C
Component	Cuba	Jamaica
Fingerprint	$f_t = \hat{s}_t - \tilde{s}_t$	$f_t = \hat{s}_t - \tilde{s}_t$
Stochastic	$\hat{s}_t = x_t - \widehat{P}_t$	$\hat{s}_t = x_t - \widehat{T}_t$
Trend	None	$\widehat{T}_t = -0.4998 + 0.00270t$
Periodic component	$\widehat{P}_t = -0.0103 \sin\left(\frac{2\pi t}{12}\right) - 0.0065 \cos\left(\frac{2\pi t}{12}\right)$	None
Stochastic	$\tilde{s}_t = \left(\frac{1}{1 - 0.4248B}\right) \tilde{a}_t$	$\tilde{s}_t = \left(\frac{1 - 0.3753B}{1 - 0.9071B}\right) \tilde{a}_t$
standard deviation of $\tilde{a}_t$	0.47 C	0.32 C

Table 4. Calculations of the ARMA fingerprint for the major Caribbean islands.

As in the previous islands, the first thirty years were used to estimate climatology of the surface air temperature of Cuba. The reference data include monthly observations from January 1948 to December 1978. The anomalies of reference-data of air temperature from Cuba exhibit a no significant trend. The harmonic analysis and the autocorrelation function show that there is a significant periodic component with period equal 12 months. The autocorrelation function of the stochastic component shows that this process can be represented by an AR(1) process. Estimates of a sinusoidal function and the stochastic component are given in Table 4. Figure 18c shows that there was a significant climate

change occurred on Cuba and becomes evident during 2007 to 2010. This figure also shows that there are some false alarms that occurred on 1981, and 1991.
In Table 4 the variable x_t represents the anomaly temperatures for the corresponding country, B is back shift operator; m = 360 and n = 758.
The reference data of monthly anomalies for Jamaica exhibit a linear trend with an increasing rate of temperature of about 0.00270 °C per month. The trend was removed and periodicity component was not significant component. The autocorrelation function shows that the stochastic component can be represented by and ARMA(1,1). The fingerprint procedure is outlined in Table 4 and estimates are also given in this table. The sequential statistical test was implemented to detect an addition deviation from the mean in the testing data. Figure 18d shows that Jamaica air temperatures indicate there is a possible reduction with respect to the linear trend; i.e., there was an over estimation of the increasing rate given for a reference data. It should be noted that Jamaica during the first two decades (1948 to 1968) shows a significant increasing rate of temperature and caused an over estimation of trend during the reference data. Jamaica increasing rate is about three times higher than Puerto Rico and four times than Dominican Republic.
In summary, Jamaica exhibits the largest increasing rate of surface air temperature and this phenomenon occurred during 1948 to 1968. Cuba shows no increasing rate during the reference data; however, exhibits a significant increment during 2007 to 2010. Puerto Rico shows a linear trend during the reference data in addition to a significant increment of temperature in 2010. Dominican Republic shows a significant trend during the reference data and no more changes during the testing data. In general, the climate change exhibited in the Caribbean islands is marginal compared to a global scale.

4.2.3 Global cloud cover

The cloud cover monthly time series was obtained from the International Satellite Cloud Climatology Project (ISCCP). The cloud products were generated from sensors located on seven satellites, and the D2 product provides the properties of the clouds observed at every three hours and presented in monthly time series during the period of July 1983 to June 2005. Some of the included variables in this data set are cloud cover, top-cloud temperature, top-cloud pressure, optical thickness, and water path. The clouds are classified based on optical thickness and on top pressure. More information can be found in the following site: http://iridl.ldeo.columbia.edu/SOURCES/.NASA/.ISCCP/.D2/.all/. Quispe (2006) developed a user friendly computer program to read and manage the cloud data files.
The global cloud cover file includes 6,596 grids and the cloud cover was selected at global and at Caribbean scales. Figure 19 shows the global cloud cover.
Data from July 1983 to June 1994 were used as reference and from July 1994 to June 2005 as the testing data; the following values were selected for n = 264 months, and m = 132 months. A significant linear trend was identified and the harmonic analysis shows that there are three significant harmonics at 12, 132 and 66 months. Thus, the linear trend and the sinusoidal functions were removed from the original data and the stochastic process follows an AR(1) model. The estimates of the trend, the sinusoidal function, and stochastic model are given in Table 5. The left panel of Figure 20 shows the performance of the sequential test, which shows and an additional reduction of cloud cover that occurs during 1997 to 2001; however, increment was also detected from 2004 to 2005 and a possible false alarm occurs on 1994. With the purpose of better understanding the behaviour of the clouds, a second analysis was conducted without trend; i.e., in this case we want to measure the deviation

with respect to the constant average of the reference data. The right panel of Figure 20 indicates that a significant reduction occurs with respect to the reference mean during 1995 to 2004. In summary, it can be concluded that a significant reduction of global cloud cover occurs and especially during the period of 1995 to 2004.

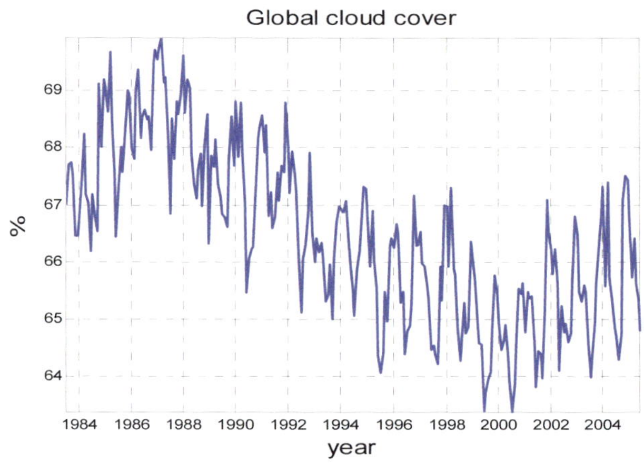

Fig. 19. Global cloud cover.

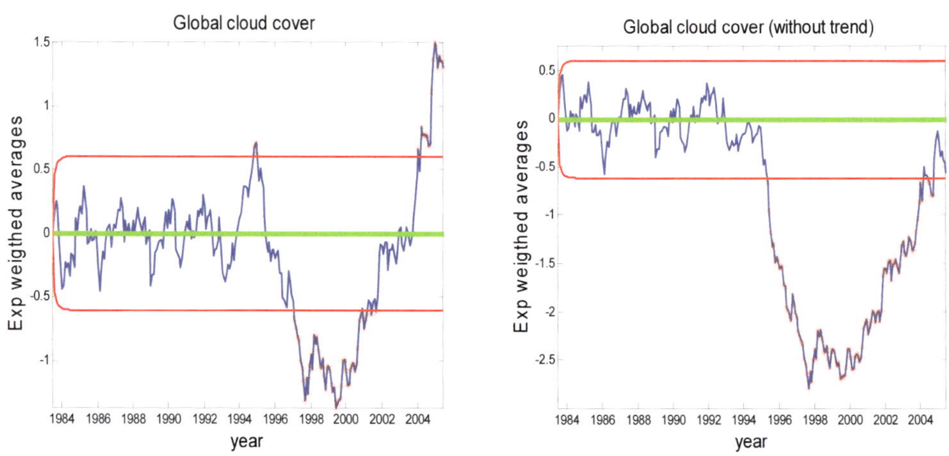

Fig. 20. The left panel shows that a significant reduction was detected during 1997 to 2001. The right panel shows an analysis without trend, and confirmed that a significant reduction of global cloud cover occurred during 1995 to 2004.

4.2.4 Caribbean cloud cover

The selected Caribbean area includes the following geographical location: latitude from 17 N to 24 N and longitude from 87 W to 64 W, as shown in Figure 21. The data was organized as follows: the reference data include from July 1983 to June 1994 and the testing data are from July 1994 to June 2005. The reference data exhibit a significant reduction rate (-0.0259 per month) of cloud cover and the harmonic analysis shows that there is a significant periodic function with period equal to twelve. The remaining stochastic component can be modelled by a moving average of order one; i.e., MA(1). This model is also a particular case of the ARMA model, in which the model does not contain the autoregressive part. The

Component	Global cloud cover			
Fingerprint	$f_t = \hat{s}_t - \tilde{s}_t$			
Stochastic	$\hat{s}_t = x_t - \hat{T}_t - \hat{P}_t$			
Trend	$\hat{T}_t = 68.49 - 0.0137t$			
Periodic component	$\hat{P}_t = \sum_{i=1}^{3} \hat{b}_i \sin\left(\frac{2\pi t}{p_i}\right) + \hat{c}_i \cos\left(\frac{2\pi t}{p_i}\right)$			
	i	$\hat{b}_i$	$\hat{c}_i$	p_i (months)
	1	-0.3819	-0.5707	12
	2	0.7968	-0.625	132
	3	0.0189	-0.4304	66
Stochastic	$\tilde{s}_t = \left(\frac{1}{1 - 0.2327B}\right)\tilde{a}_t$			
Standard deviation of $\tilde{a}_t$	0.605			

Table 5. Calculations for the ARMA fingerprint for the global cloud cover.

procedure to compute the fingerprint and the estimation of parameters are given in table 6. The sequential test indicates that no additional change was found in the Caribbean cloud clover. Table 6 shows that Caribbean cloud cover during the reference data exhibit a reduction rate of -0.0259 % per month. Figure 22 Shows that Caribbean clouds cover no additional climate change is detected.

Component	Caribbean cloud cover
Fingerprint	$f_t = \hat{s}_t - \tilde{s}_t$
Stochastic	$\hat{s}_t = x_t - \hat{P}_t$
Trend	$\hat{T}_t = 52.76 - 0.0259t$
Periodic component	$\hat{P}_t = 5.4444 \sin\left(\frac{2\pi t}{12}\right) + 2.869 \cos\left(\frac{2\pi t}{12}\right)$
Stochastic	$\tilde{s}_t = (1 + 0.2263B)\tilde{a}_t$
Standard deviation of $\tilde{a}_t$	5.99

Table 6. Estimation of the fingerprint for the Caribbean cloud cover.

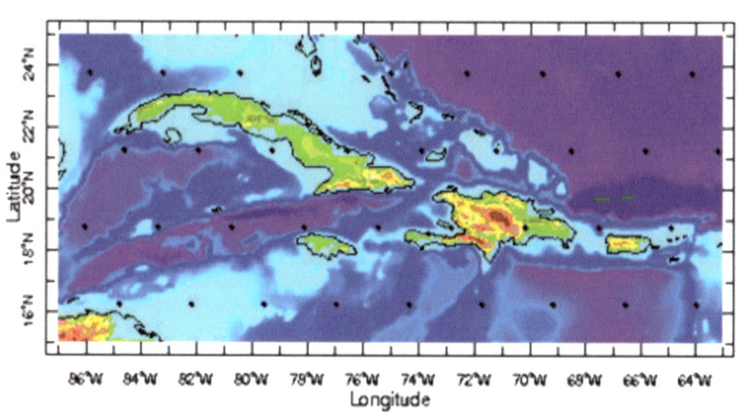

Fig. 21. The selected Caribbean region to be studied.

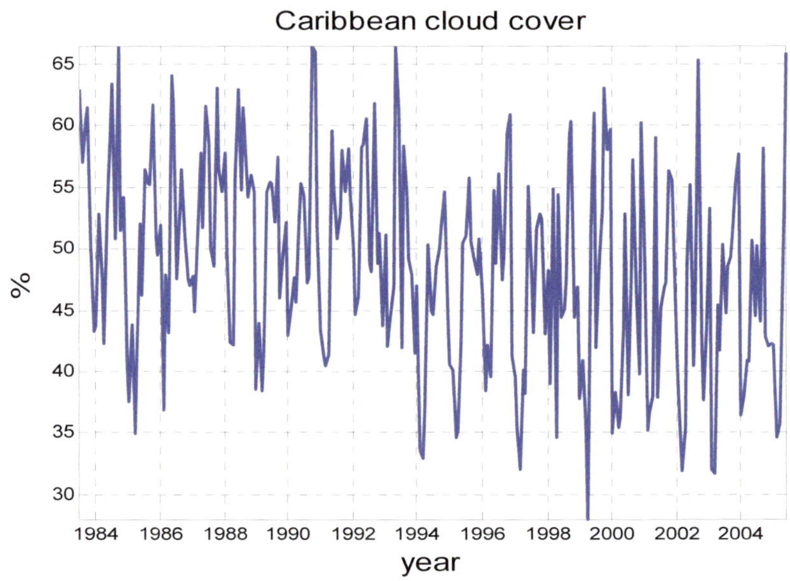

Fig. 22. The Caribbean clouds cover from July 1983 to June 2005. This data were extracted from ISCCP D2 using the Quispe (2006) program.

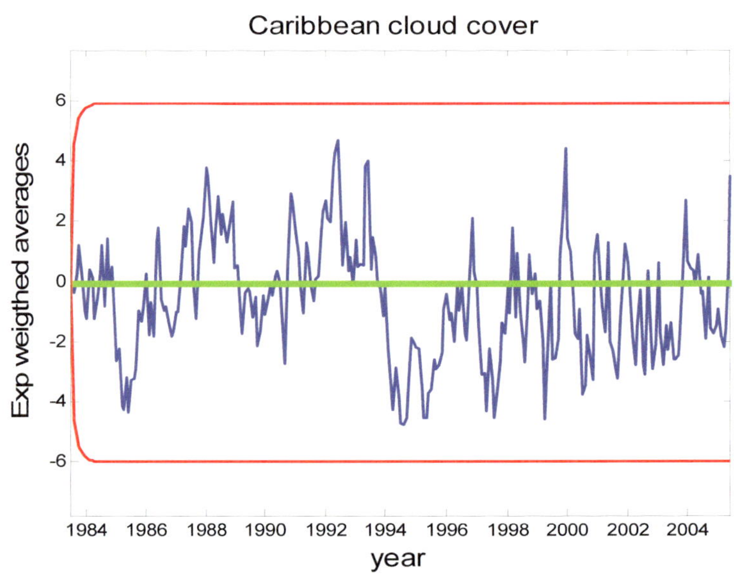

Fig. 23. Caribbean clouds cover shows a no additional reduction of cloud cover.

5. Conclusion

The major factors that have forced climate change are due to natural and anthropogenic causes such as solar radiation, volcano eruptions, increasing concentration of greenhouse gases, etc. A statistical test is proposed to detect when a significant climate change has

occurred. Usually, a climate indicator can be decomposed into three major features: trend, seasonal, and stochastic component. A climate change can be exhibited in any of the components of a given climate indicator. The introduced statistical test detects a climate change that can be observed in any of the three component of the process. The test consists on dividing the underlying time series in two parts. The first part of the observations is used for identifying trend, seasonality and stochastic components and these components are removed from the entire time series. The test consists of determining whether or not the remaining part of the series exhibits a significant deviation from the white noise. Thus, if a significant deviation from the white noise process occurs a climate change is detected; otherwise, no change has recorded. The proposed test was implemented to detected climate changes at global and Caribbean levels. The studied variables were: sunspots, concentration of carbon dioxide, surface air temperature, and cloud cover. At the Caribbean levels, it was found that cloud cover exhibits a significant reduction rate whereas the air temperature shows a significant increasing rate. Rainfall processes across the main Caribbean islands shows no significant trend during the studied period, which suggests that heavy rainfall events are being concentrated in small areas to maintain the rainfall process with no trend. The global cloud cover also shows a significant decreasing trend whereas the land-ocean surface temperature shows an increasing trend. Smaller clouds cover areas and high temperatures across the world also suggests that heavy rainfall processes will be concentrated in small continental areas causing flooding, landslide, and human and economic catastrophic impacts.
A sequential statistical test has been introduced to detect when as significant climate change has occurred. The major contribution of this work is to introduce a statistical tool to determine without ambiguity when a climate change has occurred. One of the advantages of the proposed procedure is its simplicity; however, it requires of a large sequence of a reliable climate indicator, where a climate indicator is a meteorological variable that reveals the intrinsic climatic characteristic of a given region of the Earth. It is very important to understand the interaction of the physical processes and how the climate indicators are related. One of the major limitations of the proposed detection test is that some climate indicators after removing the trend and the periodic components still retain characteristics of a nonstationary process and difference operators to induce stationary behaviour are not applicable, since the difference operator not only removes the nonstationary behaviour but also remove the climate change.

6. Acknowledgments

This research has been supported by National Aeronautics and Space Administration (NASA) EPSCoR program with grant NCC5-595 and also by the University of Puerto Rico. Authors want to recognize the technical support from several institutions that provided the climate indicator variables.

7. References

Angeles-Malaspina, Moisés E. 2005. An Assessment of Future Caribbean Climate Change Using "Business as Usual" Scenario by Coupling GCM Data and RAMs. Thesis of Master of Science in Mechanical Engineering, University of Puerto Rico-Mayagüez Campus.

Angeles, M.E. Gonzalez, E. J., Ericsson III, D. J., and Hernandez, J.L. (2006). Predictions of future climate change in the Caribbean region using global general circulation models. *International Journal of Climatology*, DOI: 10.1002/joc.1416

Barnett, T.P. Hasselman, K., M. Chelliah, T. Delworth, G. Hergel, P. Jones, E. Ramsmusson, E. Roeckner, C Ropelewski, B. Santer and S. Tett (1999), Detection and Attribution of Recent Climate Change: A Status Report, *Bull. Am. Meteorol. Soc.* Vol. 80 No. 12, December 1999.

Battisti D., Bitz M., Moritz R. (1997) Do General Circulation Models Underestimate the Natural Variability in the Arctic Climate? *Journal of Climate*, Vol. 10: 1909 - 1920.

Box G., and Jenkins (1976). *Time Series Analysis Forecasting and Control.* Golden-Day, California.

Brockwell, P. and Davis R. (2002), *Introduction to Time Series and Forecasting*, 2nd ed., Springer-Verlag New York, Inc.

Burroughs, W.J. (2001), *Climate Change: A Multidisplinary Approach*, 1st Edition, Cambridge University Press

Changnon S. A., Pielke R. A. Jr., Changnon D., Sylves R. T., and Pulwarty R., (2000) Human Factors Explain the Increased Losses from Weather and Climate Extremes; *Bull. Am. Meteorol. Soc.*; 81, 437 - 442.

Easterling D. R., Meehl G. A., Parmesan C., Changnon S. A., Karl T. R. and Mearns L. O. (2000) Climate Extremes: Observations, Modeling, and Impacts; *Science*, 289, 2068 - 2074.

Feldstein, S. B. (2002). The Recent Trend and Variance Increase of the Annular Mode, *Journal of Climate*, Vol. 15, 88-94.

Hansen, J., R. Ruedy, M. Sato, and K. Lo (2010), Global surface temperature change, *Rev. Geophys.*, 48, RG4004, doi:10.1029/2010RG000345.

Hansen, J (2005). A slippery slope: How much global warming constitutes "dangerous anthropogenic interference"? *Climatic Change*, 68: 269–279.

Huntingford C., Stott P., Allen M., and Lambert H., (2006) Incorporating model uncertainty into attribution of observed temperature change, *Geophys. Res. Lett.*, Vol. 33, L05710.

IPCC, *Climate Change 2001: The Scientific Basic*, Cambridge University Press, 697 - 738 pps.

IPCC - *Climate Change: The Physical Science Basis* 2007. Edited by Solomon, S., Qin, D., and Manning, M. Co-Chair, Co-Chair, Head, Technical Support Unit, IPCC Working Group I Melinda Marquis Kristen Averyt Melinda M.B. Tignor Henry LeRoy Miller, Jr.

Keeling, D. and Whorf T. (2005). Atmospheric CO_2 records from sites in the SIO air sampling network. In Trends: A Compendium of Data on Global Change. Carbon Dioxide Information Analysis Center, Oak Ridge National Laboratory, U.S. Department of Energy, Oak Ridge, Tenn., U.S.A. Available in wet site: http://cdiac.ornl.gov/trends/co2/sio-mlo.htm

Leuliette, E. , R. Nerem, and G. Mitchum, (2004). Calibration of TOPEX/Poseidon and Jason altimeter data to construct a continuous record of mean sea level change. *Marine Geodesy*, 27(1-2), 79-94.

MathWorks, 2000: *Optimization Toolbox for use with Matlab: User's Guide.* The MathWorks, Inc.

Meehl G., Washington W., Ammann C., Arblaster J., Wigley T., and Tebaldi C., (2004), Combinations of Natural and Anthropogenic Forcings in Twentieth-Century Climate, *Journal of Climate*, Volume 17: 3721 - 3727.

Menne, Matthew J., (2005), Abrupt Global Temperature Change and the Instrumental Record, *18th Conference on Climate Variability and Change*, P4.3.

Montgomery, D. (2001), *Introduction to Statistical Quality Control*, 4th Edition, John Wiley & Sons, Inc.

Nicolay, S., Mabille, G., Ferrweis, X., and Erpicum, M., (2010). Multi-months cycles observed in climatic data. *Climate Change and Variability* Edited by Suzanne W. Simard and Mary E. Austin Published by Sciyo Janeza Trdine 9, 51000 Rijeka, Croatia. ISBN 978-953-307-144-2.

Pandit, S.M. and Wu, S.M 1983. *Time Series and System Analysis with Applications*, 1st Edition, John Wiley & Sons, Inc.

Peterson, T.C., and D.R. Easterling, 1994: Creation of homogeneous composite climatological reference series. *International journal of climatology*, 14 (6), 671-679.

Quispe, W. (2006). *Sieve Bootstrap en Series de Tiempo de Nubosidad en el Caribe*. Thesis Master of Science in Mathematics, University of Puerto Rico, Mayagüez, 165 pages.

Ramirez-Beltran, N. D., and Sastri, T. (1997), Transient Detection With An Aplication To A Chemical Process, *Computer ind. Engng* Vol. 32, No. 4, pp. 891-908.

Roberts, S. W. (1959). Control Chart Tests Based on Geometric Moving Average. *Technometrics*, Vol 1.

Robock A. (2000). Volcanic Eruptions and Climate, *Reviews of Geophysics*, 38, 2: 191-219.

Salinger, M. J. (2005), "Climate Variability and Change: Past, Present and Future - an overview", *Climate Change* 70: 9-29.

Santer B. D., Wigley T. M. L., Mears C., Wentz F. J, Klein S. A., Seidel D. J., Taylor K. E., Thorne P. W., Wehner M. F., Gleckler P. J., Boyle J. S., Collins W. D., Dixon K. W., Doutriaux C., Free M., Fu Q., Hansen J. E., Jones G. S., Ruedy R., Karl T. R., Lanzante J. R., Meehl G. A., Ramaswamy V., Russell G. and Schmidt G. A., (2005) Amplification of Surface Temperature Trends and Variability in the Tropical Atmosphere, *Science*, 309: 1551 - 1556.

Schar C, Vidale Pl, Luthi D, Frei C, Haberli C, Liniger MA, and C. Appenzeller, (2004); The role of increasing temperature variability in European summer heatwaves; *Nature*, 427: 332-335.

Smith R., Wigley T. and Santer B. (2002), A Bivariate Time Series Approach to Anthropogenic Trend Detection in Hemispheric Mean Temperatures. *Journal of Climate*, 16: 1228 - 1240.

Stott P. and Kettleborough J., (2002). Origins and estimates of uncertainty in predictions of twenty-first century temperature rise, *Nature*, Vol. 416: 723 - 726.

She C., Krueger D., (2004), Impact of natural variability in the 11-year mesopause region temperature observation over Fort Collins, CO (41_N, 105_W), *Advances in Space Research*, Vol. 34, 330–336.

Tomé, A. R., and P. M. A. Miranda (2004), Piecewise linear fitting and trend changing points of climate parameters, Geophys. *Res. Lett.*, 31, L02207, doi:10.1029/2003GL019100

Wei, W.W.S (1990), *Time Series Analysis: Univariate and Multivariate Methods*, 1st Edition, Addison-Wesley Publishing Company

http://www.statgraphics.com/downloads.htm

http://www.minitab.com/en-US/default.aspx

http://www.mathworks.com/

6

Climate Changes of the Recent Past in the South American Continent: Inferences Based on Analysis of Borehole Temperature Profiles

Valiya M. Hamza and Fábio P. Vieira
National Observatory – ON/MCT
Brazil

1. Introduction

A detailed understanding of the nature of past climate changes is important in assessment of the effects of global warming trends identified in meteorological records (Hansen and Lebedeff, 1987). Nevertheless, there are large uncertainties in the reconstruction of the climate history of times prior to the period of instrumental records, there being considerable difficulties in experimental determination of past climate changes. In this context geothermal methods, based on results of temperature logs in boreholes stands out as one of the few methods that allow direct measurement of thermal signals in the subsurface induced by climate changes of the past.

Geothermal methods have been employed during the last few decades in extracting information on climate changes of the recent past for several regions of the northern hemisphere (e.g. Cermak, 1971; Lachenbruch et al, 1982; Beltrami et al, 1992; Bodri and Cermak, 1995; Duchkov and Sokolova, 1998; Harris and Chapman, 2001; Majorowicz and Safanda, 2001; Golovanova et al, 2001; Beltrami and Burlon, 2004). Nevertheless, very few attempts have been made in using geothermal data for examining climate variations in low latitudes of the southern hemisphere. Among the studies in this category are the works carried out in Australia (Cull, 1979, 1980; Torok and Nicholls, 1996; Taniguchi et al, 1999a, 1999b; Appleyard, 2005), Brazil (Hamza, 1991; Hamza, 1998; Hamza et al, 1991; Cavalcanti and Hamza, 2001; Cerrone and Hamza, 2003) and South Africa (Tyson et al, 1998; Jones et al, 1999). Much of the work carried out in Brazil remain as publications of limited access, such as internal reports (Hamza et al, 1978; Eston et al, 1982; Hamza et al, 1987), academic theses (Vitorello, 1978; Araújo, 1978; Santos, 1986; Ribeiro, 1988; Del Rey, 1989; Cavalcanti, 2003) and meeting proceedings (Ribeiro, 1991; Souza et al, 1991; Hamza and Cavalcanti, 2001; Cerrone and Hamza, 2003; and Conceição and Hamza, 2006). In the present work we provide a synthesis of these earlier works, with emphasis on progress obtained during the period 2006 – 2011.

As prelude to the discussion of results presented in this work we provide first a brief description of the sources and characteristics of the geothermal data employed and the criteria used for data selection. Details of the methods used for extracting information on past climate are set out in the next section. The results obtained in model simulations of temperature-depth profiles are classified into groups, representative of the major geographic zones. Finally, the climate history of South America, deduced from geothermal data, is compared with results of geothermal climate reconstructions from other continental areas.

2. Characteristics of the data base

According to the recent compilations carried out by the National Observatory (Observatório Nacional - ON/MCT) in Brazil geothermal measurements have been carried out in over 5000 localities in South America (Hamza et al, 2010; Vieira and Hamza, 2010). Most of the earlier data were acquired as parts of basic research projects for heat flow determinations and also as parts of applied research projects for oil exploration and geothermal energy assessments. The focus of data acquisition in the earlier works has been on determining temperature gradients in the deeper parts of the boreholes.

The characteristics of these data sets are variable, depending on the methods used for primary data acquisition. Of these, only the ones acquired using the so-called conventional (CVL) method provide direct information on the vertical distribution of temperatures at shallow depths and hence are potentially suitable for climate related investigations. The conventional method has been employed for geothermal studies in 134 localities, which is slightly more than 10% of the overall data set. It includes mainly temperature logs in bore holes and wells and thermal property measurements on samples representative of local geologic formations. In some cases estimation of radiogenic heat production was also carried out. The details of the experimental techniques employed for temperature and thermal conductivity measurements have been discussed in academic theses (e.g. Vitorello, 1978; Araújo, 1978; Del Rey, 1989) and in publications dealing with heat flow measurements (Hamza et al, 1987; Hamza and Muñoz, 1996; Gomes and Hamza, 2005). A direct evaluation of the quality of data acquired in the earlier works is a difficult task since the experimental techniques used for temperature and thermal conductivity measurements have undergone substantial changes over the last few decades.

The sources of conventional data sets employed in the present work may be considered as falling into five main groups:

- New data acquired in ten different localities in the Amazon region, during the period of 2006 to 2008 (Hamza, 2006; Hamza, 2007);
- Results of recent geothermal measurements in the cordilleran region of Colombia (Hamza et al, 2009; Alfaro et al, 2009).
- Data acquired during the period of 2000 - 2005, as part of geothermal projects for mapping heat flow variations in the coastal area of southeast Brazil (Gomes, 2003; Gomes and Hamza, 2005; Hamza et al, 2005). Some of these have been employed in studies of climate change the state of Rio de Janeiro (Cerrone and Hamza, 2003; Hamza et al, 2003);
- Data acquired during the decade of 1980, mainly in the state of São Paulo, as parts of hydrocarbon and geothermal energy exploration programs. Some of these data are reported as parts of academic theses of the 1980s (Santos, 1986; Ribeiro, 1988; Del Rey, 1989) while some are parts of related publications (Santos et al, 1986; Del Rey and Hamza, 1989; Hamza et al, 1987);
- Data acquired during the decade of 1970 in the southern and eastern parts of Brazil. Most of the results have been published in the Brazilian Geothermal Data Collection – Volume 1 (Hamza et al, 1978). Some are part of academic theses of the late 1970s (Vitorello, 1978 and Araújo, 1978) and related publications (Vitorello et al, 1978; Hamza,

1982). Data acquired for four boreholes in the cordilleran region of Peru is reported in the IHFC data collection by Huang and Pollack (1998). The geographic distribution of the overall data sets is illustrated in the map of Figure (1).

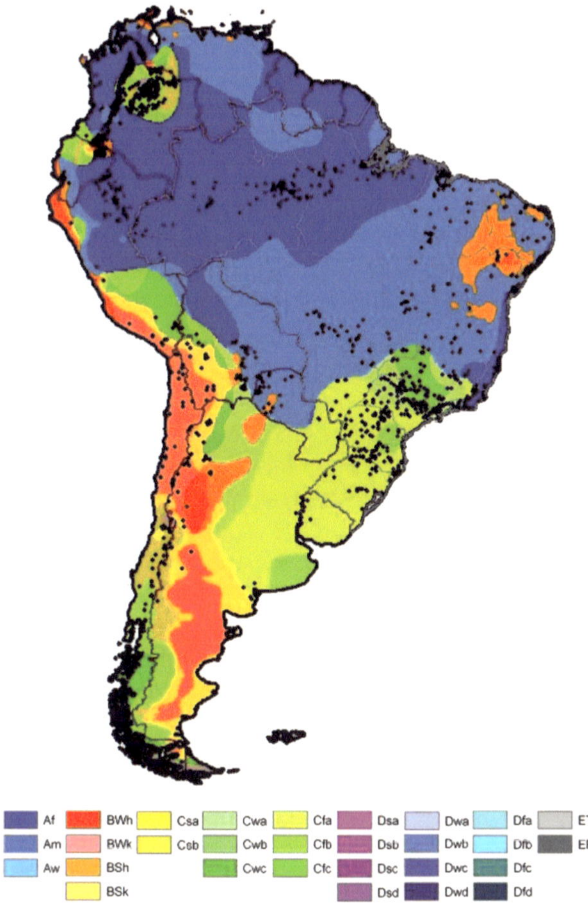

Fig. 1. Localities of geothermal measurements in South America (Vieira and Hamza, 2010). Letters in the legend refer to Koppen climate classification: A – Tropical; B – Dry; C – Temperate; D – Continental; E - Polar (Koppen, 1936; Pidwirny, 2006).

2.1 Selection criteria for climate studies

The characteristics of the conventional (CVL) data set were examined carefully to screen out records with indications of possible perturbations arising from non-climatic effects. Also, it was necessary to eliminate those which do not provide fairly reliable determinations of both the steady and the transient components of the subsurface thermal field. In an attempt to guarantee the reliability of the data set the following quality assurance conditions were imposed:

a. The depth of borehole is sufficiently large that the lower section of the thermal profile allows a reliable determination of the geothermal gradient, free of the effects of recent climate changes. Order of magnitude calculations indicate that surface temperature changes of the last centuries would penetrate to depths of nearly 150meters, in a medium with a thermal diffusivity of 10^{-6} m^2/s. Thus boreholes of at least 200 m deep are necessary for a reliable determination of the local geothermal gradient. The choice of this depth limit is rather arbitrary since the possibility that low amplitude climate signals of earlier periods are present at larger depths of up to several hundreds of meters cannot entirely be ruled out. However it is a reasonable compromise for examining subsurface thermal effects of ground surface temperature (GST) variations of the last few centuries;

b. The temperature-depth profile is free from the presence of any significant non-linear features in the bottom parts of the borehole, usually indicative of advection heat transfer by fluid movements, either in the surrounding formation or in the borehole itself;
c. The time elapsed between cessation of drilling and measurements in boreholes is at least an order of magnitude large compared to the duration of drilling, minimizing thereby the influence of eventual thermal perturbations generated during the drilling activity;
d. The lithologic sequences encountered in the borehole have relatively uniform thermal properties and are of sufficiently large thickness that the gradient changes related to variations in thermal properties does not lead to systematic errors in the procedure employed for extracting the climate related signal; and
e. The elevation changes at the site and in the vicinity of the borehole are relatively small so that the topographic perturbation of the subsurface temperature field at shallow depths is not significant.

The sites of these selected boreholes are distributed in the eastern parts of Brazil (in the states of Santa Catarina, Paraná, São Paulo, Minas Gerais, Rio de Janeiro and Bahia), in the Amazon region, central cordillera in Colombia and eastern cordillera in Peru. The majority of the selected temperature logs are from boreholes with depths greater than 200 meters. Some log data for depths less than 200 m were also considered, as these are found to provide complementary information on subsurface temperature fields at shallow depths which may be compared with those encountered in areas where deeper boreholes are situated. Most of the data sets acquired during the decades of 1970 and 1980 have temperature measurements at depth intervals of ten to twenty meters. In more recent logs measurements have been made at intervals of two meters. Such recent logs may be considered as capable of providing more robust estimates of the background temperature gradients. Data from boreholes with depths shallower than 150 meters were not considered in the present work in view of the potential uncertainties in the determination of the local undisturbed gradient and consequent difficulties in extraction of the climate signal. On the other hand, information on climate changes of the recent past available in such logs may be used in obtaining qualitative estimates of GST changes. Data acquired at shallow depths of less than 20 meters were excluded from analyses for climate changes, avoiding thereby eventual perturbing effects of diurnal and seasonal variations in the reconstruction of surface temperature history.

In an earlier study Hamza et al (2007) reported results of geothermal measurements for a number of sites in the Brazilian territory. In the present we have included results of additional studies carried out during the period of 2006 to 2011. The data sets have been classified into groups, designated as subtropical highlands, subtropical humid zones (of the interior and of the coastal areas), tropical Amazon region, semi-arid zones and cordilleran regions in western parts of the continent. The area extents of these geographic sectors are in large part similar to the prevailing climate zones indicated in Figure (1). Typical examples of temperature profiles encountered in boreholes in Brazil, Peru and Colombia are illustrated in Figure (2). In this figure some of the temperature-depth profiles have been shifted laterally to convenient positions along the temperature axis, to avoid overlap and to allow for easy visualization. Consequently, the temperature axis in Figure (2) displays only relative values.

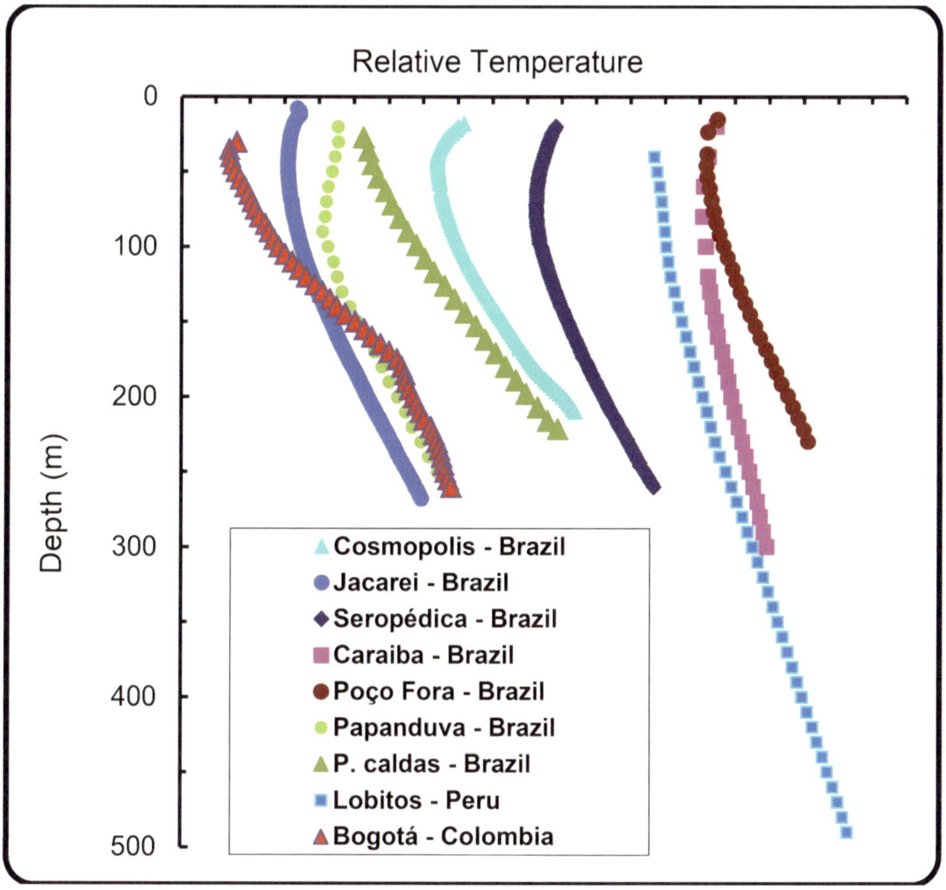

Fig. 2. Vertical distributions of temperatures in selected localities in Brazil, Colombia and Peru. The climate changes of the recent past are often considered responsible for the non-linear features in temperature logs.

The vertical distributions of temperatures in boreholes, some examples of which are illustrated in Figure (2) reveal several remarkable features. For example, the temperature depth profiles at shallow depths are consistently concave towards the depth axis, which is indicative of surface warming events of relatively recent times. The widespread occurrence of such temperature-depth profiles in almost all of the major geographic zones, irrespective of the local geological complexities and changes in soil type, is considered a clear indication that the observed features are generated by surface warming events of large spatial dimensions. Temperature profiles that are convex towards the depth axis, and hence characteristic of cooling events, were not encountered.

3. Methods employed in data analysis

Three different methods have been employed in data analysis: forward models, inversion methods and signal backstripping methods. Brief descriptions of these methods are provided in the following sections.

3.1 Forward models

The basics of forward modeling approach has been discussed extensively in the literature (Birch, 1948; Cermak, 1971; Vasseur et al, 1983; Lachenbruch et al, 1986, 1988). For the case where surface temperature variation can be represented by a power law relation, analytic solutions are readily available. Thus, for a linear (or ramp type) change in surface temperature the relation between the amplitude of the climate signal (ΔT) and the time elapsed (t) at any depth (z) is given by the relation (Carslaw and Jaeger, 1959):

$$\Delta T(z) = 4\Delta T\, i^2\, erfc\left(z / \sqrt{4\kappa t}\right) \tag{1}$$

where $i^2\ erfc$ is the second integral of the complementary error function and κ the thermal diffusivity of the medium. The best fitting ramp function is obtained by inverting the above relation using iterative procedures such as linearized Newton's method. The iterative procedure for this model, referred to as Ramp Inversion (Chisholm and Chapman, 1992; Golovanova et al, 2001; Roy et al, 2002), allows simultaneous determination of the magnitude of surface temperature change and the period of its occurrence. The main limitation of the forward model approach is that it resolves mainly the first-order features in the GST history. This is a consequence of the implicit assumption that the bottom parts of the log data, employed in determination of background temperature gradients, are free of transient perturbations.

The basic steps in forward model approach include identification and separation of the steady and transient components present in temperature profiles at shallow depths. The steady state component is determined by the flow of heat from the interior of the Earth while the transient component is induced by downward propagation of a climate related thermal signal induced at the surface. Usually a linear fit to the deeper portion of the log data, where the climate perturbation is practically absent, allows determination of the local temperature gradient. However some care is necessary in selecting the depth interval for determination of the gradient. If the temperature gradient is calculated using a small subset of data from the lowermost part of the borehole its standard deviation (σ_G) is likely to be relatively large, a consequence of the large root mean square (rms) deviation associated with the small number of data points. Progressive inclusion of data from the overlying parts in least square analysis leads to a steady initial decrease in σ_G, as the estimation of gradient becomes more robust. However, as more data are included from shallower depths (where non-linear features are present) this tendency is reversed and σ_G increases. In the present work, the depth corresponding to the minimum value of σ_G is considered as indicative of the top of the unperturbed zone. The background gradient determined for the depth interval below this zone is used for calculating the steady component of the temperature field. Subtracting it from the observed temperatures allows determination of the transient component. A similar procedure has also been employed by Roy et al (2002) in the separation of steady and transient components of temperature profiles in the Indian subcontinent.

3.2 Inverse models

In the inverse problem approach (Tarantola and Valette, 1982) a priori information is explicitly incorporated in constraining the solution. The functional space inversion (FSI) method discussed by Shen and Beck (1991, 1992) and Shen et al (1992) makes use of the non-linear least squares theory in solving the one dimensional heat conduction equation in a layered half space. The algorithm employed finds the model that minimizes the misfit function:

$$S(m)=\frac{1}{2}\left\{\left[(d-d_0)^t C_d^{-1}(d-d_0)\right]+\left[(m-m_0)^t C_m^{-1}(m-m_0)\right]\right\} \tag{2}$$

where d and d_0 are respectively the calculated and observed temperatures, m and m_0 the calculated and a priori model parameters and C_d and C_m the covariance matrices of d_0 and m_0. The term C_d indicates the uncertainty in the observed temperature-depth data while C_m indicates uncertainty in the a priori model. The selection of appropriate values of a priori standard deviations for the temperature (σ_{d0}) and thermal conductivity (σ_{k0}) data are important in determining the solutions.

The main advantage of FSI formulation is that it does not predetermine the steady state temperature profile. Instead, both the steady state and transient profiles are estimated simultaneously. In addition, it allows consideration of the vertical distribution of thermo-physical properties and their uncertainties as model parameters, allowing thereby determination of a more detailed GST history where it is possible to identify second order features. Also, FSI formulation includes as model parameters all variables that govern the conductive thermal regime (background heat flow density, the GST history as a function of time and thermal properties as functions of depth).

As prelude to the presentation of the results obtained by the functional space inversion method we provide brief descriptions of the steps taken in data processing and analysis. These include specifying the depth intervals and time periods used for inversion, setting a priori standard deviations of temperature and thermal conductivity data sets and measures taken to minimize the undesirable consequences of the null hypothesis in the inversion scheme.

The estimates of the depth at which the thermal regime is supposedly untouched by the GST variations and the time limit beyond which GST variations cannot be recovered from the given temperature log data, were chosen in accordance with the depth extent of the available temperature log data. In particular, the depth estimate is set to be greater than the deepest data point because the calculated data are projected (interpolated) from the finite element solution. As for the time limit there is no harm in setting a value compatible with or greater than the depth extent of the borehole (Shen and Beck, 1992). On the other hand, use of a shorter-than-necessary time span would end up in "telescoping" the GST history. Unless there are independent evidences indicating a rapid return to unperturbed conditions it seems prudent to assume that this return take place gradually. Use of shorter time spans leads to slight reductions in the magnitude and duration of the cooling events of the earlier periods. As pointed out by Shen and Beck (1991; 1992) the results of GST history, determined by functional space inversion, is sensitive to a priori standard deviations of thermal conductivity. The preferred values for the standard deviations are based on considerations of the trade-off between consistency of the solution and data resolution. In the present work we have used 50mK for standard deviation of the temperature data (σ_{d0})

and 1W/m/K for standard deviation of the thermal conductivity (σ_{k0}). These values are of the same order of magnitude as those adopted by Safanda and Rajver (2001) and Golovanova et al (2001).
FSI inversion makes use of an a priori null hypothesis in obtaining robust estimates of the prior steady state. However, this built-in feature can potentially lead to undesirable results when inversions are attempted for determining GST history of periods not correlated with the subsurface temperature data. Thus, in the absence of suitable temperature data for shallow depths the inversion scheme generates artificial values for the late part of GST history. It is clear that acquisition of reliable temperature data for shallow depths is important in determining GST history of the last few decades. On the other hand, the results also show that occurrence of artificial cooling trends for the decades prior to 1970 are possible only in cases where temperature measurements are restricted to depths greater than 100m.

3.3 Signal back stripping approach

Both the conventional and the Bayesian inversion methods have inherent difficulties in identification of individual thermal signals originating from climate variations that are episodic. In such cases methods based on signal back stripping approach are more convenient. Following the standard practice we also assume that the residual temperature profile represents is a superposition of individual perturbations. The characteristics of such perturbations vary, mainly as a result of differences in the magnitudes and time periods of individual GST episodes. The essence of the procedure adopted in the present work can be understood by considering magnitudes of temperature perturbations at two conveniently selected depths z_1 and z_2 in the residual profile:

$$dT_1 = A\,erfc\left(z_1 / \sqrt{4\kappa t'}\right) \tag{3a}$$

$$dT_2 = A\,erfc\left(z_2 / \sqrt{4\kappa t'}\right) \tag{3b}$$

Dividing (3a) by (3b) and designating the ratio dT_1/dT_2 by δ we have:

$$\frac{dT_1}{dT_2} = \delta = \frac{erfc\left(z_1 / \sqrt{4\kappa t'}\right)}{erfc\left(z_2 / \sqrt{4\kappa t'}\right)} \tag{4}$$

Note that equation (4) does not depend on the magnitude of the temperature perturbation but only on the selected value of the time period. Iterative methods may now be employed for determining the appropriate value of δ for the depth interval. The magnitude of this perturbation can be obtained from the relation:

$$A = \frac{dT_1}{erfc\left(z_1 / \sqrt{4\kappa t'}\right)} \tag{5}$$

In applying this procedure it is necessary to start with results of the lowermost section of the residual temperature profile. If the borehole is sufficiently deep it is fairly reasonable to assume that the residual temperatures of the deeper parts retain the effects of only the earliest perturbation. The magnitude and period of this earliest perturbation can be determined through the use of equations (3) and (5). It also opens up the possibility of removing the effects of this particular perturbation from the original residual temperature profile. The result is a back stripped profile free of the effects of the earliest perturbation. The procedure is repeated successively for identifying and removing the perturbations arising from later climate episodes.

4. Estimates of surface temperature changes

4.1 Forward model results

A summary of the results obtained in fitting forward models to the observational data discussed in this work, is presented in Table (1). It includes magnitudes of the GST change (ΔT) and their duration (t) as well as the values of the root mean square (rms) misfit between the model and the observational data. For reasons of brevity, we present here only the results for the ramp function model. Climate changes inferred on the basis of this model are grouped together for the five major geographic zones: subtropical highlands (elevations >400 meters) subtropical lowlands (elevations <400m), humid regions, semi arid zones, tropical rain forest and cordilleran regions.

The data in table (1) reveal differences in GST between the major geographic zones. For example, the highland areas seem to be characterized by changes in GST in the range of 2 to 3.8°C, the mean value being 3.3°C. The lowlands of the interior also seem to have nearly similar GST changes. On the other hand, the humid zones have relatively low GST values, in the range of 0.4 to 1.8°C, with a mean of 1.1°C. Low GST values are also found in the semiarid zones in the northeast, in the interval of 1 to 2°C. Also, the times of duration of climate changes are different for the geographic zones. Thus, the subtropical highlands, subtropical lowlands and humid zones are characterized by climate episodes with durations varying from 40 to 120yrs, the mean value being 80 yrs. The semi arid zones on the other hand have climate variations with relatively larger duration, the mean value being 150yrs.

The vertical distributions of transient components derived from forward model results (which are considered as indicative of temperature signals resulting from climate changes) are illustrated in Figure (3). Note that the magnitudes of the transient components decrease rapidly with depth. At depths greater than 150 meters these values fall below the experimental detection limits of temperature changes in boreholes. The period of GST changes fall in the range of 130 to 260 years. Hence the beginning of the climate change seems to have taken place during the period of approximately 1750 to 1850. A closer examination of the results illustrated in Figure (3) reveals some marked differences in the GST values within individual geographic zones. The primary reason for the occurrence of such intra-zonal variations is unknown at the moment, but it is likely that they are related to microclimatic histories of individual sites.

Also, the depth distribution of the transient components indicates that the magnitude of warming event is relatively smaller (in the range of 1.4 to 2.2°C) for the semi-arid zone in north eastern parts of Brazil. The close agreement between the results for the localities in Table (1) is considered as indication that local changes in vegetation cover and soil types have only a minor influence on the surface thermal budget in semi-arid zones. The GST change in this region seems to have had its beginning during the time period of 1850 to 1900, significantly earlier than the corresponding periods for other geographic zones.

Climate Zone	Locality	Coordinates	ΔT (°C)	Climate variation		rms (mK)
				Duration	Year of Onset	
Subtropical Highlands	Á. Lindóia	22° 29′/46° 38′	3.2	105	1877	4.4
	Amparo	22° 43′/46° 46′	2.8	115	1852	5.5
	Araras	22° 21′/47° 22′	3.4	40	1942	6.4
	Cosmópolis	23° 43′/47° 12′	3.8	75	1907	5.5
	Itapira	22° 28′/46° 43′	3.8	40	1942	10.2
	Rafard	23° 00′/47° 31′	3.8	60	1922	6.8
	Jacarei	23° 18′/45° 57′	2.4	105	1880	3.5
	São Paulo	23° 34′/46° 44′	2.1	105	1905	7.2
	Serra Negra	22° 36′/46° 32′	2.0	105	1877	3.9
	P. Caldas	21° 55′/46° 25′	2.0	50	1926	13.7
	Teresópolis	22° 26′/ 42° 57′	2.0	55	1945	5.5
	L. Muller	28° 40′/49° 30′	2.6	80	1895	10.0
Subtropical Lowlands	Seropédica	22° 46′/43° 39′	3.8	145	1860	23.1
	Miracema	22° 01'/41° 06'	3.6	30	1969	7.3
	Campos	21° 46'/41° 17'	3.0	20	1980	6.7
	Itapemirim	19° 06′/41° 04′	3.6	65	1910	14.1
	S. Sebastião	23° 48′/45° 25′	2.7	110	1880	7.8
Subtropical Humid	Itu	23° 15′/47° 19′	1.2	60	1922	3.2
	Jundiaí	23° 10′/46° 52′	1.2	50	1932	7.2
	Papanduva	26° 23′/50° 08′	1.2	90	1885	15.0
	Cach. Sul	30° 00′/52° 55′	1.8	70	1905	12.0
	Maricá	22° 54′/42° 45′	0.4	15	1985	2.9
	Rio Bonito	21° 25′/42° 12′	1.8	40	1960	7.7
Semi-Arid	Arraial	12° 32′/42° 50′	1.4	85	1890	8.4
	Caraiba	09° 28′/39° 50′	1.4	110	1865	9.3
	Jacobina	11° 11′/40° 31′	1.9	150	1828	8.7
	Poço Fora	09° 41′/39° 51′	1.8	105	1870	4.9
Tropical Rain Forest	Manaus	02° 57′/60° 01′	1.3	40	1960	15.0
	Belém	01° 27′/48° 27′	2.2	55	1953	9.4
	Salinópolis	00° 38′/47° 20′	2.1	65	1943	8.6
	Dom Eliseu	04° 17′/47° 34′	1.9	60	1948	9.2
Cordilleran Regions	Pen 742	81° 10′/04° 17′	4.0	200	1779	7.4
	Lobitos	81° 16′/04° 27′	2.5	200	1779	8.3
	Lomitos	70° 39′/17° 16′	2.0	200	1779	7.2
	Bogotá	74° 05′/04° 38′	2.8	150	1860	9.5

Table 1. Results of Ramp Inversions of GST changes, for selected localities in South America. ΔT is the magnitude of ramp change, and *rms* the root mean square misfit.

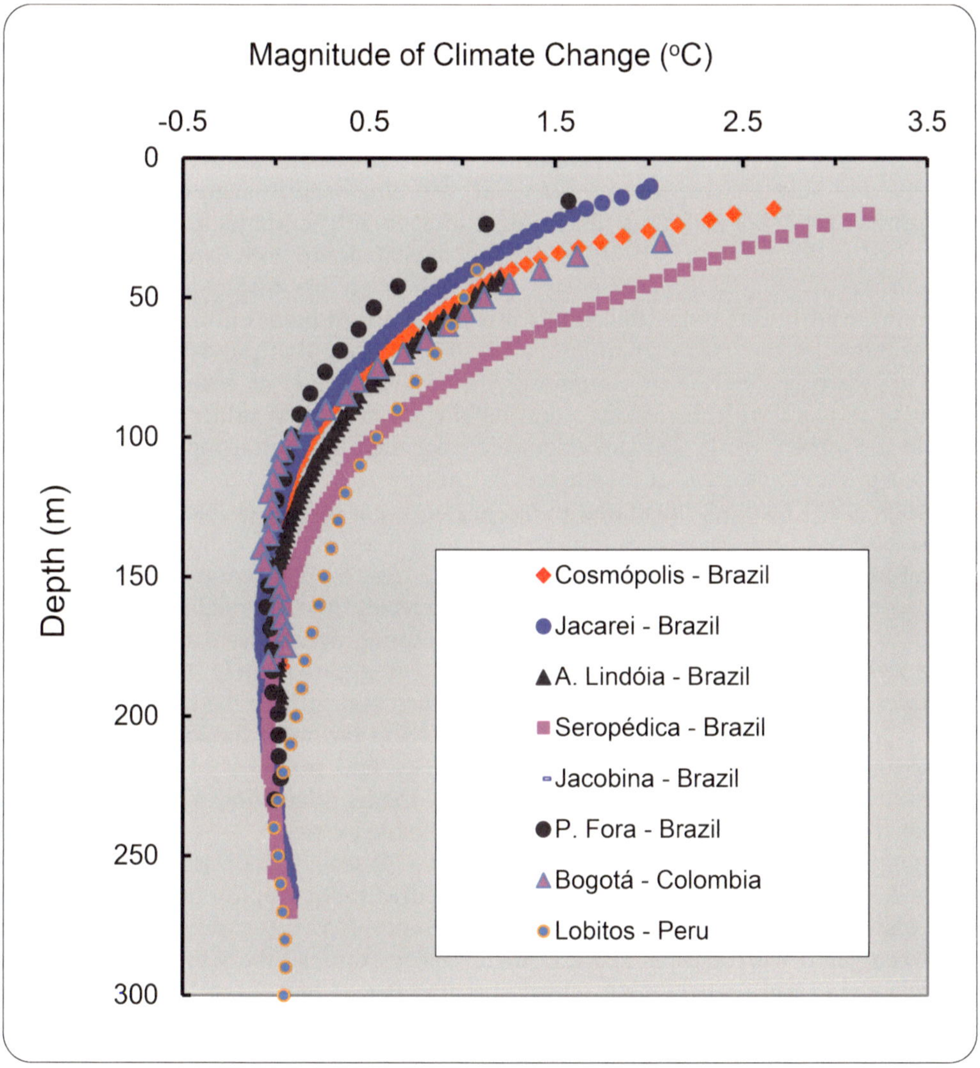

Fig. 3. Magnitudes of climate changes and its vertical distributions, deduced from forward model fits to borehole temperature profiles at selected localities in South America.

A careful examination of figure (3) reveals that some of the transient temperature profiles do have small negative values for depth intervals corresponding to the lower parts of the transient sections. The negative residuals are usually considered as arising from cooling events prior to recent warming episodes. However, the magnitudes of such events appear as subdued features, because of the implicit assumption in forward model approach that the bottom parts of the borehole are free of transient perturbations. Results of numerical simulations indicate that occurrence of negative residuals is quite sensitive to subtle changes in the value adopted for the background temperature gradient. This is the main limitation of the forward models, which resolves mainly for the first-order features in the GST history. We conclude by noting that residual temperature profiles similar to those obtained in the present work were also reported by Golovanova et al (2001) for the Urals region and Roy et al (2002) for the Indian Peninsula. A complementary analysis of this problem is provided in section (4.2) below, where we discuss the residual temperature profiles derived using the inversion method.

4.2 Results of functional space inversion

The method of Functional space inversion (FSI) was employed in analysis of temperature-depth profiles for 14 localities distributed over three main geographic zones of Brazil. The criteria used in the selection of profiles included availability of both thermal property data of subsurface layers and supplementary information on the history of changes in the vegetation cover. In discussing the results it is important to point out that the FSI method provides more detailed information on the GST history than that provided by the forward model approach. However, in comparing the GST histories of several localities it is convenient to work with deviations from the site specific mean rather than the absolute value. In the present work, GST deviations are calculated by subtracting the model results from the a posteriori estimate of the site specific mean.

The results of FSI method, illustrated in figure (4), reveal several characteristic features in the GST history of the study area. Foremost among these are the indications that surface temperatures have increased by as much as 1 to 4°C, over the last century. This observation is in reasonable agreement with the results of the ramp function model discussed in the previous section. However, in all localities the warming events seem to be preceded by cooling episodes occurring over the time period of approximately 1700 to 1900. The amplitudes of the cooling events are much less, falling generally in the range of 0.5 to 1°C. For time periods prior to the 17th century the resolving power of the inversion method is poor, a consequence of the limitations in sensitivity and precision of sensors used for temperature measurements in boreholes. Hence FSI model calculations for periods prior to 1700 may not necessarily be representative of true climate history.

Even though the data set is poor the results of Figure (4) seem to indicate that the warming trends are less pronounced in semi-arid regions relative to the tropical humid regions. The results also reveal a small time shift in the occurrence of climate warming in highlands regions compared with that for semiarid regions. For example, the warming event in the highlands region had its beginning during the time period of 1850 to 1900 while that for semi-arid regions seems to have had its beginning during the period of 1670 to 1860. Also the durations of both warming and cooling episodes appear to be relatively smaller in highland regions when compared with those for the semi-arid regions. Such differences have important implications for understanding evolutionary trends in climate history of eastern Brazil.

The vertical distributions of transient components derived from FSI inversions are similar to those found for the forward model results, illustrated in Figure (3). The magnitudes of the transient components decrease rapidly with depth. However, the residual temperatures become significantly negative at depth intervals corresponding to the lower parts of the transient sections. Such negative residuals are indicative of the occurrence of cooling events prior to recent warming episodes. We conclude that the vertical distributions of residual temperatures in FSI method are better representations of subsurface transient thermal regimes than those derived from the forward model approach.

Another conspicuous feature of the results by the FSI method in Figure (4) is the presence of brief cooling episodes for the recent decades, since 1970. There is a possibility that short period cooling episodes are spurious, a consequence of the null hypothesis employed in the computational process of the FSI method (Shen and Beck, 1992). The potential undesirable effects of the null hypothesis may be minimized by limiting the GST history to periods compatible with reliable subsurface temperature data. In the present work temperature data are available for depths less than 50 meters in all of the data sets employed in inversion schemes. Consequently the cooling trends observed in our results, for the decades prior to 1970, cannot be attributed to the spurious effects of the null hypothesis. Examples of cooling trends similar to those found in the present work can also be seen in the results reported for several localities in Europe and North America (Bodri and Cermak, 1995; Rajver et al, 1998; Stulc et al, 1998; Golovanova et al, 2001).

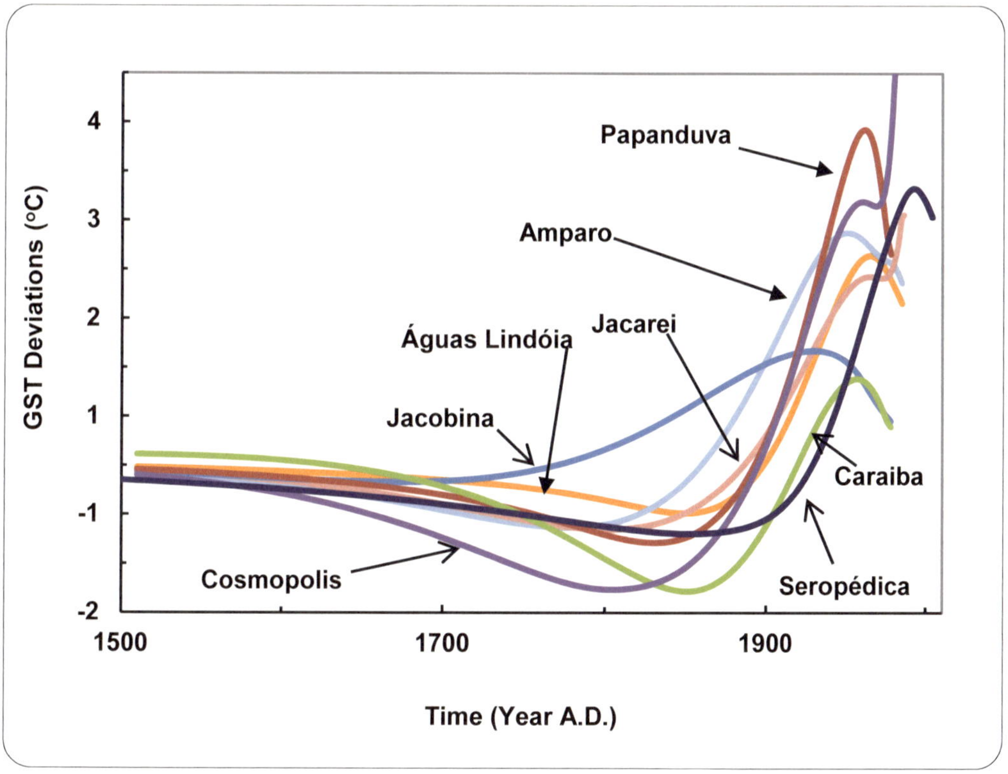

Fig. 4. History of ground surface temperature variations associated with climate changes of the recent past, deduced on the basis of functional space inversion models fits to borehole temperature profiles, for selected localities in South America (Adapted with modifications from Hamza et al, 2007).

The uncertainties in the estimates of GST in the inversion method can somewhat be improved by carrying out simultaneous inversion of temperature profiles of several sites in the same geographical province. In the present work we have carried out simultaneous inversions only for repeat measurements at Seropédica, in the state of Rio de Janeiro. In view of notable differences in the local soil conditions and vegetation cover, no attempt has been made for carrying out simultaneous inversions of temperature profiles from different locations discussed in the present work. Also, most of these sites are separated by large distances and fall within areas with distinctly different microclimate conditions and geographic characteristics.

A summary of the results of inverse modelling is presented in Table (2). The summary includes maximum and minimum values of ground surface temperatures and their respective times of occurrence. Also given in this table are the difference between the maximum and minimum values of GST, the time elapsed between the maximum and minimum and aposteriori estimates of undisturbed GST. The values of the differences in magnitudes are in the range of 1.3 to 3°C, which is slightly lower than the range indicated by the results of the forward model approach. On the other hand, the values of the differences in the time periods are in the range of 62 to 147 years, which is comparable to the ranges indicated by the results of the forward model approach.

Locality	Maximum		Minimum		Difference		T_0 (°C)	Log Year
	°C	Age	°C	Age	Mag.	Yr		
Amparo	21,1	1949	18.1	1802	3.0	147	18.9	1982
Itu	20.7	1982	19.4	1920	1.3	62	19.9	1982
Araras	22.8	1982	19.8	1910	3.0	72	20.9	1982
Campos	25.7	1999	22.7	1936	3.0	64	23.9	2000
C. Itapemirim	23.7	1971	20.0	1883	3.7	88	21.2	1976
Arraial	28.8	1967	27.2	1863	1.6	104	27.7	1976
Caraíbas	30.1	1946	28.6	1812	1.6	134	29.0	1976
Poço de Fora	30.4	1976	28.2	1822	2.1	154	28.7	1976

Table 2. Results of functional space inversions for selected data sets. Maximum and minimum values of GST variations and their respective years of occurrences are given in columns 2 - 5. Also given are model results for GST deviations and the respective periods. T_0 is the aposteriori steady state temperature.

4.3 Results of signal back-stripping approach

Techniques of signal back-stripping were used in analysis of temperature log data from 15 localities in south east Brazil. In the present work we limit the discussion to the results obtained for the locality of Seropédica (Rio de Janeiro). The vertical distributions of temperatures obtained during the various stages of the back stripping process are illustrated in the set of upper and lower panels of Figure (5).

The segment on the left side of the upper panel in this figure refers to results of the first stage of the back stripping process. Here the blue curve indicates the initial reduced temperatures and the red curve the first signal extracted by the back stripping method. The extracted signal points to an episode of climate change that took place 115 years back in time and had a magnitude of 2.9°C. Similarly, the segments on the right side of the upper panel and that on the left side of the lower panel of this figure illustrate, respectively, the results of the second and third stages of the back stripping process. According to the results obtained the signal extracted in the second stage points to an episode of climate change that took place 93 years back in time and had a magnitude of 0.2°C. On the other hand, the signal extracted in the third stage points to an episode of climate change that took place 21 years back in time and had a magnitude of 0.8°C. The segment on the right side of the lower panel illustrates the residual temperatures, after extracting the above mentioned three climate signals. The back stripping process is terminated at this third stage since the magnitudes of the residual temperatures are less than 0.05°C, below the sensitivity limit of experimental system used in data acquisition. The overall history of climate change derived from the back stripping method is illustrated in Figure (6).

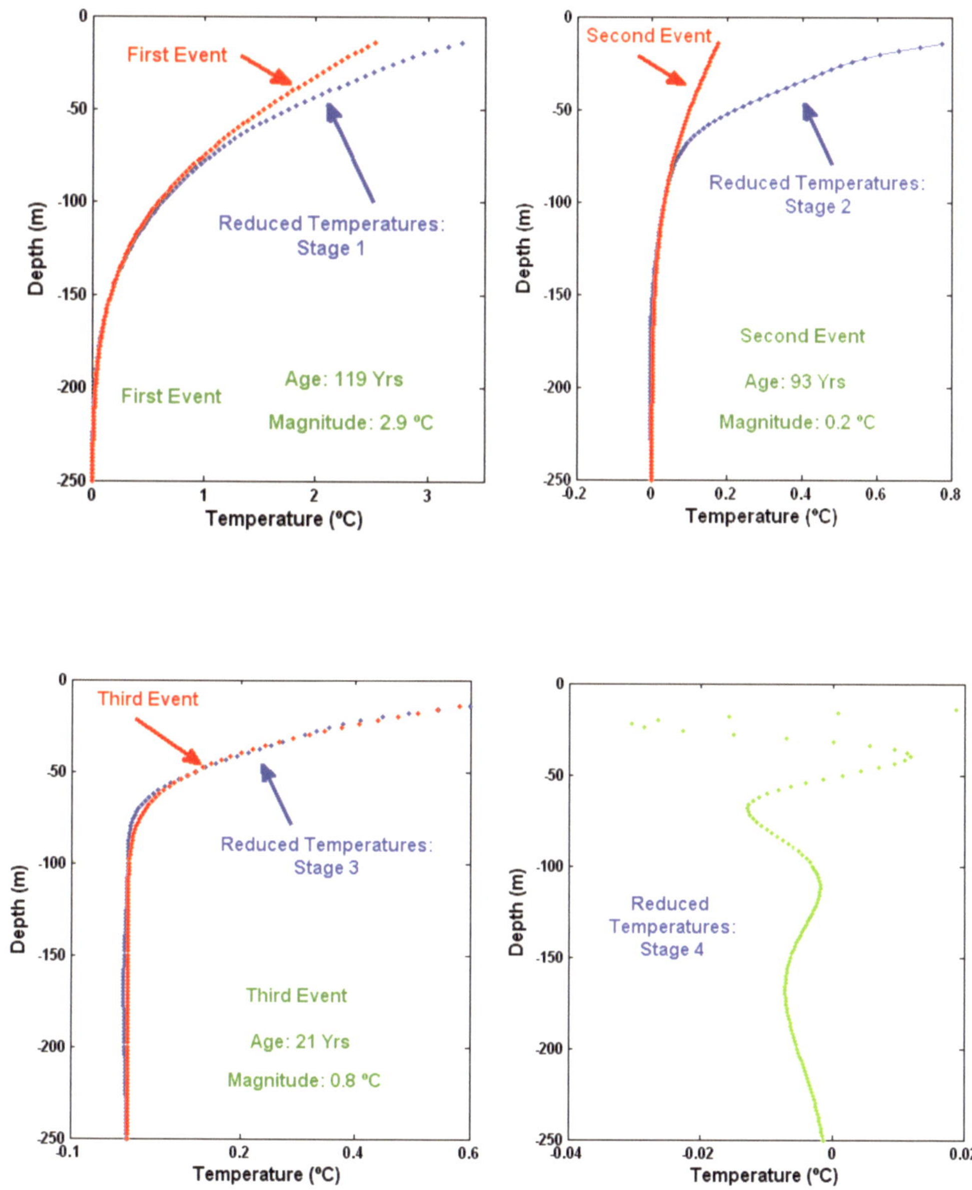

Fig. 5. Results of signal back stripping method employed in disentangling the climate history of Seropédica (Rio de Janeiro).

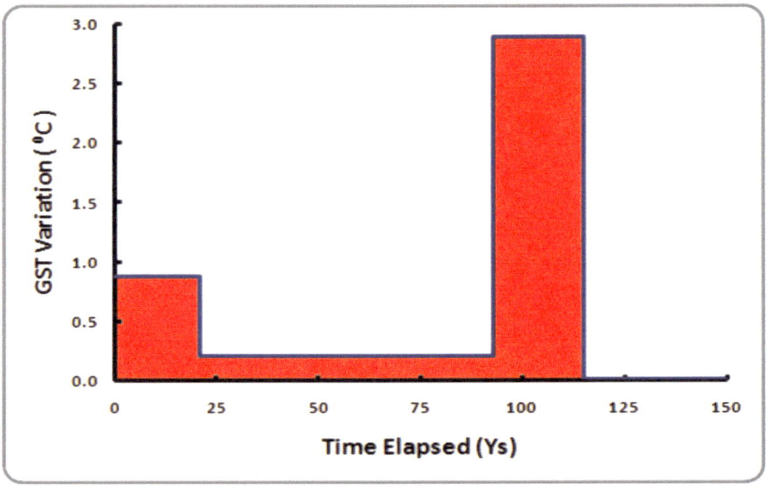

Fig. 6. Climate history of Seropedica (Rio de Janeiro) derived from results of the signal back stripping method.

5. Discussion

5.1 Climate changes in South America

The results obtained in the present work have contributed to substantial improvements in data base for climate change in South America. The area covered in the data base includes several geographic zones in Brazil, Colombia and Peru. Classical inverse models were employed in the analysis of temperature logs from over 30 localities and, in addition, Bayesian inverse modelling was carried out for data from 20 selected sites. The model results have allowed determination of the magnitude as well as the duration of ground surface temperature changes in the major geographic zones of South America. The map of recent climate changes derived from geothermal data is presented in Figure (7).
According to the results obtained the magnitude of GST changes are in the range of 2 to 3.5°C but have had their beginning during the early decades of the 20th century. Nearly similar trends are seen in temperature-depth profiles of bore holes in tropical as well as subtropical zones of the interior and coastal areas. The data from semi arid zones also indicate occurrence of surface warming events but the magnitudes are in the range of 1.4 to 2.2°C while the duration of the warming event is larger, extending back into the last decades of the 19th century. The magnitudes of GST variations are relatively large in localities which have undergone recent changes in vegetation cover. Also there are indications that GST changes are practically insignificant in areas of tropical rain forest.

5.2 Global climate changes

The improved data base on GST changes in the South American continent has contributed to a better understanding of climate changes in the southern hemisphere. In this context, it is convenient to examine also its implications for global variations. It is clear that improved assessments of global changes require an integrated analysis of both the geothermal data sets as well as the meteorological data sets (such as HadCRUT3) reported for oceanic and continental regions (Brohan et al, 2006; Jones and Moberg, 2003; Rayner et al, 2003). Major obstacles in such a venture springs from the wide disparities in data density, and it is

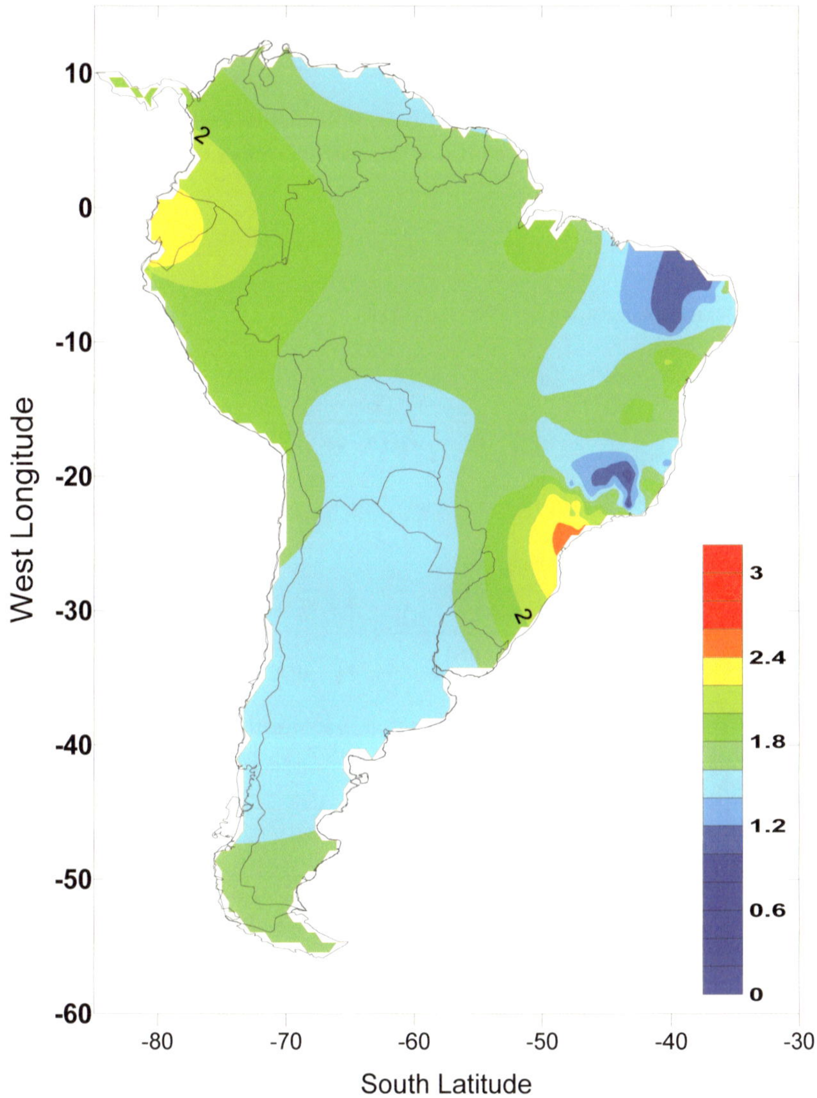

Fig. 7. Map of recent changes in ground surface temperatures derived from geothermal data sets, and supplemented with ground data for selected localities.

convenient to adopt procedures that minimize problems arising from the non-homogeneous distribution. In the present case, the surface area of the globe is divided into a regular grid system composed of 5 x 5 degree grid cells and average values of surface temperature variation in the grid elements calculated. Experimental data are available for almost all of the grid elements covering the continental regions and a significant part of the grid system for the oceanic regions. Following the common practice employed in deriving maps of global representations interpolated values were used for grid elements without data.

Techniques of spherical harmonic representation (Hamza et al, 2008) were employed in analysis of global variations. The harmonic representation of surface temperature variations (q) may be represented as:

$$q(\theta,\phi)=\sum_{n=0}^{N}\sum_{m=0}^{n}\left[A_{nm}\cos(m\phi)\;P'_{nm}(\cos\theta)+B_{nm}\;\sin(m\phi)\;P'_{nm}(\cos\theta)\right] \tag{7}$$

where ϕ is the longitude $\theta = 90 - \psi$, is the colatitude, $P'nm(\cos\theta)$ is the associated Legendre function that is fully normalized and A_{nm} and B_{nm} the coefficients of the harmonic expansion. The expression for evaluation of P'_{nm} is:

$$P'_{nm} = P_{nm}\sqrt{K_n^m} \tag{8}$$

where P_{nm} is the associated Legendre function given by:

$$P_{nm}(\cos\theta)=\frac{sen^m\theta}{2^n}\sum_{t=0}^{Int\left(\frac{n-m}{2}\right)}\frac{(-1)^t\,(2n-2t)!}{t!\,(n-t)!\,(n-m-2t)!}\cos^{(n-m-2t)}\theta \tag{9}$$

and

$$K_n^m=\frac{H(2n+1)(n-m)!}{(n+m)!}\;,\quad\begin{cases} if\;\; m=0 \Rightarrow H=0 \\ if\;\; m\neq 0 \Rightarrow H=2\end{cases} \tag{10}$$

In equation (9) $Int\,(n-m/2)$ is the largest integer that is lower than $(n-m)/2$.
Full normalization of associated Legendre functions (P_{nm}) requires that the following equations be satisfied:

$$\int_0^{2\pi}\int_0^{\pi}\left[P'_{nm}(\cos\theta)sen(m\phi)\right]^2 sen\theta\;d\theta\;d\phi=4\pi \tag{11a}$$

$$\int_0^{2\pi}\int_0^{\pi}\left[P'_{nm}(\cos\theta)\cos(m\phi)\right]^2 sen\theta\;d\theta\;d\phi=4\pi \tag{11b}$$

The coefficients A_{nm} and B_{nm} are evaluated by fitting the harmonic expansion to the set of experimental data, which are the values of surface temperature changes (q) and their respective geographic coordinates (ϕ and θ).
The results obtained on the basis of harmonic representation of the mixed data set are illustrated in the global map of surface temperature changes, in Figure (8). According to features discernible in this figure the thermal signals of climate change in continental areas of North America, Europe, Asia, West Africa and Eastern South America have magnitudes in excess of one degree centigrade, while those in most of the oceanic regions is less than one degree centigrade. On the other hand most of the oceanic regions are characterized by surface temperature variations of less than 0.8^{0}C.

6. Conclusions

Much of the information on surface temperature changes of the recent past in South America are derived from historical records of local meteorological stations. Some of the

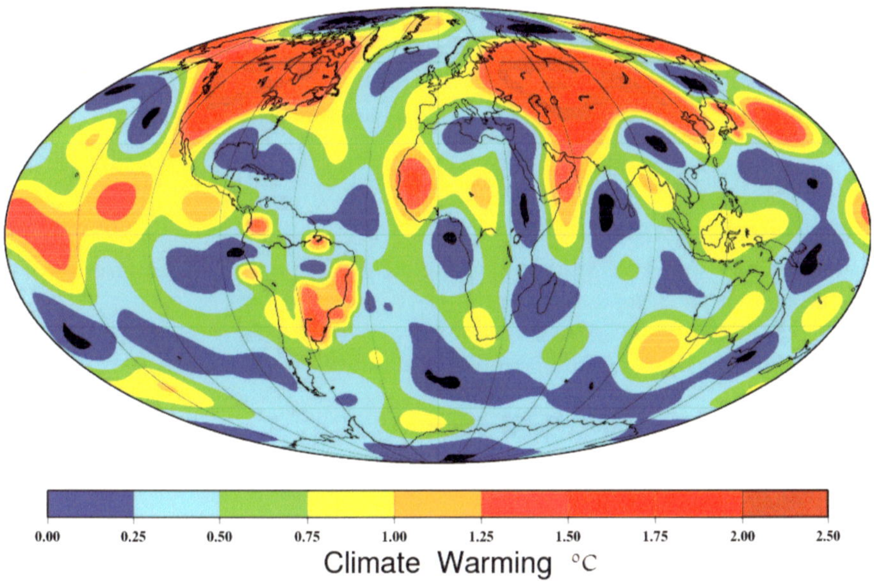

Fig. 8. Spherical harmonic representation of surface temperature variations associated with climate changes of the recent past. Note the systematic differences in magnitudes of climate change between continental and oceanic areas.

main problems in the use of such records include non-uniformity in the accuracy and precision of sensors employed in data acquisition, gaps in the time series of observations and influence of urban warming effects (most of the meteorological observatories are located in or close to towns or cities). The results reported in the present work, based on geothermal methods, are in general free of such problems.

An important result emerging from studies based on geothermal methods is that the climate was relatively cooler during the 17th and 18th centuries. The climate histories, deduced from geothermal data, are found to be consistent with the results of available meteorological records in Brazil, Colombia and Peru. Comparative studies indicate that the magnitudes and duration of recent climate changes in South America are similar to those found in other continental areas such as North America, Asia and Europe.

Analysis of surface temperature changes on global scale reveal that climate warming effects in the northern hemisphere are more pronounced because of the relatively large areas of continental blocks in these regions. On the other hand, the climate warming effects are relatively subdued in the southern hemisphere, where relatively larger proportions of oceans dominate. It is possible that systematic difference in magnitudes of global warming between continental and oceanic regions is a consequence of large scale mixing of near surface waters with those of deep ocean circulation systems. This mixing process is considered to be responsible for pronounced mitigation of the effects of global warming in the southern hemisphere.

7. Acknowledgments

This work was carried out as part of a research project with funding from Conselho Nacional de Desenvolvimento Científico - CNPq (Project No. 301865/2008-6; Produtividade de Pesquisa - PQ). We thank the topic editor and reviewers for critical comments and suggestions. Dr. Andres Papa of the Geophysics Department (ON-MCT, Rio de Janeiro) provided institutional support. The source code for functional space inversion program was provided by Dr. Paul Shen, University of Western Ontario, Canada.

8. References

Alfaro, C., Alvarado, I., Quintero, W., Hamza, V.M., Vargas, C., Briceño, L.A., (2009). Preliminary map of geothermal gradients of Colombia., XII Congreso Colombiano de Geologia, 7-11 September, Paípa - Boyacá, Columbia.

Appleyard, S.J. (2005). Late Holocene temperature record from southwestern Australia: evidence of global warming from deep boreholes, Australian J. of Earth Sciences, 52, 161-166.

Araújo R. L. C., (1978). Heat flow measurements in the Alkaline intrusive complex of Poços de Caldas (in Portuguese). Unpublished M.Sc. Thesis, University of São Paulo, São Paulo, Brazil.

Beltrami, H. and Bourlon, E., (2004). Ground warming patterns in the Northern Hemisphere during the past five centuries, Earth Plnet. Sci. Lett., 227, 169-177.

Beltrami, H., Jessop, A.M. and Mareschal, J.C., (1992). Ground temperature histories in eastern and central Canada from Geothermal measurements: Evidence of climatic change, Palaeogeogr. Palaeoclimatol. Pallaeoecol., 98, 167-184.

Birch, F., (1948). The effects of Pleistocene climatic variations upon geothermal gradients, Am. J. Sci., 246, 729–760.

Bodri, L. and Cermak, V., (1995). Climate changes of the last millennium inferred from borehole temperatures: results from the Czech Republic - Part I, Global and Planetary Change, 11, 111-125.

Brohan, P., J.J. Kennedy, I. Harris, S.F.B. Tett and P.D. Jones, (2006). Uncertainty estimates in regional and global observed temperature changes: a new dataset from 1850. J. Geophysical Research 111, D12106, doi:10.1029/2005JD006548.

Carslaw, H. S. and Jaeger, J. C., (1959). Conduction of Heat in Solids, 386pp, Oxford University Press, New York.

Cavalcanti, A. S. B., (2003). Palaeoclimate variations in Brazil based on the geothermal method (in Portuguese), Unpublished M.Sc. Thesis, Observatório Nacional, Rio de Janeiro, Brazil.

Cavalcanti, A. S. B. and Hamza, V. M., (2001). Climate changes of the recent past in southern parts of Brazil (Extended abstract), 7th International Congress of the Brazilian Geophysical Society, Salvador (BA), Brazil, 1-4.

Cermak, V., (1971). Underground temperature and inferred climatic temperature of the past millennium, Palaeogrogr., Palaeoclimatol., Palaeoecol., 10, 1-19.

Cerrone, B. N. and Hamza, V. M., (2003). Climate changes of the recent past in the state of Rio de Janeiro, based on the geothermal method (Extended abstract in Portuguese), 8th International Congress of the Brazilian Geophysical Society - SBGf, Rio de Janeiro, Brazil, 1-4.

Chisholm, T.J. and Chapman, D.S., (1992). Climate change inferred from analysis of borehole temperatures: an example from western Utah. J. Geophys. Res., 97, 14155–14175.

Conceição, E. and Hamza, V. M., (2006). Subsurface thermal signals of the greenhouse effect in the state of São Paulo (Extended Abstract in Portuguese), 2nd International Symposium of the Brazilian Geophysical Society - SBGf, Natal (RN), Brazil, 1-6.

Cull, J. P., (1979). Climatic corrections to Australian heat flow data. BMR Journal of Australian Geology and Geophysics, 4, 303-307.

Cull, J.P., (1980). Geothermal records of climate change in New South Wales, Search, 11(6), 201-203.

Del Rey, A. C., (1989). Hydrogeothermal studies of the regions of Águas de Lindoia, Amparo e Socorro - Northeastern parts of the state of São Paulo (in Portuguese), Unpublished M.Sc. Thesis, University of São Paulo, São Paulo, Brazil.

Del rey, A. C. and Hamza, V. M., (1989). Terrestrial heat flow variations in the northeastern part of the state of São Paulo: A case for transport of geothermal heat by inter-fracture fluid flows. In: A.E. Beck, G. Garven and L. Stegena (Eds.), Hydrogeological regimes and their subsurface thermal effects. Geophysical Monograph, 47, IUGG volume 2, American Geophysical Union, Washington, 137-148.

Duchkov, A.D. and Sokolova, L.S., (1998). Investigation of temperature changes at the Earth's surface via borehole geothermometry, In Problems of climatic reconstruction and environment of the Holocene and Pleistocene in Siberia (in Russian), 1, 151-157, Nauk, Novosibirsk, Russia.

Eston, S. M., Hamza, V. M., Becker, E. A. and Furumoto, S., (1982). Geothermal research in exploration of hydrocarbons in the Paraná basin (in Portuguese), Internal Report, No. 18106, Instituto de Pesquisas Tecnológicas, São Paulo, Brazil.

Golovanova, I.V., Harris, R.N., Selezniova, G.V. and Stulc, P., (2001). Evidence of climate warming in the southern Urals region derived from borehole temperatures and meteorological data, Global and Planetary Change, 29, 167-188.

Gomes, A. J. L., (2003). Assessment of Geothermal Resources of the state of Rio de Janeiro (in Portuguese), Unpublished M.Sc. Thesis, National Observatory, Rio de Janeiro (Brazil). 138pp.

Gomes, A. J. L. and Hamza, V. M., (2005). Geothermal gradients and heat flow in the state of Rio de Janeiro, Brazilian J. of Geophysics, 23, 4, 325-348.

Hamza, V. M., (1982). Terrestrial heat flow in the alkaline intrusive complex of Poços de Caldas, Brazil. Tectonophysics, 83, 45-62.

Hamza, V. M., (1998). A proposal for continuous recording of subsurface temperatures at the sites of Geomagnetic field Observatories. Rev. Geofísica, 48, 183-198.

Hamza, V. M., (1991). Recent climate changes in the southern hemisphere: the geothermal evidence (In Portuguese), Proceedings 2nd Congress of the Brazilian Geophysical Society, Salvador (BA), 971-973.

Hamza, V.M., (2006). Thermal Signals of Global Warming: Evaluation using geothermal methods (in Portuguese), IV Symposium of the Amzon Region, 12-14 September, Belém (PA), Brazil.

Hamza, V.M., (2007). Climate changes in the eastern segment of the Amazon region: Inferences based on geothermal methods (in Portuguese), V Symposium of the Amzon Region, 30 October to 1 November, Belém (PA), Brazil.

Hamza, V.M., Alfaro, C., Alvarado, I., Quintero, W., López, J., Monsalve, M.L., Pulgarín, B. and Madrid, C., (2009). Surface Temperature Variations of the Last Two Centuries in Bogotá, Columbia., XII Congreso Colombiano de Geologia, 7-11 September, Paípa - Boyacá, Columbia.

Hamza, V.M., Cardoso, R.R., Vieira, F.P. and Guimaraes, S.N., (2010). Geothermal Maps of Brazil (in Portuguese), National Atlas of Brazil - 2010, IBGE.

Hamza, V. M. and Cavalcanti, A. S. B., (2001). Thermal imprints of changes in vegetation cover and climate patterns in borehole temperature profiles in Brazil (Abstract), IAGA-IASPEI Assembly, Hanoi, Vietnam, Book of Abstracts, 413-414.

Hamza, V. M., Cavalcanti, A. S. B. and Benyosef, L.C., (2007). Surface thermal perturbations of the recent past at low latitudes-Inferences based on borehole temperature data from Eastern Brazil, Climates of the Past, 3, 1–13.

Hamza, V. M., Cerrone, B. N., Gomes, A. J. L., Cardoso, R. A., (2003). A Geothermal Climate Change Observatory in the Southern Hemisphere (Abstract), Proceedings of XXIII General Assembly, IUGG, Sapporo, Japan, v. A, p197.

Hamza, V. M., Eston, S. M., Araujo, R. L. C., Vitorello, I and Ussami, N., (1978). Brazilian Geothermal Data Collection - Series-1, Publication IPT No. 1109, p.1-316.

Hamza, V. M., Frangipani, A. and Becker, E. A., (1987). Geothermal Maps of Brazil (In Portuguese) Internal Report No. 25305, Instituto de Pesquisas Tecnologicas, São Paulo, Brazil.

Hamza, V. M. and Muñoz, M., (1996). Heat flow map of South America, Geothermics, 25(6), 599-646.

Hamza, V. M.; Ribeiro, F. B.; and Becker, E. A., (1991). Recent Climatic Changes in the Southern Hemisphere. XX General Assembly, IUGG, Vienna (Austria).

Hamza, V. M., Silva Dias, F. J. S., Gomes, A. J. L. and Terceros, Z. G. D., (2005). Numerical and functional representations of regional heat flow in South America. Physics of the Earth and Planetary Interiors, 152, 223-256.

Hansen, J.E., and S. Lebedeff, (1987). Global trends of measured surface air temperature. *J. Geophys. Res.*, 92, 13345-13372, doi:10.1029/JD092iD11p13345.

Harris, R.N. and Chapman, D.S., (2001). Mid-Latitude (30-60N) climatic warming inferred by combining borehole temperatures with surface air temperatures, Geophys. Res. Lett., 28, 747-750.

Huang, S. and Pollack, H.N., (1998). Global Borehole Temperature Database for Climate Reconstruction. IGBP PAGES/World Data Center-A for Paleoclimatology Data Contribution Series #1998-044. NOAA/NGDC Paleoclimatology Program, Boulder CO, USA.

Jones, P. D. and A. Moberg, (2003). Hemispheric and Large-Scale Surface Air Temperature Variations: An Extensive Revision and an Update to 2001, J. Clim., 16, 206–23.

Jones, W.Q.W., Tyson, P.D. and Cooper, G.P.R., (1999). Modeling climate change in South Africa from perturbed borehole temperature profiles, Quaternary International, 57/58, 185-192.

Koppen, W., (1936). Das geographisca System der Klimate, in: Handbuch der Klimatologie, edited by: Koppen, W. and Geiger, G., 1. C. Gebr, Borntraeger, 1–44.

Lachenbruch, A. H., Claudouhos, T. T. and Saltus, R. W., (1988). Permafrost temperature and the changing climate, Proc. Fifth International Conference on Permafrost, Trondheim, Norway, 1-9.

Lachenbruch, A. H., Sass, J. H. and Marshall, B. V., (1986). Changing Climate: Geothermal Evidence from Permafrost in the Alaskan Artic, Science, 234, 689-696.

Lachenbruch, A. H., Sass, J. H., Marshall, B. V. and Moses, T. H. Jr., (1982). Permafrost, heat flow and geothermal regime at Prudhoe Bay, Alaska, J. Geophys. Res., 87, 9301-9316.

Majorowicz, J.A. and Safanda, J., (2001). Composite surface temperature history from simultaneous inversion of borehole temperatures in western Canadian plains, Global and Planetary Change, 29, 231-239.

Pidwirny, M. (2006). "Climate Classification and Climatic Regions of the World". *Fundamentals of Physical Geography, 2nd Edition.*

Rajver, D, Safanda, J. and Shen, P.Y., (1998). The climate record inverted from borehole temperatures in Slovenia, Tectonophysics, 291, 263-276.

Rayner, N.A., Parker, D.E., Horton, E.B., Folland, C.K., Alexander, L.V, Rowell, D.P., Kent, E.C. and Kaplan, A., (2003). Globally complete analyses of sea surface temperature, sea ice and night marine air temperature, 1871-2000. J. Geophysical Research 108, 4407, doi:10.1029/2002JD002670.

Ribeiro, F., (1988). Study of the problem of determination of equilibrium temperatures in boreholes (in Portuguese), Unpublished Ph.D. Thesis, University of São Paulo, São Paulo, Brazil, 187pp.

Ribeiro, F., (1991). Reconstruction of the climate change from geothermal profiles: Advances in Theory (in Portuguese), Proceedings 2nd Congress of the Brazilian Geophysical Society, Salvador (BA), 959-964.

Roy, S., Harris, R.N., Rao, R.U.M. and Chapman, D.S., (2002). Climate change in India inferred from geothermal observations, J. Geophys. Res., 107, B7, 5, 1-16.

Safanda, J. and Rajver, D., (2001). Signature of last ice age in the present subsurface temperatures in the Czech Republic and Slovenia, Global and Planetary Change, 29, 241-257.

Santos, J., (1986). Heat flow density measurements in the northern parts of the Paraná basin (in Portuguese), Unpublished M.Sc. Thesis, University of São Paulo, São Paulo, Brazil, 124pp.

Santos, J., Hamza, V.M. and Shen, P.Y., (1986). A method for measurement of terrestrial heat flow density in water wells, Brazilian J. Geophysics, 4, 45-53.

Shen, P. Y. and Beck, A. E., (1991). Least square inversion of borehole temperature measurement in functional space. J. Geophys. Res., 96, 19965-19979.

Shen, P. Y. and Beck, A. E., (1992). Palaeoclimate and heat flow density inferred from temperature data in the Superior Province of the Canadian Shield. Global and Planetary Change, 98, 143-165.

Shen, P. Y. Wang, K., Beltrami, H. and Mareschal, J.-C., (1992). A comparative study of inverse methods for estimating climatic history from borehole temperature data, Palaeogeography, Palaeoclimatology, Palaeoecology, 98, 113-127.

Souza, J. R. S., Araújo, R. L. C. and Makino, M., (1991). Transient variations in shallow thermal profiles associated with microclimatic changes in the Amazon (in Portuguese), Proceedings 2nd Congress of the Brazilian Geophysical Society, Salvador (BA), 965-970.

Stulc, P., Golovanova, I.V. and Selezniova, G.V., (1998). Climate change record in the Earth – Example of borehole data analysis in the Urals region, Russia, Phys. Chem. Earth, 23, 1109-1114.

Taniguchi, M., Williamson, D.R. and Peck, A.J., (1999a). Disturbances of temperature-depth profiles due to surface climate change and subsurface water flow: 2. An effect of step increase in surface temperature caused by forest clearing in southwestern Australia, Water Resources Research, 35, 1519-1529.

Taniguchi, M., Williamson, D.R. and Peck, A.J., (1999b). Estimations of surface temperature and subsurface heat flux following forest removal in the south-west of Western Australia, Hydrological Processes, 12, 2205-2216.

Tarantola, A. and Valette, B., (1982). Inverse problems = quest for information. J. Geophys., 50, 159-170.

Torok, S. and Nicholls, N., (1996). An historical temperature record for Australia. *Aust. Met. Mag.* 45, 251-260.

Tyson, P. D., Mason, S. J., Jones, M. Q. and Cooper, G. R., (1998). Global warming and geothermal profiles: the surface rock-temperature response in South Africa, Geophys. Res. Lett., 25, 2711-2713.

Vasseur, G., Bernard, Ph., van de Mentebrouch, J., Kast, Y. and Jolivet, J., (1983). Holocene palaeotemperatures deduced from geothermal measurements, Paleogeography, Paleoclimatology, Palaeontology, 43, 237-259.

Vieira, F.P. and Hamza, V.M., (2010). Global Heat Loss: New Estimates Using Digital Geophysical Maps and GIS Techniques. IV Symposium of Brazilian Geophysical Society, Brasília, 14 a 17 de novembro de 2010.

Vitorello, I., (1978). Heat flow and radiogenic heat production in Brazil, with implications for thermal evolution of continents. Ph. D. dissertation, Univ. of Michigan, Ann Arbor, U.S.A.

Vitorello, I., Hamza, V. M., Pollack, H. N. and Araújo, R. L. C., (1978). Geothermal Investigations in Brazil, Rev. Bras. Geof., 8(2), 71-89.

7

Climate Change Impacts on Atmospheric Circulation and Daily Precipitation in the Argentine Pampas Region

Olga C. Penalba[1] and María Laura Bettolli[1,2]
[1]*Departamento de Ciencias de la Atmósfera y los Océanos, FCEN, UBA*
[2]*Consejo Nacional de Investigaciones Científicas y Técnicas*
Argentina

1. Introduction

The relevant findings of the Intergovernmental Panel on Climate Change (IPCC, 2007) highlight the increment of the global mean temperature and the need to understand how this increment will affect the climate variability and change of the regional environment. The changes in the climate are a consequence of both internal variability of the climate system and external factors, the latter being both natural and anthropogenic. The Fourth Assessment Report of IPCC describes the scientific progresses that have been achieved by researchers in the understanding the observed changes of the climatic system, the processes involved, and the establishment of future climate change projections (IPCC, 2007). The number of studies that discuss this problematic have increased considerable during the last years. However, there are many issues that need further investigation, in particular for developing countries.

Extreme climate anomalies have a negative impact on the population and economy of the affected regions. The climate has a fundamental role for regions where the economy is based on agriculture. The process of growing crops can be seriously affected by extreme temperatures, and the precipitation can be a limiting factor which conditions the success or failure of the production.

The region of interest in this study is the Pampas region, which comprises the most productive agricultural lands of Argentina. The most important grains of the country, like soybean, corn, wheat and sunflower are grown in this region. Together with their by-products, these crops promote the social and productive system of the region, and are one of the principal sources of fiscal incomes. In the campaign of 2008/09, more than 24 million hectares, compared to of the country's total of 28 million cultivated hectares of these grains, corresponded to the Pampas region (http://www.sagpya.mecon.gov.ar/). This region, located in the center east of Argentina, southeastern South America, have an extension of more than 600.000 km2. Since the grains are cultivated extensively without artificial irrigation, the precipitation is one of the climatic variables of main influence for the production, and is also a condition for the management of the crops. Therefore, the spatial and temporal distributions of the precipitation in the region, and its surplus or deficit, are of extreme importance for the successful harvests.

From the 1960s, the Pampas region was favored by an increase of precipitation on both the annual and seasonal scales (Fernández et al., 2006; Liebmann et al., 2004). This increase showed a non-stationary variability (Penalba &Vargas, 2004). Depending on the region and the time of the year, the observed cycles were: inter-decadal variability, trends, jumps or discontinuities (Barros et al., 2008; Boulanger et al., 2005; Penalba & Vargas, 2008). This hydrological condition displaced the agricultural border with around 200 km to the west, which favored substantially the agricultural activity, especially in the semiarid subregions.

However, these inter-decadal and inter-annual variations were observed in the extreme precipitation, on an annual scale and during the months of maximum precipitation (Penalba & Vargas, 2008). On a daily scale, the frequency of rainy days and of days of extreme precipitation, showed the same temporal variability (Penalba & Robledo, 2010). Furthermore, the temporal variability was greater, increasing the risk of droughts and their consequent negative impacts (Penalba et al., 2010). During 2008 and during almost the entire 2009, a severe rainfall deficit occurred in the region (Bidegain et.al., 2010), impacting strongly the gross domestic product. In Entre Ríos, which is a province within the Pampas region, the producers lost more than 50 millions of dollars in the corn harvest (Riani, 2009).

Rainfall events depend on, among other factors, the large-scale atmospheric fields. Therefore, the study of these circulation structures, their frequency, distribution and temporal variability are important elements for diagnosis and forecast, particularly in the context of future climate change.

Global Circulation Models (GCMs) are fundamental tools for climate change studies. Various studies shows that the GCMs have good capacity of representing characteristics of the South American circulation climatology, on temporal scales from monthly to decadal (Di Luca et al., 2006; Solman & Le Treut, 2006; Solman & Pessacg, 2006). On the other hand, the intermodel variability in the representation of monthly and seasonal characteristics of the precipitation and temperature in different regions of South America is high (Gulizia et al., 2009; Marengo et al., 2010; Rusticucci et al., 2010; Silvestri & Vera, 2008; Vera et al., 2006).

The skill of the GCMs in representing the circulation that conducts or conditions the precipitation events has an important role in the evaluation of future climate projections. However, the capacity of the GCMs in representing the circulation on a synoptic scale has been little explored up to now. The usefulness of the GCMs in local studies is restricted by their poor spatial resolution. Techniques of scale reduction have been developed as bridges between the large scale information generated by the GCMs and the local scale information, with the purpose of performing short- to midrange forecasts and to study the potential impacts of future climate change. In this way, the use of daily results from the GCMs for studies of local climate is subject to their aptitude in representing the atmospheric systems on a regional scale.

In this context, the present chapter is structured to fulfill the following objectives: to characterize the rainfall conditions and their probability of occurrence in the Pampas region; to identify daily circulation patterns in southern South America and to associate them with different rainfall conditions in the Pampas region; to evaluate the representation of the daily circulation patterns as simulated by a set of 12 GCMs; and to analyze the projected changes of the same patterns at different time horizons of the 21th century.

In the second section of this chapter, the area, data and methods of this study are described. In the third section, the results are analyzed and discussed and in the last section the conclusions are presented.

2. Data and methods

2.1 Data

The following data-sets were used in this study:

a. Daily mean sea level pressure (SLP) fields from NCEP reanalysis 2, provided by the NOAA/OAR/ESRL PSD, Boulder, Colorado, USA: http://www.cdc.noaa.gov/, were used to represent observed circulation for the period 1979-1999. The domain extends from 15°S to 60°S and from 42.5°W to 90°W on a 2.5° latitude-longitude grid, including the Pacific and the Atlantic Oceans and the Andes Mountains, geographical features that have a significant influence on the circulation over South America (Figure 1).

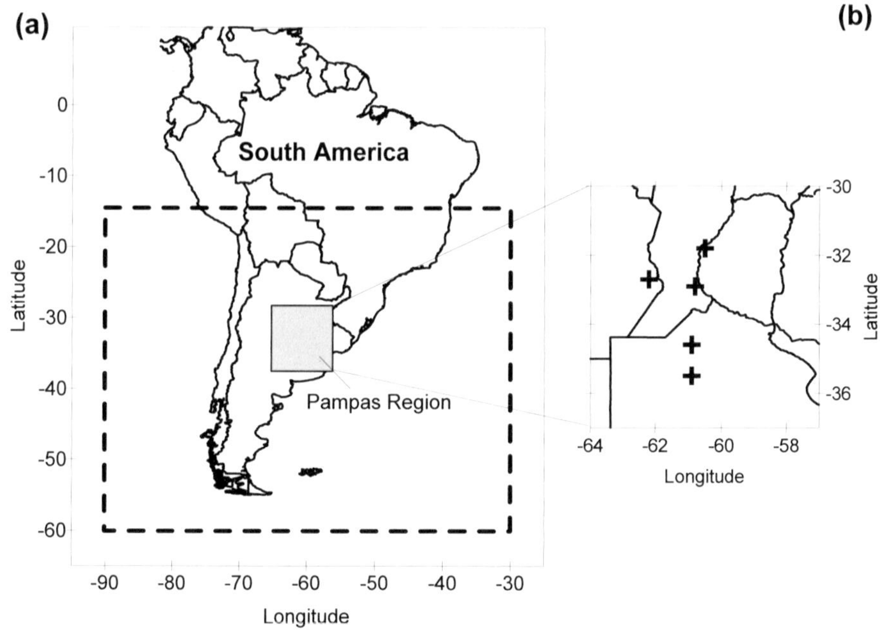

Fig. 1. (a) The Argentine Pampas Region (shaded rectangle) and the domain chosen for the atmospheric circulation fields (dashed line). (b) Locations of the five meteorological stations used in this chapter.

The focus of this analysis is to examine the model veracity with respect to sea level pressure. Previous studies (e.g. Bettolli et al., 2010) observed that the upper level patterns are less variable than the surface patterns. Furthermore, the upper level patterns can be associated with different surface synoptic types. The surface fields with their embedded synoptic systems provide a first-order control on spatial and temporal variations in precipitation.

b. Observed daily rainfall series from stations in the Pampas region were provided by the Argentine National Meteorological Service. The five stations that had both less than ten per cent of missing data and a continuity of their records were chosen for analysis (Figure 1) Although this dataset goes further back in time, we only analyze the period that coincide with the NCEP data period.

c. A set of 12 GCM SLP daily fields was used to describe present and future low level circulation (Table 1). The 20C3M experiment was used for the period 1979-1999 and the

SRES A1B 720 ppm stabilization scenario was used for the periods 2046-2065 and 2081-2099. These simulations are available from the Program for Climate Model Diagnosis and Intercomparison (PCDMI) and from the ENSEMBLES CERA archives.

The SLP fields of the models were interpolated to the NCEP reanalysis grid with an inverse distance weighting method in order to facilitate comparisons.

Identification	Model	Original Grid Resolution
A	BCCR-BCM2.0	2.7905° x2.8125 °
B	CNRM-CM3	2.79° x 2.8125°
C	CSIRO-Mk3.0	1.865° x 1.8750°
D	ECHAM5/MPI-OM	1.865° x 1.8750°
E	EGMAM	3.71059° x 3.75°
F	GFDL-CM2.0	2° x 2.5°
G	GFDL-CM2.1	2° x 2.5°
H	GISS-EH	3° x 5°
I	GISS-ER	3° x 5°
J	INGV-SXG	1.1215° x 1.125°
K	IPSL-CM4	2.5352° x 3.75°
L	UKMO-HadCM3*	2.5° x 3.75°

* Available period for present climate: 1979-1989

Table 1. List of GCMs used for the study.

2.2 Methods

In order to identify the dominant spatial structures and their degree of contribution to the total variance, Principal Component Analysis (PCA) was performed on observed NCEP SLP field (Jolliffe, 2002; Richman, 1986). The method was applied in the T-mode, withthe correlation matrix as input and the SLP daily fields as variables and the gridpoints as observations. Only the principal components (PCs) that correspond to large eigenvalues are expected to contain an interpretable signal and are retained for further analysis. Craddock & Flood (1969) suggest plotting the log eigenvalue diagram (LEV diagram) and cutting the number of PCs just behind a section where the graph approximates to a line with a relatively small slope. Following their criterion, the first 6 PCs for summer and the first 8 PCs for winter, accounting for 94.3% and 94.6% of the total variance respectively, were retained for further analyses. The first unrotated PC of the raw data can be identified with the time mean pattern (Compagnucci & Vargas, 1986; Huth, 2000). The first PC was calculated for each GCM in order to analyze the ability of the GCMs to reproduce the basic characteristics of the daily circulation at low levels.

The cluster analysis was coupled with PCA to determine the dominant circulation types (CT) of NCEP (as in Romero et al., 1999a; Rodrigues Chaves & Cavalcanti, 2001). The analysis was carried out in the subspace given by the leading unrotated PCs. As shown by Gong & Richman (1995), this combination of methods provides the most separable cluster system. The clustering algorithm used in this study is the 'k-means' method, which is a partitioning method that classifies all days into a predefined, optimal number of clusters (MacQueen, 1967). The method minimizes the variability within each cluster and maximizes the variability between clusters.

The choice of optimal number of clusters was established by the pseudo-F statistic (Calinski & Harabasz, 1974). This statistics assess the among- and within-cluster sum of squares relationship. The number of maximum local peaks in its plot indicates an appropriate number of clusters (Romero et al., 1999b). A progressive number of k clusters, from 2 to 30 were tested, and the pseudo-F statistic suggested a five-cluster solution for summer and a seven-cluster solution for winter.

The SLP fields from the GCMs, were classified using the cluster centroids from the NCEP original typing. Each GCM SLP daily field was assigned to the NCEP circulation type that correlated best with the daily field.

In order to outline and summarize the model-data comparisons, Taylor diagrams were constructed (Gleckler et al., 2008). These diagrams convey information with clarity and they quantify the degree of statistical similarity between two fields (in this study, between the observed NCEP circulation and the simulated by the GCMs) considering the correlation coefficient, the standard deviation and the root mean squared error (RMSE). The shape of the configurations can be compared, to a certain degree, through the correlation coefficient. The spatial patterns are compared directly from the values of atmospheric pressure through the standard deviation and the RMSE. The GCMs are considered to characterize and estimate the spatial patterns better, when the cloud of points is more concentrated and closer to the reference point.

The probability of occurrence of a given rainfall event conditioned to a specific circulation type (CT) was compared to the climatological probability of its occurrence (Bettolli et al., 2010). The Z statistic was used to quantify the difference between the probabilities (Infante Gil & Zárate de Lara, 1984).

3. Results

3.1 Precipitation

The Pampas region has a humid temperate climate and a flat relief. The mean annual rainfall is around 900 mm. The annual rainfall is characterized by a spatial variation in the NE-SW direction, with a significant decrease from east to west from the meridian of 65° (Penalba & Vargas, 2008). Of the annual amount, only 20% reaches the sea as water and the remaining 80% evaporates, runs off or changes the soil water amount (Berbery & Mechoso, 2001). The mean annual cycle of the region is characterized by a wet season with around 110 mm per month from October to April (warm months). During the transition months, the monthly rainfall decreases, reaching its lowest values during austral winter (around 30 mm/month) (Bettolli et. al., 2010).

In some recent application studies, the number of rain days appeared to be the key to fluctuations in total rainfall amounts; in some, variation depended on rainfall intensity; and in others, on both variables. It has also been found that the lack of water can affect crop production.

Robledo & Penalba (2008) analyzed the climatology of the different components that affect the monthly rainfall of Argentina. They calculated the amount and the frequency of the mean daily intensity and the daily extreme rainfall, using different thresholds according to the regions. In the Pampas, the spatial patterns and seasonal variation of daily rainfall above the 75th percentile show similar behavior. From October to April, the value of this threshold is around 16 mm/day, meanwhile during winter this value decreases to 5 mm/day. The frequency of rain days during the wet months is around 25%, decreasing to 16% during the austral winter months.

Considering the dry condition, the average length of a dry sequence is 5 days or 8 days in summer and winter, respectively. The maximum dry period length presents more spatial variability with values around 20 days in summer and 40 days in winter (Llano & Penalba, 2011).

In this chapter, we are interested in the characteristics of the rainfall that condition the production of crops. Winter and summer coincide with key stages of the growing season of main crops of the region (Pascale & Damario, 2003). Dry or wet conditions are defined at a regional scale, analyzing the joint information of the five stations. We analyze days with no rainfall in the five stations of the region (dry days), days with rainfalls in at least one station in the region greater than 0.1mm (rain days, R0.1) and greater than the 75th percentile (heavy rain days). As mentioned above, the 75th percentile for winter is 5mm (R5) and 16mm (R16) for summer.

The annual cycle of the percentage of days corresponding to different rainfall conditions is shown in Figure 2. The annual cycle pattern of the three humid conditions is conserved in the region, with low variability on a monthly time scale for both winter and summer seasons. The highest variability between the months of these seasons is found in the R0.1 condition. During summer, in general, the probability of rainfall at least at one station is around 50% and of intense rainfall 20%. For winter, these probabilities are 30% and 10% respectively. Although the region is classified as humid, due to its high annual precipitation amount, the annual percentage of dry days is more than 50 per cent.

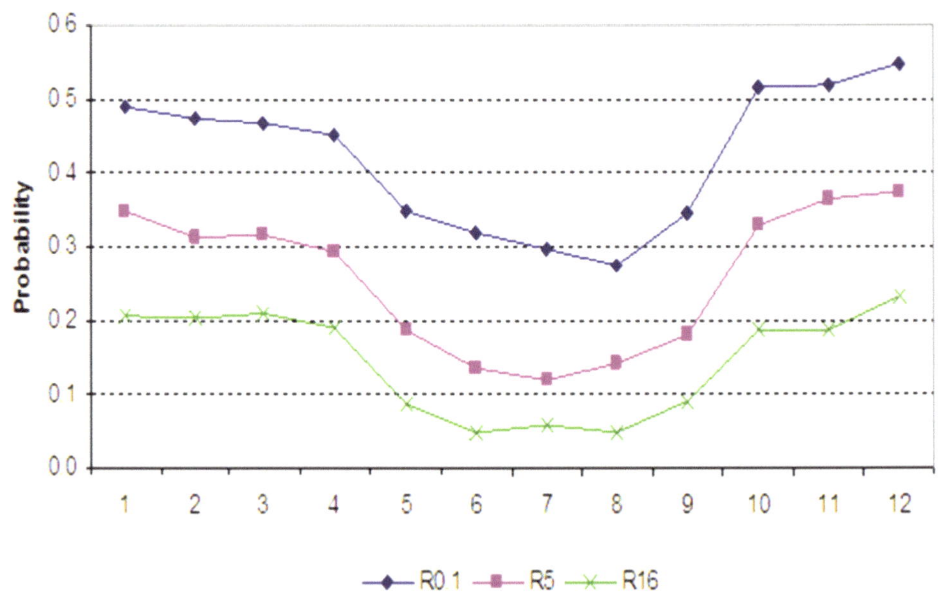

Fig. 2. Annual cycle of the probability of days per months corresponding to different rainfall conditions: 0.1 mm (R0.1), 5mm (R5), 16mm (R16) in at least one station in the region.

The daily rainfall of 10mm plays an important role for the hydrological balance of the summer months, since this amount approximates the daily evaporation (R10) (Vargas, 1979). Due to the hydrological characteristics of the Pampas region, the analysis focus on winter (JJA) and summer (DJF), coinciding with key stages of the growing season of different main crops in the region (wheat, corn and soybean). The thresholds of study are set 1 to R0.1 and no rain for both seasons and to R5 for winter and to R10 and R16 for summer.

3.2 Climatic characteristics of the atmospheric circulation

One important aspect when comparing the properties of the GCMs and NCEP dataset is the reproduction of the PCs spatial structures. In particular, the first PC spatial pattern approximates the time mean pattern; whereas the remaining PCs spatial patterns can be interpreted as deviations from the time mean (Huth, 2000). This feature is used to analyze the climatology of the GCMs.

The climatic characteristics of the mean SLP patterns, represented by the first PC of NCEP datasets, show a notorious seasonality (Figure 3). This seasonality is of great importance in determining the low-level circulation and its associated moisture advection to the region. During summer, the high pressure cells over the eastern South Pacific and western South Atlantic are positioned in their southernmost location, limitating the Westerlies to the south of 50ºS. A clear thermo-orographic low is located in the center of the continent over the eastern Andes mountain range. During winter, both the semi-permanent high systems and the Westerlies are in their northernmost position and the thermo-orographic low is absent. These climatic characteristics seem to be better captured by the GCMs in winter than in summer (Figure 3). The scatter plot diagrams of Figure 3 show that the cloud of points is more concentrated and closer to the point of high correlation (between the spatial patterns of the PC1 of NCEP and of the models), and lesser RMSE.

The spatial patterns of the best and worst performing GCMs, that is, the GCMs located at the extreme points of the scatter plot diagrams, are also shown in Figure 3. During winter, the GCMs tend to displace the Westerlies equatorward, attenuating the contribution of the subtropical highs, whereas in summer, both semi-permanent high systems are more extended to the south. For some GCMs, the difficulty in representing the Andes orographic effect on the circulation is noteworthy. As examples, the model GISS-ER and EGMAM are shown in Figure 3. GISS-ER extends the thermo-orographic low of summer towards a larger region over the mountain range, while the model EGMAM represents both semi permanent anticyclones over the continent, penetrating the range.

The variance explained by each PC can be interpreted as a measure of the strength of each spatial pattern. Therefore, an analysis of the variance values explained by the first PC is used as a simple indicator of the ability of the GCMs in representing the mean fields of the low level circulation over the region. The percentages of variance explained by the first PC are shown in Figure 4. For the NCEP dataset, values reach 70.6% and 54.6% for summer and winter respectively, indicating that the cold season is more perturbed than the warm season. During summer, the percentage of variance explained by the first PC of 10 out of the 12 GCMs is below than observed. Thus, most models tend to represent a lesser incidence of the mean pattern and, therefore, a higher presence of perturbations. During winter, 6 models keep the circulation closer to its mean than is observed (i.e., overestimation in the percentages of variance explained by the first PC in BCCR-CM2.0, CNRM-CM3, CSIROMk3.0, GFDL-CM2.0, UKMO-HadCM3 and IPSL-CM4). However, the inter-model dispersion is lower when compared with summer. It is worth mentioning that most models are capable of reproducing the seasonality of the percentage of variance explained by the first PC, which is lower in winter. The exception is IPSL-CM4 with a lower percentage in summer.

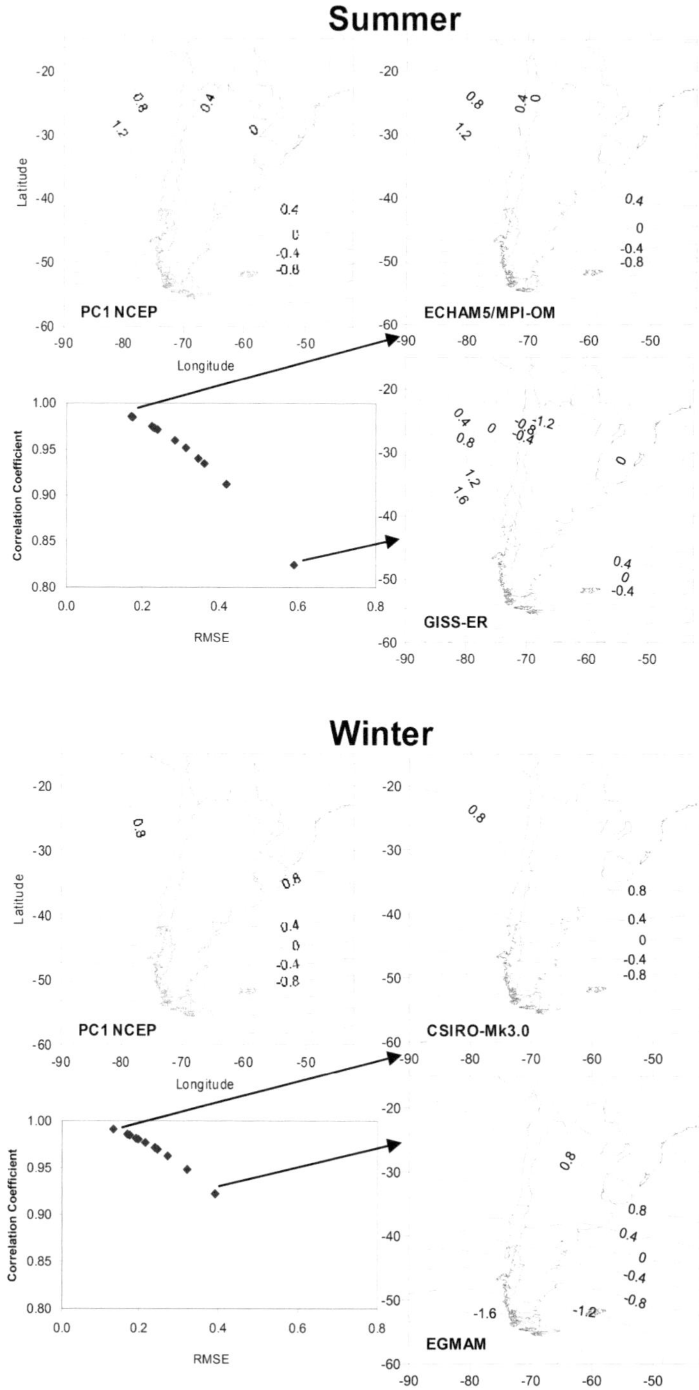

Fig. 3. Spatial pattern of the PC1 of the NCEP SLP data for summer and winter. Scatterplot of the correlation coefficients versus the RMSEs between PC1 of NCEP and of the GCMs. Spatial pattern for some selected models are also shown.

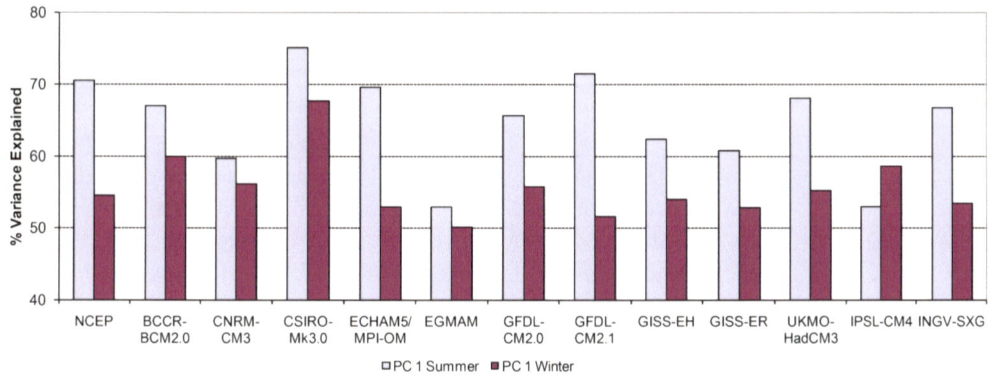

Fig. 4. Percentage of variance explained by the PC1 for summer and winter.

3.3 Observed circulation types

The circulation types for summer (CTi_S, i=1,...,5) are shown in Figure 5.

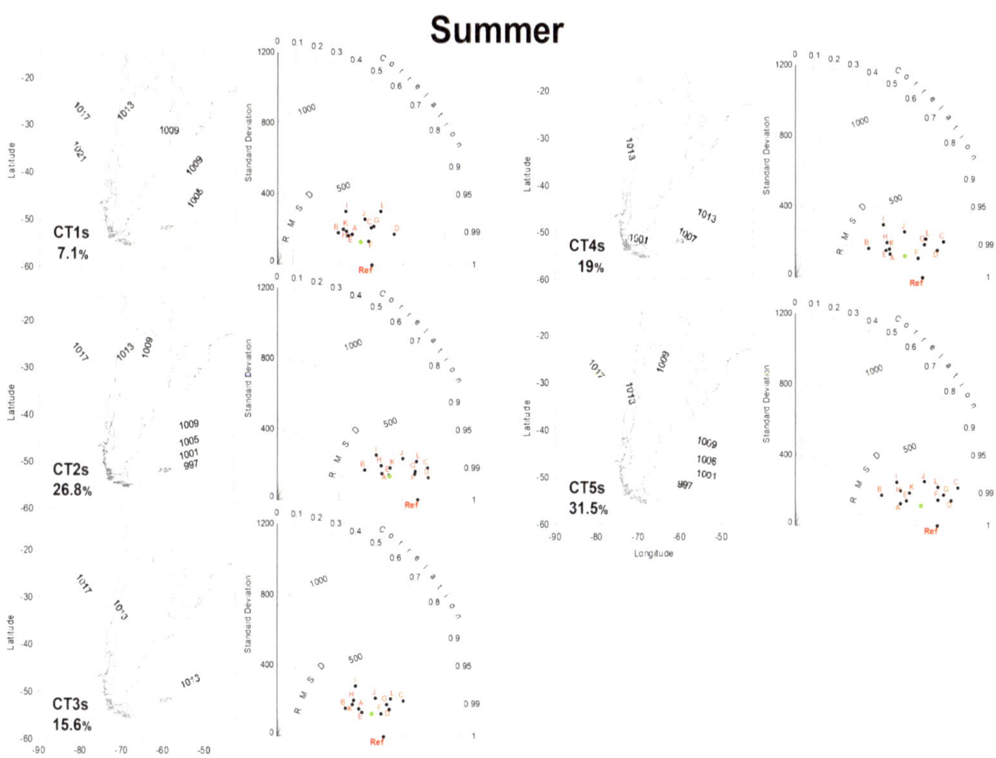

Fig. 5. Observed CTs and percentage of days corresponding to each CT for summer. The dashed (solid) lines represent sea level pressure values lower (higher) than 1013 hPa. The contour interval is 2 hPa. Taylor diagrams of the observed CTs of NCEP and the CTs of the GCMs (red letters) and of the model ensemble (green point).

$CT1_S$ is characterized by an intensification of the southern Pacific anticyclone associated with a trough axis in the northwest-southeast direction. This structure could be connected with a post-frontal anticyclone that moves forward on the continent inducing an anomalous flow from the east-southeast over the Pampas region. $CT2_S$ is characterized by a perturbation over the continent and a weakening of the Atlantic anticyclone that could be related to a cold front affecting the region. In $CT3_S$, a belt of high pressures is extended to the South, reaching around 45°S. This CT is accompanied by a centre of low pressure values to the north. $CT4_S$ shows an intensification and expansion of the southern Atlantic anticyclone, which interrupts the passage of the eastern perturbations and diverts them to the south. $CT5_S$ is the pattern that is most similar to the mean SLP field of summer (compare with the spatial pattern of the NCEP PC1 in Figure 3). This is the most frequent summer pattern, with 31.5% of the studied cases.

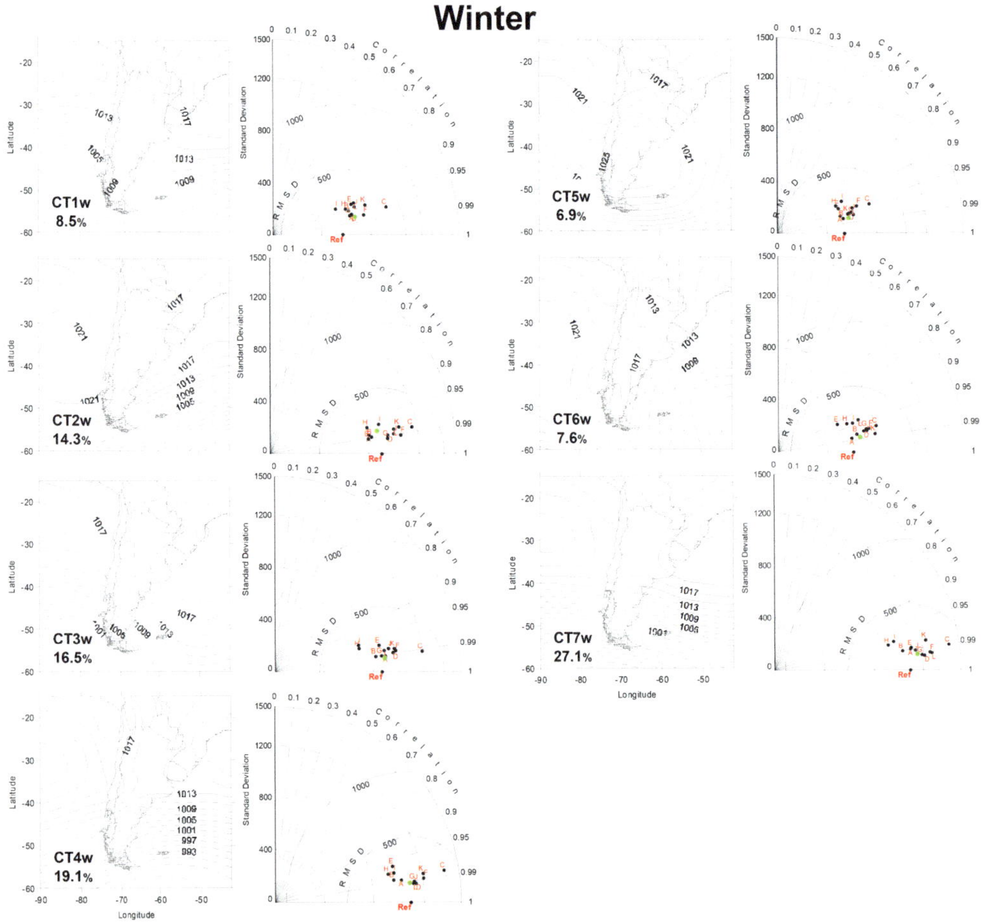

Fig. 6. Idem Figure 5 for winter.

The winter patterns (CTi_W, i=1,...,7) show more variable spatial structures than the summer patterns (Figure 6). This is due to the higher baroclinicity of the winter season and therefore, the greater contribution of synoptic perturbations. In $CT1_W$, a cyclonic perturbation dominates the circulation over the southern South Pacific Ocean, while the opposite occurs in $CT2_W$ for which a high pressure system extends towards south, entering over the continent. $CT3_W$ shows an intensification of the Atlantic anticyclone, inducing an anomaly of the northern-northeastern flow at the southern tip of the continent. $CT3_W$ shows an extension towards the north of the Westerlies, which restricts the action of both anticyclones to act to the north of 40ºS. $CT5_W$ corresponds to an intense high pressure centered over Patagonia in the south of Argentina that extends over almost the whole southern region of the continent and over adjacent oceans. $CT6_W$ can be associated with a cold front that advances towards northeast with its postfrontal anticyclone generating southern advection when getting in over the continent. Finally, $CT7_W$ can be linked with the mean SLP field of winter, similarly to summer, with a frequency of 27.1% (compare with the spatial pattern of the NCEP PC1 in Figure 3).

3.4 Observed circulation types and daily rainfall

This section quantifies the relationship between the CTs for each season and rainfall amount and persistence over the Pampas region. The purpose is to evaluate how much rainfall information for the core crop-producing region is contained in the circulation structures at a regional scale. Then, the probability of occurrence of a rainy day conditioned to a specific CT is compared with the probability of occurrence of a rainy day for the rest of the data by means of the Z-statistic. Values and significance from the Z-statistics are shown in Table 2.

Summer				
	Dry Days	R0.1	R10	R16
CT1s	-2.65	2.65	-0.06	-1.13
CT2s	-2.22	2.22	2.25	2.29
CT3s	0.67	-0.67	-0.99	-1.51
CT4s	3.68	-3.68	-3.05	-2.16
CT5s	-0.05	0.05	1.00	1.15

Winter			
	Dry Days	R0.1	R5
CT1w	-1.20	1.20	0.75
CT2w	1.21	-1.21	-2.26
CT3w	0.33	-0.33	-0.71
CT4w	1.78*	-1.78*	-1.36
CT5w	-4.72	4.72	3.06
CT6w	-1.56	1.56	0.44
CT7w	1.99	-1.99	0.34

Table 2. Z-statistics of the comparison between the conditional probability of occurrence of a day with a certain rainfall condition in each CT for summer and winter and the climatological probability of that day. If the Z-statistic value is positive (negative) and is significant, the specific circulation pattern has (does not have) a significant contribution to the rainfall event. In red, significant values at 95% and 90% (*).

For summer, $CT4_S$ has the highest contribution to the dry days, showing positive and significant values of the Z-statistic for this condition (first column in Table 2). The configuration of SLP of $CT4_S$ corresponds to an intensification of the southern Atlantic anticyclone, which interrupts the passage of the eastern perturbations and diverts them to the south. This anticyclone induces stability at low levels and can be significantly associated to the dry days of the region. $CT4_S$ is the most persistent pattern, with the 19% of the events in sequences lasting from four to seven-day (Table 3). For rainy days, positive and significant values of the Z-statistic are observed for the $CT2_S$ and $CT1_S$ patterns (R0.1, R10 and R16 columns in Table 2). Rainy days (R0.1) are significantly benefited by patterns that could be related to a post-frontal intense anticyclone that induces east-southeast anomalous flow and consequently increases moisture advection over the region (CT1s). This pattern is the less frequent NCEP pattern of the season (7.1%, Figure 5) and is also one of the less persistent patterns, with 90% of the events in sequences of one to three days (Table 3). Heavy rainy days are significantly related with a cyclonic disturbance at the centre of the continent associated with a cold front passage ($CT2_S$ in Figure 5).

	Summer					Winter						
D	CT1s	CT2s	CT3s	CT4s	CT5s	CT1w	CT2w	CT3w	CT4w	CT5w	CT6w	CT7w
1	54.3	41.1	60.5	34.8	46.7	36.6	50.3	36.8	52.5	59.5	51.3	48.3
2	24.3	23.7	22.2	31.2	22.2	33.8	25.5	27.8	20.1	21.5	30.0	24.6
3	11.4	17.8	9.0	12.1	12.8	11.3	14.5	16.5	12.3	12.7	6.3	12.7
4	4.3	8.7	4.8	8.5	7.4	7.0	5.5	9.0	7.3	2.5	10.0	3.8
5	1.4	3.2	0.6	4.3	4.7	5.6	2.1	5.3	2.8	3.8	2.5	5.5
6		2.7	0.0	4.3	1.9	2.8	1.4	1.5	2.8			0.8
7	4.3	0.9	2.4	2.1	0.8	1.4	0.0	0.0	1.1			1.7
8		1.4		1.4	1.6	1.4	0.7	2.3	1.1			0.4
9			0.6	0.7	1.2							0.8
10												0.4
11		0.5		0.7	0.4							0.4
12												
13					0.4							
14												
15												0.4

Table 3. Probability of the persistence (in days, D) of the different CTs.

	Summer		Winter
CT1s	Rainy Days Least Frequent	**CT1w**	
CT2s	Heavy Rainy Days	**CT2w**	
CT3s	Least persistent	**CT3w**	Most persistent
CT4s	Dry Days Most persistent	**CT4w**	Dry Days
CT5s	Most frequent	**CT5w**	Rainy and heavy rainy Days Least persistent Least frequent
		CT6w	Rainy Days
		CT7w	Dry Days Most frequent

Table 4. Schematic summary of results.

In Table 4, a schematic summary of the results described above is found, which will serve as a basis for the comparison with the GCMs in the next section.

Winter dry days are significantly favored by a high pressure system that extends from the Atlantic Ocean to the centre of the continent ($CT4_W$) and also by $CT7_W$, the pattern that resembles the mean pattern for winter (positive and significant values of the Z-statistic for this condition in the first column of Table 2). Rainy days and heavy rainy days are significantly benefited by structures with a high pressure system at the south of the continent, enhancing an anomalous flow from the east-southeast to the central region of Argentina and a corresponding moisture advection at low levels ($CT5_W$). This CT is the less persistent pattern with 93.7% of the events in sequences of one- to three-day lasting (Table3) and it is also the less frequent one (6.9%, Figure 6). $CT3_W$ is the most persistent pattern of winter (Table 3), coinciding with what was found for the summer $CT4_S$.

3.5 Comparison between NCEP and GCMs

3.5.1 Present climate

A diversity of aspects should be taken into account when comparing the ability of the GCMs in representing the synoptic patterns of NCEP, given that the surface climate depends on the representation of these characteristics.

The comparison of the mean spatial patterns is summarized in the Taylor diagrams of Figures 5 and 6. Although the correlation is expected to be high, since the projection of the GCM fields was defined over the centriods of the observed NCEP fields, a certain dispersion is found.

In summer, $CT2_S$ and $CT5_S$ shows the smallest dispersion, bounded between the values 0.95 and 0.99 of the Taylor diagrams. This means that the GCMs are able to reproduce the structure and position of these atmospheric systems. In particular, the accurate representation of $CT2_S$ is essential for the generation of the heavy rainfall events in the region.

In winter, the greatest correlations are close to 0.99 and are found for the $CT2_W$, $CT3_W$, $CT4_W$ and $CT7_W$. Unlike what occurs for summer, the CTs that are best represented by the GCMs are the ones associated with dry days ($CT4_W$ y CT7w). CT3w is the most persistent structure with an intensification of the Atlantic anticyclone (Table 3) that could be linked to the blocking events occurring in the Atlantic ocean (around 40°W) that are more frequent during winter (Alessandro, 2003). From the Taylor diagram, it is clear that the spatial structure of CT3w is well represented, which is key for the location of the blocking and its consequent effect on the surface variables (Figure 6).

The comparison of the standard deviation and the root mean squared error indicate an inter model dispersion according to the CT and the time of the year. For summer, the standard deviations are distributed around the observed NCEP value for all cases except for CT2s, for which most models underestimate the standard deviation (points to the left of the reference point in the Taylor diagram of Figure 5). This indicates that the GCMs tend to underestimate the amplitude of the variation of the SLP of CT2s, and consequently underestimate the depth of the systems that are associated directly with heavy rainfall.

During winter, the dispersion of the standard deviations is uniform for $CT2_W$, $CT3_W$, $CT4_W$ y $CT7_W$, which also are the CTs with best estimations of the spatial structures (Figure 6). Most GCMs overestimate the standard deviation of the types $CT1_W$, $CT5_W$ and $CT6_W$,

increasing the depth of the systems, and in particular of those that are associated with rainfall or heavy rainfall of winter ($CT5_W$ and $CT6_W$, Table 4).
In all cases, the root mean squared errors are lower than 400 hPa. Considering that the standard deviations of NCEP vary between 536 and 1088 hPa, the model errors are lower than the proper variability of the observed mean fields. Also, the majority of the errors do not reach higher values than 50% of the standard deviation.
Another aspect to take into account is the relative frequency of each CT estimated by the GCMs. Figure 7.a shows that the ensemble is able to reproduce these frequencies, although the model dispersion is considerable, especially in summer for CT3s and CT5s. In summer, the frequencies of the models overestimate the frequencies of CT3s while the frequencies of CT5s are mostly underestimated. The latter coincides with what was found in the analysis of the dominant summer pattern, PC1. CT5s is similar to the summer mean field and an underestimation of its frequencies implicates a higher contribution of the other CTs, representing the perturbations. The pattern that represents an intensification of the southern Atlantic anticyclone and its stability (CT4s) is the pattern that is best represented by the GCMs in terms of frequency. This pattern is significantly associated to the dry days of the region.
For winter, the dispersion among the models is lower than for summer. $CT5_W$ and $CT6_W$ are best represented with frequencies close to observed values. This implies that the models are capable of a quite good representation of the frequencies of the structures that are significantly associated with rain days and heavy rain days. The ensemble reproduces the observed frequencies very well in all cases.

3.5.2 Future climate

The future frequency changes of each CT show a considerable dispersion among the GCMs (Figures 7.b and 7.c), especially for the warm season. Nevertheless, the signs of the tendencies are equal for all CTs and for the two time horizons (2046-2065 and 2081-2099). It is important to point out that the future changes of the CT frequencies are smaller than the 20th century observed frequency dispersion (Figure 7). In this sense, it is difficult to quantify the uncertainty.
In the previous section, it was found that the $CT2_S$ and $CT4_S$ are significantly associated with heavy rainy days and dry days respectively (Table 4). In the projections of future CTs, the majority of the GCMs and the ensemble show an increment of $CT2_S$ frequencies and a decrease of the $CT4_S$ frequencies. This is interpreted as a trend of reduction of the dry day circulation patterns and in an increase of the frequencies of the patterns associated with rainfall of the region. These results coincides with other climate change results over the region that shows positive trends in the total seasonal precipitation and changes in the precipitation variability (Barros et al., 2005; Silvestri & Vera, 2008; Vera et al., 2006).
For winter, even if there is a slight dispersion in the frequencies of the future projections, this does not seem to indicate mayor changes in the frequencies of the patterns. The main changes are given by an increase of the frequencies of $CT2_W$ and $CT7_W$, and a decrease of $CT1_W$ and $CT3_W$.
These results would imply an increment in the variability of daily rainfall, especially during summer, and also a change in the distribution of rainfall over the Pampas region.

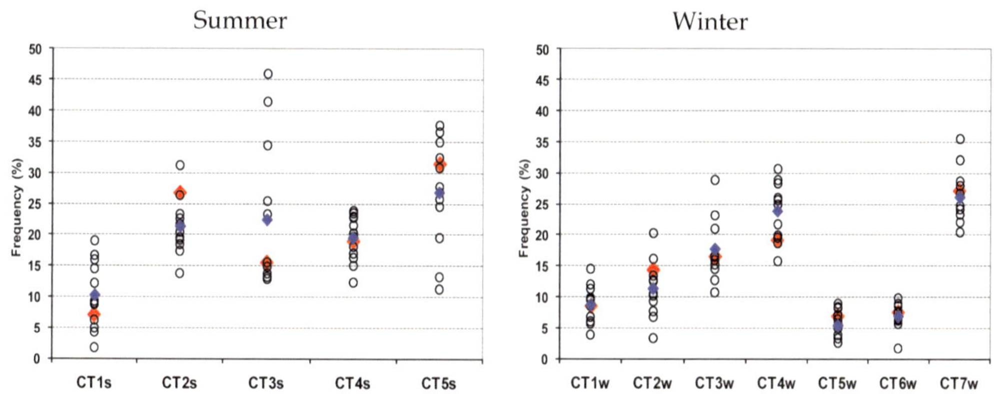

(a) 20th Century

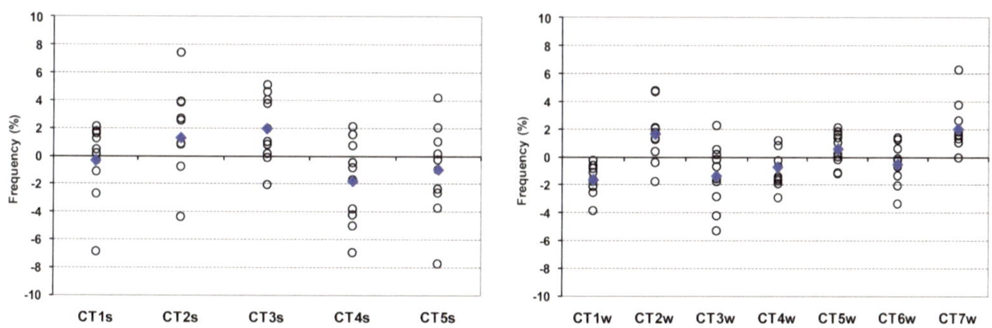

(b) (2046-2065) minus 20th Century

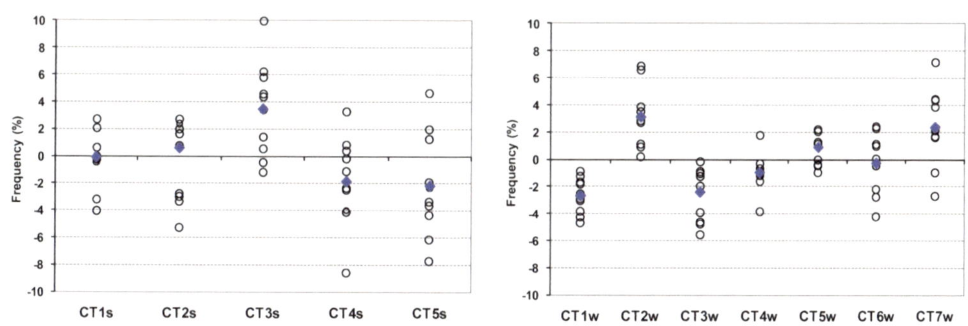

(c) (2081-2099) minus 20th Century

Fig. 7. Frequency (%) of circulation types for summer and winter for NCEP (red diamond), GCMs (circles) and ensemble of GCMs (blue diamond) for the 20th Century (a). Future changes of the frequencies of the CTs for the period 2046-2065 (b) and 2081-2099 (c).

4. Conclusion

This chapter analyzed the relation between the daily low level circulation over Southern South America and adjacent oceans and the precipitation over the region Pampas. Also, the capacity of 12 GCMs in representing the low level synoptic circulation for the period 1979-1999 was evaluated and their future projection of the XXI century was discussed.

The synoptic structures identified in this work can be associated with daily rainfall over the region of study. The classification scheme is effective not only in discriminating dry and rainy days, but also in differentiating between different thresholds of rainfall intensities. In this sense, the findings of this research help to improve our understanding of the relationship between rainfall variability and atmospheric circulation as defined by an objective classification of circulation types.

Summer dry days are related to the most persistent circulation type, corresponding to an intensification of the southern Atlantic anticyclone, which interrupts the passage of the eastern perturbations and diverts them to the south. Rainy days are significantly benefited by patterns that could be related to a post-frontal intense anticyclone that induces east-southeast anomalous flow and moisture advection over the region Heavy rainy days are significantly related with a cyclonic disturbance at the centre of the continent associated with a cold front passage.

Winter dry days are significantly favored by a high pressure system that extends from the Atlantic Ocean to the centre of the continent. Rainy days and heavy rainy days are significantly benefited by structures with a high pressure system at the south of the continent, enhancing an anomalous flow from the east-southeast to the central region of Argentina and a corresponding moisture advection at low levels.

The principal climatic characteristics of the atmospheric circulation of summer and winter are reasonably well captured by the GCMs, although the seasonality is exaggerated. For example, during winter, the Westerlies are displaced towards the equator, attenuating the contribution of the subtropical highs. During summer, both semi-permanent high systems more extended to the south than the observed fields. Most models tend to show a lower contribution of the mean pattern in comparison to NCEP and, therefore, a higher presence of perturbations.

With respect to the circulation types, the models are able to reproduce the full range of summer and winter circulation types found in the NCEP climatology. For present climate, an inter-model variability of the representation of the summer patterns is observed. The GCMs estimate reasonably well the frequency of atmospheric situations that favor dry days. For the two future time horizons analyzed, a trend of reduction of the circulation patterns associated with dry days and an increment of the frequencies of the patterns associated with rainfall of the region is observed. This is in agreement with other climate change studies over the region that shows positive trends in the total seasonal precipitation and changes in the precipitation variability.

Contrary to what is observed for summer, during winter the GCMs estimate reasonably well the frequency of the circulation types, especially those that favor heavy rain and dry conditions over the region. The majority of the GCMs indicate an increment of these patterns for future climate, principally during the second half of the 21th century.

The results above imply an increment in the variability of daily rainfall, especially during summer, and also a change in the distribution of rainfall over the Pampas region. From

these projections, favorable consequences for the trends of the agricultural production in a climate change context can be derived, especially for the agricultural border.
However, it should be stressed that although the GCMs are capable of reproducing the circulation types and their main characteristics, there is an important intermodel dispersion. This limits the reliability of the GCMs for the study of future circulation changes, and consequently of precipitation changes. As explained above the quantification of the uncertainty of the projections is a complex issue. With this in mind, the findings of this research provide insight into the possible future climate change context.
One of the important aspects related to climate change, including climate variability, is the comprehension of extreme events and the skill of the models in representing their occurrence. The impact of climate adversities in agricultural activities makes it necessary to determine in what measure the spatial and temporal variability of climate is responsible for the yields of the crops and to generate tools that permit the supervision, the estimation of impacts and the design of alert systems. The world is facing a water crisis, but improved water management in rain-fed agriculture can build resilience to cope with future water related risks and uncertainties.

5. Acknowledgment

This work was supported by the grant X170 of the Universidad de Buenos Aires; and CLARIS LPB Project (European Community's Seventh Framework Programme under Grant Agreement No. 212492). Authors thank Pablo Krieger for his assistance in calculations and graphs.

6. References

Alessandro, A.P. (2003). Blocking action situations in the south of South America during the 1990s. *Meteorologica*, Vol. 28, No.1-2, pp 23-38, ISSN 0325-187X

Barros, V.R.; Doyle, M.E. & Camilloni, I.A. (2008). Precipitation trends in southeastern South America: Relationship with ENSO phases and with low-level circulation. *Theoretical and Applied Climatology*, Vol. 93, No. 1-2, (June 2008), pp. 19-33, ISSN 0177-798X

Barros, V.; Menéndez, A. & Nagy, G. (2005). *El cambio climático en el Río de la Plata*. Buenos Aires: Consejo Nacional de Investigaciones Científicas y Técnicas, ISBN 950-692-0664-4, Buenos Aires

Berbery, E. H. & Mechoso C.R. (2001), Climatology and Hydrology of the La Plata basin, Document of the VAMOS scientific study group on La Plata basin, http://meto.umd.edu/~berbery/lpb/laplata.html, 2001

Bettolli, M.L.; Penalba, O.C. & Vargas, W.M. (2010). Synoptic Weather Types in the South of South America and their Relationship to Daily Rainfall in the Core Production Region of Crops in Argentina. *Australian Meteorological and Oceanographic Journal*, 60, No. 1, (January 2010), pp. 37-48, ISSN 1836-716X

Bidegain, M.; Skansi, M.; Penalba, O. C.; Quintana, J. & Aceituno, P. (2010). Southern South America in "State of the Climate in 2009", *Bulletin of the American Meteorological Society*, 91, S150-S152

Boulanger, J.P.; Leloup, J.; Penalba, O.C.; Rusticucci, M.; Lafon, F. & Vargas, W.M. (2005). Observed precipitation in the Paraná-Plata hydrological basin: Long-term trends, extreme conditions and ENSO teleconnections. *Climate Dynamics,* Vol. 24, No. 4, (March 2005), pp. 393-413, ISSN 0930-7575

Calinski, R.B. & Harabasz, J. (1974). A dendrite method for cluster analysis. *Communications in Statistics.*, Vol. 3, No. 1, (January 1974), pp. 1-27, ISSN: 0361-0926

Compagnucci, R.H. & Vargas, W.M. (1986). Patterns of surface pressure field during July 1972-1983 in Southern South America and the Antarctic peninsula. *Proceedings of Third International Conference on Statistical Climatology,* Vienna, Austria, 23-27 June 1986

Craddock, J.M. & Flood, C.R. (1969). Eigenvectors for representing the 500 mb geopotential surface over the Northern Hemisphere. *Quaterly Journal of the Royal Meteorological Society,* Vol.95, No. 405, (July 1969), pp. 576-593, ISSN: 1477-870X

Di Luca, A.; Camilloni, I. & Barros, V. (2006). Sea-level pressure patterns in South America and the adjacent oceans in the IPCC-AR4 models, *Proceedings of 8th International Conference on Southern Hemisphere Meteorology and Oceanography,* ISBN 85-17-00023-4, Foz de Iguazú, Brasil, 24-28 April 2006

Fernández, A.; Ciappesoni, H. & Nuñez, M. (2006). Observed trends in precipitation and temperature over Argentina. *Proceedings of 8th International Conference on Southern Hemisphere Meteorology and Oceanography,* ISBN 85-17-00023-4, Foz de Iguazú, Brasil, 24-28 April 2006

Gleckler, P.J.; Taylor, K. E. & Doutriaux, C. (2008). Performance metrics for climate models. *Journal of Geophysical Research,* 113, D06104, doi:10.1029/2007JD008972, 2008

Gong, X. & Richman, M.B. (1995). On the application of cluster analysis to growing season precipitation data in North America east of the Rockies. *Journal of Climate,* Vol. 8, No. 4, (April 1995), pp. 897-931, ISSN 0894-8755

Gulizia, C.; Camilloni, I. & Doyle, M. (2009). Evaluation of the moisture transport and convergente in southern South America in the WCRP-CMIP3 multimodel dataset. MOCA-09, Montreal, Canada, 19-29 July 2009

Huth, R. (2000). A circulation classification scheme applicable in GCM studies. *Theoretical Applied Climatology,* Vol. 67, No. 1-2, (October 2000), pp. 1-18, ISSN: 0177-798X

Infante Gil, S. & Zárate de Lara, G. (1984). *Métodos Estadísticos,* Trillas, ISBN: 9682414229, México

IPCC (2007). Solomon, S., D. Qin, M. Manning, Z. Chen, M. Marquis, K.B. Averyt, M. Tignor and H.L. Miller (Eds.). (2007). *Climate Change 2007: The Physical Science Basis. Contribution of Working Group I to the Fourth Assessment Report of the Intergovernmental Panel on Climate Change,* Cambridge University Press, ISBN: 978 0521 70596-7, Cambridge, United Kingdom and New York, NY, USA

Jolliffe, I.T. (2002). *Principal Component Analysis,* Springer-Verlag, ISBN 978-0-387-95442-4, New York, USA

Liebmann, B.; Vera, C.; Carvalho, L.; Camilloni, I.; Hoerling, M.; Allured, D.; Barros, V.; Báez, J. & Bidegain, M. (2004). An observed trend central S. American precipitation. *Journal of Climate,* Vol. 17, No. 22, (November 2004), pp 4357-4367, ISSN 1520-0442

Llano, M.P. & Penalba O.C. (2011). A climatic analysis of dry sequences in Argentina. *Internacional Journal of Climatology* , Vol. 31, No. 4, (March 2011), pp 504-513, ISSN 0899-8418

MacQueen, J. (1967). Some methods for classification and analysis of multivariate observations. *Proceedings of Fifth Berkeley Symposium on Mathematical Statistics and Probability*. Berkeley CA, University of California Press, USA

Marengo, J.; Rusticucci, M.; Penalba, O. C. & Renom, M. (2010). An intercomparison of model-simulated in extreme rainfall and temperature events during the last half of the XX century. Part 2: Historical trends. *Climatic Change*. Vol. 98, No 3-4, (February 2010), pp. 509-529, ISSN 0165-0009

Pascale, J.Y. & Damario, E.A. (2004). *Bioclimatología agrícola y agroclimatología*, Universidad de Buenos Aires, ISBN 9789502908229, Buenos Aires, Argentina

Penalba, O.C.; Rivera, J. A. & Bettolli, M.L. (2010). Trends and periodicities in the annual amount of dry days over Argentina. Looking towards the climatic change. *Second International Conference on Drought Management: Economics of Drought and Drought Preparedness in CIHEAM Journal Options Méditerranéennes*, 4 - 6 March 2010, Istanbul, Turkey

Penalba, O. C. & Robledo, F. (2010). Spatial and temporal variability of the frequency of extreme daily rainfall regime in the La Plata Basin during the 20th century. *Climatic Change*. Vol. 98, No.3-4, (February 2010), pp 531-550, ISSN 0165-0009

Penalba, O.C. & Vargas, W. M. (2004). Interdecadal and Interannual variations of annual and extreme precipitation over central-northeastern Argentina. Changes in the extreme precipitation seasonal cycle. *International Journal of Climatology*, Vol. 24, No 12, (October 2004), pp. 1565-1580, ISSN 0899-8418

Penalba, O.C. & Vargas, W.M. (2008). Variability of low monthly rainfall in La Plata Basin. *Meteorological Applications*, Vol. 15, No. 3, (September 2008), pp. 313 - 323, ISSN 1469-8080

Riani, J. (2009). Importantes pérdidas por la sequía en Entre Ríos, La Nación, 22 May 2009.

Richman, M.B. (1986). Rotation of Principal Components. *International Journal of Climatology*, 6, pp. 293-335, ISSN 0899-8418

Robledo, F. & Penalba O.C. (2008). Análisis estacional de la frecuencia diaria y la intensidad de los eventos extremos de precipitación sobre el sudeste de Sudamérica, *Meteorologica*, Vol 32 (33), No 1-2, pp. 31-49, ISSN 0325-187X

Rodrigues Chaves, R. & Cavalcanti, I. (2001). Atmospheric Circulation Features Associated with Rainfall Variability over Southern Northeast Brazil. *Monthly Weather Review*, 129, No 10, (October 2001), pp. 2614-2626, ISSN 1520-0493

Romero, R.; Sumner, G.; Ramis, C. & Genovés, A. (1999a). A classification of the atmospheric circulation patterns producing significant daily rainfall in the Spanish Mediterranean area. *International Journal of Climatology*, Vol. 19, No. 7, (June 1999), pp. 765–785, ISSN 0899-8418

Romero, R.; Ramis, C.; Guijarro, J. & Sumner, G. (1999b). Daily rainfall affinity aeras in mediterranean Spain. *International Journal of Climatology*, Vol. 19, No. 5, (April 1999), pp. 557-578 ISSN 0899-8418

Rusticucci, M.; Margeno, J.; Penalba, O.C. & Renom, M. (2010). An intercomparison of model-simulated in extreme rainfall and temperature events during the last half of the XX century. Part 1: Mean values and variability. *Climatic Change*, Vol. 98, No. 3-4, (February 2010), pp. 493-508,. ISSN 0165-0009

Silvestri, G. & Vera, C. (2008). Evaluation of the WCRP-CMIP3 model simulations in the La Plata Basin. *Meteorological Application,* Vol. 15, No. 4, (December 2008), pp. 497–502, ISSN 1469-8080

Solman, S. & Le Treut, H. (2006). Climate change in terms of modes of atmospheric variability and circulation regimes over southern South America. *Climate. Dynamics*, Vol. 26, No. 7-8, (June 2006), pp. 835–854, ISSN 1432-0894

Solman, S. & Pessacg, N. (2006). Analysis of low frequency variability patterns and circulation regimes over southern South America and their response to global warming as depicted by IPCC AR4 AOGCMs. *Proceedings of 8th International Conference on Southern Hemisphere Meteorology and Oceanography,* Foz de Iguazú, Brasil, 24-28 April 2006

Vargas, W.M. (1979). *Atlas. Excesos y Déficits de Humedad en la Región Húmeda y Semiárida Argentina. Tomo 1.* Instituto Nacional de Ciencia y Técnica Hídricas, Buenos Aires, Centro de Investigaciones Biometeorológicas

Vera, C.; Silvestri, G.; Liebmann, B. & Gonzalez, P. (2006). Climate change scenarios for seasonal precipitation in South America from IPCC-AR4 models. *Geophysical Ressearch Letters*, 33, L13707 doi:10.1029/2006GL025759

8

Holocene Vegetation Responses to East Asian Monsoonal Changes in South Korea

Sangheon Yi
Korea Institute of Geoscience and Mineral Resources
Korea

1. Introduction

The Korean Peninsula, surrounded by the sea on three sides (east, west, and south), is located on the eastern end of the Asian continent adjacent to the West Pacific and belongs to the temperate zone with four distinct seasons, which are largely controlled by the East Asian monsoon. During the winter, from December to February, it is cold and dry due to the establishment of the strong Siberian anticyclone on the Tibetan Plateau. Meanwhile, the summer, from June to August, is hot and wet, with frequent heavy rains (An, 2000; Nakagawa et al., 2006; Yancheva et al., 2007). The modern climate of Korea is characterized by a mean annual temperature of 12.2℃, ranging from 5.1℃ to 13.6℃, with a monthly mean daily maximum temperature of up to 19.4℃ and a monthly mean daily minimum temperature of 0℃ over the past 30 years (1971-2000). Precipitation is relatively high (mean, 1299 mm), and about 70% of the annual precipitation falls in summer, especially from June to August (Korean Meteorological Administration, http://kma.go.kr).

Pollen studies are well suited to examining the impact of rapid climate change on terrestrial ecosystems because the reponse of vegetation to climate fluctuation is pronounced and can occur on decadal time scales (Tinner & Lotter, 2001). Pollen analysis provides information that is unavailable from other sources and offers a unique and invaluable perspective on natural, climate-induced vegetation changes and environmental reconstruction (Birks & Birks, 1980; Davis, 1994) despite its limitations compared with marcropaleontology. Therefore, among the various terrestrial paleoclimate proxies, pollen has proven to be a most useful tool.

Pollen studies were carried out in South Korea with a focus mainly on reconstructing vegetation and climate from the Quaternary sediments of wetlands (e.g., Chang et al., 1987; Choi et al., 2005; Jang et al., 2006; Jun et al., 2010; Park & Yi, 2008; Yi et al., 2004, 2005, 2008a, 2008b; Yoon, 1997), lakes (e.g., Chang & Kim, 1982; Fujiki & Yasuda, 2004; Yasuda et al., 1980), and archaeological sites (e.g., Chung et al., 2006, 2010; Yi, 2011; Yi & Kang, 2009; Yi & Kim, 2009; Yi et al., 2011; Yoon et al., 2005). Early, non-dated pollen studies were conducted to interpret local vegetation history. Recently, pollen investigations have reconstructed vegetation and climate changes with geologic ages using radiocarbon dates.

The age-controlled pollen data are used herein. The response of vegetation in South Korea to East Asian monsoon climate change is discussed based on the pollen datasets.

2. East Asian monsoon

The East Asian monsoon is an integral part of the global climatic system. Monsoon climates, especially monsoon-derived precipitation, are important to the maintenance of the environments of East Asia. This monsoon regime is a sub-system of the Asian monsoon circulation (An, 2000). The East Asian monsoon is formed by thermal differences between the Asian landmass and the Pacific Ocean; the area to the northwest of the front is under the strong influence of the continental Siberian air mass, which is dry and has large seasonal temperature variations, whereas the area to the southeast of the front is controlled by the oceanic Pacific air mass, which is high in moisture and has small seasonal temperature variations (Yancheva et al., 2007). The coastal East Asian regions are characterized by prominent seasonality due to the seasonal migration of the monsoon front across these regions (Nakagawa et al., 2006). As the seasonal temperature variability on the continent is greater than that in the ocean, the temperature gradient between the two air masses creates a surface-air pressure gradient that seasonally changes its direction, forcing NW and SE surface winds in winter and summer, respectively (Fig. 1).

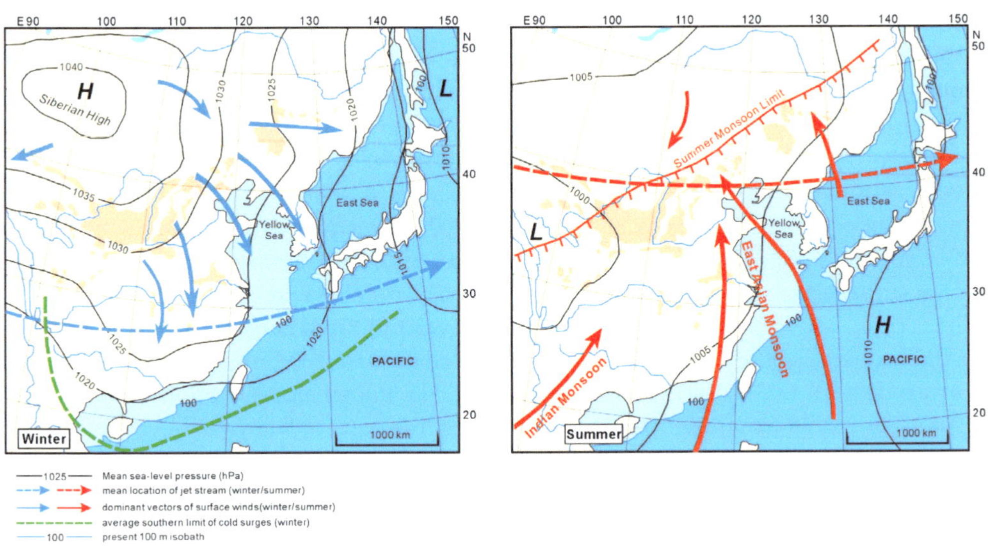

Fig. 1. Seasonal changes in the wind system of the East Asian monsoon area (modified from Yancheva et al., 2007).

2.1 Evolution of the East Asian monsoon

The evolution of the East Asian monsoon is a principal factor controlling paleoenvironmental changes in the East Asian region (An, 2000). Cenozoic uplift of the Himalayan-Tibetan Plateau is thought to contribute to the Asian Monsoons, including the East Asian and Indian monsoons, and the Northern Hemispheric ice ages (Qiang et al., 2001). Qiang et al.'s (2001) study of the magnetostratigraphy of the Jiaxian red clay section of the Chinese Loess Plateau documented that the onset of the East Asian monsoon occurred in the Late Miocene (8.35 Ma). Moreover they recognized long-term changes in the East Asian monsoon since the Late Miocene based on accumulation rate and grain size analysis of the eolian dust deposited in the central Chinese Loess Plateau. For example, the strengthening of the East Asian winter monsoon occurred between 3.5 Ma and 3.1 Ma and intensified further after 2.6 Ma. However, Sun and Wang (2005), based on compilation of paleobotanical and lithological data from China, suggested that an initiation of monsoon climate system in East Asia was further back to the latest Oligocene. Subsequently, paleomonsoonal studies have used multi-proxies of loess-paleosols (e.g., An, 2000; An et al., 1990), caves (e.g., Dykoski et al., 2005; Wang et al., 2001), lake sediments (e.g., Nakagawa et al., 2006; Wu et al., 2006; Wűnnemann et al., 2006; Yamada, 2004), ice cores (e.g., Yao et al., 2001), and pollen (e.g., Feng et al., 2006; Makohonienko et al., 2008; Yi et al., 2003a, 2003b).

2.2 Modern climate of Korea

South Korea is in the temperate zone controlled by the East Asian monsoon, which is characterized by distinct seasonal changes, with a warm, wet summer and a cool, dry winter. However, the present climate conditions are trending toward increases in extreme precipitation and drought.

Baek and Kwon (2005) showed a decreasing trend in April precipitation and an increasing trend in August precipitation for 1954–2002. Chang and Kown (2007) investigated the spatial patterns of trends in summer precipitation for 1973–2005 and pointed out a significant increase in June precipitation for the northern and central-western part of Korea. Bae et al. (2008) reported that the long-term trends in annual precipitation and runoff were increasing in the northern part of Korea and decreasing in the southwestern part of Korea. Furthermore, a recent study on climate trends in Korea reported annual precipitation increases and increases in the number of severe precipitation events (Jun et al., 2010). It showed that the increase in annual precipitation is mainly associated with increases in the frequency and intensity of heavy precipitation during the summer season (June–September), whereas precipitation during the spring and winter seasons showed a decreasing trend. This variation in precipitation is likely to increase flood and drought risk.

3. Vegetation of Korea

The first studies of Korean plants were made by Japanese researchers (e.g., Nakai, 1952; Uyeki, 1911, 1933), followed by Korean botanists (e.g., Chung & Lee, 1965; Kong, 2007; Lee, 1985; Lee & Yim, 2002; Yim, 1977; Yim & Kira, 1975). Yim and Kira (1975) first established a vegetation map of the Korean Peninsula, which consists of conifer forest (subalpine zone), deciduous broadleaved forest (temperate zone), and evergreen forest (subtropical zone). The deciduous broadleaved forest is further divided into three zones at different latitudes: the northwest temperate zone, the central temperate zone, and the southern temperate zone. Moreover, the vertical vegetation zone is divided on the basis of the elevation of mountain ranges. Local vegetation is primarily controlled by climate, soil, geomorphology, and artificial factors. In all, the distribution of Korean forests is band-shaped and changes with variations in temperature depending on latitude and elevation (Fig. 2).

Temperature is an important factor in the growth and distribution of plants. The mean annual temperature in Korea is 2.5–10.0°C in the northern region (39°N–43°N), 10.0–12.5°C in

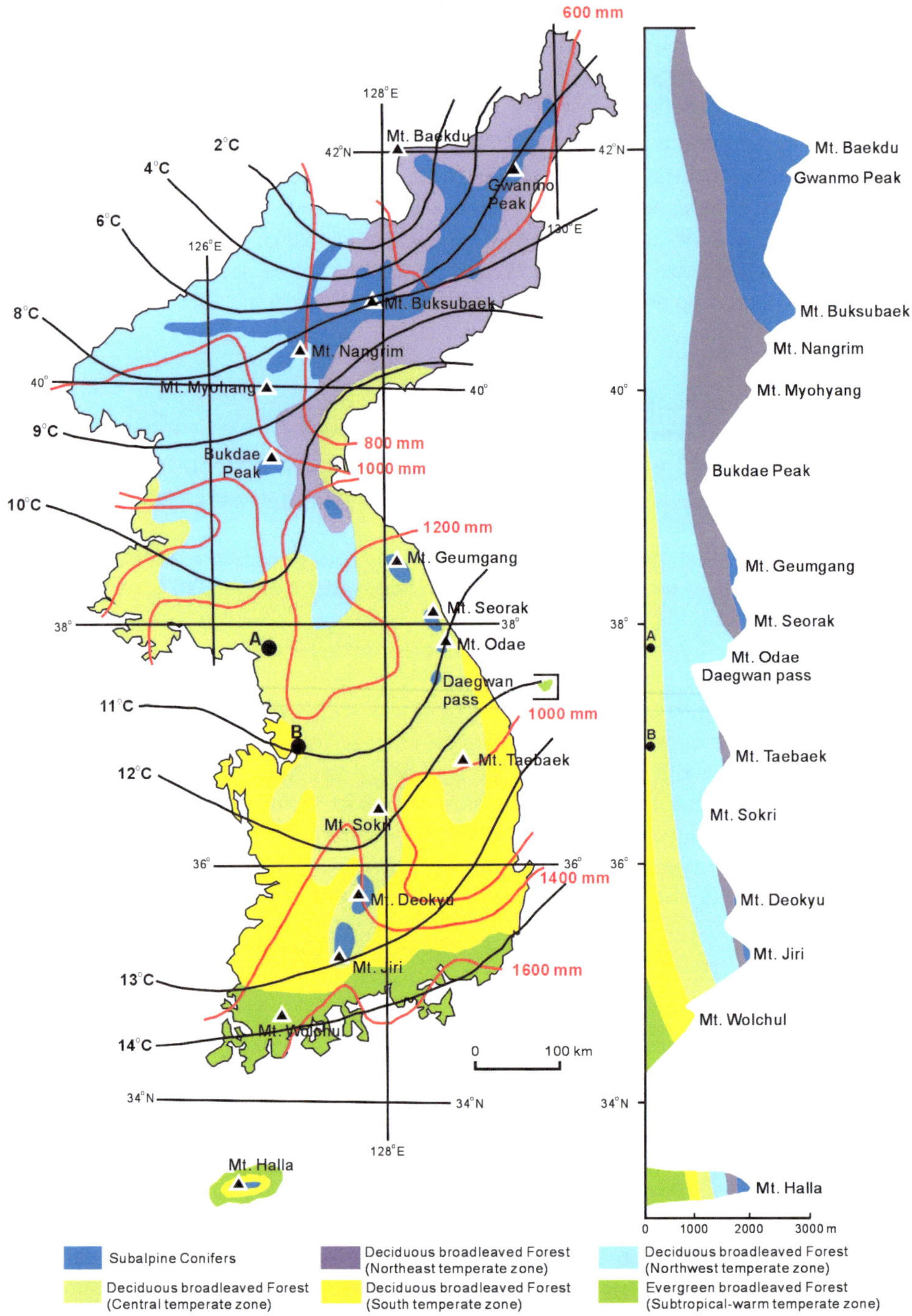

Fig. 2. Vertical and latitudinal modern vegetation map with an isothermal and an isobaric line (modified from Yim & Kira, 1975). N–S cross-section showing the forests distributed across the peninsula with elevation. Pollen records discussed in the text are from the Paju-Unjeong site (A) and Pyeongtaek site (B).

the central region (37°N–39°N), and 12.5–15.0°C in the southern region (33°N–37°N). The mean monthly daily maximum temperature is 32.5–35°C in the northern region and 37.5–40.0°C in the southern region; the mean daily minimum temperature ranges from -5°C to -15°C in the subtropical region, -15°C to -20°C in the warm-temperate south central region, -20°C to -25℃ in the warm-temperate central region, -25℃ to -30℃ in the warm-temperate north central region, and -35℃ to -45℃ in the subalpine northern region (Kong, 2007). Subalpine conifer forest is mainly distributed in North Korea and consists of evergreen conifers, such as a fir (*Abies holophylla, A. koreana, A. nephrolepsis*), spruce (*Picea jezoenisis*), pine (*Pinus koraiensis, P. pumila*), and yew (*Taxus cuspidate*), and deciduous broadleaved trees, such as birch (*Betula costata,* and *B. platyphylla* var. *japonica*). These hardwood trees grow under subalpine climate conditions, with a mean annual temperature of 5°C (Fig. 3a, 3b).

Mixed conifer and deciduous broadleaved forests are dominated by pines (*Pinus desiflora*) and oaks (*Quercus mongolica*), with other hardwood trees such as elm (*Ulmus parvifolia, U. davidiana* var. *japonica*), *Carpinus laxiflora*, hazel (*Corylus teterophylla* var. *thunbergii*), lime (*Tilia amurensis*), and maple (*Acer palmatum*) (Fig. 3c).

Deciduous broadleaved forest (DBF) is distributed between 35°N and 43°N, except in the subalpine area. The main trees are maples (*Acer palmatum*), oaks (*Quercus dentate, Q. aliena* and *Q. serrata*), birches (*Betula platyphylla* var. *japonica*), *Zelkova serrate, Styrax japonica, Carpinus tschnonoskii, Lindera erythrocarpa, Lindera obtusiloba,* and *Acer mono* (Fig. 3d). This forest is further divided into three zones, the north temperate, central temperate, and south temperate, based on floral components. The north temperate zone of the DBF is distributed from the north central region to the borders of China and Russia. This zone is composed of *Tilia amurnesis* var. *glabrata, Acer tegmentosum, Acer okamotoanum, Betula schmidtii, Quercus mongolica, Corylus sieboldiana, Abies holophylla,* and *Pinus koraiensis*. The central temperate zone of the DBF is distributed south to 40°N on the east coast, to 39°N on the west coast, and to 38°N in the central area and consists of *Zelkova serrata, Strax japonica, Quercus mongolica, Q. serrata, Q. aliena, Lindera obtusiloba, Juniperus chinensis, Abies holophylla,* and *Pinus densiflora*. The south temperate zone of the DBF develops between 35°N and 36°N and ranges to 38°N on the east coast and to 37°30′N on the west coast. This zone is characterized by the predominance of deciduous broadleaved trees of *Carpinus tshonoskii, Meliosma oldhamii, Pourthiaea villosa, Acer palmatum, Pterocarya strobilacea,* and *Celtis sinensis*, evergreen broadleaved trees of *Euonymus japonica, E. fortune* var. *radicans,* and *Daphniphyllum macropodum*, and conifers of *Pinus densiflora, P. thunbergii,* and *Cephalotaxus koreana*.

Subtropical evergreen forest is located along the south coast and is limited to 35°N in inland areas and 35°30′N in coastal areas, including several islands. The main components are trees such as *Quercus acuta, Q. glauca, Q. myrsinaefolia, Castanopsis cuspidate* var. *sieboldii, Cinnamomum camphora, Machilus thunbergii,* and *Euonymus japonica* and shrubs including *Camellia japonica, Ilex integra, Aucuba japonica,* and *Eurya japonica* (Fig. 3e).

Additionally, coastal conifers such as *Pinus thunbergii* grow along the west, south, and east coasts (Fig. 3f). Salt marshes composed of *Sueda japonica, S. glauca, Salicornia europaea, Salosa komarovii,* and *Phragmites communis* are found in patches along the west and south coasts (Fig. 3g). Some recreational forests, run by the national and local governments, were designed by the afforestation of conifers (Fig. 3h).

(a) Conifer forest (*Pinus densiflora*)

(b) Subalpine conifer (*Abies koreana*)

(c) Mixed coniferous & deciduous broadleaved forest (*Pinus densiflora* & *Quercus mongolica*)

(d) Deciduous broadleaved forest (*Quercus aliena* & *Ulmus davidiana*)

(e) Evergreen deciduous broadleaved forest (*Machilus thunbergii*)

(f) Coastal conifer forest (*Pinus thunbegii*)

(g) Salt marsh (*Phragmites communis* & *Suaeda japonica*)

(f) Afforestation (*Chamaecyparis obtusa*)

Fig. 3. Modern forest types of South Korea.

4. Palynological studies in South Korea

Quaternary palynological studies in Korea have mainly focused on the reconstruction of vegetation and climate change since the last glacial maximum (LGM). In the middle to late 20th century, Korean pollen investigations were carried out in peat or organic-rich soil from coastal wetlands or lagoons (e.g., ., Jo, 1979; Oh, 1971; Park, 1990; Tsukada, 1977; Yi et al., 1996; Yoon, 1997). Subsequently, a number of palynological studies were performed due to increased peat layer recovery from excavations of inland wetland and archaeological sites (e.g., Chung & Lee, 2006; Kim et al., 2001; Seo & Yi, 2001; Yi et al., 2006, 2008a). Attempts were made to reconstruct the natural vegetation history in response to climate change from the peat and wetland samples and to intrepret human-induced changes based on age-controlled pollen profiles, (agriculture and land-use in forests). The palynological studies are shown in Table 1.

Two Holocene pollen analyses are introduced herein: the Paju-Unjeong site is located in the west central area, and the Pyeongtaek site is located in the western coastal area on the Korean Peninsula (Fig. 2).

4.1 Pollen assemblages of the Paju-Unjeong area

Zone UJ10-I (elevation 17.954–18.164 m a.s.l., ca. 8425–4700 cal yr BP) was quantitatively dominated by *Quercus* and *Alnus* together with a few broadleaved deciduous trees of *Fraxinus, Ulmus/Zelkova, Magnolia,* and *Castanea* and the conifer *Pinus*. The dominant herbs were members of the Cyperaceae, with Gramineae and *Artemisia*. Additionally, accessory trees and shrubs and herbaceous taxa were present throughout this zone. Subzone UJ10-Ia (elevation 17.954–17.984 m a.s.l., ca. 8425–7520 cal yr BP) was characterized by the predominance of hardwood trees of *Quercus* and *Alnus* and herbs of Cyperaceae. The common pollen grains were from trees of *Fraxinus, Ulmus/Zelkova, Magnolia, Salix,* and *Castanea* and herbs of Gramineae and *Ambrosia*. Subzone UJ10-Ib (elevation 17.984–18.164 m a.s.l., ca. 7520–4700 cal yr BP) was dominated by *Quercus, Alnus,* and Cyperaceae, which gradually declined, in combination with *Pinus, Magnolia, Ulmus/Zelkova,* Gramineae, and *Artemisia*. The occurrence of *Pinus* somewhat increased upward within this subzone. Other accessory taxa occurred sporadically.

Zone UJ10-II (elevation 18.164 –18.354 m a.s.l., ca. 4700–2170 cal yr BP) was characterized by the dominance of *Pinus, Quercus,* Gramineae (≥35 μm), Gramineae, and Cyperaceae. Common elements included hardwood trees of *Magnolia* and *Buxus* and herbs of *Artemisia* associated with a few *Myrica, Carpinus, Juglans, Ulmus/Zelkova, Fraxinus, Castanea,* and *Ambrosia*. This zone was defined by a distinct decrease in *Quercus* and *Alnus* deciduous broadleaved trees and an apparent increase in *Pinus* conifers and Gramineae (≥35 μm) and Gramineae herbs toward the top of the zone. *Tilia* was found only within this zone.

Zone UJ10-III (elevation 18.354–18.764 m a.s.l., ca. 2170 cal yr BP–Modern) was conspicuously marked by the growing dominance of *Pinus*, Gramineae (≥35 μm), Gramineae, and Cyperaceae, replacing *Quercus* and *Alnus*, which decreased dramatically. Common taxa were the pollen grains derived from *Magnolia* and *Artemisia*, together with a few *Ulmus/Zelkova, Myrica, Carpinus, Juglans, Salix, Fraxinus,* and *Castanea*. Pollen grains of *Fagopyrum* and Ericaceae appeared only within this zone. The boundary of the subzone was designated by a sudden increase in *Pinus, Fagopyrum,* and *Artemisia* and the first appearance of Ericaceae. Subzone UJ10-IIIa (elevation 18.354–18.534 m a.s.l., 2170 to ca. 440 cal yr BP) featured the predominance of herbs, Gramineae (≥35 μm), and Cyperaceae associated with

Archive	Site	Location	Reference	Age
Lake	Yongrangho	Socho-shi, Gangwon Province	Tsukada (1977), Chang & Kim (1982)	Holocene
	Hyangho	Gangneung-shi, Gangwon Province	Fujiki & Yasuda (2004)	Holocene
	Bangeojin	Ulsan-shi, Gyeongsang Province	Jo (1979), Chang & Kim (1982)	Holocene
	Wolhamji	Buyeo-gun, Chungcheong Province	Chang & Kim (1982)	Holocene
Moor, Bog	Mt. Daeam	Inje-gun, Gangwon Province	Chang et al. (1987), Choi & Koh (1989)	Holocene
	Moojaechi	Ulsan-shi, Gyeongsang Province	Park & Chang (1998), Choi (2001)	Holocene
Wetland	Youngyang Basin	Yongyang-gun, Gyeongsang Province	Yoon & Jo (1996)	Late Pleistocene-Holocene
	Hanam	Hanam-shi, Gyeonggi Province	Yi et al. (2008b)	Late Pleistocene-Holocene
	Paju-Unjeong	Paju-shi, Gyeonggi Province	Yi et al. (2011)	Holocene
Coastal wetland	Imja-do	Shinan-gun, Jeolla Province	Yi et al. (2004)	Holocene
	Cheollipo	Taean-gun, Chungcheong Province	Park (1990), Jang et al. (2006)	Holocene
	Ilsan	Goyang-shi, Gyeonggi Province	Yoon (1997), Yi et al. (2005)	Holocene
	Pyeongtaek	Pyeongtaek-shi, Gyeonggi Province	Oh (1971), Jun et al. (2010)	Holocene
Archaeo-logical site	Sorori	Cheongwon-gun, Chungcheong Province	Kim (2001)	Late Pleistocene
	Unjeonri	Cheonan-shi, Chungcheong Province	Park (2004)	Holocene
	Anyoungri	Tacheon-myeon, Chungcheong Province	Seo & Yi (2003)	Late Pleistocene
	Poonggi	Asan-shi, Chungcheong Province	Yi et al. (2006)	Late Pleistocene
	Yongdong	Naju-shi, Jeolla Province	Chung & Lee (2006)	Late Pleistocene
	Yeanri	Gimhae-shi, Gyeongsang Province	Yi & Saito (2003c)	Holocene
	Cheonggye-cheon	Seoul	Yi et al. (2008a)	Holocene
	Jinju	Jinju-shi, Gyeongsang Province	Chung et al. (2006)	Late Pleistocene
	Piseori	Muan-gun, Jeolla Province	Chung et al. (2005)	Late Pleistocene
Island	Hanon	Seogyuipo-shi, Jeju Province	Chung (2007)	Late Pleistocene

Table 1. List of palynological studies in South Korea.

Pinus, *Quercus*, and Gramineae. *Quercus* pollen gradually decreased, whereas *Pinus*, Gramineae (≥35 μm), and *Artemisia* pollen grains gradually increased toward the upper boundary. Moreover, the pollen grains of *Quercus* and *Alnus* decreased, whereas Gramineae (≥35 μm) and Gramineae increased compared with frequencies in the preceding zone. Subzone UJ10-IIIb (elevation 18.534–18.764 m a.s.l., ca. 440 cal yr BP to Modern) was characterized by *Pinus* and Gramineae (≥35 μm), which were the prominent contributors and occurred in very abundant amounts. This zone was marked by an increase in *Pinus* and an apparent decrease in Gramineae, Cyperaceae, and *Artemisia*. *Quercus* and Gramineae (≥35 μm) were continuously present (Yi, 2011) (Fig. 4).

4.2 Pollen assemblages of the Pyeongtaek area

Zone HS-I (depth 192–187 cm, ca. 10 600 to ca. 10 400 cal yr BP) was dominated by *Quercus* (up to 22%) together with a few broadleaved deciduous trees of *Alnus* and *Betula* and the conifer, *Pinus*. The dominant herbs were members of the Cyperaceae, with a few Gramineae and *Typha*. Additionally, accessory trees and shrubs and herbaceous taxa were present throughout this zone.

Zone HS-II (depth 187–122 cm, ca. 10 400 to ca. 8000 cal yr BP) was defined by a distinct increase in *Quercus* deciduous broadleaved trees and an apparent decrease in Cyperaceae herbs. Arboreal pollen indicated that conifers of *Pinus* and *Larix* and deciduous broadleaved trees such as *Carpinus*, *Juglans*, *Ulmus/Zelkova*, and *Alnus* increased slightly from the preceding zone. Chenopodiaceae showed an increasing trend, whereas Cyperaceae decreased throughout the zone. Gramineae was still common in this zone. The pollen concentration of all taxa decreased compared with that in the preceding zone.

Zone HS-III (depth 122–77 cm, ca. 8000 to ca. 6000 cal yr BP) was characterized by the sudden expansion of T-C-C (Taxaceae-Cephalotaxaceae-Cupressaceae) and Cyperaceae compared with zone HS-II. *Quercus* declined in frequency and gradually decreased toward the top of the zone. Chenopodiaceae decreased to the point of being rare in this zone. Gramineae slightly increased. The pollen concentration varied among samples (average 380 000 grains/g) but was relatively high, reaching 1 000 000 grains/g.

Zone HS-IV (depth 77–44 cm, ca. 6000 to ca. 4500 cal yr BP) was marked by the predominance of *Alnus* and T-C-C associated with *Quercus*, Gramineae, and *Typha*. Fern spores increased in frequency, with greater numbers toward the top of the zone. Freshwater algae and aquatic pollen were common. This zone showed the highest pollen concentration (average 500 000 grains/g) among the pollen zones (Jun et al., 2010) (Fig. 5).

5. Vegetation change and the East Asian monsoonal fluctuation in South Korea during the Holocene Period

5.1 Vegetation changes

The available age-controlled pollen datasets allow us to infer the vegetational history of South Korea. The vegetation changes in the eastern and western parts of South Korea are discussed. During the early Holocene (10 000–7000 cal yr BP), subalpine conifer forest was replaced by broadleaved deciduous forest dominated by hardwood oak trees due to climatic amelioration. Moreover, the forest components evidenced by the pollen records are greater than those of the preceding period. During the mid-Holocene optimum, the former forest was replaced by a mixed subtropical and warm-temperate broadleaved forest, which was characterized by evergreen oak and thermophilous hardwood trees. These trees were composed mainly of oak

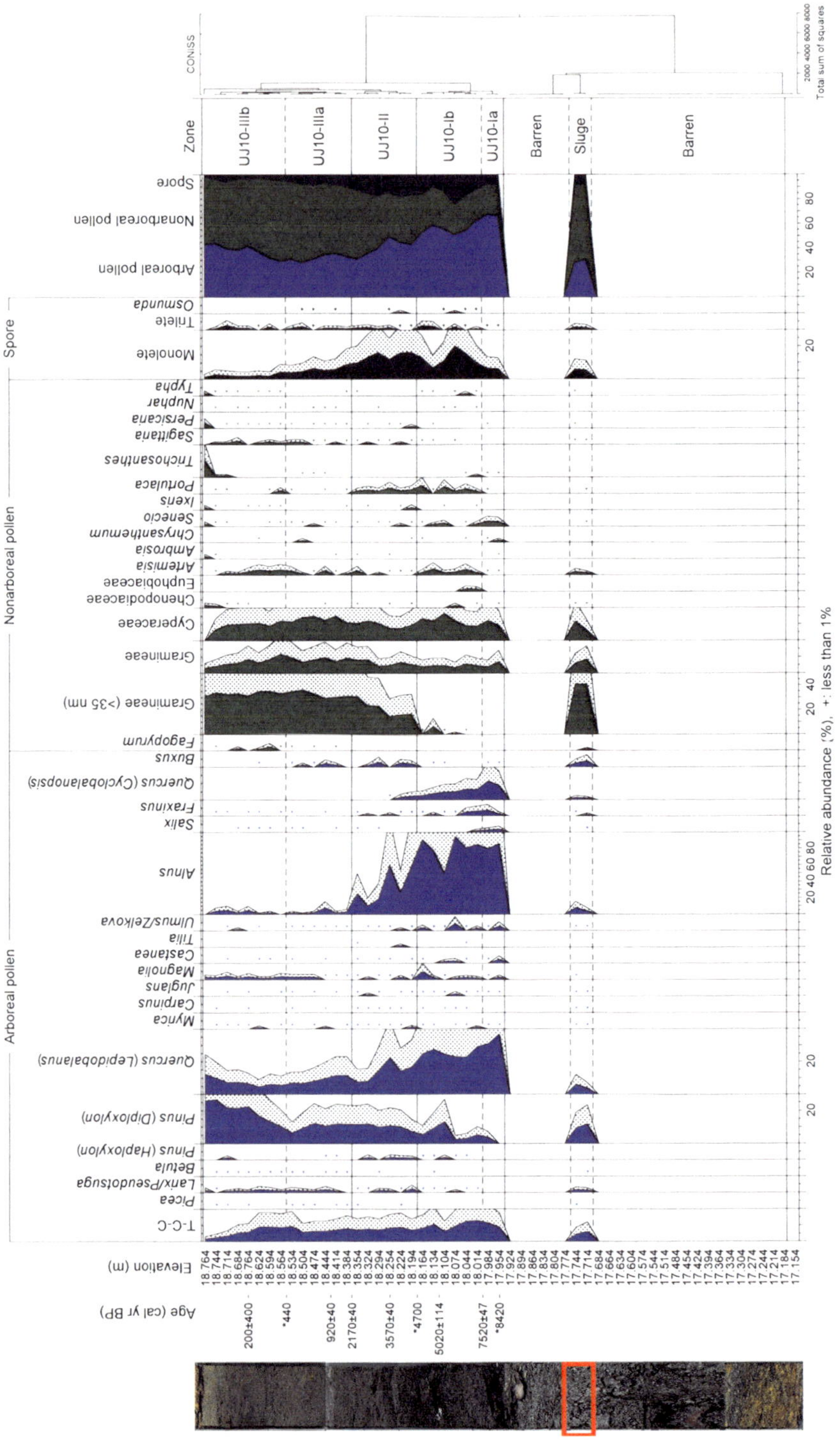

Fig. 4. Pollen diagram of selected taxa with the pollen zones of core UJ10 of the Paju-Unjeong area (modified from Yi, 2011).

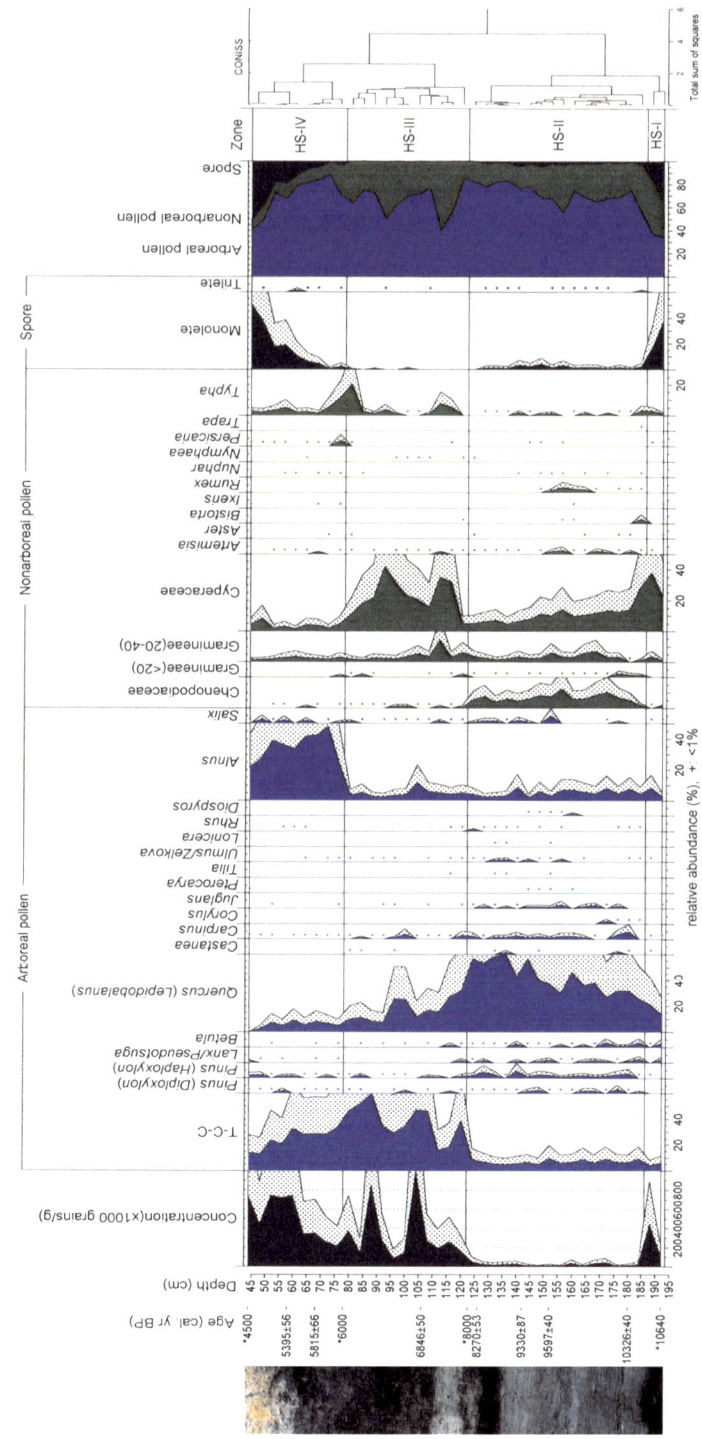

Fig. 5. Pollen diagram of selected pollen taxa with pollen zones of Hwangsan Trench in the Pyeongtaek wetlands (modified from Jun et al., 2010).

[*Quercus* (*Cyclobalanopsis*)] and *Q.* (*Lepidobalanus*)], hornbeam (*Carpinus*), hazel (*Corylus*), alder (*Alnus*), zelkova (*Zelkos*), and elm (*Ulmus*). The mixed subtropical and warm-temperate broadleaved forest flourished under the favorable warm and wet climatic conditions. After the mid-Holocene optimum, the subtropical and warm-temperate forest shrank and was replaced by a temperate forest due to climatic deterioration. Beginning about 2000 cal yr BP, the forest was affected by human impacts, such as cultivation, slash-and-burn agriculture, and deforestation, recorded in the pollen by the first appearance of the agricultural indicator buckwheat (*Fagopyrum*) in association with sudden increases in synanthropogenic indicators [*Ambrosia, Plantago, Artemisia,* and Gramineae (≥35 μm)] and secondary pine trees (Fujiki & Yasuda, 2004; Jo, 1979; Tsukada, 1977).

Compared with the east coast, pollen studies from the west coast are more numerous (e.g., Jang et al., 2006; Jun et al., 2010; Oh, 1971; Park, 1990, Yi et al., 2005; Yi et al., 2010; Yoon, 1997) because there are a plenty of wetlands along this coast. From 8000 to 5000 cal yr BP, subtropical evergreen and warm-temperate forest occupied this area, especially the hills and low mountainous areas, resulting in a high proportion of pollen from trees and shrubs with a smaller proportion from herbs. The evergreen and warm-temperate forest consisted mainly of oak [*Quercus* (*Cyclobalanopsis*)] and *Q.* (*Lepidobalanus*)], alder (*Alnus*), willow (*Salix*), hornbeam (*Carpinus*), hazel (*Corylus*), zelkova (*Zelkos*), and elm (*Ulmus*). However, Jun et al. (2010) pointed out that salt marsh (*Suaeda*) pollen appeared only at certain periods that were comparable to transgression periods of the Yellow Sea (Chough et al., 2004; Park, 1992). The favorable conditions characterized by high moisture and warmer temperatures during the mid-Holocene optimum and the transgression of the Yellow Sea accelerated the flourishing of forests along the western region. Later, the forest was replaced by conifer-dominated forest with an herb-dominated understory until about 2000 cal yr BP. Additionally, the components of the hardwood forests showed a sudden decline in alder (*Alnus*) and an increase in birch (*Betula*) and hazel (*Corylus*) owing to climatic deterioration. Beginning about 2000 cal yr BP, anthropogenic indicators, including *Fagopyrum, Ambrosia, Plantago, Artemisia,* and Gramineae (≥35 μm), and pine trees indicate that human activity played an important role in disturbing the forest and in secondary forestation.

Korea Meteorological Administration (KMA) reports that there was a mean annual temperature of 12.2°C and 1255 mm in a mean annual precipitation during the 1973–1980, but the temperature and precipitation increased up to 12.9°C and 1469 mm, respectively, between 2001 and 2007. This meteorological phenomena show the climate conditions of Korea are changing to be subtropical zones caused by global warming. With such climate conditions, types and communities of Korean forest can be expected to change. For example, a Korean fir (*Abies koreana*), which is a sensitive to temperature, cannot be survived any longer in high mountains, such as Mt. Halla in Jeju Island, Mt. Duckyoo in central region and Mt. Seolak in eastern region. Moreover, north limit of vegetation distribution will migrate northward. Evergreen broadleaved forests may widely distribute and expand north to about 37°N as they did during the mid-Holocene pollen record (e.g., Jang et al., 2006; Yi et al., 2008). The distribution of both a coniferous forest and a mixed forest of conifer and deciduous broadleaved tree will be reduced. South temperate zone of deciduous broadleaved forest (DBF) will replace the central temperate zone of DBF in South Korea (Fig. 2).

In summary, the Holocene pollen records reflect differences in forest plant assemblages between the western and eastern regions during the early to middle Holocene. In the eastern coastal area, dominance alternated between oak and pine over time, reflecting climate changes during the early to middle Holocene. However, in the western coastal area, oak and alder were co-dominant taxa during the early to middle Holocene. This is a reflection of the geomorphic features of and the marine environmental influence over the Korean Peninsula. The Korean Peninsula is geomorphologically highly mountainous in the east and flat, low, and wide in the west (Fig. 2). During the early to middle Holocene, the west coast experienced wetter conditions for a longer period of time than did the east coast during sea-level rise. From the late Holocene (ca. 2000 cal yr BP), pine trees and agricultural indicators increased over South Korea, reflecting the intensity of human impact since that time (Fig. 6). Due to global warming, no subalpine conifer, especially a Korean fir (*Abies koreana*), will be exist in South Korea in near future. Also evergreen broadleaved forest will further spread north to 37°N, and south temperate zone of DBF will occupy the region of central temperate zone of DBF.

5.2 Vegetation responses to turbulent East Asian monsoonal changes during the early to middle Holocene

About 9000 cal yr BP, the concentration of CO_2 (up to 380 ppm) reached a maximum in the atmosphere along with increased solar radiation in the Northern Hemisphere in the summer (Berger, 1978; Neftel et al., 1982). In July, solar radiation was 7% higher than at present. As a result, the seasonal range of temperatures was considerably increased. The difference in warming between the continent and ocean was higher, leading to monsoons (Kutzbach, 1981; An et al., 2000; Shi et al., 2011). The majority of thermophiles, particularly evergreen trees, did not grow well because of cold winters, even though summers were relatively warm. This is why deciduous oak forests were so common in South Korea during the early Holocene.

From the mid-Holocene optimum, the percentage of broadleaved deciduous components gradually decreased, whereas the percentage of pollen grains from evergreen oak [*Quercus* (*Castanopsis*)] increased (Fig. 7). During that time, deciduous broadleaved forests were replaced by evergreen forests mostly composed of *Q.* (*Castanopsis*). The climate became milder, with warmer winters. More proportional solar radiation per season (Fig. 7) led to a decreased annual range of temperatures: July temperatures slightly decreased, and January temperatures increased. It is possible that there was related adaptation to a slightly warmer winter climate. The range of deciduous oak forests shrank in this region, probably due to increased moisture. Mean annual temperature in Korea (Sohn, 1984) was 2–3°C higher than presently. Temperature differences between the northern and southern parts of the Korean Peninsula remarkably increased toward the end of this phase (Sohn, 1984). An enhanced summer monsoon provided favorable conditions to an evergreen oak–dominated forest.

Pollen studies in China indicate that forest vegetation occupied a larger area during the mid-Holocene (Yu et al., 1998, 2000) due to warm and wet climate together with greater than present summer insolation and stronger Pacific monsoon activity (Winkler & Wang, 1993). The oak-pine woodland and steppe were replaced by a dry steppe vegetation in northeast China region after late Holocene of ca. 3500 cal. yr BP (Liu et al., 2002; Tarasov et al., 2006).

Takahara et al. (2000) concluded that the vegetation distribution at mid-Holocene time (6000 yr BP) was rather similar to present. From their pollen study, they pointed out the broadleaved evergreen and warm mixed forest may have been present at higher elevations in the mountains of central Japan. But, the northern limit of the biome was apparently similar to present as a consequence of northward migration of the biome under warmer and wetter conditions.

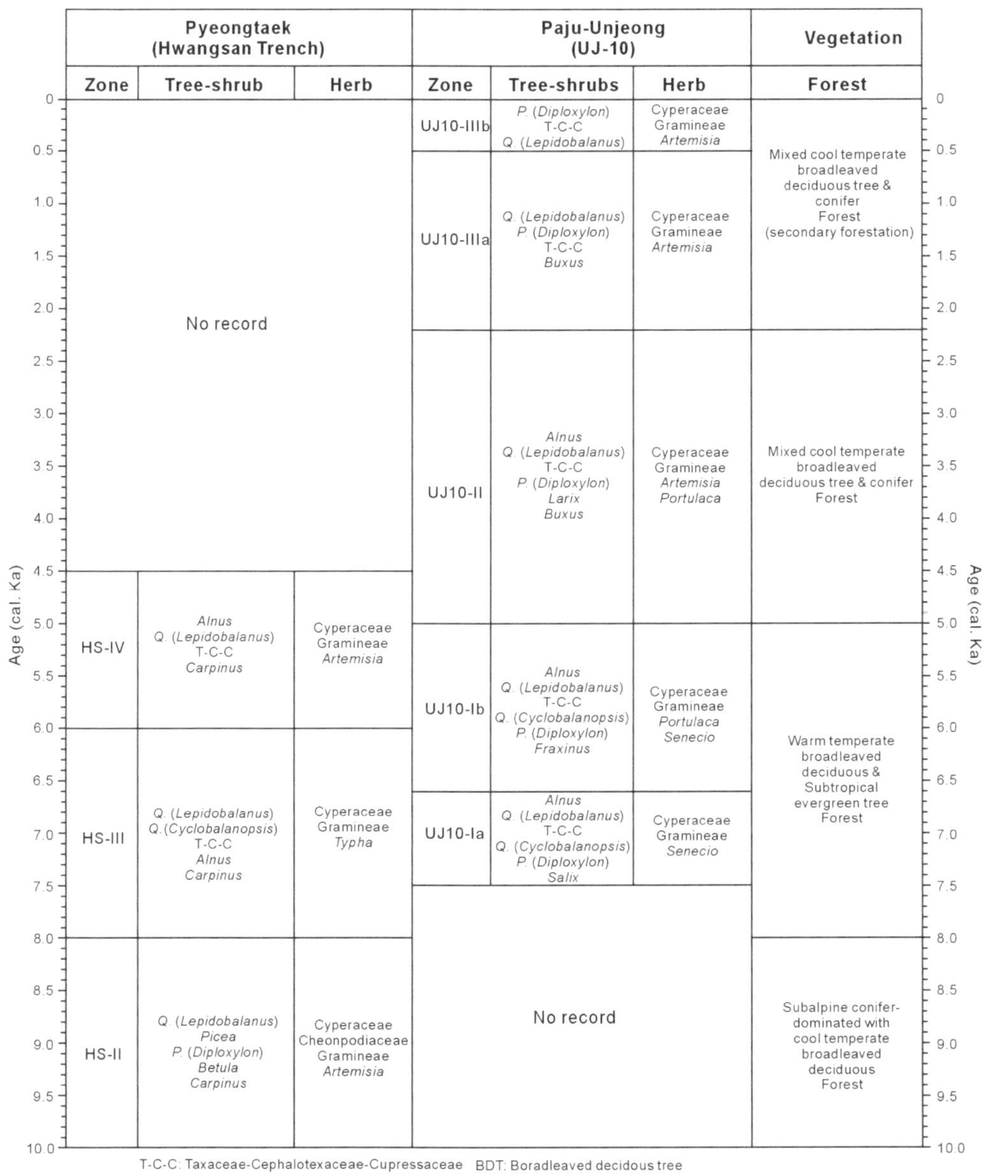

Fig. 6. Correlation of pollen zones, with main trees and herbs, between the Pyeongtaek and Paju-Unjeong areas and forest history during the Holocene period (compiled from Jun et al., 2010 and Yi et al., 2011).

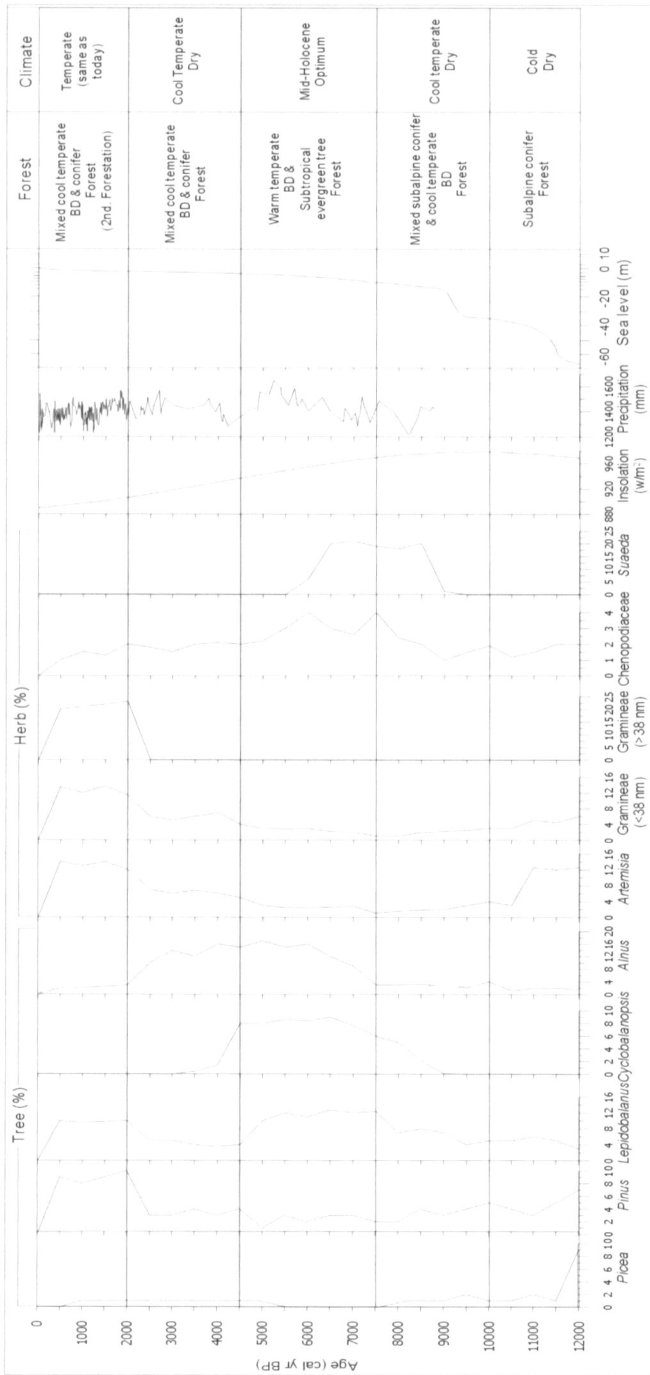

Fig. 7. Vegetation and climate changes based on pollen indicators in South Korea during the Holocene period. Insolation (July), precipitation, and sea level are from Berger (1978), Hu et al. (2008), and Park (1992) and from Chough et al. (2004), respectively. BD: Broadleaved deciduous

6. Conclusion

The Korean Peninsula, surrounded by the sea on three sides (east, west, and south), is located on the eastern end of the Asian continent and belongs to the temperate zone with four distinct seasons largely controlled by the East Asian monsoon. During the summer, the Korean Peninsula is occupied by a subtropical high pressure system and experiences warm, wet conditions with frequent, heavy rainfalls. During the winter, it is cold and dry under the dominant influence of the northwesterly Siberian high air mass. The Korean Peninsula is an area sensitive to climate changes. Therefore, well-preserved records from the Korean Peninsula that provide a continuous climate history are a source of valuable information of the East Asian monsoonal system.

Age-controlled pollen stratigraphy was obtained from several organic-rich sediments in wetlands and archaeological sites in South Korea. Holocene vegetation and climate were deduced from pollen records. During the Early Holocene (ca. 10 400–8000 cal yr BP), dry, cool-temperate conditions encouraged *Quercus*-dominated deciduous broad-leaved forests with conifers (*Pinus*) and cool-tolerant birch (*Betula*) in the hills and mountainous areas as post-glacial warming began. A mid-Holocene climate optimum occurred between 7000 cal yr BP and 5000 cal yr BP, when evergreen broadleaved oak [*Quercus* (*Cyclobalanopsis*)] and deciduous broadleaved hardwood trees flourished and migrated northward to the central Korean Peninsula (N 37-38°), while conifers and cool-tolerant birch (*Betula*) retreated higher up the mountains. After the mid-Holocene (ca. 5000 to ca. 3000 cal yr BP), evergreen and deciduous broadleaved *Quercus*-dominated forests were replaced by mixed conifer and deciduous broadleaved forests due to climatic deterioration to dry, cool-temperate conditions. During the late Holocene (ca. 3000–2000 cal yr BP), mixed coniferous and deciduous broadleaved forests were continuous in the mountainous areas under wet, cool-temperate climatic conditions. Since 2000 cal. yr BP, forests were primarily affected by human disturbance in Korea. That is, sterilized mountain soil conditions caused by human activity such as deforestation and cultivation accelerated the expansion of coniferous forest, replacing the former mixed-vegetation forest. Human impact is indicated by the occurrence of cultivated plants, Gramineae (≥35 μm), and buckwheat (*Fagopyrum*) combined with synanthropogenic *Ambrosia* and *Artemisia*.

Owing to global warming, no subalpine conifer, especially a Korean fir (*Abies koreana*), will be exist in South Korea in near future. Also evergreen broadleaved forest will further spread north to N 37°, and south temperate zone of deciduous broadleaved forest will occupy the region of central temperate zone of deciduous broadleaved forest.

7. Acknowledgments

This study was supported by the basic research project of the Korean Institute of Geosciences and Mineral Resources (KIGAM). The pollen datasets were a partly result of this project.

8. References

An, Z.; Liu, T. S.; Lu, Y. C.; Porter, S. C.; Kukla, G.; Wu, X. H. & Hua, Y. M. (1990). The long term paleomonsoon variation recorded by the loess-paleosol sequence in central China. *Quaternary International*, Vol.7/8, (January 1990), pp.91-95, ISSN 1040-6182

An, Z. (2000). The history and variability of the East Asian paleomonsoon climate. *Quaternary Science Reviews*, Vol.19, (January 2000), pp.171-187, ISSN 0277-3791

Bae, D. H.; Jung, I. W. & Chang, H. (2008). Long-term trend of precipitation and runoff in Korean river basins. *Hydrological Processes*, Vol.22, No.14, (July 2008), pp.2644-2656, ISSN 1099-1085

Baek, H. J. & Kwon, W. T. (2005). The climatological characterstics of monthly precipitation over the Han- and Nakdong-river basins: part I. Variability of area averaged time series. *Journal of Korean Water Resources Association*, Vol.38, (July 2005), pp.111-119 (in Korean with English abstract)

Berger, A. L. (1978). Long-term variations of caloric insolation resulting from the Earth's orbital elements. *Quaternary Research*, Vol.9, No.2, (March 1978), pp.139-167, ISSN 0033-5894

Birks, H. H. & Birks, H. J. B. (1980). *Quaternary Palaeoecology*. The Blackburn Press, USA, ISSN 1-930665-56-3

Chang, C. H. & Kim, C. M. (1982). Late Quaternary vegetation in the Lake of Korea. *Korean Journal of Botany*, Vol.25, No.1, (March 1982), pp.37-53, ISSN 0583-421X

Chang, H. & Kwon, W. T. (2007). Spatial variations of summer precipitaion trends in South Korea, 1973-2005. *Environmental Research Letter*, Vol.2, No.4, (November 2007), pp.1-9, ISSN 1748-9326

Chang, N. K.; Kim, Y. P.; Oh, I. H. & Son, Y. H. (1987). Past vegetation of the moor at Mt. Daeam in terms of pollen analysis. *Korean Journal of Ecology*, Vol.10, (December 1987), pp.195-204, ISSN 2093-4521 (in Korean with Enlgish abstract)

Choi, K. R. (2001). Palynological study of Moojechi Bog. *Korean Journal of Quaternary Research*, Vol.15, No.1, (June 2001), pp.13-20, ISSN 1226-8448 (in Korean with English abstact)

Choi, K. R. & Koh, J. K. (1989). Studies on moor vegetation of Mt. Daeam, east-central Korea. *Korean Journal of Ecology*, Vol.12, No.4, (December 1989), pp.237-244, ISSN 1975-020X (in Korean with English abstact)

Choi, K. R.; Kim, K. H.; Kim, J. W.; Kim, J. C.; Lee, G. K.; Yang, D. Y. & Nahm, W. H. (2005). Vegetaion history since the Mid-Late glacial from Yeongsan River Basin, Southwestern Korea. *Korean Journal of Ecology*, Vol.28, No.1, (Marh 2005), pp.37-43, ISSN 1975-020X (in Korean with English abstact)

Chough, S. K.; Lee, H. J.; Chun, S. S. & Shinn, Y. J. (2004). Depositional processes of late Quaternary sediments in the Yellow Sea: a review. *Geosciences Journal*, Vol.8, No.2, (June 2004), pp.211-264, ISSN 1226-4806

Chung, C. H. (2007). Vegetation response to climate change on Juju Island, South Korea, during the last deglaciation based on pollen record. *Geosciences Journal*, Vol.11, No.2, (June 2007), pp.147-155, ISSN 1226-4806

Chung, C. H. & Lee, H. J. (2006). Palynological study from the Pleistocene sediments of the Yongdong archaeological site, Naju area. *Journal of Korean Paleolithic Society*, Vol.13, pp.1-6 (in Korean with English abstract)

Chung, C. H.; Lee, H. J.; Lim, H. S. & Kim, J. B. (2005). Palynological study of the Late Quaternary sediments at Piseo-ri, Muan, Korea. *Journal of Korean Earth Science Society*, Vol.26, No.6, (August 2005), pp.597-602, ISSN 1225-6692 (in Korean with English abstract)

Chung, C. H.; Lim, H. S. & Yoon, H. I. (2006). Vegetation and climate changes during the Late Pleistocene to Holocene inferred from pollen record in Jinju area, South Korea. *Geosciences Journal*, Vol.10, (December 2006), pp.423-431, ISSN1226-4806

Chung, C. H.; Lim, H. S. & Lee, H. J. (2010). Vegetation and climate changes during the Late Pleistocene to Holocene inferred from pollen record in Gwangju area, South Korea. *Quaternary International*, Vol.227, (November 2010), pp.61-67, ISSN 1040-6182

Chung, T. H. & Lee, W. C. (1965). A study of the Korean woody plant zone and favorable reion for the growth and proper sepcies. *Journal of Sunkyunkwan University*, Vol.10, (September 1965), pp.329-366 (in Korean with English abstract)

Davis, K. O. (1994). Aspects of archaeological palynology: methodology and application. *American Association of Stratigraphic Palynologists Contributions*, No.29, (December 1994), pp.1-5, ISSN 0160-8843

Dykoski, C. A.; Edwards, R. L.; Cheng, H.; Yuan, D.; Cai, Y.; Zhang, M.; Lin, Y.; Qing, J.; An, Z. & Revenaugh, J. (2005). A high-resoultion, abolute-date Holocene and deglacial Asian monsoon record from Dongge Cave, China. *Earth and Planetary Science Letters*, Vol.233, (April 2005), pp.71-86, ISSN 0012-821X

Evstigneeva, T. A. & Naryshkina, N. N. (2010). The Holocene climatic optimum at the Southern coast of the Sea of Japan. *Paleontological Journal*, Vol.44, No.10, (December 2010), pp.1262-1269, ISSN 0031-0301

Feng, Z. D.; Tang, L. Y.; Wang, H. B.; Ma, Y. Z. & Liu, K. B. (2006). Holocene vegetation variations and the associated environmental changes in the western part of the Chinese Loess Plateau. *Palaeogeography, Palaeoclimatology, Palaeoecology*, Vol.241, No.3-4, (November 2006), pp.440-456, ISSN 0031-0182

Fujiki, T. & Yasuda, Y. (2004). Vegetation history during the Holocene from Lake Hyangho, northeastern Korea. *Quaternary International*, Vol.123-125, (December 2004), pp.63-69, ISSN 1040-6182

Hu, C.; Henderson, G. M.; Huang, J.; Xie, S.; Sun, Y. & Johnson, R. (2008). Quantification of Holocene Asian monsoon rainfall from spatially separated cave records. *Earth and Planetary Science Letters*, Vol.266, No.3-4, (February 2008), pp.221-232, ISSN 0012-821X

Jang, B. O.; Yang, D. Y.; Kim, J. Y. & Choi, K. R. (2006). Postglacial vegetation history of the central western region of the Korean Peninsula. *Journal of Ecology and Field Biology*, Vol.29, (December 2006), pp.573-580, ISSN 2093-4521 (in Korean with English abstract)

Jo, K.; Woo, K. S.; Lim, H. S. & et al. (*in press*). Holocene and Eemian climatic optima in the Korean Peninsula based on textural and carbon isotopic records from the stalagmite of the Daeya Cave, South Korea. *Quaternary Science Reviews*, doi:10.1016/j.quascirev.2011.02.012, ISSN 0277-3791

Jo, W. R. (1979). Palynological studies on postglacial age in eastern coastal reigon, Korean Peninsula. *Annals of the Tohoku Geographic Association*, Vol.31, pp.23-35 (In Japanese with English abstract)

Jun, C. P.; Yi, S. & Lee, S. J. (2010). Palynological implication of Holocene vegetation and environment in Pyeongtaek wetland, Korea. *Quaternary International*, Vol.227, (November 2010), pp.68-74, ISSN 1040-6182

Jung, I. W.; Bae, D. H. & Kim, G. (2010). Recent trends of mean and extreme precipitation in Korea. *International Journal of Climatology*, Vol.31, No.3, (March 2010), pp.359-370, ISSN 08998418

Kim, J. Y.; Yang, D. Y.; Bong, P. Y.; Lee, Y. J. & Park, J. H. (2001). A study on vegetation history of organic muds of Sorori archaeological site, Oksan-myeon, Cheongwon-gunm, Korea. *Korean Journal of Quaternary Research*, Vol.15, No.2, (December 2001), pp.75-82, ISSN 1226-8448 (in Korean with English abstact)

Kong, W. S. (1994). The vegetation history of Korea during the Holocene Period. Korean *Journal of Quaternary Research*, Vol.8, No.1, (December 1994), pp.9-22

Kong, W. S. (2007). *Biogeography of Korean Plants*. Geobook, ISBN 978-89-959394-0-6 (in Korean)

Kutzbach, J. E. (1981). Monsoon climate of the Early Holocene: Climate experiment using the Earth's orbital parameters for 9000 years ago. *Science*, Vol.214, No.4516, pp.59-61, ISSN 00368075

Lee, T. B. (1985). *Illustrated flora of Korea*. Hyangmunsa Publisher, Seoul (in Korean)

Lee, W. C. & Yim, Y. J. (2002). *Plant Geography with special reference to Korea*. Kangwon Naitonal Unviersity Press, ISBN 89-7157-137-3 (in Korean)

Liu, H.; Xu, L. & Cui, H. (2002). Holocene history of desertification along the woodland-steppe border in northern China. *Quaternary Research*, Vol. 57, No. 2, (March 2002), pp. 259–270, ISSN 0033-5894

Lorius, C.; Jouzel, J. & Ritz, R. (1985). A 150,000 year climatic record from Antarctic Ice. *Nature* 316, pp.591-596, ISSN 0028-0836

Makohonienko, M.; Kitagawa, H.; Fujiki, T.; Liu, X.; Yasuda, Y. & Yin, H. (2008). Late Holocene vegetation changes and human impact in the Changbai Mountains area, Northeast China. *Quaternary International*, Vol.184, (June 2008), pp.94-108, ISSN 1040-6182

Nakagawa, T.; Tarasov, P. E.; Kitagawa, H.; Yasuda, Y. & Gotanda, K. (2006). Seasonally specific responses of the East Asian monsoon to deglacial climate changes. *Geology*, Vol.35, No.7, (July 2006), pp. 521-524, ISSN 0091-7613

Nakai, T. (1952). *Synoptical Sketch of Korean flora*. Bulletin of Natural Science Museum, Tokyo

Naryshkina, N. N. & Evstigneeva, T. A. (2009). Sculpture of pollen grains of Quercus L. From the Holocene of the South of the Sea of Japan. *Paleontological Journal*, Vol.43, No.10, (December 2009), pp.1309-1315, ISSN 0031-0301

Neftel, A.; Oeschger, H. & Schwander, J. (1982). Ice core sample measurements give atmosphere CO_2 content during the past 40,000 yr. *Nature*, Vol.295, pp.391-394, ISSN 0028-0836

Oh, C. Y. (1971). A pollen analysis in the peat sediments from Pyungtaek county. *Korean Journal of Botany*, Vol.14, (March 1971), pp.126-133, ISSN 0583-421X

Park, I. K. (1990). A pllen analytical study of the peat sediments from the Chollipo arboretum in Southwestern, Kora. *Korean Journal of Ecology*, Vol.13, No.4, (December 1990), pp.311-320, ISSN 1975-020X (in Korean with English abstact)

Park, J. H. & Yi, S. (2008). Postglacial environments of the Chungnam Province inferred from pollen analysis: with emphasis of change in climate and vegetation together with human impact. *Journal of the Paleontological Society of Korea*, Vol.24, No.1, (June 2008), pp.55-75, ISSN 1225-0929(in Korean with English abstract)

Park, J. H. (2004). Vegetation history of deposits at Unjeong-ri archaeological site, Mokcheon-eub, Cheonan-si, Korea. *Journal of the Geomorphological Association of Korea*, Vol.11, No.4 (December 2004), pp.61-68, ISSN 1226-4293 (in Korean with English abstract)

Park, J. K. & Chang, N. K. (1998). Past vegetaion of Moojaechi on Mt. Jungjok by pollen analysis. *Korean Journal of Ecology*, Vol.21, No.5, (June 1998), pp.427-433, ISSN 1975-020X (in Korean with English abstact)

Park, Y. A. (1992). The changes of sea level and climate during the late Pleistocene and Holocene in the Yellow Sea region. *Korean Journal of Quaternary Research*, Vol.6, No.1, (December 1992), pp.3-19

Qiang, X. K.; Li, Z. X.; Powell, C. M. & Zheng, H. B. (2001). Magnetostratigraphic record of the Late Miocene onset of the East Asian monsoon and Pliocene uplift of northern Tibet. *Earth and Planetary Science Letters*, Vol.187, No.1-2, (April 2001), pp.83-93, ISSN 0012-821X

Seo, K. S. & Yi, M. S. (2003). A study on spore from the archaeological site of Anyoungri, Kongju, Korea. *Journal of Cultural Science and Technology*, Vol.2, No.(1), (February 2003), pp.29-35 (in Korean with English abstract)

Shi, Z.; Liu, X.; Sun, Y.; An, A.; Liu, Z. & Kutzbach, J. (2011). Distinct response of East Asian summer and winter monsoons to orbital forcing. *Climate of the Past Discussion*, Vol.7, No.2, (March 2011), pp.943-964, ISSN 18149340

Sohn, P. K. (1984). The paleoenvironment of Middle and Upper Pleistocene Korea, *in* R. P. Whyte (ed.), *The Evolution of the East Asian Environment*, University of Hong Kong, Hong Kong, Vol.4, pp.877-893

Sun, X. & Wang, P. (2005). How old is the Asian monsoon system?–Palaeobotanical records from China. Palaeogeography, Palaeoclimatology, Palaoecology, Vo.222, No.3-4, (July 2005), pp.181-222, ISSN 0031-0182

Takahara, H.; Sugita, S.; Harrison, S. P.; Miyoshi, N.; Morita, Y. & Uchiyama, T. (2000). Pollen-based reconstructions of Japanese biomes at 0,6000 and 18,000 ^{14}C yr BP. *Journal of Biogeography*, Vol.27, No.3, (May 2000), pp.665-683, ISSN 1365-2699

Tarasov, P.; Jin, G. & Wagner, M. (2006). Mid-Holocene environmental and human dynamics in northeastern China reconstructed from pollen and archaeological data. Palaeogeography, Palaeoclimatology, Palaaeoecology, Vol.241, No.2, (November 2006), pp.284-300, ISSN 0031-0182

Tinner, W. & Lotter, A. F. (2001). Central European vegetation response to abrupt climate change at 8.2 ka. *Geology*, Vol.29, No.6, (June 2001), pp.551-554, ISSN 00917613

Tsukuda, M. (1977). The environment change history in Korea. I. The vegetaiton change history in Sogcho. *The Quaternary Research Program and Abstract*, Vol.6, pp.21 (in Japanese)

Tsukuda, M. (1988). Japan *in* B. Huntley & T. Webb (ed.), *Vegetation History*, Kluwer Academic Publishers, Dordrecht, pp.459-518

Uyeki, K. (1911). On the forest zones in Korea. *Bulletin of Korean Agricultural Society*, Vol.6 (in Japanese)

Uyeki, H. (1933). O the forest zones of Korea. *Acta Phytotax & Geobotany*, Vol.2, No.2, (1933), pp.73-85 (in Japanese)

Wang, Y. J.; Cheng, H.; Edwards, R. L.; An, Z.; Wu, J. Y.; Shen, C. C. & Dorale, J. A. (2001). A high-resolution absolute-dated late Pleistocene monsoon record from Hulu Cave, China. *Science*, Vol.294, (December 2001), pp.2345-2348, ISSN 0036-8075

Winkler, M. G. & Wang, P. K. (1993). The Late-Quaternary Vegetation and Climate of China. *In* H. E. Wright, J. E. Kutzbach, T. Webb III, W. F. Ruddiman, F. A. Street-Perrott & P. J. Bartlein (eds.), *Global Climates since the Last Glacial Maximum*. University of Minnesota Press, Minneapolis, pp. 265–293

Wu, Y.; Lűcke, A.; Jin Z.; Wang, S.; Schleser, G. H.; Battarbee, R. H. & Xia, W. (2006). Holocene climate development on the central Tibetan Plateau: a sedimentary records from Cuoe Lake. *Palaeogeography, Palaeoclimatology, Palaeoecology*, Vol.234, No.2-4, (May 2006), pp.328-340, ISSN 0031-0182

Wűnnemann, B.; Mischke, S. & Chen, F. (2006). A Holocene sedimentary record from Bosten Lake, China. *Palaeogeography, Palaeoclimatology, Palaeoecology*, Vol.234, No.2-4, (May 2006), pp.223-238, ISSN 0031-0182

Yamada, K. (2004). Last 40 ka climate changes as deduced from the lacustrine sediments of Lake Biwa, central Japan. *Quaternary International*, Vol.123-125, (February 2004), pp.43-50, ISSN 1040-6182

Yancheva, G.; Nowaczyk, N. R.; Mingram, J.; Dulski, P.; Schettler, G.; Negendank, J. F.W.; Liu, J.; Sigman, D. M.; Peterson, L. C. & Haug, G. H. (2007). Influence of the intertropical convergence zone on the East Asian monsoon. *Nature*, Vol.445, (January 2007), pp. 74-77, ISSN 0028-0836

Yang, D. Y.; Kim, J. Y.; Nahm, W. H.; Ryu, E.; Yi, S.; Kim, J. C.; Lee, J. Y. & Kim, J. K. (2008). Holocene wetland environmental change based on major element concentrations and organic contents from the Chellipo coast, Korea. *Quaternary International*, 176-177, (January 2008), pp.143-155, ISSN 1040-6182

Yao, T.; Xu, B. & Pu, J. (2001). Climatic changes on orbital and sub-orbital time scale recorded by the Guliya ice core in Tibetan Plateau. *Science in China (Series D)*, Vol.44, (December 2001), pp.360-368

Yasuda, Y. (1978). Prehistoric environment in Japan. Palynological approach. *Science Report of Tohoku University, Series 7, Geography*, Vol.28, No.2, pp.117-281

Yasuda, Y.: Tsukada, M.; Kim, J. M.; Lee, S. T. & Yim, Y. J. (1980). Environmental changes and the origin of agriculture in Korea. *Oversease Research Report*, Ministry of Education, Japan, pp.1-19 (in Japanese)

Yi, S. (2011, *in press*). Pollen as an indicator of possible land-use regimes since the late Neolithic Age in South Korea: a cast study of a Paju Unjeong site. *Vegetation History and Archaeobotany*, ISSN 0939-6314

Yi, S.; Saito, Y.; Oshima, H.; Zhou, Y. & Wei, H. (2003a). Holocene environmental history inferred from pollen assemblages in the Huanghe (Yellow River) delta, China: climatic change and human impact. *Quaternary Science Reviews*, Vol.22, (March 2003), pp.609-628, ISSN 0277-3791

Yi, S.; Saito, Y.; Zhao, Q. & Wang, P. (2003b). Vegetation and climate changes in the Changjiang (Yantze River) Delta, China, during the past 13,000 years inferred from pollen records. *Quaternary Science Reviews*, Vol.22, (June 2003), pp.1501-1519, ISSN 0277-3791

Yi, S. & Saito, Y. (2003c). Palynological evidence for late Holocene environmental change on the Gimhae fluvial plain, southern Korean Peninsula: reconstructing the rise and fall of Golden Crown Gaya State. Geoarchaeology, Vol.18, No.8, (December 2003), pp.831-850, ISSN 1520-6548

Yi, S.; Nam, S. I.; Chang, S. W. & Chang, J. H. (2004). Holocene environmental changes in the tidal sediments of west coast of South Korea inferred from pollen records. *Journal of the Geological Society of Korea*, Vol.40, No.2, (June 2004), pp.213-225, ISSN 0435-4036 (in Korean with English abstract)

Yi, S.; Ryu, E.; Kim, J. Y.; Nahm, W. H.; Yang, D. Y. & Shin, S. C. (2005). Late Holocene paleoenvironmental changes inferred from palynological and diatom assemblages in Isanpo area, Ilsan, Gyeonggi-do, Korea. *Journal of the Geological Society of Korea*, Vol.41, (September 2005), pp.295-322, ISSN 0435-4036 (in Korean with English abstract)

Yi, S.; Kim, J. Y.; Oh, K. C.; Yang, D. Y.; Ryu, E. & Oh, K. J. (2006). Late Pleistocene paleoenvironments of the Poonggi-dong area, Asan, inferred from pollen analysis. *Journal of the Geological Society of Korea*, Vol.42, No.1, (March 2006), pp.57-68, ISSN 0435-4036 (in Korean with English abstract)

Yi, S.; Kim, J. Y.; Yang, D. Y.; Oh, K. C. & Hong, S. S. (2008a). Mid- and Late-Holocene palynofloral and environmental change of Korean central region. *Quaternary International*, Vol.176-177, (January 2008), pp.112-120, ISSN 1040-6182

Yi, S.; Kim, J. Y.; Yang, D. Y.; Kim, J. C.; Nahm, W. H. & Yun, H. S. (2008b). Palynological implication for environmental changes in the Hanam area, Gyeonggi Province since the Last Glacial Maximum. *Journal of the Geological Society of Korea*, Vol.44, No.5, (October 2008), pp.673-684, ISSN 0435-4036 (in Korean with English abstract)

Yi, S. & Kang, B. W. (2009). Application of pollen, non-pollen palynomorphs and phytolith analyses to the archaeology: case study from the Chunghyo-dong archaeological site, Gyeongju. *Journal of Paleontological Society of Korea*, Vol.25, No.1, (June 2009), pp.77-102, ISSN 1225-0929(in Korean with English abstract)

Yi, S. & Kim, J. Y. (2009). Pollen indication of Holocene vegetation and environments in the Sacheon-dong archaeological site, Cheongju, Chungbuk Province. *Journal of Paleontological Society of Korea*, Vol.25, No.1, (June 2009), pp.63-76, ISSN 1225-0929(in Korean with English abstract)

Yi, S.; Kim, J. Y. & Jia, H. (2011, *submitted*). Pollen record of agricultural cultivation in the west-central Korean Peninsula since the Neolithic Age. *Quaternary International*, ISSN 1040-6182

Yim, Y. J. (1977). Distribution of forest vegetation and climate in the Korean Peninsula IV. Zonal distribution of forst vegetation in relation to thermal climate. *Japanese Journal of Ecology*, Vol.27, (March 2007), pp.269-278

Yim, Y. J. & Kira, T. (1975). Distribution of forest vegetation and climate in the Korean Peninsula I. Distribution of some indices of thermal climate. *Japanese Journal of Ecology*, Vol.25, No.2, (June 1975), pp.77-88, ISSN 00215007

Yoon, S. O. (1997). The Holocene environmental change and reconstruction of the Palaeogeography at Ilsan area, with the special reference to pollen analysis. *Journal of the Korea Geographical Society*, Vol.32, No.1, (March 1997), pp.15-30, ISSN 1225-6633 (in Korean with English abstract)

Yoon, S. O. & Jo, W. R. (1996). The late Quaternary environmental change in Youngyang Basin, southeastern part of Korean Peninsula. *The Korean Geographical Society*, Vol.31, No.3, (September 1996), pp.447-468, ISSN 1225-6633 (in Korean with English abstract)

Yoon, S. O.; Kim, H. R.; Hwang, S. & Choi, J. M. (2005). Environmental change and agricultural acitivities during the Late Holocene in Geumcheon-ri, Milyang City, Korea. *Journal of the Korean Archaeological Society*, Vol.56, (June 2005), pp.7-48, ISSN 1015-373X (in Korean with English abstract)

Yu, G.; Prentice, I. C.; Harrison, S. P. & Sun, X. (1998). Pollen-based biome reconstruction for China at 0 and 6000 years. *Journal of Biogeography*, Vol.25, No.6, (November 1998), pp.1055–1069, ISSN 1365-2699

Yu, G.; Chen, X.; Ni, J.; Cheddadi, R.; Guiot, J.; Han, H.; Harrison, S. P.; Huang, C.; Ke, M.; Kong, Z.; Li, S.; Li, W.; Liew, P.; Liu, G.; Liu, J.; Liu, K.-B.; Prentice, I. C.; Qui, W.; Ren, G.; Song, C.; Sugita, S.; Sun, X.; Tang, L.; Van Campo, E.; Xia, Y.; Xu, Q.; Yan, S.; Yang, X.; Zhao, J. & Zheng, Z. (2000). Palaeovegetation of China: a pollen data-based synthesis for the mid-Holocene and last glacial maximum. *Journal of Biogeography*, Vol.27, No.3, (May 2000), pp.635–664, ISSN 1365-2699

9

Climate Signals from ^{10}Be Records of Marine Sediments Surrounded with Nearby a Continent

Kyeong Ja Kim[1] and Seung-Il Nam[2]
[1]*Korea Institute of Geoscience and Mineral Resources*
[2]*Korea Polar Research Institute*
Korea

1. Introduction

Climate signals from ^{10}Be records in marine environments have been studied for last two decades (Aldahan et al., 1997, Bourlès et al., 1989, Christl et al., 2003, Eisenhauer et al., 1994, Horiuchi et al., 2000, 2001, Kim and Nam, 2010, Knudsen et al., 2008, McHargue et al., 2010, McHargue and Donahue, 2005). Understanding of regional climate signals is feasible through not only ^{10}Be but also ^{9}Be from the sediment. This is because ^{9}Be is terrigenous origin while ^{10}Be signal is affected by climatic condition and production at the top of atmosphere. Recent study from the East Sea of Korea (05-GCRP-21) indicated that climate signals from ^{10}Be records of Korean marine sediment are generally representing global climate variations during warm and cold periods from Last Glacial Maximum to Holocene and also MIS (marine isotope stage) 6 to Eemian. The ^{10}Be records of the East Sea are well compared with those from the oxygen isotopic record of this marginal sea (Kim and Nam, 2010). During the warm interglacial periods the ^{10}Be concentrations per sediment mass have significantly increased while during the cold glacial periods those have decreased (Aldahan et al., 1997, Eisenhauer et al., 1994). This result also shows that a vivid record of ^{10}Be/^{9}Be indicates a significant increase of ^{10}Be at a time of 120 kyr, which might be an indication of the paleomagentic excursion.

Interestingly, it was found that the ^{10}Be concentrations per 1g sediment of this region were about 30% lower than other ^{10}Be records of largely open marine environment. We also found that ^{10}Be concentrations of the Blake Outer Ridge were similar to those from Korean marine sediments (McHargue et al., 2000). Two study areas are located nearby large continents: the East Asia (the East Sea) vs North America (the Atlantic). This could be caused by sediment inflow to the marine environment which is close enough to the continent. Therefore, local marine environmental influence is revealed through the beryllium isotopes. Both cases would have similar climatic signals due to their geographical locations nearby continent. The lower ^{10}Be concentrations for these regions could be also involved in ocean current and circulation. Relatively deep sea water of these regions may not be well mixed rapidly with the surface water and old sea water with relatively lower ^{10}Be concentration remains in the sediment records. For this chapter, we investigated climatic signals from Be isotope records of the East Sea of Korea and the Mendeleev Ridge of the Arctic Ocean and compared with the records from the Blake Outer Ridge studied by McHargue et al., 2000. In addition, global ^{10}Be records of marine sediments for various regions will be briefly discussed. This chapter would provide a new insight guide into understanding climate signals through ^{10}Be records of various marine environments.

2. Beryllium isotopes in terrestrial environments

2.1 Cosmic ray induced ^{10}Be production rate

Production rates of cosmogenic isotopes depend on geomagnetic latitude, altitude, and flux of incoming cosmic rays to the earth (Lal, 1988). The Earth's geomagnetic field deflects incoming cosmic rays and has an effect on the production rate of *in situ* cosmogenic isotopes. This deflection affects the incident angle and the rigidity of cosmic rays. The rigidity is defined as $r = pc/q$, where p is the momentum, q is the charge of the particle, and c is the velocity of light (O'Brien, 1979). The vertical cutoff rigidity is the lowest at the geomagnetic poles and highest at the equator. Therefore, greater cosmic rays reach to the poles and attenuation length at low latitude is greater than at high latitude (Simpson and Uretz, 1953). Since geomagnetic latitude affects the production rate of cosmogenic isotopes, understanding of the secular variation of the Earth's geomagnetic field is important. It is known that variations of geomagnetic field intensity cause changes in the flux of cosmic rays, in solar activity, and in shielding by the Earth's magnetosphere. Laj et al., 1996 describes geomagnetic intensity and ^{14}C abundance in the atmosphere and ocean during the past 50 kyr. This paper shows geomagnetic change effects on the ^{14}C production, which is increased by the decrease of the Earth's dipole moment. Similarly, the relationship between ^{10}Be production rate and geomagnetic field intensity was studied using deep-sea sediments. These results also demonstrate the importance of the relationship between cosmogenic nuclide production and intensity of geomagnetic dipole moment variation (Frank et al., 1997). Thus, production of cosmogenic isotope should be corrected with the variation of geomagnetic dipole moment variation. It has been found that the Earth's geomagnetic pole is essentially the geographic pole for periods greater than about 2 kyr (Champion, 1980, Ohno and Hamano, 1992). Therefore, a correction from geographic reading to geomagnetic reading is required. For most cases of ^{10}Be or ^{26}Al surface exposure dating samples, the working range of age is from several thousand years to a few million years. Thus, in this case, the correction for geomagnetic reading may not be required, but the correction of production related to secular variation of geomagnetic dipole moment intensity is required. Masarik et al., 2001 demonstrated the correction of *in situ* cosmogenic nuclide production rates for geomagnetic field intensity variations during the past 800 kyr. This paper indicated that at the equator integrated production rates for exposure ages between ~40 and 800 kyr are 10 to 12% higher than the present day value, whereas at latitudes greater than 40 degree, geomagnetic field intensity variations have hardly influenced *in situ* cosmogenic nuclide production.

Production rates as a function of both latitude and altitude have been studied in the past. For a few decades, models from Lal and Peters, 1967 and Lal, 1988, have been widely used for the scaling factors for production rates of *in situ* produced cosmogenic nuclides. A third degree polynomial equation found in Lal, 1991, enables one to calculate the production rate of ^{10}Be and ^{26}Al with respect to geomagnetic latitude and altitude. A reevaluation of scaling factors for these production rates has been attempted recently using non-dipole contributions of the geomagnetic field to the cosmic ray flux and observed attenuation lengths (Dunai, 2000). The scaling factors for the nuclear disintegration with respect to geomagnetic latitude and altitude from Lal's work are involved in the geocentric axial dipole hypothesis and this is appropriate for time scales exceeding 200 kyr. The non-dipole components of the Earth's magnetic field contribute up to 20% to the total field; therefore, they must be considered for short-term effects of cosmic rays. It is known that the new scaling factors and those of Lal are significantly different, by up to about 30%, especially at high altitude and at low latitude. Currently, a few other research groups have been involved in studying production rates of cosmogenic nuclide or measurement of neutron flux as a function of geomagnetic latitude and altitude. This additional research on this field may provide a confirmation of these scaling factors for the production of *in situ* cosmogenic isotopes. (Stone, 2000, Graham et al., 2005a,b,c, 2000, Lifton et al., 2001, Desilets et al., 2006).

2.1.1 Production rate in the atmosphere

^{10}Be is produced in the atmosphere by nuclear interactions with oxygen and nitrogen (Peters, 1955, Goess and Phillips, 2001). The intensity of the cosmic ray flux depends on galactic and solar sources, and modulation by the heliomagnetic and geomagnetic fields. Both ^{10}Be is produced by spallation reactions in the atmosphere, and then ^{10}Be is well mixed up (Ueikkila et al., 2009) and removed from the atmosphere by precipitation scavenging of aerosol particles to land and sea. Eventually, these nuclides are deposited within ocean sediment. The ^{10}Be concentration at 10.7 km of stratosphere and at 19.2 km in the tropospheric concentration are known to be 7 x 10^6 atoms/m^3, and 1.3 x 10^7 atoms/m^3, respectively. The global average ^{10}Be production rate is found to be (1.21±70) x 10^6 atoms/cm^2/yr (Monaghan et al., 1985). Estimates of the ^{10}Be production rate derived from measurements on ice cores, lake sediments, and deep-sea sediments range from 0.35 x 10^6 atoms/cm^2/yr to 1.89 x 10^6 atoms/cm^2/yr. (Monaghan et al., 1985). Castagnolie et al., 2003 demonstrated reconstruction of the modulation parameter M from the open solar magnetic flux proposed by Solanki et al., 2000, and experimental values calculated from the GCR spectra measured with balloons and spacecraft are compared well with ^{10}Be concentration measured at the Dye3 ice core, assuming constant accumulation rate during the period of 1810-1997. The production rate of ^{10}Be ranged from 0.015 to 0.025 atoms/cm^2/s (Castagnolie et al., 2003).

The precipitation onto the surface of the Earth and the deposition of ^{10}Be in soils is influenced by climate. In turn, the production of ^{10}Be in the atmosphere is influenced by the magnetic dipole field of the Earth to which it is inversely related. This relationship between the production of ^{10}Be and the geomagnetic field has been shown by the correlation between the variations of ^{10}Be and those of the measured paleo-inclination data of the dipole field in sediments (Frank et al., 1997, Frank, 2000, Masarik et al., 2001, Laj et al., 2000), and the concentrations of ^{10}Be in marine sediments and the measured paleointensity (Carcaillet et al., 2004, McHargue and Donahue, 2005).

The influence of climate on the deposition of ^{10}Be, otherwise is problematic for interpretations of the cosmic-ray flux, in itself is a worthy subject for study. For example, variations in the deposition rates of ^{10}Be and sediments affect the ^{10}Be/^{9}Be ratio due to the uneven mixing of the two isotopes in the hydrological cycle. That, ^{10}Be, produced largely in the atmosphere, is transported to the surface of the earth by rain and dry precipitation to the sea. In contrast, ^{9}Be, derived from terrigenous materials, is transported to the sea largely by rivers, and to a lesser extent by atmospheric deposition.

Location	^{10}Be	Source	Reference
USA continent	(1.38±0.36) ~(3.96 ± 0.35) x 10^6 atoms/cm^2/yr	rain	Monagahan et al., 1985
USA Hawaii	1.9 x 10^3 ~ 8.94 x 10^4 atoms/g	rain	Monagahan et al., 1985
Tropical region	Avg. 1.53 x 10^4 atoms/g	rain	Monagahan et al., 1985
Illinois, USA	2 ~7 x 10^7 atoms/g	soil	Brown et al., 1989
Japan, Kikari Is	(0.80 ~ 7.17) x 10^9 atoms/g	soil	Maejima et al., 2005
Japan, Kikari Is	(2.0 ~ 3.5) x 10^6 atoms/cm/yr	rain	Maejima et al., 2005
New Zealand	(2.1 ~ 2.9) x 10^4 atoms/g	rain	Graham et al., 2003
Korea, Masanri	(0.67 ~ 1.47) x 10^8 atom/g	soil	Kim et al., 2011a
India	(0.43~3.34) x 10^7 atoms/l	rain	Somayajulu et al., 1984
Global	(1.21 ±0.70) x 10^6 atoms/cm^2/yr	rain	Monagahan et al., 1985

Table 1. Production rate and concentration of ^{10}Be in the rain and soil.

2.1.2 ^{10}Be in land surface

Precipitation was collected for approximately one year during 1980 at seven localities in the continental U.S.A. (Monagahan et al., 1985). The ^{10}Be flux ranged from (1.38±0.36) x 10^6 to (3.96±0.35) x 10^6 atoms/cm^2/yr (Monagahan et al., 1985). In the case of Hawaii, ^{10}Be concentration ranges from 1.9 x 10^3 to 8.94 x 10^4 atoms/g in rain water. The mean ^{10}Be deposition rate in temperate latitude is determined to be 1.53 x 10^4 atoms/g in rain water (20% error). Monagahan et al., 1985 indicated that the concentration of ^{10}Be in surface soils and river sediments varies between 10^7 and 10^9 ^{10}Be atoms/g soil with the modal concentration of ^{10}Be lying between 4 x 10^8 and 6 x 10^8 atom/g. The relationship between annual rainfall and ^{10}Be deposition rate is plotted to be linearly proportional to each other (Maejima et al., 2005). This study shows that ^{10}Be fluxes (cm/day) for the two rain collection sites are relatively higher during a collection period of January 22 to April 22 than other collection period (Maejima et al., 2005).

Seasonal variations for ^7Be and ^{10}Be concentration in Tokyo and Hachijo-Island during a period of 2002 to 2003 were similar to each other. The peak value for ^7Be and ^{10}Be concentration appeared in April and October, respectively. Especially, in April when stratosphere-troposphere exchange occurs, peak values for the atomic ratio $^{10}Be/^7Be$ appeared. Low ^7Be and ^{10}Be concentrations and the atomic ratio of $^{10}Be/^7Be$ appeared in summer, July to August. Because the composition of the aerosol of Tokyo was almost same to the nearby soil, it is considered that Tokyo was strongly influenced by re-suspended soil contamination. Yamagata et al., 2005 indicated that using Al concentration in the aerosols, the enrichment of ^{10}Be concentration by re-suspended soil contamination was estimated to be about 30%.

In the case of Southern Hemisphere, Graham et al., 2003 demonstrated ^7Be and ^{10}Be fluxes at 36 to 45°S were determined to be (1.2~14) x 10^7 atoms/kg and (2.1~2.9) x 10^7 atoms/kg, respectively. These results are similar to those for rain sampled at mid-latitude sites across the USA from 1986 to 1994. The annual ^7Be and ^{10}Be flux rates are ~15 and ~27 x 10^9

atoms/m^2, respectively, at the northern sites of Leigh and Gracefield, and are significantly lower at ~9 and ~19 x 10^9 atoms/m^2, respectively, at the southern site of Denidin, because of the lower average rainfall there. Graham et al., 2003 indicated that ^{7}Be/^{10}Be in New Zealand ranged 0.47 to 0.61 and this is significantly lower than the ratio in USA (0.69~0.78). This is due to re-suspended dust to the primary atmospheric ^{10}Be in the rain sample in New Zealand. Interestingly, the ratio of ^{7}Be/^{10}Be at three sites are 0.70 (Leigh), 0.65 (Gracefield) and 0.50 (Dunedin). These results suggest an overall reduction in the ^{7}Be/^{10}Be ratio from north to south, due to increasing residence time for Be isotopes in the atmosphere above New Zealand. The mean residence time for ^{7}Be and ^{10}Be in the atmosphere above New Zealand range from 77 to 109 days and are lower in the summer than the winter due to transfer of older stratospheric air to the tropopause in late spring-early summer (Graham et al., 2003).

Maejiam et al., 2005 demonstrated that ^{10}Be concentrations of six soil samples on the raised coral reef terraces of Kikari Island, southwest Japan ranged from 0.80 to 7.17 x 10^9 atoms/g. The annual deposition rate of ^{10}Be from the atmosphere to Kikari Island from 2000 and 2002 ranged from 2.0 to 3.5 atoms/cm^2/y. The minimum absolute age was calculated from the inventory of meteoric ^{10}Be in the soil, and the annual deposition rates of ^{10}Be are ranged from 8 to 136 kyr (Maejima et al., 2005). A 36 cm of soil depth profile from the Roberts Massif, Antarctica was studied to obtain the age of soil by Graham et al., 1997. The sampling site is located in the edge of the nearby East Antarctic Ice Sheet at an altitude of 2700 m. This site is considered to have been ice-free for an extremely long period of time, of the order of several million years. The results of Graham et al., 1997 determined its minimum soil age of 12 million years which is much older than other ^{40}Ar/^{38}Ar dating result of 8 million years for volcanic deposit, Scoria associated with soils on the tills laid down by the Meserve Glacier, Antarctica.

2.1.3 ^{10}Be in the ocean

Using radiocarbon, the sedimentation rates during glacial periods and deglacial periods for the western Arctic Ocean were found to be 0.5 cm/kyr and 1-2 cm/kyr, respectively (Darby et al., 1997). A recent study shows that the concentration of ^{10}Be in the authigenic fraction of the sediment normalized to the total sediment mass is indirectly correlated to the oxygen isotope curve (McHargue and Donahue, 2005). For example, a low ^{10}Be/^{9}Be ratio in sediments would imply that terrestrial source of ^{9}Be has increased compared to the more oceanic ^{10}Be. Correlation of ^{10}Be with δ^{18}O recorded in marine sediment from the Blake Outer Ridge (DSDP site 72) shows a climatic effect on the ^{10}Be record in addition to cosmogenic effects. Age-corrected ^{10}Be variability in the sediment cores studied in Aldahan et al., 1997 and the oxygen isotope stratigraphy with the climatic stages numbered from 1 to 10, is generated from Aldahan et al., 1997. ^{10}Be from sediments of the Arctic Ocean covering the past 350 kyr shows well defined trends of Be isotopes coincident with interglacial/glacial climatic cycles and demonstrates that the sedimentation rates are higher during glacial periods and lower generally due to low sedimentation/accumulation rate during interglacial periods (Aldahan et al., 1997).

The ^{10}Be records of four sediment cores, taken along a transect from the Norwegian Sea via the Fram Strait to the Arctic Ocean, demonstrate that high ^{10}Be concentration are related to interglacial stages and that sediment sequences with low ^{10}Be concentration are related to glacial stages. This study confirms that the sharp contrast of high and low ^{10}Be concentrations at climatic stage boundaries are an independent proxy for climatic and sedimentary change, and can be applied for ^{10}Be stratigraphic dating of sediment cores (Eisenhauer, 1994).

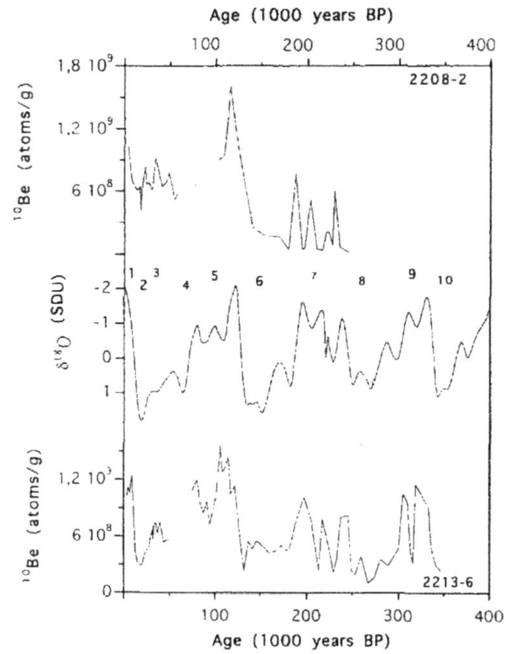

Fig. 1. Age corrected ^{10}Be as a function of age (Aldahan et al., 1997).

Also, Carcaillet et al., 2004 produced high resolution authigenic $^{10}Be/^{9}Be$ records over the last 300 kyr from sedimentary cores off the Portuguese coast. Comparison of $^{10}Be/^{9}Be$ and benthic $\delta^{18}O$ records from the two cores suggested that dipole moment lows may be associated with the end of interglacial episodes, and have a quasi-period of 100 kyr (Carcaillet et al., 2004). In a recent study, McHargue and Donahue, 2005 showed a strong correlation between ^{10}Be and oxygen isotope stages from the Blake Outer Ridge in the Atlantic Ocean. This relationship between climate and ^{10}Be deposition suggest that ^{10}Be could be used, in addition to, or as proxy for $\delta^{18}O$ in the studies of the climatic influences on marine sedimentation.

Other considerations are the carbonate flux in sediment which is strongly correlated to ^{10}Be flux, and carbonate-free sediments from which $\delta^{18}O$ is difficult to obtain from foraminifera. In addition, the two isotopes of beryllium, as stated above, are source dependent, thus the relationship of ^{10}Be to ^{9}Be in the sediments is a function of the relative contributions from atmospheric and terrestrial sources, and their mixing time in the sea.

Paleomagnetic intensity obtained from deep sea cores is well described in recent publications (Valet, 2001, Guyodo et al., 2000, Guyodo and Richter, 1999). However, it was found that climatic influence on ^{10}Be deposition can be significant, obscuring those variations from its production in the atmosphere, and thus must be addressed (Frank et al, 1997, Kok, 1999). Scavenging corrected ^{10}Be records compared to calculated variation of the global ^{10}Be production based on paleomagnetic intensity records (Christl et al., 2003) (Figure

2) show the variation of ^{10}Be flux in each location with general agreement of inversely proposal to the paleomagentic intensity from Mazaud et al., 1994, Yamazaki and Iokyr et al., 1994, and Guyodo and Valet, 1999.

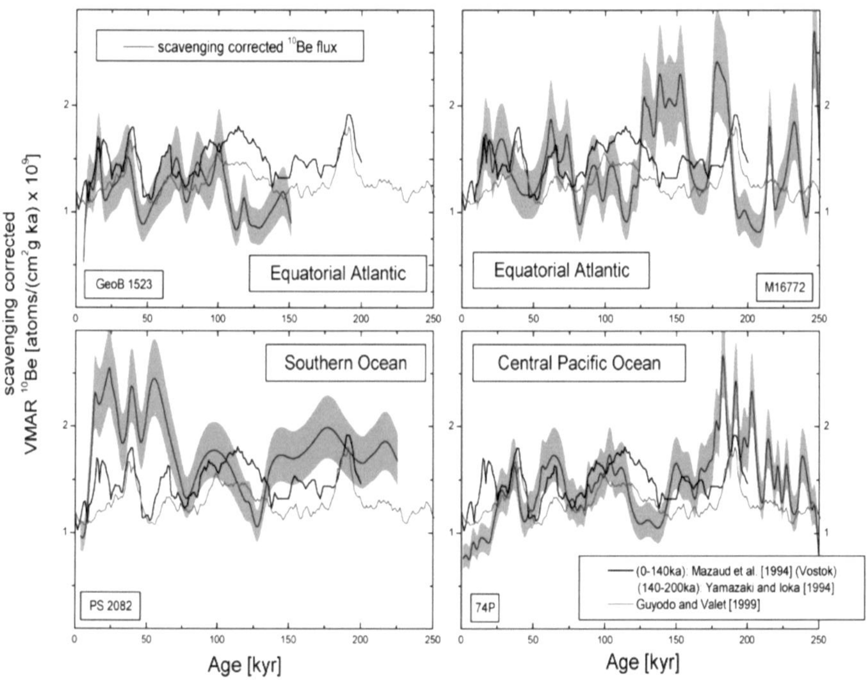

Fig. 2. Scavenging corrected ^{10}Be records compared to calculated variation of the global ^{10}Be production based on paleomagnetic intensity records (Christl et al., 2003).

2.2 ^{10}Be chemistry

Generally, to extract authigenic beryllium isotopes from sediments, the procedure of Bourles et al., 1989 is used. About one gram of sediment is leached in a solution of 25% acetic acid and hydroxlyamine-HCl to separate the "authigenic" fraction of the sediment from the "terrigenous" fraction. Most samples had more than 1 g of dry sediment; however, in the case of less than 1 g, two or three neighbouring samples were combined for the analysis. When ^{10}Be is normalized to the mass of the authigenic fraction, it should more accurately reflect its concentration in ocean water than ^{10}Be normalized to the total mass of the sediment (McHargue and Donahue, 2005, McHargue et al., 2010). This fraction is mostly composed of exchangeable ions, carbonates, and Fe-Mn hydroxides. Two aliquots of the leachate are prepared, one for the elemental analysis with ICP-MS/ICP-AES, and one for the preparation of AMS samples.

Figure 3 shows the flow chart of ^{10}Be chemistry to extract authigenic beryllium from sediment. This chemistry includes two steps of purification procedures using perchloric acid and nitric/hydrochloric acid. These steps are important to extract authigenic brylllium isotopes. Sometimes, this step is repeated to remove unwanted organic materials. When the unwanted organic materials are not completely removed, the residue sample is often difficult to dissolve in weak acidic solution for ICP analysis. This also causes a further problem in the step of Be separation using Na-EDTA. The concentration level for Be is mostly in ppb range; therefore, Be analysis was performed using ICP-MS. For AMS, the Be fraction is precipitated as $Be(OH)_2$ and combusted to BeO. $^{10}Be/^{9}Be$ ratios for chemical blank are found to be less than ~ 1×10^{-14} with 2 mg of ^{9}Be carrier. ^{9}Be and other elements can be measured by ICP-MS and ICP-AES.

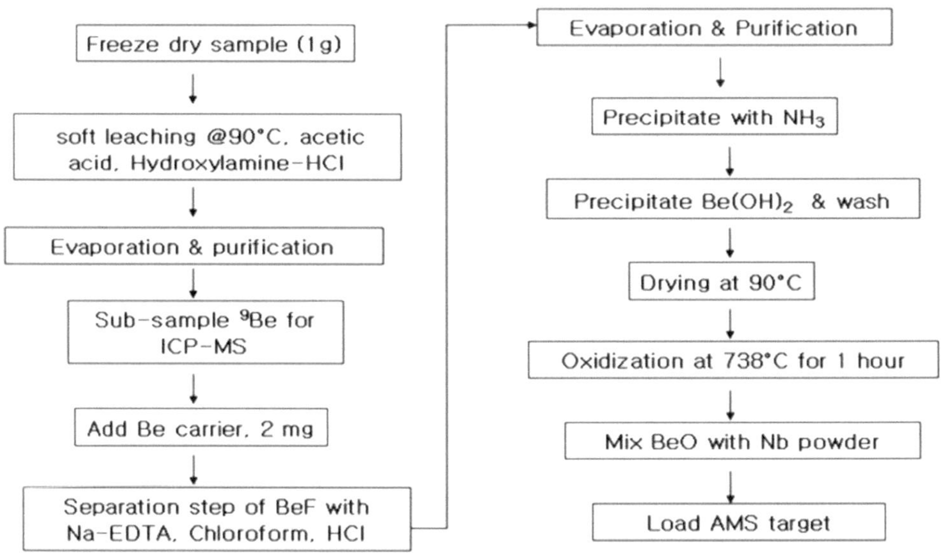

Fig. 3. Description of authigenic Be isotope extraction method.

2.3 The climate signal of ^{10}Be from nearby a continent

The signal of ^{10}Be from the East Sea in the Pacific Ocean and the Black Outer Ridge in the North Atlantic Ocean may give similar climatic influence because of its proximity to continents (Kim and Nam, 2010, McHargue et al., 2000). The depths of basins where sediment core collected were about 3,700 m and 3,818 m in the East Sea and Blake Outer Ridge, respectively. Because both cores were collected in the basin of the ocean, we might expect water circulation could be weaker. This could allow rather older waters can remain at the bottom of the basin. In this case, we may expect lower concentration of ^{10}Be in sediment compared to samples collected from other open seas (Bourlès et al., 1989, Knudsen et al., 2008). Also, the influence from the continent would be similar in both regions. The terrestrial origin of ^{10}Be over glacial/deglacial time period may similarly appear.

As shown in Figure 4, three locations were examined with respect to ^{10}Be concentration as a function of time and any related proxies for each site. Mostly, maximum ^{10}Be concentrations in various marine sediment samples appear to be above 1×10^{9} atoms/g sediment (Bourlès et al., 1989, Knudsen et al., 2008). Both the East Sea and Blake Outer Ridge, ^{10}Be concentrations are reached at 8×10^{8} atoms/g. This value is at least 30 percent lower than the most maximum value of ^{10}Be in each marine sediment core. Also, when ^{9}Be is investigated with ^{10}Be, ^{9}Be signal may be another indicator as a signal of sediment input from the land to the offshore. In the case of the study of the East Sea, the ^{9}Be values also show similar trend to those of ^{10}Be. his shows that both warmer periods of the Holocene and the Eemian,

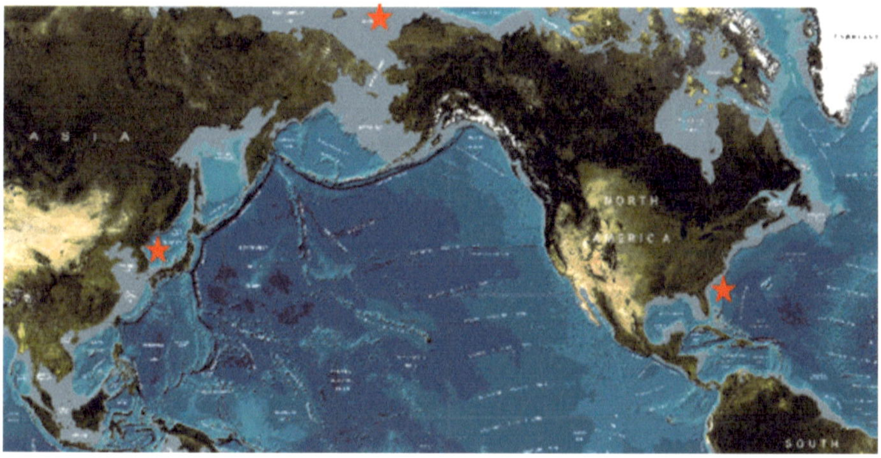

Fig. 4. Locations of the coring sites of the East Sea, Black Outer Ridge, and Mendeleev Ridge in the western Arctic Ocean.

wetter and warmer climate influenced ^{9}Be to be transported from the land to the ocean. ^{10}Be is transported from the both land and atmosphere. The signal of ^{10}Be/^{9}Be especially stands for lack of either lack of ^{9}Be transport or higher production rate of ^{10}Be, possibly associated with paleomagnetic intensity. In Figure 5, A, B, and C are associated with relatively higher ^{10}Be/^{9}Be compared to neighbouring ^{10}Be/^{9}Be values. A could be partially due to lowered value of ^{9}Be; B and C could be due to production rate of ^{10}Be. This study shows high production rate of Be at 15.5 and near 120 kyr. Also, lower ^{10}Be production rate is shown at 130.6 kyr during MIS 6 (Figure 5) (Kim and Nam, 2010). These three points can be identified easily with ^{10}Be/^{9}Be rations. In the Figure 5, the regions associated with climatic influence are clearly shown as the ^{10}Be/^{9}B ratios to be within the value between 2 and 3. This observation could be useful in future analysis. Based on Wagner et al., 2000, lower paleomagnetic intensities are associated with the ages at 1.5, 2.5, 4.0, and 6.5 kyr (Christl et al., 2003). Therefore, B could be likely involved in higher production rate of ^{10}Be at 1.5 kyr (Kim and Nam, 2010) (Figure 5).

Figure 6 shows the ^{10}Be concentration and M/Mo as determined from measured NRM/ARM of core CH88-10P with respect to depth and time. This figure shows that ^{10}Be concentration is inversely proportional to the relative paleomagnetic intensity. The production rate of ^{10}Be occurred at about 40 and 65 kyr. The peak values of ^{10}Be/^{9}Be reached at maximum at about 40 kyr. This time period is named as Laschamp paleomagnetic excursion where the expected ^{10}Be reaches at a maximum value. These paleomagnetic excursions are well compared with GRIP records.

A recent investigation on ^{10}Be and ^{9}Be from the Mendeleev ridge in the Arctic Ocean shows that ^{10}Be record at 75 kyr reveals production rate decrease evidently for at least 35 kyr of duration (Kim et al., 2011b). At this time the paleomagnetic intensity is found to be at maximum (Figure 2) (Christl et al., 2003; Flank et al., 1997). Interestingly, the values of magnetic susceptibility (Guyodo and Valet, 1996) obtained from a lake (Lac du Bouchet, France) are high as well as δ^{18}O. Also, ^{9}Be is relatively high which stands for a warm climate. The results of this study confirm that ^{10}Be reveals predominantly paleomagnetic features over the δ^{18}O at the extreme point of paleomagnetic intensity. This situation brings us to have precaution in a misuse of ^{10}Be as a climatic tracer. This study confirms the fact that the ^{10}Be record for climatic tracer, comparison with ^{9}Be is essential. When both beryllium isotopes behaves similarly, the pattern of Be can be used to determine whether the record of Beryllium isotopes is associated with colder or warmer climate based on their consistent concentration trend. The total authigenic ^{10}Be and ^{9}Be ($^{9}Be \gg {}^{10}Be$) can be referred as ^{9}Be because of their amount ratio in terrestrial environment. The ^{9}Be can be used as another climatic indicator like Sr, Ca, opal, TOC. The study at The Mendeleev Ridge confirms that ^{9}Be generally has a positive correlation with opal, TOC, $\delta^{13}C_{org}$ and negative correlation with $CaCO_3$ (Nam unpublished) (Figure 7). General trend of ^{9}Be clearly show the anti-correlation between ^{9}Be and Ca or Sr (Boulrès et al., 1989, Kim et al., 2011b). Therefore, we can conclude that ^{10}Be has a positive correlation with $\delta^{18}O$ which gives ^{10}Be to be used as a climate indicator, however, this is only true when ^{9}Be reveals similar climatic pattern with ^{10}Be. Both ^{10}Be and ^{9}Be show lower concentration at a cold/dry climate and higher concentration at a warm/wet climate period (Kim and Nam, 2010).

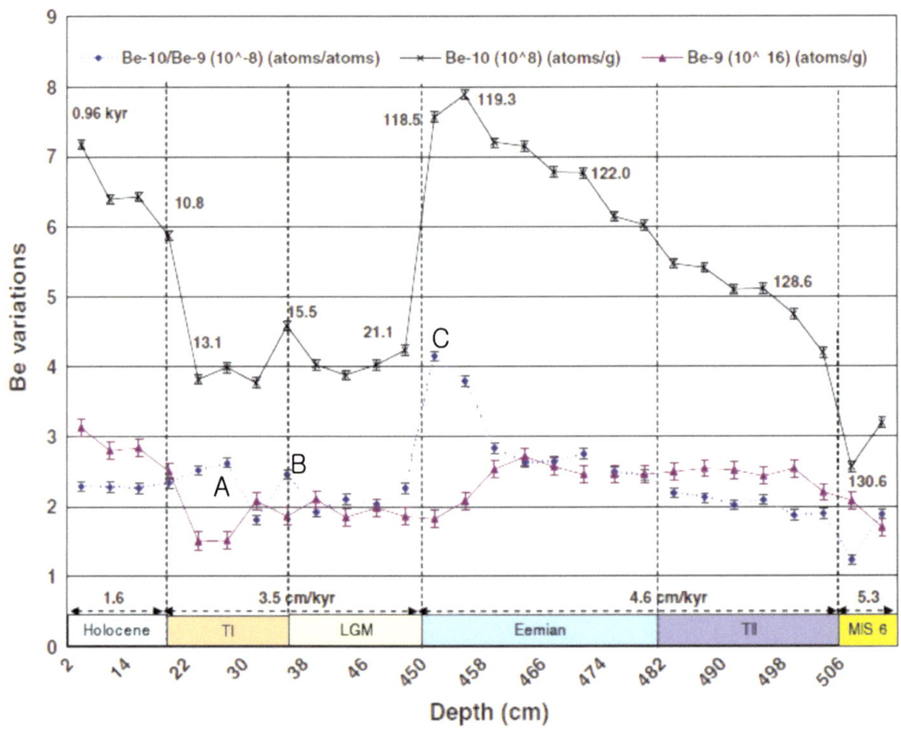

Fig. 5. Authigenic beryllium isotope records from the East Sea, Korea (Kim and Nam, 2010).

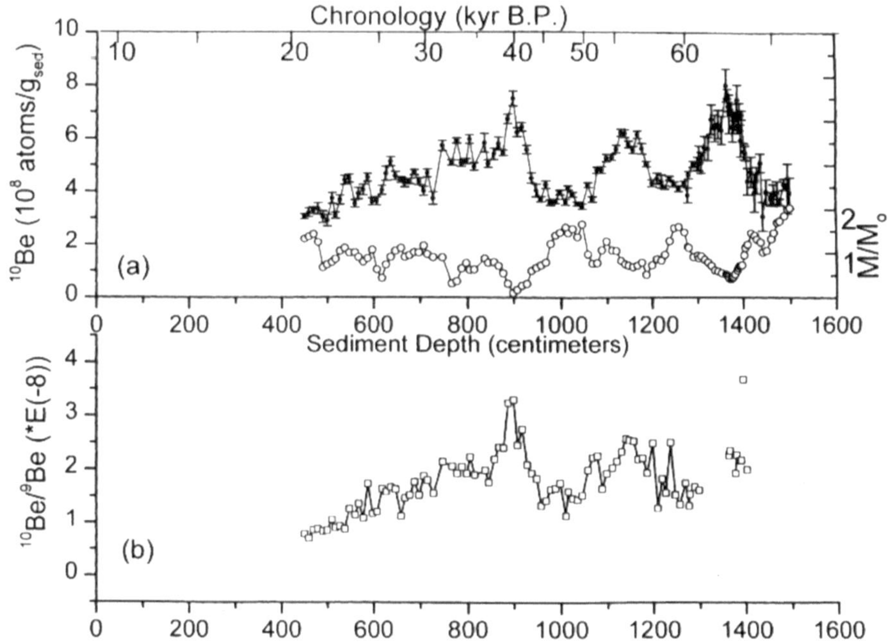

Fig. 6. The ^{10}Be concentration and M/Mo as determined from measured NRM/ARM (Schwartz et al., 1998) of core CH88-10P versus depth and time (McHargue et al., 2000).

2.4 Current problem and future research

A number of investigations show that there has been positive correlation between oxygen isotope and ^{10}Be concentration (Aldahan et al., 1997). Also, a positive correlation between oxygen isotopes and paleomagnetic intensity and also magnetic susceptibility is investigated (Carcaillet et al., 2004). During the Holocene, paleomagnetic intensity was gradually increased since the time of Laschamp excursion. This confirmed that production rate of 10Be at present is the lowest value since Laschamp excursion. However, the 10Be values recent years are higher than the 10Be concentration and the trend of 10Be is similar to that of δ18O value. This implies that 10Be is closely related to climatic and temperature variation. Because of this contradictory fact, confining the cause of climate change using nuclides which sun's activity related became important.

Although there have been a number of investigation on climate study using above parameters, obscurity in finding the cause of climate change is still remain. The relationship among production rate of 10Be, paleomagenetic intensity, and Sun-climate connection was studied (Sharma, 2002). This study estimated changes in 10Be production rate and the geomagnetic field intensity, variations in solar activity were calculated for the last 200 kyr., and confirms that the production of ^{10}Be in the Earth's atmosphere depends on the galactic cosmic ray influx that is affected by the solar surface magnetic activity and the geomagnetic dipole strength. However, large variations in the solar activity are evident. The marine δ^{18}O record and solar modulation are strongly correlated at the 100 kyr timescale. This proposes that variation in solar activity control the 100 kyr glacial-interglacial cycles. Sharma, 2002 suggested that the ^{10}Be production rate variations may have been under-estimated during the interval between 115 kyr and 125 kyr, and may have biased the results (Sharma, 2002).

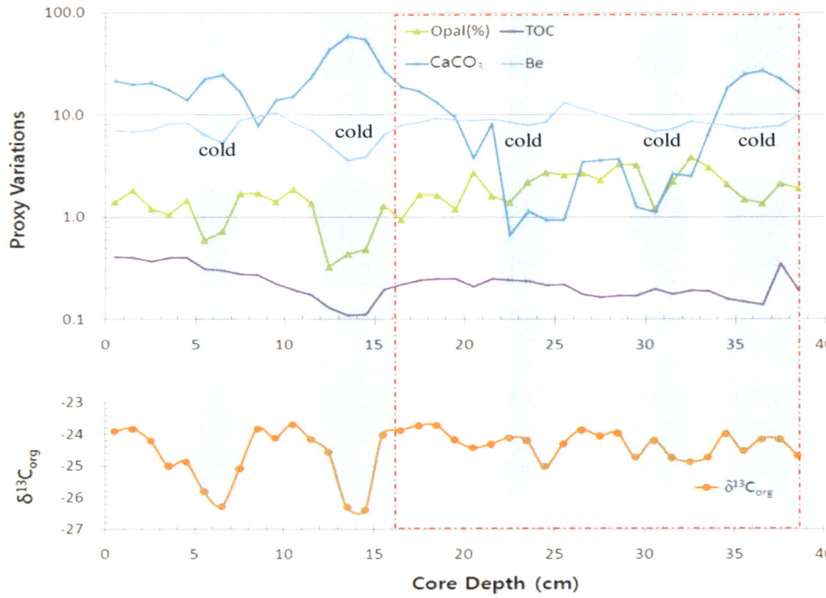

Fig. 7. Multi-proxy record from the core (PS72/396-3) from Mendeleev Ridge, the Arctic Ocean (Kim et al., 2011b).

Usoskin et al., 2004 indicated that the reconstructed sunspot record exhibits a prominent period of about 600 years, in agreement with earlier observations based on cosmogenic isotopes. Also, there is evidence for the century scale Gleissberg cycle and a number of shorter quasi-periodicities whose periods seem to fluctuate on millennium time scale. This invalidates the earlier extrapolation of multi-harmonic representation of sunspot activity over extended time intervals and the present high level of sunspot activity is unprecedented on the millennium time scale (Usoskin et al., 2004). Accepting solar forcing of Holocene and galacial climatic shift implies that the climate system is far more sensitive to small variation in solar activity than generally believed. In order to fully understand how sensitive climate really is for variations in solar activity, we need to look for additional evidence and to quantify such evidence, both in paleorecords and in observations of present climate with models to estimate climate change in the future (Geel et al., 1999).

3. Conclusions

As a climate indicator, ^{10}Be has been frequently investigated because of its property associated with rainfall, dust fallout and its production mechanism in the atmosphere by cosmic-rays. Similar patterns of ^{10}Be and δ^{18}O, magnetic susceptibility records show climatic influence, however, ^{10}Be is incorporated with the production rate which is inversely proportional to the paleomagenetic intensity. The cyclic orbital forcing effect toward ^{10}Be, δ^{18}O, and paleomagenetic intensity are connected, ^{10}Be signal is clearly mixed with climatic component and earth's paleo magnetic strength. Scrutinizing authigenic ^{10}Be with ^{9}Be, a region either climatic or production related zone can be evidently identified by looking at the ^{10}Be/^{9}Be ratios. Investigation of ^{9}Be is another useful tracer to examine climatic influence of marine environments with other multi-proxies which have positive or anti-correlated with ^{9}Be. Marine environments like the East Sea of Korea, Blake Outer Ridge, and Mendeleev Ridge are associated with significant terrigeous input during deglacial periods. Examining both beryllium isotopes together with other multi-proxy stratigraphy will provide understanding the pattern of environmental change at various glacial/interglacial events much more evidently.

4. Acknowledgment

This study is partially supported by Korean-IODP at Korea Institute of Geoscience and Mineral Resources and also funded by K-Polar (PP11070) at Korea Polar Research Institute.

5. References

Aldahan, A.; Ning, S.; Possnert, G.; Backman, J. & Bostrom, K. (1997) ^{10}Be records from sediments of the Arctic Ocean covering the past 350 kyr. *Marine Geology* 144, 147-162.

Brown, L.; Stensland, G.; Klein, J., Middleton, R. (1989) Atmospheric deposition of ^{7}Be and ^{10}Be, *Geochimica et Cosmochimica Acta* 53, 135-142.

Bourlès, D.; Raisbeck, G. M. & Yiou, F. (1989) ^{10}Be and ^{9}Be in marine sediments and their potential for dating, *Geochimica et Cosmochimica Acta* 53, 443-452.

Carcaillet, J.; Bourlès, D.; Thouveny, N. & Arnold, M. (2004) A high resolution authigenic ^{10}Be/^{9}Be record of geomagnetic moment variation over the last 300 kyr from sedimentary cores of the Portuguese margin, *Earth and Planetary Science Letters* 219, 397-412.

Castagnoli, G. C.; Cane, D.; Taricco, C. & Bhandari, N. (2003) GCR flux decline during the last three centries : Extra terrestrial and terrestrial evidences, *28th Int. Cosmic Ray Conference,* 4045-4048.

Champion, D. E. (1980) Holocene geomagnetic secular variation in the western United States: implications for the global geomagnetic field. *U.S. Geological Survey Open-file Report,* 80-824 , 1-314.

Christl, M.; Strobl, C. & Mangini, A. (2003) Beryllium-10 in deep-sea sediments; a racer for the Earth's magnetic field intensity during the last 200,000 years, *Quaternary Science Reviews* 22, 725-739.

Darby, D. A.; Bischof, J. F. & Jones, G. A. (1997) Radiocarbon chronology of depositional regimes in the western Arctic Ocean, *Deep-Sea Research II* 44, 1745-1757.

Desilets, D.; Zreda, M. & Ferre,´ T. (2007) Scientist water equivalent measured with cosmic rays at 2006 AGU Fall Meeting, *EOS Transactions AGU,* 88(48), 521, doi:10.1029/2007EO480001.

Dunai, T. J. (2000) Scaling factors for production rates of in situ produced cosmogenic nuclides: a critical reevaluation, *Earth and Planetary Science Letters* 176, 157-169.

Eisenhauer, A. (1994) ^{10}Be records of sediment cores from high northern latitudes: Implications for environmental and climatic changes, *Earth and Planetary Science Letters* 124, 171-184.

Frank, M.; Shwarz, B.; Baumann, S.; Kubik, P. W.; Suter, M. & Mangini, A. (1997) A 200 kyr record of cosmogenic radionuclide production rate and geomagnetic field intensity

from ^{10}Be in globally stacked deep-sea sediments, *Earth and Planetary Science Letters* 149, 121-129.
Frank, M. (2000) Comparison of cosmogenic radionuclide production and geomagnetic field intensity over the last 2000,000 years, The *Philosophical Transactions of the Royal Society, London* A 358 , 1089-1107.
Geel., B. van; Raspopov, O. M.; Renssen, H.; Plicht, J. van der; Dergachev, V. A. & Meijer, H. A. J. (1999) The role of solar forcing upon climate change, *Quaternary Science Reviews* 18, 331-338.
Goess, J. C. & Phillips, F. M. (2001) Terrestrial in situ cosmogenic nuclides: theory and application. *Quaternary Science Reviews, 20*, 1475-1560.
Graham, I. J.; Barry, B. J.; Hemmingsen, I. D.; Hutchinson, E. F.; Pohl, P. K. & Zondervan, A. (2005) *GNS science report*, 2005/02, 43 p., Institute of Geological and Nuclear Sciences, Lower Hutt, New Zealand.
Graham, I. J.; Barry, B. J.; Ditchburn, R. G. & Zondervan, A. (2005) *GNS science report* 2005/03, 29 p.
Graham, I. J.; Barry, B. J.; Ditchburn, R. G.; Shea, M. A.; Smart, D. F. & Whitehead, N. E. (2005) *GNS science report* 2005/04, 32 p.
Graham, I. J.; Barry, B. J.; Ditchburn, R. G. & Whitehead, N. E. (2000) *Nuclear Instruments and Methods B* 172, 802-805.
Graham, I. J.; Ditchburn, R. G. & Whitehead, N. E. (1998) ^{10}Be spikes in Plio-Pleistocene cyclothems, Wanganui Basin, New Zealand: identification of the local flooding surface (LFS), *Sediment Geology*, 122, 193-215.
Graham, I.; Ditchburn, R.; & Barry, B. (2003) Atmospheric deposition of ^{7}Be and ^{10}Be in New Zealand rain (1996-98), *Geochimica et Cosmochimica Acta,* 67 (3), 361-373.
Graham, I. J.; Ditchburn, R. G.; Sparks, R. J. & Whitehead, N. E. (1997) ^{10}Be investigations of sediments, soil, and loess at GNS, *Nuclear Instruments and Methods B*123, 307-318.
Guyodo, Y. & Valet, J.-P. (1996) Relative variations in geomagnetic intensity from sedimentary rcords: the past 200,000 years, *Earth and Planetary Science Letters* 143, 23-36.
Guyodo, Y.; Gaillot, P. & Channell, J. E. T (2000) Wavelet analysis of relative geomagnetic paleointensity at ODP Site 983, *Earth and Planetary Science Letters* 184, 109-123.
Guyodo, Y. & Richter, C. (1999) Paleointensity record from Pleistocene sediments (1.4-0 Ma) off the California Margin, *Journal of Geophysical Research,* 104, B10, 22,953-22,964.
Guyodo, Y. & Valet, J.-P. (2010) Relative variations in geomagnetic intensity from sedimentary records:the past 200,000 years, *Earth and Planet ary Science Letters,* 143, 23-36.
Heikkilla, U.; Beer, J. & Feichter, J. (2009) Meridional transport and deposition of atmospheric ^{10}Be, atmos. *Chemical Physics,* 9, 515-527.
Kim, K. J.; Zhou, L. P., Kim, J. H.; Kim, J. Y.; Park, Y. A. (2011a) Dating of aeolian sand deposits in Korea using OSL and Be-10, *the 12th International Conference on Accelerator Mass Spectrometry*, March 20-25, Wellington, New Zealand.
Kim, K. J.; Nam, S.-I.; Stein, R.; Mattiessen, J. (2011b) Glacial history and paleo-oceanographic changes of the western Arctic Ocean (Mendeleev Ridge) using beryllium isotopes , *Arctic Science Summit Workshop*, March 29-31, 2011, Seoul, Korea.

Kim, K. J. & Nam, S.-I. (2010) Climatic signals from the ^{10}Be records of the Korean marine sediments, *Nuclear Instruments and Methods*, B 268, 1248-1252.

Kim, K. J.; Jull, A. J. T.; Kim, J.-H.; Matsuzaki, H.; Ohlendorf, C. & Zolitschka, B. (2010) Tracing Environmental Change of Potrok Aike, Argentina using Beryllium Isotopes, *Geophysical Research Abstracts Vol. 12, EGU2010-3788-1*, 2010 EGU General Assembly.

Kok, Y. S. (1999) Climatic influence in NRM and Be-derived geomagnetic paleointensity data, *Earth and Planetary Science Letters*, 166, 105-119.

Knudsen, M. F.; Henderson, G. M., Frank, M.; Niocaill, C. M. & Kubik, P. W. (2008) In-phase anomalies in Beryllium-10 production and palaeomagnetic field behaviour during the Iceland Basin geomagnetic excursion, *Earth and Planetary Science Letters* 265, 588-599.

Laj, C.; Kissel, C.; Mazaud, A.; Channell, J. E. T. & Beer, J. (2000) North Atlantic palaeointensity stack since 75 kyr (NAPIS-75) and the duration of the Laschamp event, *Philosophical Transactions of Royal Society, London, A* 358 , 1009-1025.

Laj, C. (1996) Alain Mazaud and jean-Claude Duplessy, Geomagnetic intensity and ^{14}C abundance in the atmosphere and ocean during the past 50 kyr, *Geophysical Res. Lett.*, vol 23, No. 16 , 2045-2048.

Lal, D. & Peters, B. (1967) Cosmic Ray Produced Radioactivity on the Earth, *Handbuck der Physik*, XLVI/2 , 551.

Lal, D. (1988) In situ-produced cosmogenic isotopes in terrestrial rocks, *Annual Reviews of Earth and Planetary Science*, 16, 355-388.

Lal, D. (1991) Cosmic ray labeling of erosion surfaces: in situ nuclide production rates and erosion models, *Earth and Planetary Science Letters* 104, 424-439.

Lifton, N. A.; Jull, A. J. T. & Quade, J. (2001) A new extraction technique and production rate estimate for in situ cosmogenic ^{14}C in quartz, *Geochimica et Cosmochimica Acta*, 65, 1953-1969.

Horiuchi, K.; Kobayashi, K.; Oda, T.; Nakyrmura, T.; Fujimura, C.; Matsuzaki, H. & Shibata, Y. (2000) Climate-induced fluctuations of ^{10}Be concentration in Lake Baikyrl sediments, *Nuclear Instruments and Methods in Physics Research* B172, 562-567.

Horiuchi, K.; Goldberg, E.L.; Kobayashi, K.; Oda, T.; Nakyrmura, T. & Kyrwai, T. (2001) Climate-induced variations of cosmogenic beryllium-10 in the sediments of Lake Baikyrl of the last 150 ky from AMS, SRXRF and NAA data, *Nuclear Instruments and Methods in Physics Research. A*470, 396-404

Mazaud, A.; Laj, C. & Bender, M. (1994) A geomagnetic chronology for Antarctic ice accumulation. *Geophysical Research Letters* 21, 337–340.

McHargue, L. & Donahue, D. J. (2005) Effects of climate and the cosmic-ray flux on the ^{10}Be content of marine sediments, *Earth and Planetary Science Letters*, 232, 193-207.

McHargue, L. R.; Donahue, D.; Damon, P. E.; Sonett, C. P.; Biddulp, D. & Burr, G. (2000) Geomagnetic modulation of the late Pleistocene cosmic-ray flux as determined by ^{10}Be from Blake Outer Ridge marine sediments, *Nuclear Instruments and Methods in Physics Research B* 172, 555-561.

McHargue, L.; Jull, A. J. T. & Cohen, A. (2010) Measurement of ^{10}Be from Lake Malawi (Africa) drill core sediments and implications for geochronology. *Palaeogeography, Palaeoclimatology, Palaeoecology*, Doi. 10.1016/j.palaeo.2010.02.012.

Masarik J. & Beer, J. (1999) Simulation of Particle fluxes and Cosmogenic Nuclide Production in the Earth's Atmosphere. *Journal of Geophysics Research*, 104, 12099 (1999).

Masarik, J.; Frank, M.; Shafter, J. M. & Wieler, R. (2001) Correction of in situ cosmogenic nuclide production rates for geomagnetic field intensity variations during the past 800,000 years, *Geochimica et Cosmochimica Acta* 65 (17), 299-3003.

Maejima, Y.; Matsuzaki, H. & Higashi, T. (2005) Application of cosmogenic ^{10}Be to dating soils on the raised coral reef terraces of Kikyri Island, southwest Japan, *Geoderma,* 126 389-399.

Monaghan, M. C.; Krishnaswami, S. & Turekian, K. K. (1985/86) The global-average production rate of ^{10}Be, *Earth and Planetary Science Letters,* 76 (1985/86) 279-287.

O'Brien, K. A. (1979) Secular variations in the production of cosmogenic isotopes in the earth's atmosphere. *Journal, of Geophysical Research,* 84, 423-31.

Ohno, M. & Hamano, Y., (1992) Geomagnetic poles over the past 10,000 years. *Geophysical Reseach Letters,* 19, 1715-18.

Peters, B. (1955) Radioactive beryllium in the atmosphere and on the earth, *Proceedings of Indian Academy of Sciences,* 41, 67-71.

Sharma, P. & Middleton, R. (1989) Radiogenic production of ^{10}Be and ^{26}Al in uranium and thorium ores: Implications for studying terrestrial samples containing low levels of ^{10}Be and ^{26}Al, *Geochimica et Cosmochimica Acta,* 53 , 709-716.

Sharma, P. & Somayajulu, B. L. K (1982) ^{10}Be dating of large manganese nodules from world oceans, *Earth and Planetary Science Letters,* 59, 235-244.

Simpson, J. & Uretz, A. R. B. (1953) Cosmic-Ray Neutron Production in Elements as a Function of Latitude and Altitude, *Physics Review,* 90, No. 1 , 44-50.

Sharma, M. (2002) Variations in solar magnetic activity during the last 200,000 years: is there a Sun-climate connection? *Earth and Planetary Science Letters* 199, 459-472.

Somayajulu, B., L., K.; Sharma, P.; Beer, J.; Bonani, G.; Hofmann, H.-J.; Morezoni, e., Nessi, M., Suter, M. & Wolfli, W. (1984) ^{10}Be annual fallout in rains in India, *Nuclear Instruments and methods in Physics Research* B5, 398-403.

Stone, J. O. (2000) Air pressure and cosmogenic isotope production. *Journal of Geophysics Research,* 105, 23,753-23,759.

Strack, E.; Heisinger, B. B.; Dockhom, B.; Hartmann, F. J.; Korshinek, G. & Noler, E. (1994) Determination of erosion rates with cosmogenic ^{26}Al, *Nuclear Instruments and Methods,* B 92, 317-320.

Usoskin, I. G.; Mursular, K.; Solanki, S.; Schussler, M. & Alanko, K. (2004) Reconstruction of solar activity for the last millennium using ^{10}Be data, *Astronomy and Astrophysics,* 413, 745-751.

Valet, J.-P. (2001) Time Variations in Geomagnetic Intensity, *Reviews of Geophysics*, 41/1, 4-44.

Yamagata, T.; Saito, T.; Nagai, H. & Matsuzaki, H. (2005) Seasonal variation for ^{7}Be and ^{10}Be concentrations in the atmosphere at Tokyo and Hachijo-Island during the period of 2002 and 2003, *The 10th International Conference on Accelerator Mass Spectrometery,* Sept. 5-10, Berkeley, CA. USA.

Yamazaki, T. & Iokyr, N. (1994) Long-term secular variation of the geomagnetic field during the last 200 kyr recorded in sediment cores from the western equatorial Pacific. *Earth and Planetary Science Letters,* 128: 527–544.

10

Drought Analysis Based on SPI and SAD Curve for the Korean Peninsula Considering Climate Change

Minsoo Kyoung[1], Jaewon Kwak[2], Duckgil Kim[2], Hungsoo Kim[2] and Vijay P. Singh[3]
[1]*Samsung Loss Control Center, Samsung Fire & Marine Insurance. CO. Seoul*
[2]*Department of Civil Engineering, Inha University, Incheon*
[3]*Department of Biological and Agricultural Engineering Texas A & M University, College Station, TX 77843*
[1,2]*South Korea*
[3]*USA*

1. Introduction

In recent years, Korea has been experiencing serious drought and water scarcity problems. Korea is classified as a water-deficient country by the United Nations(UN). These problems are further compounded by the rapidly growing population, especially in the urban areas. Attaining water security is one of the major concerns and top priorities of the Korean government. Further complicating these issues is the global climate change, which is currently at the forefront of scientific research. With the projected global temperature increase due to increases in greenhouse gas emissions, scientists generally agree that the global hydrological cycle will intensify and suggest that extremes (e.g. droughts, floods) will become more common. Therefore, one major concern arising from climate change is its potential effects on water resources in terms of (increases in) droughts, and its impacts on different health, environmental, economic, and social sectors. Changes in the frequency and magnitude of droughts will have enormous impacts on water management, agriculture, and aquatic ecosystems. According to Korea Water Resource Association (KWRA), material damage recorded was 522 million USD in 1967, 584 million USD in 1968, 216 million USD in 1981, and 287 million USD in 1982 (http://kwra.or.kr/news/en_04.html). Researches on drought have, however, lagged behind in terms of quality and quantity.

This paper addresses the assessment of the drought characteristics in the Korean Peninsula considering climate change. Researches on temporal and spatial characteristics of drought are necessary to evaluate potential impact of drought and to carry out rational management of water resources. The most common method for interpreting drought relies on the drought index, which considers the impacts of drought severity, frequency, its affecting area, and duration. It is because the drought index explains the current drought situation by meteorological factors such as precipitation and past meteorological conditions close to that of the present.

Henriques and Santos (1999) proposed the use of thiessen method to produce SAF (Severity-Area-Frequency) curve. Hisdal and Tallaksen (2003), and Mishra and Singh (2009) also relied on SAF curve to analyze droughts. Furthermore, Kim et al. (2002) took advantage of geostatic technique used by Matheron (1963) to generate Intensity-Areal extent-Frequency curve. In one instance, SAD (Severity-Area-Duration) curves, in which the frequency in SAF curve is replaced by duration, were used to analyze the relationship between droughts and their areal extent (Andreadis et al., 2005). Precipitation data necessary for computing the SPI were obtained from a 32-years (1973-2004) dataset that originates from 58 stations operated by the Korea Meteorological Administration (KMA) and 130-years(1970~2099) dataset that was downscaled from GCM data to 58 stations of KMA. To produce SAD curves, it is necessary to expand SPI values for each of precipitation stations into the affected area. To do so, EOF analysis was carried out to contract data into core spatial information. Then, kriging method was used for spatial expansion. Climate change effect on drought was assessed by comparing SAD curves which were constructed from observed and downscaled precipitation respectively.

2. Climate change scenario for South Korea

2.1 Review of GCMs

Since the General Circulation Model (GCM) typically simulates the earth by putting a grid on it, peninsula states like Korea and some island countries tend to be depicted as part of sea, depending on location or the size of a grid. In this regard, reviewing whether Korea is described as land or not can allow this study to find an appropriate model for Korea. The reason is as follows: Since the GCM has disparate simulation models for land and sea, and once a land is presented as a sea in the simulation, consequent meteorological variables can present far distant results from reality. Figure 1 shows how GCMs differently suggested by countries describe the Korean peninsula.

As showed in Figure 1, the Korean Peninsula is simulated as a land in a few models, such as MIHR(Japan), CNCM3(France), and FGOALS(China) and other GCMs depict only some part of Korea as a land. In this study, therefore, such GCMs are excluded from reviews as they are assumed not suitable for the evaluation of climate change impacts on the Korean peninsula.

GCMs are known for their relatively precise simulation of temperatures. This study, thus, selected a CGM that properly presents the temperatures of the Koreans peninsula. The selection was done by comparing the grid-type temperature data of the central part of Korea, which is provided by the IPCC DDC, with the temperature data observed at Seoul weather station.

Among the simulation results from each GCM of 20c3m which simulates the 1990s, the statistics of monthly mean temperature (mean, variance, and distortion) of the center of the Korean peninsula was compared with that observed at Seoul station. Since 20c3m generally simulates the time period from 1985 to 1999, the statistics of monthly mean temperatures from 1961 to 1999 were firstly calculated and compared by taking into account the observation period. Several GCM models in Figure 1, such as MIHR, CNCM3, FGOALS which assume Korea as land, and other models, which describe some part of Korea as sea, were additionally reviewed for comparison as well.

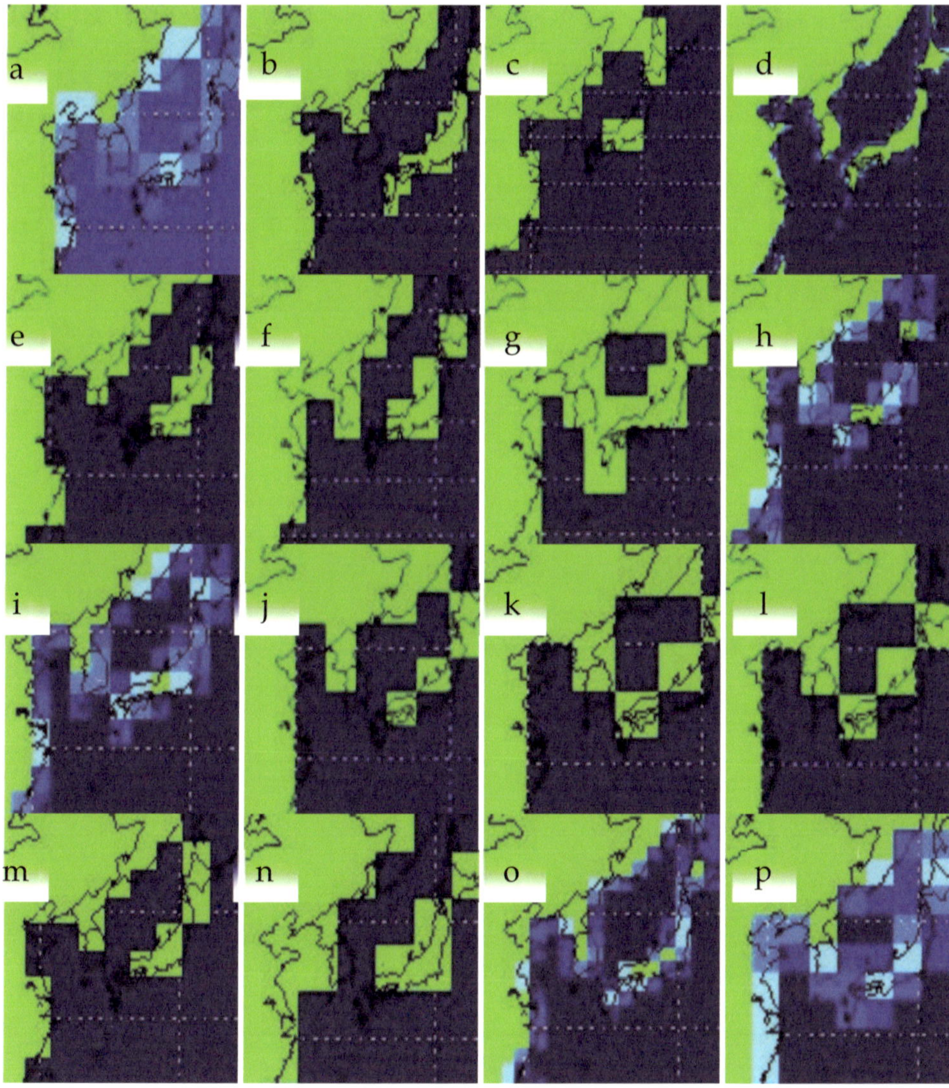

Fig. 1. GCMs grid around the Korea peninsular (a-BCM2, b-CGHR, c-CGMR, **d-MIHR**, e-CSMK3, **f-CNCM3**, **g-FGOALS**, h-CFCM20, i-CFCM21, j-GIAOM, k-GIEH, l-GIER, m-HADCM3, n-HADGEM, o-INCM3, p-IPCM4)

Overall, three models - MIHR, CNCM3, and FGOALS -, which assume Korea as land, simulate monthly mean temperatures of Korea relatively well. In the case of FGOALS, proposed meteorological variables have small variances. Thus, considering the nature of SK-NN method that detects and downscales the pattern of similar meteorological variables, it was expected that a number of locations would have several same precipitations. Therefore, this study had to eliminate FGOALS. Among CNCM3 and MIHR left, CNCM3, which was suggested as the most appropriate model for Korea in Kyoung (2010), was chosen, as the GCM for the analysis of drought on the Korean Peninsula.

	Jan	Feb	Mar	Apr	May	Jun	Jul	Aug	Sep	Oct	Nov	Dec
Seoul	-3.0	-0.7	5.0	11.6	18.0	21.0	25.6	25.5	20.8	14.3	6.8	-0.1
CNCM3	-3.3	-2.5	2.4	7.7	13.3	17.8	20.9	21.8	18.7	12.5	5.9	0.1
FGOALS	-7.0	-4.4	1.3	8.2	14.5	18.8	21.3	22.2	18.0	10.7	2.4	-4.0
MIHR	-3.3	-2.1	3.3	10.3	17.1	21.6	24.2	24.3	20.0	12.7	5.2	-0.4
GIAOM	18.0	18.6	20.0	22.0	24.2	25.8	26.4	26.3	25.0	23.0	20.7	19.5
GIEH	13.3	13.9	16.2	18.9	21.9	24.5	26.5	27.0	26.0	23.2	20.9	16.5
GIER	12.8	13.0	15.2	17.1	19.5	22.3	24.8	25.9	24.8	22.1	18.6	15.2
INCM3	4.4	5.8	10.3	15.5	20.8	24.8	26.8	26.8	24.5	19.3	12.8	6.6
IPCM4	14.9	14.2	15.6	18.5	21.9	24.9	26.7	27.1	26.1	23.6	20.6	17.4
MRCGCM	-1.6	-0.1	4.6	9.5	14.3	18.7	21.6	21.7	18.2	12.6	6.4	0.9

Table 1. Comparison of monthly mean temperature between observation and simulation from GCMs

	Jan	Feb	Mar	Apr	May	Jun	Jul	Aug	Sep	Oct	Nov	Dec
Seoul	5.1	3.6	2.4	1.7	0.8	0.7	1.6	1.2	0.7	1.2	2.3	3.5
CNCM3	5.2	3.5	2.4	1.0	1.3	1.4	0.7	0.7	0.8	0.7	1.1	1.3
FGOALS	4.9	5.7	4.4	1.6	0.5	0.5	0.3	0.3	0.6	1.9	3.2	4.5
MIHR	2.5	1.7	1.3	0.8	0.5	0.6	0.5	0.5	0.5	0.6	1.1	1.3
GIAOM	0.6	0.8	1.1	1.5	1.2	0.4	0.1	0.2	0.9	1.5	1.7	0.4
GIEH	1.9	1.9	1.5	0.6	0.5	0.5	0.2	0.2	0.4	0.5	1.0	1.4
GIER	1.7	1.9	0.8	0.4	0.3	0.4	0.3	0.3	0.5	0.9	1.2	0.7
INCM3	4.4	6.0	3.8	1.6	1.3	0.4	0.2	0.2	0.5	1.0	1.5	3.6
IPCM4	1.0	1.0	1.5	1.0	0.7	0.5	0.3	0.3	0.3	0.6	0.7	0.9
MRCGCM	1.5	1.1	1.1	0.8	0.4	0.3	0.4	0.7	0.7	1.0	1.5	1.0

Table 2. Comparison of variance of monthly mean temperature between observation and simulation from GCMs

	Jan	Feb	Mar	Apr	May	Jun	Jul	Aug	Sep	Oct	Nov	Dec
Seoul	-0.4	0.2	-0.6	0.9	0.2	0.0	0.7	-0.5	0.8	0.5	-0.1	-0.7
CNCM3	-2.2	-0.6	0.1	-0.6	-0.9	0.0	-1.4	0.1	0.4	-0.2	1.0	-0.1
FGOALS	-0.3	0.0	0.2	-0.4	0.0	-0.2	-0.6	-0.4	0.3	0.6	-0.5	-0.2
MIHR	-0.4	0.3	-0.8	0.1	0.3	-0.4	0.9	-0.5	0.5	0.3	0.7	-0.8
GIAOM	-0.4	-0.5	-0.2	0.0	-0.1	-0.3	-0.1	-0.5	-0.2	0.0	-0.2	-0.9
GIEH	-0.2	-0.1	-0.2	-0.4	-0.1	-0.1	0.2	0.1	0.2	0.2	-0.3	0.4
GIER	-0.3	0.4	-0.3	0.3	0.2	0.2	0.2	0.1	-0.2	-0.6	0.2	0.4
INCM3	0.2	-0.3	-0.2	-0.3	-0.1	-0.5	0.0	0.2	0.5	-0.4	0.0	-0.6
IPCM4	0.1	0.4	0.4	0.3	0.6	-0.2	-0.4	-0.9	-0.4	0.5	-0.2	-0.3
MRCGCM	-0.2	0.4	0.3	-0.3	-0.6	0.4	0.8	0.5	0.1	-0.1	-0.2	-0.1

Table 3. Comparison of skewness of monthly mean temperature between observation and simulation from GCMs

2.2 Application of downscaling technique

1) Selecting a climate change scenario

In this study, the SRES A1B scenario was selected as a socioeconomic scenario to predict the precipitation changes in Korea. This is because this study recognized the A1B scenario as the best match for circumstances under which the entire world is exerting great efforts to maximize the efficiency of energy resources and find alternative energy sources. And this study also examined how drought and probability precipitation in Korea can change, should the A1B scenario be considered on the basis of the 20c3m scenario used as the initial conditions of the SRES scenario.

In order to predict the impact of climate change on Korea in the short, medium, and long term, the time periods were divided as the followings:

CASE 1 (1980s) : 1970 ~ 1999 (20c3m, Reference duration)
CASE 2 (2020s) : 2010 ~ 2039 (A1B, Projection duration)
CASE 3 (2050s) : 2040 ~ 2069 (A1B, Projection duration)
CASE 4 (2080s) : 2070 ~ 2099 (A1B, Projection duration)

2) Adopting a spatial downscaling technique using the SK-NN method

A general K-NN technique is mainly used for short term prediction after estimating parameters($\widehat{\alpha_1}, \widehat{\alpha_2}, \ldots \ldots \widehat{\alpha_\alpha}$) necessary for prediction by considering delay time(τ) and embedding dimensions(m) in a single time series. This K-NN method, which was proposed by Casdagli (1992), and Casdaglli and Weigend (1994), is a method primarily applied to Chaos time-series under the assumption that past time-series patterns would be repeated in the future after converting the time series data of a single variable into a set of vectors with the consideration of its delay time and embedding dimension. This method was also used in studies of Young(1994), Lall and Sharma(1996), Lall et al.(1996), Rajagopalan and Lall(1999), Buishand and Brandsma(2001), Yates et al. (2003), Sharif and Burn(2006).

This study downscaled data from the meteorological variables provided monthly to monthly total precipitations observed by stations by applying the SK-NN (Simplest K-Nearest Neighbor) method, a simplest method of K-NN. This was based on the assumption that under the condition of specific hydrometeorological variables, the same precipitation can occur. Therefore, vectors were formed by using temperature, specific humidity, wind speed and sea level pressure which affect precipitation not using a single time series. Delay time was excluded, because it is insignificant, given that monthly mean by averaging original time series data is used. Embedding dimension was assumed to have four dimensions of meteorological variables - temperature, specific humidity, wind speed and sea level pressure - used in the downscale method.

Above all, for application, a set of vector needs to be constructed with meteorological variables of the NCEP data, including temperature, specific humidity, wind speed and sea level pressure, capable of affecting the precipitation which is intended to be downscaled, as follows:

$$Y_t = (a_t, b_t, c_t, \ldots\ldots, n_t)$$

Here, **n**is the number of meteorological variables affecting the variables to be downscaled, and **t** shows that the variables are time-series data, implying there **t** is number of vector Y_t. In addition, in order to calculate the optimal number of nearest (k), provided vector Y_t was divided by Training set, Y_{ttra} and Prediction set, Y_{tpre}. In order to investigate the similarity between Y_{tpre} and Y_{ttra}, this study calculated the distance between them, or $\|Y_{tpre} - Y_{ttra}\|$. When the most similar Y_{ttra} with Y_{tpre} is identified through the process, the precipitation which occurred under the condition of Y_{ttra} is inferred to occur again in Y_{tpre}. Here, the number of nearest is determined by the number of which leads to a optimal result, while it is changing the number of similar Y_{ttra} with Y_{tpre} in order of its similarity. With the number of the nearest determined, Y_t is identified as many as the number of the nearest, then, precipitation at $Y_t^{'}$.is calculated by getting arithmetic means of precipitation at Y_t,

This study selected appropriate meteorological variables like average temperature, humidity, southerly wind and sea level pressure for the downscale method, and downscaled precipitation data of 58 stations country-wide relevant to 20c3m and A1B scenarios from CNCM3 model.

This study initiated calibration and validation of the SK-NN model by using NCEP data and total precipitation data from stations in order for the application of SK-NN. To this end, the observed data was divided into three parts - Training set, Calibration set and Validation set. The first step for calibration is to find a duration that has the meteorological variable of NCEP value similar to the meteorological variables of NCEP - average temperature, humidity, southerly wind and sea level pressure - relevant to a calibration period. Then, the precipitation in locations is designated as the precipitation for the calibration period. In this case, the estimation of errors by increasing the number of meteorological variable groups of Training set from one to twenty, which are similar to the meteorological variables of NCEP relevant to calibration period, may produce K, the number of optimal meteorological variable groups for SN-NN. In order to evaluate the application of estimated value of K, the calibration was proceeded as follows: (See Figure 2)

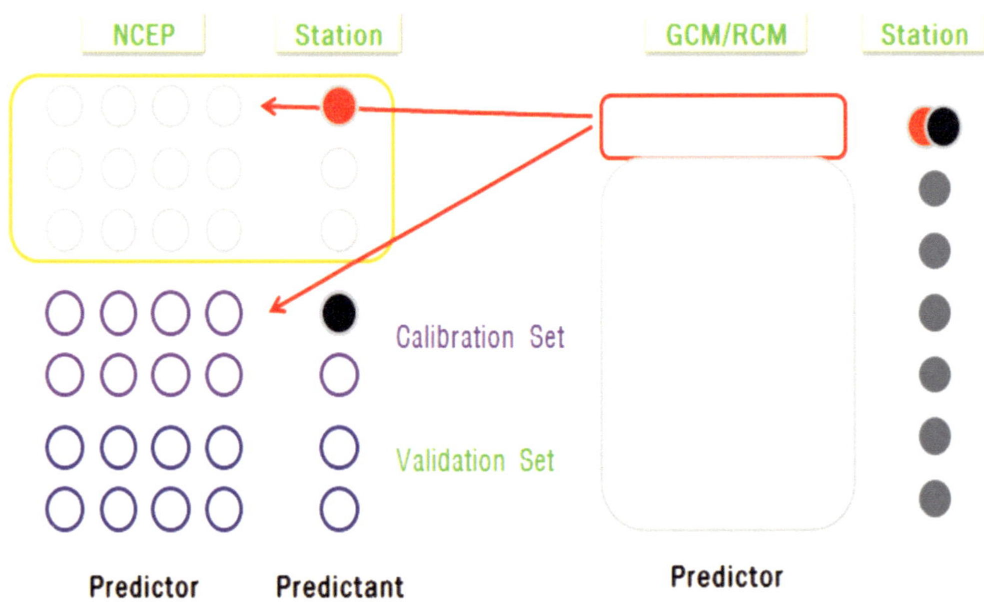

Fig. 2. Basic concept of Simplest K-NN method as a downscaling technique

When the application of the model is estimated through calibration and validation, it may be possible to get the precipitation of target locations of GCM by using the same meteorological variables with those of NCEP - average temperature, humidity, southerly wind and sea level pressure - suggested by GCM. In other words, the precipitation values for GCM may be estimated by averaging the results of extracted K number of meteorological variable groups of NCEP - average temperature, humidity, southerly wind and sea level - which have the same values with those of GCM.

3) Bias Correction by using Quantile Mapping

Those downscaled results from GCM tend to show some level of bias against observed outcomes in real. The most commonly used method to correct such bias is Quantile mapping suggested by Panofsy and Brire (1963). The Quantile mapping approach used as a way to correct runoff (Wood et al, 2004; Hamlet et al, 2003; Hashino et al., 2007), and in regard to climate change, GCM data in Palmer et al(2004), Fowler et all(2007b), Durman et al.(2001), Kim et al(2008), Kyoung et al(2009a, 2009b). The Quantile mapping's general procedure is presented in Figure 3.

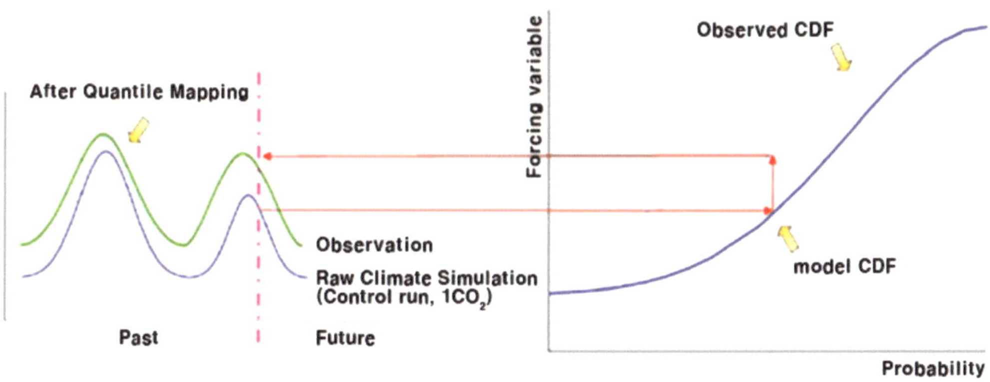

Fig. 3. Quantile mapping.

With the Quantile mapping approach, this study estimated the bias of observed data and GCM , with respect to both the monthly total precipitation of 20c3m(CASE1) scenario, which was downscaled by SK-NN, and that of 58 stations operated by the Korea Meteorological Administration (KMA). Then, the results were applied to those downscaled cases of A1B, CASE2, CASE3 and CASE4, to correct monthly total precipitation bias for each case.

4) KMA RCM

The National Institute of Meteorological Research (NIMR) in Korea is providing downscaled RCM for the Eastern Asia region. It uses the IPCC SRES A1B, A2 and B1, and generates a global climate change scenario by applying anthropogenic forcing to ESHO-G. It also generates a regionally downscaled climate change scenario for Korea by proceeding dynamical downscaling of a global climate change scenario with a regional climate model, MM5. Finally, it offers a high resolution climate change scenario by using a statistical downscaling technique.

3. Standardized precipitation index considering climate change

3.1 Standardized precipitation index

In this study, downscaled data for 58 stations were used to construct SPI for each of stations. SPI, which was developed by Mckee et al. (1993), is a relatively simple drought index based only on precipitation. Table 4 shows the status of droughts in accordance with the scope of index.

SPI	Moisture
more than 2.00	extremely wet
1.50 ~1.99	severe wet
1.00 ~ 1.49	wet
-0.99 ~ 0.99	normal
-1.49 ~ -1.00	dry
-1.99 ~ -1.50	severe dry
less than -2.00	extremely dry

Table 4. Classification of moisture by SPI

The SPI is calculated from the long term record of precipitation in each location (generally at least 30 years). Values of SPI are derived by comparing the total cumulative precipitation for a particular station or region over a specific time interval with the average cumulative precipitation for that same time interval over the entire length of the record. The SPI was

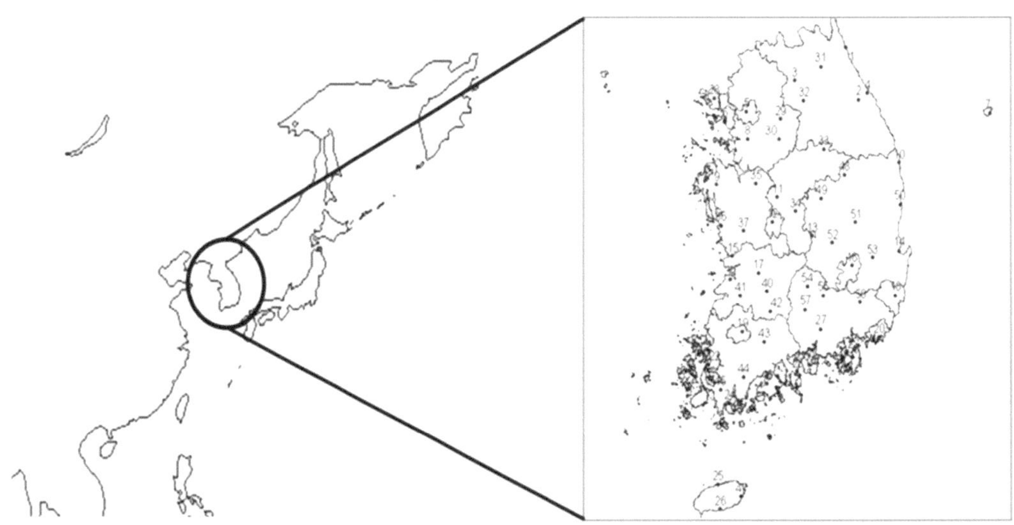

Fig. 4. Locations of Rainfall station, Korea

designed to enhance the detection of onset and monitoring of drought conditions on a variety of time scales. A key feature of the SPI is the flexibility to measure drought at different time scales. The data will be fitted to normal distribution and be normalized to a flexible multiple time scale such as 3-, 6-, 12-, 48-month and etc. This temporal flexibility allows the SPI to be useful in both short-term agricultural and long-term hydrological applications.

Figure 4 and Table 5 show the location and geographical designation of precipitation stations of Korea Meteorological Administration that provided data for the calculation of the SPI. Finally, Figure 5 ~ Figure 8 show the calculated SPI in conjunction with major locations.

No.	Station	No.	Station	No.	Station	No.	Station
1	Sokcho	16	Daegu	31	Inje		
2	Daegwallyong	17	Jeonju	32	Hongcheon	46	Sungsanpo
3	Chuncheon	18	Ulsan	33	Jecheon	47	Yeongju
4	Gangneung	19	Gwangju	34	Boeun	48	Mungyeong
5	Seoul	20	Pusan	35	Cheonan	49	Youngdong
6	In Cheon	21	Tongyeong	36	Boryeong	50	Uiseong
7	Ulleungdo	22	Mokpo	37	Buyeo	51	Gumi
8	Suwon	23	Yeosu	38	Buan	52	Yongcheon
9	Seosan	24	Wando	39	Imsil	53	Gochang
10	Uljin	25	Jeju	40	Jeongeup	54	Hapcheon
11	Chongju	26	Seogwipo	41	Namwon	55	Miryang
12	Daejeon	27	Jinju	42	Suncheon	56	Sancheong
13	Chupungryong	28	Ganghwa	43	Jangheung	57	Geoje
14	Pohang	29	Yangpyong	44	Haenam	58	Namhae
15	Gumsan	30	Icheon	45	Goheung		

Table 5. Rainfall stations

3.2 Assessment of climate change effect on drought occurrence and frequency

This study calculated SPI for precipitation at downscaled 58 stations in order to assess climate change effect on drought, and drought spell based on the estimated SPI. The drought spell may be assessed by SPI and using average precipitation index implies that SPI is constantly less than zero and its minimum value is less than -1(Dubrovsky et al, 2004). Building on droughts selected by such basis, climate change inducing a drought occurrence behavior was analyzed, and for this, droughts with the duration of 6 months and 12 months were selected.

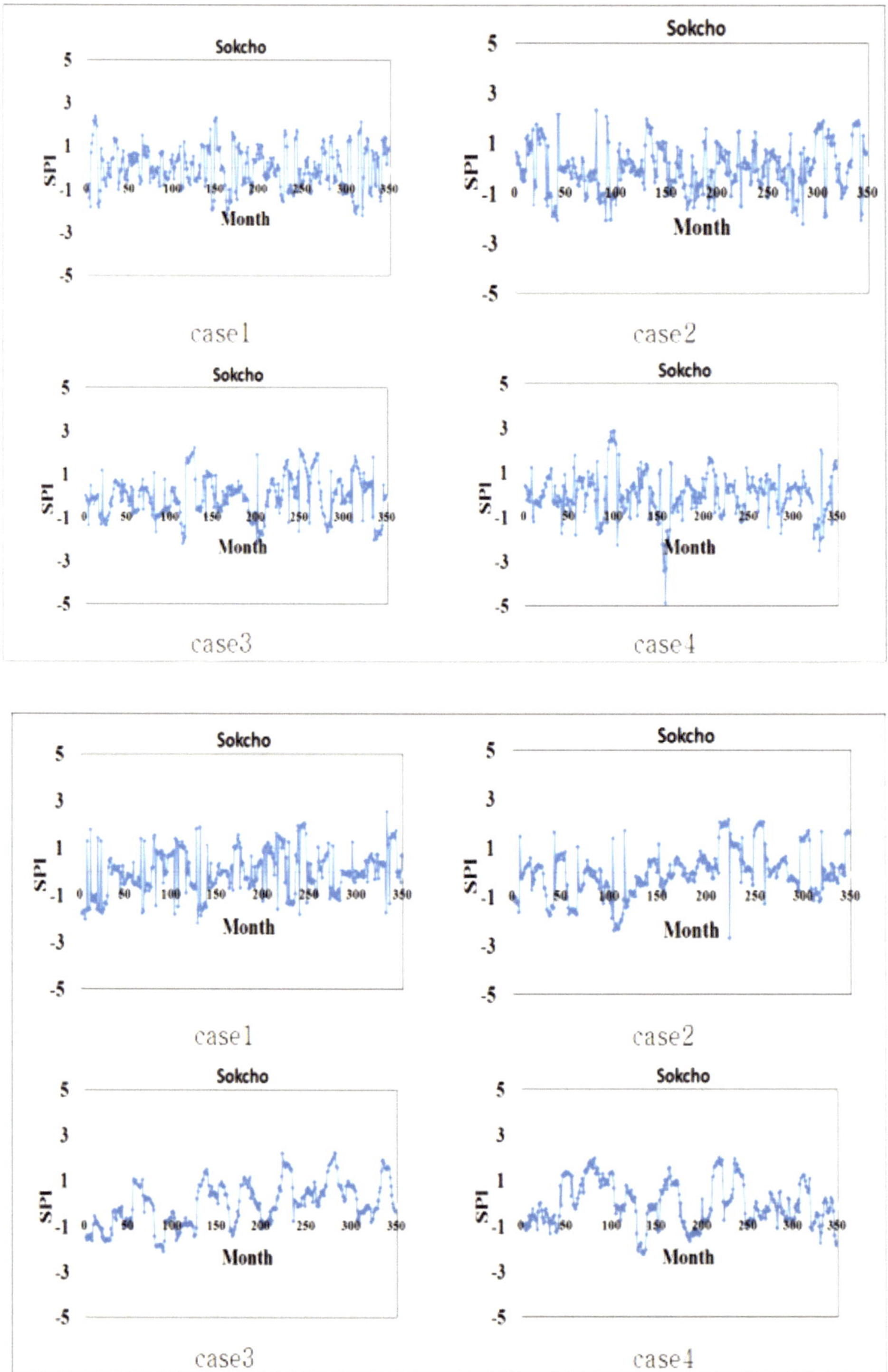

Fig. 5. SPI results of Sokcho (upper part: CNCM3 & downer part : KMA RCM)

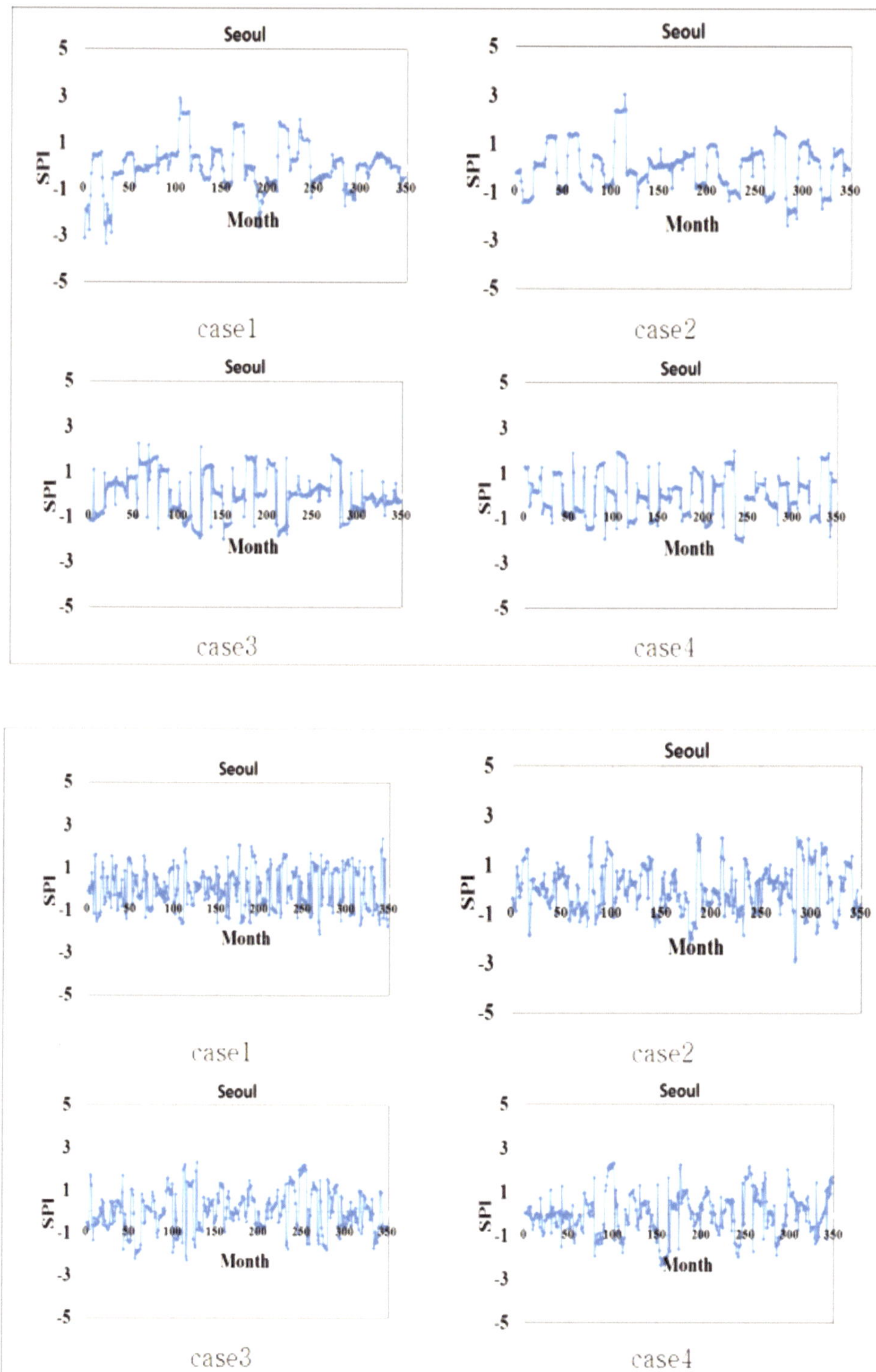

Fig. 6. SPI results of Seoul (upper part: CNCM3 & downer part : KMA RCM)

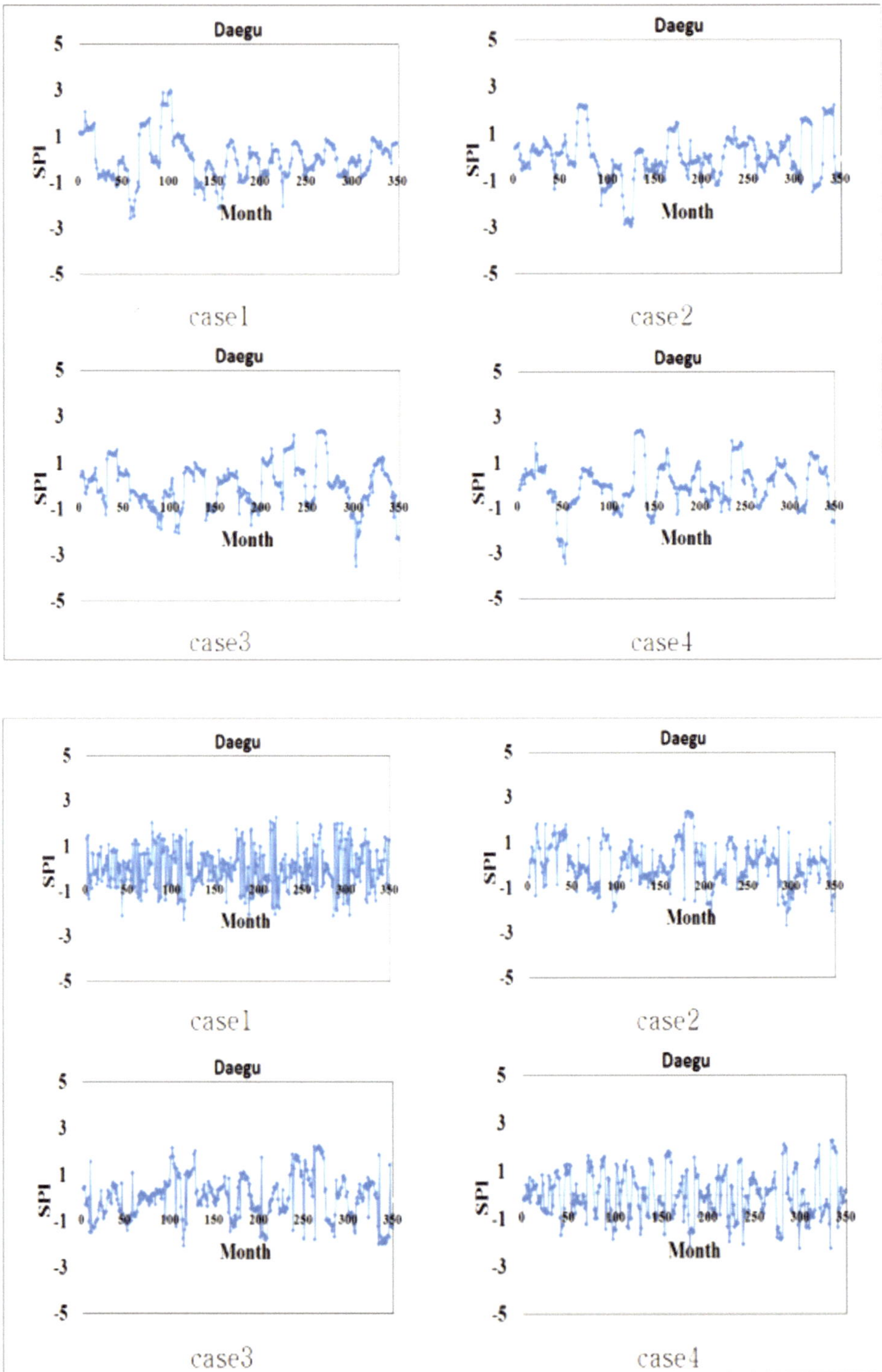

Fig. 7. SPI results of Daegu (upper part: CNCM3 & downer part : KMA RCM)

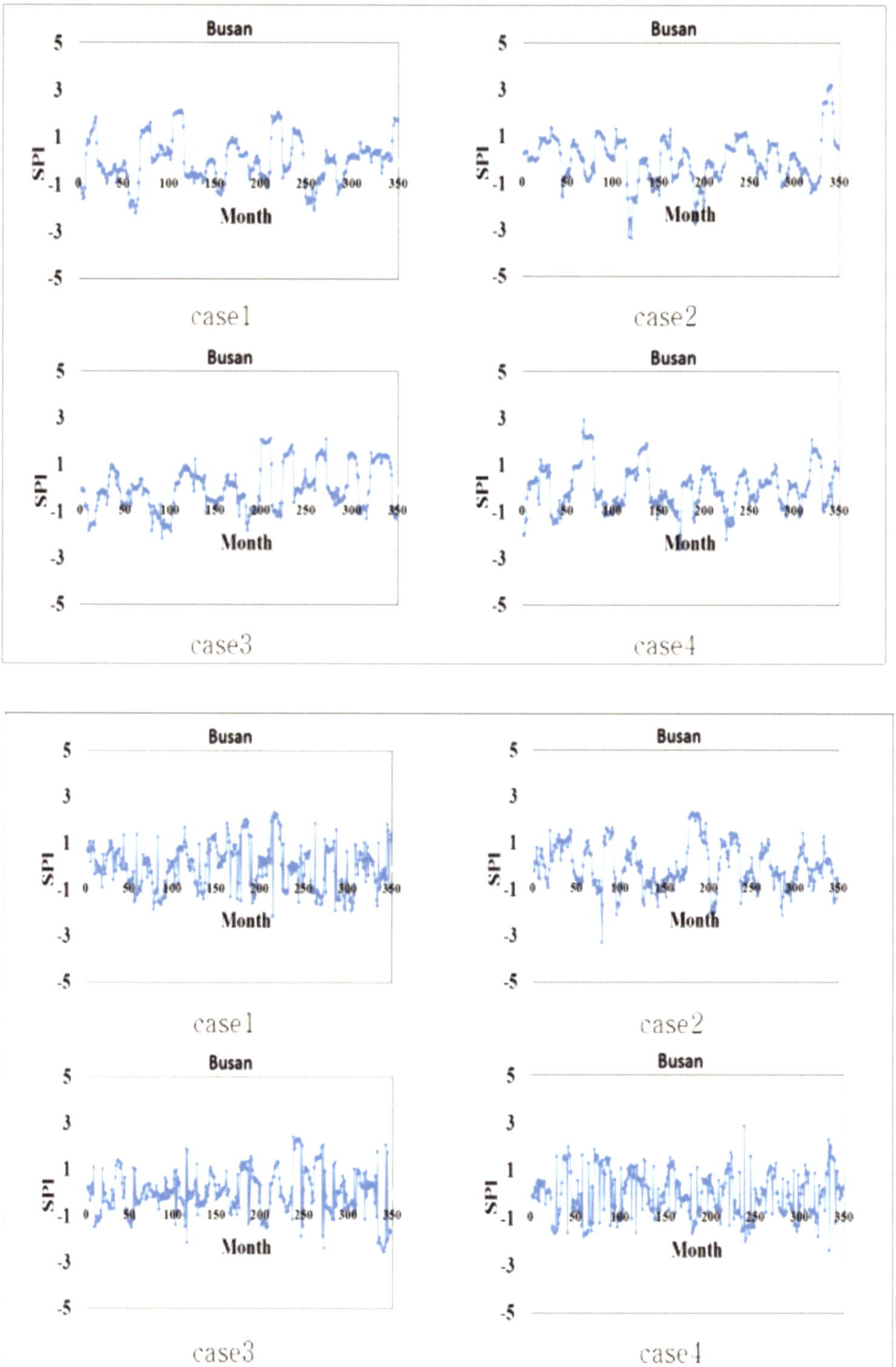

Fig. 8. SPI results of Busan (upper part: CNCM3 & downer part : KMA RCM)

GCM	Station	CASE1	CASE2	CASE3	CASE4	CASE1	CASE2	CASE3	CASE4
		more than 6 month				more than 12 month			
KMA RCM	Sokcho	1	2	5	3	0	0	0	0
	Seoul	2	2	2	4	0	0	0	0
	Daegu	0	4	4	2	0	0	0	0
	Busan	3	1	4	3	0	0	0	0
CNCM3	Sokcho	3	3	5	2	0	0	2	2
	Seoul	4	4	5	5	0	1	0	0
	Daegu	3	3	3	4	0	1	1	1
	Busan	4	3	2	2	1	2	1	0

Table 6. Frequencies of drought

Across those 58 stations, there are 138 droughts with the six-month duration in CASE1, 68 in CASE2, 171 in CASE3 and 121 in CASE4. In other words, the number of drought occurrences decreases in CASE2, and increases in CASE3, and then drops again. As for the CNCMs model, there are 185 droughts with the six-month duration in CASE1, 221 in CASE2, 190 in CASE3, and 186 in CASE4 across 58 stations, implying that the number of droughts is likely to increase in the future.

KMA RCM showed that there are five droughts with the 12-month duration in CASE1, zero in CASE2, seven in CASE3 and three in CASE$, representing that droughts with a longer duration may not return. Meanwhile, CNCM3 showed 50, 60, 38 and 35 times of drought for each case. It is possible to predict that the drought with the 12-month duration and above may steadily decrease duc to the effect of climate change. This may be explained by the fact that the effect of gradually increasing precipitation reduces the number of drought occurrences with relatively a longer duration.

Table 7 shows the average severity of drought in major locations by drought duration (six and twelve months).

The average severity of the droughts with the duration of more than six months was -1.16, -0.95, -1.31, and -1.41 for each CASE in KMA RCM. For CASE4, as showed in Table 6, although the number of drought occurrences decreases, their severity increases, representing a possibility of much severe drought in the future. As for CASE2 of CNCM3, it was expected that relatively severe drought is likely to occur in the south central area, particularly Dae-gu, and some part of the southern coastal area. For CASE3 and CASE4, some parts of the southern coastal area and nearby inland areas were expected to be the main area affected by droughts though the potential drought areas may be reduced.

KMA RCM failed to simulate drought with the duration of over 12 months. So, when analyzing the average severity of such drought based on the CNCM3 model that successfully simulated the number of observed drought occurrences, it was predicted that droughts would occur mainly in the southern and eastern coastal area and Gangwon province. It was also predicted that such expected areas of drought would slightly decline through CASE3 and CASE4. In general, it can be predicted that drought with the duration of 12 months or more would slowly decrease because of climate change effects and this may be because drought with relatively a longer duration decreases as precipitation gradually increases.

4. Establishment of SAD curve

The empirical orthogonal function (EOF) is mainly used for indentifying major patterns of spatial variability. EOF analysis is widely applied to water resources management field

(Sauquet et al., 2000; Benjamin, 2002; Hisdal and Tallaksen, 2003; Yoo and Kim, 2004; Viron et al., 2006; Zveryaev and Arkhipkin, 2008; Perry and Niemann, 2008).

GCM	Station	CASE1	CASE2	CASE3	CASE4	CASE1	CASE2	CASE3	CASE4
		more than 6 month				more than 12 month			
KMA RCM	Sokcho	-1.7	-1.5	-1.6	-1.9	0	0	0	0
	Seoul	-1.4	-1.5	-1.8	-1.6	0	0	0	0
	Daegu	0.0	-1.6	-1.7	-1.7	0	0	0	0
	Busan	-1.4	-1.7	-1.6	-1.5	0	0	0	0
CNCM3	Sokcho	-1.5	-1.8	-1.5	-1.6	0.0	0.0	-1.61	-1.6
	Seoul	-1.4	-1.5	-1.4	-1.4	0.0	-1.8	0.0	0.0
	Daegu	-1.5	-1.9	-1.6	-1.7	0.0	-2.6	-1.9	-2.5
	Busan	-1.3	-1.6	-1.6	-1.4	-1.6	-1.9	-1.6	0.0

Table 7. Average severity for the occurred drought

In this study EOF analysis is applied to SPI in order to analyze quantitatively the regional behavior of drought. The individual EOFs are obtained by performing the SVD (single value decomposition) for the original data matrix A and the equation is as follows;

$$A = U \times S \times V^T$$

where columns of the matrix U are the EOFs of A, and each EOF is a mutually independent component. Diagonal components of the matrix S are the eigen values of A, and the sum of them is the variance of the original data. Each column of the matrix V is the coefficient of time series and represents the temporal behavior of the corresponding EOF. The matrix A is consisted of the number of locations (M) and the temporal length of SPI (N), so its size is $M \times N$. The size of matrix U has $M \times M$ and that of matrix S is $M \times N$. The matrix V is consisted of the size of $N \times N$.

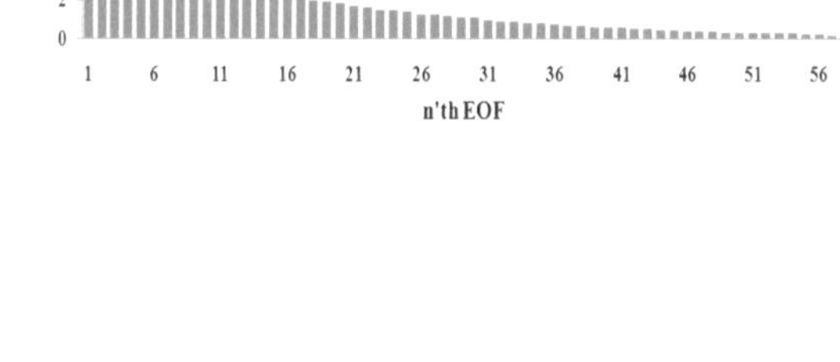

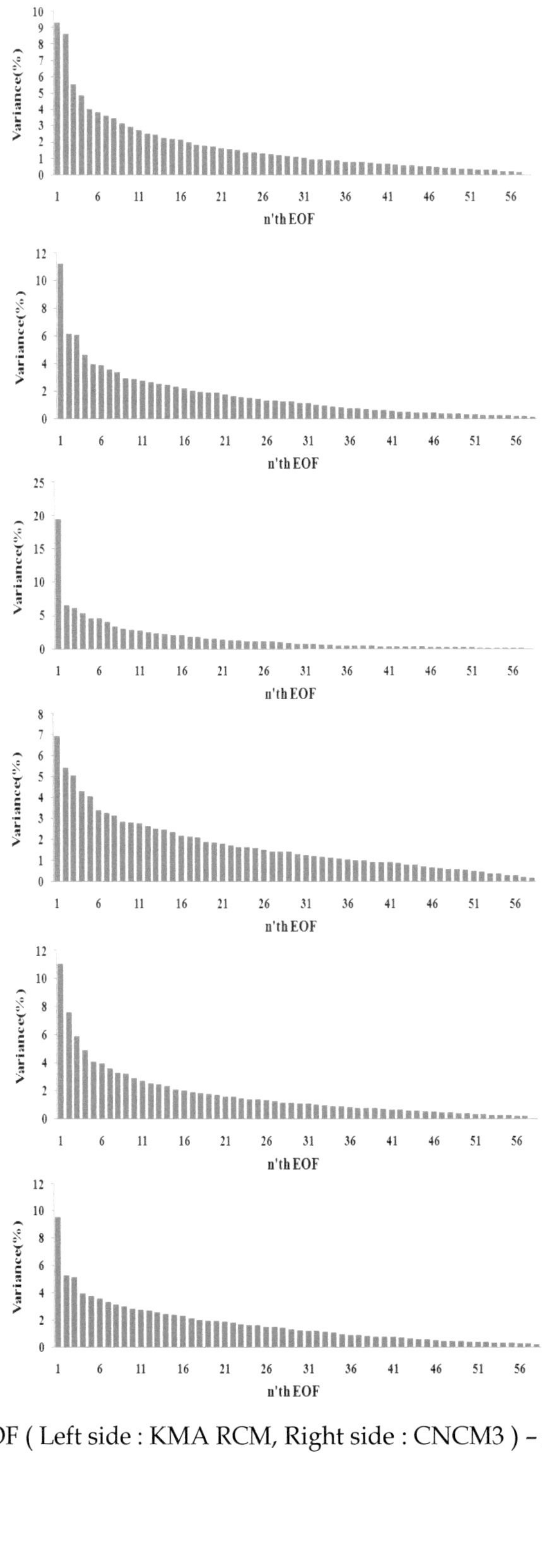

Fig. 9. Variance of EOF (Left side : KMA RCM, Right side : CNCM3) – from top to down is Case 1 to Case 4

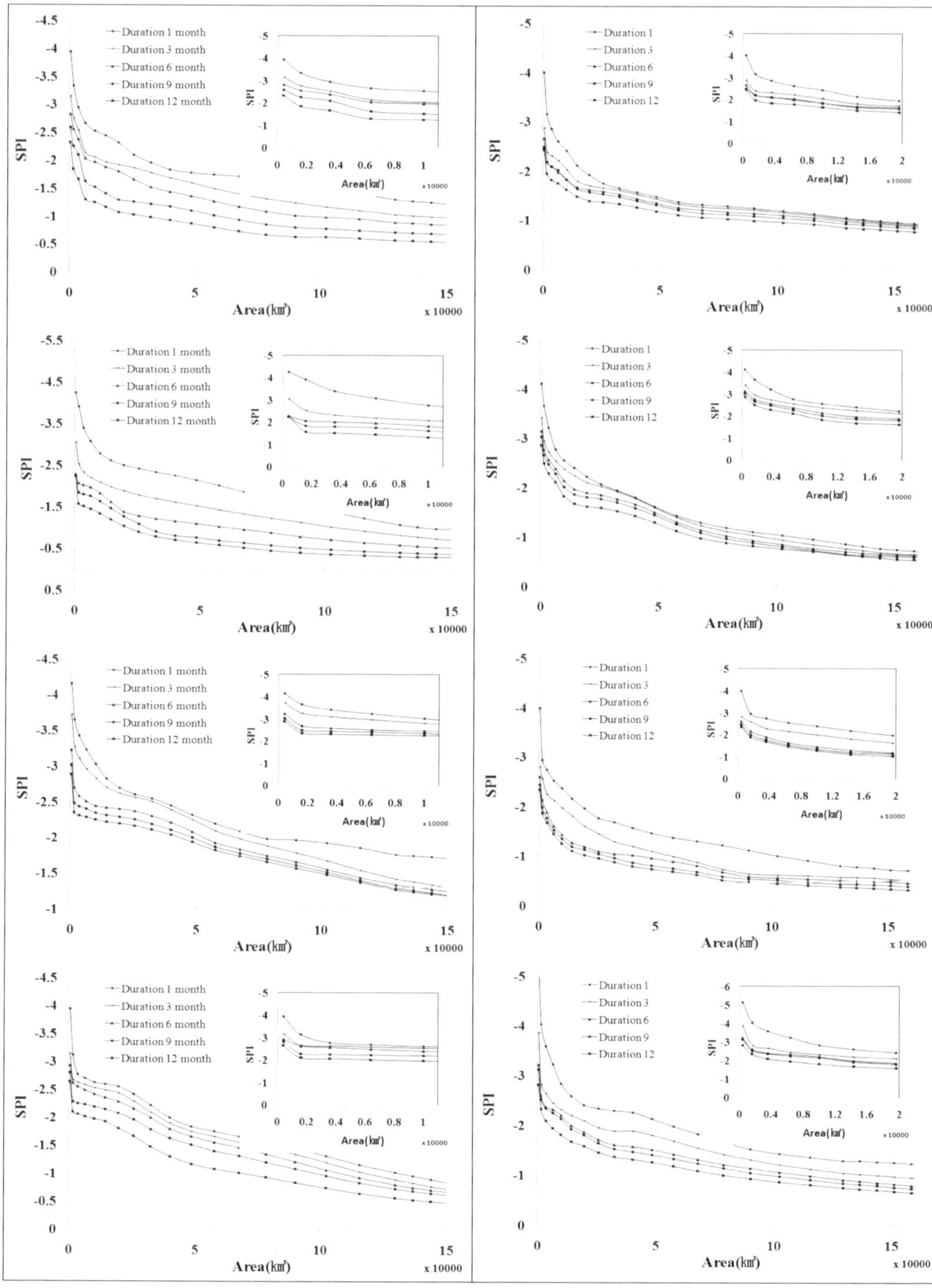

Fig. 10. SAD curves for whole South Korea (Left side : KMA RCM, Right side : CNCM3) – from top to down is Case 1 to Case 4

The number of EOF is as many as 58, which is the same number with the stations used. In this study, every variance value from each CASE was arranged by all models presented here. Those variance values were marked as the percentages(%) of the total variance values to present the variance values of each EOF.

As for KMA RCM, it was expected that two strong spatial patterns would appear in CASE2 followed by one pattern of spatial variability with a strong explanatory power in CASE3. Then, CASE4 showed various types of partial variability. In contrast, the CNCM3 model maintained the pattern of spatial variability with an apparent explanatory power from CASE1 to CASE2 but overall, the spatial variability pattern did not seem to be significantly different in CASE3. However, the variances of the first EOF are expected to show apparently different characteristic.

In order to draw the SAD curve, it is necessary to expand spatially each EOF. For the spatial expansion of each EOF, the kriging method can be used. The whole Korea is divided into 10,000 (100×100) grids with a width and length of six kilometers. Hisdal and Tallaksen (2003) also applied the kriging method to assign an EOF's value at each grid point. Such spatially expanded EOFs are then inversely transformed to calculate SPI time series at each grid point. Finally, averaging spatially and temporally, one can calculate the maximum spatio-temporally averaged SPI value at a specified temporal and spatial scale to draw SAD curve. Figure 10 shows drought severity-affected area curves for the duration of 1, 3, 6, 9 and 12 months.

SAD curve serves as an indicator showing how drought severity, which is estimated by using SPI to assess the drought prone area, changes by area and duration. As a result of generating the SAD curve for each model, it was predicted that the smaller the affected area is, the higher the severity is likely to be in CASE2 representing the near future while except for the drought with the three month duration, behaviors similar to CASE1 were seen in CASE3. Since overall drought severity in CASE4 became higher, drought is expected to cause the most severe damage.

5. Summary and conclusions

In this study, the impact of climate change on drought in Korea was evaluated. In order to consider such climate change impact, French GCM, CNCM3 and RCM by National Institute of Meteorological Research (NIMR) under the Korea Meteorological Administration were used. In addition, downscaled monthly precipitation was estimated by using SK-NN, and NIMR's RCM was used to calculate SPI for 58 stations in Korea. The SPI was used to compare the number of drought occurrences and average severity of the total drought, and droughts both with the duration of over six months and over twelve months. Drought severity by spatial extent in accordance with duration was calculated, generating SAD curve by utilizing SPI. The analysis of this study shows that the risk of drought in Korea is growing. Therefore, it is important to come up with responsive measures for drought from a mid-and-long term perspective.

6. Acknowledgement

This work was supported by the National Research Foundation Grant funded by the Korean Government (MEST) (**NRF-2009-220-D00104**)."

7. References

Andreadis, K. M., Clark, E. A., Wood, A. W., Hamlet, A. F., Lettenmaier, D. P. (2005) Twentieth-Century Drought in the Conterminous United States. Journal of Hydrometeorology Vol. 6, No. 6, pp. 985-1001

Benjamin, L. H. (2002) The Long-Range Predictability of European Drought. Degree of doctor of philosophy. University college London.

Buishand, T. A., Brandsma, T. (2001) Multisite simulation of daily precipitation and temperature in the Rhine Basin by nearestneighbor resampling, Water Resources Research, Vol. 37, No. 11, pp. 2761–2776

Casdagli, M. (1992) Chaos and deterministic versus stochastic nonlinear modeling, Journal of the Royal Statistical society, Statistics in society, Series B 54, pp. 303–324

Casdagli, M. and Weigend, A. (1994) Exploring the Continuum Between Deterministic and Stochastic Modelling, Forecasting the Future and Understanding the Past, Eds. A. S. Weigned and N. A. Gershenfeld, SFI Studies in the Sciences of Complexity, Proc. Vol. XV, Addison-Wesley, 993

Dubrovsky M., Buchtele, J., Zalud, Z. (2004) High-frequency and lowfrequency variability in stochastic daily weather generator and its effect on agricultural and hydrologic modelling, Climatic Change, Vol. 63, pp. 145–179

Durman, C.. F., Gregory, J. M., Hassell, D. C., Jones, R. G. and Murphy, J. M. (2001) A comparison of extreme European daily precipitation simulated by a global and a regional climate model for present and future climates. Quarterly Journal of the Royal Meteorological Society, Royal Meteorological Society, Vol. 127, No. 573, pp. 1005-1015

Fowler, H. J., Kilsby, C. G. and Stunell, J. (2007b) Modeling the impacts of projected future climate change on water resources in north-west England. Hydrologic & Earth System Sciences, EGU, Vol. 11, No. 3, pp. 1115-1126

Hamlet, A.F., Lettenmaier, D.P. and Snover, A. (2003) Climate change streamflow scenarios for critical period water planning studies:A technical methodology. Journal of Water Resources Planning and Management, ASCE, in review

Hashino, T., Bradley, A. A., and Schwartz, S. S. (2007) Evaluation of bias-correction methods for ensemble streamflow volume forecasts. Hydrology and Earth System Science, EGU, Vol. 11, pp. 939-950

Henriques AG, Santos MJJ. 1999. Regional Drought Distribution Model. Phys. Chem. Earth (B), Vol. 24, No. 1-2, pp. 19–22

Hisdal, H., Tallaksen, L. M. (2003) Estimation of regional meteorological and hydrological drought characteristics : a case study for Denmark. Journal of Hydrology, Vol. 281, No. 3, pp. 230-247

Kim T. W., Valdés J. B., Aparicio J. (2002) Frequency and Spatial Characteristics of Droughts in the Conchos River Basin, Mexico. Water International, Vol. 27, No. 3

Kim B. S., Kim B. K., Kyoung M. S., Kim H. S. (2008) Impact Assessment of Climate Change on Extreme Rainfall and I-D-F Analysis. Journal of Korea Water Resources Association. Korea Water Resources Association. Vol. 41, No. 4, pp. 379-394

Kyoung M. S., Lee Y. W., Kim H. S., Kim B. S., (2009a) Assessment of Climate Chanage Effect on Temperature and Drought in Seoul : Based on the AR4 SRES A2 Senario. Korean Society of Civil Engineers Journal of Civil Engineering. Korean Society of Civil Engineers. Vol. 29, No. 2B, pp. 267-276

Kyoung M. S., Lee J. K., Kim H. S. (2009b) Downscaling Technique of Monthly GCM Using Daily Precipitation Generator. Korean Society of Civil Engineers Journal of Civil Engineering. Korean Society of Civil Engineers. Vol. 29, No. 5B, pp. 441-452

Kyoung M. S. (2010) Assessment of Climate Change Effect on Drought and Frequency Based Precipitation, Department of Civil Engineering, INHA University, Incheon, Korea. pp. 44-48

Lall, U., Rajagopalan, B., Torboton, D. G. (1996) A nonparametric wet/dry spell model for resampling daily precipitation, Water Resources Research, Vol. 32, No. 9, pp. 2803–2823

Lall, U., Sharma, A. (1996), A nearest neighbour bootstrap for time series resampling, Water Resources Research, Vol. 32, No. 3, pp. 679–693

Matheron G. 1963. Principles of Geostatistics. Econom. Geol., Vol. 58, No. 8, pp. 1246–1266

McKee, T. B., Doesken, N. J. and Kleist, J. (1993) The relationship of drought frequency and duration to time scales. Preprints, 8th Conference on Applied Climatology, 17-22 January, Anaheim, California, pp.179-184.

Mishra AK, Singh VP. 2009. Analysis of drought severity-area-frequency curves using a general circulation model and scenario uncertainty, J. Geophys. Res., Vol. 114

Palmer, R., Wiley, M., Kameenui, A. (2004) Will Climate Change Impact Water Supply and Demand In the Puget Sound?, Department of Civil and Environmental Engineering University of Washington, Seattle WA.

Panofsy, H. A., Brire, G. W. (1963) Some application of Statistics to Meteorology, Pennsylvania State University, University Park, Pennsylvania, pp. 224.

Perry and Niemann, (2008) Generation of soil moisture patterns at the catchment scale by EOF interpolation. *Hydrol. Earth Syst. Sci.* 12: 39–53.

Rajagopalan, B., Lall, U., (1999) A k-nearest neighbour simulator for daily precipitation and other variables, Water Resources Research, Vol. 35, No. 10, pp. 3089–3101

Sauquet E, Krasovskaia I, Leblois E. (2000) Mapping mean monthly runoff pattern using EOF analysis. *Hydrology and Earth System Sciences* 4(1): 79-93.

Sharif, M., Burn, D. H. (2006) Simulating climate change scenarios using improved K-nearest neighbor model, Journal of Hydrology, Vol. 325, pp. 179-196

Viron O, Panet I, Diamaent M. (2006) Extracting low frequency climate signal from GRACE data. *eEarth* 1: 9-14.

Wood, A. W., Leung, L. R., Sridhar. V. and Lettenmaier, D. P. (2004) Hydrologic implications of dynamical and statistical approaches to downscaling climate model outputs. Climatic Change, Vol. 62, Issue 1-3, pp. 189-216.

Yates, D., Gangopadhyay, S., Rajagopalan, B., Strzepek, K. (2003) A technique for generating regional climate scenarios using a nearest-neighbor algorithm, Water Resources Research, Vol. 39, No. 7, pp. 7-14

Young, K.C., (1994) A multivariate chain model for simulating climatic parameters with daily data, Journal of Applied Meteorology, vol. 33, pp. 661–671

Yoo CS, Kim SD. (2004) EOF Analysis of surface soil moisture field variability. *Advance in Water Resources* 27(8): 831-842, DOI: 10.1016.

Zveryaev II, Arkhipkin AV. (2008) Structure of climate variability of the Mediterranean sea surface temperature. Part II. Principle modes of variability. *Russian Meteorology and Hydrology* 33(7): 446–452, DOI: 10.3103/S1068373908070066.

Part 2

Changes in Fauna and Flora

11

Review of Long Term Macro-Fauna Movement by Multi-Decadal Warming Trends in the Northeastern Pacific

Christian Salvadeo[1], Daniel Lluch-Belda[1], Salvador Lluch-Cota[2] and Milena Mercuri[1]
[1]*Centro Interdisciplinario de Ciencias Marinas del Instituto Politécnico Nacional*
[2]*Centro de Investigaciones Biológicas del Noroeste*
La Paz, B.C.S.,
Mexico

1. Introduction

Worldwide marine ecosystems are continuously responding to changes in the physical environment at diverse spatial and temporal scales. In addition to the seasonal cycle, other natural patterns occur at the interannual scale, such as El Niño-La Niña Southern Oscillation (ENSO) with a period of about three to five years (Wang & Fiedler, 2006). When ocean conditions stay above or below the long-term average for periods of 10 to 20 years we recognize decadal fluctuations (Mantua et al., 1997), and those with periods longer than 50 years are known as regime (Lluch-Belda et al., 1989). On the ocean, marine populations respond to these variations in different ways, such as changes in their distribution and abundance. Evidence suggests that this multi-decadal scale climate variations are cyclic, which generates recurrent changes in the production level of marine ecosystems in ways that may favor one species or a group over another.
Abrupt changes between multi-decadal phases are known as regime shifts (Overland et al., 2008). The best documented regime shift in the North Pacific occurred in the mid-1970, with strong physical and biological signals, including ocean productivity (Ebbesmeyer, et al., 1991; Roemmich & McGowan, 1995), strong biomass and distribution changes in sardine and anchovy populations (Kawasaki, 1983; Lluch-Belda et al., 1989), and several other fish populations (Beamish et al., 1993; Mantua et al., 1997; Holbrook et al., 1997). These changes impacted marine food webs and ultimately affected the distribution and survival of marine top predators such as seabirds and marine mammals (Trites & Larkin, 1996; Veit et al., 1997; Trites et al., 2007). In this work we review published reports on long term macro-fauna (nekton) movements as related to multi-decadal temperature trends in the Northeastern Pacific.

2. Long term ocean surface variability on the southern California current system

The study area (Fig 1) is under the influence of the California Current System, where, several authors have observed environmental and biological multi-decadal climate signals

(Lluch-Belda et al., 1989; Ware, 1995; Mantua et al., 1997). To describe the environmental conditions on the California Current System, monthly gridded (2x2 degree) sea surface temperatures (from January 1900 to December 2010) were analyzed for the area limited by the 20-42°N latitude and 102-140°W longitude (Fig. 1). The data base is known as "Extended Reconstructed Sea Surface Temperature" and was obtained from the National Oceanic and Atmospheric Administration (NOAA) web site (http://lwf.ncdc.noaa.gov/oa/climate/research/sst/ersstv3.php).

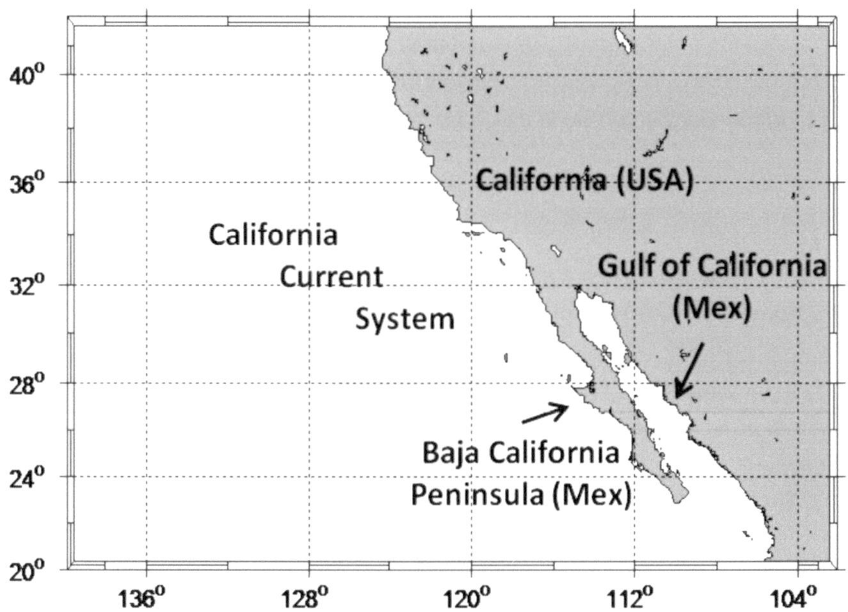

Fig. 1. Study area; USA: United States of America; Mex: Mexico.

To isolate scales of variability from the SST time series, we computed the long term mean and the seasonal signal by fitting annual and semiannual harmonics to the 110-year monthly mean time series (Ripa, 2002). Then we computed SST anomalies as residuals containing sub-seasonal (meso-scale) and low frequency variability (interannual and large scales) after extracting the long term mean and seasonal signals at each grid point. To analyze the regional modes of SST anomalies over the study area (Fig. 1), an empirical orthogonal functions analysis (EOF) was conducted using SST anomalies. The EOF decomposes the variability of the anomalies in a set of N uncorrelated orthogonal functions; each n-function represents an independent "mode of variability" (Björnsson & Venegas, 1997; Venegas, 2001).

The first EOF mode of SST anomalies explains 48% of the total variance over the study area. The spatial pattern shows a typical distribution of a global mode, where the surface temperature increase (decrease) in the whole area at the same time and according to the sign of the EOF time series, which explains up to 50% of the unseasonal SST variability off California and Baja California Peninsula (Fig. 2, upper panel). This mode shows a great interannual and multi-decadal variability in its time series (Fig. 2, lower panel). Two long warming trends and two long cooling trends are evident. Warming trends occurred between the late 1910s and the end of the 1930s, and from 1975 until the end of the 1990s, while the cooling trend occurred from the beginning of the twentieth century to the late 1910s, and between the early 1940s and 1975. The strong warming event at the end of the 1950s was not considered as a long term trend, because this was caused by the strong El Niño 1958-59 event, and a few years later the SST recovered its cooling trend until 1975. Also, our results suggested a new cooling trend beginning with the new century. The origin of these multi-decadal trends is subject to debate. In this regard, several studies and hypothesis have been developed to explain the physic mechanisms that are underlying this multi-decadal variability, but are not the matter of this work.

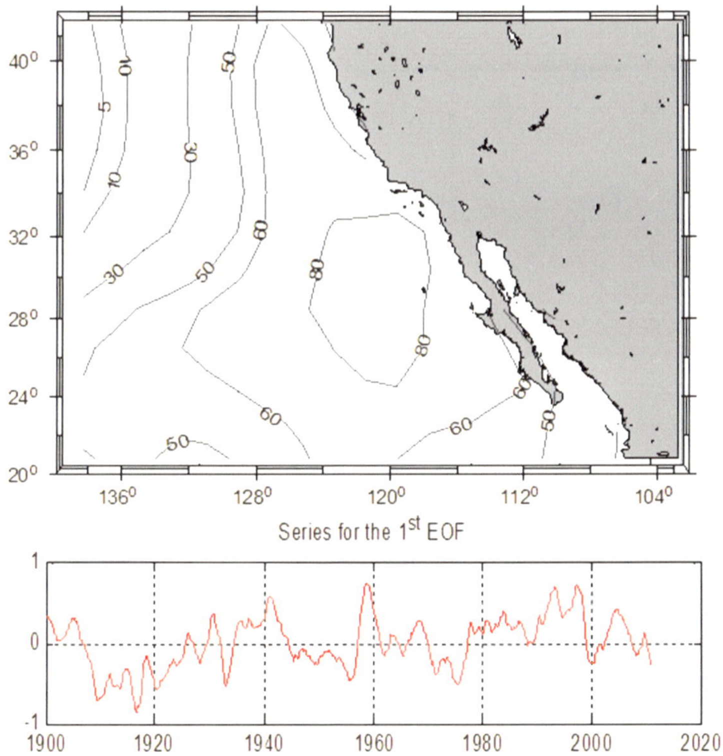

Fig. 2. Local explained variance (%) and temporal patterns of the first EOF mode of SST anomalies.

3. Long term macro-fauna movement

The California sardine (*Sardinops sagax caeruleus*) is the most abundant fish species in the northeast Pacific. It is a key component of the California Current pelagic ecosystem, being the main prey of several pelagic species such as seabirds, marine mammals, predatory fishes and squid (Bakun et al., 2010). This sardine has two core centers of distribution, one in the west coast of the Baja California Peninsula, and the other inside the Gulf of California. From these centers, schools may expand into the surrounding waters when environmental conditions are suitable. This species tends to have large interannual fluctuations in its abundance, due to strong variations in recruitment related primarily to environmental variability in their spawning areas (Lluch-Belda et al., 1986; Hammann et al., 1998). In addition to these interannual fluctuations, this group has a not yet totally understood regime shift time scale (~60 years) of global alternation between sardine and anchovy populations, due to the expansion and contraction of their populations (Fig. 3; Kawasaki, 1983; Lluch-Belda et al., 1989; Baumgartner et al., 1992; Chavez et al., 2003; Bakun et al., 2010). These can be seen in the commercial landings of California state (USA) waters (Fig. 4) and in fossil records over the last 2000 years (Baumgartner et al., 1992). Chavez et al (2003) related this regime shift to the SST variability in the northeast Pacific. This relationship is evident in the sardine landings (Fig. 4), where increases are evident during warming trends (1920-1940 & 1975-2000) and a decrease during the cooling trend (1940-1975).

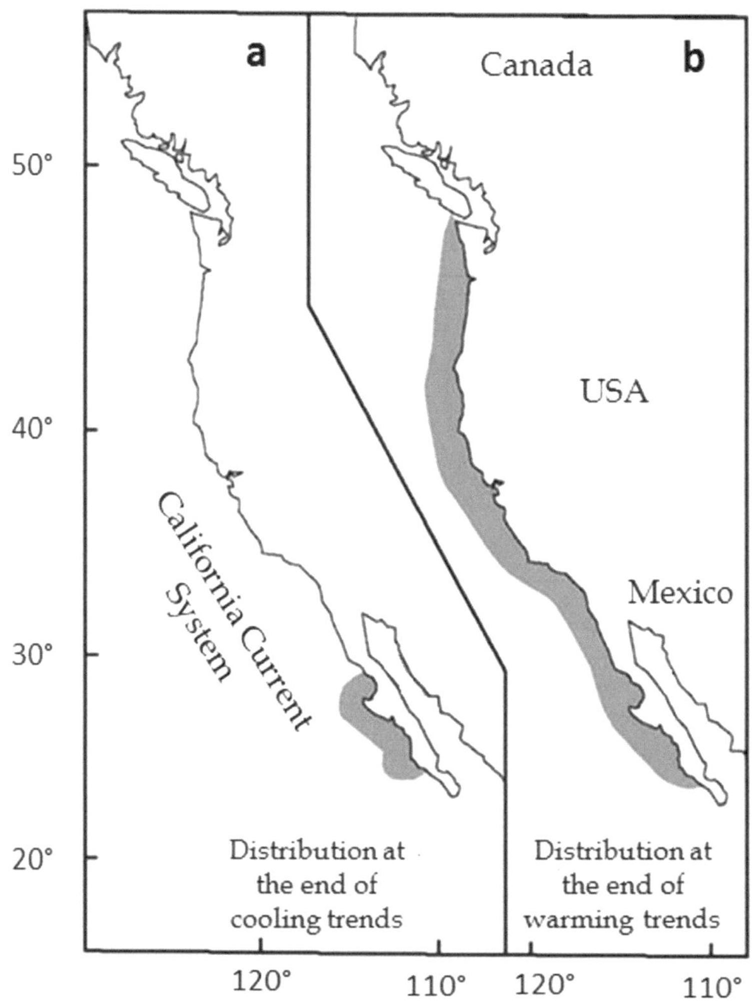

Fig. 3. Contraction (a) and expansion (b) of California sardine populations in the Northeast Pacific at the end of cooling and warming periods respectively (Bakun et al., 2010).

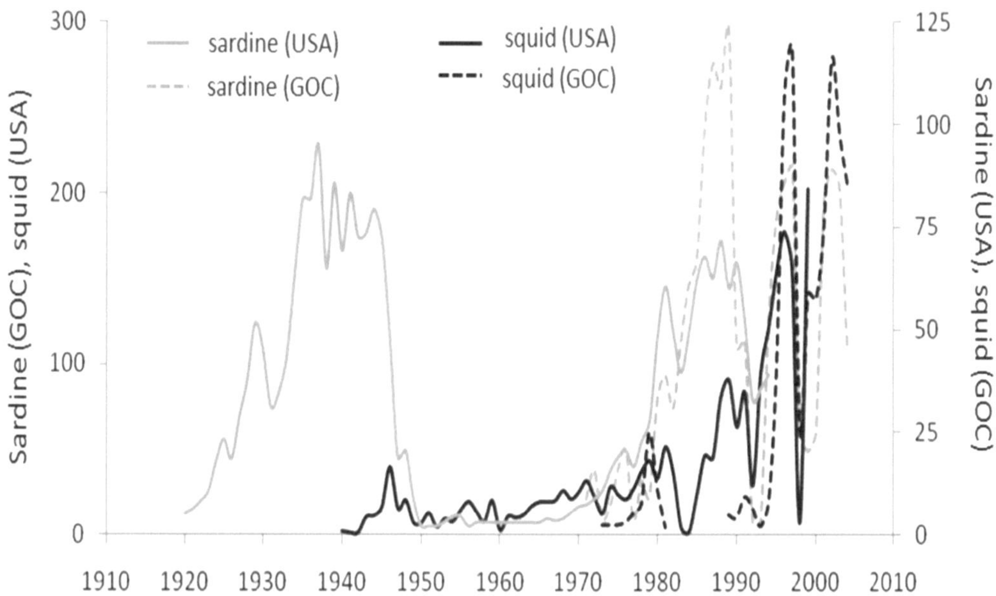

Fig. 4. California sardine landings at California waters (USA; thousands of tons) from FAO (1997), and for Gulf of California waters (GOC; thousands of tons) from SAGARPA; jumbo squid landings at California waters (USA; millions of pounds) from NOAA web page (http://www.pfeg.noaa.gov/research/climatemarine/cmffish/cmffishery.html), and Gulf of California (GOC; thousands of tons) from SAGARPA.

The Jumbo squid (*Dosidicus gigas*) is a large ommastrephid (up to 50 kg mass and overall length of 2.5 m) endemic to the Eastern Tropical Pacific. This squid is an important component of the marine food web that prey on small pelagic and mesopelagic fishes, crustaceans and squids (Markaida & Sosa-Nishizaki, 2003; Armendáriz-Villegas, 2005; Field et al., 2007); being an energy transfer from the mesopelagic food web to higher trophic level species as tunas, billfish, sharks, and marine mammals (Galván-Magaña et al., 2006; Field et al., 2007). The jumbo squid maintain the largest squid fishery in the world, which operates off the coasts of Peru, Chile and Central America, and in the Gulf of California (Morales-Bojórquez et al., 2001; Waluda & Rodhouse 2006). Recent scientific publications, anecdotal observations and fisheries landings pointed out a range expansion of jumbo squid throughout the California Current and southern Chile over the past decade (Fig. 4 & 5; Cosgrove, 2005; Chong et al., 2005; Wing, 2006; Zeidberg & Robinson, 2007). This sustained range expansion has generated hypotheses related to changes in climate-linked oceanographic conditions and reduction in their competing top predators (Zeidberg & Robinson, 2007; Waters et al., 2008). However, the coincidence of the recent poleward range expansions in both hemispheres, and the reports of the increases in the abundance off the west coasts of North and South America in the late 30s (Rodhouse, 2008), (just at the end of the 1910-1940 warming trend), suggests a physically-induced forcing mechanism. This may be related with long term warming trends and the poleward expansion of their primary habitat (Bazzino, 2008).

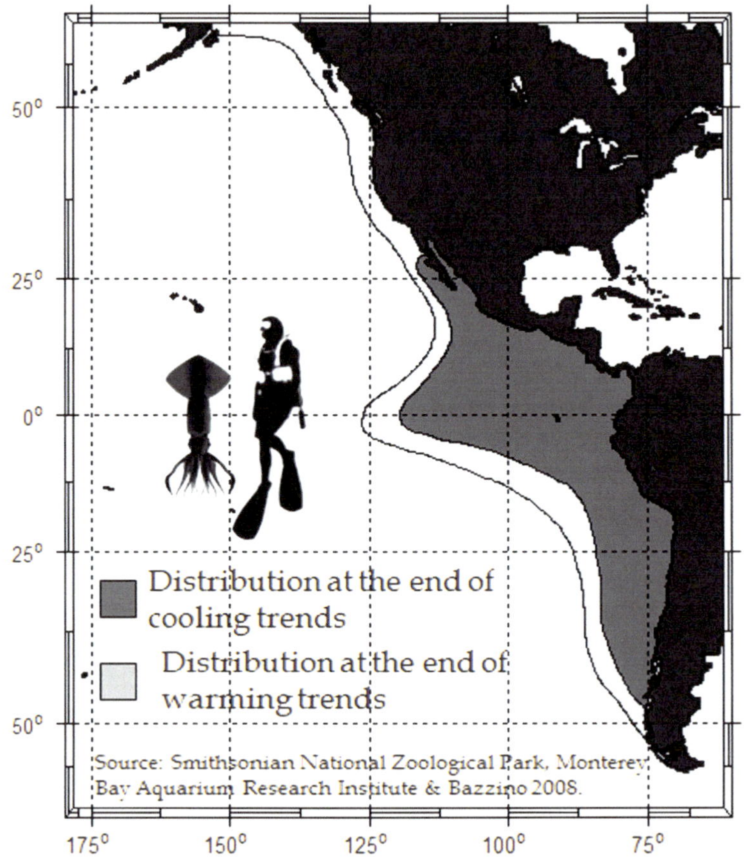

Fig. 5. Jumbo squid expansion during multi-decadal environmental trends.

The sperm whale *(Physeter macrocephalus)* is the largest odontocete, or toothed whale. This predator can be found in all world oceans in deeper waters, feeding largely on epi- and mesopelagic squid species (Whitehead, 2003). Groups of females and immatures are distributed on tropical and temperate waters, while solitary males are distributed on polar waters and only go to lower latitudes to breed. In the California Current System, Barlow & Forney (2007) showed that the abundance of sperm whales is temporally variable, and the two most recent estimates (2001 and 2005) were markedly higher than the estimates for 1991−96. Related to this increased in whales abundance, Jaquet et al. (2003) noted that few sightings of sperm whales were reported during the 1980s along the Baja California Peninsula; then their abundance appeared to increase since 1992. Actually these whales occur into the Gulf of California year-round and the high proportion of mature females and first-year calves suggests that this area is an important breeding and feeding ground for the sperm whale (Jaquet et al., 2003). As sperm whales are known to forage on jumbo squid, these authors coincided that the increased in the presence of sperm whale in both regions could be related with the expansion of jumbo squid in the California Current System and in the Gulf of California during the past two decades. Concurrently, a decrease in sperm whale abundance in the Galapagos Islands since the early 1990s has been observed (Whitehead et al., 1997), as well as animals from Galapagos have been spotted inside the Gulf (Jaquet et al., 2003), suggesting a northward shift in their distribution.

The Pacific white-sided dolphin (*Lagenorhynchus obliquidens*) is an average-sized oceanic dolphin (from 2 to 2.5 m) found in temperate waters of the North Pacific Ocean, feeding on small pelagic and mesopelagic fish and squid. In the eastern Pacific, large groups of this species are frequently seen in the California Current System (Leatherwood et al., 1984; Stacey & Baird, 1990; Keiper et al., 2005). The southern boundary of the distribution of Pacific white-sided dolphins is the Gulf of California, where the species has been observed only in its southwest area during the winter and spring (Aurioles et al., 1989). During the last 3 decades, Salvadeo et al. (2010) documented a decline in the presence of this dolphin species in the southwest Gulf of California, just during the end of the last warming trend in the California Current System (Fig. 2). Considering that the thermal environment is physiologically important to animals, the authors listed three evidences consistent with a poleward shift in their range: 1) The occurrence of this dolphin has decreased by approximately 1 order of magnitude per decade since the 1980s, (Table 1); 2) their monthly contraction to cooler months of the year (Fig. 6); and 3) the occurrence of this dolphin has increased on the west coast of Canada from 1984 to 1998 (Morton, 2000).

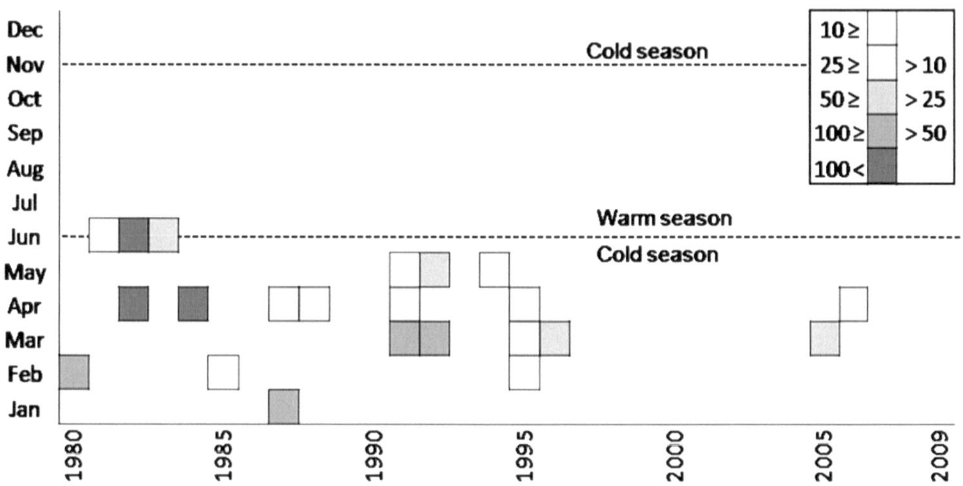

Fig. 6. Historical numbers of animals per month of Pacific white-sided dolphin from the southwest Gulf of California (Salvadeo et al. 2010).

Period	Effort	Sightings	Animals	Mean	Min.	Max.	SD	Sightings/hrs	Animals/hrs
1980s	252	10	647	65	2	200	67	0.039	2.56
1990s	1659	16	316	20	1	45	12	0.010	0.19
2000s	1986	2	50	25	20	30	7	0.001	0.03

Table 1. Pacific white-sided dolphin: accumulated historical data from the southwest Gulf of California for the last 3 decades. Effort (h); sightings: number of occasions when the species was observed; mean, minimum (min.), maximum (max.), and SD for group size; sightings h–1 and animals h–1: abundance relative to effort; 1980s: 1978–1988; 1990s: 1989–1999; 2000s: 2000–2009 (Salvadeo et al., 2010).

The gray whale (*Eschrictius robustus*) is a medium sized baleen whale reach 14 m in length and weigh of 45 metric tons. Some pods of gray whales breed every boreal winter at three lagoons along the Baja California Peninsula. At the end of the breeding season, the whales migrate to the feeding grounds in the Bering and Chukchi Seas, where they feed on benthic fauna (Rice & Wolman, 1971). The population of gray whales seems to have reached carrying capacity, with population size fluctuating between 20,000 and 22,000 animals (Rugh et al., 2008). As the Pacific white-sided dolphin, the evidences pointed out a possible poleward shift of the gray whale distribution related to the last warming SST trend. These evidences are: 1) there is an apparent long term tendency in the use of breeding lagoons, increasing at the northern lagoon and decreasing at the southern lagoon (Urbán et al., 2003a); 2) the decrease in the numbers of whales at the breeding lagoons during the last years, also observed from shore-based surveys at Piedras Blancas during the northbound migration (Urbán et al., 2010); 3) an increase in calf sightings at California (USA) correlating with warmer sea surface temperature anomalies (Shelden et al., 2004); 4) a range expansion into Arctic waters (Moore and Huntington, 2008); 5) during warming El Niño years the whales tend to use northern areas more intensively than in normal years (Gardner & Chávez-Rosales, 2000; Urbán et al., 2003b); 6) the unusual sighting of a gray whale in the Mediterranean Sea, it is another possible effect of their expansion to the north, which allows them to cross the Arctic to the Atlantic (Scheinin et al., 2011); and 7) in spite of having an increasing population of gray whales in the eastern Pacific, the observations of individuals inside the Gulf of California has been consistently declining (Salvadeo et al., 2011).

4. Conclusions

Two well defined long term climate warming trends were observed in the SST anomalies, these appear to be part of cyclical changes that include cooling trends over the study area (Fig 2). Changes in the SST are indicators of more complex ocean processes related to alterations in oceanic and atmospheric circulations, which ultimately affect the enrichment of superficial waters. The biological responses to those ocean processes are complex and not well understood.

There are evidences which indicate that distribution shifts related to long term ocean warming had occurred for some species, including poleward shifts (gray whale and Pacific white-sided dolphin), range expansions (California sardine and jumbo squid) and redistribution (sperm whale). The distributions of most species are defined by interactions between available environmental conditions and the ecological niches that they occupy on the ecosystem (Macleod, 2009). For gray whales and Pacific white-sided dolphins the cause of their range shift is apparently driven by the importance of thermal environment for the species. This poleward shift caused by thermal niche was also recorded in stranding records of dolphin species in the north-eastern Atlantic Ocean (Macleod et al., 2005). For the sperm whale it seems to be related with a trophic link, because their redistribution appears to be coupled with the range expansion of their primary prey, the jumbo squid. Multi-decadal range shift related with trophic interactions was also observed in the north-eastern Atlantic Ocean, from the subpolar gyre variability via plankton, to marine top predators (Hátún et al., 2009)

For the California sardine and the jumbo squid, their range expansions appear to be related with the extension of suitable habitat for their reproduction and recruitment. These range shifts seems to be cyclical, where their populations retract to subtropical areas during cooling trends and expand to temperate areas during warming trends. For cetacean species, this cycle was not observed yet, possibly due to the lack of information, so maybe this could also happen. These recurrent populations' changes also were observed on small pelagic fish and squids in other world oceans current systems (Fig. 7), and show the links between multi-decadal global ocean climate variability and regional fish and squid populations (Lluch-Belda et al., 1989; Schwartzlose et al., 1999; Sakurai et al., 2000; Tourre et al., 2007). These synchronous population shifts are consequence of cyclic changes on the environment that affect the production level of marine ecosystems in ways that may favor one species or group of species over another, affecting the marine food web structure and function.

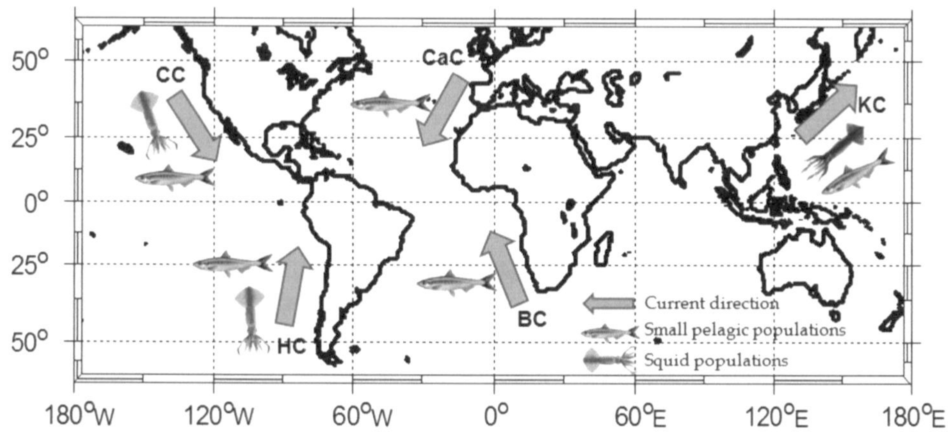

Fig. 7. Oceans current systems, where distribution shift were recorded on small pelagic fish and squid populations; ocean currents: California (CC), Canary (CaC), Kuroshio (KC), Humboldt (HC) and Benguela (BC); source: Lluch-Belda et al., 1989; Schwartzlose et al., 1999; Sakurai et al., 2000; Tourre et al., 2007, Bazzino 2008.

In conclusion, there are evidences that distribution shift occurred for some species due to long term ocean warming. Future scientific studies need to focus on understand the mechanisms of these long term cyclic variations and their effects on marine fauna, and incorporate this knowledge into the management and conservation approaches of the living marine resources.

Finally, the first EOF mode of SST anomalies showed a cooling trend for the last 10 years (Fig. 2). If the observed trends during the past are replicated, we should expect the beginning of a new ecological cycle, forced by climate tendencies that will restrict the distribution of California sardine to the west coast of the Baja California peninsula; and will move the jumbo squid range southward, forcing lower squid population levels at the west coast of the Baja Peninsula and the Gulf of California; related with this, a subsequent movement of sperm whales to other areas of the Pacific would occur, and the return of white-sided dolphins and gray whales as seasonal visitors of the Gulf of California.

5. Acknowledgment

We acknowledge the Consejo Nacional de Ciencia y Tecnología (CONACyT) and the Programa Institucional de Formación de Investigadores from the Instituto Politécnico

Nacional (PIFI-IPN) for the scholarships given to C.S. This work was done under the project "Patrones de cambio climático en el océano y sus efectos ecológicos", financed by "SEP-CONACyT". We also thank Emilio Beier for his help in the SST analysis, and Dr. German Ponce with the SEMARNAT-CONACYT projetc No 108270 for his support for the publication of this chapter.

6. References

Armendáriz-Villegas, J. (2005). Hábitos alimenticios del Calamar Gigante (Dosidicus gigas; Orbigny, 1835), en el Centro del Golfo de California durante los años 2002 y 2003, Bachelor Thesis, Universidad Autónoma de Baja California Sur, La Paz, B.C.S., México

Aurioles D. G., Gallo-Reynoso J. P., Muñoz E. L. & Ejido J. V. (1989). El delfín de costados blancos (Lagenorhynchus obliquidens, Gill 1865) (Cetacea: Delphinidae) residente estacional en el suroeste del Golfo de California, México. *Anales del Instituto de Biología, Universidad Nacional Autónoma de México, Serie Zoología* Vol.60, pp.459–472, ISSN 0368-8720

Bakun, A., Babcock, E. A., Lluch-Cota, S. E., Santora, C. & Salvadeo C. J. (2010). Issues of ecosystem-based management of forage fisheries in "open" non-stationary ecosystems: the example of the sardine fishery in the Gulf of California. *Reviews in Fish Biology and Fisheries* Vol.20, No.1, pp. 9-29. ISSN 1573-5184

Barlow, J. & Forney K. A. (2007). Abundance and density of cetaceans in the California Current ecosystem. Fishery *Bulletin* Vol.105, pp. 509–526. ISSN 0090-0656

Baumgartner, T. R., Soutar, A. & Ferreira-Bartrina V. (1992). Reconstruction of the history of Pacific sardine and Northern Pacific anchovy populations over the past two millennia from sediments of the Santa Barbara basin. CalCOFI Reports Vol.33, pp. 24-40. ISSN 0575-3317

Bazzino, G. (2008). Estructura poblacional, movimientos horizontales y migraciones verticales del calamar gigante en el Golfo de California y en el Océano Pacífico frente a la Península de Baja California. PhD. Thesis, Centro de Investigaciones Biológicas del Noroeste, La Paz, B.C.S., México. 126 pp.

Beamish, R. J. & Bouillon, D. R. (1993). Pacific salmon production trends in relation to climate. Canadian Journal of Fisheries and Aquatic Sciences Vol.50, pp. 1002–1016. ISSN 1205-7533

Björnsson, H. & Venegas S. A. (1997) A manual for EOF and SVD analyses of climatic data, McGill University, Retrieved from http://www.geog.mcgill.ca/gec3/wp-content/uploads/2009/03/Report-no.-1997-1.pdf

Chavez F. P., Ryan J., Lluch-Cota S. E. & Niquen C. M. (2003). From anchovies to sardines and back: multi-decadal change in the Pacific Ocean. Science Vol.299, pp. 217–221. ISSN 1095-9203

Chong, J., Oyarzun, C., Galleguillos, R., Tarifeño, E., Sepúlveda R. & Ibáñez, C. (2005). Fishery biology parameters of jumbo squid, Dosidicus gigas (Orbigny, 1835) (Cephalopoda: Ommastrephidae), in Central Chile coast (29°S-40°S) during 1993-1994. Gayana Vol.69, No.2, pp. 319-328. ISSN 0717-652X

Cosgrove, J. A. (2005). The first specimens of Humboldt squid in British Columbia. PICES Press Vol.13, No.2, pp. 30–31. ISSN 1195-2512

Ebbesmeyer, C. C., Cayan, D. R., McLain, D. R., Nichols, F. H., Peterson, D. H. & Redmond K. T. (1991). 1976 step in the Pacific climate: forty environmental changes between 1968-1975 and 1977-1984. Seventh annual Pacific climate (PACLIM) workshop, Asilomar, California, USA, April 1990

FAO. (1997). Review of the state of world fishery resources: marine fisheries. FAO Fisheries Circular No.920 FIRM/C920, ISSN 0429-9329, Rome, Italy.

Field, J. C., Baltz, K., Phillips, A. J. & Walker, W. A. (2007). Range expansion and trophic interactions of the jumbo squid, Dosidicus gigas, in the California Current. CalCOFI Reports Vol.48, pp. 131-146. ISSN 0575-3317

Galván-Magaña, F., Olson, R. J., Bocanegra-Castillo, N. & Alatorre-Ramirez, V G. (2006). Cephalopod prey of the apex predator guild in the epipelagic eastern Pacific Ocean In: The role of squid in open ocean ecosystems, Olson, R. J. & Young, J. W. (Eds.), pp. 45-48, GLOBEC international project, ISSN 1066-7881, Honolulu, Hawaii, USA

Gardner, S. C. & Chávez-Rosales, S. (2000). Changes in the relative abundance and distribution of gray whales (Eschrichtius robustus) in Magdalena Bay, Mexico during an El Niño event. Marine Mammal Science Vol.16, pp. 728–738, ISSN 0824-0469

Hammann M. G., Nevárez-Martínez M. O. & Green-Ruíz, Y. (1998). Spawning habitat of the Pacific sardine (Sardinops sagax) in the Gulf of California: Egg and larval distribution 1956-1957 and 1971-1991. CalCOFI Reports Vol.39, pp. 169-179, ISSN 0575-3317

Hátún, H., Payne, M. R., Beaugrand, G., Reid, P. C., Sandø A. B., Drange, H., Hansen, B, Jacobsen, J. A. & Bloch D. (2009). Large bio-geographical shifts in the north-eastern Atlantic Ocean: From the subpolar gyre, via plankton, to blue whiting and pilot whales. Progress in Oceanography Vol.80, pp. 149–162, ISSN 0079-6611

Holbrook, S., Schmitt, R. & Stephens, J. (1997) Changes In An Assemblage of Temperate Reef Fishes Associated With A Climate Shift. Ecological Applications Vol.7, pp. 1299-1310, ISSN 1051-0761

Jaquet, N., Gendron, D. & Coakes, A. (2003). sperm whales in the gulf of California: residency, movements, behavior, and the possible influence of variation in food supply. Marine Mammals Science Vol.19, No.3, pp.545-562, ISSN 0824-0469

Kawasaki, T. (1983). Why do some pelagic fishes have wide fluctuations in their numbers? - biological basis of fluctuation from the viewpoint of evolutionary ecology, In: Reports of the Expert Consultation to Examine Changes in Abundance and Species Composition of Neritic Fish Resources. Sharp, G. D. & Csirke J. (eds.), pp. 1065-1080, FAO Vol.91, No.2-3, Rome, Italy

Keiper, C. A., Ainley, D. G., Allen, S. G. & Harvey J. T. (2005). Marine mammal occurrence and ocean climate off central California, 1986 to 1994 and 1997 to 1999. Marine Ecology Progress Series Vol.289, pp. 285–306. ISSN 1616-1599

Leatherwood, S., Reeves, R. R., Bowles, A. E., Stewart, B. S. & Goodrich, K. R. (1984). Distribution, seasonal movements and abundance of Pacific white-sided dolphins in the eastern North Pacific. Science Report Whales Research Institute Vol.35, pp. 129–157

Lluch-Belda, D., Magallón, B. F. J. & Schwartzlose, R. A. (1986). Large fluctuations in the sardine fishery in the Gulf of California: possible causes. CalCOFI Reports Vol.27, pp. 136-140, ISSN 0575-3317

Lluch-Belda, D., Crawford R. J. M., Kawasaki, T., MacCall, A. D., Parrish, R. H., Schwartzlose, R. A. & Smith P. E. (1989). Worldwide fluctuations of sardine and anchovy stocks: the regime problem. South Africa Journal of Marine Science Vol.8, pp. 195–205, ISSN 1814-2338

MacLeod, C. D. (2009) Global climate change, range changes and potential implications for the conservation of marine cetaceans: a review and synthesis. Endangered Species Research Vol.7, pp. 125-136, ISSN 1613-4796

MacLeod, C. D., Bannon, S. M., Pierce, G. J., Schweder, C., Learmonth, J. A., Reid, R.J. & Herman, J. S. (2005) Climate change and the cetacean community of northwest Scotland. Biological Conservation Vol.124, pp. 477–483, ISSN 0006-3207

Mantua, N. J., S. R. Hare, Y. Zhang, J. M. Wallace, and R. C. Francis. 1997. A Pacific decadal climate oscillation with impacts on salmon. Bulletin of the American Meteorological Society Vol.78, pp. 1069-1079, ISSN 0003-0007

Markaida U, Sosa-Nishizaki O (2003) Food and feeding habits of jumbo squid *Dosidicus gigas* (Cephalopoda: Ommastrephidae) from the Gulf of California, Mexico. Journal of the Marine Biological Association of the United Kingdom Vol.83, pp. 507–522, ISSN 0025- 3154

Moore, S. E., & Huntington, H. P. (2008). Arctic marine mammals and climate change: impacts and resilience. Ecological Applications Vol.18, pp. 157–165, ISSN 1051-0761

Morales-Bojórquez, E., Cisneros-Mata, M. A., Nevarez-Martínez, M. O. & Hernández-Herrera, A. (2001). Review of stock assessment and fishery biology of Dosidicus gigas in the Gulf of California, Mexico. Fisheries Research Vol.54, pp. 83-94, ISSN 0165-7836

Morton, A. (2000). Occurrence, photo-identification and prey of Pacific white-sided dolphins (Lagenorhynchus obliquidens) in the Broughton Archipelago, Canada 1984–1998. Marine Mammals Science Vol.16, pp. 80-93, ISSN 0824-0469

Overland, J., Rodionov, S., Minobe, S., & Bond, N. (2008). North Pacific regime shifts: Definitions, issues and recent transitions. Progress In Oceanography Vol.77, No.2-3, pp. 92-102, ISSN 0079-6611

Rice, D. W., & Wolman, A. A. (1971). The life history and ecology of the gray whale (Eschrichtius robustus). American Society of Mammalogists, Special Publication No.3 Stillwater, Oklahoma, USA

Ripa, P. (2002). Least squares data fitting. Ciencias Marinas Vol.28, pp.79-105,ISSN: 0185-3880

Roemmich, D. & McGowan, J. (1995). Climate warming and the decline of zooplankton in the California Current. Science Vol.267, pp. 1324-1326. ISSN 1095-9203

Rodhouse P. G. (2008). Large-scale range expansion and variability in ommastrephid squid populations: a review of environmental links. CalCOFI Reports Vol.49, pp. 83-89, ISSN 0575-3317

Rugh, D. J., Breiwick, J., Muto, M. M., Hobbs, R. C., Sheldon, K. W., D'Vincent, C., Laursen, I. M., Rief, S. L., Maher, S. L., & Nilson, S. D. (2008). Report of the 2006–2007 census of the eastern North Pacific stock of gray whales. Alaska Fisheries Science Center, Retrieved from www.afsc.noaa.gov/Publications/ProcRpt/PR2008-03.pdf

Sakurai, Y., Kiyofui, H., Saitoh, S., Goto, T. & Hiyama, Y. (2000). Changes in inferred spawning areas of Todarodes pacificus (Cephalopoda: Ommatrephidae) due to changing environmental conditions. ICES Journal of Marine Science Vol.57, pp. 24-30, ISSN 1095-9289

Salvadeo C. J., Lluch-Belda, D., Gómez-Gallardo, A., Urbán-Ramírez, J. & MacLeod, C. D. (2010). Climate change and a poleward shift in the distribution of the Pacific white-sided dolphin in the northeastern Pacific. Endangered Species Research Vol.11, pp. 13-19, ISSN 1613-4796

Salvadeo C. J., Lluch-Cota, S., Maravilla-Chavez, M. O., Alvarez-Castañeda S. T., Mercuri, M. & Ortega-Rubio, A. (2011). Impact of climate change on gray whale populations, whale watching, and conservation. Conservation Science Symposium, Loreto, BCS, Mexico, May 25-28, 2011

Scheinin, A. P., Kerem, D., MacLeod, C. D., Gazo, M., Chicote C. A. & Castellote, M. (2011). Gray whale (Eschrichtius robustus) in the Mediterranean Sea: anomalous event or early sign of climate-driven distribution change?. Marine Biodiversity Records Vol.4, e28, ISSN 1755-2672

Schwartzlose, R. A., Alheit, J., Bakun, A., Baumgartner, T. R., Cloete, R.,. Crawford, R. J. M., Fletcher, W. J., Green-Ruiz, Y. Hagen, E., Kawasaki, T., Lluch-Belda, D., Lluch-Cota, S., MacCall, A. D., Matsuura, Y., Nevarez-Martinez, M. O., Parrish, R. H., Roy, C., Serra, R., Shust, K. V., Ward, M. N. & Zuzunaga, J. Z. (1999). Worldwide large-scale fluctuations of sardine and anchovy populations. South Africa Journal of Marine Science, Vol.21, pp. 289–347, ISSN 1814-2338

Shelden, K. W., Rugh, D. J., & Schulman-Janiger, A. (2004). Gray whales born north of Mexico: indicator of recovery or consequence of regime shift? Ecological Applications Vol.14, pp. 1789–1805, ISSN 1051-0761

Stacey, P. J. & Baird, R. W. (1990). Status of the white-sided dolphin, Lagenorhynchus obliquidens, in Canada. Canadian *Field-Naturalist* Vol.105, pp. 219–232, ISSN 0008-3550

Tourre, Y. M., Lluch-Cota, S. E. & White, W. B. (2007). Global multi-decadal ocean climate and small-pelagic fish population. Environmental Research Letters Vol.2, doi:10.1088/1748-9326/2/3/034005, ISSN 1748-9326

Trites, A. W. & Larkin, P. A. (1996). Changes in the abundance of Steller sea lions (Eumetopias jubatus) in Alaska from 1956 to 1992: how many were there? Aquatic Mammals Vol.22, pp. 153-166, ISSN 0167-5427

Trites, A. W., Deecke, V. B., Gregr, E. J., Ford, J. K. B. & Olesiuk P. F. (2007). Killer whales, whaling and sequential megafaunal collapse in the North Pacific: a comparative analysis of dynamics of marine mammals in Alaska and British Columbia following commercial whaling. Marine Mammal Science Vol.23, pp. 751-765, ISSN 0824-0469

Urbán, R. J., Rojas, L. B., Pérez-Cortéz, H., Gómez-Gallardo, A. U., Swartz, S. Ludwing, S. & Brownell, L. (2003a). A review of gray whales on their wintering grounds in Mexican waters. Journal of Cetacean Research and Management Vol.5, pp. 281–295, ISSN 1561-0713

Urbán, R. J., Gómez-Gallardo, U. A. & Ludwig, S. (2003b). Abundance and mortality of gray whales at Laguna San Ignacio, Mexico, during the 1997–98 El Niño and the 1998–99 La Niña. Geofisica Internacional Vol.42, pp. 439-446, ISSN 0016-7169

Urbán, R. J., Gómez-Gallardo U. A., Rojas-Bracho, L. & Swartz S. L. (2010). Historical changes of gray whales abundance in San Ignacio and Ojo de Liebre breeding lagoons, Mexico. IWC Scientific Committee Paper, SC/62/BRG36, ISSN-0255-2760

Venegas S. A. (2001). Statistical methods for signal detection in climate. Danish Center for Earth System Science Report No.2, Retrieved from www.atmos.colostate.edu/~davet/AT655/notes/VenegasNotes.pdf

Veit, R. R., Mcgowan, J. A., Ainley, D. G., Wahls, T. R., & Pyle, P. (1997). Apex marine predator declines ninety percent in association with changing oceanic climate. Global Change Biology Vol.3, pp. 23–28, ISSN 1365-2486

Waluda, C. M. & Rodhouse, P. G. (2006). Remotely sensed mesoscale oceanography of the central Eastern Pacific and recruitment variability in *Dosidicus gigas*. Marine Ecology Progress Series Vol.310, pp. 25–32. ISSN 1616-1599

Wang, C. & Fiedler, P. C. (2006). ENSO variability and the eastern tropical Pacific: A review. Progress in Oceanography Vol.69: 239-266. ISSN 1369-9350

Ware, D. M. (1995). A century and a half of change in the climate of the NE Pacific. Fisheries Oceanography Vol.4, pp. 267-277, ISN 1365-2419

Waters, G. M., Olson, R. J., Field, J. C. & Essington T. E. (2008). Range expansion of the Humboldt squid was not caused by tuna fishing. PNAS Vol.105, No.3, E5, ISSN 1091-6490

Whitehead, H. (2003). "*Sperm Whales: Social Evolution in the Ocean*." University of Chicago Press, ISBN 9780226895178, Chicago USA

Whitehead, H., Christal, J. & Dufault , S. (1997). Past and distant whaling and the rapid decline of sperm whales off the Galápagos Islands. *Conservation Biology* Vol.11, pp. 1387-1396, ISSN 0888-8892

Wing, B. L. (2006). Unusual invertebrates and fish observed in the Gulf of Alaska, 2004–2005. PICES Press Vol.14, No.2, pp. 26–28

Zeidberg L. D. and B. H. Robinson. 2007. Invasive range expansion by the Humboldt squid, Dosidicus gigas, in the eastern North Pacific. PNAS Vol.104, No.31, pp. 12948-12950, ISSN 1091-6490

12

Global Heating Threatens the `I`iwi (*Vestiaria coccinea*), Currently a Common Bird of Upper Elevation Forests in Hawaii

Anthony Povilitis
Life Net Nature, Willcox, Arizona
USA

1. Introduction

The `I`iwi is one of 17 surviving Hawaiian honeycreepers (Fringillidae: Drepanidinae) of 37 species known historically and 55 extant prior to human arrival on Hawaii (Pratt 2009). Its closest relative is the extinct Hawaii Mamo (*Drepanis pacifica*) (Pratt 2005). Disease and habitat loss are primary reasons for the decline of Hawaiian honeycreepers and other native forest birds. Extinctions continue to this day, with the most recent being the Poo-uli (*Melamprosops phaeosoma*) in 2004.

The `I`iwi, a scarlet bird with black wings and tail, and a long curved, salmon-colored bill, is generally placed in the monotypic genus *Vestiaria*. It is a largely nectarivorous species that occurs commonly in closed canopy, high-stature native forests above 1500 m elevation (Fancy and Ralph 1998). `I`iwi breed and winter primarily in mesic and wet forests dominated by native 'ōhi'a (*Metrosideros polymorpha*) and koa (*Acacia koa*) trees (Scott et al. 1986). They often travel widely in search of ʻōhiʻa flowers and are important ʻōhiʻa pollinators (Mitchel et al. 2005). The birds respond to seasonal flowering patterns, often moving to lower elevations where they are exposed to deadly disease (Pratt 2005). The `I`iwi uses its long bill to extract nectar from decurved corollas of Hawaiian lobelioids, which have become far less common on Hawaii over the past century (Smith et al. 1995).

Female `I`iwi typically lay two eggs, and they alone are thought to incubate eggs and brood young (Mitchel et al. 2005). But males provision females with food off the nest. Breeding takes place predominantly from February to June, and is usually associated with peak flowering of 'ōhi'a (Fancy and Ralph 1998).

For native Hawaiians, the `I`iwi and other forest birds have a spiritual nexus. Feathered objects represented gods, ancestors, and divine lineage (Amante-Helweg and Conant 2009). Red feathers of clothing, such as cloaks, capes, and helmets, were predominantly from `I`iwi. Once a familiar sight on all main Hawaiian Islands, the `I`iwi remains an icon of Hawaii's native forests.

Today `I`iwi occur in higher elevation habitats largely free of avian disease, to which the species is highly susceptible. With climate change, these refugia may be lost entirely as pathogens and vectors advance upslope in response to higher ambient temperatures. This prognosis points to the needs for swift remedial action by responsible U.S. federal and State of Hawaii authorities to prevent the `I`iwi from joining the tragically long list of extinct or feared extinct Hawaiian birds (Banko and Banko 2009).

2. Population status

The `I`iwi occurs on the Hawaiian islands of Kauai, Oahu, Maui, Molokai, and Hawaii (Gorresen et al. 2009). Once widely distributed in native forests on all major Hawaiian Islands, it is now mostly restricted to elevations above 1250 m because of avian diseases and habitat loss elsewhere (Warner 1968, Scott et al. 1986, Fancy and Ralph 1998, Pratt 2005).

Kauai -- `I`iwi numbers decreased by 62%, from 26,000 ± 3,000 to 9,985 ± 960, between the 1970s and 2000 (Foster et al. 2004, Gorresen et al. 2009);. `I`iwi range contracted from 140 to 110 sq km, consistent with a shift in its low elevation boundary from ~900 m to >1,100 m.
Oahu - Few, if any, birds remain; 8 individuals dispersed in 3 isolated locations were reported in 1994-1996 (VanderWerf and Rohrer 1996).
Molokai -- Few birds (1-3) were detected from 1988-2004 (Reynolds and Snetsinger 2001, Gorresen et al. 2009), contrasting with 12 in 1979 (Scott et al. 1986).
Maui -- About 19,000 ± 2,000 individuals occurred in restricted upper elevation habitats of east Maui (Scott et al. 1986); ~ 180 ± 150 birds were reported in isolated west Maui prior to 1980 (Scott et al. 1986); the west Maui population persists today at a very low number (Gorresen et al. 2009).
Hawaii Island - 340,000 ± 12,000 birds were estimated in higher elevation range; ~1,000 birds in lower elevation Kohala and Puna areas (Scott et al. 1986); overall downward trends are evident in recent decades (Camp et al. 2009a, Gorresen et al. 2009); of 10 study locations, `I`iwi appear now absent at one, declining at 5, stable at 3, with no estimate for 1 (Gorresen et al. 2009).
Regional breakout of data for Hawaii Island:
Northeast area: For the Hakalau Forest National Wildlife Refuge (Hakalau Unit; 1,500-2,000 m elevation) population trend data vary from stable (over a 21-year period) to declining (during a recent 9-year period), except for increasing numbers in limited newly restored upper elevation habitat (Camp et al. 2009a). Recent `I`iwi numbers were estimated at ~61,000 birds.
Central windward area: `I`iwi frequency decreased 54% between late 1970s and 1986-2000 periods in National Park and Hamakua areas, with specific study area declines and evidence of upward range contraction (Gorresen et al. 2005, Camp et al. 2009b);`I`iwi showed pronounced decline at lower elevations (East Rift, <1,000 m elev., and `Ōla`a, ~1,200-1,400 m, 1977-1994 data); modest declines (Kūlani-Keauhou, 1,500-2,000 m, 1977-2003 data) or stability (Mauna Loa Strip, ~1,500-2,000 m, 1977-1994 data) at higher elevations.
Southeast area: Lower `I`iwi density in the Ka`û area (2002 and 2005 data) than previously (1976 and 1993 data) (Gorresen et al. 2009); recent estimate of ~ 78,000 birds, with 60% occurring above 1,500 m (Gorresen et al. 2007).
Leeward (western) area: `I`iwi densities have dramatically declined in the Hualâlai and Kona regions (1997-2000); they are decreasing at lower elevations (<1,500 m; Kona Forest Unit, Hakalau Forest National Wildlife Refuge); stable only at upper elevations (Gorresen et al. 2009); `I`iwi range is contracting upslope, with few occurrences below 1,100 m during the breeding season (Camp et al. 2002).

Table 1. `I`iwi population estimates for Hawaiian islands.

`I`iwi are declining everywhere in Hawaii, except at high elevation on east Maui Island and northeast Hawaii Island (Gorresen et al. 2009) (Table 1), and population extinctions are impending throughout the Islands (Banko and Banko 2009).

On Kauai, in the western portion of the species' range, `I`iwi numbers have declined sharply (Table 1). Risk of extirpation from the island is of immediate concern because of severely diminished disease-free habitat. Oahu, Molokai, and the isolated western area of Maui have small remnant `I`iwi populations at high risk of extinction (Gorresen et al. 2009). `I`iwi are gone from nearby Lanai. These four areas comprise the central portion of the species' geographic range.

On east Maui and the Island of Hawaii, forming the eastern part of the species' range, `I`iwi populations are restricted to high elevations (Table 1). While some populations are still large, they are at risk of fragmentation and decimation resulting from the spread of avian disease driven by climate warming (Pratt et al. 2009).

The `I`iwi population of Hawaii has been estimated at 360,000 birds (Pratt et al. 2009), with the vast majority of birds (~90%) occurring on Hawaii Island (Scott et al. 1986). Declines in `I`iwi abundance corresponding with reduced lower elevation range since the early 1970s are consistent with anticipated impacts of mosquito borne disease (Foster et al. 2004). The population trend is downward on all islands, with some stability in high elevation areas (Pratt et al. 2009). Climate change is now setting the stage for widespread disease transmission at the highest elevations on Maui and Hawaii Island (Benning et al. 2002; LaPointe et al. 2005).

`I`iwi pairs are reported to produce on average only 1.33 chicks per year, reflecting low productivity characteristic of Hawaiian forest birds in general (Woodworth and Pratt 2009). However, the `I`iwi has the lowest annual survivorship reported (55% ± 12 SE for adults and 9% ± 5 for juveniles) for any extant species of honeycreeper, reflecting the impact of malaria and avian pox and/or low re-sighting probabilities (Fancy and Ralph 1998; Pratt 2005).

'I'iwi populations have suffered from fragmentation as well as reduced size and range. Small population units are at risk of extinction from random demographic fluctuations, localized catastrophes (severe storms, wild fire, disease outbreaks, volcanism, etc.), inbreeding depression, and genetic drift (Primack 2006).

3. Forest habitat

Most of the `I`iwi's original forest habitat has been cleared for food crops, livestock grazing, tree plantations, and land development, with habitat losses since human settlement ranging from 52% on Hawaii Island to 85% on O'ahu (Fancy and Ralph 1998). The amount of habitat available to the 'I'iwi and other forest birds has declined over the past few decades as many areas become dominated by invasive non-native species (Price et al. 2009). On the island of Hawaii, additional forest habitat loss results from land development, logging, and conversion to livestock pasture.

`I`iwi habitat across Hawaii is primarily threatened by destruction and adverse modification by feral pigs and other exotic ungulates (goats, sheep, mouflon, deer, cattle) (USFWS 2006, Pratt et al. 2009). Alien animals destroy forest understory vegetation, eliminate food plants for birds, create mosquito breeding sites through ground disturbances, provide openings on the forest floor for weeds, transport weed seeds to native forests, cause soil erosion, disrupt seedling regeneration of native plants, and girdle young trees (Fancy and Ralph 1998; Pratt 2005; USFWS 2006). Spread of exotic ungulates that are especially difficult to contain (i.e., axis deer on Maui and Molokai, black-tailed deer on Kauai, and mouflon sheep on Hawaii Island) represent a growing threat to `I`iwi habitat as these high-jumping species invade areas even with fencing designed to exclude feral pigs and goats (Price et al. 2009). Browsing and soil compaction by feral pigs, goats, and deer in Molokai has reduced 'ōhi'a forest to grassy scrubland (Hess 2008).

Hawaiian forests are severely modified by invasive alien plant species that displace native plants used by foraging and nesting birds (Scott et al. 1986; Foster et al. 2004) and increase the frequency of forest fires (Pratt et al. 2009). Herbivory by the introduced black rat on the flowers and fruits of native plants may also reduce food resources for native birds and impact regeneration of native plants (Banko and Banko 1976). Introduced predatory insects also may reduce or eliminate specialized native insects that are needed for pollination of plants important to `I`iwi.

Introduced species of insects and birds can compete with native birds for food and other resources. `I`iwi may face competition from Japanese White-eye (*Zosterops japonicas*) (Mountainspring and Scott 1985), a malaria resistant species, whose numbers have increased at least on Kauai over the past 30 years (Foster et al. 2004). Negative correlations between `I`iwi and Japanese White-eye densities may stem from competition for limited nectar resources (Fancy and Ralph 1998). There are no current efforts to control competing species within the recovery areas of endangered forest birds (USFWS 2006). Habitat degradation by non-native mammals, plants, and invertebrates will likely continue to result in loss, modification, and curtailment of `I`iwi habitat and range.

4. Climate change

The `I`iwi survives in habitat largely free of avian malaria (*Plasmodium reluctum*) and bird pox (*Aviapoxvirus*). Such habitat is currently limited to 8.9 ha (22 acres) on Kauai, 2,632 ha (6,500 acres) on Maui, and 6,478 ha (16,000 acres) on Hawaii Island, with virtually none on Oahu and Molokai (Pratt et al. 2009). The elevational advance of these pathogens driven by climate change immediately endangers the smaller `I`iwi populations on Kauai, Oahu, Molokai, and west Maui, and threatens the larger ones on east Maui and Hawaii Island.

4.1 `I`iwi is highly vulnerable to disease

Avian disease is a primary reason for the decline of I`iwi and other Hawaiian honeycreepers (Pratt 2005, Atkinson and LaPointe 2009). Warner (1968) demonstrated high susceptibility of honeycreepers that died from avian malaria and bird pox after experimental exposure to mosquito infested lower elevations where the birds were absent. Van Ripper et al. (1986) also provided experimental evidence of high susceptibility of I`iwi to avian malaria. More recently, Atkinson et al. (1995) experimentally exposed several species of honeycreepers to a single bite of a malaria infected mosquito and found that effects were most severe in `I`iwi with significantly higher mortality and clear manifestations of malaria disease at death. `I`iwi were infected by either single (low-dose) or multiple (high-dose) mosquito bites. Mortality in both groups was significantly higher than in uninfected controls, reaching 100% of high-dose birds and 90% (9 of 10) in low-dose birds.

While some individual I`iwi are known to have at least temporarily survived malaria, there is no evidence of population level tolerance or resistance to the disease. Atkinson et al. (1995) found that the one `I`iwi that survived malaria after a single experimental bite from an infected mosquito did not develop new parasitemia after multiple bites from infected mosquitoes. This indicated that `I`iwi are capable of an immunological response at least to the administered strain of malaria. Freed et al. (2005) discovered tolerance to malaria in two wild `I`iwi that successfully bred 2-years post infection. However, broken head feathers in these birds suggested physiological costs of malarial tolerance that could reduce survivorship of wild birds. Studies of experimentally infected birds indicate that tolerant birds likely retain chronic infection for life (Atkinson et al. 2001, Valkiunas 2005). Challenges to the immune system by stress or excessive energy expenditure can result in recrudescence of a chronic infection to higher parasitemia levels (Freed et al. 2005). Infected birds lose weight and suffer malaria related pathologies (Atkinson et al. 2001), and would be expected to be more susceptible than healthy birds to predation, competition, avian pox, unfavorable weather, and other stressors. A comparison of infection incidence in `I`iwi and other Hawaii forest birds suggests that few `I`iwi survive exposure in the wild (Atkinson et al. 2005).

It is uncertain if exiting larger populations of `I`iwi on Maui and Hawaii Island could evolve tolerance rapidly enough to avoid extinction from increased malaria parasitism. This would depend on exposure rapidity, the extent of current disease tolerance, if any, the virulence of *Plasmodium* strains, patterns of selection and genetic drift, rates of evolution in hosts, vector, and pathogen, and other factors. The avian disease system on Hawaii would be further complicated if new reservoir hosts or vectors enter the picture (Atkinson and LaPointe 2009).

Lethal effects of avian poxvirus have also been experimentally demonstrated in Hawaiian honeycreepers (Jarvi et al. 2008). Freed et al. (2005) found a dead `I`iwi in the field with massive poxvirus sores on its ankles. The bird also tested positive for malaria. A significantly high proportion of Hawaiian forest birds with avian pox also had chronic malaria, suggesting interaction between the two diseases (Atkinson et al. 2005).

The downward trajectory of `I`iwi populations (Table 1) indicates a pattern of decline similar to Hawaiian forest birds already acknowledged to be endangered and very vulnerable to disease, and dissimilar to populations of the unlisted Amakihi (*Hemignathus* spp.) (Shehata et al. 2001, Woodworth et al. 2005) and Apapane (*Himatione sanguine*) (Atkinson et al. 2005) which have shown some disease resistance and population persistence at lower elevations.

Among the most endangered Hawaiian bird species, the `Ō`ū, (*Psittirostra psittacea*), like the `I`iwi, was widespread on all main islands across a wide range of habitats a century ago (USFWS 2006). However, Ō`ū primarily inhabited the lower to mid-elevation forests where the impact of introduced mosquito-borne diseases was first manifested. Today, the `Ō`ū is probably extinct. Similar widespread exposure of `I`iwi to avian diseases can be expected in coming decades as a consequence of climate change.

4.2 Disease will spread over `I`iwi range as ambient temperatures rise

Avian malaria in Hawaii has been mostly confined to elevations below 1500 m (van Riper et al. 1986) where cool temperatures limit mosquito presence and development of the malaria parasite (LaPointe 2000). Recent climate modeling, however, has projected avian malaria to reach elevations up to or beyond 1900 m within this century, affecting most if not all remaining forest bird habitat (Benning et al. 2002).

Benning et al. (2002) modeled changes in malaria prevalence for Hawaiian honeycreepers at high quality habitat sites, assuming a 2° Celsius (C) increase in regional temperatures (based on International Panel on Climate Change 2007 projections; see Meehl et al. 2007). Current low-risk habitat diminished by 57% (665 to 285 ha) at the Hanawi Natural Area Reserve, Maui. Low-risk habitat at the Hakalau National Wildlife Refuge on Hawaii Island declined by 96% (3,120 to 130 ha). On Kauai (the Alakai Swamp), currently with little or no malaria free habitat, a 2° C warming placed most habitat (84%) at highest risk for malaria infection in native birds. Current mean ambient temperatures are believed to already allow limited disease transmission throughout Kauai as all `I`iwi habitat occurs below 1600 m elevation (LaPointe et al. 2005).

The effects of a 2° C warming would almost certainly eliminate the small `I`iwi populations from the lower-elevation islands of Kauai, Molokai, and Oahu, and from West Maui. Larger populations on East Maui and Hawaii Island would be expected to decline severely in a manner corresponding to decreases (~60-96%) in high elevation, disease-free refuges (Atkinson and LaPointe 2009).

The prognosis for `I`iwi and many other native forest birds appears worse than indicated by the Benning et al. (2002) model. The model assumed an increase of 2° C above current temperature, corresponding to ~2.7° C increase above pre-industrial levels. However, recent analysis of global heating indicates that temperature increases in Hawaii and elsewhere are unlikely to be limited to 2° C in this century. Increases in global temperature are currently on a trajectory to reach 2° C (above pre-industrial levels) by mid-century and about 5° C by 2100 (Meinshausen et al. 2009, Sokolov et al. 2009). Global greenhouse gas emissions would need to be halved by 2050 (from 1990 levels) to keep near the 2° C level with a high probability (55-88%) (Meinshausen et al. 2009). Unfortunately, under current multi-national policies regarding greenhouse gas emissions, there is virtually no chance of limiting heating to 2° C even with full policy implementation (Rogelj et al. 2009). For Hawaii, only a low global emissions scenario would likely keep temperature increases to 2° C (Karl et al. 2009).

An added concern is the risk of abrupt increases in global temperature unaccounted for in most modeled climate projections (Lovelock 2009). For example, a global climate model used by Sokolov et al. (2009) did not fully incorporate positive feedbacks that may occur, for example, if increased temperatures cause a large-scale melting of permafrost in arctic regions and subsequently release large quantities of methane, a very potent greenhouse gas (Rice 2009). If these positive feedback loops should occur, and evidence in mounting that they will (McCarthy 2010), temperatures are likely to increase to an even greater degree in Hawaii.

For Hawaii, Giambelluca et al. (2008) document a long-term increase in temperature and an accelerated rate of increase over the past few decades consistent with global trends (0.04° C C/decade over an 88-year period, and about 0.2° C/decade since 1975). Moreover, since 1975 higher elevation temperatures exceeded average warming (a 0.27° C/decade increase) with steepest increases in minimum (night time) temperature (near 0.5° C/decade), which is likely the most limiting for malaria transmission. The recent surface temperature trend in Hawaii is only slightly lower than the overall global trend. Similar surface warming has been detected elsewhere in the Pacific, and is associated with an increase in sea surface temperatures, upper ocean heat content, and sea level height (Richards and Timmermann 2008).

In Hawaii, the upper limit of mosquito presence appears to have increased substantially, from about 600 m in the late 1960s to 1100-1500 m in recent decades (Pratt 2005). Freed et al. (2005) reported that prevalence of malaria in Hawaiian forest birds at 1900 m on the island of Hawai'i more than doubled over a decade. A highly significant increase of malaria in `I`iwi was associated with much warmer summertime air temperatures. The 13° C threshold for malaria development projected for 1900 m sites by the conservative Benning et al. (2002) model was surpassed in 2001 by a wide margin (4.4° C; Freed et al. 2005). Measured temperatures were believed to exceed model expectations because the site was strongly affected by the island's trade wind inversion layer related to tropical air circulation. The altitude of the inversion has averaged 1900 m, above which cooler, drier conditions prevail (*Atlas of Hawaii*, 3rd edition). The response of the inversion layer to climate heating is uncertain (Pounds et al. 1999, Loope and Giambelluca 1998). If the inversion layer rises, disease epizootics could become commonplace at higher elevations with devastating short-term consequences for `I`iwi. If the inversion falls, and higher temperatures become associated with high-elevation drought, the effects would be very damaging to upper elevation Hawaiian forests and ultimately to surviving honeycreepers including the `I`iwi (Benning et al. 2002). Given that scenario, or if the inversion layer remains stable, high-elevation forest bird populations may be squeezed between expanding disease transmission from lower elevations and the upper limits of suitable habitat (Atkinson and LaPointe 2009). Hawaii may see an increased frequency of heavy rain events and increased rainfall during summer months (Karl et al. 2009), conditions that, along with increased temperature, are likely to facilitate breeding of malaria-carrying mosquitoes (Ahumada et al. 2004). At the same time, overall annual precipitation for the Hawaiian Islands may decline (Chu and Chen 2005) thereby affecting habitat quality (e.g., 'ōhi'a forest) for the `I`iwi.

4.3 Confounding population stressors and threats

Ectoparasites, particularly chewing lice (Phthiraptera), may impact `I`iwi by increasing morbidity and reducing the ability of birds to survive environmental challenges. Freed et al. (2008) documented an explosive increase in the prevalence of chewing lice in all bird host species at a study site on Hawaii Island. The number of major fault bars in wing and tail feathers, a sign of nutritive stress, was correlated with intensity of infection, suggesting an indirect cost to parasitized birds. Poorer body condition preceded the outbreak indicating the synergistic effect of multiple stressors on forest birds. At a minimum, chewing lice will increase food requirements of hosts. This indirect cost may be especially relevant because it can affect the ability of birds to mount a sufficient immune defense against diseases like avian malaria and pox. Chewing lice may also directly contribute to bird mortality (Freed et al. 2008).

Additional risks to `I`iwi from disease include potential introductions of West Nile virus, new avian malaria vectors (such as temperate varieties of *Culex quinquefaciatus*), or biting midges (Culicoides) that transmit avian diseases.

Introduced rats are serious predators on adults and nests of Hawaiian forest birds, and are abundant in high elevation habitats (Atkinson 1977, Scott et al. 1986, Fancy and Ralph 1998, VanderWerf and Smith 2002). Feral cats, introduced small Indian mongoose, and the native Short-eared Owl and introduced Barn Owl may also impact native Hawaiian birds (Scott et al 1986; Kowalsky et al. 2002). Predator control efforts generally have not been conducted over areas large enough to result in significant improvement in the status of imperiled forest birds (USFWS 2006). Logistical and other obstacles to predator control can be great, especially in rugged bird habitat.

Epizootics involving avian malaria or other pathogens could quickly eliminate remaining `I`iwi from the lower elevation islands of Kauai, Oahu, and Molokai, and from west Maui in the near term, and could diminish and fragment 'I'iwi populations on higher elevation east Maui and Hawaii Island. There is currently no habitat on Kauai, Oahu, and Molokai where mean ambient temperature entirely restricts malaria development (Benning et al.

2002). These islands are vulnerable to avian malaria at all elevations on a more or less ongoing basis. A recent avian malaria outbreak on Hawaii Island was associated with increases in summertime temperatures related to tropical inversion layer conditions (Freed et al. 2005). Outbreaks of malaria can be triggered by warm periods linked to inversion layer dynamics or El Niño events, and will likely intensify and persist longer with ongoing climate change.

Hurricanes are known for their devastating effects on island birds (Foster et al. 2004). They reduce habitat by blowing down trees and by creating forest openings that facilitate the spread of invasive alien plants. The `I`iwi decline on Kauai after a 1992 hurricane may have partially resulted from the birds seeking substitute nectar resources at lower elevations where risk of malaria transmission is highest (Foster et al. 2004).

Hurricanes are likely to intensify in a warmer climate (Meehl et al. 2007) in terms of wind speeds and precipitation, though the number of storms may be fewer (Bengtsson et al. 2007). Infectious mosquitoes can be carried upslope in strong winds, a probable factor in malaria outbreaks on Hawaii above 1900 m elevation (Freed et al. 2005).

On Hawaii Island, volcanism presents a potential threat to substantial acreage of forest bird habitat. For example, a large portion of the Upper Waiākea Forest Reserve, location of some of the last observations of `Ō`ū and considered prime habitat for the species, was inundated by the 1984 Mauna Loa lava flow which destroyed thousands of acres of forest and created a treeless corridor over 1 km wide (USFWS 2006).

5. Conservation

Current regulations by the U.S. government and the State of Hawaii are inadequate to conserve high elevation forests needed to buffer the `I`iwi and other susceptible forest birds against the upslope advance of avian diseases driven by global heating. While some progress has been made to re-forest former upper elevation habitat areas with native trees and reduce or eliminate harmful alien species from existing ones, huge tracts of land needed for forest bird conservation in Hawaii remain degraded or without native tree cover (USFWS 2009). A preponderance of lands intended for forest bird recovery are not managed conservation lands (Pratt et al. 2009). Management actions identified in existing forest bird conservation plans have not been implemented at ecologically relevant scales, and successful efforts to restore higher elevation forests must occur across tens of thousands of areas, not hundreds (Scott 2009). On the Island of Maui, for example, more than half of the lands identified for forest bird recovery remain without native forests, have only remnant forest patches, or are dominated by introduced tree species and other alien vegetation (personal observation). Yet restoration of high elevation koa/`ōhi`a forest to protect native birds is clearly a stated conservation priority (Scott et al.1986, USFWS 2006)

At current rates, reforestation and forest enhancement efforts for Hawaiian forest birds will not achieve habitat conservation goals in time to build and expand populations robust enough to withstand avian malaria and other consequences of climate change. Of over 140 actions for forest bird recovery relating to reforestation and securing recovery areas (USFWS 2006), 61% have not begun, 37% are ongoing, and only 2% are complete or partially so (USFWS 2009). Likewise, of more than 160 actions designed to reduce or eliminate exotic ungulates and mammalian bird-predators, 71% are not yet underway, 27% are ongoing, and less than 2 % are complete or partially complete.

Poor political and policy decisions are responsible for the current inadequacy of management to prevent forest bird extinctions. The problem includes conflicting management goals and policies, most notably involving state forest lands (USFWS 2006, 2009), and failure to provide necessary funding (Leonard 2008).

Leonard (2009) discusses political obstacles to saving Hawaiian forest birds, including a state mandate to provide public hunting opportunities of exotic ungulates even where incompatible with conservation of native birds. Actions such as fencing and ungulate control for bird conservation may result in the loss of hunting areas, which is very controversial within his state agency (Leonard 2008). Even proposals for protecting limited forest in areas of little or no public access receive fierce opposition from local hunters (San Nicolas 2010). Native forest restoration is also hampered by agency decisions favoring exotic tree species or leasing for livestock (USFWS 2006).

In terms of addressing climate change, existing international and U.S. regulatory goals to reduce global greenhouse gas emissions are clearly inadequate to safeguard the `I`iwi against climate related extinction. As discussed, severe shrinkage of habitat absent of or at low risk of avian disease is expected with a 2° C rise in ambient temperature. While the 2009 U.N. Climate Change Conference in Copenhagen called on countries to hold the increase in global temperature below 2° C, the *non-binding* "Copenhagen Accord" that emerged from the conference fell way short of that goal. A summary by the Pew Center (2010) of four analytical reviews of the Accord found that collective national pledges to cut greenhouse gas emissions are inadequate to achieve the 2° C goal, and instead suggest emission scenarios leading to a 3 to 3.9° C warming.

Economic growth is the most significant factor driving projected increases in carbon dioxide emissions, as the world continues to rely on fossil fuels for most of its energy use (USEIA 2009). Yet the prevailing economic and political framework for the U.S. and most other countries is to maximize growth as a priority. In a high growth scenario, world carbon dioxide emissions increase at an average rate of 1.8 percent annually from 2006 to 2030, as compared with 1.4 percent under standard assumptions (USEIA 2009).

The United States is responsible for over 20% of worldwide carbon dioxide emissions (USEIA 2004). While the U.S. Environmental Protection Agency (EPA) currently has some authority to regulate greenhouse gas emissions, the agency bends under political pressure (Bravender and Samuelsohn 2010) and has not set targets or standards to protect the 'I'iwi or other wildlife. Prospects for regulations within the foreseeable future adequate to stem the climate-change threat to the `I`iwi are very poor. For example, the U.S. Congress has failed to pass climate change legislation and, as of early 2011, is considering bills to block EPA's limited authority to regulate greenhouse gas emissions (New York Times 2011).

The nation's top wildlife agency, the US Fish and Wildlife Service (USFWS), is focused on a climate adaptation strategy for wildlife in general but with little, if any, emphasis on regulation of greenhouse gas emissions (USFWS 2010). The agency has been urged to promote reductions in emissions while expediting upper elevation habitat restoration in conservation plans for endangered Hawaiian forest birds (Povilitis and Suckling 2010).

The 'I'iwi is not included on the USFWS list of endangered species and therefore does not merit the conservation provisions of the U.S. Endangered Species Act (ESA), such as the protection of its "critical habitat." Also, like other Hawaiian honeycreepers, it is not protected under the U.S. Migratory Bird Treaty Act. In 2010, a formal request was made to the USFWS to list the 'I'iwi under the ESA and designate and protect critical habitat for the species (Center for Biological Diversity 2010). While the listing of species endangered by climate change is controversial because of an overall backlog of listing requests, new administrative strategies or procedures will be needed if the U.S. is to fulfill its commitment to safeguard endangered wildlife (Woody 2011).

The ESA defines the term "critical habitat" to mean specific areas essential to the conservation of the species which may require special management considerations or protection (ESA 1973). Critical habitat designation would legally ensure that U.S. government actions avoid jeopardizing the species and promote its conservation. Designation would alert federal and state agencies and private landowners to the need for habitat management and restoration actions in areas essential for the 'I'iwi, and require remedies to institutional conflicts that undermine habitat conservation. Specific measures to reduce the climate change/disease threat include reforestation, elimination or control of alien species inimical to the survival of `I`iwi, and special measures to monitor and reduce (or eliminate) occurrence of avian malaria vectors. Programs to re-establish native forests, reduce rat depredation, control weeds, and fence out and remove ungulates are essential for forest bird recovery in high elevation habitats that serve as native bird refugia (Gorresen et al. 2005). Reducing mortality in key habitat areas, such as that caused by rodent predation, may lessen the threat from disease by improving survival and reproduction of any birds with disease tolerance or natural immunity (VanderWerf and Smith 2002). The evolutionary acceleration of disease resistance through rodent control is possible (USFWS 2006).

Critical habitat should include all areas needed to provide sufficient forested habitat to support viable or potentially viable `I`iwi populations on Kauai, Oahu, Molokai, Maui, and Hawaii Island, as each island represents a significant portion of the species' natural range. This should include areas on Maui and Hawaii Island above the current limit of tree growth to accommodate any forest expansion resulting from climate change. Critical habitat designation for `I`iwi would extend habitat protection to other listed endangered Hawaiian birds, where ranges overlap since most currently listed forest birds do not have critical habitat designations.

6. Conclusion

The best available science indicates that global warming will allow avian diseases to spread throughout most or all of the `I`iwi's geographic range. The `I`iwi is highly vulnerable to avian diseases and cannot sustain itself where disease prevails. `I`iwi in the central portion of the species' range (Oahu, Molokai, and west Maui) are critically endangered because of small population sizes and exposure to malaria. Those to the west (on Kauai) are severely threatened as disease free habitat is fast disappearing. `I`iwi in the eastern portion of the range (east Maui and Hawaii Island) face further population declines and eventual extinction with ongoing climate change. The `I`iwi merits protection under the US Endangered Species Act that includes designation and conservation of critical habitat.

7. References

Ahumada, J. A., D.A. LaPointe and M. D. Samuel. 2004. Modeling the population dynamics of Culex quinquefasciatus (Diptera: Culicidae) along an elevational gradient in Hawaii. Journal of Medical Entomology 41(6):1157-1170.

Amante-Helweg, V.L.U. and S. Conant. 2009. Hawaiian culture and forest birds in Conservation Biology of Hawaiian Forest Birds: Implications for Island Avifauna. T. K. Pratt, C.T. Atkinson, P. C. Banko, J. D. Jacobi, B. L. Woodworth (eds.). Yale University Press. New Haven.

Atkinson, I. A. E. 1977. A reassessment of factors, particularly Rattus rattus L., that influenced the decline of endemic forest birds in the Hawaiian Islands. Pacific Science 31:109-133.

Atkinson, C. T., R. J. Dusek, and J. K. Lease. 2001. Serological responses and immunity to superinfection with avian malaria in experimentally-infected Hawai'i 'Amakihi. Journal of Wildlife Diseases 37:20–27.

Atkinson, C. T. and D. A. LaPointe. 2009. Ecology and pathogenicity of avian malaria and pox. Chapter 9 in Conservation Biology of Hawaiian Forest Birds: Implications for Island Avifauna. Carter T. Atkinson, Paul Christian Banko, James D. Jacobi, Bethany Lee Woodworth. Yale University Press. New Haven.

Atkinson, C.T., J. K. Lease, R. J. Dusek, and M. D. Samuel. 2005. Prevalence of pox-like lesions and malaria in forest bird communities on leeward Mauna Loa volcano, Hawaii. Condor 107:537-546.

Atkinson, C. T., K. L. Woods, R. J. Dusek, L. Sileo, and W. M. Iko. 1995. Wildlife disease and conservation in Hawaii: Pathogenicity of avian malaria (Plasmodium relictum) in experimentally infected 'I'iwi (Vestiaria coccinea). Parasitology 111:S59-S69.

Banko, W. E. and P. C. Banko. 1976. Role of food depletion by foreign organisms in historical decline of Hawaiian forest birds. Pages 29-43 in Proceedings of 1st conference in Natural sciences, Hawai'i Volcanoes National Park, Hawai'i.

Banko, W. E. and P. C. Banko. 2009. Historic decline and extinction in Conservation Biology of Hawaiian Forest Birds: Implications for Island Avifauna. T. K. Pratt, C.T. Atkinson, P. C. Banko, J. D. Jacobi, B. L. Woodworth (eds.). Yale University Press. New Haven.

Bengtsson, L., K. I. Hodges, M. Esch, N. Keenlyside, L. Kornblueh, J.-J. Luo, and T. Yamagata. 2007. How may tropical cyclones change in a warmer climate? Tellus 59A:539-561.

Benning, T. L., D.A. LaPointe, C.T. Atkinson, and P. M. Vitousek. 2002. Interactions of climate change with biological invasions and land use in the Hawaiian Islands: Modeling the fate of endemic birds using a geographic information system. Proceedings of the National Academy of Sciences 99(22):14246-14249.

Bravender, R. and D. Samuelsohn. 2010. EPA's gradual phase in of GHG Regs garners qualified praise from Senators. New York Times. Feb. 23.

Camp, R. J., P. M. Gorresen, B. L. Woodworth, and T. K. Pratt 2002. Preliminary analysis of forest bird survey data for national parks in Hawaii: Hawaii Volcanoes National Park 1977-2000. Interim Report, May 2002. Hawaii Forest Bird Interagency Database Project, Pacific Island Ecosystems Research Center, Biological Resources Division, U.S.G.S.

Camp, R. J., T. K. Pratt, P.M. Gorresen, J. J. Jeffrey, and B. L. Woodworth. 2009a. Passerine bird trends at Hakalau Forest National Wildlife Refuge, Hawai'i. Hawai'i

Cooperative Studies Unit Technical Report HCSU-011. University of Hawai'i at Hilo.

Camp. R. J, B. L. Woodworth, T. K. Pratt, C. Collins, and H. Howitt. 2009b. The Hawaii bird interagency database project. http://biology.usgs.gov/pierc/HFBIDPSite/HCC_2000_Poster.pdf Accessed 9-22-09

Center for Biological Diversity. 2010. Petition to list the 'I'iwi (*Vestiaria coccinea*) as threatened or endangered under the U.S. Endangered Species Act. Tucson, Arizona. U.S.A.. August 24.

Chu, P.-S., and H. Chen, 2005: Interannual and interdecadal rainfall variations in the Hawaiian Islands. Journal of Climate 18:4796-4813.

Fancy, S. G. and C. J. Ralph. 1998. `I`wi (Vestiaria coccinea) in A. Poole and F. Gill, eds. The Birds of North America, No. 327. The Birds of North America, Inc., Philadelphia, PA, U.S.A.

Foster, J. T., E. J. Tweed, R. Camp, B. L. Woodworth, C. Adler, and T. Telfer. 2004. Long-term population changes of native and introduced birds in the Alaka'i Swamp, Kaua'i. Conservation Biology 18(3):716-725.

Freed, L. A., R. L. Cann, M. L. Goff, W. A. Kuntz, and G. R. Bodner. 2005. Increase in avian malaria at upper elevation in Hawaii. Condor 107:753–764.

Freed, L. A., M. C. Medeiros, and G.R. Bodner. 2008. Explosive increase in ectoparasites in Hawaiian forest birds. Journal of Parasitology 94(5):1009–1021.

Giambelluca, T. W., H.F. Diaz, and M. S. A. Luke. 2008. Secular temperature changes in Hawaii. Geophysical Research Letters 35:L12702.

Gorresen, P.M., R. J. Camp, and T. K. Pratt. 2007. Forest Bird Distribution, Density and Trends in the Ka`ū Region of Hawai`i Island. Open-File Report 2007–1076.

Gorresen, P.M., R.J. Camp, T. K Pratt, and B. L. Woodworth. 2005, Status of forest birds in the central windward region of Hawai`i Island: population trends and power analyses: U.S. Geological Survey, Biological Resources Discipline, Open-File Report 2005-1441.

Gorresen, P. M., R. J. Camp, M. H. Reynolds, B. L. Woodworth, and T. K. Pratt. 2009. Status and trends of native Hawaiian songbirds. Chapter 5 in Conservation Biology of Hawaiian Forest Birds: Implications for Island Avifauna. T. K. Pratt, C.T. Atkinson, P. C. Banko, J. D. Jacobi, B. L. Woodworth (eds.). Yale University Press. New Haven.

Harvell, C. D., C. E. Mitchell, J. R. Ward, S. Altizer, A. P. Dobson, R. S. Ostfeld, and M. D. Samuel. 2002. Climate warming and disease risks for terrestrial and marine biota. Science 296:2158–2162.

Hess, S.C. 2008. Wild sheep and deer in Hawaii - a threat to fragile ecosystems. USGS, Hawaii National Park, Hawaii.

Jarvi, S. I., D. Triglia, A. Giannoulis, M. Farias, K. Bianchi, and C. T. Atkinson. 2008. Diversity, origins and virulence of Avipoxviruses in Hawaiian Forest Birds. Conservation Genetics 9:339–348.

Karl, T. R., J. M. Melillo, and T. C. Peterson (editors). 2009. Global climate change impacts in the United States. Cambridge University Press, New York.

Kowalsky, J. R., T. K. Pratt, & J. C. Simon. 2002. Prey taken by feral cats (Felis catus) and barn owls (Tyto alba) in the Hanawi Natural Area Reserve, Maui, Hawai'i. 'Elepaio 62:129-131.

LaPointe, D. A. 2000. Avian malaria in Hawai'i: the distribution, ecology, and vector potential of forest-dwelling mosquitoes. Ph.D. dissertation, University of Hawaii at Manoa, Honolulu, HI.

LaPointe, D., T. L. Benning, and C. Atkinson. 2005. Avian malaria, climate change, and native birds of Hawaii. In Climate change and biodiversity (T.E. Lovejoy and L. Hannah eds). Yale University Press, New Haven.

Leonard, D. L. 2008. Recovery expenditures for birds listed under the US Endangered Species Act: the disparity between mainland and Hawaiian taxa. Biological Conservation 141:2054–2061.

Leonard, D. L. 2009. Social and political obstacles to saving Hawaiian birds: Realities and remedies. Chapter 24 in Conservation Biology of Hawaiian Forest Birds: Implications for Island Avifauna. T. K. Pratt, C.T. Atkinson, P. C. Banko, J. D. Jacobi, B. L. Woodworth (eds.). Yale University Press. New Haven.

Loope, L.L. and T.W. Giambelluca. 1998. Vulnerability of island tropical montane cloud forests to climate change, with special reference to East Maui, Hawaii. Climatic Change 39:503-517.

Lovelock, J. 2009. The vanishing face of Gaia. Basic Books. New York.

McCarthy, J. 2010. Reflections on: our planet and its life, origins, and futures. Science 326:1646-1655.

Meehl, G. A., T. F. Stocker, W. D. Collins, P. Friedlingstein, A. T. Gaye, J. M. Gregory, A. Kitoh, R. Knutti, J. M. Murphy, A. Noda, S. C. B. Raper, I. G. Watterson, A. J. Weaver, and Z.-C. Zhao, 2007: Global climate projections. In: Climate Change 2007: The Physical Basis. Contribution of Working Group I to the Fourth Assessment Report of the Intergovernmental Panel on Climate Change [Solomon, S., D. Qin, M. Manning, Z. Chen, M. Marquis, K.B. Averyt, M. Tignor, and H.L. Miller (eds.)]. Cambridge University Press, Cambridge, UK, and New York, pp. 747-845.

Meinshausen, M., N. Meinshausen, W. Hare, S. C. B. Raper, K. Frieler, R. Knutti, D. J. Frame, and M.R. Allen. 2009. Greenhouse-gas emission targets for limiting global warming to 2 degrees C. Nature 458:1158-1162.

Mitchell, C, C. Ogura, D. W. Meadows, A. Kane, L. Strommer, S. Fretz, D. Leonard, and A. McClung. 2005. Hawaii's Comprehensive Wildlife Conservation Strategy. Department of Land and Natural Resources. Honolulu, Hawai'i.

Mountainspring, S., and J. M. Scott. 1985. Interspecific competition among Hawaiian forest birds. Ecological Monographs 55:219-239.

New York Times. 2011. Global Warming. Science Topics. Updated: Jan. 13, 2011
http://topics.nytimes.com/top/news/science/topics/globalwarming/index.html

Pew Center on Global Climate Change. 2010. http://www.pewclimate.org/copenhagen-accord Accessed May 5.

Pounds, J. A., M. P. L. Fogden and J. H. Campbell. 1999. Biological response to climate change on a tropical mountain. Nature 398:611-615.

Povilitis, A. and K. Suckling. 2010. Addressing climate change threats to endangered species in U.S. recovery plans. Conservation Biology 24(2):372-376.

Pratt, D. H. 2005. The Hawaiian honeycreepers. Oxford University Press. Oxford.

Pratt, T. K. 2009. Origins and evolution. Chapter 1 in Conservation Biology of Hawaiian Forest Birds: Implications for Island Avifauna. T. K. Pratt, C.T. Atkinson, P. C. Banko, J. D. Jacobi, B. L. Woodworth (eds.). Yale University Press. New Haven.

Pratt, T. K., C. T. Atkinson, P. C. Banko, J. D. Jacobi, B. L. Woodorth, and L. A. Mehrhoff. 2009. Can Hawaiian forest birds be saved? Chapter 25 in Conservation Biology of Hawaiian Forest Birds: Implications for Island Avifauna. T. K. Pratt, C.T. Atkinson, P. C. Banko, J. D. Jacobi, B. L. Woodworth (eds.). Yale University Press. New Haven.

Price, J.P., J. D. Jacobi, L.W. Pratt, F.R. Warshauer, and C.W. Smith. 2009. Protecting forerst bird populations across landscapes. Chapter 16 in Conservation Biology of Hawaiian Forest Birds: Implications for Island Avifauna. T. K. Pratt, C.T. Atkinson, P. C. Banko, J. D. Jacobi, B. L. Woodworth (eds.). Yale University Press. New Haven.

Primack, R. B. 2006. Essentials of Conservation Biology. 4th Edition. Sinauer Associates Sunderland, MA.

Reynolds, M. H. and T. J. Snetsinger. 2001. The Hawai`i rare bird search 1994-1996. Pp. 133-143 in J. M. Scott, S. Conant, and C. van Riper, III, eds. Evolution, ecology, conservation, and management of Hawaiian birds: a vanishing avifauna. Studies in Avian Biology No. 22. Cooper Ornithological Society. Allen Press, Lawrence, KS, U.S.A.

Rice, D. 2009. Global warming may be twice as bad as previously expected. USA Today. May 21.

Richards, K. and A. Timmermann. 2008. Climate change and biodiversity in Melanesia. Regional Climate Change Projections for the Southwestern Pacific with a focus on Melanesia International Pacific Research Center, School of Ocean and Earth Science and Technology, University of Hawaii CCBM Paper 1. Bishop Museum Technical Report 42(1).

Rogelj, J., B. Hare, J. Nabel, K. Macey, M. Scheffer, K. Markmann, and M. Meinshausen. 2009. Halfway to Copenhagen, no way to 2 C. Nature Reports 3:81-83.

San Nicolas, C. 2010. Hunters take aim at plan to preserve forest. The Maui News. April 30. http://www.mauinews.com/page/content.detail/id/531038.html?nav=10

Scott, J. M. 2009. Forward. in Conservation Biology of Hawaiian Forest Birds: Implications for Island Avifauna. T. K. Pratt, C.T. Atkinson, P. C. Banko, J. D. Jacobi, B. L. Woodworth (eds.). Yale University Press. New Haven.

Scott, J. M., S. Mountainspring, F. L. Ramsey, and C.B. Kepler. 1986. Forest bird communities of the Hawaiian Islands: their dynamics, ecology, and conservation. Studies in Avian Biology No. 9. Cooper Ornithological Society. Allen Press. Lawrence.

Shehata, C. L., L. A. Freed, and R. L. Cann. 2001. Changes in native and introduced bird populations on O'ahu: infectious diseases and species replacement. Studies in Avian Biology 22:264–273.

Smith, T. B., L. A. Freed; J. Kaimanu Lepson and J. H. Carothers. 1995. Evolutionary Consequences of Extinctions in Populations of a Hawaiian Honeycreeper. Conservation Biology 9(1):107-113.

Sokolov, A. P., P. H. Stone, C. E. Forest, R. Prinn, M. C. Sarofim, M. Webster, S. Paltsev, C. A. Schlosser, D. Kicklighter, S. Dutkiewicz, J. Reilly, C. Wang, B. Felzer, and H. D. Jacoby. 2009. Probabilistic forecast for 21st century climate based on uncertainties in emissions (without policy) and climate parameters. Journal of Climate DOI: 10.1175/2009JCLI2863.1.

U.S. Endangered Species Act (ESA) 1973. U.S. Code 16, chap. 35.

U.S. Energy Information Administration (USEIA). 2004. International energy outlook. http://www.eia.doe.gov/oiaf/ieo/emissions.html. Accessed May 5.

U.S. Energy Information Administration (USEIA). 2009. International energy outlook. http://www.eia.doe.gov/oiaf/ieo/emissions.html. Accessed May 5.

U.S. Fish and Wildlife Service (USFWS). 2006. Revised Recovery Plan for Hawaiian Forest Birds. Region 1, Portland.

U.S. Fish and Wildlife Service (USFWS) 2009. USFWS website. Accessed 16 September 2009. https://ecos.fws.gov/roar/pub/planImplementationStatus.action?documentId=1001509

U.S. Fish and Wildlife Service (USFWS). 2010. National Fish and Wildlife Climate Adaptation Strategy. http://www.fws.gov/nfwcaspvgp.html Accessed 5 May 2010.

Valkiunas, G. 2005. Avian malaria parasites and other haemosporidia. CRC Press, Boca Raton.

VanderWerf, E. A., and J. L. Rohrer. 1996. Discovery of an 'I'iwi population in the Ko`olau Mountains of O`ahu. `Elepaio 56(4):25-28.

VanderWerf, E. A., and D. G. Smith. 2002. Effects of alien rodent control on demography of the O`ahu `Elepaio, an endangered Hawaiian forest bird. Pacific Conservation Biology 8:73-81.

van Riper, C. III, S. G. van Riper, M. L. Goff, and M. Laird. 1986. The epizootiology and ecological significance of malaria in Hawaiian land birds. Ecological Monographs 56:327–344.

Warner, R. E. 1968. The role of infectious diseases in the extinction of the endemic Hawaiian avifauna. Condor 70:101–120.

Woodworth, B. L., C. T. Atkinson, D. A. LaPointe, P. J.Hart, C. S. Spiegel, E. J. Tweed, C. Henneman, J. LeBrun, T. Denette, R. DeMots, K. L. Kozar, D. Triglia, D. Lease, A. Gregor, T. Smith, and D. Duffy. 2005. Host population persistence in the face of introduced vector-borne diseases: Hawaii amakihi and avian malaria. Proceedings of the National Academy of Sciences (USA) 102:1531–1536

Woodworth, B.L. and T. K. Pratt. 2009. Life history and demography. Chapter 8 in Conservation Biology of Hawaiian Forest Birds: Implications for Island Avifauna.

T. K. Pratt, C.T. Atkinson, P. C. Banko, J. D. Jacobi, B. L. Woodworth (eds.). Yale University Press.

Woody, G. 2011. Wildlife at Risk Face Long Line at U.S. Agency. New York Times. 20 April.

13

Possible Effects of Future Climate Changes on the Maximum Number of Generations of *Anopheles* in Monsoon Asia

Shunji Ohta and Takumi Kaga
Faculty of Human Sciences, Waseda University
Tokorozawa, Saitama 359–1192,
Japan

1. Introduction

The predicted increase in global temperatures is expected to affect ecosystem, human health and society. In particular, a longer growing period due to climate warming is expected to enhance the growth and reproduction of some kinds of organisms (Kiritani, 2006; Ohta & Kimura, 2007). Northern and altitudinal shifts in vector species that cause infectious diseases have been observed worldwide (Barker & Lindsay, 2000; Hales et al., 2002). Conversely, the endemicity of vector-borne diseases such as malaria has decreased during the past century (Gething et al., 2010). The reason for this decrease is the economic development and disease control during this period (Hay et al., 2009; Gething et al., 2010). However, the recession of malaria during the last century did not correspond to a reduction in the ranges of vector species. Because of the complex relationship between malaria and climate (Martin & Lefebvre, 1995), the potential distribution of vectors and climate conditions of their habitats without human interference need to be elucidated.

Malaria is the chief disease caused by vectors breeding on shallow surface waters. Because survival and reproduction rates of mosquitoes are mainly determined by the temperature and humidity of their habitats, climate factors determine whether a location is suitable for the transmission of a wide range of infectious diseases. Projected changes in temperatures and the hydrological cycle will cause changes in the geographical distributions and population dynamics of vectors, thus altering the patterns of infectious disease transmission (Martin & Lefebvre, 1995; Martens et al., 1999). In particular, one of the most interesting aspects is how climate change could affect the geographical distributions and incidence patterns of the diseases caused by mosquitoes.

However, most investigators assessing the impacts of climate change on the incidence and geographical range of malaria assume that mosquito distribution would not change under future climate conditions (Martens et al., 1999; van Lieshout et al., 2004). These investigators assessed the risk of malaria transmission based on the geographical distribution of particular mosquito species by country or administrative unit (World Health Organization, 1989; Jetten & Takken, 1994). However, mosquitoes are not always uniformly distributed within a country, and their range typically crosses national borders (Kashiwada & Ohta, 2010). Martin & Lefebvre (1995) developed the Malaria Potential Occurrence Zone model, which was applied with outputs derived from 5 General Circulation Models (GCMs) to provide first-order estimations of the changes in potential malaria transmission suitable for vectors; they revealed an increased incidence of seasonal malaria at the expense of perennial malaria. Rogers & Randolph (2000) assessed the current and future global distributions of malaria by using parasite dynamics with a GCM output and the estimated changes in human population living in malaria-transmission zones.

Most entomological studies are unable to incorporate a detailed distribution of the vector mosquito species, even though it is the basis for the risk assessment of malaria transmission. Monsoon Asia has a few observation sites of *Anopheles* mosquitoes (Kashiwada & Ohta, 2010). Thus, mechanistic model estimations of *Anopheles* distribution are necessary to enhance the existing data regarding their distribution. In addition, although vector distribution in Africa has been extensively studied, research on *Anopheles* in Asia, consisting of more than the half the entire human population, is sparse. Bhattacharya et al. (2006) studied the impacts of climate on *Anopheles* and malaria in India. Their study is the meaningful one of very few researches intended for Asian regions.

Therefore, we studied the effects of future climatic changes on the maximum number of generations per year of *Anopheles* mosquitoes in Monsoon Asia by using our model (Kashiwada & Ohta, 2010) and climate data in the absence of human interference. First, we studied the characteristics of changes in the energy and water balance of mosquito habitats owing to future climate changes. Second, we estimated the changes in the seasonal and geographical distribution patterns of *Anopheles* mosquitoes and discussed the changes in the potential distribution of malaria risk in Monsoon Asia under future climatic conditions.

2. Recent models for estimating vector distribution

Various types of approaches are used for estimating a range of meteorological variables for the vector species, depending on the purpose such as establishing an early warning system for malaria risk and making a high-resolution global map of the disease. Most studies use meteorological data, which are essential for estimating the vector distribution and malaria risk. However, meteorological records are usually obtained from urban areas, and coverage can be relatively sparse or inappropriate in less-developed countries (Martens & Thomas, 2005).

2.1 Integrated process-based epidemic model

An integrated and process-based model for estimating the "transmission or epidemic potential" index of a malaria-mosquito population (Craig et al., 1999; Martens et al., 1999; van Lieshout et al., 2004) improves the previous approaches by incorporating female mosquito density (a calculation of the basic reproduction rate), associated favorable climate conditions, and human population densities. This index approach is very useful for determining the risk of malaria transmission in a community where no mosquito observations have been conducted. However, climate conditions suitable for adult female mosquitoes are not always optimal for the immature stages of mosquitoes. Bayoh & Lindsay (2003) concluded that the optimal temperature range for *Anopheles* development narrows as the mosquito develops. Furthermore, the immature stages of mosquitoes live in water environments such as puddles, pools, or streams (Hoshen & Morse, 2005). Thus, the water temperature range suitable for the growth and development of the mosquitoes should be considered (Hopp & Foley, 2001; Kashiwada & Ohta, 2010).

Although some issues relate to the life history of mosquitoes, these epidemiological methods are useful and valuable for rapidly assessing malaria risk. Recently, Billingsley et al. (2005) successfully integrated these entomological techniques and compared malaria risk in different ecological and epidemiological settings. Conner & Mantilla (2008) attempted to develop and test integrated early warning systems that aim to provide early warnings regarding changes in epidemic risk in southern Africa based on seasonal climate forecasts. However, Ebi (2009) suggested that increased climate variability render early warning systems based on these variables more unreliable.

2.2 High-resolution map model

Hay et al. (2009) focused largely on malaria parasites and created a map of malaria endemicity within previously defined stable spatial limits of *Plasmodium falciparum* transmission by using a model-based geostatistical procedure and human population data. On the other hand, Martin & Lefebvre (1995) developed the Malaria Potential Occurrence Zone model with climate data that estimates malaria risk on the basis of the minimum and maximum temperatures and humidity required for the vector species. In addition, some species of mosquitoes in Africa have been mapped using simple climate data (Lindsay & Martens, 1998; Lindsay et al., 1998; Hay et al., 2002). Recent studies mapping vector species attempted to explain the geographic distribution of *Anopheles* and *Aedes* mosquitoes by analyzing climate variables for the mosquito observation sites by using niche-based distribution models (Foley et al., 2008, Medley, 2010) or a fuzzy logic model (Craig et al., 1999; Ebi et al., 2005). These studies have generated high-resolution maps of the vectors over large areas by using climate data; these maps are good indicators of the present distributions of the vectors. However, these maps do not represent the temporal occurrence of the vectors, because the calculation time intervals in these studies were typically at least 1 month. Mosquito development and the life cycle during the immature life stages occur at time scales ranging from several days to a few weeks (Hopp & Foley, 2001; Depinay et al., 2004).

2.3 Biological model

The most popular biological model expresses malaria parasites dynamics by using climate factors such as temperature (Martens et al., 1999; Rogers & Randolph, 2000). Other ecophysiological and entomological approaches have been employed to explain the temporal occurrence of vectors by describing their life histories. These models calculate development according to air temperature at each developmental stage of the insect, based on the assumption that the growth of an insect depends on the temperature of its habitat (Hopp & Foley, 2001; Ikemoto, 2003; 2005; 2008; Depinay et al., 2004; Pascual et al., 2006). Temperature affects the survival of the parasite only during the life cycle of *Anopheles* mosquitoes. According to experimental data, the optimal air temperature is between 22°C and 30°C; temperatures exceeding this value affect the activity of *Anopheles* mosquitoes. Although such ecophysiological models can generally explain the temporal population dynamics of pest insects at a specific site, they are unable to estimate the geographical distribution of the *Anopheles* vector (Depinay et al., 2004; Ikemoto, 2005). Therefore, the quantities of the available water and temperatures of the available aquatic habitats, which vary rapidly, must be incorporated into the model, because immature mosquitoes develop in natural or artificial water bodies as mentioned above. However, the volumes and temperatures of water in the small puddles in which immature mosquitoes live have not been measured over a broad area.

2.4 Habitat-oriented model

Besides simple climatic variables such as air temperature and precipitation, the activities of the mosquitoes depend on the variables of their native habitats such as water and soil conditions. Lindsay et al. (1998) showed that the activities of *Anopheles gambiae* and *Anopheles arabiensis* depend on the humidity index (the ratio of precipitation to potential evaporation during the optimum 5 months). The Malaria Potential Occurrence Zone model (Martin & Lefebvre, 1995) is a numerical model based on physics and biology and estimates whether environmental conditions are favorable for both the malaria parasite and its vectors. A characteristic of this model is that it uses a parameter for minimum atmospheric moisture, which is defined as the ratio of precipitation to potential evaporation obtained from simple meteorological data. The water conditions in the abovementioned studies are monthly or annual mean values. Hopp & Foley (2001) expressed the time variation of the surface water balance and soil moisture content by using a simple bucket model. In addition, air temperature instead of water temperature has been used in the model; this variable has been used to obtain the distribution of *Aedes* mosquitoes (Hopp & Foley, 2001). To determine the distributions of mosquitoes, the simple bucket model of Hopp & Foley is a very effective tool for developing accurate data for the daily water budget at the soil surface. This water balance model has been developed mainly in the fields of boundary-layer meteorology and agricultural meteorology.

3. Model description

The abovementioned review study revealed that many types of model construction require different data to estimate the distributions of mosquitoes. In this study, we selected the *Anopheles* mosquito, because some species live particularly in Monsoon Asia. Although various *Anopheles* species are located at a given site, we used the physiological data of some species available from references to consider their broad-scale distribution. The life history of *Anopheles* mosquitoes is described as a function of climate variables on the basis of an ecophysiological approach coupled with models of the energy balance of surface water and soil.

We constructed a model consisting of 3 parts: the primary part of the model calculates the growth of the mosquito (section 3.1), and the 2 subprocesses of the model determine the moisture conditions of the mosquito habitat (section 3.2) and calculate water temperature (section 3.3). The calculation time interval for all 3 parts of the model is 1 day. Methods using these models provide simultaneous spatial and temporal distributions of *Anopheles* mosquitoes at a fine resolution.

3.1 Mosquito growth model

The daily development of *Anopheles* mosquitoes was calculated using the temperature-dependent growth rate when the immature mosquitoes had an available water body to live in. This developmental value was summed each day until the mature-adult stage when the *Anopheles* mosquitoes are capable of laying eggs (Fig. 1a). The soil moisture content at root depth was determined before developing the growth model of the mosquito, as described in the next section. The threshold of soil moisture content was set at 70% of the value of the holding capacity (Dunne & Willmott, 1996) to exclude the emergence of mosquitoes during the dry season from the validation process. When the soil moisture content exceeded the threshold, the developmental rate of the mosquitoes was determined by the temperatures of the habitat. Water temperature was used to calculate the development of the egg, larval, and pupal stages; air temperature was used to calculate the development of the adult mosquito (Fig. 1b). The development at a given time (in this study, this value was assumed as 1 day) can be described as follows:

$$\Delta d_k = \frac{1}{D_{temp}} \times \Delta t_k \tag{1}$$

where Δd_k is the development at kth day, D_{temp} is the duration for the completion of each developmental stage (day) at a given constant temperature (°C), and Δt_k is the time unit of development. The inverse of the D_{temp} for each stage ranges from 0 to 1, with 0 representing no development and the amount of development increasing until the value approaches 1 (Lardeux et al., 2008). Values for D_{temp} were obtained from the results of laboratory studies and field observations of some *Anopheles* species (Mogi & Okazawa, 1996; Bayoh & Lindsay, 2003; Hu et al., 2003; Depinay et al., 2004; Lardeux et al., 2008). The relationships between the temperature threshold and mosquito development at each developmental stage were determined by the results obtained from these studies (Kashiwada & Ohta, 2010). With reference to these relationships, the growth begins when the temperature exceeds 15°C (Fig. 1a). The value of Δd_k was summed on a day-to-day basis to determine the amount of development. Cumulative Δd_k was determined separately in each developmental stage ("*stage*" expressed in Fig. 1a)—egg, larva, pupa, and adult—as their developmental rates are calculated differently. Each stage of development is completed when the accumulated development is 1. The developmental stage then proceeds with the next one. The value of Δd_k for each developmental stage was summed continuously as follows to determine the duration of mosquito development in the life history of the mosquito:

$$C_{total} = \sum_{k=l}^{m} \Delta d_{egg,k} + \sum_{k=m+1}^{n} \Delta d_{larva,k} + \sum_{k=n+1}^{o} \Delta d_{pupa,k} + \sum_{k=o+1}^{p} \Delta d_{adult,k} \tag{2}$$

where C_{total} is the cumulative development; l, m+1, n+1, and o+1, denote the start day of each developmental stage; m, n, o, and p denote the end day of each developmental stage; and $\Delta d_{egg,k}$, $\Delta d_{larva,k}$, $\Delta d_{pupa,k}$, and $\Delta d_{adult,k}$, are the development of eggs, larvae, pupae, and adults at the kth day. The value of each Δd_k ranges from 0 to 1; hence, C_{total} ranges from 0 to 4 (Fig. 1c). Values between 0 and 1 were attributed to embryonic development. Larval and pupal development was assigned values between 1 and 2, and between 2 and 3, respectively. Values exceeding 3 were assigned after the emergence of adults. Values exceeding 4 were attributed when adult mosquitoes oviposited eggs. The alternation of generations was counted over a 1-year period (G_q: qth generation), and the maximum number was recorded as the maximum number of generations (G_{max}) to estimate the frequency of occurrence (Kiritani, 2006; Kashiwada & Ohta, 2010).

3.2 Water balance model for moisture conditions of the mosquito habitat

The growth and development of immature mosquitoes can occur in small water bodies such as puddles or artificial water containers. To consider the potential distribution of *Anopheles* mosquitoes, the availability of a "natural" water surface was assumed in this study. There was no consideration of human activities such as irrigation or sewerage systems. Under natural conditions, the mosquito habitat would be small pools in marshes or wetlands with

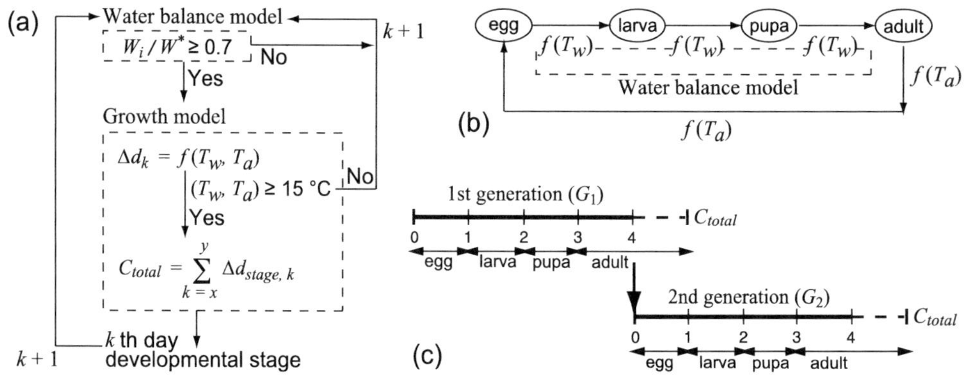

Fig. 1. Model procedures. (a) Determination of moisture conditions and mosquito growth; (b) Relationship between life history and temperatures; (c) Calculation of mosquito growth.

some plant cover. Therefore, in the present study, *Anopheles* mosquitoes were assumed to develop in a small water body with soil, which is an area of several tens of square meters. The soil moisture content in the mosquito habitat must be estimated to exclude areas that are too arid for an immature mosquito to survive. Subsequently, soil moisture content at the depth of plant roots was calculated using the water budget approach proposed by Hopp & Foley (2001) and Tao et al. (2003). The soil moisture content was calculated on the *i*th day (W_i) by using the following equation:

$$W_i = min\ (W_{i-1} + P_i + M_i - AE_i \quad or \quad W^*) \tag{3}$$

where W_{i-1} is the soil moisture content at the end of the previous (i – 1) day, P_i is the daily precipitation (mm), M_i is the daily snow melt on the *i*th day, AE_i is the actual evapotranspiration on the *i*th day, and W^* is the holding capacity of soil moisture, which reflects the types of soil texture as well as organic content and plant root depth in soil, according to Dunne & Willmott (1996). When the W_i value obtained from these calculations was larger than W^*, W_i was equal to W^*. A similar procedure by Tao et al. (2003) was adopted to define the value of AE_i as a ratio of the available moisture content to the potential evapotranspiration calculated using the FAO Penman–Monteith method (Allen et al., 1998). Calculation of the potential evapotranspiration requires data for net radiation, air temperature, air humidity, and wind speed. Net radiation was calculated using an energy balance equation as described in the next section (Ohta et al., 1993a; 1995; Ohta & Kimura, 2007), with basic climate factors and geographical data such as albedo and altitude. The values of air temperature and humidity were directly used in the model. In the present study, wind speed was assumed to be constant at 2 m s^{-1}, according to Allen et al. (1998), because the wind speed in modeled plant populations is uncertain.

3.3 Estimation of the temperature of water bodies inhabited by mosquitoes

Although the temperatures of the water bodies in which *Anopheles* mosquitoes live directly affect immature mosquito development, water temperature has unfortunately not been measured with fine resolution either spatially or temporally, in contrast to what has been performed for air temperature and precipitation. Many researchers (e.g., Hopp & Foley, 2001) often substituted air temperature for water temperature. However, the daily average water temperature in temperate regions is 2-4°C lower than the daily average air temperature during the period of insufficient net radiation in the winter (Ohta et al., 1993a; Ohta & Kimura, 2007). Conversely, the daily average water temperature is 1-2°C higher than the daily average air temperature in the spring (Ohta et al., 1993a; Ohta & Kimura, 2007). Because these seasonal changes in the differences between air and water temperatures are complex, it cannot be assumed that air temperature is equivalent to water temperature. If these daily differences in temperatures are summed during the growing season of the mosquitoes, the calculation error will gradually increase.

The contribution of the energy fluxes in the soil, stored per unit water column, to the energy balance of shallow waters can be omitted if the energy balance is estimated daily or for a prolonged period. Because the energy stored in plant biomass is very small compared to the other components of the energy balance equation, the energy balance of shallow waters without the percolation and flux of irrigation (Ohta et al., 1993a; 1995; Ohta & Kimura, 2007) can be simply expressed using the following equation:

$$R_n = 15h\,(e_{sat} - e_a) + h\,(T_w - T_a) \tag{4}$$

where R_n is the net radiation at the water surface (W m^{-2}), h is the transfer coefficient for sensible heat, e_{sat} is the saturation vapor pressure at the water temperature (kPa), e_a is the water vapor pressure in air (kPa), T_w is the water temperature at a depth of 0.05–0.10 m (K), and T_a is the air temperature (K). It is assumed that the value of h without the plants is nearly constant at 8.36×10^{-5} W m^{-2} K^{-1}. In the presence of a relatively large difference between T_w and T_a, the net radiation affecting T_w can be given by

$$R_n = R_{na} - h_R\,(T_w - T_a) \tag{5}$$

where R_{na} is the net radiation (W m^{-2}) to be calculated assuming that T_a equals T_w, h_R (= $4\sigma T_a^3$) is the radiative heat transfer coefficient (W m^{-2}), and σ is the Stefan-Boltzmann constant (W m^{-2} K^{-4}). From equations (4) and (5), the daily mean of T_w can be obtained numerically by using the daily means of the basic climate observation values as described by Ohta et al. (1993a; 1995) and Ohta & Kimura (2007).

This modeled shallow water condition is nearly the same as the ideal habitat for the immature stages of *Anopheles* mosquitoes considered in the present study. Therefore, T_w was estimated using the simple energy balance model as well as with the methods developed by Ohta et al. (1993a; 1995) and Ohta & Kimura (2007). Calculation of T_w requires basic climate observation values, including air temperature, solar radiation, cloud cover, and vapor pressure.

3.4 Validation of the model

Our model as described above was validated using many observations in the study area obtained from published data (Ono, 1992; Dev, 1996; Konradsen et al., 1998; Lee et al., 2002; Chen et al., 2002; Overgaard et al., 2002; Toma et al., 2002; Singh et al., 2004; Yeom et al., 2005; Chen et al., 2006; Rueda et al., 2006). The main results of the model validation are summarized below.

If the value of the cumulative development (C_{total}) calculated from the kth day was greater than 3, we defined the kth day as the day of mosquito emergence. This calculation was then conducted in all cases where the developmental start day was 1st–365th day of the year. As the mosquito could have emerged across different years, the calculation period included 2 years. The first date of mosquito emergence was defined as the mosquito appearance date, and the end date of mosquito emergence was defined as the mosquito disappearance date (Kashiwada & Ohta, 2010). The predicted emergence or disappearance dates obtained from the model were concordant with the observation data. Although the root mean square errors between the observed and modeled values for these dates were approximately 25 days, the values were acceptable for simulating the potential mosquito distribution, because the mosquito occurrence observations were repeated over several weeks to a few months, as described previously (Dev, 1996; Konradsen et al., 1998; Overgaard et al., 2002).

During comparison of the more detailed observation data at representative sites, which ranged from temperate to subtropical zones (Ono, 1992; Konradsen et al., 1998; Lee et al., 2002; Toma et al., 2002), the model was able to estimate seasonal patterns of occurrence of mosquitoes with reasonable accuracy (Kashiwada & Ohta, 2010). In addition, the geographical distributions of the northern limit of the mosquitoes in Monsoon Asia obtained from this model were corroborated by the observed distributions in summer (Kashiwada & Ohta, 2010).

4. Data used for calculation

4.1 Current climate data

Data for the Asian monsoon region (70–150° E, 10° S–50° N) were used in the model. All meteorological data had a spatial resolution of 0.5° longitude × 0.5° latitude. The Climatic Research Unit — Global Climate Dataset, available through the IPCC Data Distribution Center (New et al., 1999) was used for the model calculations. Monthly data for air temperature, precipitation, shortwave radiation, vapor pressure, and cloud cover were included in the dataset for 1961–1990 climate normals. If daily data were not available for the model, monthly values were converted to daily data by using linear interpolations or cubic spline, ensuring consistency of the daily values with monthly average or total values. Elevation data derived from GTOPO30 (U.S. Geological Survey) were used to calculate vapor pressure and longwave radiation in the present study, as well as correction methods described in Ohta et al. (1993b). Elevation data in GTOPO30 are regular at a spatial resolution of 30 arcseconds (approximately 1 km); we averaged and converted them to a spatial resolution of 30 arcminutes. The monthly surface albedo (0–1) in each 0.5° longitude × 0.5° latitude resolution (Ohta et al., 1993a) was used. When the air temperature was less than 0°C, the value of surface albedo was set to 0.7.

4.2 Future climate data

The future climate variables used in this study were obtained from the climate projections of the Coupled General Circulation Model (CGCM3.1) by the Canadian Centre for Climate Modelling & Analysis, which contributed to the IPCC 4th Assessment Report (Flato et al., 2000). The climate variables of the CGCM3.1/T63 involved daily mean air temperature, daily total precipitation, and daily accumulated shortwave radiation on a spatial resolution of 2.81° longitude × 2.81° latitude. These values are outputs from the IPCC 720 ppm stabilization experiment (A1B scenario) using the CGCM3.1 for the years 2001–2100 initialized from a simulation experiment starting from the end of the 20th century. In the present study, the spatial resolution of the output data from CGCM3.1/T63 was converted into a spatial resolution of 0.5° longitude × 0.5° latitude by using the interpolation method.

5. Energy and water balances of the habitat of mosquitoes

5.1 Net radiation at the water surface

Soil moisture content and water temperature were determined by the net radiation obtained from equations (4) and (5). In addition, because net radiation is the only energy source on which all life depends, it is essential to draw the geographical distribution of net radiation to estimate the life history of mosquitoes. Fig. 2 compares the seasonal changes in the geographical distribution of net radiation between the current and future climates.

In the current climate, the amount of monthly mean net radiation in warm- and cool-temperate and subarctic climate zones changes seasonally, reflecting the difference in the characteristics of geographical location and the effects of the Asian monsoon. In subtropical and tropical climate areas at latitudes below 20°N, hardly any seasonal changes in net radiation occur throughout the year. In summer, the amount of net radiation at latitudes above 20°N as well as in subtropical and tropical climate areas increases rapidly.

The amount of net radiation in areas with alpine, ice, and desert climates is negative, except in summer; this implies the emission of thermal radiation from underlying surfaces into the atmosphere. In the future climate conditions, areas with negative net radiation in autumn, winter, and spring will decrease gradually. A noteworthy difference between the current and future climates in Eastern China is the increase in the amount of net radiation from June to October during which the average amplitude of the increase is approximately 30 W m^{-2}. This is mainly because the amount of shortwave radiation during the same period obtained from the CCGM3.1 increases independently of increases in air temperature.

5.2 Moisture conditions of the mosquito habitat

Fig. 3 represents the seasonal changes in geographical distribution of soil moisture conditions for the mosquito habitat under the present and future climates. This value is expressed as the ratio of W_i / W^*. The calculation was conducted daily; the value of the 15th day of each month is indicated on the map. These maps indicate that many of the areas of the Asian monsoon region generally have humid conditions. The values of W_i / W^* in humid regions (e.g., the whole areas in subarctic, temperate, and subtropical or tropical zones), in alpine regions (e.g., the Himalayas), and in the arid belt (i.e., The Gobi, Takla Makan, and Great Indian Desert, and the Mongol Plateau) do not change seasonally, and water conditions in these areas are constant throughout the year (Fig. 3a). In contrast, the values of W_i / W^* in some areas, including in the entire area of the Indian Peninsula and excluding the west coast and Indochina (i.e., Myanmar, Thailand, Laos, Cambodia, and Vietnam), change seasonally (Fig. 3a). Especially, from December to April, the soil moisture contents in these areas are remarkably limited, reflecting the dry season with small precipitation.

These characteristics of soil moisture content in the mosquito habitat under the current climate conditions would not change under the future climate conditions (Fig. 3a and 3b). Thus, the increases in evapotranspiration from the soil surface due to the increases in air temperature, shortwave radiation, and net radiation in the future climate condition do not affect the soil moisture content in the mosquito habitat.

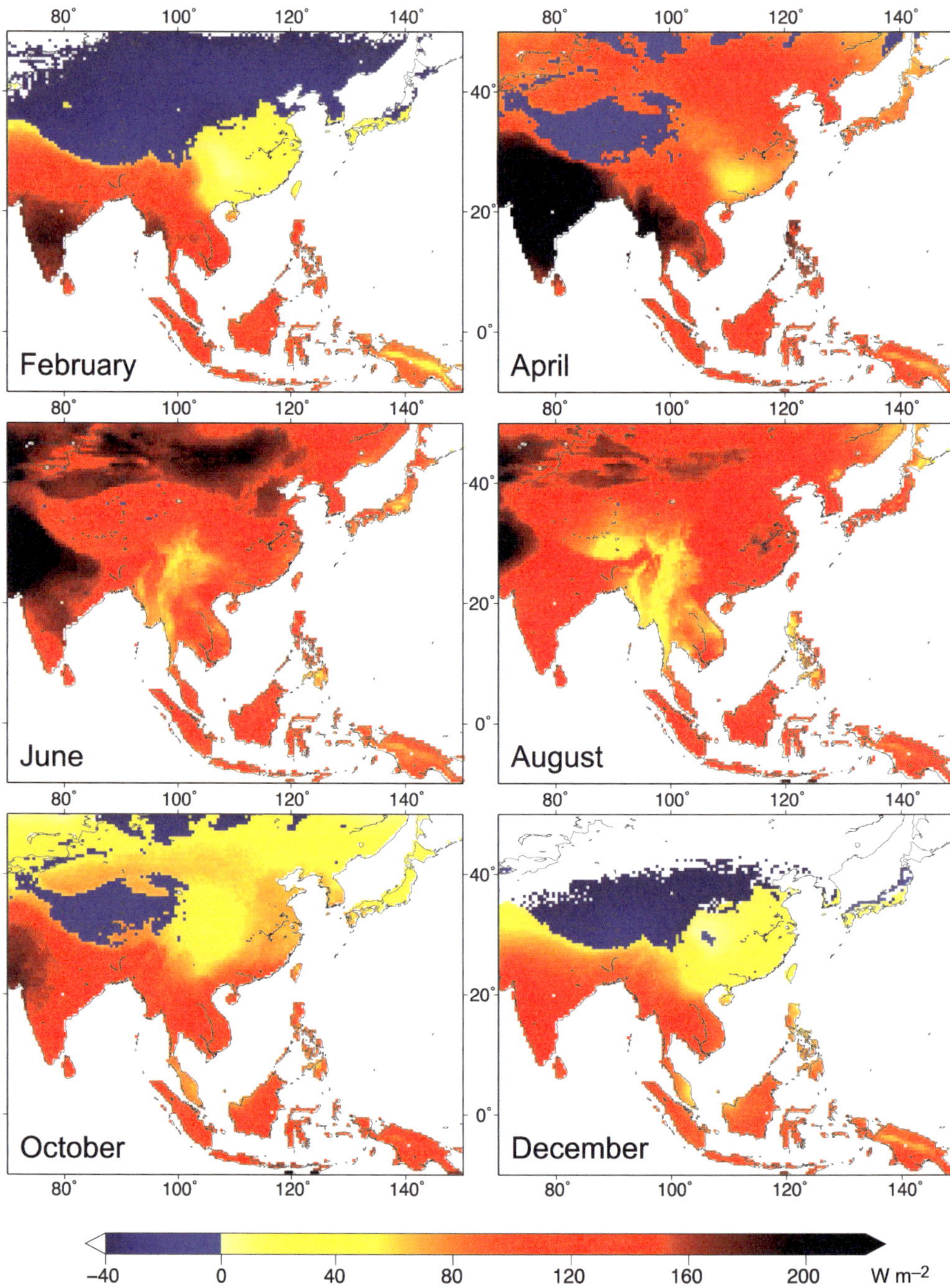

Fig. 2a. Seasonal patterns of geographic distribution of monthly net radiation at water surfaces under current climate (W m-2).

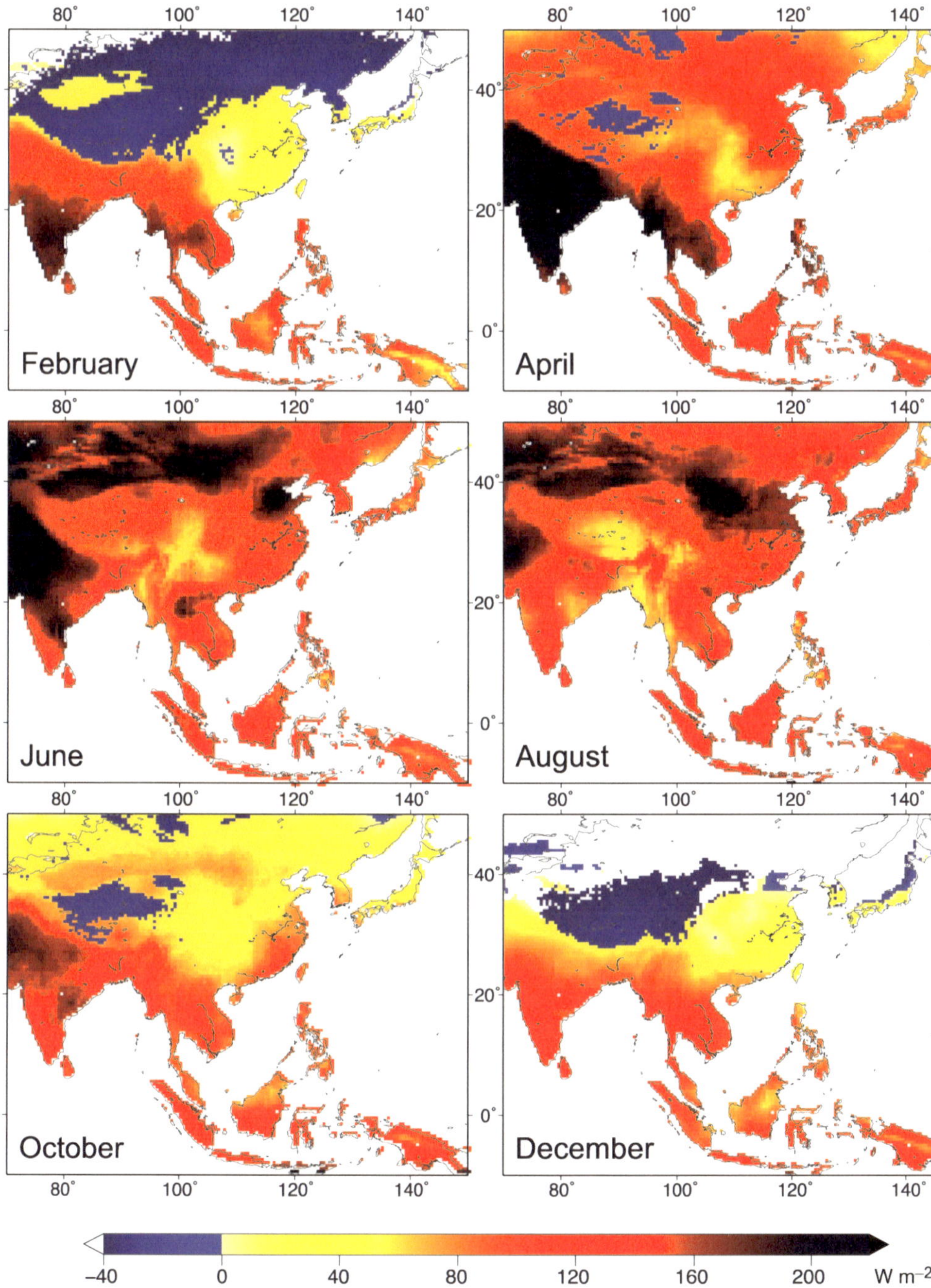

Fig. 2b. Seasonal patterns of geographic distribution of monthly net radiation at water surfaces under future climate (W m^{-2}).

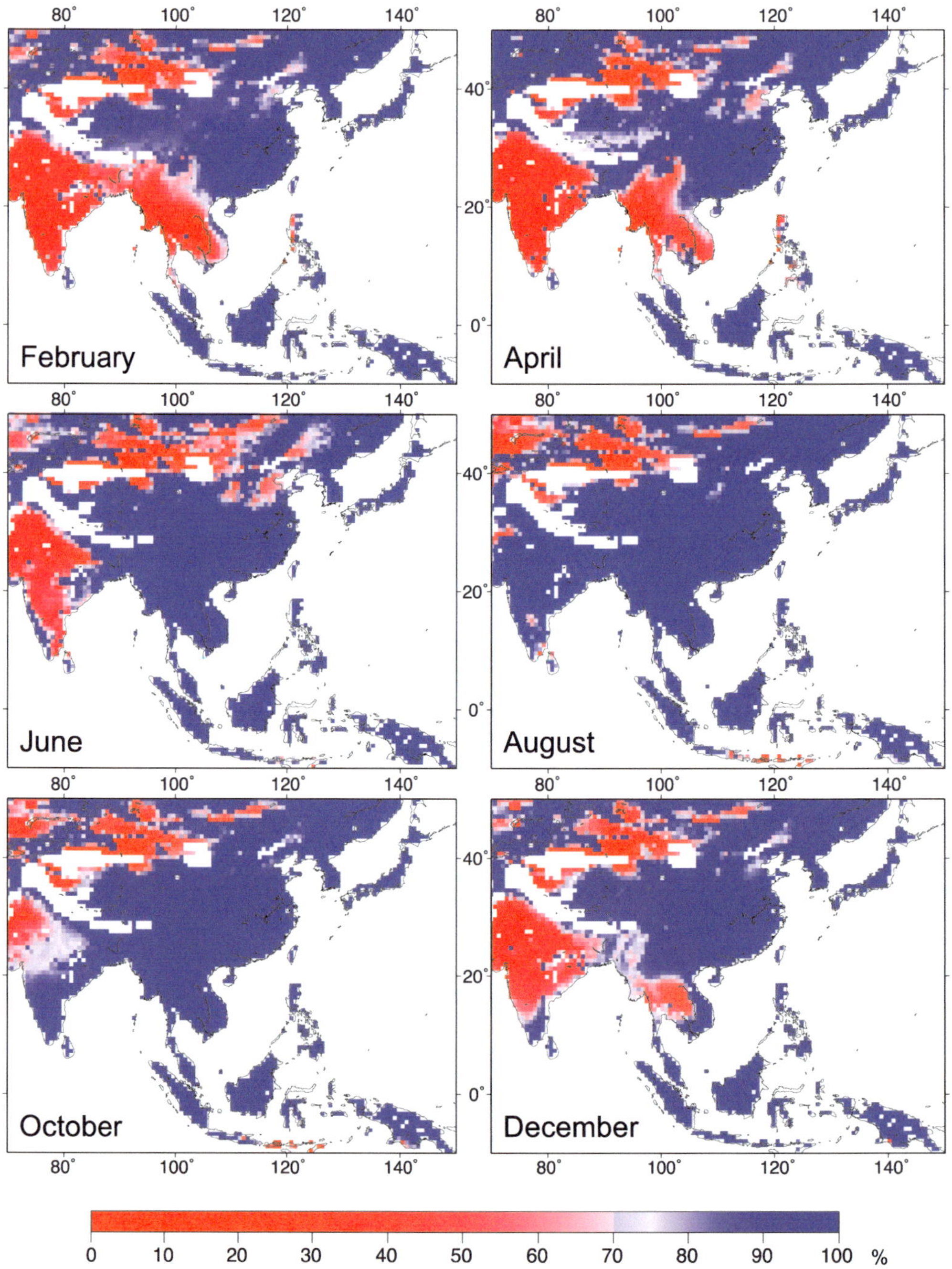

Fig. 3a. Seasonal patterns of geographic distribution of the monthly mean soil moisture under current climate (%).

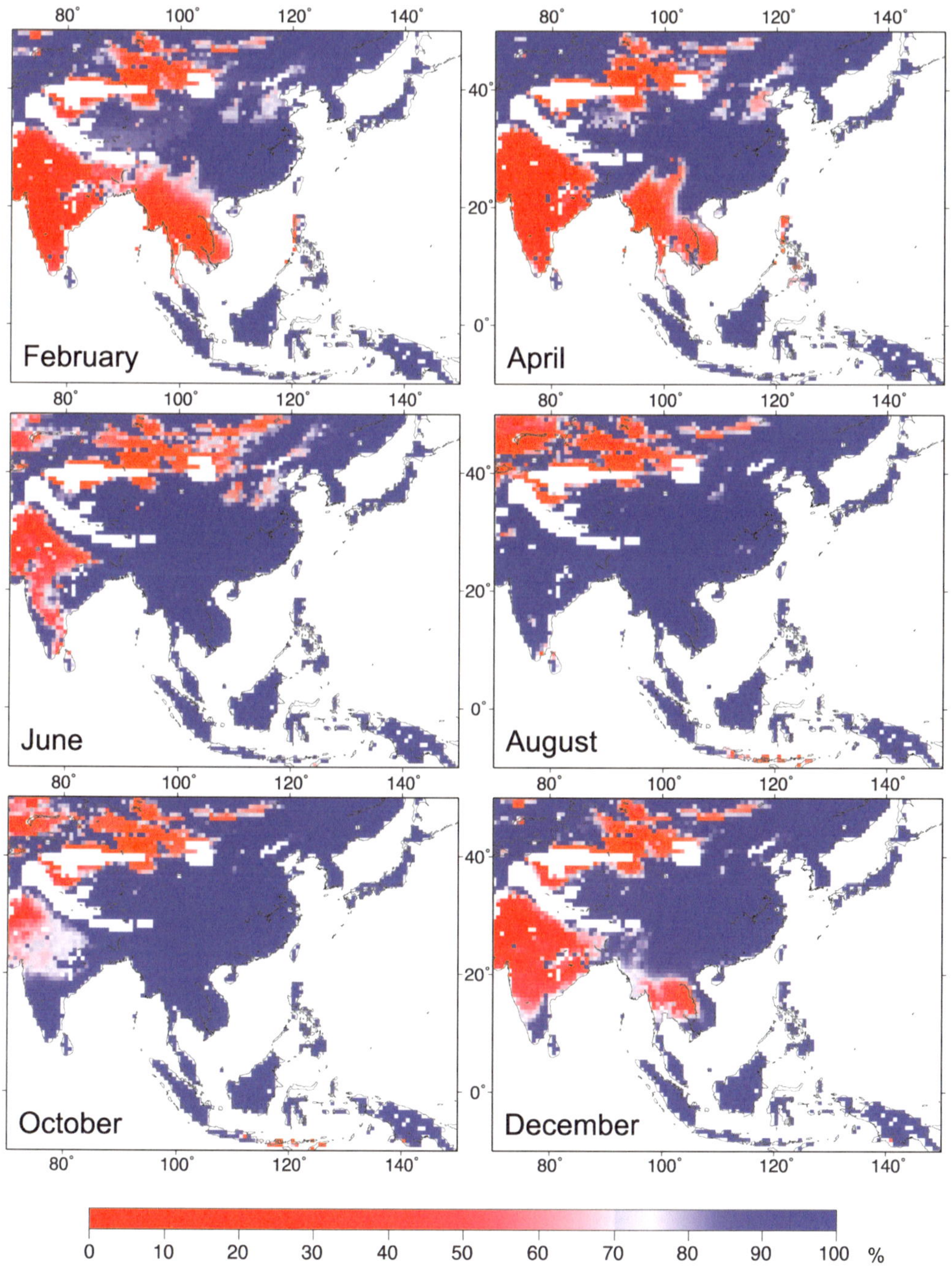

Fig. 3b. Seasonal patterns of geographic distribution of the monthly mean soil moisture under future climate (%).

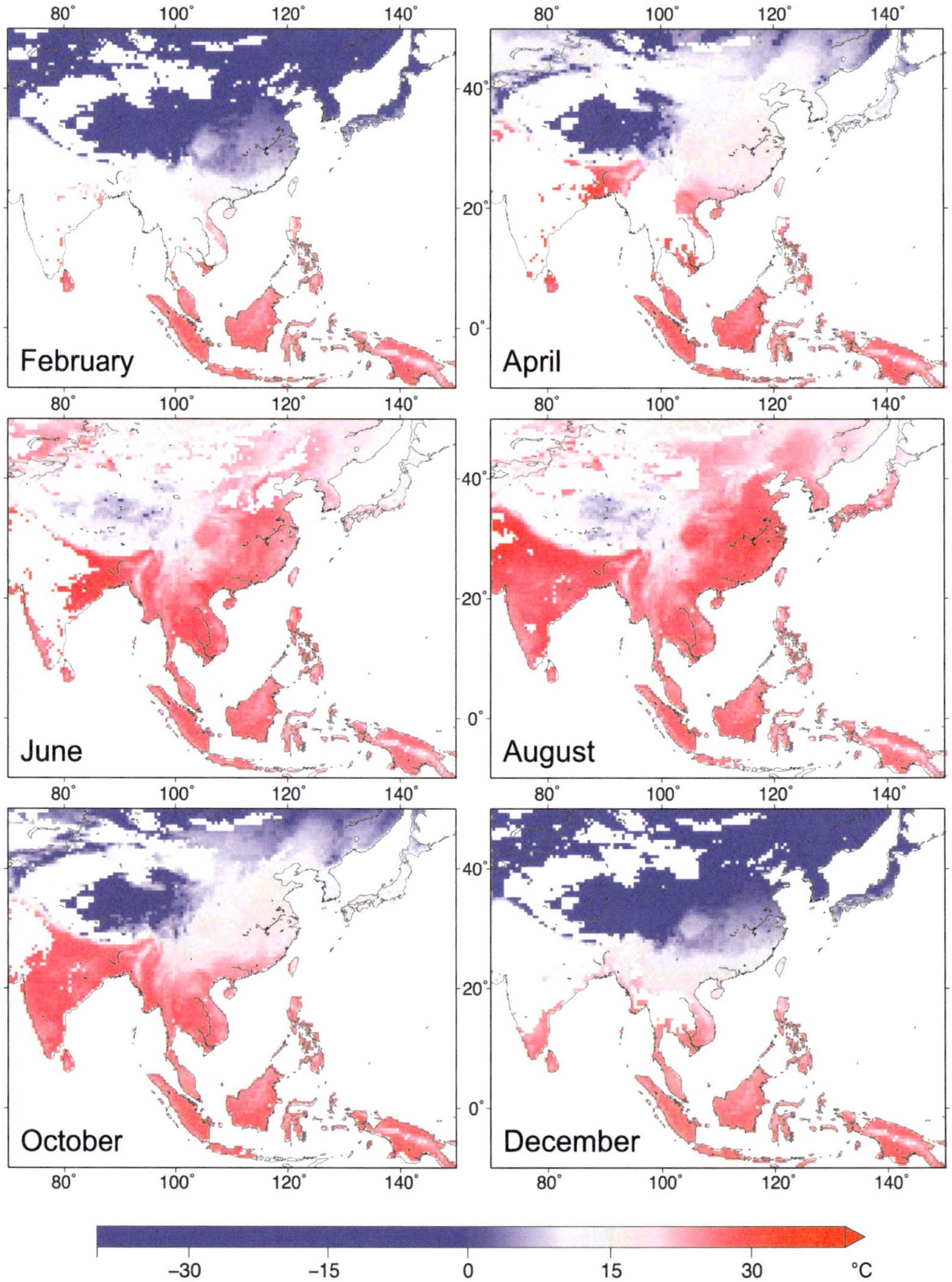

Fig. 4a. Seasonal patterns of geographic distribution of the daily mean water temperature under current climate (°C). The white areas denote the scarcity of surface water of soil unsuitable for mosquitoes.

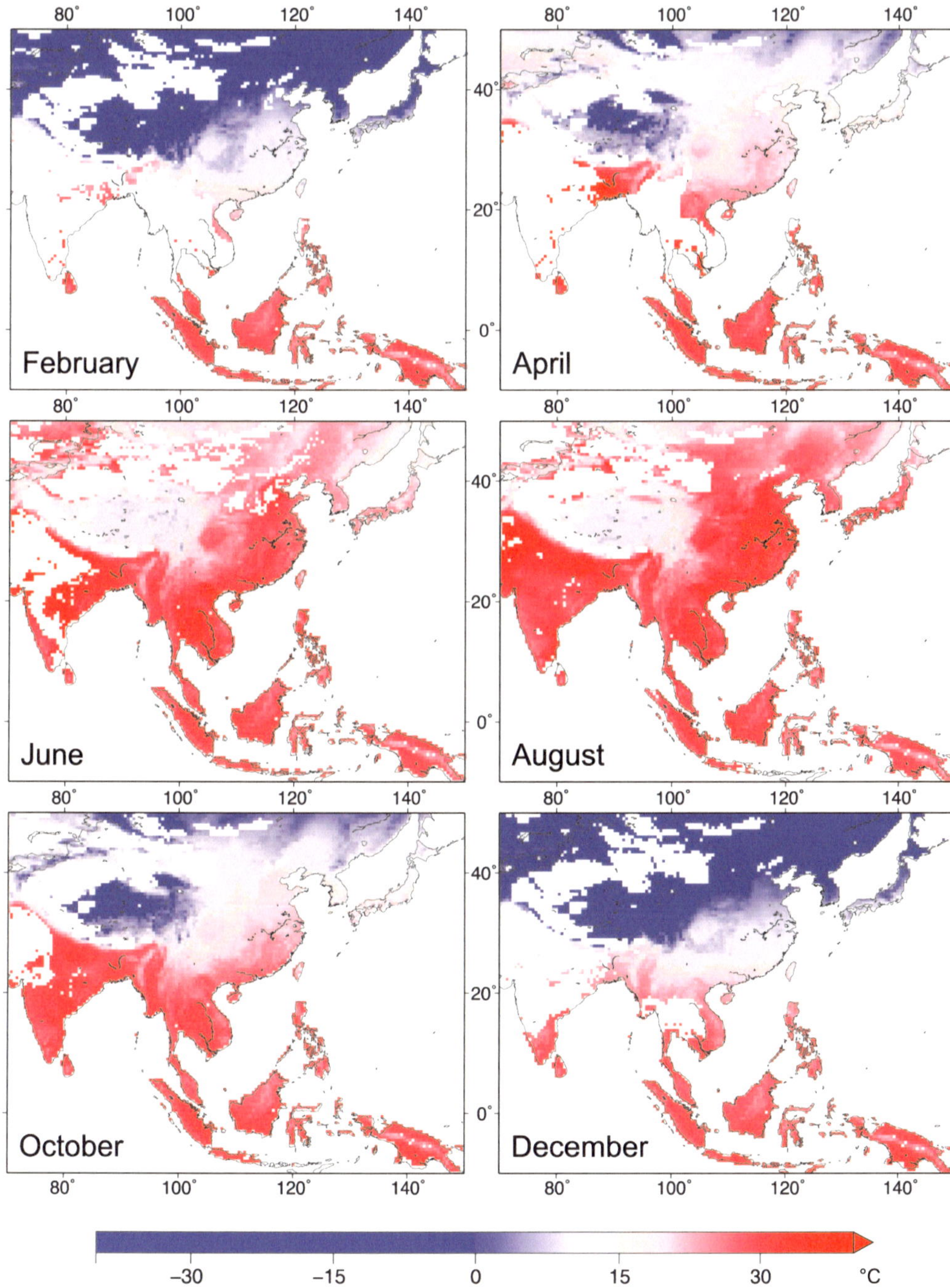

Fig. 4b. Seasonal patterns of geographic distribution of the daily mean water temperature under future climate (°C). The white areas denote the scarcity of surface water of soil unsuitable for mosquitoes.

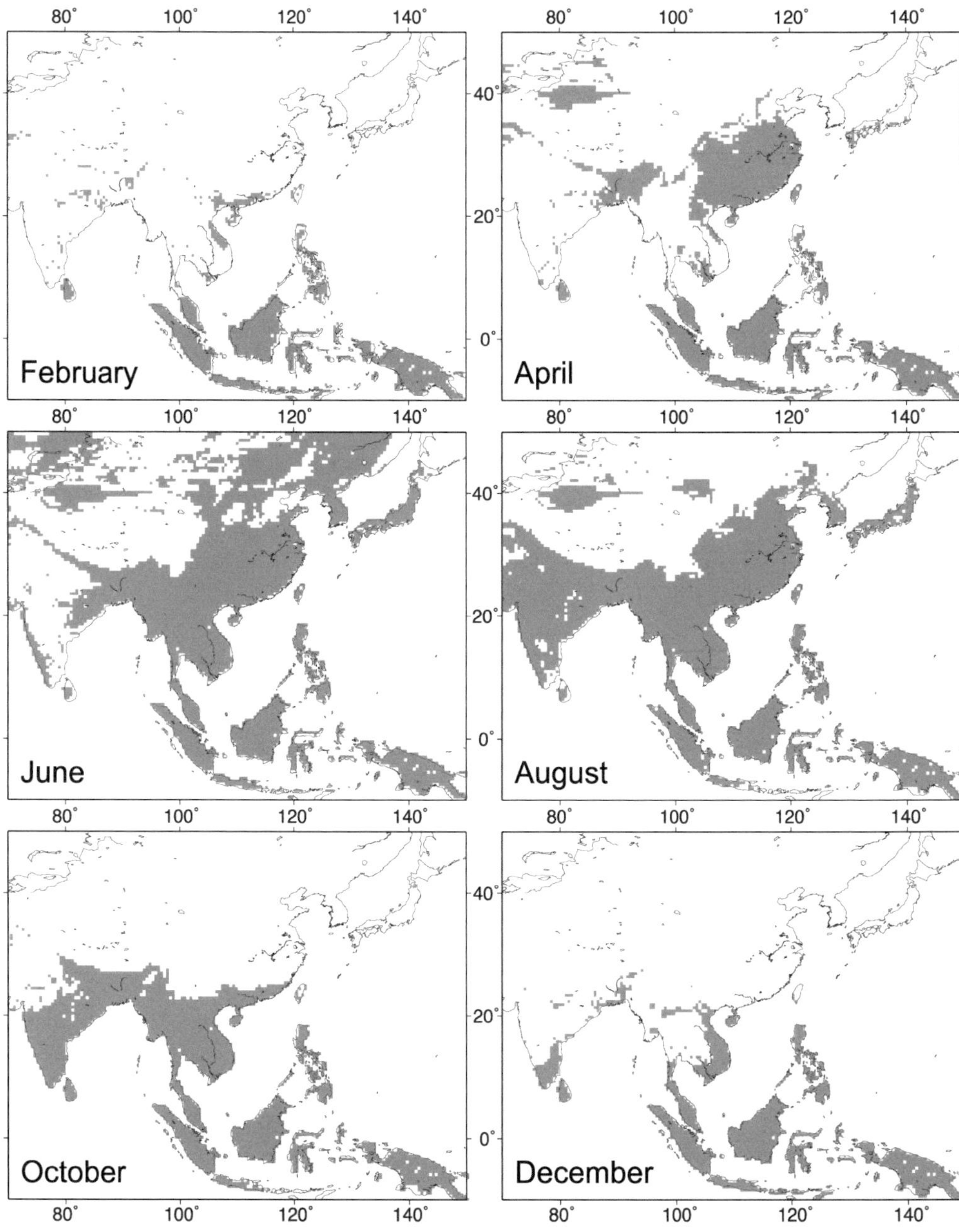

Fig. 5a. Geographic distribution of bimonthly adult mosquito occurrence under current climate. Gray shading denotes the area in which the maximum development reached the adult stage (when the cumulative developmental stage (C_{total}) exceeded 4).

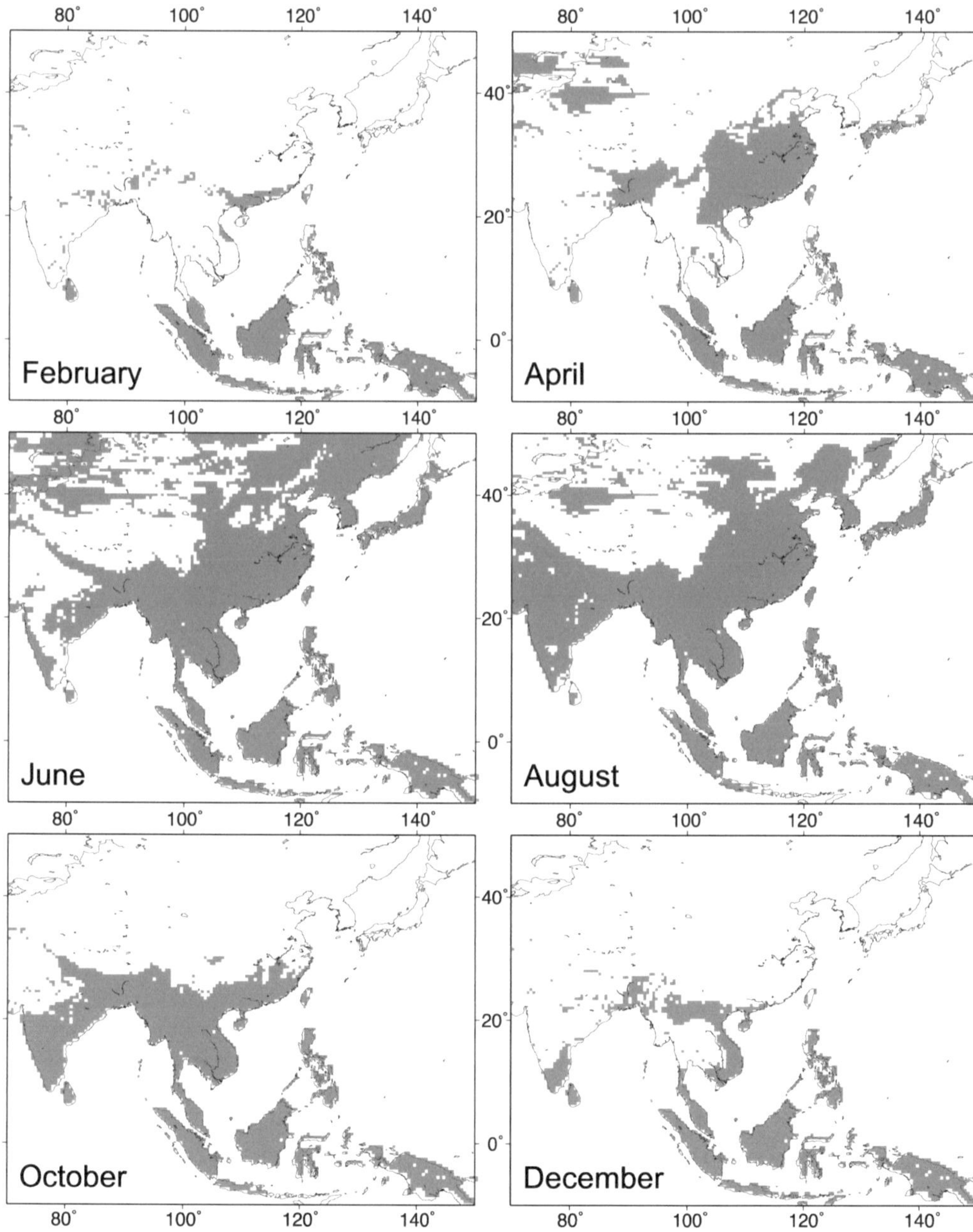

Fig. 5b. Geographic distribution of bimonthly adult mosquito occurrence under future climate. Gray shading denotes the area in which the maximum development reached the adult stage (when the cumulative developmental stage (C_{total}) exceeded 4).

5.3 Temperature of the water bodies inhabited by mosquitoes

Fig. 4 shows the changes in the temperature of the water in which immature mosquitoes live in the future climate conditions. The smallest difference between the water temperatures in tropical and subarctic regions is found in mid-summer; the difference gradually becomes larger, peaking in winter (Fig. 4). The seasonal patterns of the geographical distribution of water temperatures in Monsoon Asia show a trend similar to those of air temperatures. The temperatures of ponded shallow water increase by approximately 1.0–2.5°C in the future climate (Fig. 4), synchronizing the changes in air temperatures obtained from the CGCM3.1. However, the magnitudes of increase in water temperature in the future climate are approximately 0.5–1.0°C smaller than the magnitudes of increase in air temperature. This is because the temperature rise is suppressed by the energy characteristics of water.

The white areas in Fig. 4 denote the scarcity of surface water of soil unsuitable for mosquitoes. As described in section 3.1, because mosquito development begins when water temperatures exceed 15°C, the blue areas in Fig. 4 denote the shortage of thermal resources for mosquitoes, and the red areas in Fig. 4 represent the growing period of mosquitoes in terms of water temperatures. On comparing Fig. 4a and 4b, it is apparent that the red areas in southeastern China along the Tropic of Cancer extend in April and December. Furthermore, the red areas in eastern China, especially those including densely populated cities, extend in October. The water temperatures in the southern and eastern parts of China in the future climate increase by approximately 3°C compared with those in the current climate. When considering future mosquito growth, it is important to consider that this value in China is greater than that in temperate regions, the eastern part of Indochina (approximately 2°C), and inland India (approximately 1.5°C).

6. Spatiotemporal distribution of *Anopheles* mosquitoes

6.1 Shift of seasonal changes in geographical distribution

To clarify seasonal variations in the geographical distribution of adult mosquito abundance, the developmental stage of the mosquitoes was determined using the model described in section 3. Fig. 5 shows the abundance of mosquitoes when the cumulative developmental stage (C_{total}) reached 4. The most notable observation in Fig. 5a is that mosquito emergence in the present climate is limited seasonally in most areas of Monsoon Asia, including subarctic regions, cool- and warm-temperate regions, and the Indian and Indochina Peninsula. However, in large areas of tropical regions in Southeast Asia, including Sumatra, Kalimantan, Sulawesi, New Guinea, and the Philippines, the adult emergence of *Anopheles* occurs year round. Adult mosquitoes in semi-arid tropical countries such as Thailand and Myanmar appeared only from June to October. In inland India, mosquitoes only appeared from August to October. The appearance of adult mosquitoes in temperate regions such as Japan, Korea, and northeastern China was limited to summer.

In the future climate (Fig. 5b), although there is little change in adult emergence in temperate regions in spring, the northern distribution limit of adult mosquitoes in mid-summer could shift approximately 200–300 km north. In autumn, the northern distribution limit of adult emergence in subtropical regions such as southern China could shift approximately 100 km north. In addition, *Anopheles* could still potentially appear even in winter. This suggests that the areas where mosquitoes are able to appear year round could expand in the subtropical regions of southern China. On the other hand, the emerging pattern of *Anopheles* in tropical regions, excluding the southwestern coastal areas of the

Indian Peninsula, in the future climate are unchanged, compared to that in the current climate.

6.2 Changes in the average developmental length and geographical distribution of the maximum number of generations of *Anopheles* mosquitoes

The duration of mosquito development from egg hatching to adult was counted from the day when the egg stage began to the day of oviposition activity. This duration, which is required for the development of the adult mosquito, was defined as the developmental length in the present study. The average developmental length ($D_{average}$, day) during a year was calculated as follows:

$$D_{average} = \left(\sum_{q=1}^{G_{max}} D_q \right) \times \left(\frac{1}{G_{max}} \right) \quad (6)$$

where D_q is the developmental length (day) of the qth generation from egg hatching to adult emergence, q is the number of generations, and G_{max} is the maximum number of generations as described in section 3.1. $D_{average}$ indicates the necessary duration of development from an egg to the adult mosquito. Model estimations of $D_{average}$ in the current climate obtained using equation (6) range from approximately 2 weeks to more than 1 month (Fig. 6). The $D_{average}$ value in the regions with the shortages of water and thermal resources is higher than that in the regions suitable for the growth of mosquitoes. When the $D_{average}$ in the current climate is less than 20 days, the $D_{average}$ in the future climate is approximately 2 days shorter than that in the current climate; when the $D_{average}$ in the current climate is more than 20 days, the $D_{average}$ in the future climate is approximately 3 days shorter than that in the current climate (Fig. 6).

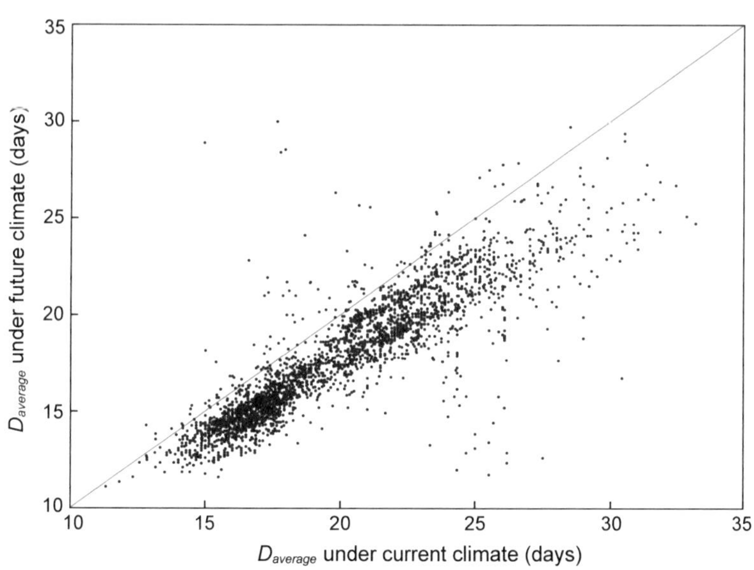

Fig. 6. The average duration of mosquito development (day) under current and future climates.

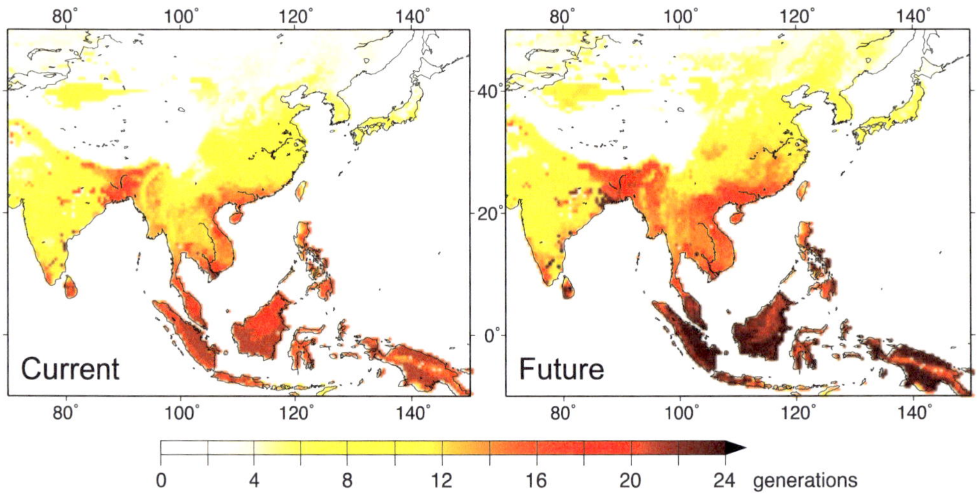

Fig. 7. Geographic distribution of the maximum number of generations under current and future climates.

These smaller $D_{average}$ values in the future climate increase the maximum number of successive generations (G_{max}) in most areas and increase the frequency of mosquito occurrence (Fig. 7). Fig. 7 indicates that G_{max} depends on the availability of moisture (Fig. 3) and on water temperature (Fig. 4). The values of G_{max} in temperate regions and inland India were approximately 6 to 8 generations in the current climate. The G_{max} in the eastern part of Indochina and southern China were approximately 14 to 16 generations. The G_{max} values in tropical regions were more than 20 generations.

In the future climate conditions, the G_{max} in tropical regions and southern and southeastern China would increase by approximately 3 generations under warmer climate conditions. The G_{max} in temperate regions would increase by approximately 2 generations, and the G_{max} in inland India would increase by approximately 1-2 generations, although the growing period of the mosquitoes is still limited to the period from August to October. However, the increase in the G_{max} in eastern Indochina would be slightly less than 1 generation in the future climate conditions. In addition, the G_{max} in southern Thailand would decrease by 1 generation.

7. Discussion and conclusions

The present model, driven by simple meteorological data, was able to predict the effects of future climate change on the temporal variations and the potential distribution of the *Anopheles* mosquito that possibly transmits malaria in Monsoon Asia. Four main conclusions can be drawn from the results.

First, although the G_{max} values in the future climate would not increase in regions where water resources are a limiting factor for the growth of mosquitoes, the future G_{max} values would generally increase by approximately 2-3 generations (Fig. 7). This is because the future climate projected by the CGCM3.1 would improve the current thermal environment in regions with shortage of thermal resources (Fig. 2), especially increased water temperature suitable for the growth of mosquitoes (Fig. 4). However, the water resources in semi-arid regions would remain unchanged in the future climate (Fig. 3). Consequently, the G_{max} values in the future climate increase by the function of smaller $D_{average}$ values (Fig. 6) owing to the increase in water temperature (Fig. 4). Accordingly, the southern part of China will experience the most remarkable changes in vector distribution in future climate conditions. The northern limit of the current distribution of tropical *Plasmodium falciparum* malaria in Monsoon Asia could shift approximately 100 km north, especially in autumn (Fig. 5). Another region with distinct changes in vector distributions in future climate conditions is the eastern part of China, which currently has a low density of the current vector species. The northern limit of the current distribution of temperate *Plasmodium vivax* malaria in Monsoon Asia could shift approximately 200–300 km north, especially in spring (Fig. 5). The survival of the mosquitoes in the southwestern coastal regions of India in early summer would be likely in the future climate conditions (Fig. 5), reflecting improved soil moisture conditions (Fig. 3).

Second, the shifts in the seasonal patterns of the geographical distributions of vectors projected in this study are concordant with those of the malaria potential obtained from other studies (Martens et al., 1999; Rogers & Randolph, 2000; van Lieshout et al., 2004; Bhattacharya et al., 2006). The integrated model with socioeconomic processes predicts the expansion of tropical malaria in southern China (Rogers & Randolph, 2000; van Lieshout et al., 2004). According to Martens et al. (1999), the temperate regions where malaria can occur would extend in the future climate in eastern China; this prediction is concordant with our results. In addition, the future geographical distributions of vectors in the southwestern coastal regions of India mentioned above conform to the projected endemic regions obtained from the simulation with a regional climate model (Bhattacharya et al., 2006). Interestingly, the northern limit of the geographical distribution of the vectors estimated in the present study is less than that estimated by Martin & Lefebvre (1995) who studied the annual potential. Also, it must be noted that the geographical distributions obtained from these studies indicate the epidemic potential of malaria (Hay et al., 2009; Gething et al., 2010) and not that of the vector. Although long-term trends for malaria transmission were not significantly associated with a recent climate warming (Gething et al., 2010), it was found from this study that the potential ranges of vector species could change due to future climatic conditions.

Third, the present model that described the developmental processes of a representative *Anopheles* mosquito is unable to adequately or precisely indicate the temporal patterns of the mosquito emergence, especially in tropical regions. Using the original concept of "intrinsic optimum temperature for development" and a thermodynamics model, Ikemoto (2005) concluded that the growth of *Anopheles* mosquitoes is suppressed and stressed by temperatures exceeding 30°C. According to our projections, the future G_{max} values in the all the tropical regions would increase by 2 generations. However, considering that high temperatures injure mosquitoes (Depinay et al., 2004; Ikemoto, 2005; 2008), it is possible that these increases in G_{max} in tropical regions will not occur. The possibility of a subsequent generation is only one of the factors for determining the magnitude of a population. This might lead to overestimations in tropical regions. Therefore, a population dynamics model that can simulate the growth of many individual mosquitoes stochastically should be applied to predict the future distribution of mosquitoes. The model based on population dynamics provides useful data regarding the magnitudes of a mosquito population in future climate conditions as well as the maximum number of generations for the risk assessment of malaria. Pascual et al. (2006) suggests that the observed temperature changes would be significantly amplified by the mosquito population dynamics with a difference in the biological response at least 1 order of magnitude larger than that in the environmental variable.

Finally, the probability of our results regarding changes in G_{max} may increase or decrease the range of the mosquitoes owing to different climate models or future scenarios for greenhouse gas emissions. To elucidate the probability of the projection range, mosquito abundance must be simulated using our model with a variety of future emission scenarios and climate models.

8. Acknowledgements

We wish to thank Prof. Yasushi Morikawa, Prof. Hiroshi Koizumi, and Prof. Shoji Nishimura (Waseda University) for their useful comments on this study. We are also grateful to Miss Momoyo Kashiwada for her technical and devoted support on handling the climate data and collecting the related articles. This study was funded in part by the Ministry of Education, Science and Culture of Japan, Grant-in-Aid for Scientific Research (C-21510022) and a Waseda University Grant for Special Research Projects (2008B-244; 2010A-088).

9. References

Allen, R. G.; Pereira, L. S.; Raes, D. & Smith, M. (1998). *Crop Evapotranspiration: Guidelines for Computing Crop Water Requirements.* United Nations Food and Agriculture Organization, Irrigation and Drainage Paper 56. Rome, 300 p.

Barker, I. K. & Lindsay L. R. (2000). Lyme borreliosis in Ontario: determining the risks. *Canadian Medical Association Journal*, 162, 1573–1574.

Bayoh, M. N. & Lindsay S. W. (2003). Effect of temperature on the development of the aquatic stages of *Anopheles gambiae* sensu stricto (Diptera: Culicidae). *Bulletin of Entomological Research*, 93, 375–381.

Bhattacharya, S.; Sharma, C.; Dhiman, R. C. & Mitra, A. P. (2006). Climate change and malaria in India. *Current Science*, 90, 369–375.

Billingsley, P. F.; Charlwood, J. D. & Knols, B. G. J. (2005). Rapid assessment of malaria risk using entomological techniques: taking an epidemiological snapshot, In: *Environmental Change and Malaria Risk,* Takken, W.; Martens, P. & Bogers, R. J., (Eds.), 41–50, Springer, Netherlands.

Connor, S. J. & Mantilla G. C. (2008). Integration of seasonal forecasts into early warning systems for climate-sensitive diseases such as malaria and dengue, In: *Seasonal Forecasts, Climatic Change and Human Health,* Thomson, M. C.; Garcia-Herrera, R. & Beniston, M. (Eds.), 71–84, Springer, New York.

Craig, M. H.; Snow, R. W. & le Sueur, D. (1999). A climate-based distribution model of malaria transmission in sub-Saharan Africa. *Parasitology Today*, 15, 105–111.

Chen, B.; Harbach, R. E. & Butlin, R. K. (2002). Molecular and morphological studies on the *Anopheles minimus* group of mosquitoes in southern China: taxonomic review, distribution and malaria vector status. *Medical and Veterinary Entomology*, 16, 253–265.

Chen, B.; Butlin, R. K.; Pedro, P. M.; Wang, X. Z. & Harbach, R. E. (2006). Molecular variation, systematics and distribution of the *Anopheles fluviatilis* complex in southern Asia. *Medical and Veterinary Entomology*, 20, 33-43.

Dev, V. (1996). *Anopheles minimus*: its bionomics and role in the transmission of malaria in Assam, India. *Bulletin of the World Health Organization*, 74, 61-66.

Depinay, J. M. O.; Mbogo, C. M.; Killeen, G.; Knols, B.; Beier, J.; Carson, J.; Dushoff, J.; Billingsley, P.; Mwambi, H.; Githure, J.; Toure, A. M. & McKenzie, F. E. (2004). A simulation model of African *Anopheles* ecology and population dynamics for the analysis of malaria transmission. *Malaria Journal*, 3, 29.

Dunne, K. A. & Willmott, C. J. (1996). Global distribution of plantextractable water capacity of soil. *International Journal of Climatology*, 16, 841-859.

Ebi, K. L.; Hartman, J.; McConnell, J. K.; Chan, N. & Weyant, J. (2005). Climate suitability for stable malaria transmission in Zimbabwe under different climate change scenarios. *Climatic Change*, 73, 375-393.

Ebi, K. L. (2009). Managing the changing health risks of climate change. *Current Opinion in Environmental Sustainability*, 1, 107-109.

Flato, G. M.; Boer, G. J.; Lee, W. G.; McFarlane, N. A.; Ramsden, D.; Reader, M. C. & Weaver, A. J. (2000). The Canadian Centre for Climate Modelling and Analysis global coupled model and its climate. *Climate Dynamics*, 16, 451-467.

Foley, D. H.; Rueda, L. M.; Peterson, A. T. & Wilkerson, R. C. (2008). Potential distribution of two species in the medically important *Anopheles minimus* complex (Diptera: Culicidae). *Journal of Medical Entomology*, 45, 852-860.

Gething, P. W.; Smith, D. L.; Patil, A. P.; Tatem, A. J.; Snow, R. W. & Hay, H. I. (2010). Climate change and the global malaria recession. *Nature*, 465, 342-345.

Hales S.; de Wet, N.; Maindonald, J. & Woodward, A. (2002). Potential effect of population and climate changes on global distribution of dengue fever: an empirical model. *Lancet*, 360, 830-834.

Hay, S. I.; Cox, J.; Rogers, D. J.; Randolph, S. E.; Stern, D. I.; Shanks, G. D.; Myers, M. F. & Snow, R.W. (2002). Climate change and the resurgence of malaria in the East African highlands. *Nature*, 415, 905-909.

Hay, S. I.; Guerra, C. A.; Gething, P. W.; Patil, A. P.; Tatem, A. J.; Noor, A. M.; Kabaria, C. W.; Manh, B. H.; Elyazar, I. R. F.; Brooker, S.; Smith, D. L.; Moyeed, R. A. & Snow, R. W. (2009). A world malaria map: *Plasmodium falciparum* endemicity in 2007. *PLOS Medicine*, 6, 0286-0302.

Hopp, M. J. & Foley, J. A. (2001). Global-scale relationships between climate and the dengue fever vector *Aedes aegypti*. *Climatic Change*, 48, 441-463.

Hoshen, M. B. & Morse, A. P. (2005). A model structure for estimating malaria risk, In: *Environmental Change and Malaria Risk,* Takken, W.; Martens, P. & Bogers, R. J., (Eds.), 41-50, Springer, Netherlands.

Hu, X. M.; Tsuda, Y. & Takagi, M. (2003). Survival and development of larvae of three tropical malaria vectors (Diptera: Culicidae) under a seasonally changing temperature condition in Nagasaki, Japan. *Medical Entomology and Zoology*, 54, 371-379.

Ikemoto, T. (2003). Possible existence of a common temperature and a common duration of development among members of a taxonomic group of arthropods that underwent speciational adaptation to temperature. *Applied Entomology and Zoology*, 38, 487-492.

Ikemoto, T. (2005). Intrinsic optimum temperature for development of insects and mites. *Environmental Entomology*, 34, 1377-1387.

Ikemoto, T. (2008). Tropical malaria does not mean hot environments. *Journal of Medical Entomology*, 45, 963-969.

Jetten, T. H. & Takken, W. (1994). *Anophelism without malaria in Europe: a review of the ecology and distribution of the genus Anopheles in Europe*. Wageningen Agricultural University Press, Wageningen. 69 p.

Kashiwada, M. & Ohta, S. (2010). Modeling the spatio-temporal distribution of the *Anopheles* mosquito based on life history and surface water conditions. *Open Ecology Journal*, 3, 29–40.

Kiritani, K. (2006). Predicting impacts of global warming on population dynamics and distribution of arthropods in Japan. *Population Ecology*, 48, 5–12.

Konradsen, F.; Stobberup, K. A.; Sharma, S. K.; Gulati, O. T. & van der Hoek, W. (1998). Irrigation water release and *Anopheles culicifacies* abundance in Gujarat, India. *Acta Tropica*, 71, 195–197.

Lardeux, F. J.; Tejerina, R. H.; Quispe, V. & Chavez, T. K. (2008). A physiological time analysis of the duration of the gonotrophic cycle of *Anopheles pseudopunctipennis* and its implications for malaria transmission in Bolivia. *Malaria Journal*, 7, 141.

Lee, J. S.; Lee, J. W.; Cho, S. H. & Ree, H. I. (2002). Outbreak of vivax malaria in areas adjacent to the demilitarized zone, South Korea, 1998. *American Journal of Tropical Medicine and Hygiene*, 66, 13–17.

Lindsay, S. W. & Martens, W. J. (1998). Malaria in the African highlands: past, present and future. *Bulletin of the World Health Organization*, 76, 33–45.

Lindsay, S. W.; Parson, L. & Thomas, C. J. (1998). Mapping the ranges and relative abundance of the two principal African malaria vectors. *Proceedings of the Royal Society of London. Series B. Biological Sciences*, 265, 847–854.

Martens, P.; Kovats, R. S.; Nijhof, S.; de Vries, P.; Livermore, M. T. J.; Bradley, D. J.; Cox, J. & McMichael, A. J. (1999). Climate change and future populations at risk of malaria. *Global Environmental Change*, 9, S89–S107.

Martens, P. & Thomas, C. (2005). Climate change and malaria risk: complexity and scaling, In: *Environmnetal Change and Malaria Risk*, Takken, W.; Martens, P. & Bogers, R. J., (Eds.), 3–14, Springer, Netherlands.

Martin, P. H. & Lefebvre, M. G. (1995). Malaria and climate: sensitivity of malaria potential transmission to climate. *Ambio*, 24, 200–207.

Medley, K. A. (2010). Niche shifts during the global invasion of the Asian tiger mosquito, *Aedes albopictus* skuse (Culicidae), revealed by reciprocal distribution models. *Global Ecology and Biogeography*, 19, 122–133.

Mogi, M. & Okazawa, T. (1996). Development of *Anopheles sinensis* immatures (Diptera: Culicidae) in the field effects of temperature and nutrition. *Medical Entomology and Zoology*, 47, 355–362.

New, M.; Hulme, M. & Jones, P. (1999). Representing twentieth-century space–time climate variability. Part 1: development of a 1961–90 mean monthly terrestrial climatology. *Journal of Climate*, 12, 829–856.

Ohta, S.; Uchijima, Z.; Seino, H. & Oshima, Y. (1993a). Probable effects of CO_2-induced climatic warming on the thermal environment of ponded shallow water. *Climatic Change*, 23, 69–90.

Ohta, S.; Uchijima, Z. & Oshima, Y. (1993b). Probable effects of CO_2-induced climatic changes on net primary productivity of terrestrial vegetation in East Asia. *Ecological Research*, 8, 199–213.

Ohta, S.; Uchijima, Z. & Oshima, Y. (1995). Effects of 2 × CO_2 climatic warming on water temperature and agricultural potential in China. *Journal of Biogeography*, 22, 649–655.

Ohta, S. & Kimura, A. (2007). Impacts of climate changes on the temperature of paddy waters and suitable land for rice cultivation in Japan. *Agricultural and Forest Meteorology*, 147, 186–198.

Ono, M. (1992). *Research on expansion forecast of animal mediation infection.* Global Environment Research Fund, B-13, pp. 342–351., Tokyo (in Japanese)

Overgaard, H. J.; Tsuda, Y.; Suwonkerd, W. & Takagi, M. (2002). Characteristics of *Anopheles minimus* (Diptera: Culicidae) larval habitats in northern Thailand. *Population Ecology*, 31, 134–141.

Pascual, M.; Ahumada, J A.; Chaves, L. F.; Rodó, X. & Bouma, M. (2006). Malaria resurgence in the East African highlands: Temperature trends revisited. *Proceedings of the National Academy of Sciences of the United States of America*, 103, 5829–5834.

Rogers, D. J. & Randolph, S. E. (2000). The global spread of malaria in a future, warmer world. *Sicence*, 289, 1763–1766.

Rueda, L. M.; Kim, H. C.; Klein, T. A.; Pecor, J. E.; Li, C.; Sithiprasasna, R.; Debboun, M. & Wilkerson, R. C. (2006). Distribution and larval habitat characteristics of *Anopheles hyrcanus* group and related mosquito species (Diptera: Culicidae) in South Korea. *Journal of Vector Ecology*, 31, 198–205.

Singh, O. P.; Chandra, D.; Nanda, N.; Raghavendra, K.; Sunil, S.; Sharma, S. K.; Dua, V. K. & Subbarao, S. K. (2004). Differentiation of members of the *Anopheles fluviatilis* species complex by an allele-specific polymerase chain reaction based on 28S ribosomal DNA sequences. *American Journal of Tropical Medicine and Hygiene*, 70, 27–32.

Tao, F.; Yokozawa, M.; Hayashi, Y. & Lin, E. (2003). Future climate change, the agricultural water cycle, and agricultural production in China. *Agriculture Ecosystems & Environment*, 95, 203–215.

Toma, T.; Miyagi, I.; Malenganisho, W. L. M.; Murakami, H.; Nerome, H. & Yonamine, M. (2002). Distribution and seasonal occurrence of *Anopheles minimus* in Ishigaki Island, Ryukyu Archipelago, Japan, 1998–1999. *Medical Entomology and Zoology*, 2, 29–42.

U.S. Geological Survey (February 2010). Global Topographic Data, Available from http://www1.gsi.go.jp/geowww/globalmap-gsi/gtopo30/gtopo30.html

van Lieshout, M.; Kovats, R. S.; Livermore, M. T. J. & Martens, P. (2004). Climate change and malaria: analysis of the SRES climate and socio-economic scenarios. *Global Environmental Change*, 14, 87–99.

World Health Organization (1989). *Geographical distribution of arthropod-borne diseases and their principal vectors.* WHO, Geneva. 134 p.

Yeom, J. S.; Ryu, S. H.; Oh, S.; Lee, W. J.; Kim, T. S.; Kim, K. H.; Kim, Y. A.; Ahn, S. Y.; Cha, J. E. & Park, J. W. (2005). Status of *Plasmodium vivax* malaria in the Republic of Korea during 2001-2003. *American Journal of Tropical Medicine and Hygiene*, 73, 604-608.

14

Climate Change and Shifts in the Distribution of Moth Species in Finland, with a Focus on the Province of Kainuu

Juhani H. Itämies[1], Reima Leinonen[2] and V. Benno Meyer-Rochow[3,4]

[1]*Kaitoväylä 25 A 6; SF-90570 Oulu;*

[2]*Centre for Economic Development, Transport and the Environment for Kainuu, Kajaani,*

[3]*Faculty of Engineering and Sciences, Jacobs University Bremen, Research II, D-28759*

[4]*Bremen and Department of Biology; Oulu University; SF-90014 Oulu,*

[1,2,3]*Finland*

[4]*Germany*

1. Introduction

Distributions and abundances of insect species depend on a variety of factors, but whether we focus on food plants and availability, environmental niches and shelters, predators or parasites, by far the most important limiting factor is climate. Shifts in insect community structure have successfully been correlated with glacial and inter-glacial periods (Coope 1995; Ashworth 1997; Morgan 1997), but have also attracted the attention of researchers concerned with current climate trends. Parmesan (2001) and Forester et al. (2010) examined examples from North America and Europe and emphasized that predictions of responses to a warmer climate must incorporate observations on habitat loss or alteration, land management and dispersal abilities of the species in question.

For the United Kingdom, Hill et al. (2001) have summarized data on changes in the distribution of specifically three butterfly species (*Pararge aegeria, Aphantopus hyperantus,* and *Pyronia tithonus*). These authors report that the three species have been shifting northward since the 1940s and they present maps of simulated butterfly distributions for the period 2070-2099, based on the changes seen since the 1940s. According to that scenario Iceland will see some colonies of these three species in less than a hundred years. Monitoring specifically moth distributions has been less popular (Chen et al. 2009; Park et al. 2009), and especially microlepidopteran species have not received the attention they deserve, so that few detailed observations on their distribution pattrerns, covering longer periods, exist.

Söderman (1994) reports that a programme to monitor moth abundance and species composition has been developed and run as a Nordic project since the middle of the 1990´s. At the moment the following countries are involved in this Moth Monitoring Scheme (MMS): Denmark, Finland, Iceland, Sweden, (at regional level), Estonia, Latvia, Lithuania and Russia (restricted to Karelia, the St.Petersburg region, and the Pskov region). The methods recommended and explained further below are meant to be used by all participants of the project. To our disappointment, however, moth monitoring in the other Nordic countries has not progressed at the same level that it has in Finland, which is why we focus on the Finnish data in this article.

Collecting butterflies and moths has a long tradition in Finland and the basic knowledge of the distribution of the Finnish macrolepidopteran species is rather good, especially thanks to the comprehensive works by Kaisila (1962), Mikkola (1979) and Huldén et al. (2000). Systematic, regular monitoring of moths in Finland started in 1993 (Somerma et al 1993) and has involved a network of light traps located all over the country.

The contribution of the Province of Kainuu to that project has been significant (see further below) and has helped us to create a solid base to notice changes in the Finnish moth fauna. In the Kainuu area we have not only carried out some extra collecting, employing methods additional to those used elsewhere in Finland (cf., Material and Methods), but we have also been paying particular attention to the microlepidoptera. Based on the two sets of data (from Finland generally and Kainuu Province in particular), we were able to detect changes in the composition of the moth fauna at local county level as well as the wider, biogeographical region, namely the entire territory of Finland.

1.1 Characteristics of the study area

The biogeographical province of Kainuu lies in the eastern part of Middle Finland (Fig. 1) and includes Lake Oulujärvi and a stretch of the river Oulujoki. The lake basin and smaller rivers are defining characters of the province's two easternmost areas. The middle parts of Kainuu, i.e., Puolanka, Hyrynsalmi and Sotkamo, are rich in hills reaching heights of 200-300 m above sea level. In the vicinity of water there are also many eskers running from northwest to southeast and the most remarkable esker series is one that passes through Sotkamo-Vuokatti- Oulujärvi -Rokua.

The natural surroundings of Kainuu are characterized by numerous bodies of water, wide forest areas, eskers, hills and bogs. The large forests give the area a stamp of pristine wilderness. Over 80% of the land area of Kainuu is covered by forests. In the southwestern and western parts of the province bogs dominate the landscape. Especially large boggy areas occur in the vicinity of Vaala. The total area of Kainuu is 24 456 km^2, of which the land area alone amounts to 21 571 km^2. The province borders on the following territories: Karelia borealis in the southeast, Savonia borealis in the south, Ostrobothnia borealis in the west, Kuusamo in the northeast and Russia in the east. The number of human inhabitants is around 85 000 (www.Kainuu.fi).

1.2 The Kainuu moth fauna

The macrolepidopterous fauna of the biogeographical province of Kainuu has been well documented thanks to studies in that area that began in the middle of the 19 th century (Tengström 1869). Additional information, including faunal composition of Kainuu moths, has come from investigations by Aro (1900), Valle (1910), Heikinheimo (1939) and Mikkola (1955). In 1993 Leinonen (1993) gathered together all the information available up to that date and added his own observations. Consequently, information specifically on the marolepidopterous moth fauna of Kainuu has been kept at a high level until the beginning of the last decade of the 20 th century. An updated compilation of the macrolepidoptera for the whole of Finland up to 1997 was published by Huldén et al (2000), also including the then latest available data from the Kainuu area.

Fig. 1. Location of Kainuu biogeographical province in eastern Middle Finland from approximately 64 - 66 degree latitude. 15 mm correspond to ca. 100 km.

On the other hand, there were no published Finland-wide data on the microlepidopterous fauna even though lists of captures or sightings of microlepidopteran species were available from Finland at county level. Basically the most up-to-date work on Finnish microlepidoptera is that by Kyrki (1978), whose study material came from museums and several private collections. At that time 433 microlepidopterous species were known from the Kainuu area alone. Two additions supplemented the earlier publication (Kyrki 1979; Kyrki & Tabell 1984) and even more recent updates were published by Kerppola et al (1995) and Kullberg et al. (2002). Thus, material to compare the present Kainuu microlepidopterous fauna with older compilations does exist, but a thorough investigation into the question of the compositional origins of the microlepidopterous fauna had not been carried out until we started our research in 1987. We have recently published results on the moth fauna of the Friendship Nature Reserve, a joint park between Russia and Finland (Leinonen et al. 1997). The Finnish parts of the park lie near the eastern border of Kainuu province and most of the finds that this report deals with were made by us in the same area and are included in the above mentioned catalogues.

2. Material and methods

2.1 Equipment used and areas visited

One of the largest monitoring exercises in Finland is the Nocturnal Moth Monitoring (NMM) project, which started in 1993 (Somerma et al 1993, Söderman 1994) and still continues (Söderman et al 1994, 1995, 1997, 1999, Nieminen 1996, Leinonen et al 1998a, 1999, 2000, 2003). In this project the numbers of species and individuals of the so-called macrolepidoptera are recorded. Furthermore the distribution of the sexes, number of generations per year and any noteworthy changes have been recorded. From the Kainuu study sites we also have identified and counted the so-called microlepidoptera (Leinonen & Itämies, unpublished data).

For Finland's NMM project, altogether 205 light traps around the country have been in operation for at least one collecting period ever since the project started (Fig. 2). Basic automatic and passive collecting has involved Jalas light-traps (Jalas 1960). The traps were initially equipped with 160 W mixed light lamps and later with 125 W mercury vapour lamps. The traps were emptied once a week and all so-called macrolepidopterous species were identified, counted and sexed. As with some butterflies (e.g., *Pieris napi*: Meyer-Rochow and Järvilehto, 1997) in some species of moths different colour morphs were noticed and recorded. Identifications were made by the members of the Finnish Lepidopterological Society and the data were fed into the Hertta-database and preserved in separate files for each year at the Finnish Environment Institute. At this moment the database includes almost six million records.

The number of traps used over the years has fluctuated (Table 1). There were 5 sites where a trap was in operation every year since 1993 and in the town area of Kajaani one trap was in operation for 21 years between 1984-2005. All light trap material is included into the data of the NMM project. Additionally, bait-trapping has been used extensively in connection with the Oulu-model trap (Fig. 3), according to Laaksonen et al. (2006). More randomly we have also made use of malaise, pheromone and colour traps. During the first years of monitoring the lepidopteran faunas in the Friendship Park area of Kuhmo and Suomussalmi, we used baited traps of the "Jalas type", but later we used the "Oulu-model". The largest bait

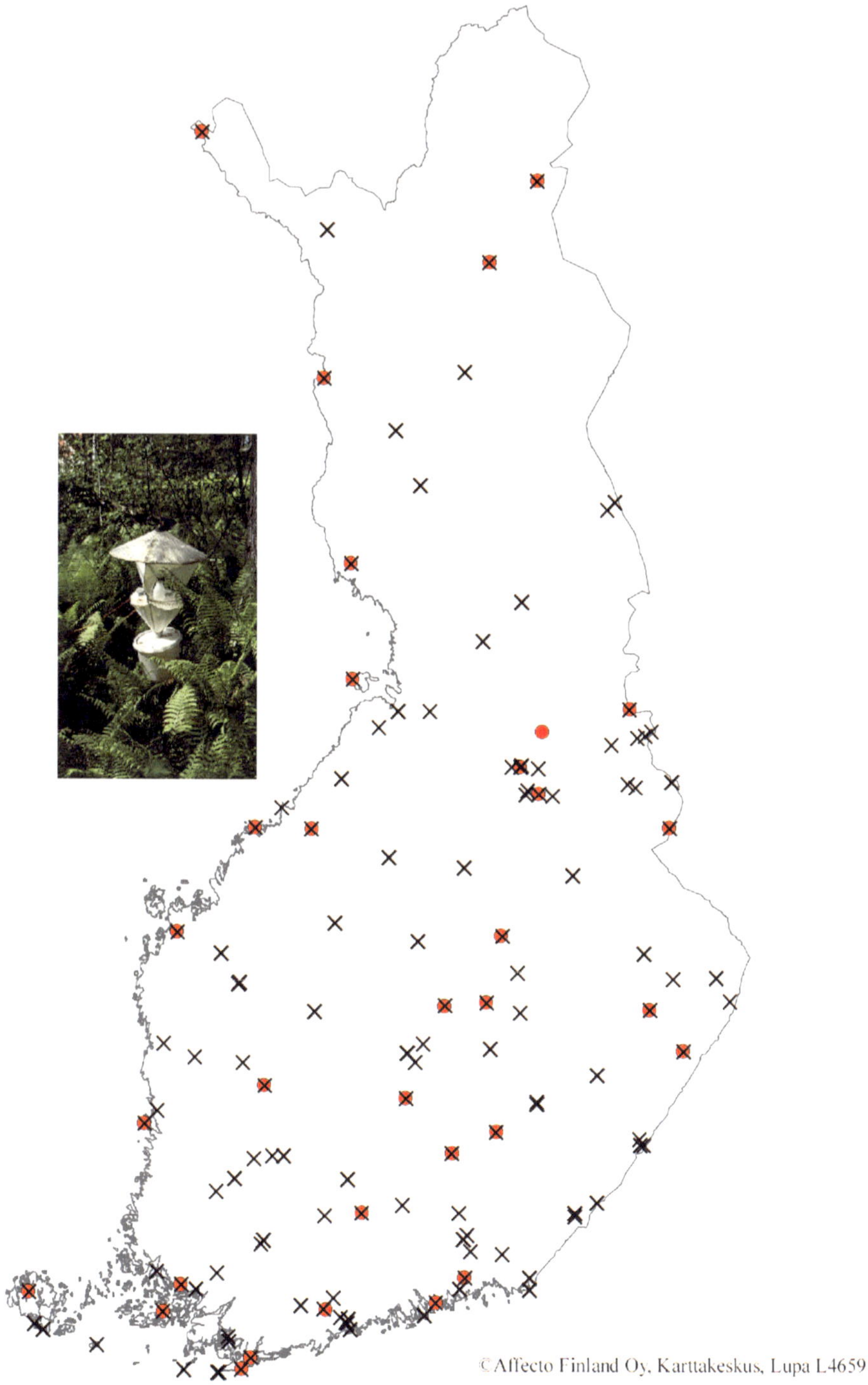

Fig. 2. The light trapping network of the Finnish Nocturnal Moth Monitoring project. Key to symbols: black cross = trap has been in use <15 years; red circle = trap has been in use > 15 years.

Fig. 3. The Oulu-model bait trap (cf. Laaksonen et al. 2006) used in the moth mapping of the Kainuu area.

	light	bait	malaise	pheromone	colour
1987	1				
1990	1				
1991	2	3			
1992	4	20			
1993	12	18			
1994	15	19			
1995	17	29		30	
1996	18	1		17	18
1997	15	31			27
1998	10	5	1	15	18
1999	11	15			23
2000	10	6	1	18	21
2001	10	112		29?	21
2002	10	74		17	18
2003	11	19	4		21
2004	6	1	4	1	18
2005	7	9	5		10
2006	6	2	2		
2007	6	3	3		1
2008	6	2	2	6	2
2009	6	1	2	15	3
2010	6	1	3	15	3

Table 1. List of yearly numbers and types of traps used in the Kainuu area in the years 1987 and 1990-2010.

trapping was carried out in the years 2001-2003, when over one hundred bait traps were used in the joint area of Kainuu and Russian Karelia. This study on the impact of the primeval forest area of Vienan Karelia was carried out to assess the animal populations of old Kainuu forests (including Landscape Ecology of moths in Russian Karelia and the adjacent Finnish province Kainuu), with special attention given to the influence of human activity upon the local moth fauna (Várkonyi & Leinonen 2004).

Our thorough mapping-exercises were based on active searching, either by visiting new sites or by collecting moths from certain food plants. Emphasis has been on species not previously reported from the Kainuu area.

2.2. Groups of Lepidoptera investigated

We have kept micros (Micropterigidae-Pterophoridae) and macros (Papilionidae-Noctuidae) separate, but the general level of knowledge regarding species in these groups varies from region to region. In Finland there are many people who occupy themselves with macros, but much fewer enthusiasts, perhaps one tenth, involve themselves with micros. The first task, therefore, was to identify the new arrivals to the region. We use two different terms and distinguish: a) new provincial finds (npf), which means a species occurring in the province of Kainuu, but not having been reported earlier from there, and b) species that can be considered new to the region and therefore represent new arrivals (nar). Judgments on what is and is not a new provincial find or a new arrival are based largely on published reports.

The interpretation of what constitutes a new arrival is dependent on our experience and also on reports of events from other provinces of Finland. In many cases identification was not at all difficult, but there were also controversial cases. The first difficulty stems from the huge yearly fluctuations of moth catches, obviously increasingly pronounced as we move further northwards (see e.g., Marttila 1992, Itämies & Pulliainen 2006). We had to examine data covering periods of five to six years in order to obtain a basic idea of the moth fauna in the study area. However, there were some rather clear cases that could immediately be recognized as new arrivals even during the first two years of study, based on our own experience and the information available from numerous publications and catalogues.

2.3 Additional information

Another problem presented the group of species that was retrieved through active searching. We could not be absolutely certain, if a species collected from the Kainuu area had already become established there or represented a recent and fleeting visitor. Problematic species such as these were routinely left out of the group of new arrivals. We can illuminate these problems by some exemplary species. *Notocelia roborana* was not included as a new arrival, although it was not recorded from Kainuu until 2003. It depends on rose bushes, which means that it is more or less bound to cultural surroundings. There is of course a small possibility that it had arrived recently, but events in other counties do not support this idea. It seems to have been overlooked earlier. Rather similar is the case with *Eucosma hohenwartiana*, first reported in 2002. It is an inhabitant of meadows with *Centaurea* species, which in the Kainuu area are extremely rare. The species seems to expand its range, but we left it out of the group of arrivals, because we believe it must have been overlooked earlier.

In order to establish which kinds of species are indeed invading the northern regions of Finland, we grouped species together according to their habitat preferences (see Appendix).

One species could be listed in more than one habitat, if it was able to exist in several biotopes. This means that the total number of species deemed to represent new arrivals, may be somewhat larger than the real number. We therefore assigned "habitat points" to each species. The habitat classification used was as follows: df = deciduous forest; cf= coniferous forest; c= cultural surroundings, i.e. habitats made by man like yards, gardens, parks etc.; m= meadows; dm =dry meadows; r= ruderal areas; b= bogs, fens and marshes; dh= extremely sun exposed heaths and slopes; bu= bushes, all kind of border areas of forests, meadows and roads with rich deciduous bushes; cc= clear cuts.
In addition, two columns were added that provided information on the type of targeted searching, i.e., d = actively searched, and ph=collected by pheromones. In the last column the number of migratory moth species is given. These were added to the arrivals, but they were not assigned to any of the habitat groups. Finally, in our comprehensive table (Appendix) we have used a three-step-scale to evaluate the present status of the arrivals as to their success in settling in the Kainuu province.
Another aspect for which we wanted to obtain information, was related to the arrivals' hibernation characteristics. Towards this end, groupings of macrolepidoptera were based on the results of Mikkola & Jalas (1977, 1979), and Mikkola et al. (1985, 1989) while those involving microlepidoptera were based on Schütze (1931), Emmet (1979) and Svensson (1993). Nomenclature follows that of Kullberg et al. (2002).

3. Results

During the years 1987-2010 altogether 258 species of moths were determined as new, i.e., not recorded until then from the biogeographical province of Kainuu. Of these species 187 were representatives of micro and 71 of macrolepidoptera. Altogether 38 species were the results of targeted searching and, as could be expected, this pattern of new records occurred at the beginning of our study period (Fig. 4) and on the other hand most of them (37) were microlepidoptera. It should not surprise that with the kind of massive collecting that took place at that time, increasing numbers of new provincial finds happened mainly during the first few years.
When we arbitrarily divide the series into three periods (1987-95, 1996-2000 and 2001-2010), more interesting features can be revealed. The huge peak of new species is still obvious for the first period (108 species), but one also notices that the last period, especially the first seven years of it, were almost equally profitable in that 97 new species were recorded over that time. There seems to be some kind of collapse during the last three years of the third period, but only future observations will be able to tell us whether this signifies a permanent feature or has to be regarded as a temporary event. In total 77 species were identified as new arrivals and the greatest bulk of these arrived during the last period (Fig. 4).

3.1 Habitat distribution

We distinguished 11 habitat groups to which the species could be assigned (see Material and Methods). Most species were found to belong to the deciduous forest group. When we added to that group species living in various bushy habitats, we see that together they reach 123 points. Meadows and dry meadows follow with 98 points. A rather large number of species also occurred in cultural surroundings (49) and on ruderates (22). Very few new species, however, were collected from bogs (14) and from sun-exposed dry heaths (7).

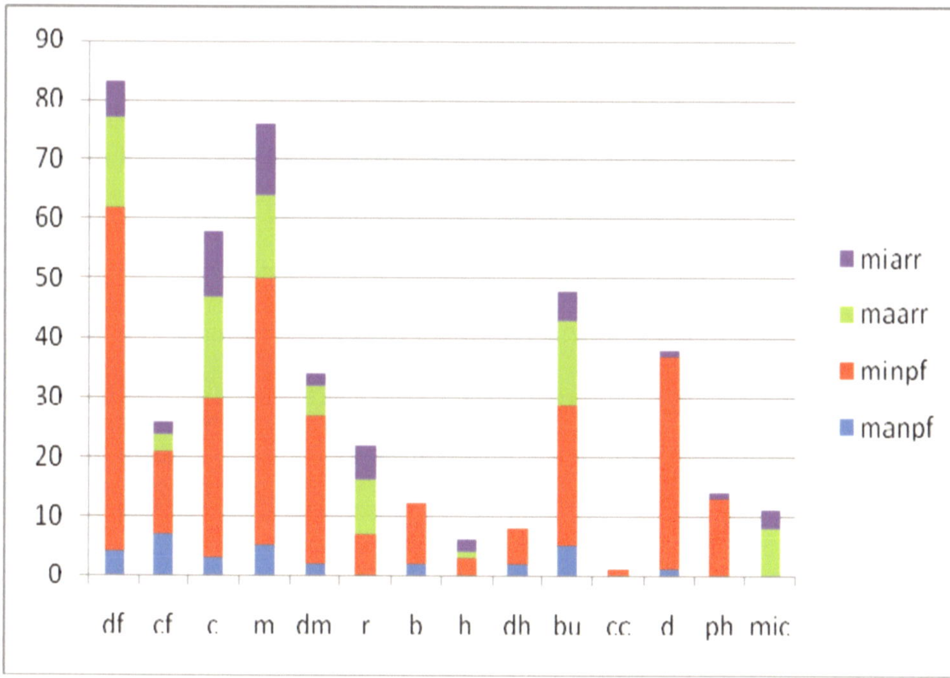

Fig. 4. Habitat distribution of new provincial finds (npf) and new arrivals (arr) of microlepidoptera (mi) and macrolepidoptera (ma) in the province of Kainuu for the years 1987 and 1990-2010 combined. For abbreviations of habitats on the abscissa see Material and Methods. The ordinate shows the number of species.

When we inspect more closely the habitat distribution of the arrivals versus other new provincial finds, we can extract some features from the material (Fig. 5). Species living in man-made surroundings (c and r) seem to be most actively expanding into the Kainuu region. Meadow habitats are also well inhabited by new arrivals, but the proportion of deciduous forest species, for instance, is much lower than in other new finds. Bogs and dry heaths did not contain any new arrivals, although other new finds were present in these habitats. Through active searching we could hardly find any newcomers, but species hitherto overlooked were readily discovered with this kind of planned searching. A similar result was obtained when pheromone-traps were used. However, it can be expected that if we had kept these traps year after year in the same place, we should have been able to show some newcomers also by this method, but this was unfortunately not done.

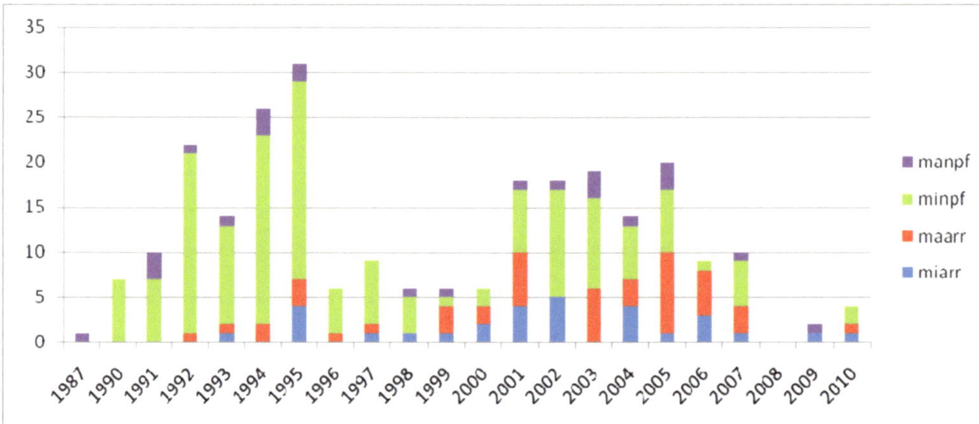

Fig. 5. Yearly catch of new provincial finds (npf) and new arrivals (arr) of microlepidoptera (mi) and macrolepidoptera (ma) in the biogeographical province of Kainuu for the years 1987 and 1990-2010 combined. The ordinate shows number of species; the abscissa depicts the year of monitoring.

3.2 Hibernation phases

Hibernation as larvae was most common amongst all of the newcomers (Fig. 6) irrespective as to whether we singled out micro and macrolepidoptera . Among the macrolepidopterous species, however, there were also many that hibernated in the egg stage and also wintering over as pupae was not exactly rare either. Of the new arrival microlepidopteran species, three were hibernating as imagines, but none of the new arrival macros were in the group of adult overwinterers. The hibernation phases of the latter were rather evenly distributed between egg, larva and pupa, but for the microlepidopteran new arrival species larval hibernation was the rule, although egg hibernators were moderately represented as well.

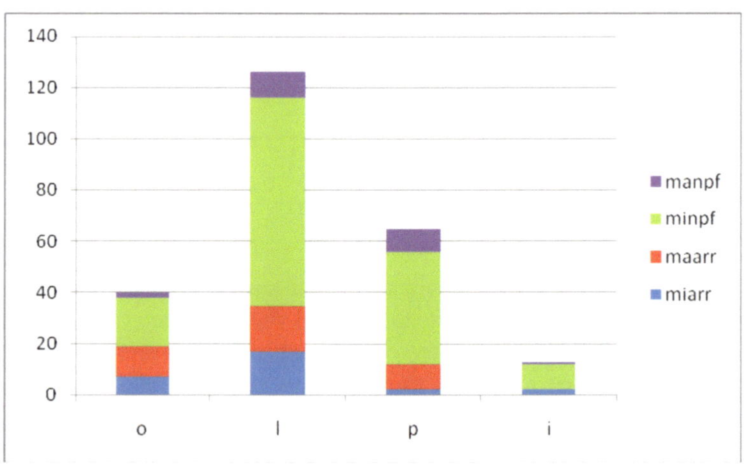

Fig. 6. Hibernating phases of new provincial finds and new arrivals of moths in the Kainuu area, based on data from 1987 and 1990-2010 combined. Explanation of symbols on the abscissa: o = ovum; l = larva; p = pupa; i = imago. The ordinate shows number of species; npf = new provincial find; arr = new arrival; mi = microlepidoptera; ma = macrolepidoptera. The ordinate shows number of species.

3.3 Settling down and becoming established

None of the eleven migratory species appear to have become regulars in the area, although some of them were collected in more than just one year, like, for example, *Agrius convolvuli* (Fig. 7). Settlement history of other newcomers has varied greatly. Many of them were collected only once, but some have expanded their range considerably. We can name as the most successful newcomers *Gracillaria syringella, Oncocera semirubella, Cryptocala chardinyi, Araschnia levana* and *Argynnis paphia.* To date altogether 17 new arrivals seem to have more or less successfully colonised the Kainuu area and become established there (Appendix).

Fig. 7. *Agrius convolvuli*, a typical migratory hawk moth reported a couple of times also from the Kainuu area during the last few years. Photo Reima Leinonen.

3.3.1 Observed changes in the moth fauna Finland-wide

The available general information on the Finnish lepidopterous fauna stems from the large network of amateurs and their observations and from professional research efforts at various Finnish universities. The Finnish Lepidopterological Society has over a thousand members, which gives a good idea about the breadth of the hobby of butterfly and moth collectors in Finland with its 5.4 million inhabitants. Moreover, children learn already at school facts about Finnish butterflies and moths (Meyer-Rochow 2008). During the last few years increasingly more attention has been paid to changes in the natural surroundings. One of the largest monitoring exercises in Finland is the Nocturnal Moth Monitoring project (abbreviated NMM), which started in 1993 (Somerma et al 1993, Söderman 1994) and some conclusions presented in this chapter are based on that material, representing a short review of the results of 17 years of monitoring. Year-by-year summaries of this project have been presented quite regularly (Söderman et al 1994, 1995, 1997, 1999, Nieminen 1996, Leinonen et al 1998a,1999, 2000, 2003) and specific surveys based on those many years of data have also been presented at various seminars, in conference abstracts and review articles like the most recent one by Leinonen et al (2011).
For collecting of proper material see chapter 2.3. above.

3.3.2 Observed changes in the moth fauna of northern Finland

In the results of the NMM we find evidence in support of the northward spread of several species, suggesting that this northward expansion could be the result of an increasingly warmer climate. Numerous new arrivals have been recorded from Finland and many southern species have expanded their ranges northward. Examples of southern species spreading into northern regions are the following ones: *Allophyes oxyacanthae* (Fig. 8), *Protodeltote pygarga* and *Cosmia trapezina*. *Cryptocala chardinyi* is one of the most illuminating species when it comes to the rapid expansion of distribution area in Finland (Fig. 9).

The opposite phenomenon to the above mentioned scenario is that there are indications that certain species have started to withdraw from southern Finland or at least their population sizes have been dwindling. We can mention, for example, the following species: *Entephria caesiata*, *Xanthorhoe annotinata* and *Dasypolia temple*. Maps based on the situation at the beginning of NMM and after fifteen years show the obvious trend happening in the area of this moth (Fig. 10).

Fig. 8. *Allophyes oxyacanthae*, one of numerous moth species presently extending its range northward in Finland. Photo Reima Leinonen.

The structure of the species composition has also undergone change. Populations of species earlier known to be abundant have visibly decreased. Species compositions have started to resemble more and more those typical of Central Europe, where the number of species is high but the numbers of individuals are relatively low. At the same time the populations of some species have increased dramatically and we can talk about pest risks. Here we can take *Lymantria monacha* as an example of a species which may in the near future be so abundant as to be able to cause forest disasters even in Finland like it has done in Central Europe.

In the northernmost part of Finland, in Lapland, the geometrid moth *Epirrita autumnata* (Fig. 11) is known to have mass outbreaks with intervals of approximately ten years, when it severely affects mountain birch forests. The trees will stand a one-year-disaster, but if the catastrophe goes on for several years, the trees are likely to die. The eggs of *E. autumnata* are

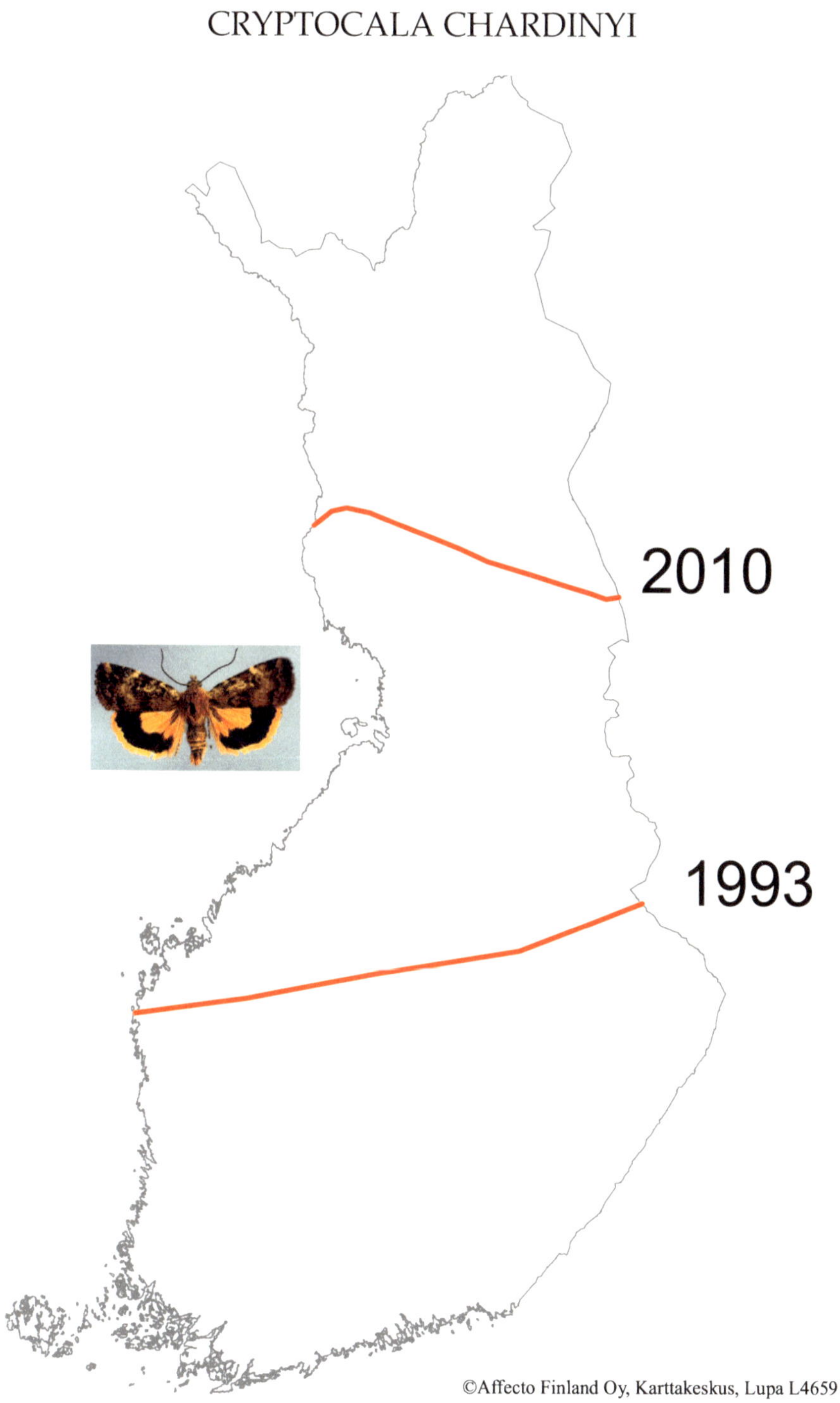

Fig. 9. Map showing the increase in the distribution area of *Cryptocala chardinyi* in Finland based on data from the Finnish Nocturnal Moth Monitoring project.

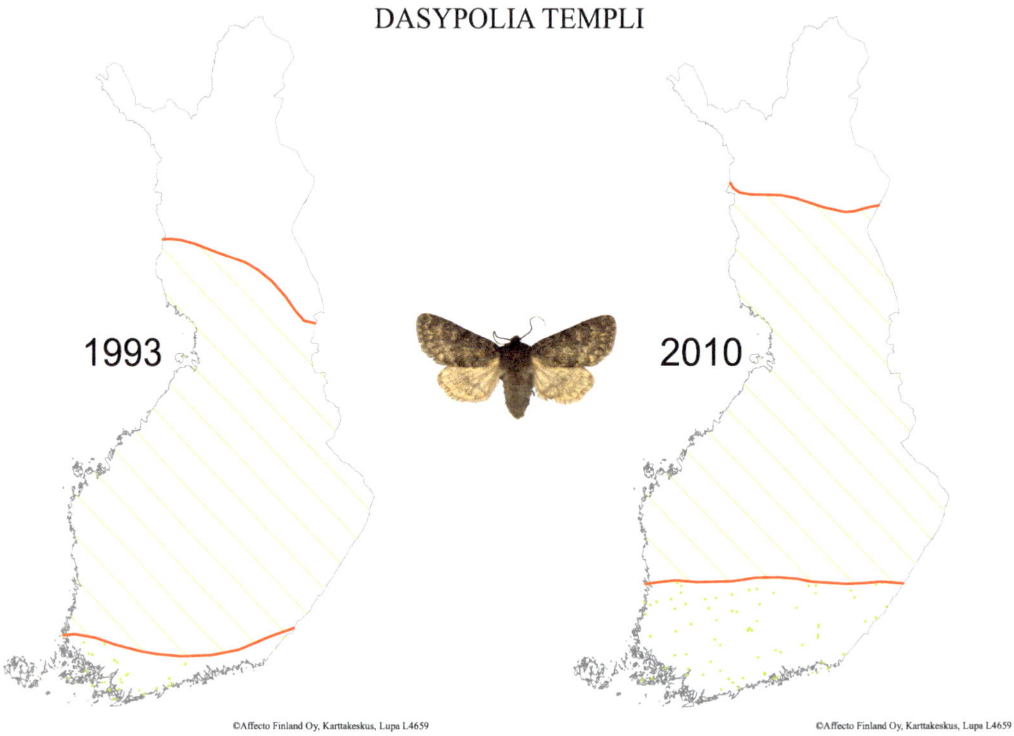

Fig. 10. Maps showing the northward withdrawal of *Dasypolia templi* in Finland based on data from the Finnish Nocturnal Moth Monitoring project.

Fig. 11. *Epirrita autumnata*, a geometrid moth which in northernmost Finland has massive outbreaks in mountain birch forests (*Betula pubescens*) and which in the future may occur more often, if the winters get milder due to global climate warming.

vulnerable to very low temperatures, i.e. if they encounter temperatures of -40°C, all but approximately 2 % will die. However, if temperatures stay above -35°C, then up to 96% of the eggs will survive. Mild winters due to global climate warming may thus promote disasters in northern mountain birch forests.

One of the clearest indications of climate warming in our NMM data is the increase of second annual generations. This has been observed in a growing number of species and in more and more northern localities. *Gymnoscelis rufifasciata* is an extreme example of this phenomenon, because it has nowadays even a third generation quite regularly in some areas of southern Finland (Fig. 12). Having second or even third generations in regions as far north as Lapland or northern Finland generally is not necessarily beneficial to the species, for it can be a risk to the population. If the species does not get ready for hibernation, that part of population will be lost and in the long run the species may even become endangered.

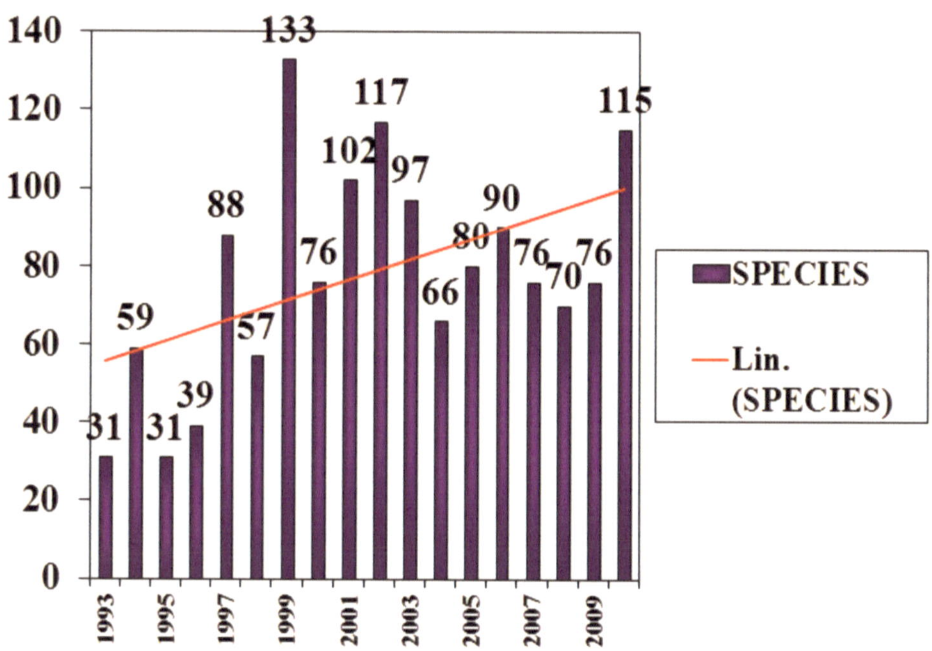

Fig. 12. The number of species with second generations per annum has steadily been increasing with *Gymnoscelis rufifasciata*, a geometrid moth representing the ultimate example of an increase in annual generations in Finland to two or even three generations during the last few years, based on the material of the Finnish Nocturnal Moth Monitoring project.

3.4 Alpha-diversities

As to the diversity index, we used the data of the FMM and calculated the so called alpha-index. At the beginning of monitoring activities in 1993 the value was a little above 60, while at the same time for Lithuania it was, for instance, close to one hundred. At present the alpha-indices are close to a hundred in certain parts of southern Finland. This reflects the increase of species numbers and that the observed specimen numbers are closer to each other, i.e. the huge mass occurrences of individual species are lacking. In northern Finland the alpha-values stay still low mainly due to two reasons. The effectiveness of light traps is weak around midsummer in Lapland due to the continuous summer daylight and the species numbers are originally low to start with. Secondly, in the north we still have species that exhibit quite regular mass outbreaks, thus keeping alpha-values low.

4. Discussion

It is not unusual to find that insect populations peak and decline in cycles and that the cycles can have periodicities varying from one to many years. This is one major difficulty in trying to draw conclusions on insect abundances in relation to climate change (Meyer-Rochow 2008). Therefore, it is essential that one has data on population densities and species compositions over long time spans for a defined geographic region; another equally important requirement is that one has reliable weather records over many years for the region in question. Our investigation meet both of these criteria.

The longest continuous set of Finnish temperature data exists from the area of Helsinki, Kaisaniemi, from where we have almost two hundred years of recordings. Statistics preserved at the Finnish Meteorological Institute show that the average daily temperatures and especially the lowest daily temperatures have been rising most clearly during the last decade. This means that nighttime temperatures have risen, thus enabling species demanding higher minimum temperatures to spread and survive even in the northern regions of Finland. At the same time the winters have become shorter and milder. On the other hand, increasing summer rains, observed in the last few years, could present a problem for some species.

The results of the Finnish NMM project demonstrate convincingly that there has been a dramatic increase over the last ten years in the number of new records for species of moths in Finland generally and the biogeographical province of Kainuu in particular. This is not, however, a new phenomenon, for Kaisila (1962) had already shown how the rising average summer temperature around the turning point of the 1940s, had caused many species in Finland to expand their ranges northward. He named several species as expansive ones. Yet, many of the species Kaisila (1962) had identified as those with northward expanding ranges, later withdrew again from their northernmost expanses when cooler summers returned (see for instance Huldén et al. 2000).

The period studied by Kaisila (1962) ended in 1961, but Mikkola (1997) followed up Kaisila's work and analysed migrations of Lepidoptera until 1996. The overlap of three years with the NMM (which had started in 1993) and the availability of the early NMM data were, thus, a bonus to Mikkola's investigation. According to Mikkola's analyses, 30 recent invaders and 40 older expansive species were once again spreading into northern Finland. These basic and important investigations, however, were almost totally lacking information on microlepidoptera, because the general knowledge with regard to the latter was so poor that they had to be left out of the analyses. At province level, however, similar expansion trends in this moth group as compared to the macrolepidoptera were noticed. Since species belonging to the microlepidoptera are frequently very specific with regard to their larval food plants, one could have expected that this would have presented a hindrance to their northward spread and therefore function as a severe obstacle in range expansion.

Yet, our results show that changes in the moth fauna of Finland are not only continuing, but are possibly taking place at a greater pace than ever before. We must, of course, keep in mind that the number of collectors has increased remarkably and that the traps and trapping efforts have become vastly more efficient than what they were some 50 years ago. In spite of that we dare say, however, that the moth fauna, especially in northern Finland, has been undergoing a dramatic change over the last few years.

4.1 Possible reasons of the northward expansion

What could possibly be the cause of this situation? Moths and in particular their larvae depend on live, green plants, some very strictly limited to one, or at least a few, food plants; others with a somewhat wider spectrum of acceptable food plants (see for instance Seppänen 1954). Although relevant data are incomplete, we have good reason to assume that microlepidopteran species contain more food specialists than macrolepidoptera do. Against this biological background, we can easily understand how a change in the environment and in the way the land is used will influence the floral composition, vegetation, and thus the welfare of numerous species of moths. Restructuring the Finnish agriculture has resulted in huge changes from small units to much larger ones. This has especially led to a decrease in various meadow habitats (e.g. Vainio et al. 2001, Kivinen et al. 2008) and is reflected by the dwindling numbers of certain moth species that depend on the presence of such meadows (Rassi et al. 2010).

One very large scale change in the environment has been due to reforms in the forestry industry. Here the splitting of large forest areas into much smaller units has to be mentioned as an ongoing process. As a consequence, the age structure of the forests has become much younger (Várkonyi & Leinonen 2004). Yet the changes mentioned above should speed up species declines and lead to the disappearance of some species, not an increase and range expansion. Therefore the reason for the observed northward expansion of many species must be sought elsewhere and most likely climate change, i.e. global warming, is involved.

4.2 The role of global warming

What kinds of species are invading Finland now? Our results from Kainuu show that many new arrivals prefer meadows and cultural surroundings. Many of them are also living on decorative plants or plants of ruderate areas. This seems to be in disagreement with what had been stated above, e.g., the decrease of meadow habitats, age-structural changes in forest trees, etc. However, we need to remember we were referring to mostly dry habitats. Wetter meadows have in fact been increasing, at least locally (Kivinen et al. 2008). Furthermore, the changes in forestry practices have resulted in increased clear cuts, which in the first steps of succession act like meadows and after some years turn into thick bush, both not only acceptable, but frequently preferred to other habitats, like for instance dense forests, by new arrivals. Mikkola (1997) stated that expansive species are mostly associated with bushes and trees. Although Finland is still largely covered by forests, especially along the eastern and northern parts, it is quite clear that many species cannot unlimitedly migrate northward, where suitable habitats (and to a lesser extent the required food plants) are missing. This holds true in spite of the fact that the changing climate would suit some of these expanding species. Virtanen & Neuvonen (1999) state that the northern limit of the macrolepidopteran distribution in Finland is principally determined by climatic factors, with food plants probably determining distributions in no more than 3 % of the species. This may apply to permanent and long-established species, but in the case of new arrivals the situation could be different. We can only speculate how many of the new arrivals might have been able to settle down in Finland and Kainuu, if the environment had been the same as, say, some 50 or 60 years ago!

Species overwintering as eggs or adults would be expected to benefit from a rise in winter temperature (Virtanen & Neuvonen 1999), because in their analyses from different geographical zones in Finland these authors were able to show, how in southern areas the proportion of adult and egg hibernators was large and, on the other hand, in northern areas larval hibernation was dominating. In our results from the Kainuu area there is some support for this view: in the new microlepidopterous arrivals there were rather many species wintering as eggs. Noteworthy is, however, that the proportion of imago hibernators has not yet significantly increased in the Kainuu area.

4.3 Summary

In summary, drastic changes of the moth fauna are presently taking place in Finland both at state level and the local scale, for instance the Kainuu area. We can only speculate how huge this change may be on a global scale or at least with regard to the northern hemisphere alone. Volney and Fleming (2000), cited in http://www.eoearth.org/article/Climate_change_and_insects_as_a_forest_disturbance_in_the_arctic, have attempted to predict what impact the global change might have on insects of boreal forests. Huang et al. (2010) have tried to do the same for the subtropical island nation of Taiwan. Not surprisingly, emphasis has been on possible effects of destructive moths and their range expansions (Vanhanen et al. 2007; Huang et al. 2010), but climate change and its consequences on lepidopteran diseases have also come under scrutiny, for example at the Centre for Ecology and Hydrology of the United Kingdom (http://www.ceh.ac.uk) and in the United States (Altizer and de Roode 2010).

That moths not only expand their ranges into higher latitudes but also into areas of higher elevation has been shown for tropical regions (Chen et al. 2009) and that a warmer climate, in the case of the winter moth *Operophtera brumata*, can disrupt the synchrony of oak and insect phenology has been examined by Visser and Holleman (2001). Because of the usefulness of Lepidoptera as an indicator of climate change (Forester et al. 2010), studies on distributions and abundances of these insects are being conducted on a worldwide scale. It is clear that dramatic changes are taking place, but that moths can also adapt to recent climate changes show observations by Park et al. (2009) from islands in southern Korea. What exactly the ecological consequences of the shifts affecting moth populations (on a global scale) might be with regard to entire ecosystems and biomes, given the roles caterpillars and adult Lepidoptera play in them, is impossible to predict just yet - or perhaps never.

5. Appendix

List of new moth (Lepidoptera) arrivals to Kainuu area in 1987, 1990-2010. Explanation to columns: A = name of the species; B = lepidopterous group (1= microlepidoptera and 2 = macrolepidoptera); C = arrival year; D = habitat classification (see material and methods for the abbreviations); E = state of settlement into Kainuu area (s = a couple of finds, ss = regurlarly found in some places and sss = widespread nowadays at Kainuu); F = found by targeted searching = d and with pheromone = ph; G = hibernating phase (o = ovum, l = larva, p = pupa and i = imago).

A	B	C	D	E	F	G
Gracillaria syringella	1	1993	c	sss		p
Arghyresthia goedartella	1	1995	df	s		l
Eidophasia messingiella	1	1995	m,c			e
Croesia holmiana	1	1995	c	s		e
Loxoterma siderana	1	1995	c	s		l
Exaeretia allisella	1	1997	r,c	ss		l
Trachysmia advenella	1	1998	df,bu			l
Elachista adscitella	1	1999	m,df	s		l
Oncocera semirubella	1	2000	m			l
Limnaecia phragmitella	1	2000	h			l
Echromius ocelleus	1	2001	Mi			i
Loxostege sticticalis	1	2001	Mi			p
Phycitodes lacteellus	1	2001	Mi			
Isoprichtis striatella	1	2001	m,r			p
Synanthedon flaviventris	1	2002	df,bu		d, ph	l
Lobesia abscisana	1	2002	m,r	s		o
Coleophora tamesis	1	2002	h			l
Dipleurina lacustrata	1	2002	cf			l
Nyctegretis lineana	1	2004	m			l
Dichrorampha simpliciana	1	2004	r,c,m			l
Acrolepiopsis assectella	1	2004	c			i
Sophronia semicostella	1	2004	dm			l
Psoricoptera speciosella	1	2005	df,bu			o
Coleophora artemisicolella	1	2006	r,c,m	s		l
Caryocolum blandelloides	1	2006	c,m			o
Cymolomia hartigiana	1	2006	cf			o
Depressaria daucella	1	2007	m,c,r			i
Choreutis pariana	1	2009	c			o
Endothenia ericetana	1	2010	m,bu			l
Xestia sexstrigata	2	1992	c,m,			l
Amphipyra perflua	2	1993	df			e
Colotois pennaria	2	1994	df			e
Fabriciana adippe	2	1994	dm			l
Timandra griseata	2	1995	dm	s		l
Erannis defoliaria	2	1995	df			e
Blepharita amica	2	1995	bu,bu,c			e
Ipimorpha subtusa	2	1996	df			e
Allophyes oxyacanthae	2	1997	bu,r,c			e
Lomographa temerata	2	1999	df,bu	s		p
Cryptocala chardinyi	2	1999	m,c	sss		l
Agrius convolvuli	2	1999	Mi			p
Argynnis paphia	2	2000	m,dm	ss		l
Pyrrhia umbra	2	2000	c,bu			p
Acronicta alni	2	2001	df			p
Photedes fluxa	2	2001	m			l
Schranckia costaestrigalis	2	2001	m,r			l

Idaea biselata	2	2001	df	s	l
Mesoligia literosa	2	2001	m		l
Polia nebulosa	2	2001	bu,cf		l
Photedes captiuncula	2	2002	Mi		l
Staurophora celsia	2	2003	bu,r,df, c	s	e
Catocala sponsa	2	2003	Mi		e
Xestia triangulum	2	2003	bu,c,cf,m		l
Eilema lutarellum	2	2003	m, dm		l
Eupithecia linariata	2	2003	m,r		p
Lomographa bimaculata	2	2003	df,bu		p
Cerastis leucographa	2	2004	bu,r,c		p
Araschnia levana	2	2004	m,c	sss	p
Apamea ophiogramma	2	2004	bu,m,c		l
Macroglossa stellatarum	2	2005	Mi		p
Agrotis ipsilon	2	2005	Mi		i
Cosmia pyralina	2	2005	Mi		e
Apamea scolopacina	2	2005	m,c,bu		l
Catocala nupta	2	2005	Mi		e
Catocala fulminea	2	2005	Mi		e
Arenostola semicana	2	2005	reed bed		e
Paracolax tristalis	2	2005	df		l
Trisateles emortualis	2	2005	df		p
Gortyna flavago	2	2006	c,r,		e
Mniotype satura	2	2006	bu,df,r		e
Ipimorpha retusa	2	2006	df		e
Protodeltote pygarga	2	2006	bu,r,c	s	p
Herminia tarsipennalis	2	2006	df,bu,c		l
Atolmis rubricollis	2	2007	cf		p
Scopula incanata	2	2007	dm,m,c		l
Xestia xanthographa	2	2007	m		l
Larentia clavaria	2	2010	c,r	s	e

6. References

Altizer, S., de Roode, J. 2010: When butterflies get bugs: The ABC of lepidopteran disease. American Butterflies 18 (2): 16-27.

Aro, J.E. 1900: Suomen perhoset. Helsinki 290 pp.

Ashworth, A.C. 1997: The response of beetles to Quaternary climate change. In: Huntley, B. et al. (eds.) Past and future rapid environmental changes: the spatial and evolutionary responses of terrestrial biota. Springer Verlag, Berlin. 119-127 pp.

Chen, I-C., Shiu, H-J., Benedick, S., Holloway, J.D., Chey, V.K., Barlow, H.S, Hill, J.K., Thomas, C.D. 2009: Elevation increases in moth assemblages over 42 years on a tropical mountain. Proceedings of the National Academy of Sciences (U.S.A.) 106(5): 1479-1483.

Coope, G.R. 1995: Insect faunas in ice age environments: why so little extinction? In: Lawton, J.H., May, R.M. (eds) Extinction rates. Oxford University Press, Oxford (U.K.). 55-74 pp.

Emmet, A.M. 1979: A field guide to the smaller British lepidoptera. The British Entomological & Natural History Society. London. 271 pp.

Forester, M.L., McCall, A.C., Sanders, N.J., Fordyce, J.A., Thorne, J.H., O'Brien, J., Waetjen, D.P., Shapiro, A.M. 2010: Compound effects of climate change and habitat alteration shift patterns of butterflyn diversity. Proceedings of the National Academy of Sciences (U.S.A.) 107(5): 2088-2092.

Heikinheimo, O. 1939: A stipendiate report of the moth faunistic reasearch done in the areas of the communes of Kuhmo and Pielisjärvi. In Finnish. Preserved in the Archives of the Finnish Lepidopterological Society. Helsinki.

Hill, J.K., Thomas, C.D., Fox, R., Moss, D., Huntley, B. 2001: Analysis and modeling range changes in UK butterflies. In: Woiwood, I.P., Reynolds, D.R., Thomas, C.D. (eds.) Insect movement: mechanisms and consequences.CABI Publishing, Wallingford (UK). 415-441.

Huang, S-H., Cheng, C-H., Wu, W-J. 2010: Possible impacts of climate change on rice insect pest and management tactics in Taiwan. Crop, Environment, and Bioinformatics 7: 269-279.

Huldén, L., Albrecht, A., Itämies, J., Malinen, P. & Wattenhovi, J. 2000: Suomen suurperhosatlas. Atlas of Finnish Macrolepidoptera. Finnish Lepidopterological Society. Natural History Museum. Helsinki. Viestipaino Oy. 328 pp.

Itämies, J. & Pulliainen, E. 2006. Sallan Värriötunturin suurperhosfauna valorysäsaaliiden perusteella vuosina 1978-2004. (Macrolepidoptera fauna at Värriötunturi fell area, northeastern Finnish Lapland, as indicated by light trap samples during 1978-2004) (In Finnish with English summary). -Baptria 31:63-93.

Jalas, I. 1960: Eine leichtgebaute, leichttransportable Lichtreuse zum Fangen von Schmetterlingen. - Ann.Entomol. Fennici 26:44-50.

Kaisila. J. 1962: Immigration und Expansion der Lepidopteren in Finnland in the Jahren 1869-1960. (English summary). -Acta Entomologica Fennica 18: 1-452

Kerppola, SW., Albrecht, A. & Huldén, L. 1995: Suomen pikkuperhosten levinneisyyskartasto (Lepidoptera). Distribution maps of Microlepidoptera in Finland (Lepidoptera). -Baptria 20 (2a): 1-79.

Kivinen, S., Kuussaari, M., Heliölä, J., Luoto, M., helenius, J. & Härjämäki, K. 2008: Maatalousmaiseman rakenteen muutokset ja niiden merkitys lajiston monimuotoisuudelle. (In Finnish). -Suomen ympäristö 4: 112-127.

Kullberg, J., Albrecht, A., Kaila, L. & Varis, V. 2002: Checklist of Finnish Lepidoptera - Suomen perhosten luettelo. - Sahlbergia 6:45-190.

Kyrki, J. 1978: Suomen pikkuperhosten levinneisyys. I. Luonnontieteellisten maakuntien lajisto (Lepidoptera:Micropterigidae- Pterophoridae). (Distribution of Microlepidoptera in Finland. I. Fauna of the biogeographical provinces. In Finnish.). -Notulae Entomologicae 58: 37-67.

Kyrki, J. 1979: : Suomen pikkuperhosten levinneisyys. II. Lisäyksiä luonnontieteellisten maakuntien lajistoon (Lepidoptera: Micropterigidae- Pterophoridae). (Distribution of Microlepidoptera in Finland. II.Additions to the Fauna of the biogeographical provinces. In Finnish.). -Notulae Entomologicae 59: 125-131.

Kyrki, J. & Tabell, J. 1984: Lisäyksiä Suomen luonnontieteellisten maakuntien lajistoon (Lepidoptera: Micropterigidae- Pterophoridae). (Additions to the Fauna of the biogeographical provinces in Finland. In Finnish.). -Notulae Entomologicae 64: 134-144.

Laaksonen, J., Laaksonen, T., Itämies, J., Rytkönen, S. & Välimäki, P. 2006: A new efficient bait-trap model for Lepidoptera surveys - the "Oulu" model. -Entomologica Fennica 17: 153-160.

Leinonen, R. 1993: Kainuun suurperhoset. The Macrolepidoptera of Kainuu. -Baptria 18 (2a): 1-73.

Leinonen, R., Söderman, G., Lundsten, K-E. & Grönholm, L. 1998: Valtakunnallisen yöperhosseurannan tulokset 1996, Results of the Finnish National Moth Monitoring Scheme 1996. -Baptria, 23, (2),79-82.

Leinonen, R., Söderman, G. & Lundsten, K-E. 1999: Valtakunnallisen yöperhosseurannan tulokset 1997, Results of the Finnish National Moth Monitoring Scheme 1997. - Baptria 24, (1), 47-54.

Leinonen, R., Söderman, G. & Lundsten, K-E. 2000: Valtakunnallisen yöperhosseurannan tulokset 1998, Results of the Finnish National Moth Monitoring Scheme 1998. - Baptria 25, (4), 163-170.

Leinonen, R., Lundsten, K-E., Söderman, G. & Tuominen-Roto, L. 2003: Valtakunnallisen yöperhosseurannan tulokset 1999, Results of the Finnish National Moth Monitoring Scheme 1999. -Baptria 28, (1),16-22.

Leinonen, R., Itämies, J., Söderman, G. & Pöyry, J. 2011: The Finnish Moth Monitoring Scheme: patterns and trends with respect of recent climate change. The 20th anniversary celebration publication of Finnish-Russian Friendship Park. Suomen Ympäristö. Manuscript submitted.

Marttila, O. 1992: Yöperhosten (Lepidoptera: Drepanoidea, Geometroidea, Bombycoidea, Sphingoidea, Noctuoidea) lennon ajoittuminen ja sen yhteys tehoisaan lämpösummaan. Kymmenen vuoden (1981-1990) yöperhosseuranta Joutsenossa. (In Finnish). Etelä-Karjalan Allergia- ja Ympäristöinstituutti, tammikuu 1992. 49 pp. + 4 appendix.

Meyer-Rochow, V.B. 2008: Insects and their uses in a cold country: Finland. Entomol. Res. 38, S28-S37.

Meyer-Rochow, V.B. & Järvilehto, M. 1997: Ultraviolet colours in *Pieris napi* from northern and southern Finland: Arctic females are the brightest! Naturwissenschaften 84, 165-168.

Mikkola, A.V.V. 1955: Lepidopterous observations at Kainuu. A manuscript in Finnish, preserved in the archives of the Finnish Lepidopterological Society. Helsinki.

Mikkola, K. & Jalas, I. 1977: Suomen perhoset. Yökköset 1. Otava Keuruu. 256 pp.

Mikkola, K., Jalas, I. 1979: Suomen perhoset. . Yökköset 2. Otava Keuruu. 304 pp.

Mikkola, K., Jalas, I & Peltonen, O. 1985: Suomen perhoset. Mittarit 1. Tampereen Kirjapaino Oy Tamprint. 260 pp.

Mikkola, K., Jalas, I. & Peltonen, O. 1989: Suomen perhoset. Mittarit 2. Hangon Kirjapaino . 280 pp.

Mikkola, K. 1997: Population trends of Finnish Lepidoptera during 1961-1996. - Entomologica Fennica 8 (3): 121-143.

Morgan, A.V. 1997: Fossil coleopteran assemblages in the Great Lakes region of North America: past changes and future prospects. In: Huntley, B. et al. (eds.) Past and future rapid environmental changes: the spatial and evolutionary responses of terrestrial biota. Springer Verlag, Berlin. 129-142 pp.

Nieminen, M. (ed.) 1996: International Moth Monitoring Scheme - proceedings of a seminar Helsinki, Finland 10. April 1996. Tema Nord 1996:630, 84 pp.

Park, M., An, J-S., Lee, J., Lim, J-T., Choi, S-W. 2009: Diversity of moths (Insecta: Lepidoptera) on Bogildo Island, Wando-Gun, Jeonnam, Korea. Journal of Ecological Field Biology 32 (2): 129-135.

Parmesan, C. 2001: Coping with modern times? Insect movement and climate change. In: Woiwood, I.P., Reynolds, D.R., Thomas, C.D. (eds.) Insect movement: mechanisms and consequences.CABI Publishing, Wallingford (UK). 387-413.

Rassi, P., Hyvärinen,E., Juslén, A. & Mannerkoski,I. (eds.) 2010: Suomen lajien uhanalaisuus - Punainen kirja (The endangered Finnish species - Red data book) (In Finnish)- Ympäristöministeriö ja Suomen Ympäristökeskus. 685 pp.

Schütze, K.T. 1931: Die Biologie der Kleinschmetterlinge. Frankfurt am Main. 235 pp.

Seppänen, E. 1954: Suomen suurperhostoukkien ravintokasvit. Die Nahrungspflanzen der Grossschmetterlings-raupen Finnlands. Animalia Fennica 8:1-268.

Somerma, P., Söderman, G. & Väisänen, R. 1993: Valtakunnallisen yöperhosseurannan opas. Vesi-ja Ympäristöhallituksen Monistesarja. Helsinki. 58 pp.

Svensson, I. 1993: Fjärilkalender. Kristianstad Publ., Sverige. 124 pp.

Söderman, G. (ed.) 1994: Moth Monitoring Scheme. A handbook for field work and data reporting. Environmental Report No. 8.Environment Data Centre, National Board of Waters and the Environment. Helsinki. 63 pp.

Söderman, G., Väisänen, R., Leinonen, R. & Lundsten, K-E. 1994: Valtakunnallisen yöperhosseurannan 1. vuosiraportti, Finnish Moth Monitoring Newsletter 1993. Helsinki. 78 pp.

Söderman, G., Lundsten, K-E., Leinonen, R. & Grönholm, L. 1995: Valtakunnallisen yöperhosseurannan 2. vuosiraportti, Finnish Moth Monitoring Newsletter 1994. Helsinki. 74 pp.

Söderman, G., Lundsten, K-E., Leinonen, R. & Grönholm, L. 1997: Valtakunnallisen yöperhosseurannan 3. vuosiraportti, Nocturna Annual Newsletter 1995. Suomen Ympäristö- sarja, luonto ja luonnonvarat nr. 62. Helsinki. 68 pp.

Söderman, G., Leinonen, R., Lundsten, K-E. & Tuominen-Roto, L. 1999: Yöperhosseuranta 1993-97. Suomen Ympäristö-sarja, Luonto ja luonnonvarat, nr. 303. Helsinki. 64 s.

Tengström, J.M.J. 1869: Catalogus lepidopterorum Faunae Fennicae praecursorius. 43 pp.

Vainio, M., Kekäläinen, H.., Alanen, A. & Pykälä, J. 2001. Suomen perinnebiotoopit. Perinnemaisemien valtakunnallinen loppuraportti. Suomen Ympäristö 527: 1-163.

Vanhanen, H., Veteli, T.O., Päivinen, S., Kellomäki, S., Niemelä, P. 2007: Climate change and range shift in two insect defoliators: gypsy moth and nun moth - a model study. Silva Fennica 41(4): 621-638.

Várkonyi, G. & Leinonen, R. 2004: Yöperhoskantojen ekologia Kainuun ja Vienan Karjalan metsämaisemissa. Teoksessa: Heikkilä, R. & Várkonyi, G. (eds.), Vienan Karjalan erämaa-alueiden vaikutus Kainuun vanhan metsän eläinpopulaatioihin, Kuhmo,7-46.

Virtanen, T. & Neuvonen, S. 1999: Climate change and macrolepidopteran biodiversity in Finland. –Chemosphere- Global Change Science 1:439-448.

Visser, M.E., Holleman, L.J.M. 2001: Warm springs disrupt the synchrony of oak and winter moth phenology. Proceedings of the Royal Society, London B 268: 289-294.

Volny, W.J.A., Fleming, R.A. 2000: Climate change and impacts of boreal forest insects. Agriculture, Ecosystems and Environment 82: 283-294.

Webpages:

http://www.ceh.ac.uk , cited 25.05.2011

http://www.eoearth.org/article/Climate_change_and_insects_as_a_forest_disturbance_in_the_arctic , cited 25.05.2011

http://www.Kainuu.fi/Matkailu/Luonto , cited 23.02.2011.

15

Effects and Consequences of Global Climate Change in the Carpathian Basin

János Rakonczai
University of Szeged, Department of Physical Geography and Geoinformatics
Hungary

1. Introduction

The consequences of global climate change are quite versatile on Earth. The rise of surface temperature is more or less general, however its degree can be different in space. Its effect can mostly be proved by the well-measurable changes observed in ice-covered areas. Significant alterations are observed also in the distribution of precipitation which means excess rainfall at certain places and water-shortage elsewhere. In the last century a 0.8°C rise in surface temperature and a 60–80 mm decrease in precipitation were detected in the Carpathian Basin. Since both temperature values and precipitation show considerably high standard deviation, it is very difficult to prove the trend character of these changes. Nevertheless, based on our research, there are some environmental indicators (groundwater, soil, vegetation, and biomass) which are suitable for revealing the true consequences and tendency of climate change.

2. Climate changes of the geologic recent past

Several researches in the field of climatology, geology, archaeology and environmental history inform us about the climatic changes of the last centuries and millennia. The volumes of the IPCC published in the last two decades have revealed in detail that in the tertiary period, surface temperature on Earth was even 10°C higher than today, and during the glacials (e.g. several ten thousand years ago during the last ice age) relatively rapid fluctuations (even 10°C) occurred in mean annual temperature (Broecker 1997). For the last thousand years, evaluable records and observations are available. Based on these data, main temperature tendencies can be drawn for the globe but at the same time more significant spatial differences occur. This statement can be confirmed by the temperature changes of the last millennium in the world and in the Carpathian Basin (Fig. 1). It can be observed that the little ice age in the middle ages caused a more significant fall in temperature in the basin and lasted longer than globally.

Precipitation also showed significant alterations in the last thousand years. Paleobotanical researches showed that in the boreal phase (8–9 ka ago) dry steppe was characteristic in the Carpathian Basin with around 300 mm mean annual precipitation. Archaeological data indicate a significantly drier period around the end of the Sarmatian age (1–5th century A. D.), and for example along with these, even 3–5 metres groundwater-table fall was also revealed near the River Danube (Knipl & Sümegi 2011). The changes in precipitation can also be designated by the alterations in the water level of Lake Balaton in the last millennium (Fig. 2), however it was also artificially modified occasionally because of military purposes.

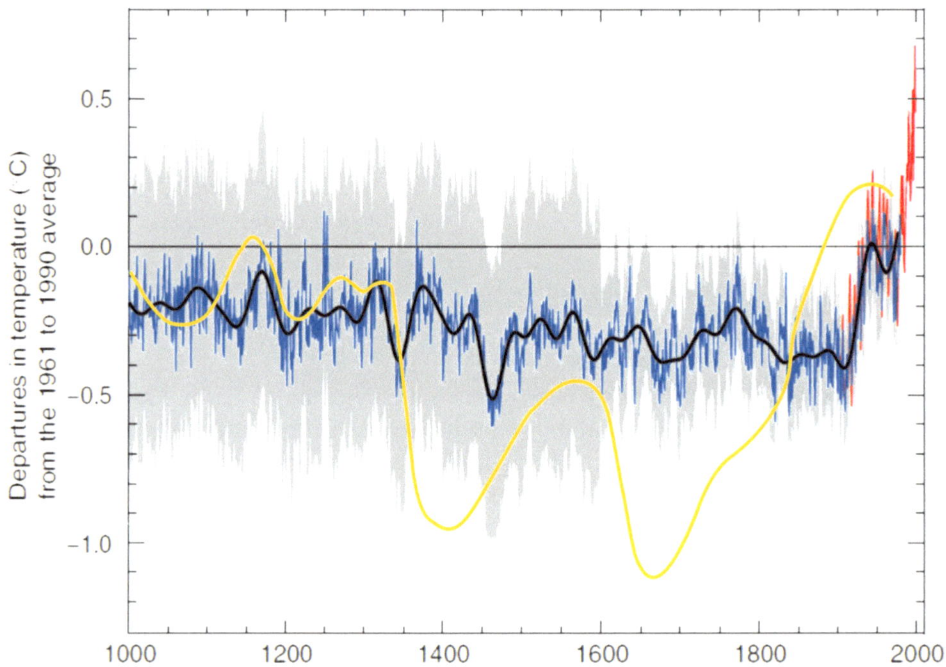

Fig. 1. Mean annual temperature in the Northern Hemisphere (black) and in Eastern-Europe (yellow) (based on IPCC 2001 and Varga-Haszonits 2003).

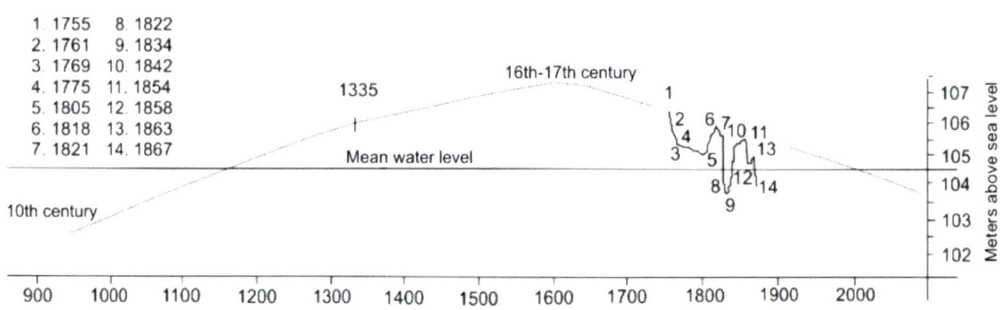

Fig. 2. The water level of Lake Balaton in the last thousand years (Rácz 2011).

More and more detailed environmental historical researches have also revealed that the Carpathian basin, which has a small territory on an Earth scale (around 100.000 km^2), provided quite variable conditions (many times just because of its basin characteristics) for the ecosystem (Sümegi 2011). Thus, the effects of climate change can occur diversely in the landscape. Therefore, in the research of the consequences of recent climatic changes, climate historical analogues are advisable to be taken into consideration.

3. Groundwater as an indicator of climate change

The primary water source of near-surface aquifers is precipitation, changes of which can be reflected by these reservoirs. Where the groundwater* can be refilled from surface waters or by subsurface flow from distant areas, this connection is not strong. In those areas where there is no opportunity for external water supply, significant alterations in the water-table can be formed due to climatic changes showing the trends of precipitation relations.

The territory located between the two large rivers of the Carpathian Basin (the River Danube and the River Tisza) is an ancient alluvial fan and appears as an elevated ridge (about 60–80 metres high) (Fig. 3). It has no surface or underground water-influx from the surroundings, so precipitation is the only source of groundwater, except of the areas with lower elevation along the rivers. In this maximum 10–20 km wide zone, the characteristic annual fluctuation of groundwater-table remains invariable (maximum in spring, minimum in autumn) but longer periods with lower precipitation causes decrease in its amplitude (Fig. 4)

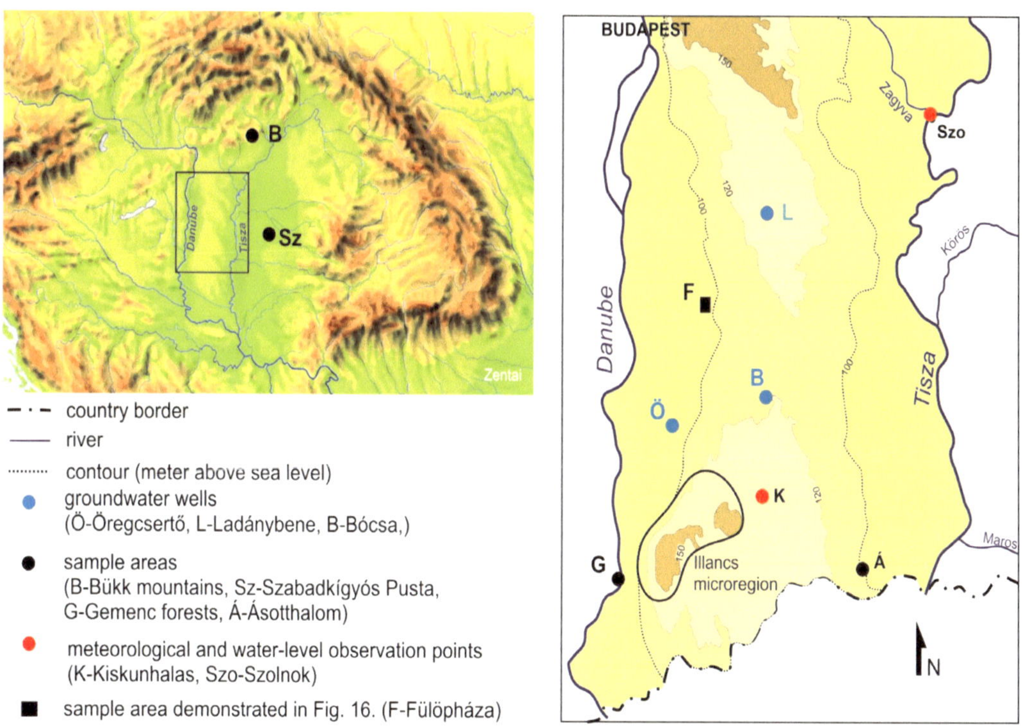

Fig. 3. Terrain model of the Carpathian Basin and the Danube–Tisza Interfluve, displaying the areas under investigation and demonstration in this study.

*shallow groundwater which means the water of the porous aquifer above the uppermost water-retaining layer. As these aquifers have priority for agricultural purposes, well network has already been built in the 1930s for collecting data of the groundwater table by continuous monitoring. However, the number of these observing wells changes, 1500–1700 wells working permanent (Szalai 2011) provide reliable base for spatial evaluation.

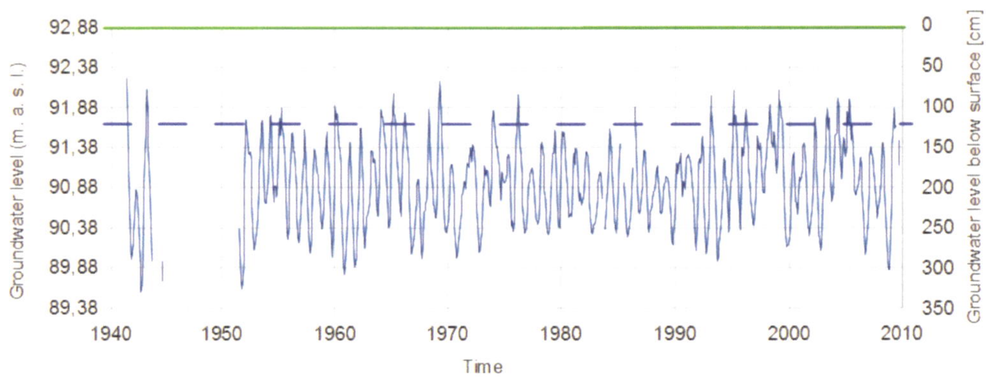

Fig. 4. The multiyear-changes in the level of groundwater-table (at Öregcsertő) on the alluvial flood plain of River Danube (20 km from the river) do not refer to decreasing precipitation (Data source: VITUKI, Hungary).

The green line represents the surface, the blue dashed line shows the average of yearly highest water-levels between 1996–2005 on River Danube.

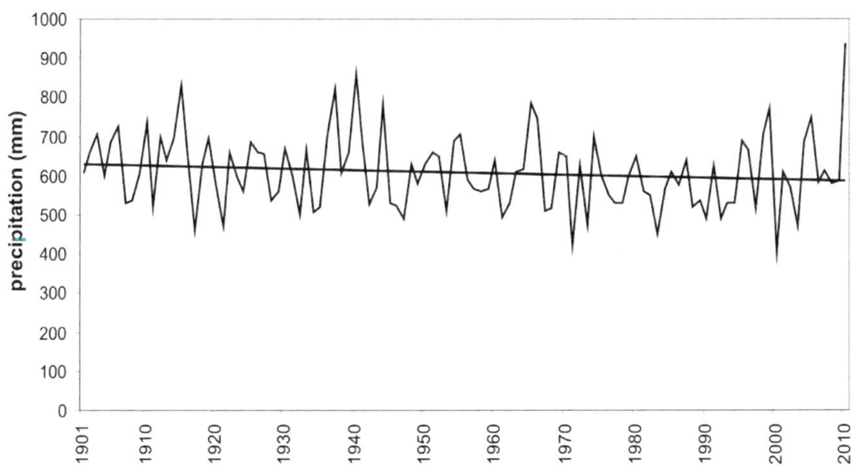

Fig. 5. Spatial averages of mean annual precipitation in Hungary between 1900 and 2010 (Data source: Hungarian Meteorological Service).

In higher areas, the decreasing annual precipitation - in spite of extremities (Fig. 5) - resulted in a continuously decreasing groundwater-table (Fig. 6). In the upper aquifers, the annual average water-levels have sunk as much as 6–8 m in the last 40 years. The decline of the groundwater-level is in close connection with the relief (height above sea level), which confirms that the main cause of sinking here is precipitation-shortage. In areas, where the decrease of the groundwater-table is less significant (related to the height above sea level too), a period with higher amount of precipitation can contribute to the rise of the water level, or even can normalise its state (Fig. 7). Where the groundwater-table fell significantly and the infiltration is limited due to the relief, humid years cause only temporary rise in water-levels, the decreasing tendency continues almost irresistibly (see Fig. 6). Moreover, certain anthropogenic causes also contribute to the process, for example the irrigation from subsurface aquifers in dryer periods.

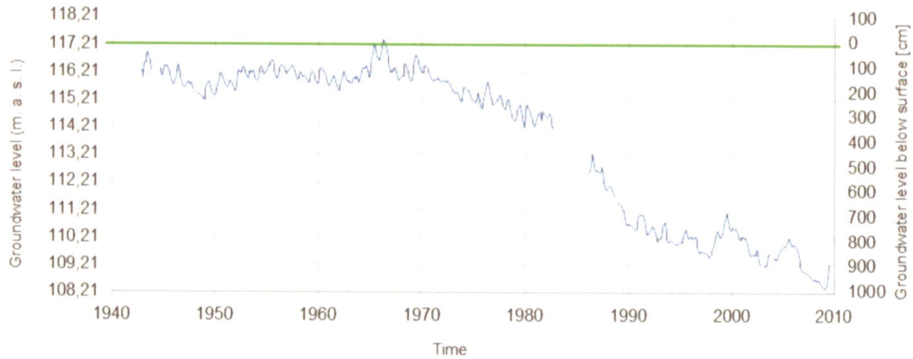

Fig. 6. The declining groundwater table in at a test site (Ladánybene) with higher elevation, where precipitation is the only source of groundwater, showing the consequences of the decreasing amount of precipitation. (Data source: VITUKI, Hungary) It can be seen that humid periods (like 1999 or 2010) can only temporarily modify the negative trend.

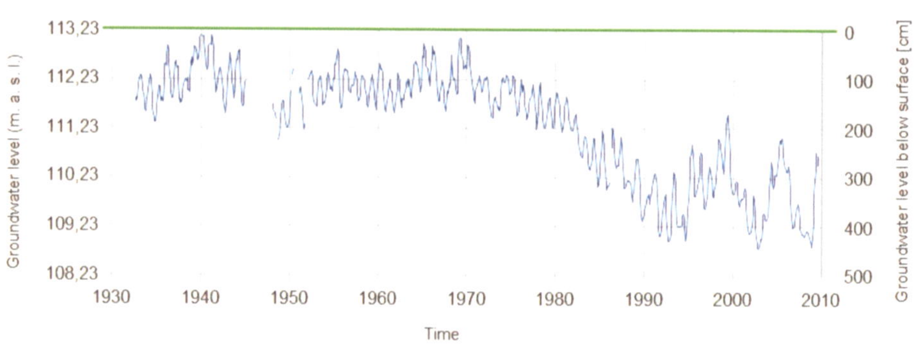

Fig. 7. Groundwater-well (Bócsa) showing the permanent effect of the decreasing precipitation but reacting more sensitively to precipitation surplus (Data source: VITUKI).

Year	Water shortage (km^3)
1980	1,15
1985	2,32
1990	4,08
1995	4.80
2000	2,84
2003	4,81
2010	~ 0,5–0,8

Table 1. The approximate quantity of water deficit in the Danube-Tisza Interfluve related to the first part of the 1970s.

By geoinformatical methods, we determined the degree of water shortage in the Danube–Tisza Interfluve, being the most affected territory in the Carpathian Basin (Fig. 8). We calculated that following arid periods, the water shortage of the region (about ten thousand km^2) is around 5 billion km^3 (Table 1), which approximately equals the total annual water

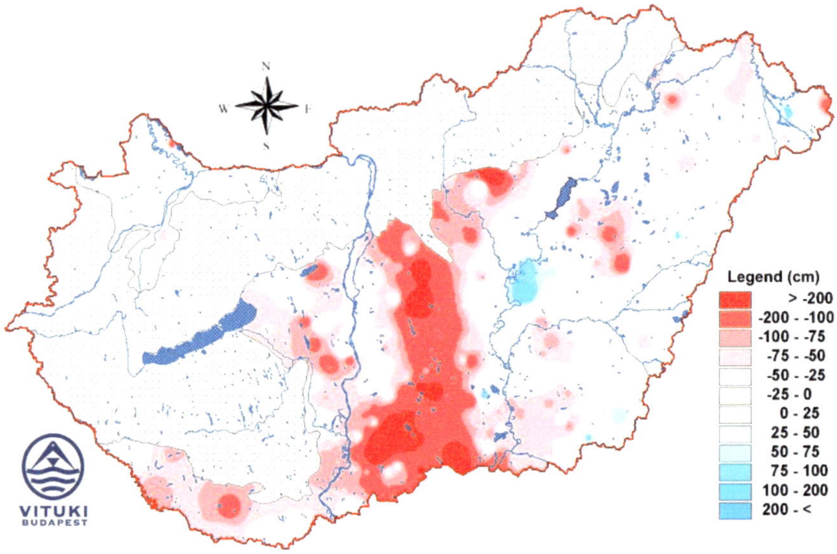

Fig. 8. Spatial distribution of differences between the average groundwater-level in the 1971–2000 period and the mean groundwater in 2009 in Hungary (Data source: VITUKI Environmental Protection and Water Management Research Institute).

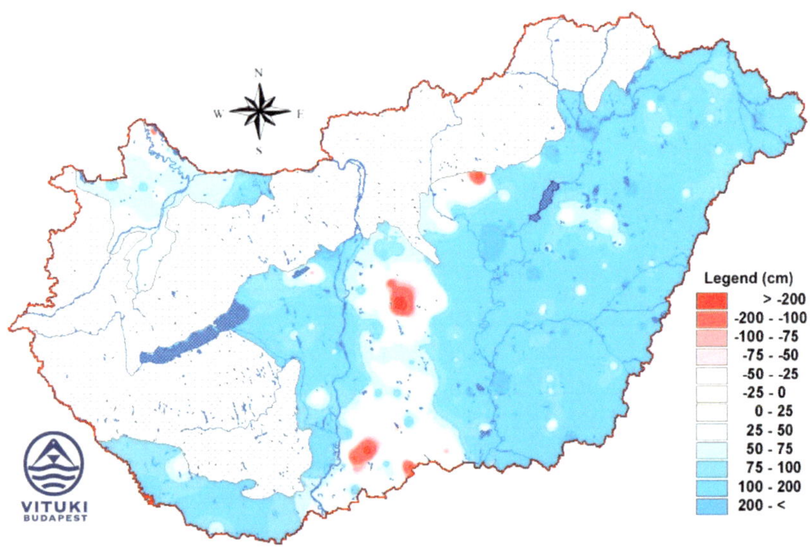

Fig. 9. Spatial distribution of differences between the average groundwater level in the 1971–2000 period and the mean groundwater level in December 2010 in Hungary* (Data source: VITUKI Environmental Protection and Water Management Research Institute).

* In the case of areas without data, there is no groundwater due to geological attributes or there are no observations allowing map representation because of small spatial extension.

consumption of Hungary. From August 2009 until the end of 2010, the most humid period has been experienced on record in the Carpathian Basin but in the case of areas with the highest fall in groundwater level, water-shortage can be still observed - however, its degree has significantly declined. These confirm our previous hypothesis (Rakonczai 2007) that on one tenth of the territory the aridification processes seems to be irreversible, since the groundwater-resources can not regenerate after an extreme humid period (Fig. 9).

The descending groundwater-table causes significant economic damage in areas mostly affected by the aridification process: fruit plantations and traditionally dug groundwater-wells dried out, farming became unpredictable, thus the aridification process has become a socio-political question as well (Ladányi et al. 2011).

4. Altering soils

The continuous sinking of the groundwater-table due to precipitation-shortage causes the alteration of certain soil types. The most spectacular changes can be observed in the case of saline soils. Since the waters of the Carpathian Basin are characterised by high salt content, the decrease of groundwater modifies vertical salt transfer processes (and its direction) in the soil profile, resulting usually the descending of the salt-accumulation zone.

Nature conservation specialists have already been registering the decreasing extent of wetlands from the 1970s and in some areas the diminishing salt content of soils related to the declining groundwater-table level as well (Kákonyi 2006). A well-observable consequence of the alterations is the decreasing extent and number of salt efflorescences being previously frequent in saline areas. In the 19th century, the salt swept and collected from the dry surface had an important economic role in the Great Hungarian Plain. E.g. according to the official registering Akasztó, a small settlement in the Danube–Tisza Interfluve, was the first on the list of "salt-producers" with 400-500 tons of salt in Hungary in 1893 (Aradi & Iványosi-Szabó 1996). According to a comprehensive soil evaluation in 2002, salt efflorescences can be observed only in the case of 32 places from the previous 164 saline areas in Hungary (Tóth 2002) and salt-collection has already been discontinued for long decades. The desalinisation process was revealed by Hungarian investigations related to the MEDALUS program* in the case of more sample areas (Kertész et al. 2001). However, the most exact alterations were detected by the author of this paper (due to some luck).

In the middle of the 1970s, detailed geomorphological and pedological investigations were performed on the Szabadkígyós Pusta (being a division of the Körös-Maros National Park at present - see Fig. 3 point Sz) as a part of the preparations for declaring the area protected. Beside the precise morphological mapping of sodified bench microforms typical in the saline landscape, sample plots were identified for joint evaluations made together with botanists (Rakonczai 1986). In the framework of the later study detailed botanic surveys were made along with the analysis of the chemical parameters of soils lying under different vegetation types. At that time nobody thought that after 25-30 years this area can be appropriate to detect the consequences of climate change. The research activity has been renewed at Szabadkígyós from 2003, when it was noticed on a field trip that during the elapsed quarter of a century the previously saline territory significantly changed, though most of the original sampling sites remained identifiable (Photo 1). We have already suspected at this point that the observed transformations were driven by the changes in the

* Mediterranean Desertification And Land Use

water cycle. Later it turned out that there were some further factors in the background as well, though the aridity experienced between the late 1980s and mid 1990s seemed to be the most important factor behind changes. During the long lasting drought period the groundwater table sank, and thus the effect of waters, having sometimes a 5000 mg/l salt content, became less and less apparent on the surface. As a result saline precipitations (salt efflorescence) started to disappear from the soil surface and the decreasing salt content enabled the gradual advance of grassy vegetation (Photo 2 and Photo 3).

Photo 1. Sample area assigned in 1979 in May 2006 (in Szabadkígyós Pusta).

Photo 2 and 3. Saline soil with salt efflorescences in 1976 become covered with steppe vegetation by 2006.

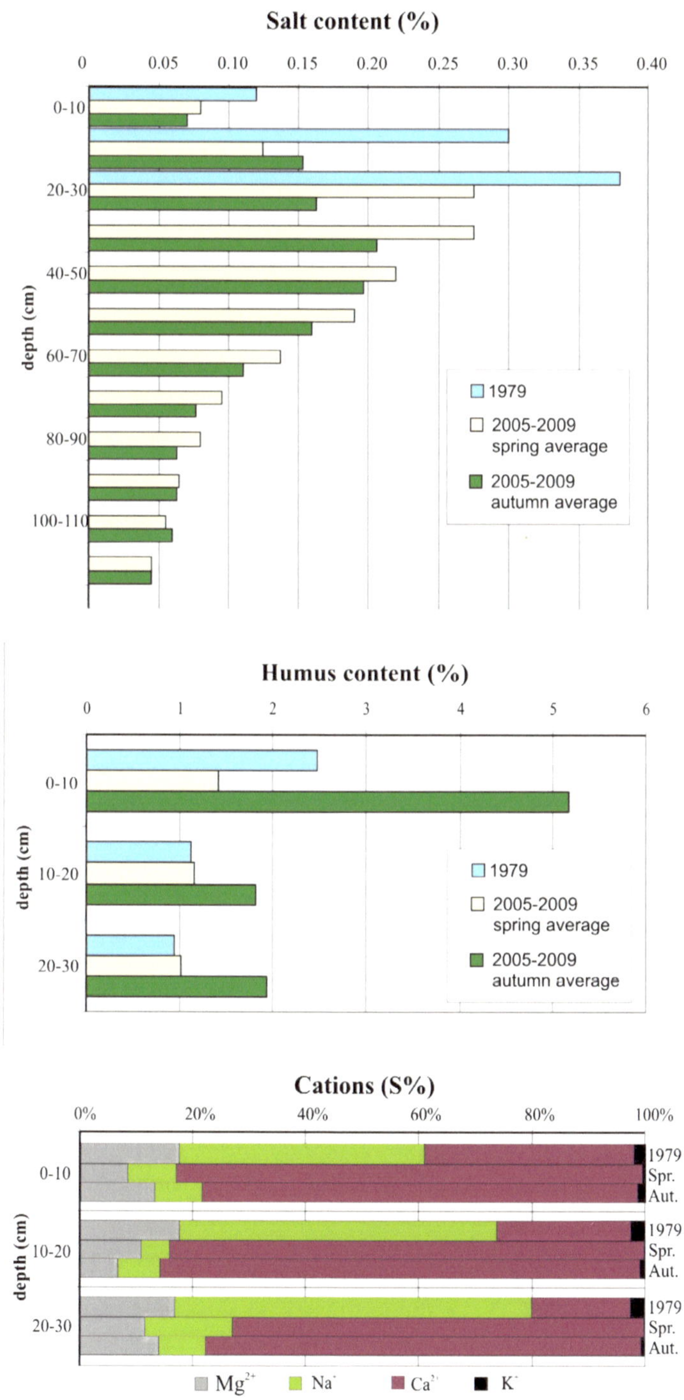

Fig. 10. Alteration of some characteristic attributes of a saline soil profile based on the measurements between 1979 and the average of 2005–2009 period in Szabadkígyós Pusta (Note: Soil sampling in 1979 was performed for the upper 30 cm).

Soil samples collected regularly since 2005 (Barna 2010) made the quantitative assessment of soil transformation possible. Results justify the physical and chemical background of modification. During the past 30 years, due to the mentioned environmental changes, the total salt content of soils has decreased significantly, especially the amount of Na dropped, thus providing more favourable conditions for plants. The gradual advance of vegetation has been followed by the considerable increase of humus content (Fig. 10).

Research prove that as a result of the aridification process, a decreasing salt content and an increasing humus content can be observed in the upper soil of saline areas in the Carpathian Basin, and at the same time steppification process can be registered. According to our experiences, this alteration can take place within 2-3 decades.

5. Vegetation as an indicator of climate change

One of the best indicators of climate change is vegetation, since it can indicate short-term alterations and it can be an important marker of extremities as well. By buffering the effects of climate extremities it can also be suitable for detecting long-term tendencies. Vegetation can sign the changes of life conditions by the degree or the lack of adaptation. Previous research revealed that in the past flora reacted to the alterations of climatic conditions by the shift of the vegetation zones. But at what extent is vegetation able to respond to the changes faster than before in landscapes fragmented by anthropogenic activities (Csorba 2008) and detached by artificial barriers?

5.1 Biomass

In terms of agriculture, a good or a bad harvest of a year is mainly determined by the consequences of climatic variability. Nevertheless, on the same territory different plants are grown year by year and water can be supplemented by irrigation if it is needed. Therefore, it is not possible to examine properly the relationship between crop yield and climatic data at the same place.

Based on our investigations, forests can be suitable for analysing the connection between biomass production and climatic conditions, since their location is permanent for many years and the effects of natural precipitation is not modified by irrigation. Increasing spatial resolution (and quality) as well as the easier availability of satellite imagery improved our investigations.

Using (NDVI and EVI) vegetation index series data of the MODIS sensor, carried by the Terra satellite, the relationship between forests and precipitation was investigated in case of several forest areas. It was shown that vegetation index data proved to be suitable for revealing the spatial differences of water up-take by plants (from precipitation and groundwater) (Ladányi et al 2009).

As a first step of our investigation, forests of a microregion (Illancs see Fig. 3) mostly affected by the groundwater depletion were analysed. In this case, landuse was influenced by the forestation in the 1900s. Even though forest management plays an important role in the area, large homogenous forest patches are rare, rather many small ones with different composition of species are characteristic. Nowadays, native stands are uncommon, mainly planted acacia and pine are the dominant tree species.

On the basis of a 10-year-period (2000–2009), vegetation dynamics of the main forest types were analysed and the relationship between precipitation and biomass production was investigated (Ladányi et al 2011). Vegetation dynamics were taken into consideration only during the vegetation period (from April to September) because this is the most active period of trees, and in winter snow and cloud cover can highly modify the evaluation. Forests were chosen as control sites, where the groundwater-sinking is not significant and where the groundwater is easily available for trees.

Our evaluation, not reviewed in detail at this time, showed the differences of the main forest types (acacia as deciduous and pine as coniferous forest) in vegetation dynamics (Fig. 11) and demonstrated the relationship between vegetation index values and precipitation distribution (Fig. 12). The tendency of annual curves for both the NDVI and EVI vegetation indices shows unambiguously the natural features of both forest types. As a consequence of the different character of the foliage, deciduous forests are characterised by lower values than coniferous forests in the first part of the year and reach higher values later. The reduced importance of precipitation at the end of the vegetation period can also be observed (September 2001).

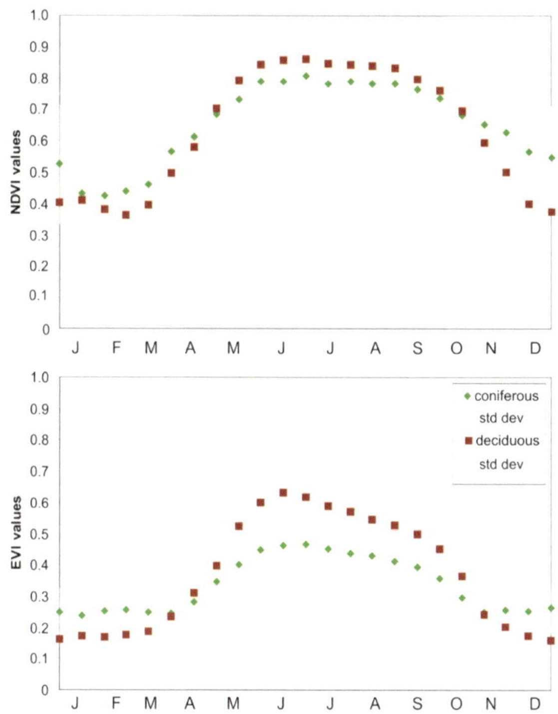

Fig. 11. Vegetation index values of coniferous and deciduous forests (2000-2009) (Ladányi 2011).

In the sample area, the relationship between biomass production (its calculation is not demonstrated in detail, it can be found in Ladányi et al. 2011), and precipitation was investigated (Table 2). In the demonstrated results the intervals are given where the correlation between the vegetation indices and the determining periods of precipitation was the strongest.

Based on the investigations, a strong relationship can be established between precipitation and biomass for both vegetation indices in the Illancs microregion. The strength of the connection slightly differs in case of the different tree species but it is significant in both cases. It can be strange for the first sight that winter precipitation has no significance in biomass production in this landscape, since (as we know) soil is impregnated by water in the winter period which supply later moisture for vegetation.

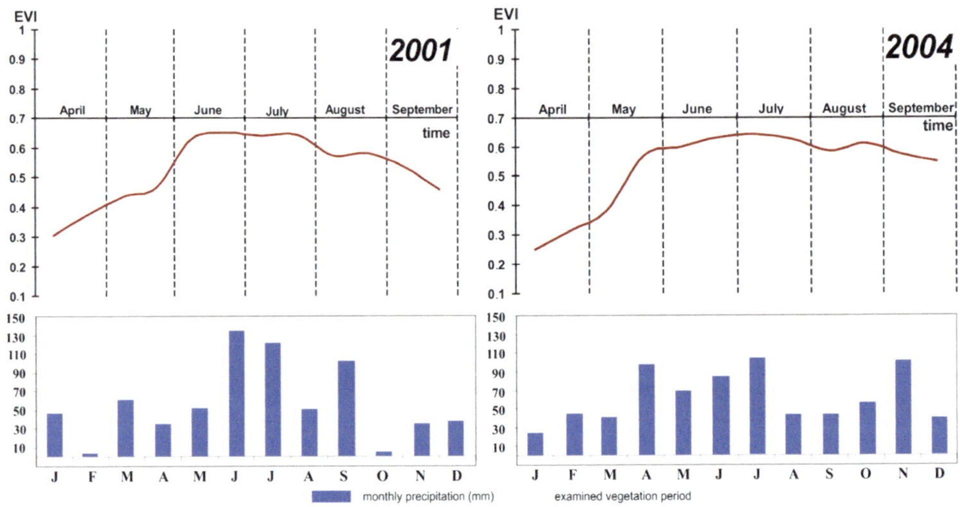

Fig. 12. Vegetation indices for deciduous forests in years with different precipitation distributions (Ladányi 2011).

Examined forests	Correlation coefficient (EVI)	Determining period (EVI)	Correlation coefficient (NDVI)	Determining period (NDVI)
Illancs microregion, coniferous forest	0.76 - 0.84	III-VI	0.85 - 0.93	III-VI
Illancs microregion, deciduous forest	0.84 - 0.95	III-VI	0.83 - 0.92	III-VI
Alluvial flood plain, Gemenc forest (control)	<0.2	-	<0.2	-
Edge of the Sandland, Ásotthalom, coniferous forest (control)	0.8	XII-IV	0.76	I-VIII

Table 2. Correlation between precipitation and biomass production based on vegetation indices. Roman number labels indicate the month range over which precipitation was accumulated. (Precipitation data source from Kiskunhalas) (Ladányi 2010).

In order to determine the appropriateness of vegetation indices in analysing "climate sensitivity" of various landscapes, control sites were selected in the neighbourhood of the sample area where vegetation is not solely dependent on rainfall variability but can use other sources of water as well. Gemenc forest (being a part of Duna-Dráva National Park - see Fig. 3 - G), is such an area on the floodplain of the Danube. Here, the regular river floodings and the continuous connection of groundwater with the river through the sandy, pebbly silt ensure stable water supply for local tree species. (The comparison can be slightly influenced by the fact that this forest stand consists of a mixture of tree species typically with larger water demand). In this area, vegetation index curves showed less differences in each year compared to the Illancs sample area. Furthermore, biomass production did not show a strong correlation with any periods of precipitation so - according to our prehypothesis - this control site does not appear to be sensitive to environmental changes from the point of view of precipitation.

As a further control site, a larger pine forest was selected on the eastern edge of the Danube-Tisza Interfluve (Ásotthalom) on soils with heavier structure, where the effect of regional groundwater-sinking process is less severe (1-2 m) and where rivers do not affect the groundwater-level substantially. The analysis of vegetation indices also showed strong relations with the precipitation (Table 2) but winter period proved to be important as well. In this case groundwater affects the water balance of vegetation so it is less exposed to the variability of precipitation.

Biomass investigations in certain parts of the Danube–Tisza Interfluve demonstrated that in areas highly affected by the groundwater table decreasing process, forests hardly depend on deep situated groundwater (since water demand is supplied from other source) so they are more sensitive to the extremities in the distribution of precipitation. Furthermore the results show that biological activity of woody vegetation can be sensitive to the declining groundwater table, so in an indirect way, it can be in connection with the long-term alterations of climate change.

5.2 Precipitation as the limiting factor of woody vegetation

Living organisms are influenced by several environmental factors (according to their own attributes). The most typical of them are water, air (oxygen, nitrogen, carbon-dioxide), nutrients and a certain range of temperature. Water supply and temperature can be directly affected by climate change.

As an effect of global climate change, the shift of forests towards the poles (and towards the peaks of mountains) can be observed in several parts of the Earth. The advancing edges of forests are investigated by researchers worldwide, but less attention is paid to the retreating low-latitude and low elevation limits of forest, however, this would be more important from both ecological and social viewpoint, as well. Beside the obvious ecological reasons of the shift of forests towards the poles (for example decreasing biodiversity of forest ecosystems), the yield-loss of forests can also result economical damages. Furthermore, the changes of landscape (the disappearance of forests) can have esthetical consequences or can affect even the quality of life (Mátyás 2010).

In the 1970–1980s, Central Europe was hit by massive forest dieback. In this period, the damage affected many commercial species, wiping out about 15 percent of oaks in a decade. The reasons were not known exactly at that time, acid rain, pests or illnesses were assigned as the main cause of the problem (Jakucs 1990). The reason could also be climate change at that time, exactly the shifting of the xeric limit, there were not enough precipitation for forests.

In the last decade the death of forests can also be observed due to the shifting xeric limit in Hungary. The spatial retreat of oak as the most sensitive species of the current climatic changes, its weakening vitality and mortality has already been obvious. It was shown that oak forests have been significantly withdrawing between the first and the last decades of the 20th century (Mátyás et al. 2010, Fig. 13). Research in the last years revealed that not the decreasing mean precipitation but the water shortage of arid years are the limiting factors of

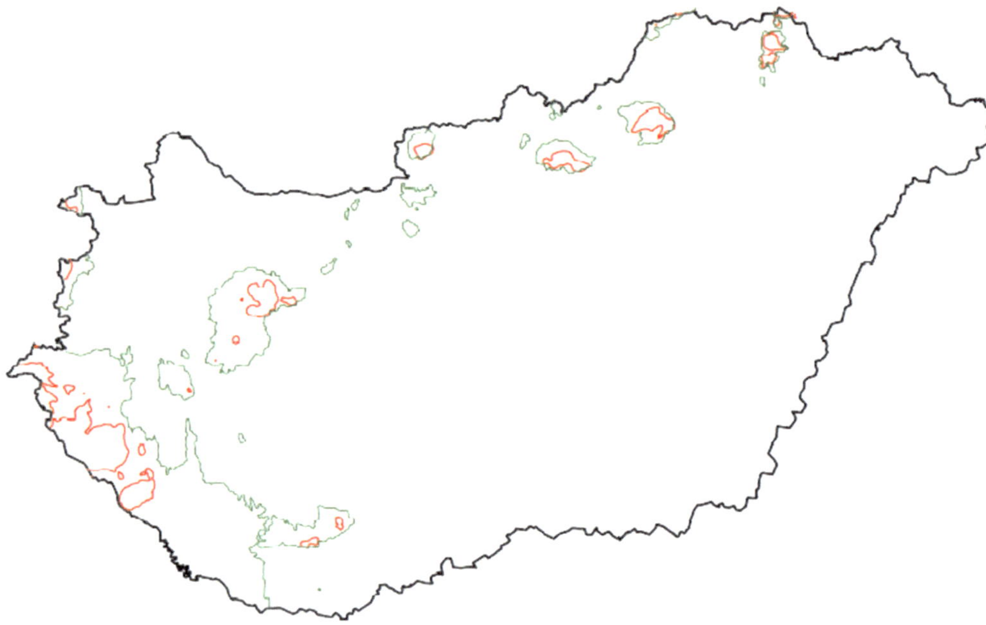

Fig. 13. Change of climate favourable for beech in course of a century: shift of contour line 24 of Ellenberg's climate quotient (=long-term minimum for beech) for Hungary, in the average of the years 1901–1930 (green) and 1975–2004 (red) (Mátyás et al 2010)

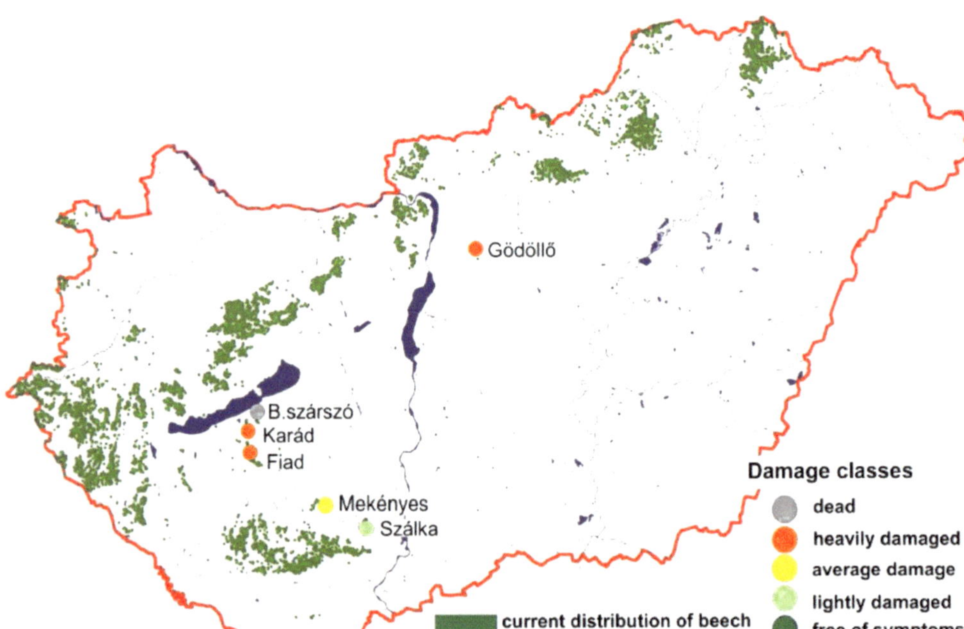

Fig. 14. Beech decline observation points with the observed damage classes in 2005 (Berki et al. 2007)

beech forest stands to subsist (Berki et al 2007). Due to the small amount of precipitation in the first decade of the 21st century, especially between 2000 and 2004 (for example the mean areal precipitation was 400 mm in Hungary but in significant territories only 200–300 mm were measured) significant damages were detected in the south-western part of the country and in case of some small forests on the xeric limit (Fig. 14). In the most endangered areas regarding climatic changes in West-Hungary, there was a need for salvage cutting in case of 30–50 percent of forests; trees had to be removed long before the economic optimum. Forest management can adapt hard to the rapid environmental changes, since the long lifespan of forest trees needs planning for 80–120 years (Mátyás 2010). As a result of the species drying out due to climate change, the composition of forests changes in an unfavourable direction. For example in the Bükk Mountains (see Fig. 3 - B), the retreating of *Quercus petraea* was followed by the expansion of shrubs and later the spread of maple (Fig. 15), and soil attributes were worsened by the composition of the changing leaf-litter (Tóth et al. 2008).

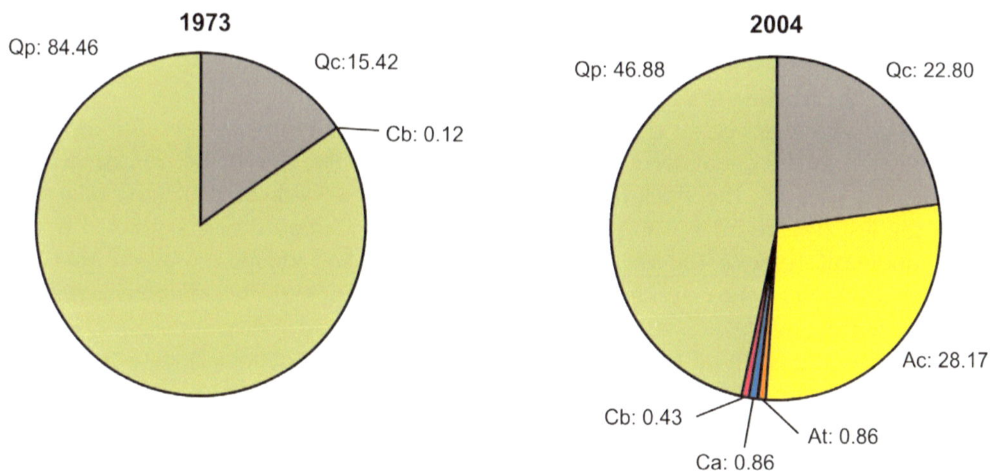

Qp=*Quercus petraea*, Qc=*Quercus cerris*, Ac=*Acer campestre*, At=*Acer tataricum*, Ca=*Cerasus avium*, Cb=*Carpinus betulus*

Fig. 15. The alteration of species composition of the forest stand in a sample area located in the Bükk Mountains (Kotroczó et al. 2007).

Similar forest- and tree damages were detected worldwide after 2000 mainly in mountain regions but these events have been treated as isolated problems. In certain areas, for example in North Africa, they are not even considered problematic, as dying trees simply make room for more goat pastures. But it has been realised that the advancing xeric limit negatively affects Ukraine, Center Asia and even North China, moreover the western part of North America. In these areas there have been sparse observations about forest mortality, however, they were considered local, transient problems and the global coherence of the process was not revealed. Forests are rarer in plain areas due to the more intensive agriculture and the dense settlement structure, thus the effects of climate change on forests can be recognised more difficult. By nowadays it has become clear that the alteration, registered in the Carpathian Basin, is not a transient and isolated phenomenon, but a part of a global process. Thus, strategic preparations are needed because of the predicted climate change in the next decades. For example, the spread of drought tolerant species has to be promoted professionally and an important task is to protect the genome of the endangered valuable populations. There is a need for changing in nature conservation practices as well: instead of the previous passive protection, active steps are necessary (Mátyás 2010).

5.3 The effect of climate change on the alteration of plain's vegetation

Landuse has changed significantly in the plain areas of the Carpathian Basin during the last 1,5–2 centuries due to human activity: e.g. as a result of the high demand for cereals at the beginning of 19th century, almost all the potential cultivatable area was converted into arable land. In the mid-19th century one of Europe's largest regulations of rivers was completed in the Great Hungarian Plain. It resulted that nature conservation research developing in the second half of 20th century in Hungary could take into account just those areas which are unsuitable, less worthy for more intensive farming on the base of „principle of remains". The extension of these remained natural areas are relatively small, they are often wetlands at lower elevation, which could have been utilized only as pastures. Thus the effects of climate change on natural vegetation can be found in a highly mosaic landscape, where the response of vegetation is strongly restricted and the climatic affects are often combined with the consequences of anthropogenic interventions. In the followings we present some typical examples for the reaction of vegetation to changing environmental conditions.

The reaction of vegetation to changing water supply and altering soil conditions can be perceived first in the decrease of the number of individuals. One of its well-conceivable consequences was e.g. the stock-decrease of the halophyte herb of *Matricaria chamomilla* during the last two decades. The reduction of its collected amount was explained first by socio-economical reasons (buying up became stricter), and nobody thought that soil-alterations resulted by the drier climate are standing in the background, which was revealed by later research.

We mentioned at point 4 (altering soils) that 5 sample plots (see Photo 1) were pointed out for a soil-vegetation connection research at the end of 1970s in a saline area in Southeast Hungary (see Fig. 3 - Sz point). Their first detailed botanical surveys were performed in the vegetation period (between April and September) of the years of 1980–1982 (Kovács & Molnár 1986). Repeated surveys occurred in these plots in 2006 and 2009 (Margóczi et al 2008, Barna 2011).

The alteration of the area is significant considering the vegetation, and the decrease of salinity can be confirmed unambiguously. The number of species has decreased: in 1980 40, but in 2009 only 33 species occurred in this area. Because of the higher productivity of soils resulted by the alterations the coverage of vegetation became higher in 2006 and in 2009 too comparing with the one in 1980. The total coverage of the plant species of strongly saline soils almost halved; while the number of plant species of slightly and moderately saline soils increased to a less degree, as they took the places of the regressive euhalopyte plant species. The number of non-salt-tolerant pseudohalopyhte species decreased, but their total coverage increased significantly. On the other hand the number and coverage of species with higher soil moisture requirement became higher to 2006 reflecting more precipitation of those periods (2005, 2006), however it decreased again because of the dry first half of the year 2009. The changes are illustrated on a sample plot covered by the strongly halophyte saline meadow community of Agrostio-Beckmannietum (Fig. 16). It can be seen well that the alteration of soils results the change of plant communities too: this strongly saline plot covered by Agrostio-Beckmannietum saline meadows in 1980 was turned to a less saline meadow type of Agrostio-Alopecuretum until 2006.

As a consequence of climatic drying and water regulations during the last decades the wetlands of the sand-ridge of the Danube–Tisza Interfluve have also begun to dry out in many places. This has also been accompanied by the degradation of vegetation. These aridification processes promote the desiccation and leaching of the saline habitats in the sand-ridge becoming uncharacteristic which results that the annual salt pioneer vegetation and *Puccinellia* swards (salty meadow-type) of solonchak soils are turned into less saline *Agrostio-Caricetum* saline meadows, then into degraded *Achilleo-Festucetum*-like sand steppe-grasslands. The less salt-tolerant species (e.g. *Festuca pseudovina* and *Agropyron repens*) have become dominant in these leaching grasslands.

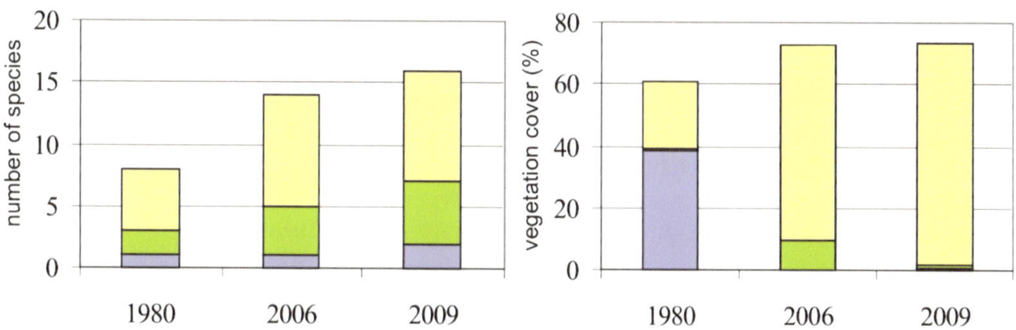

Fig. 16. Number and coverage of non-salt-tolerant pseudohalophyte species (yellow), astenohalopyte species of moderately saline soils (green) and euhalophyte species of strongly saline soils (blue) in a sample plot covered by *Agrostio-Beckmannietum* in 1980 (Barna 2011).

The former wetter *fen-woodlands* were suppressed partly because of climatic reasons, on the other hand they were cut out. Nowadays on their places mainly tussock meadows and wet fens (e.g. with *Schoenus nigricans* or *Eriophorum sp.*) can be found, but they became rare too. The above mentioned wettest fen vegetation-types were replaced by Molinia fens, sedgefields and meadows (these grassland-types were also formerly widely distributed in moderately wet depression formed by deflation), which are turned into stepping Molinia fens then sand steppe-grasslands as an effect of further dry-out. The above mentioned transformation resulted a *shift in habitat-zonation* on the edge of the deflation hollows (Deák 2011). In Hungary all the fens and bogs are maintained by local groundwater-flows and microclimatic conditions (e.g. special geomorphological position, protective buffer-zone forest maintained microclimate), so these are highly climate-sensitive communities. The groundwater-level decrease due to climate change (see point 3) caused the force back and extinction of fen habitats in large areas, and this process will probably intensify in the future if the climatic trends of the last decades continue. In the Danube–Tisza Interfluve affected by aridification the most in the Carpathian-basin, the above mentioned two typical direction of degradation processes can be pointed out (Fig. 17), but there are also such areas where the degradation of natural habitats hasn't appeared during the last 1.5 decades (Margóczi et al. 2011). These are mainly relatively deeper places in the edge-zone of the sand-ridge, where the lateral groundwater-flows maintain some moisture even in the drier seasons. There is also an example for a special rescue effect, when the former more extended fen vegetation found its (probably last) refuge in an artificial channel reach.

One of the important consequences of vegetation changes in the Great Hungarian Plain is that the present stocks of a certain vegetation-type occur not necessary in the place where they were 100–200 years ago. While the typical annual salt pioneer vegetation, loess steppe-grasslands and sandy oak forests live just there, where they existed even 200 years ago, the present stocks of sand steppe-grasslands and the willow-poplar alluvial forests were wet meadows for the most part 200 years ago, however their stocks of that time have mainly disappeared for today (Molnár & Biró 2011). The extension and place of the alluvial forests were highly influenced by the regulation of the rivers, the abandonment of pastures and the plantation of the forests. The transformation and the regression of the natural vegetation make favorable conditions for the aggressive expansion of adventive species. Though these species occupy the natural grasslands less frequently (e.g. non-treated wet meadows can be exceptions), but their expansion can be seen well on the increasing number of fallows, which cover more and more areas mainly in areas with insufficient soil conditions for arable lands. Altogether it can be stated that the effect of climate change in the case of non-woody vegetation of the plains can be very various because of the differences in water supply, soils and micro-relief. Certain vegetation-assemblages have disappeared, degradation controlled vegetation transformation rows can be registered depending on soils and water supply and the species composition of the different plant associations and their surface-coverage ratios transformed significantly too even in case of slighter external changes in the most drying areas. But it can also happen that a habitat-type is forced back to a tolerable refugium with compulsory migration. In some cases it can be observed that landscape features can buffer more or less even the unfavorable effects of climate change, so the vegetation hardly changes or doesn't change at all.

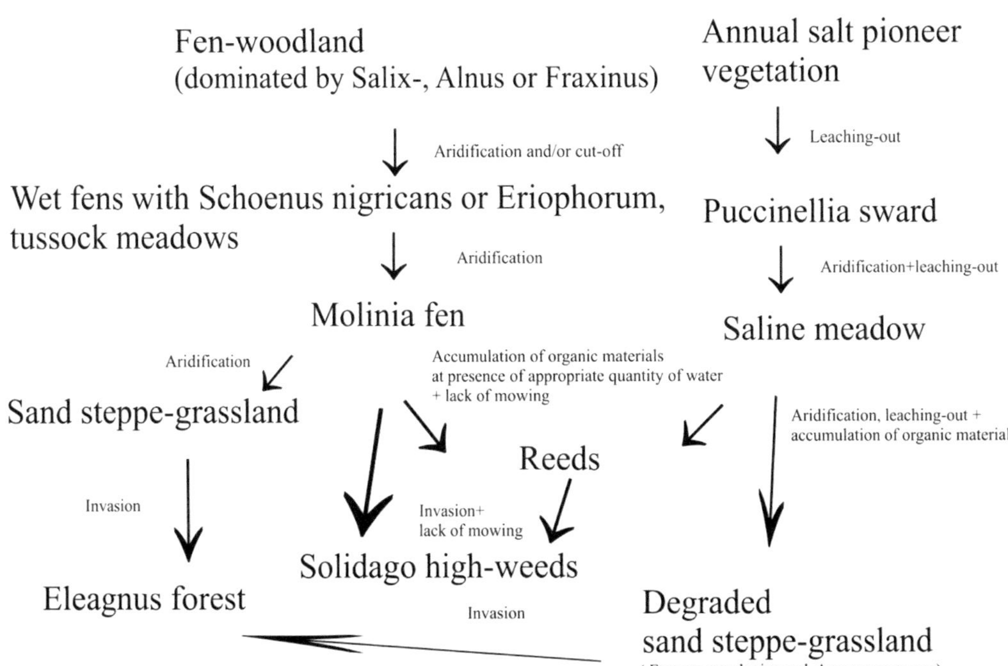

Fig. 17. Registered degradation processes of the vegetation in the Danube-Tisza Interfluve (Deák 2011).

Photo 4. Retreating fen vegetation to the channel crossing a deflation hollow in Illancs region (forced shift of the vegetation zones). (Photo: Ladányi 2010)

6. Landscape changes

Due to the decrease of near-surface water resources, the alteration of landscapes can be observed in many places. This process has been going on for a long time, however background reasons have changed in the last few decades. While in the previous 1.5-2 centuries mainly anthropogenic influence had dominated landscape change (causing alterations), in the last 30–40 years the consequences of natural alterations have taken over. The most spectacular changes are the decrease and the alteration of wetland habitats (Fig. 18). For example, only a few number of saline lakes remained from the previously numerous ones, their extent also decreased and their bed is occupied by offshore weed vegetation (*Bolboschocnus maritimus* is spreading the most agressively). In certain humid years the previous wetlands seems to regenerate but this is impossible just due to the weed vegetation (Kovács 2008). Landscape alterations related to climate change are much more complex than the decreasing extent of wetlands - however it is doubtless that most of the changes are due to the alteration in the natural water-cycle of landscapes. The simplified connection system of the changes are demonstrated in Fig. 19.

The long-term precipitation shortage (which indicates higher water-exploitation from subsurface aquifers for irrigation in lack of surface water) causes permanent decrease of the groundwater-table. This significant and permanent decline of groundwater level modifies the salt transfer in soils which might result the transformation of genetic soil type. This modification is followed by the change of natural vegetation which results in the significant alteration of natural landscapes (see Photo 2 and 3). As the ratio of agricultural fields and built-up areas in the Carpathian Basin is high, the effects of climate change on landscapes

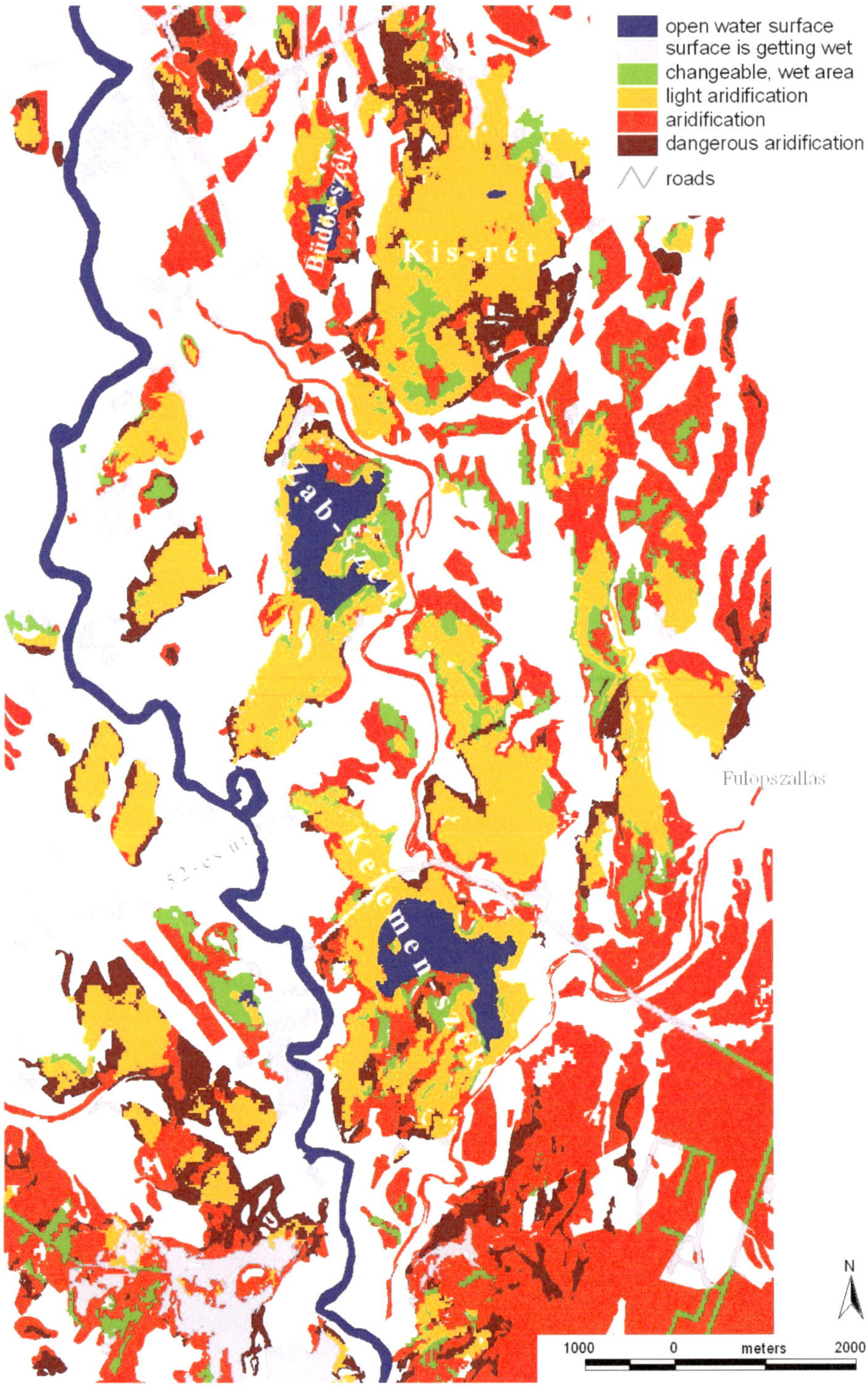

Fig. 18. Evaluation of a sample area located in the Danube–Tisza Interfluve in the viewpoint of the aridification process (Kovács 2008).

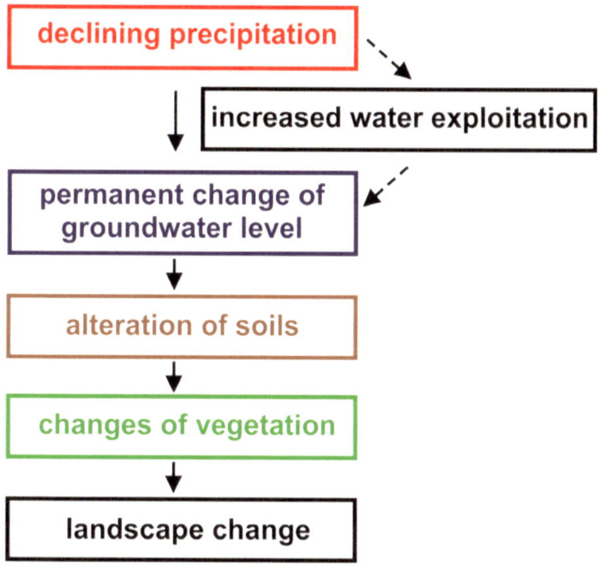

Fig. 19. The process of landscape change due to aridification.

can be observed more difficult. However, at least two of the most important consequences can be presumed:

- Due to the decreasing extent of saline areas, typical landscapes alter, for example the famous Hungarian "pusta" is under transformation at numerous locations
- The decrease in salt content and the increase in organic matter content of soils enable agricultural utilisation in several protected areas (e.g. Natura 2000) discontinuing the causes of their protection.

7. Floods and climate change

One consequence of the global climate change (due to the assessment of the IPCC) is the increase of extremities (for example in terms of precipitation). This statement – as it was shown before – is also confirmed by Carpathian Basin data. Increasing flood levels of the large rivers detected in the last decades are considered as a consequence of the climate change. However, it is not supported by the detailed evaluation of floods, since extreme precipitation events do not occur at the same time in the whole catchment, but they affect only some tributaries. In the Tisza River Basin the last flood forming in the whole catchment area occurred in 1970. Nevertheless, in the last decade (since 1999) the highest water-levels were exceeded several times. These maximum water stages were not in accordance with the maximum water discharge (Fig. 20) (and they formed only in certain parts of the river) showing that *climate change is not in the background of the increasing flood levels in case of large rivers* (Rakonczai & Kozák 2011). This is also confirmed by the fact that in the case of Hungarian rivers, the highest flood levels occurred in 20 different years (Fig. 21).
To sum up, it can be established that the increasing number of "flash floods" in smaller catchments can be connected to the extremities of precipitation due to climate change, however, in the case of large rivers, the changes in river channel capacity (for example floodplain aggradation) is the main cause of record floods.

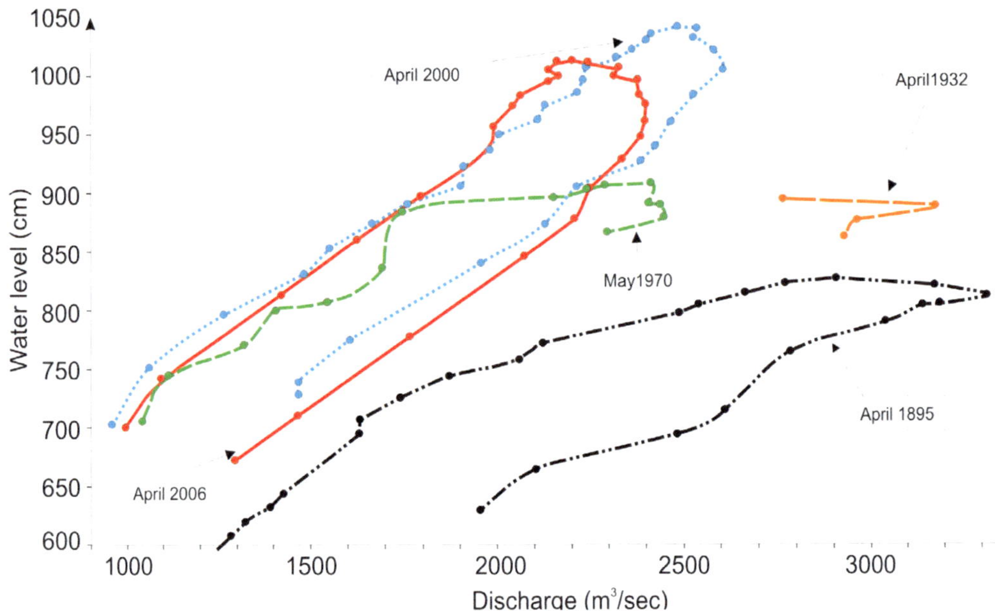

Fig. 20. Characteristic stage-discharge curves on the Szolnok reach of the Tisza (based on the data of the Middle Tisza District Water Directorate).

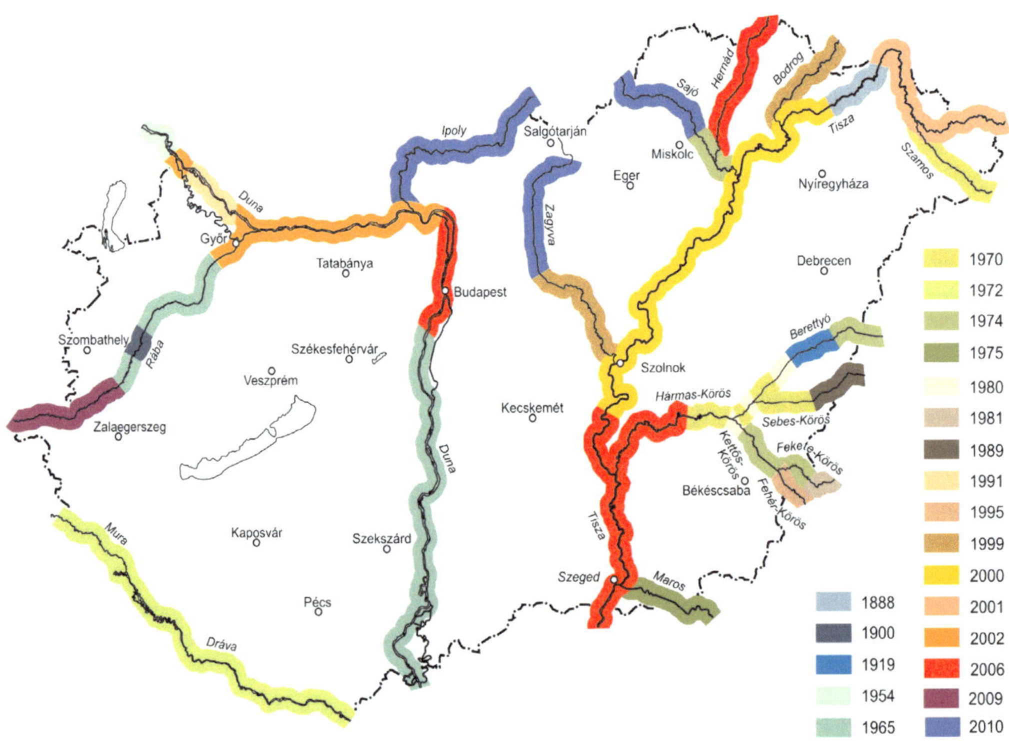

Fig. 21. Periods of record floods in Hungary.

8. Summary

Global climate change observed nowadays has preliminaries: geological and environmental historical data prove its former occurrence (having more significant consequences) in the past. The consequences of the changes - due to the differences in physical geography - cause changes with quite different scales in the landscape.

Significant territories of the Carpathian Basin are so much influenced by human activities that the effect of climate change can not be detected directly. Detailed research proved that the changes of the two main climate elements (precipitation and temperature) in the last decades have contributed to significant, trend-like alterations in the landscape. Furthermore, natural alterations are completed by the consequences of anthropogenic activities and the changes have significant social relations as well (Fig. 22). The decreasing precipitation caused groundwater-table sinking, as a result of which soil and vegetation altered, contributing to landscape changes. The detected landscape changes can serve as a base for a climate-sensitivity map in the future. However, it is important to note that every alteration in the landscape can not be explained exclusively by climate change.

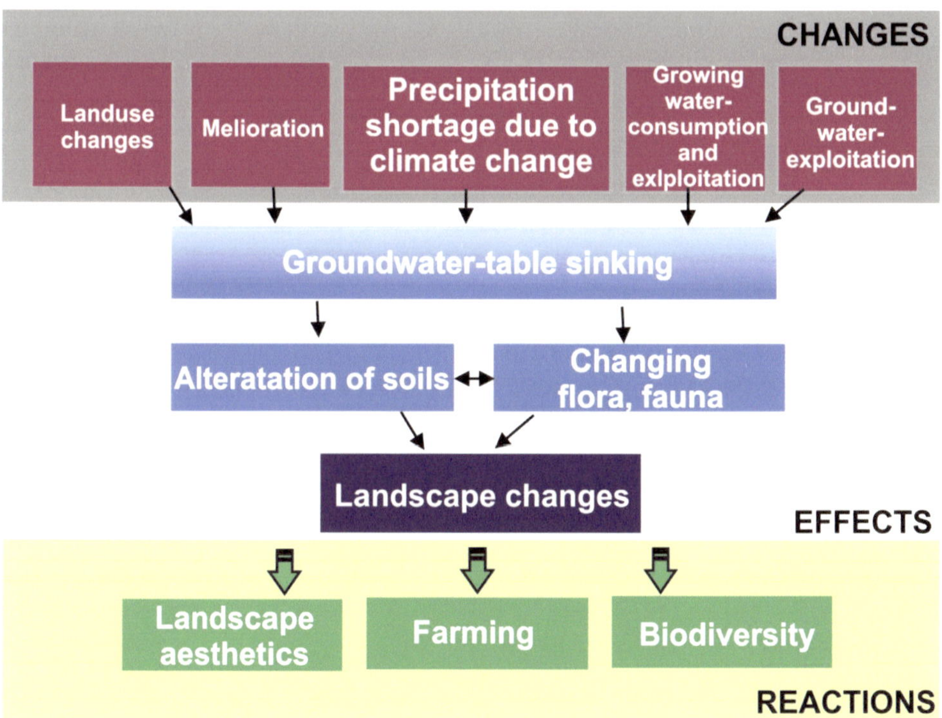

Fig. 22. The revealed relationship-system of aridification in the Danube-Tisza Interfluve (Great Hungarian Plain) (Ladányi 2010).

9. Acknowledgements

The research was funded by the Project named „TÁMOP-4.2.1/B-09/1/KONV-2010-0005 - Creating the Center of Excellence at the University of Szeged", supported by the European Union and co-financed by the European Regional Development Fund.

10. References

Aradi, Cs. & Iványosi-Szabó, A. (1996). Az Alföld természeti szépségei (Natural values of the Great Hungarian Plain). In: *A mi Alföldünk* (Our Great Hungarian Plain), Rakonczai, J. & Szabó, F. (eds.). pp. 17-36. Nagyalföld Alapítvány. ISBN 963 85437 3 6, Békéscsaba, Hungary

Barna, Gy. (2011). Tájváltozás vizsgálata a Szabadkígyósi pusztán (Analysis of landscape changes in Szabadkígyós Pusta). In: *Környezeti változások és az Alföld* (Environmental changes and the Great Hungarian Plain), Rakonczai, J. (ed.), pp. 345-354. Nagyalföld Alapítvány. ISBN 978-963 85437 8 3, Békéscsaba, Hungary

Berki, I.; Móricz, N.; Rasztovits, E. & Víg, P. (2007). A bükk szárazságtolerancia határának meghatározása (Determination of the drought tolerancelimit of beech). In: *Erdő és klíma (Forest and climate)* Vol.V., Mátyás, Cs. & Vig, P. (eds.). pp. 213-228. Sopron, Hungary

Broecker, W. S. (1997). *Will Our Ride into the Greenhouse Future be a Smooth One? GSA Today* 5. pp. 1-7.

Csorba, P. (2008). Landscape ecological fragmentation of the small landscape units (Microregions) of Hungary based on the settlement network and traffic infrastructure. *Ekológia* Vol. 27. No.1. pp. 99-116, ISSN 1335-342X, Bratislava, Slovakia

Deák, J. Á. (2011). A növényzet tájléptékű változásai a Kiskunság délkeleti részén (Changes of vegetation on choric level on the southeastern part of Kiskunság). In: *Környezeti változások és az Alföld* (Environmental changes and the Great Hungarian Plain), Rakonczai, J. (ed.), pp. 327-338. Nagyalföld Alapítvány. ISBN 978-963 85437 8 3, Békéscsaba, Hungary

IPCC (2001). Climate Change. Third Assessment Report. Intergovermental Panel on Climate Change. ISBN 0521 80767 0

Jakucs, P. (1990). A magyarországi erdőpusztulás ökológiai megközelítése (Forest dieback in Hungary from ecological aspects). *Fizikai szemle* Vol.40. No.8. pp. 225-234.

Kákonyi, Á. (2009). Szikes tavak a Duna-Tisza közén (Saline lakes on the Danube-Tisza Interfluve). In: *Magyar Hidrológiai Társaság Vándorgyűlése. ISBN 978-963-8172-23-5.* 01-03.07.2009. Available at: http://www.hidrologia.hu/vandorgyules/27/dolgozatok/12kakonyi_arpad.htm

Kertész, Á.; Papp, S. & Sántha, A. (2001). Az aridifikáció folyamatai a Duna-Tisza közén (The processes of aridification on the Danube-Tisza Interfluve). *Földrajzi Értesítő.* Vol.50. No.1-4. pp. 115-126.

Knipl, I. & Sümegi, P. (2011). Két rendszer határán. Az ember és a környezet kapcsolata a sárközi dunai allúvium és a Duna-Tisza köze peremén (On the border of two systems. Connection of humanity and the environment on the edge of the Danube alluvium and the Danube-Tisza Interfluve). In: *Környezeti változások és az Alföld* (Environmental changes and the Great Hungarian Plain), Rakonczai, J. (ed.), pp. 45-54. Nagyalföld Alapítvány. ISBN 978-963 85437 8 3, Békéscsaba, Hungary

Kotroczó, Zs.; Krakomperger, Zs.; Koncz, G.; Papp, M.; Bowden, R.D. & Tóth J.A. (2007). A Síkfőkúti cseres-tölgyes fafaj összetételének és struktúrájának hosszú-távú változása (Long-term alterations in species composition and structure of sessile oak - turkey oak forests at Síkfőkút). In: *Természetvédelmi Közlemények* Vol.13. pp. 93-100.

Kovács, F. (2008). Evaluation of landscape changes using GIS methods with special regard to aridification. In: *Proceedings Volume of 15th Congress of ISCO*, Kertész Á. (ed.). Available at: http://tucson.ars.ag.gov/isco/isco15/pdf/Kovacs%20F_Evaluation%20of%20landscape%20changes%20using.pdf

Kovács, A. & Molnár, Z. (1986). A Szabadkígyósi Tájvédelmi Körzet fontosabb növénytársulásai (Main plant communities of Szabadkígyós Landscape protection area). In: *Békés megyei Környezet- és Természetvédelmi Évkönyv* (Yearbook of nature conservation and environmental protection in Békés county), Réthy Zs. (ed.) Vol.6. pp. 165–200. Békéscsaba, Hungary

Ladányi (2010): Evaluation of landscape changes in the case of an environmentally and climate sensitive landscape located in the Danube-Tisza Interfluve, the Illancs microregion. PhD thesis. University of Szeged, Hungary.

Ladányi (2011): A természeti és társadalmi környezet hatása egy Duna–Tisza közi kistájra. Illancs környezetállapota és tájváltozásai az elmúlt évszázadban (The effect of natural and socio-environment on a microregion located in the Danube-Tisza Interfluve. The environmental state and the landscape changes of Illancs in the last century). In: *Környezeti változások és az* Alföld (Environmental changes and the Great Hungarian Plain), Rakonczai, J. (ed.), pp. 295–306. Nagyalföld Alapítvány. ISBN 978-963 85437 8 3, Békéscsaba, Hungary

Ladányi, Zs.; Rakonczai, J.; Kovács F.; Geiger, J. & Deák J. Á. (2009). The effect of recent climatic change on the Great Hungarian Plain. *Cereal Research Communication.* Vol.37. (suppl.) pp. 477-480.

Ladányi, Zs.; Rakonczai,J. & van Leeuwen, B. (2011). Precipitation vegetation interaction on the Danube–Tisza Interfluve. *Journal of Applied Remote Sensing*. Vol.5. (in press)

Margóczi, K.; Rakonczai, J.; Barna, Gy. & Majláth, I. (2008). Szikes növénytársulások összetételének és talajának hosszú távú változása a Szabadkígyósi pusztán (Long-term alterations in species composition of saline plant communities and the soils in Szabadkígyós Pusta). *Crisicum*. Vol.5. pp. 71–83, Szarvas, Hungary

Mátyás, Cs. (2010). Forecasts needed for retreating forests. *Nature*. Vol. 464. (Issue 7293.) p. 1271.

Mátyás, Cs.; Berki, I.; Czúcz, B.; Gálos, B.; Móricz, N. & Rasztovits, E. (2010). Future Beech in Southeast Europe from the Perspective of Evalutionary Ecology, *Acta Sylvatica&Lignaria Hungarica.* Vol 6. pp. 91–110. Sopron, Hungary

Molnár, Zs. & Biró, M. (2011). A Duna–Tisza köze és a Tiszántúl természetközeli növényzetének változása az elmúlt 230 évben: összegzés tájökológiai modellezések alapozásához (Changes of natural vegetation on the Danube–Tisza Interfluve and the Transtisza Region in the last 230 years: summary to establish landscape modelling). In: *Környezeti változások és az* Alföld (Environmental changes and the Great Hungarian Plain), Rakonczai, J. (ed.), pp. 75–85. Nagyalföld Alapítvány. ISBN 978-963 85437 8 3, Békéscsaba, Hungary

Rácz, L. (2011). Éghajlati változások az Alföldön a honfoglalástól a 19. század végéig (Climate changes in the Great Hungarian Plain from the Hungarian conquest until the end of the 19th century). In: *Környezeti változások és az Alföld* (Environmental changes and the Great Hungarian Plain), Rakonczai, J. (ed.), pp. 55–62. Nagyalföld Alapítvány. ISBN 978-963 85437 8 3, Békéscsaba, Hungary

Rakonczai, J. (1986). A Szabadkígyósi Tájvédelmi Körzet talajviszonyai (Pedology of the Szabadkígyós Landscape Protection Area). In: *Békés megyei Környezet- és Természetvédelmi Évkönyv* (Yearbook of nature conservation and environmental protection in Békés county), Réthy Zs. (ed.) Vol.6. pp. 19–41. Békéscsaba, Hungary

Rakonczai, J. (2007). Global change and landscape change in Hungary. Geografia fisica e dinamica quaternaria. Vol.30. No.2, 229-232.

Rakonczai, J. & Kozák, P. (2011). The consequences of human impacts on Hungarian river basins. Zeitschrift für Geomorphologie. Vol.55. (suppl.) pp. 95-107, Stuttgart, Germany

Sümegi, P. (2011). Az Alföld élővilágának fejlődése a jégkor végétől napjainkig (The development of ecosystem in the Great Hungarian Plain from the end of iceage untill nowadays). In: *Környezeti változások és az Alföld* (Environmental changes and the Great Hungarian Plain) , Rakonczai, J. (ed.), pp. 35–44. Nagyalföld Alapítvány. ISBN 978-963 85437 8 3, Békéscsaba, Hungary

Szalai, J. (2011). Talajvízszint-változások az Alföldön (Changes in groundwater-table on the Great Hungarian Plain). In: *Környezeti változások és az Alföld* (Environmental changes and the Great Hungarian Plain), Rakonczai, J. (ed.), pp. 97–110. Nagyalföld Alapítvány. ISBN 978-963 85437 8 3, Békéscsaba, Hungary

Tóth, J. A.; Krakomperger Zs.; Kotroczó Zs.; Koncz G.; Veres Zs. & Papp M., (2008). A klímaváltozás hatása a Síkfőkúti cseres-tölgyes avarprodukciójára és talajdinamikai folyamataira (The effect of climate changes on leaf-litter production and pedological processes of sessile oak-turkey oak forests at Síkfőkút). In: *Talajvédelem* (Suppl.), pp. 543-554.

Tóth, T. (2002). *Szikes talajok tér és időbeli változatossága* (Spatial and temporal variability of saline soils). MTA Phd Thesis. Budapest, Hungary.

Varga-Haszonits, Z. (2003). Az éghajlatváltozás mezőgazdasági hatásának elemzése, éghajlati szcenáriók (Analysis of the agricultural effects of climate change, climatical scenarios). *„Agro-21" Füzetek* 31. pp. 9–28.

16

Climate Change Impact on Quiver Trees in Arid Namibia and South Africa

Danni Guo[1], Renkuan Guo[2], Yanhong Cui[2],
Guy F. Midgley[1], Res Altwegg[1] and Christien Thiart[2]
[1]*Climate Change and Bioadapation Division*
South African National Biodiversity Institute, Cape Town
[2]*Department of Statistical Sciences, University of Cape Town, Cape Town*
South Africa

1. Introduction

The climate fluctuates and changes naturally, and adding the common problems of land transformation and deforestation, its impact can be very harsh on the natural environment, and cause a decline in the biodiversity of plants and animals. *Aloe dichotoma*, common name Quiver tree, is an important part of the arid regions, such as Namaqualand and Bushmanland in South Africa, and in arid parts of southern Namibia. This succulent tree species occurs in rocky areas, and it can grow quite rapidly under the right conditions. Succulents are able to survive long periods of drought conditions, due to the fact succulent plants has special water-storing tissue which makes part of the plant fleshy, and the Quiver tree has succulent leaf and stem (Van Wyk and Smith, 1996).

The Quiver tree has a 200 year life span, and can grow up to 9 meters tall, and it occurs in summer and winter rainfall regions, and can live under a variety of climatic conditions (Fig. 1). The Quiver tree is important to the ecosystem due to the fact that it is as a source of moisture for a wide variety of mammals, birds, and insects. Foden's detailed study of the demographic data of the Quiver trees show that negligible recruitment has occurred in certain populations for 50 years, and the effects of non-climatic variables, such as herbivory, competition, seed availability, fungal pathogens, plant collection... are very small (Foden, 2002).

Today, the Quiver trees are threatened by agricultural expansion, overgrazing, and mining, as well as droughts and other climate changes (Foden, 2002). Climate changes is one of the major factors affecting the existence of Quiver trees, while the Quiver tree will unlikely to be affected by small climatic fluctuations, but will be affected larger or long term climatic changes. The Quiver tree can potentially provide a good indication of long term climate changes in the arid regions (Foden, 2002).

Previous onsite observations show that Quiver trees are very sensitive to temperature changes, and does not do well under extreme hot and dry conditions. Observations has also shown that the Quiver trees might be responding to higher temperatures by shifting its distribution range towards higher and higher altitudes, showing a preference for slightly cooler regions (Midgley et al., 2009).

Fig. 1. Quiver trees in Kaukusib Koppie, Namibia, 2008

To have a complete picture of the Quiver tree's response to recent climatic changes, samples of Quiver trees were collected from their distributional range in arid Namibia and South Africa, and this is a continuing project funded by the South African National Biodiversity Institute. However, data uncertainty accompanied the sample collections as well.

In 2001 and 2002, Wendy Foden and various assistants collected Quiver trees sample data across the whole range, sampling 53 sites in detail. In 2008 Graeme Ellis re-sampled a subset of the population, sampling 41 sites. In 2009, Res Altwegg and Cory Merow did a thorough analysis and comparison of between the 2 sets of data, for 35 sites, excluding certain sites due to lack of data for comparisons.

In this chapter, we are going one step further, that is, to try to complete all the missing data values, which will allow us to do a more complete comparison. The incomplete data issue is complicated on many fronts:

1. While the 2001-2002 data are the most complete, however, there are several sites with missing variables from the 53 sample sites.
2. The 2008 data is a subset of the original 2001-2002 data, covering 41 sites, and the 2008 data variables are not the same as the 2001-2002 data variables.
3. The 2009 analysed data comparison between 2001-2002 and 2008 data is for 35 sites, with 18 sites uncalculated due to missing data.
4. With all 3 datasets, missing data is a major part of the problem, which renders the 3 datasets incomplete and only partially comparable.

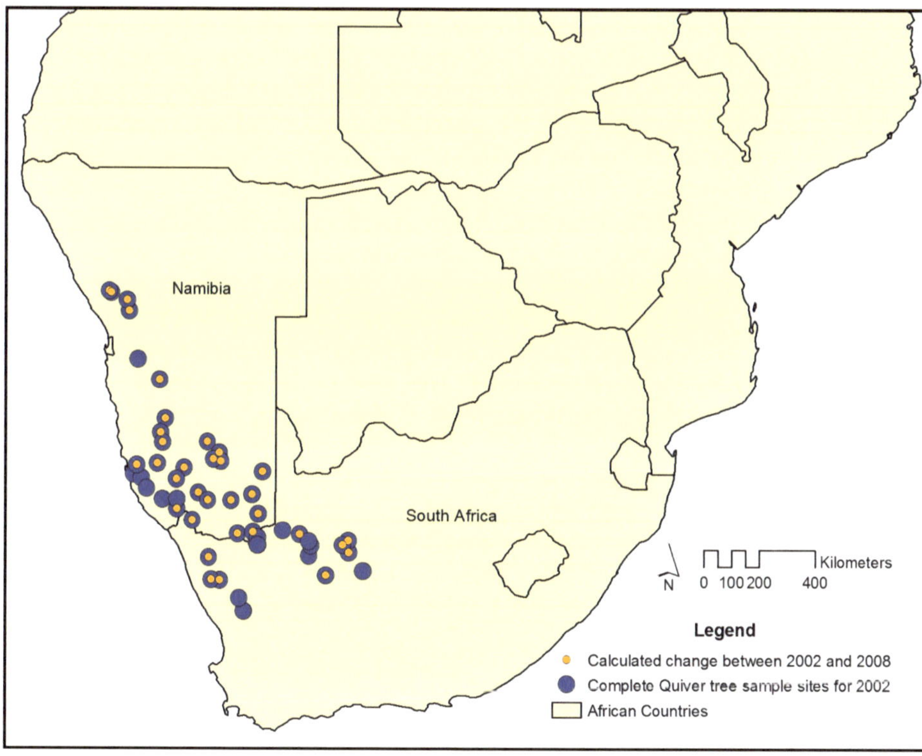

Fig. 2. Available Quiver tree sample site data in Arid Namibia and South Africa

Of course, another sample collection study would be ideal, but to due lack of funding and trained professionals, this is not possible at this stage. In any case, a full usage should be made of the 2002 and 2008 data. In Fig. 2, the Quiver trees sample sites are shown, the Quiver tree's distribution extending over the arid regions in Namibia and South Africa. As one can see from the figure, the 2008 study is a subset of the original 2002 samples.

2. Impreciseness in quiver tree's sample data

To investigate the climate change impacts on the Quiver's spatial distribution from 2002 to 2008, we must have the Quiver's population evolution information. We as biodiversity scientists need to reveal the climate change impacts on plant species based on the data collected. Only solid data-oriented analysis can provide near-true information to the public and governmental decision-making body. In this case, the Quiver tree's dataset is not complete enough to perform a solid data-oriented analysis because of the impreciseness features of the dataset.

Impreciseness is a fundamental and intrinsic feature in a solid data oriented spatial modelling exercises due to the observational data shortage and incompleteness. Facing the impreciseness reality the spatial statisticians often rely on expert's knowledge to compensate the inadequacy and in accuracy in collected observational data. Nevertheless, such a dependence on experts' knowledge engineering will still open a backdoor to pseudo-scientific believe or claim.

Impreciseness is referred to a term with an intrinsic property governed by an uncertain measure or an uncertainty distribution for each of the actual or hypothetical members of an uncertainty population (i.e., collection of expert's knowledge). An uncertainty process is a repeating process whose outcomes follow no describable deterministic pattern, but follow an uncertainty distribution, such that the uncertain measure of the occurrence of each outcome can be only approximated or calculated.

Definition 2.1: Impreciseness is an intrinsic property of a variable or an expert's knowledge being specified by an uncertainty measure.

In geo-statistics analyst communities it is seldom to mention uncertainty measure. It is true, spatial researchers are used to the term "uncertainty", which represents randomness in their eyes. Actually, this perception is wrong. Researchers have long realized the there are many forms of uncertainty, randomness is merely a member of uncertainty family. How can we differentiate different forms of uncertainty? The answer is fairly straght forward: the appropriate measure specifies a particular form of uncertainty. For example, probability measure specifies random uncertainty.

The uncertainty modelling without a measure specification will not have an rigorous mathematical foundations and consequently the modelling exercise is baseless and blindness. In other words, measure specification is the prerequisite to spatial data collection and analysis. For example, without Kolmogrov's (1950) three axioms of probability measure, randomness is not defined and thus statistical data analysis and inference has no foundation at all.

Notice that imprecise probability theory is a potential remedy to address the observational data inaccuracy and inadequacy because most the geo-statistics analysts are familiar with probability theory. However the imprecise probability based spatial modelling might be not feasible. Just as Utikin and Gurov (2000) has commented, "the probabilistic uncertainty model makes sense if the following three premises are satisfied: (i) an event is defined precisely; (ii) a large amount of statistical samples is available; (iii) probabilistic repetitiveness is embedded in the collected samples. This implies that the probabilistic assumption may be unreasonable in a wide scope of cases."

It is therefore inevitably to seek another form of uncertainty theory to meet the impreciseness challenges. In the theoretical basket, interval uncertainty theory (Moore, 1966), fuzzy theory (Zadeh, 1965, 1978), grey theory (Deng, 1984), rough set theory (1982), upper and lower provisions or expectations (Walley, 1991), or Liu's uncertainty theory (2007, 2010) may be chosen. Nevertheless, Liu's (2007, 2010) uncertainty theory is the only one built on an axiomatic uncertain measure foundation and fully justified with mathematical rigor. Therefore it is logical to engage Liu's (2007, 2010) uncertainty theory for guiding us to understand the intrinsic character of imprecise uncertainty and facilitate an accurate mathematical definition of impreciseness in order to establish the foundations for uncertainty spatial modelling under imprecise uncertainty environments.

3. Uncertain measure foundation

The Uncertainty Theory was founded by Liu in 2007 and refined in 2010 (Liu, 2007, 2010). Nowadays uncertainty theory has become a branch of mathematics.

A key concept in uncertainty theory is the uncertain measure: Let Ξ be a nonempty set (space), and $\mathfrak{A}(\Xi)$ the σ-algebra on Ξ. Each element, let us say, $A \subset \Xi$, $A \in \mathfrak{A}(\Xi)$ is called an uncertain event. A number denoted as $\lambda\{A\}$, $0 \le \lambda\{A\} \le 1$, is assigned to event $A \in \mathfrak{A}(\Xi)$, which indicates the uncertain measuring grade with which event $A \in \mathfrak{A}(\Xi)$ occurs. The normal set function $\lambda\{A\}$ satisfies following axioms given by Liu (2011):

Axiom 1: (Normality) $\lambda\{\Xi\} = 1$.

Axiom 2: (Self-Duality) $\lambda\{\cdot\}$ is self-dual, i.e., for any $A \in \mathfrak{A}(\Xi)$, $\lambda\{A\} + \lambda\{A^c\} = 1$.

Axiom 3: (σ- Subadditivity) $\lambda\left\{\bigcup_{i=1}^{\infty} A_i\right\} \le \sum_{i=1}^{\infty} \lambda\{A_i\}$ for any countable event sequence $\{A_i\}$.

Definition 3.1: (Liu, 2007, 2010) A set function $\lambda : \mathfrak{A}(\Xi) \to [0,1]$ satisfies *Axioms 1-4* is called an uncertain measure. The triple $(\Xi, \mathfrak{A}(\Xi), \lambda)$ is called an uncertainty space.

Definition 3.2: (Liu, 2007, 2010) An uncertain variable is a measurable function ξ from an uncertainty space $(\Xi, \mathfrak{A}(\Xi), \lambda)$ to the set of real numbers.

Definition 3.3: (Liu, 2010) Let ξ be a uncertainty quantity of impreciseness on an uncertainty measure space $(\Xi, \mathfrak{A}(\Xi), \lambda)$. The uncertainty distribution of ξ is

$$\Psi_{\xi}(x) = \lambda\{\tau \in \Xi \mid \xi(\tau) \le x\} \tag{1}$$

Theorem 3.4: (Liu, 2007, 2010) Let $\Psi_{\xi_1}, \Psi_{\xi_2}, \cdots, \Psi_{\xi_n}$ be uncertainty distributions for the uncertainty variables $\xi_1, \xi_2, \cdots, \xi_n$ on $(\Xi, \mathfrak{A}(\Xi), \lambda)$ respectively. Let $\Psi_{(\xi_1, \xi_2, \cdots, \xi_n)}$ be the joint distribution of uncertainty vector $(\xi_1, \xi_2, \cdots, \xi_n)$. If $\xi_1, \xi_2, \cdots, \xi_n$ are independent, then

$$\Psi_{(\xi_1, \xi_2, \cdots, \xi_n)}(x_1, x_2, \cdots, x_n) = \min_{1 \le i \le n} \Psi_{\xi_i}(x_i) \tag{2}$$

for any real numbers $x_1, x_2, \cdots, x_n \in \mathbb{R}$.

Definition 3.5: (Liu, 2007, 2010) Let ξ be a uncertainty variable on an uncertainty measure space $(\Xi, \mathfrak{A}(\Xi), \lambda)$. The expectation ξ is defined by

$$\mathrm{E}[\xi] = \int_0^{+\infty} \lambda\{\xi \ge r\}\, dr - \int_{-\infty}^{0} \lambda\{\xi \le r\}\, dr \tag{3}$$

provided that one of the two integrals exists at least.

Definition 3.6: (Liu, 2007, 2010) Let ξ be a uncertainty variable on an uncertainty measure space $(\Xi, \mathfrak{A}(\Xi), \lambda)$.with uncertainty distribution function and Ψ_{ξ} a finite expectation μ. Then the variance of ξ, denoted by $V[\xi]$, is defined by

$$V[\xi] = 2\int_0^{+\infty} (r-\mu)\big(1 - \Psi_{\xi}(r) + \Psi_{\xi}(2\mu - r)\big)dr. \tag{4}$$

Liu's uncertainty theory (Liu, 2007, 2010) does not define multivariate uncertainty distribution, (Guo, 2010; Guo et al., 2007), except for the independent uncertainty variables.

Hence it is necessary to facilitate a framework for showing the preparation to define the uncertain covariance.

Let $\Psi_{\xi_1},\Psi_{\xi_2},\cdots,\Psi_{\xi_n}$ be uncertainty distributions for the uncertainty variables $\xi_1,\xi_2,\cdots,\xi_n$ on $(\Xi,\mathfrak{A}(\Xi),\lambda)$ respectively. Let $\Psi_{(\xi_1,\xi_2,\cdots,\xi_n)}$ be the joint distribution of uncertainty vector $(\xi_1,\xi_2,\cdots,\xi_n)$. Assuming that ξ_i and ξ_j two arbitrary pair of uncertainty variables within the uncertainty vector $(\xi_1,\xi_2,\cdots,\xi_n)$ which have finite expectations μ_1 and μ_2 respectively. Denote $\Psi_{(\xi_i,\xi_j)}$ bivariate uncertainty distribution function.

Definition 3.7: Let $\eta_{ij}=(\xi_i-\mu_i)(\xi_j-\mu_j)$ be product of centered uncertainty variables ξ_i and ξ_j. The uncertainty distribution of η_{ij} is defined by

$$\Psi_{\eta_{ij}}(y)=\sup_{z_1z_2=y}\Psi_{(\xi_i,\xi_j)}(z_i,z_j),\ \forall(z_i,z_j)\in\mathbb{R}^2 \tag{5}$$

Definition 3.8: Let $\eta_{ij}=(\xi_i-\mu_i)(\xi_j-\mu_j)$ be product of centered uncertainty variables ξ_i and ξ_j. The expectation of centered product η_{ij} is called the covariance between uncertainty variables ξ_i and ξ_j, that is by

$$\gamma_{ij}\triangleq \mathrm{E}[\eta_{ij}]=\int_0^{+\infty}\left(1-\Psi_{\eta_{ij}}(r)\right)dr-\int_{-\infty}^{0}\Psi_{\eta_{ij}}(r)dr \tag{6}$$

Theorem 3.9: Let $\xi_1,\xi_2,\cdots,\xi_n$ be independent uncertainty variables on $(\Xi,\mathfrak{A},\lambda)$. Then

$$\Psi_{\eta_{ij}}(y)=\sup_{z_1z_2=y}\left(\Psi_{\xi_1}(z_1)\wedge\Psi_{\xi_2}(z_2)\right),\ \forall(z_i,z_j)\in\mathbb{R}^2 \tag{7}$$

Remark 3.10: Different from independent random variables X_1 and X_2, whose $\mathrm{E}\left[(X_1-\mathrm{E}[X_1])(X_2-\mathrm{E}[X_2])\right]=0$. In uncertainty theory, independent $\xi_1,\xi_2,\cdots,\xi_n$ do not imply $\eta_{ij}=0$.

Remark 3.11: The formation of uncertain variance-covariance does touch the detailed functional form of multivariate uncertainty joint distribution. For practical applications, the form of multivariate uncertainty joint distribution is not necessarily available, but the paired uncertainty bivariate distribution must be given.

Similar to the concept of stochastic process in probability theory, an uncertain process $\{\xi_t,\ t\ge 0\}$ is a family of uncertainty variables indexed by t and taking values in the state space $\mathbb{S}\subseteq\mathbb{R}$.

Definition 3.12: (Liu, 2007, 2010) Let $\{C_t,t\ge 0\}$ be an uncertain process.

(1) $C_0=0$ and all the trajectories of realizations are Lipschitz-continuous;

(2) $\{C_t,t\ge 0\}$ has stationary and independent increments;

(3) every increment $C_{t+s}-C_s$ is a normal uncertainty variable with expected value 0 and variance t^2, i.e., the uncertainty distribution of $C_{t+s}-C_s$ is

$$\Psi_{C_{t+s}-C_s}(z)=\left(1+\exp\left(-\frac{\pi z}{\sqrt{3}t}\right)\right)^{-1} \tag{8}$$

then $\{C_t,t\ge 0\}$ is called an uncertain canonical process.

Theorem 3.13: Assuming that $\{C_t, t \ge 0\}$ is an uncertain canonical process. Therefore, in the autocovariance and autocorrelation of uncertain canonical process $\{C_t, t \ge 0\}$ are

$$\sigma_{s,t} = s^2 + \mathrm{E}\left[C_s\left(C_t - C_s\right)\right] = s^2 + \delta_{s,t-s} \tag{9}$$

and

$$\rho_{s,t} = \frac{s}{t} + \frac{\delta_{s,t-s}}{st}, \ (s < t) \tag{10}$$

where

$$\begin{aligned} \delta_{s,t-s} &= E[C_s(C_t - C_s)] \\ &= \int_0^{\infty}\left(1 - \sup_{xy=r}\left(\Psi_{C_s}(x) \wedge \Psi_{C_{t-s}}(y)\right)\right)dr \\ &- \int_{-\infty}^{0} \sup_{xy=r}\left(\Psi_{C_s}(x) \wedge \Psi_{C_{t-s}}(y)\right)dr \end{aligned} \tag{11}$$

4. Habitat measure

Note that the Quiver tree's 2002 data subset contains percentage of dead trees, percentage of juvenile trees, the total density of trees, and the average age of dead trees. Also, Quiver tree's 2008 data subset contains population growth rate, survival reproductives, survival non-reproductives, and proportion juvenile. Considering the aim of this research: climate change impacts on the Quiver tree's population, we select total density of tree (in 2002 data subset) and population growth rate (in 2008 data subset) as the our analysis data. We will build our model based on the total tree population size

$$V_i^{(k)} = A^{(k)}\kappa_i^{(k)} \tag{12}$$

where the $V_i^{(k)}$ is the total population size at year i over area size $A^{(k)}$, and total density $\kappa_i^{(k)}$.

Notice that among 53 sites, 43 sites have total density observational values. Ten sites have missing values. That is, without total density observation {10,12,14,15,17,19,33,36,37,128} site number set.

For inexperienced go-statistics analysts, kriging method might be a choice for filling the ten missing vaules. Nevertheless, after examining the 2002 total density of tree's 43 observations, we found that geometric distance does not play a role in dertermining the value of the total density of trees. we can find out that the total density observations are not similar even for the very nearby sites geometrically within the same ecological sub-region. For example, Site , see Table 1.

From Table 1, it is obvious that for the Quiver tree's growth status, $\{Z(s), s \in \mathbb{R}^2\}$, the spatial location $s \in \mathbb{R}^2$ only partially links to the tree habitation. While the geometric distance between two locations s and $s+h$ is no longer fully reflecting the habitat closeness. The Quiver tree's growth status observations reveal a fact that the total density status is extremely sensitive to the ecological conditions of individual site area (or the area habitat).

The graphical location and the isotropic distance will not offer much information, rather the Quiver tree's growth observation $Z(s)$ gives the best indication of the tree area habitat.

Group/ site no.		Coordinates		Total density of trees	Geometric distance	Difference in total density	Density / distance
		longitude	latitude				
1	40	21.64455	-30.00486	36.41	1.1977228	65.42	54.62
	42	22.83446	-29.86828	101.83			
2	101	18.20583	-30.15250	70.00	0.285919122	6.67	23.33
	102	17.92017	-30.14033	63.33	0.688683879	16.67	24.20
	103	17.83583	-29.45683	46.67	0.787944636	23.33	29.61
3	3	19.25749	-27.49479	33.33	0.003999625	52.38	13096.47
	4	19.26062	-27.49230	54.17			

Table 1. 3 groups of sites and their total densities

The measure of habitat closeness can be defined by the habitat distance, which is a virtual distance indicating the aggregate ecological environment of an area. In such a circumstance, if we still impose conventional isotropic distance assumption to construct a kriging predictor, the prediction would be no sense at all.

It is logical to argue that in order to utilize the n observations $\{Z(s), s \in \mathfrak{N}_{s_0} \subset \mathbb{R}^2\}$ contained in the neighbourhood of s_0: $\mathfrak{N}_{s_0} = \{s_0, s_1, s_2, \cdots, s_n\}$, at s_0 without an Quiver tree's observation, to predict $Z(s_0)$, utilize the habitat distance information is far more efficient that the conventional isotropic distance.

As to how to define the habitat distance measure, it depends upon the habitat spatial distribution pattern as well as the form of uncertainty governing the spatial process $\{Z(s), s \in \mathbb{R}^2\}$.

In this study, we are going to calculate habitat distance in following three approaches:

(1) Grouping method. This approach classifies total density observations $\{Z(s), s \in \mathbb{R}^2\}$ into groups by the sizes of $Z(s)$. Then, for each group, the variance is calculated. choose the variance of group observation as habitat measure. We may face the situation at which some $Z(s)$ values may be far away from those grouped observations. Then we may group them according to geometrical distance criterion, then calculate the variances for those isolated total density observations. Table 2 lists a few observational groups and group habitat distances.

It is logical to ignore those 3 groups because of extreme group variances, which should not accepted as habitat measures.

(2) Inverse-distance methods. Notice that there are ten sites without total density observations and thus their habitat measures cannot be determined. However, later we will see the empirical habitat measure is necessary for uncertain kriging.

In this paper, we propose two schemes for addressing the habitat measure at missing total density value site. The first one is the conventional inverse distance estimation. The second one is utilizing the empirical habitat distance for uncertain inverse distance estimation.

The conventional inverse distance estimation. Table 5 summarizes the results (at $\lambda = 1.0$ because near optimal $\lambda = 0.1$ gives illogical empirical results).

Group/ Site no.		Coordinates		Total density of trees	Habitat measure
		longitude	latitude		
1	42	22.83446	-29.86828	101.83	1.1112310
	11	18.18643	-26.21253	100.00	
	113	22.18500	-29.06700	100.00	
2	20	16.23986	-25.59329	90.48	2.0052690
	13	15.15805	-21.48867	91.70	
	124	19.43333	-28.83333	93.30	
3	4	19.26062	-27.49230	33.33	3.4715580
	2	19.28217	-28.66671	33.33	
	40	21.64455	-30.00486	36.41	
	123	19.40000	-28.98333	36.70	
4	104	17.28650	-28.30430	40.00	11.1111222
	108	19.45333	-28.10333	43.33	
	103	17.83583	-29.45683	46.67	
5	107	18.56950	-27.68433	113.33	11.5226107
	29	15.47427	-26.58937	106.67	
	30	16.76122	-27.03209	111.11	
6	41	22.37461	-29.29727	86.05	11.5534900
	121	19.43333	-29.06667	80.00	
	122	19.43333	-29.06667	80.00	
	3	19.25749	-27.49479	85.71	
7	112	22.37000	-28.93500	130.00	14.9633000
	126	21.10000	-29.40000	123.30	
	26	16.31455	-27.64893	130.00	
8	23	15.62385	-26.99582	22.22	17.0993200
	9	17.01403	-26.66638	28.07	
9	102	17.92017	-30.14033	63.33	19.9113500
	1	18.78301	-28.71597	63.33	
	101	18.20583	-30.15250	70.00	
	125	20.23333	-28.61667	70.00	
	120	18.83333	-30.70000	73.30	
10	110	18.98883	-31.10050	50.00	21.5049200
	5	19.59336	-26.79313	54.17	
	24	15.78774	-27.31541	59.26	
11	106	17.98333	-26.40217	166.67	53.3333900
	18	16.21400	-23.95119	153.33	
	105	18.23967	-26.47550	153.33	
	22	15.36674	-26.88367	166.67	
	28	16.14727	-26.54360	166.67	

Table 2. Groups of total densities and habitat measures (group variances)

It is obvious that there are five sites left listed in Table 3.

Group/ site no.		Coordinates		Total density of trees	Group variance
		longitude	latitude		
12	127	21.16667	-29.08333	6.70	
13	34	16.77862	-27.64536	282.61	1843.7403440
	111	18.98567	-31.10583	343.33	
14	109	20.79000	-28.74017	200.00	12482.0000000
	21	16.31081	-25.88543	252.17	

Table 3. 3 Groups of total densities left with group variances

Site no.	Coordinates		Predicted total density of trees	Predicted habitat measure
	longitude	latitude		
15	14.57582	-21.20986	103.2131	1.1112310
14	14.63088	-21.26169	102.9913	1.1112310
12	15.21586	-21.81786	102.3649	1.1112310
17	15.51133	-23.30893	110.5426	11.5226107
19	16.39317	-25.14609	115.2105	11.5226107
33	16.77479	-27.94278	121.9919	14.9633000
36	17.47540	-27.45209	106.7434	11.5226107
10	17.78013	-25.87191	110.3903	11.5226107
37	17.80606	-27.67034	101.0099	1.1112310
128	21.10000	-28.95000	82.67477	11.5534900

Table 4. 10 Sites of Inverse Distance predicted total densities

From Table 4, we can see that since the estimated total density estimators are calculated in terms of graphical distances. Those estimators are empirical prior information. As to habitat measure, they are taking the values by comparing the estimated total density to the group total density values in Table 4. Definitely, the empirical habitat measures are not all accurate.

(3) Uncertain canonical process regression methods. Because the Quiver tree's total density is very sensitive to local area ecological environments, it is logical to consider if some uncertain relationship between the observed total density and ecological environmental variables, say, annual temperature, and annual rainfall. Table 4 lists the data, from which a linear relationship between group average total density y_{d_i} and group average annual temperature T_{d_i} and group average annual rainfall R_{d_i} at the empirical distance d_i, $i = 1,2,\cdots,11$, is intended to establish.

The uncertain regression model takes a form:

$$y_{d_i} = \alpha_0 + \alpha_1 T_{d_i} + \alpha_2 R_{d_i} + \sigma C_{d_i}, \quad i = 1,2,\cdots,11 \tag{13}$$

where errors C_{d_i}, $i=1,2,\cdots,11$ are from the uncertain canonical process $\{C_d, d\in[0,+\infty)\}$. As to $\{C_d, d\in[0,+\infty)\}$. The fitted regression is

$$\hat{y}_{d_i} = \underset{(66.6838)}{500.4016} + \underset{(6.1464)}{30.1100} T_{d_i} - \underset{(2.0621)}{8.8953} R_{d_i}, \quad i=1,2,\cdots,11 \tag{14}$$

Site no.	Coordinates		Predicted total density of trees	Predicted habitat measure
	longitude	latitude		
15	14.57582	-21.20986	110.3210	11.5226
14	14.63088	-21.26169	26.5995	17.0993
12	15.21586	-21.81786	55.0573	21.5049
17	15.51133	-23.30893	83.5152	11.5535
19	16.39317	-25.14609	8.72890	2.0582
33	16.77479	-27.94278	147.0394	53.3334
36	17.47540	-27.45209	72.2531	19.9114
10	17.78013	-25.87191	20.9684	17.0993
37	17.80606	-27.67034	29.9036	17.0993
128	21.10000	-28.95000	101.3857	1.1112

Table 5. 10 Sites of Regression Predicted total densities left with habitat measures

However, it is necessary to mention that the uncertain regression model is carried on the group average total density against group average annual temperature and group annual rainfall and thus the predictability is limited although the model goodness-of fit ($R^2 \approx 0.7$) is far better than that of classical regression model ($R^2 \approx 0.21$), whose coefficients for annual temperature and annual rainfall are both insignificant statistically. We are not going to use these regression predicted total density of trees as the basic results for evaluations of climate change impacts.

Finally, we must emphasize that for any given spatial process $\{Z(s), s\in\mathbb{R}^2\}$, which has no order, this fact would prevent geostatistical analysts from utilizing 1-dimensional statistical approaches for predictions, for example, the uncertain regression engaged. But after defining appropriate habitat distance for spatial process, then the habitat distance set can be indexed an uncertain canonical process as the desired partner process, and hence the intrinsic covariance structure of the partner process will be available and thus simplify greatly the semi-variogram computations, while in probabilistic kriging the theoretical semi-variogram must be replaced by sample or experimental semi-variogram. The uncertain semi-varogram is semi-data oriented, in contrast, the probabilistic sample or experimental semi-variogram is fully data-oriented.

5. Uncertain kriging predictor

We first point out that kriging predictor is not intrinsic to the probabilistic spatial statistics. As long as variance, covariance, and semi-variogram concepts can be established on some uncertain theory, no matter it is Zadeh's fuzzy theory, rough set theory, grey theory, random set theory, or interval theory, or Liu's uncertainty theory, new kriging predictor can

be solidly established with similar mathematical formulae as that of probabilistic ordinary kriging (Cressie, 1991). However, the geometric location s should be replaced by habitat distance $d(s)$, which is the square root of habitat measure.
An uncertain semi-variogram is defined by

$$2\gamma\left(d_{s_i}-d_{s_j}\right)=V\left(Z\left(d_{s_i}\right)-Z\left(d_{s_j}\right)\right) \tag{15}$$

where d_{s_i} and d_{s_j} are two habitat distances at location s_i and location s_j respectively.
The uncertain kriging predictor is defined by

$$Z\left(d_{s_0}\right)=\sum_{i=1}^{n}\lambda_i Z\left(d_{s_i}\right) \tag{16}$$

where $\lambda_1,\lambda_2,\cdots,\lambda_n$ are determined by minimizing the following objective function.

$$\mathrm{E}\left[\left(Z\left(d_{s_0}\right)-\sum_{i=1}^{n}\lambda_i Z\left(d_{s_i}\right)\right)^2\right]-2\psi\left(\sum_{i=1}^{n}\lambda_i-1\right) \tag{17}$$

where ψ is the Lagrange's multiplier. Then the uncertain kriging equation system is given by

$$\underline{\lambda}_U=\Gamma_U^{-1}\underline{\gamma}_U \tag{18}$$

where

$$\begin{aligned}
&\underline{\lambda}_U=\left(\lambda_1,\lambda_2,\cdots,\lambda_n,\psi\right)',\\
&\underline{\gamma}_U=\left(\gamma\left(d_{s_1}-d_{s_0}\right),\gamma\left(d_{s_2}-d_{s_0}\right),\cdots,\gamma\left(d_{s_n}-d_{s_0}\right),1\right)'\\
&\Gamma_U=\left(o_{ij}\right)_{(n+1)\times(n+1)}\\
&o_{ij}=\begin{cases}\sigma^2 & i=j=1,2,\cdots,n\\ \gamma\left(d_{s_i}-d_{s_j}\right) & i\neq j=1,2,\cdots,n\\ 1 & i=n+1,j=1,2,\cdots,n\\ 0 & i=j=n+1\end{cases}
\end{aligned} \tag{19}$$

Theorem 5.1: Let

$$\underline{\lambda}=\left(\lambda_1,\lambda_2,\cdots,\lambda_n\right)' \tag{20}$$

then,

$$\underline{\lambda}'=\left(\underline{\gamma}+\underline{1}\frac{\left(1-\underline{1}'\Gamma^{-1}\underline{\gamma}\right)}{\underline{1}'\Gamma^{-1}\underline{1}}\right)\Gamma^{-1} \tag{21}$$

where

$$\psi = -\frac{\left(1-\underline{1}^{'}\Gamma^{-1}\underline{\gamma}\right)}{\underline{1}^{'}\Gamma^{-1}\underline{1}} \tag{22}$$

Theorem 5.2: Let $d_{s_i,s_j} \triangleq d_{s_j} - d_{s_i}$, $d_{s_i} \le d_{s_j}$. Then the semi-variogram of habitat distance d_{s_i,s_j} is

$$\gamma\left(d_{s_i,s_j}\right) = \frac{1}{2}\left(d_{s_j} - d_{s_i}\right) - \delta_{d_{s_i},d_{s_j}-d_{s_i}} \tag{23}$$

Proof: It is noticed that for $\forall d_{s_i} \le d_{s_j}$

$$\begin{aligned}
&\gamma\left(d_{s_i,s_j}\right)\\
&=\frac{1}{2}\mathrm{E}\left[\left(Z\left(d_{s_i}\right)-Z\left(d_{s_j}\right)\right)^2\right]\\
&=\frac{1}{2}\mathrm{E}\left[\left(\left(Z\left(d_{s_i}\right)-m\right)-\left(Z\left(d_{s_j}\right)-m\right)\right)^2\right]\\
&=\frac{1}{2}\mathrm{E}\left[\left(Z\left(d_{s_i}\right)-m\right)^2\right]+\frac{1}{2}\mathrm{E}\left[\left(Z\left(d_{s_j}\right)-m\right)^2\right]\\
&-\mathrm{E}\left[\left(Z\left(d_{s_i}\right)-m\right)\left(Z\left(d_{s_j}\right)-m\right)\right]\\
&=\frac{1}{2}\left(\sigma^2\left(Z(s_i)\right)+\sigma^2\left(Z\left(s_j\right)\right)\right)-\mathrm{E}\left[\varepsilon\left(d_{s_i}\right)\varepsilon\left(d_{s_j}\right)\right]\\
&=\frac{1}{2}\left(d_{s_i}+d_{s_j}\right)-\sigma_{d_{s_i},d_{s_j}}\\
&=\frac{1}{2}\left(d_{s_i}+d_{s_j}\right)-\left(d_{s_i}+\delta_{d_{s_i},d_{s_j}-d_{s_i}}\right)\\
&=\frac{1}{2}\left(d_{s_j}-d_{s_i}\right)-\delta_{d_{s_i},d_{s_j}-d_{s_i}}
\end{aligned} \tag{24}$$

However, at location s_0, there is no observation $Z(s_0)$ and thus the habitat distance d_{s_0} is undefined. It is necessary to "define" the habitat distance d_{s_0} in order to carry on the uncertain kriging predictions.

Definition 5.3: If geometric location s_k is close to s_0, at which $Z(s_k)$ is observed, then the habitat distance of $Z(s_k)$ is defined as the habitat distance d_{s_0} at location s_0.

There is possibility that some $\lambda_{i_{np}} < 0$. Sort $\{\lambda_1, \lambda_2, \cdots, \lambda_n\}$ such that $\lambda_{(1)} \le \lambda_{(2)} \le \cdots \le \lambda_{(n)}$, assuming $i_p : \lambda_{(i_p-1)} < 0, \lambda_{(i_p)} > 0$ e that there are n_{np} terms which but $\sum_{i=1}^{n_{s_0}} \lambda_{(i)} = 1$, but $\sum_{i=1}^{n_{s_0}} \left|\lambda_{(i)}\right| > 1$. Let us define the adjusted coefficient $\lambda_i^{adjusted}$ as

$$\lambda_{(i)}^{adjusted} = \begin{cases} \frac{1}{\left(i_{s_p}\right)-1}\left(1-\frac{\sum_{i=(i_p)}^{(n_{s0})}\lambda_i}{\sum_{i=(1)}^{(n_{s0})}|\lambda_i|}\right) & \text{if } (i)<(i_p) \\ \frac{\lambda_{(i)}}{\sum_{(i)=1}^{(i_{n_{s0}})}|\lambda_i|} & \text{if } (i)\geq(i_p) \end{cases} \tag{25}$$

It is obvious that constraint

$$\sum_{(i)=1}^{(n_{s0})}\lambda_{(i)}^{adjusted} = 1.0 \tag{26}$$

is truly kept and also all $\lambda_i^{adjusted} \in (0,1)$.
We will calculate those uncertain kriging predictors for the ten sites {10,12,14,15,17,19,33,36, 37,128} without total density observations in following 4 steps.
Step 1: Determine the 10 neighbourhoods

$$\left\{\mathcal{N}_{s_0^{(k)}}\right\}_{k=1}^{10} \tag{27}$$

where $s_0^{(k)} \in \{10,12,14,15,17,19,33,36,37,128\}$. Each $\mathcal{N}_{s_0^{(k)}}$ contains six sites with observed total density values.
Step 2: For each individual neighbourhood $\mathcal{N}_{s_0^{(k)}}$, for example, $\mathcal{N}_{s_0^{(k)}} = \left\{s_1^{(k)}, s_2^{(k)}, \cdots, s_6^{(k)}\right\}$, determine the corresponding habitat distance from Table 3, 4, and 6, respectively, denoted by

$$\mathcal{D}^{(k)} = \left\{d_{s_1^{(k)}}, d_{s_2^{(k)}}, \cdots, ds_{6}^{(k)}, d_{s_0^{(k)}}\right\} \tag{28}$$

Step 3: Calculate those uncertain semi-variograms

$$\gamma\left(d_{s_i^{(k)},s_j^{(k)}}\right),\ \forall d_{s_i^{(k)}} < d_{s_j^{(k)}} \quad \text{and}$$

$$\gamma\left(d_{s_0^{(k)},s_i^{(k)}}\right),\ \forall d_{s_0^{(k)}} < d_{s_i^{(k)}} \ (\text{or } \gamma\left(d_{s_i^{(k)},s_0^{(k)}}\right),\ \forall d_{s_i^{(k)}} < d_{s_0^{(k)}}) \tag{29}$$

in terms of

$$\gamma\left(d_{s_i^{(k)},s_j^{(k)}}\right) = \frac{1}{2}\left(d_{s_j^{(k)}} - d_{s_i^{(k)}}\right) - \delta_{d_{s_i^{(k)}},d_{s_j^{(k)}}-d_{s_i^{(k)}}} \tag{30}$$

with

$$\delta_{d_{s_i^{(k)}}, d_{s_j^{(k)}} - d_{s_i^{(k)}}} = \int_0^{\infty} \left(1 - \sup_{xy=r} \left(\Psi_{C_{d_{s_i^{(k)}}}}(x) \wedge \Psi_{C_{d_{s_j^{(k)}} - d_{s_i^{(k)}}}}(y) \right) \right) dr - \int_{-\infty}^{0} \sup_{xy=r} \left(\Psi_{C_{d_{s_i^{(k)}}}}(x) \wedge \Psi_{C_{d_{s_j^{(k)}} - d_{s_i^{(k)}}}}(y) \right) dr, \tag{31}$$

Step 4: Calculate $\left\{ \lambda_1^{(k)}, \cdots, \lambda_6^{(k)} \right\}$ according to equation (25).

Step 5: Calculate the kriging predictor

$$Z\left(s_0^{(k)}\right) = \sum_{i=1}^{n^{(k)}} \lambda_i^{(k)} Z\left(s_i^{(k)}\right) \tag{32}$$

The upper bound $1 \le n^{(k)} \le 6$, in case of $n^{(k)} < 6$, which implies some locations have the same habitat distance value, we simply put the average of total density values with the same habitat distance to participate the uncertain kriging predictor computations.

Step 6: Repeat **Step 2** to **Step 5**, until all 10 uncertain kriging predictors are calculated. Then stop.

Table 7 summarizes uncertain kriging predictor values, and lists the no total density observation sties (similar to Table 6).

site no.	Coordinates		Uncertain kriging predicted total density of trees
	longitude	latitude	
15	14.57582	-21.20986	132.446
14	14.63088	-21.26169	155.987
12	15.21586	-21.81786	177.641
17	15.51133	-23.30893	132.567
19	16.39317	-25.14609	99.266
33	16.77479	-27.94278	62.493
36	17.47540	-27.45209	117.045
10	17.78013	-25.87191	121.982
37	17.80606	-27.67034	72.249
128	21.10000	-28.95000	44.087

Table 6. Uncertain Kriging Predictions for 10 sites

Next, we will use uncertain kriging methods to predict the missing values for the population growth rate. There are 18 sites without values: Site number {120,111,110,125,22,23,24,26,123,124,121,122,17,34,126,128,127,42}

site no.	Coordinates		Habitat measure	Uncertain kriging Predicted population growth rate
	longitude	latitude		
120	18.83333	-30.7	19.91135	1.02490171
111	18.98567	-31.1058	1843.74	1.027150902
110	18.98883	-31.1005	21.50492	1.025065854
125	20.23333	-28.6167	19.91135	1.01815189
22	15.36674	-26.8837	53.33339	0.995053254
23	15.62385	-26.9958	17.09932	1.030192199
24	15.78774	-27.3154	21.50492	1.036620495
26	16.31455	-27.6489	14.9633	1.006659104
123	19.4	-28.9833	3.471558	1.001714497
121	19.43333	-28.8333	2.005269	1.004648936
124	19.43333	-29.0667	11.55349	1.014237773
122	19.43333	-29.0667	11.55349	1.014237773
17	15.51133	-23.3089	11.5535	0.999365585
34	16.77862	-27.6454	1843.74	1.000142274
126	21.1	-29.4	14.9633	1.02687582
128	21.1	-28.95	1.1112	1.014980439
127	21.16667	-29.0833	2.0582	1.005842705
42	22.83446	-29.8683	1.111231	1.003521691

Table 7. Uncertain Kriging Predictions for 18 sites

6. Climate change impacts

Now we are ready to calculate the missing values of 2008 total density of trees, which are 25 sites. We utilize the compound growth rate formula:

$$\kappa_{2008}^{(k)} = \kappa_{2002}^{(k)}\left(1 + r^{(k)}\right)^6 \tag{33}$$

Denote the Site k population growth rate during i years' period as $\rho_{(i)}^{(k)}$, then, we can use the compound growth formula,

$$\rho_{(i)}^{(k)} = \left(1 + r^{(k)}\right)^i \tag{34}$$

Then

$$r^{(k)} = \exp\left(\frac{1}{i}\ln\left(\rho_{(i)}^{(k)}\right)\right) - 1 \tag{35}$$

Therefore, if the total density $\kappa_{2002}^{(k)}$ is available, no matter it is observed or predicted in terms of uncertain kriging approach, $\kappa_{2008}^{(k)}$ will be determined.

After the calculations, we have complete data of $\kappa_{2008}^{(k)}$ and $\kappa_{2008}^{(k)}$ for site k, then The change in site k will be

$$\delta^{(k)} = \kappa_{2008}^{(k)} - \kappa_{2002}^{(k)} \quad (36)$$

The final uncertain predicted results together with the original sampled results of 2002 and 2008 are shown in the Fig. 3. The reason that they look exactly the same with absolutely no difference at all, is because the increase and decrease in total densities are so small, that they cannot be viewed just by looking at the intervals.

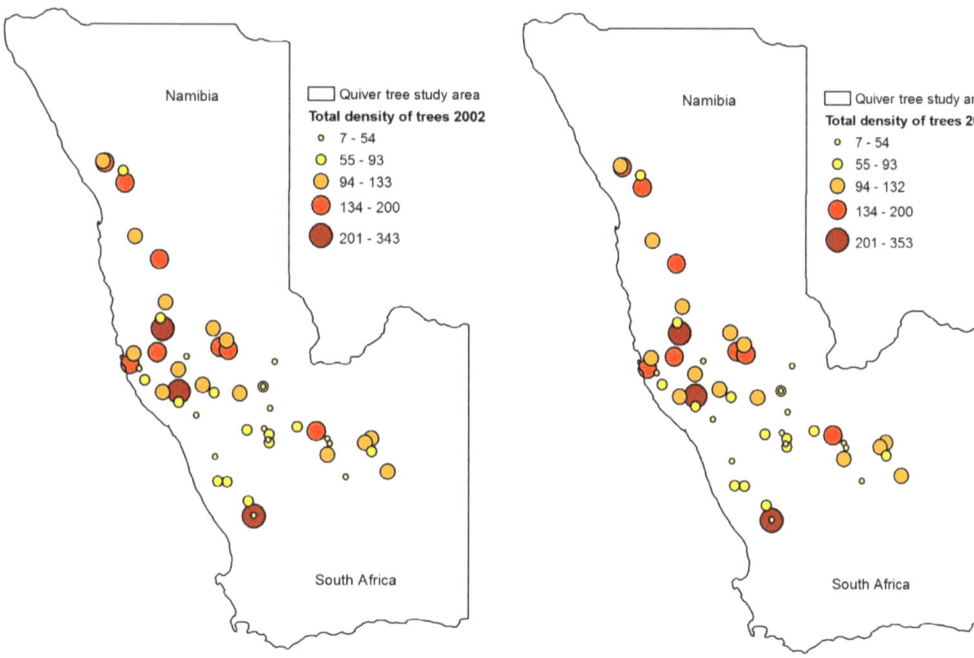

Fig. 3. Sampled and Predicted total density of Quiver trees in Namibia and South Africa, 2002 and 2008

In Fig. 4 the Ordinary kriging prediction maps of sampled and predicted total density of Quiver trees in 2002 and 2008 are shown. Since the population increase and decrease are very small, which is why they are only reflected using kriging maps. However, the kriging maps are slightly distorted in value, due to re-calculations of sample and predicted values, and the average mean values are used.

Table 8 and Fig. 5 show the change in the total density of Quiver trees over a 6 year period, between 2002 and 2008.

	Negative change ($\delta^{(k)} < 0$)	Positive change ($\delta^{(k)} > 0$)	unchanged ($\delta^{(k)} = 0$)
No. Of sites	7	18	28
Percentages	13.21%	33.96%	52.83%

Table 8. Climate change impact on total density of Quiver trees over 6 year period

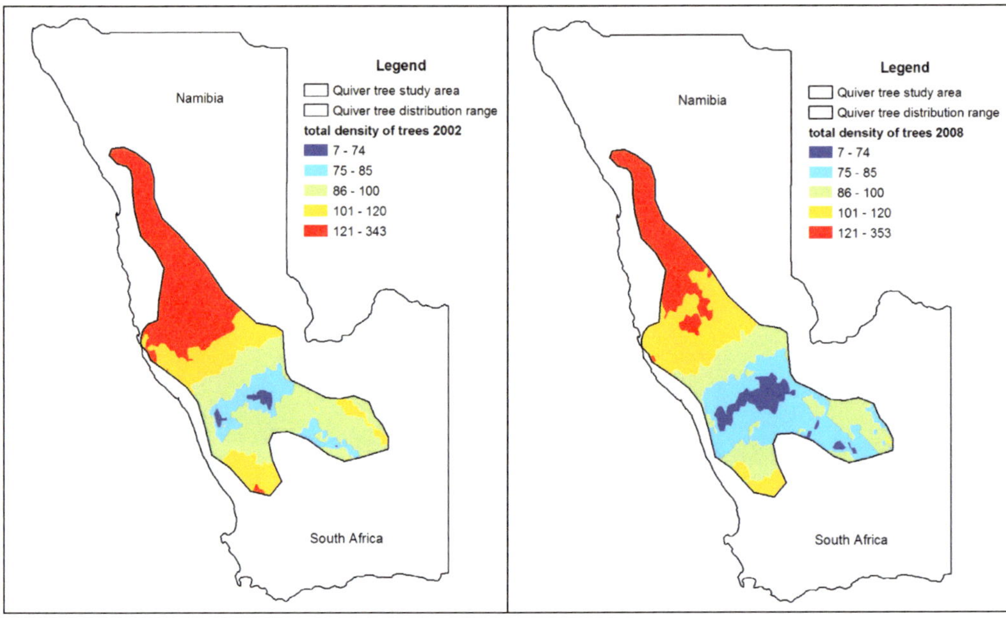

Fig. 4. Ordinary Kriging prediction maps of the total density of Quiver trees, 2002 and 2008

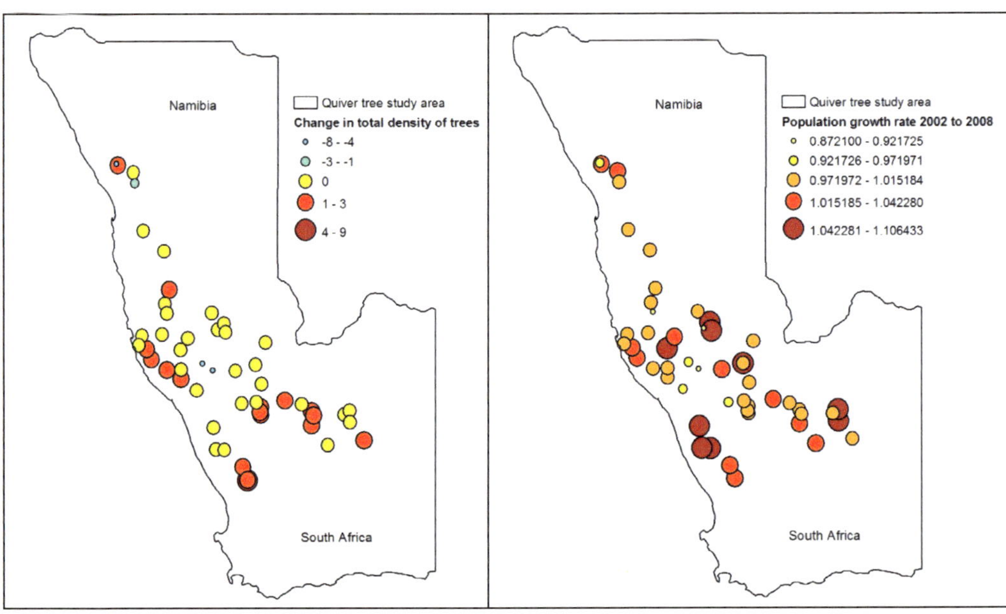

Fig. 5. Predicted change in total density of trees, and calculated population growth of Quiver trees, between 2002 and 2008

As one can see from the Fig. 5 change map, the change in total density of trees is very small. Most sample sites show no changes at all, with 18 sites show an increase of 1 to 9 trees, and 7 sites showing a decrease of 1 to 8 trees. These are small but definite changes in Quiver tree's total density over the 6 year period, between 2002 and 2008.

7. Conclusion

In this paper, we utilize the partially collected Quiver tree's sample data, 2002 and 2008 datasets, particularly, the total density of trees to evaluate the climate change impact on the Quiver tree's population. Because the data collection sites are only 53 and the Quiver tree's population size is extremely sensitive to it habitat environment, the conventional methodologies are not applicable. Based on our observations, we propose the habitat measure to catch up the closeness of the Quiver tree's population, which is later defining the habit distance of the tree's population. The habitat distance set is then regarded as the partner process index set. Therefore, we are able to engage Liu's Uncertainty theory for developing a new uncertain kriging approach to facilitate the prediction task: utilizing "complete" (53 values) data to perform the change evaluation. Once the missing values are filled, the evaluation of climate change can be performed. We are aware that our new prediction results have not had a validation process because of the time-constraint. In the future, it is necessary to engage the model accuracy and validation checking. Overall, there is a small but definite change in Quiver trees over the 6 year period, between 2002 and 2008, possibly due to climate changes over time.

Of course, monitoring the Quiver trees and using it as a climate change indicator, is a continuing process and effort for the South African National Biodiversity Institute. The results from this paper help to complete the missing data or un-sampled data, and would be useful for future comparisons, when another sample collection is made. The uncertain prediction methods and calculation process may be useful with other kinds of plant species data that displays similar problems, such as missing sample values. The predicted values together with the real sample values could be very useful in examining climate change impact over time and for studying the comparisons of plant species from different periods.

8. Acknowledgements

Sincere thanks to Wendy Foden and Graeme Ellis for their contribution to *Aloe dichotoma* data collections, and to the South African National Biodiversity Institute for their support of the Aloe Project. This book chapter is supported financially by the National Research Foundation of South Africa (Ref. No. IFR2009090800013) and (Ref. No. IFR2011040400096).

9. References

Bardossy, A.; Bogardi, I. & Kelly, E. (1990). Kriging with imprecise (fuzzy) variograms, I: Theory. *Mathematical Geology*, Vol. 22, pp. 63–79.

Cressie, N. (1991). *Statistics for Spatial Data.* Wiley-Interscience, John-Wiley & Sons Inc. New York.

Deng, J.L. (1984). *Grey dynamic modeling and its application in long-term prediction of food productions.* Exploration of Nature, Vol. 3, No. 3, pp. 7-43.

Draper, N. & Smith, H. (1981). *Applied Regression Analysis.* 2nd Edition. John Wiley & Sons, Inc. New York.

Foden, W. (2002). *A Demographic Study of Aloe dichotoma in the Succulent Karoo: Are the Effects of Climate Change Already Apparent?* MSc Thesis, Percy FitzPatrick Institute of African Ornithology, Botany Department, University of Cape Town.

Guo, D. (2010). *Contributions to Spatial Uncertainty Modelling in GIS*. Lambert Academic Publishing (online.lap-publishing.com). ISBN 978-3-8433-7388-3

Guo, D.; Guo, R.; Midgley, G.F. & Rebelo, A.G. (2007). PDEMR Modelling of Protea Species in the Population Size of 1 to 10, in Cape Floristic Region from 1992 to 2002, South Africa. *Journal of Geographic Information Sciences*, Vol. 12, No. 2, pp. 67-78.

Midgley, G.F.; Altwegg, R.; Guo, D. & Merow, C. (2009). *Are Quiver Trees a Sentinel for Climate Change in Arid Southern Africa?* South African National Biodiversity Institute. ISBN: 978-0-620-43639-7

Kolmogorov, A.N. (1950) *Foundations of the Theory of Probability*. Translated by Nathan Morrison. Chelsea, New York.

Liu, B.D. (2007). *Uncertainty Theory: An Introduction to Its Axiomatic Foundations*. 2nd Edition. Springer-Verlag Heidelberg, Berlin.

Liu, B.D. (2010). *Uncertainty Theory: A Branch of Mathematics of Modelling Human Uncertainty*. Springer-Verlag, Berlin.

Liu, B.D. (2011). *Uncertainty Theory*, 4th Edition, 17 February, 2011 drafted version.

Liu, S.F. & Lin, Y. (2006). *Grey Information*. Springer-Verlag, London.

Mase, S. (2011). GeoStatistics and Kriging Predictors, In: *International Encyclopedia of Statistical Science*. Editor: Miodrag Lovric, 1st Edition, 2011, LVIII, pp. 609-612, Springer.

Moore, R.E. (1966). *Interval Analysis*. Prentice-Hall, Englewood Cliff, NJ. ISBN 0-13-476853-1

Pawlak, Z. (1982). Rough Sets. *International Journal of Computer and Information Sciences*, Vol. 11, pp. 341-356.

Utkin, L.V. & Gurov, S.V. (1998). New reliability models on the basis of the theory of imprecise probabilities. *Proceedings of the 5th International Conference on Soft Computing and Information Intelligent Systems*, Vol. 2, pp. 656-659.

Utkin, L.V. & Gurov, S.V. (2000). New Reliability Models Based on Imprecise Probabilities. *Advanced Signal Processing Technology, Soft Computing*. Fuzzy Logic Systems Institute (FLSI) Soft Computing Series - Vol. 1, pp. 110-139, Charles Hsu (editor). Publisher, World Scientific. November 2000. ISBN 9789812792105

Van Wyk, B.E. & Smith, G. (1996). *Guide to Aloes of South Africa*. Pretoria, Briza Publications.

Walley, P. (1991). *Statistical Reasoning with Imprecise Probabilities*. London: Chapman and Hall. ISBN 0412286602

Zadeh, L. A. (1965). Fuzzy sets. *Information and Control*, Vol. 8, pp. 338-353.

Zadeh, L. A. (1978). Fuzzy sets as a basis for a theory of possibility. *Fuzzy Sets and Systems*, Vol. 1, pp. 3-28.

17

Changes in the Composition of a Theoretical Freshwater Ecosystem Under Disturbances

Ágota Drégelyi-Kiss[1] and Levente Hufnagel[2]
[1]*Bánki Donát Faculty of Mechanical and Safety Engineering, Óbuda University*
[2]*Adaptation to Climate Change Research Group of Hungarian Academy of Sciences*
Hungary

1. Introduction

Direct and indirect effects of the climate change on the terrestrial and oceanic ecosystems could also be observed in the last decades. Researches warn that there are significant changes in phenological, morphological and physiological properties of taxa and changes in spread of species, in frequency of epidemics. But the potential effects of climate change on natural ecosystems and the answers have been given indicated by the living communities which are less than the known of the complex natural ecosystems.

There is a dynamical equilibrium between the climate and ecosystems at natural systems. If the system is affected by anything, then there will be a response to maintain the equilibrium. This process can be a sudden or a gradual effect. Seems, nowadays the unpredictable, sudden changes will be the significant.

Our goal is to analyse the effect of some temperature-climate patterns on the production and community ecological relations in a strongly simplified theoretical model. This elaborated Theoretical Ecosystem Growth Model (TEGM) works as a freshwater algae ecosystem. The novelty of this modelling is characterized by a guild-specific approach at first (where competitive relationships can be manifested); on the other hand the population-dynamic model has been connected with the outputs of global circulation models. So this connection enables us to examine directly the effects of climate change.

Our expectations towards the climate-ecosystem model (TEGM) were as follows:

Depending on the adjusted constant temperature value the species which have optimum reproduction rate in that temperature let be with the largest number of specimens.

Increasing the daily random fluctuation species with narrow adaptation ability are extruded by the species with wide adaptation ability.

The diversity of the ecosystem with the increase of disturbances, to let change it according to the maximum curve, which refers the presence of Intermediate Disturbance Hypothesis.

In our earlier researches the distribution of the algae community of a theoretical freshwater ecosystem is examined by changing the temperature. The temperature was changed according to plan in order to estimate the various effects separately. The examined temperature patterns are as follows: constant temperature (293K, 294K, and 295K), the

temperature changes as a sine function over the year and historical and future climate patterns. (Drégelyi-Kiss & Hufnagel, 2009, 2010, Hufnagel et al., 2010)
In this work it was examined how the theoretical ecosystem growth model (TEGM) reaches the equilibrium in case of three signal such as unit impulse, unit step and unit ramp. The response of the theoretical ecosystem is analysed in case of these signals as disturbances. The daily random fluctuation is also examined on the basis of simulated ±1…±11 K random numbers.

2. Literature overview

The latest IPCC report (Fischlin et al., 2007) points out that a rise of 1.5-2.5 ^{0}C in global average temperature causes relevant changes in the structure and functioning of ecosystems, primarily with negative consequences for the biodiversity and goods and services of the ecological systems.
There are several consequences of the decrease in biodiversity. The most scenic is the decrease in the number of species. Secondly the decrease in genetic diversity has to be mentioned, there are a lot of cases where stands of the frequent species decrease. At the third case the contents of ecosystems change also, the various habitats allow of being and maintenance of creatures between different geological and climatic conditions. This kind of role of ecosystems is less known (Nechay 2002).
A natural system has a dynamic equilibrium between the climate and ecological systems. If the ecosystem is affected, then a response starts in order to keep the equilibrium. The degree of this reaction can be a sudden response by leaps or on the other hand gradual. Some variables such as the phenological properties follow the changing climatic conditions simply; in these cases gradual shifts could be expected (Fitter et al. 2002). In case of sudden responses there is a good example in the maritime tidal zone where the community significantly alters under small-scale temperature increase, which is caused by drastic decrease in the number of dominant predators (Sanford 1999).
According to the forecasts the probability of extreme weather conditions, the effects will be significant for the further occurrence of sudden effects. There are some quick extreme events and the given sudden responses behind the events which seemed to have experienced gradual changes (Easterling et al., 2000). In case of climate change this is not about the shift of the system being in equilibrium, but the succession could break or unhoped-for steps occur. In a mediterranean scrub regenerated after a fire the number of species does not change under artificial drought-treating while the number of species increases in the control parcel fluently (Penuelas et al., 2007). The reaction of the run-down, degrading and regenerated communities for the climate change differs significantly from the reaction of natural ecosystems. These processes are important because there are much more of these areas like natural.
The interpretation of the phenomena of disturbance has been changed with the development of the science. Earlier it is stated that the disturbance is a deviation from the equilibrium circumstances, nowadays it is as important factor to maintain the ecological integrity of the ecosystems. There are several definitions for the disturbance. (Laska, 2001) According to Grime (1973) the disturbance is such an event where the amount of the biomass decreases. It could also be stated that the disturbance differentiates in time, disturbs the life of a community, population or ecosystem, and changes the usability of the resources, environmental factors (Pickett & Parker, 1997). Summarizing it could be said that the

disturbance does not mean decrease in biomass in all cases, this is a phenomenon well-bordered in time which results dynamic patch-pattern.

The type and the intensity of the disturbance affect the succession processes therefore seconder succession starts mostly (flood, hurricane, natural catastrophes). This is because certain elements of the original community could maintain in contrast with primer successions (such as volcano explosion) (Dobson et al., 1997). The spread and the amount of the disturbance could be very different which is affected by the heterogeneity of the environment. It is stated that the processes are usually unique and related to a given area (Pickett & Parker, 1997).

After the perturbation there is usually fast succession, where the surviving species and the members of the original community participate. Complex relations are developed to adapt to their new environment, biotic and abiotic factors (MacMahon, 1998). The rhythm of the changes slows down, and then the habitat gets into quasi equilibrium through continuous adaptation.

According to the Gause's Law of competitive exclusion (Gause, 1934; Hardin, 1960) the number of the limiting factors restrict the number of species coexisted which controls the composition of the communities and populations. The competition could not be maintained in the long run. There could be three types of processes, such as the more vulnerable species disappear, adapt or drift toward other environmental factors. There is stable state if every species are restricted by different environmental factors. If more species make a competition for the same resources, then the genre with the best adaptability will exclude the others, so the succession process tends to a climax state with small diversity.

The limiting resources are different in case of various living beings. The increase of the plants is restricted by nitrogen, phosphorus, soil humidity (in case of terrestrial plants), sunlight and other biologically important elements. For example, in case of phytoplankton in temperate zone it is rare that there is more than three restriction parameter at the same time. So the phytoplankton community has to tend to his equilibrium with 1-3 dominant species during the succession process by competitive exclusion principle. But it is observed that the phytoplankton communities have much more species than expected. This phenomenon is known as "plankton paradox". It could be explained that the boundary conditions (e.g. continuous changes in environment, sunlight, turbulence) change faster than the competitive exclusion may occur. (Hutchinson, 1957). Summarizing the theory and the observations do not agree with each other. (Padisák, 1998)

Some researchers state that the ecological and environmental factors are in continuous interaction, the habitats of plankton do not reach the equilibrium state where only one genre is dominant. (Scheffer, 2006).

There are several hypotheses where the relationship between the disturbance and diversity are examined (Magura et al., 2006). The most spread theory is the Intermediate Disturbance Hypothesis (IDH). It is stated that the diversity increases in case of small or moderate disturbances (Connell, 1978; Grime, 1973). According to Increasing Disturbance Theory the smallest diversity exists in strongly disturbed areas (Gray, 1989). The Habitat-Specialist Hypothesis states that the diversity of living beings of original habitats decreases with reaching the strongly disturbed areas (Magura et al., 2004).

The species richness in tropical forests as well as that of the atolls is unsurpassable, and the question arises why the theory of competitive exclusion does not prevail here. Trees often fall and perish in tropical rainforests due to storms and landslide, and corals often perish as a result of freshwater circulation and predation. It can be said with good reason that

disturbances of various quality and intensity appear several times in the life of the above mentioned communities, therefore these communities cannot reach the state of equilibrium. The Intermediate Disturbance Hypothesis (IDH) (Connell, 1978) is based on this observation and states the following:
In case of no disturbance the number of the surviving species decreases to minimum due to competitive exclusion.
In case of large disturbance only pioneers are able to grow after the specific disturbance events.
If the frequency and the intensity of the disturbance are medium, there is a bigger chance to affect the community.
There are some great examples of IDH in the case of phytoplankton communities in natural waters (Haffner et al., 1980; Sommer 1995; Viner & Kemp, 1983, Padisák 1998, Olrik & Nauwerck, 1993). Nowadays it is accepted that diversity is the biggest in the second and the third generations after the disturbance event (Reynolds, 2006).

3. Material and methods

3.1 The elaborated Theoretical Ecosystem Growth Model (TEGM)

An algae community consisting of 33 species in a freshwater ecosystem was modelled (TEGM; Drégelyi-Kiss & Hufnagel, 2009, 2010). During the examinations the behaviour of a theoretical ecosystem was studied by changing the temperature variously. Several author draw attention to the temperature as main control factor in case of freshwater ecosystems (Christou & Moraitou-Apostolopoulou, 1995; Iguchi 2004; Dippner et al. 2000, Vadadi-Fülöp et al., 2009).
The conceptual diagram of the TEGM model (Fig. 1) describes the mathematical calculations during the modelling process (Sipkay et al., 2010). The model has two important input parameters. One is the various reproductive functions; the other is functions of the temperature patterns.
Rivalry begins among the species with the change of temperature. In every temperature interval, there are dominant species which win the competition. The increase of the population is not infinite because of the restrictive function of the model. The ecosystem reaches a dynamic equilibrium state for an input temperature. In the course of this equilibrium the following output parameters are examined: the dominant species and their numbers, the value of use of resources, the diversity of the ecosystem and the duration of reaching the equilibrium.
Algae species are characterised by the temperature interval in which they are able to reproduce. This reproductive feature depends on their temperature sensitivity. There are four types of species based on their sensitivity: super generalists (SG0, SG1), generalists (G1-G5), transitional species (T1-T9) and specialists (S1-S14). The temperature-optimum curve originates from the normal (Gaussian) distribution, where the expected value is the temperature optimum (Drégelyi-Kiss & Hufnagel, 2010). The used temperature range for the optimum values is from 277K to 301K, and the lay of the optimum curves is symmetrical in this range.
The restrictive value of reaching the sunlight (K_k) was set to 10^7 value in the first phase of the simulation studies (TEGMa model), in the second phase the intensity of the sunlight changing during a year was considered (TEGMb model):

$$K_k = d_1 \cdot \sin(d_2 \cdot k + d_3) + d_4 + \varepsilon \tag{1}$$

where d_1=4950000, d_2=0,0172, d_3=1,4045, d_4=5049998, ε: has uniform distribution in the interval of (-50000,50000).

The constant values of the K_j restrictive function is set in a way where the period of the function is 365.25, the maximum place is on 23rd June and the minimum place is on 22nd December. (These are the most and the least sunny days.)

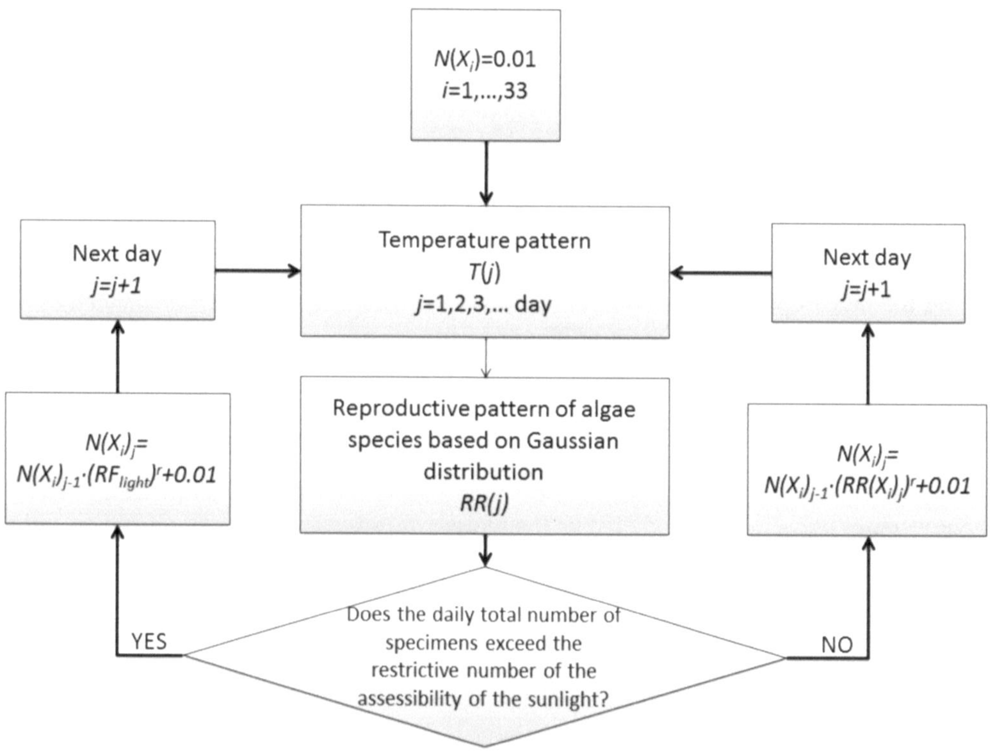

Fig. 1. Conceptual diagram of the TEGM model (RR: reproduction rate, RF: restriction function related to the accessibility of the sunlight, N(Xi): the number of the ith algae species, r: velocity parameter)

3.2 Special functions

The dynamic state of a system depends on the stationary state before the disturbance and the type and amount of the perturbation. Under special signals the composition of the theoretical ecosystem changes and new equilibrium state evolves. It is important to examine the way of reaching the new balance regarding the stability of the system.

The reaching of the equilibrium state of the theoretical ecosystem model was examined by three special functions. One of them is the Dirac delta function, which can be modelled as a large change in temperature lasted small time. The other is the step function which modelling the remaining significant change in temperature. The third one represents the slowly increasing temperature day by day. (Pokorádi, 2008)

Dirac-delta means an impulse, which has zero value always except for one moment, when it takes infinite large value:

$$\int_{-\infty}^{+\infty} \delta(t)dt = 1 \tag{2}$$

The unit impulse-like signal is modelled by an increase in the temperature in a day which throws the ecosystem off balance. The theoretical ecosystem is in equilibrium at constant temperature (293K, 294K and 295K), then on the 1001th day of the simulation the temperature has a sudden higher value, and the next day of the simulation the temperature pattern sets back the constant temperature which had before the disturbance. The magnitude of the temperature impulse is between 1K and 100K values.
The unit step function (Heaviside function) could be understand as "power-on" phenomena:

$$1(t) = \begin{cases} 0, & t < 0 \\ 1, & t > 0 \end{cases} \tag{3}$$

During the unit step-like examination the value of constant temperature is changed on the 1001th day of the simulation with 1K or 2K temperature up or down.
The unit ramp function could be described by the following equation:

$$t1(t) = \begin{cases} 0, & t < 0 \\ t, & t > 0 \end{cases} \tag{4}$$

The unit ramp-like study is modelled as the temperature increases slowly with consecutive days through 10 years. The following cases are examined:
T = 294 K - 294.365 K (the gradient of temperature is 0.0001 K/day)
T = 294 K - 297.652 K (the gradient of temperature is 0.001 K/day)
T = 268 K - 286.26 K (the gradient of temperature is 0.005 K/day)
T = 268 K - 304.52 K (the gradient of temperature is 0.01 K/day)
It is important to study the effects of daily temperature fluctuations. This was modelled as the disturbance has a uniform distribution (between ±1K… ±11K). During the simulation the given random fluctuation on the constant and increasing temperature pattern was analysed.

4. Results

4.1 Examination of impulse unit

The theoretical ecosystem is in equilibrium at constant temperature (293K, 294K and 295K), then on the 1001th day of the simulation the temperature has a sudden higher value, and the next day of the simulation the temperature pattern sets back the constant temperature which had before the disturbance. The magnitude of the temperature impulse is between 1K and 100K values.
Making the impulse unit-like simulations with TEGMa model the time of reaching the new equilibrium state depends on the magnitude of the given temperature impulse (1-100K) (Table 1). The first row of the table shows the duration of reaching the equilibrium at the beginning of the simulation. This time is 32 days and 34 days in case of faster ecosystem depending on the used constant temperature pattern. The ecosystem which has smaller reproducibility (r=0.1) the time is 151 days and 187 days, respectively.

$T_{impulse}$	*r*=1			*r*=0.1		
	T=293K	*T*=294K	*T*=295K	*T*=293K	*T*=294K	*T*=295K
at the beginning	32 days	34 days	32 days	151 days	187 days	151 days
1 K	0 day	21 days	0 day	0 day	36 days	0 day
3 K	23 days	21 days	24 days	51 days	50 days	51 days
5 K	28 days	18 days	22 days	62 days	59 days	62 days
10 K	32 days	31 days	32 days	88 days	82 days	89 days
15 K	32 days	34 days	32 days	114 days	103 days	132 days
20 K	32 days	34 days	32 days	136 days	131 days	151 days
30 K	32 days	34 days	32 days	151 days	183 days	151 days
100 K	32 days	34 days	32 days	151 days	187 days	151 days

Table 1. The time of reaching the equilibrium at various temperatures at the beginning and on the effect of the $T_{impulse}$ temperature on the 1001th day of the simulation study (different velocity parameters, TEGMa model)There are two groups according to the setting temperature.

The simulations made at 293K and 295K with r=1 parameter are similar, the duration of reaching the equilibrium is the longest in case of 10K or larger temperature impulse. In case of slower ecosystem 20K and 30K impulse is essential to get the desired time. The other group is related to the simulations made at 294K. During these simulations there must be more time to reach the new evolved equilibrium state in every case.

On the effect of small temperature impulse there are not changes in the composition of the theoretical ecosystem. In case of moderate impulses (3K, 5K, 10K, 15K) it is stated that the distribution of the species are not the same before and after the interference. At 293K and 295K the transient (T7) extrudes the specialist (S13 and S14, respectively), and the number of specimens of generalist (G4) and supergeneralist (SG1) increase. For example at 293K temperature in case of 5K impulse on the 1001th day of the simulation with TEGMb, r=1, it requires almost 30 years to reach the equilibrium state had been before the 1000th day (Figure 2).

There are similar patterns in the change of the composition of the theoretical ecosystem at 294 K temperature, also. If the system is affected by 3K impulse in case of r=1, and 3K or 5K impulse in case of r=0.1, respectively, then the specialist (S14) which has higher optimum temperature for the reproduction extrudes the other one (S13), while they share the resources on a fifty-fifty way before the interference. To take into account that the composition of the theoretical ecosystem is totally symmetric the temperature impulses are examined toward the lower temperature value (i.e. negative impulses), also. In case of -3K temperature impulse the S13 genre is the dominant. In case of faster ecosystem on the effect of 5K temperature impulse and with r=0.1 and 10K impulse parameters the T7 transient genre wins the competition, the productivity of the generalist increases.

Comparing the faster and the slower ecosystems it is stated that the change in the composition of the equilibrium state is similar, in case of *r*=1 and 3K impulse; and in case of *r*=0.1 and 10 K temperature impulse. If 15 K or more impulse is given in case of faster ecosystem, then the composition of the species are the same before and after the interference. (Fig. 3-4)

It is important to study how the diversity and the adaptability work out in the course of smaller impulses. The distribution of species and the diversity of the theoretical ecosystem

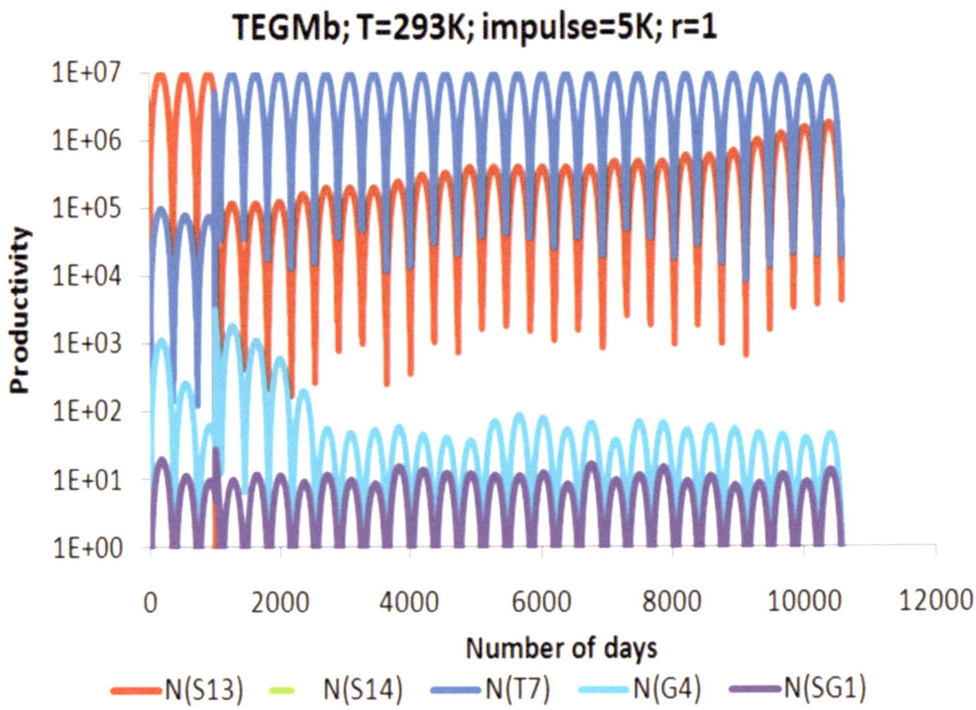

Fig. 1. The results of TEGMb simulation in case of the given 5K temperature impulse with r=1 velocity parameter

strongly depend on the setting value of the velocity parameter. The diversity of the faster ecosystem increases with smaller temperature impulses (3K, 5K) during TEGMa simulation. The diversity continuously increases during the 30 years of simulation in case of 5 K impulse and decreases in case of 3 K impulse using the TEGMb model. There are an increase in the diversity value in the course of larger impulses (10K, 15K, and 20K) for the slower ecosystem in both, TEGMa and TEGMb cases. The evolving time of the new equilibrium state is the slower where the simulation has r=0.1, T=294 parameters.
Summarizing it is stated that the new equilibrium state evolved on the effects of small and medium temperature impulses differs significantly from the state before the interference.

4.2 Examination of step unit

During the unit step-like examination the value of constant temperature is changed on the 1001th day of the simulation with 1K or 2K temperature up or down.
The value of the constant temperature function (293K, 294K, and 295K) is changed at the 1001th day of the simulation. At 293 K temperature the conditions are suited for the S13, K7, G4 and SG1 species optimum reproducibility. In this temperature the composition does not change in case of 1K temperature step. (Table 2)

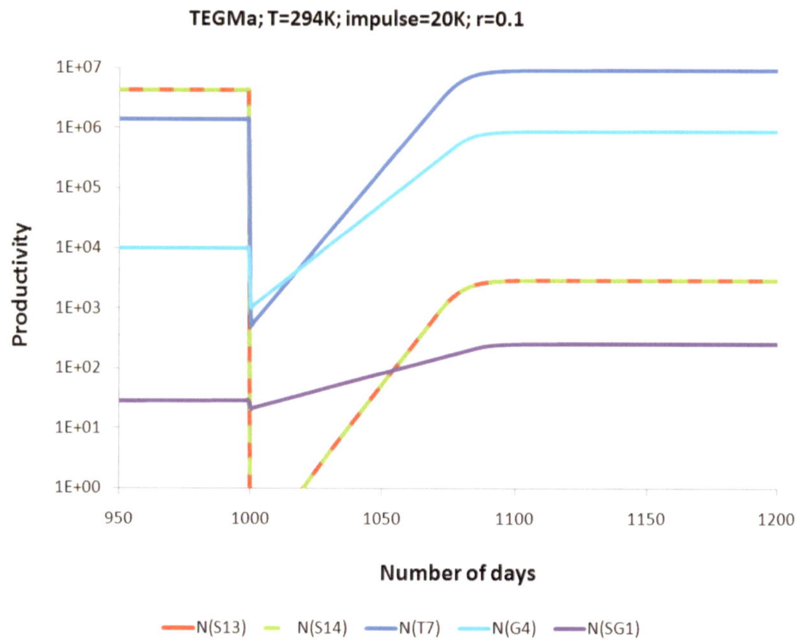

Fig. 3. The effect of 20 K temperature impulse on the 1001th day of the simulation (in case of *r*=0.1, TEGMa model, T=294K)

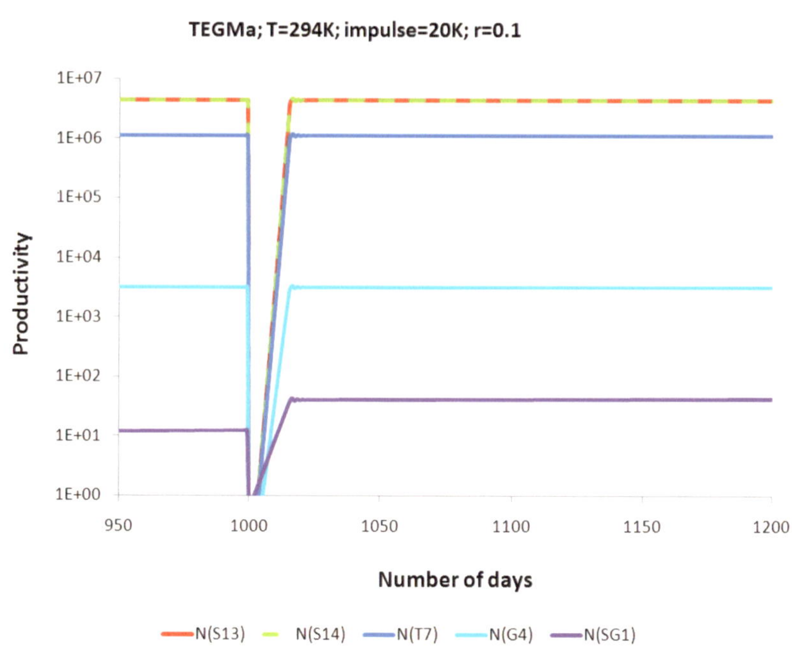

Fig. 4. The effect of 20 K temperature impulse on the 1001th day of the simulation (in case of *r*=1, TEGMa model, T=294K)

r	T1	T2	Duration of reaching the equilibrium state	Speciality after the unit step
1	293	294	-	the composition does not change
0.1	293	294	-	the composition does not change
1	294	293	20 days (and S14 disappears)	S14 disappears, S13 increases
0.1	294	293	200 days (and S14 disappears)	S14 disappears, S13 increases
1	293	295	10 days (S13 disappears in 20 days)	T7 win the competition, not the specialist as could be expected
0.1	293	295	80 days (S13 disappears after 200 days)	
1	295	293	10 days (S14 disappears in 20 days)	
0.1	295	293	80 days (S14 disappears after 200 days)	

Table 2. Examination of T1 →T2 unit step in case of ecosystems having different velocity parameters (TEGMa model)

On the effect of +1 K temperature ramp at 294 K constant temperature the composition of equilibrium does not change significantly in case of both the slower or the faster ecosystem (S14 appears with 10 number of specimens). In case of 294K→293K change the S13 specialist win the competition in 20 days, as expected.
In can be seen in Figure 5 that the productivity decreases strongly in case of 2K change. The slower ecosystem reaches the equilibrium later than the faster type. There is no noticeable change in productivity in cases of 1K temperature step.

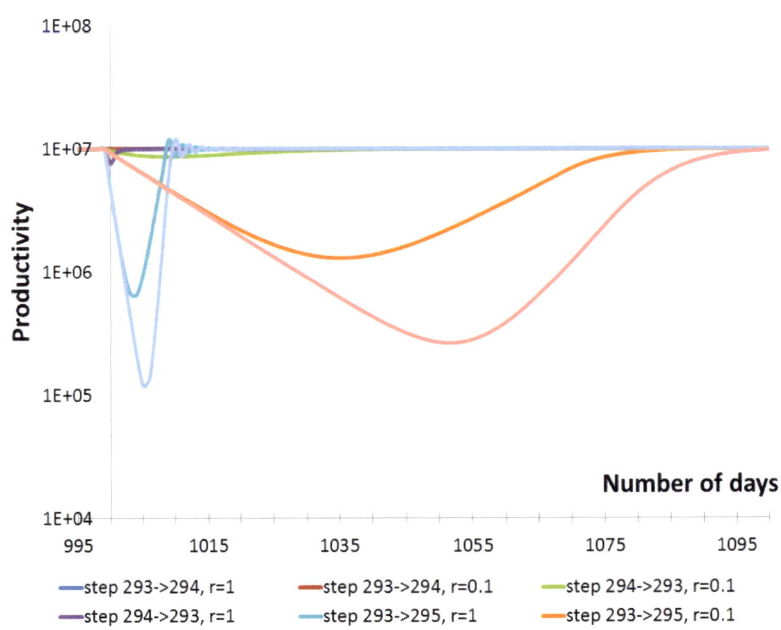

Fig. 5. The number of days of simulation versus the productivity of the theoretical ecosystem due to temperature step signal (TEGMa)

Observing the Shannon diversity values of the ecosystem it is stated (Figure 6) that the diversity value, which belongs to the new equilibrium state, moves through a local maximum value depending on the temperature.
Summarizing the effect of temperature step it is stated that the composition of the ecosystem being equilibrium at base temperature determines, what kind of the diversity of the ecosystem will be.

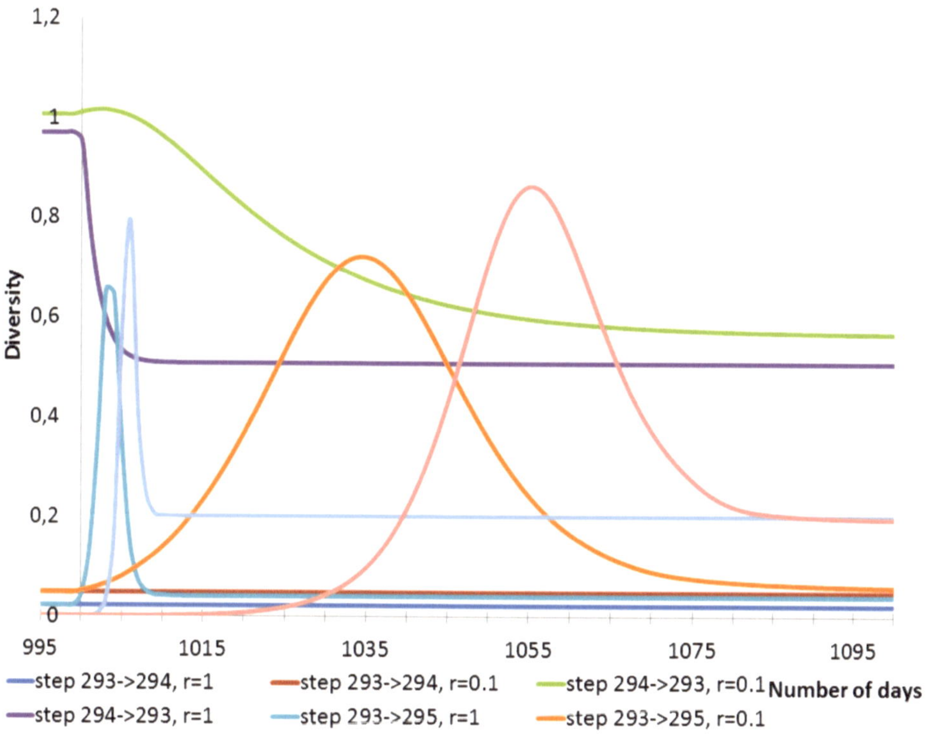

Fig. 6. Diversity of the theoretical ecosystem versus the number of days of simulation due to temperature step (TEGMa)

4.3 Examination of ramp temperature function

During ramp temperature function the value was daily increased from 268K continuously with various amounts (0.0001K...0.01K). It can be seen the appearance of some species depending on gradient. The local maximum values of the diversity exist where the specialist and the generalists have just exchanged with each other. (Figure 7-8.)

4.4 Daily random fluctuation

The daily random fluctuation was modelled as the disturbance has a uniform distribution (between ±1K... ±11K). In case of constant temperature pattern the results of the simulation study can be seen in Fig. 8, which is the part of the examinations where the random fluctuations were changed until ± 11K. The number of specimens in the community is

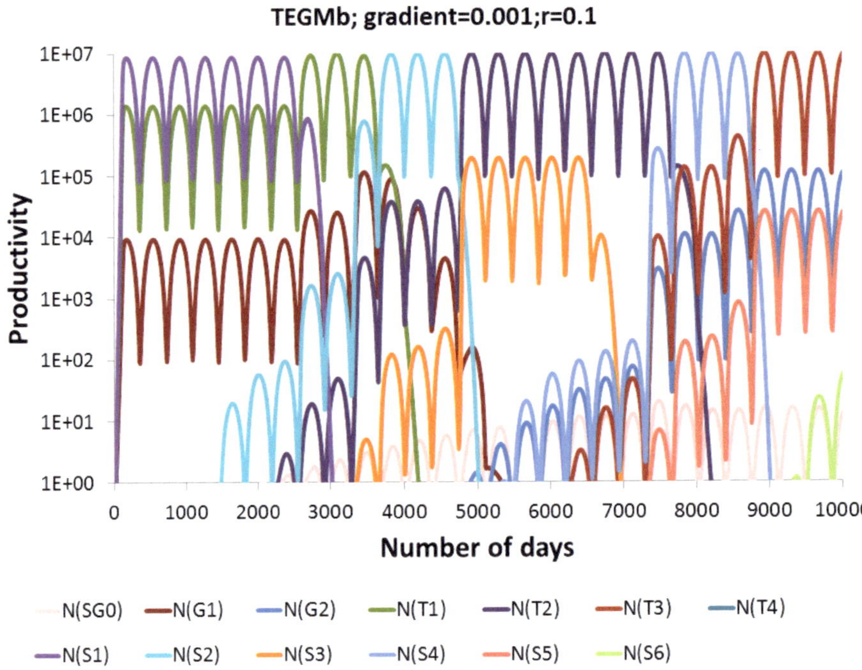

Fig. 7. The productivity results of TEGMb simulation with r=0.1 parameter in case of increasing temperature pattern (gradient=0.001K/day, T_0=268K)

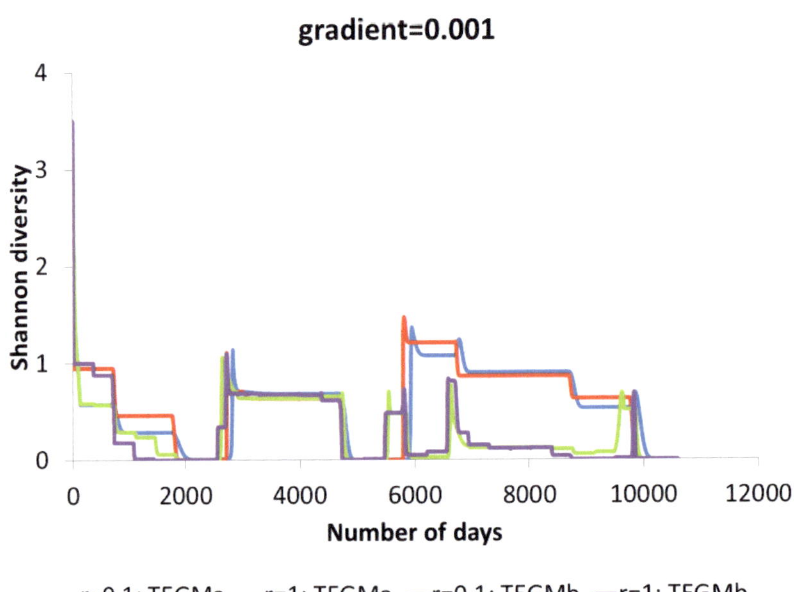

Fig. 8. The daily diversity value with TEGMb simulation and r=0.1 parameter in case of increasing temperature pattern (gradient=0.001K/day, T_0=268K)

permanent and maximum until the daily random fluctuation values are between 0 and ±2K. Significant decrease in the number of specimens depends on the velocity factor of the ecosystem which has faster reproductive ability shows lower local maximum values than the slower system in the experiments. The degree of the diversity is greater in the case of r=0.1 velocity factor than in the case of the faster system. If there is no disturbance, the largest diversity can be presented found at 294 K for both speed values. If the fluctuation is between ± 6K and ± 9K, the diversity values are nearly equally low. In case of the biggest variation (± 11K) the degree of the diversity increases strongly.

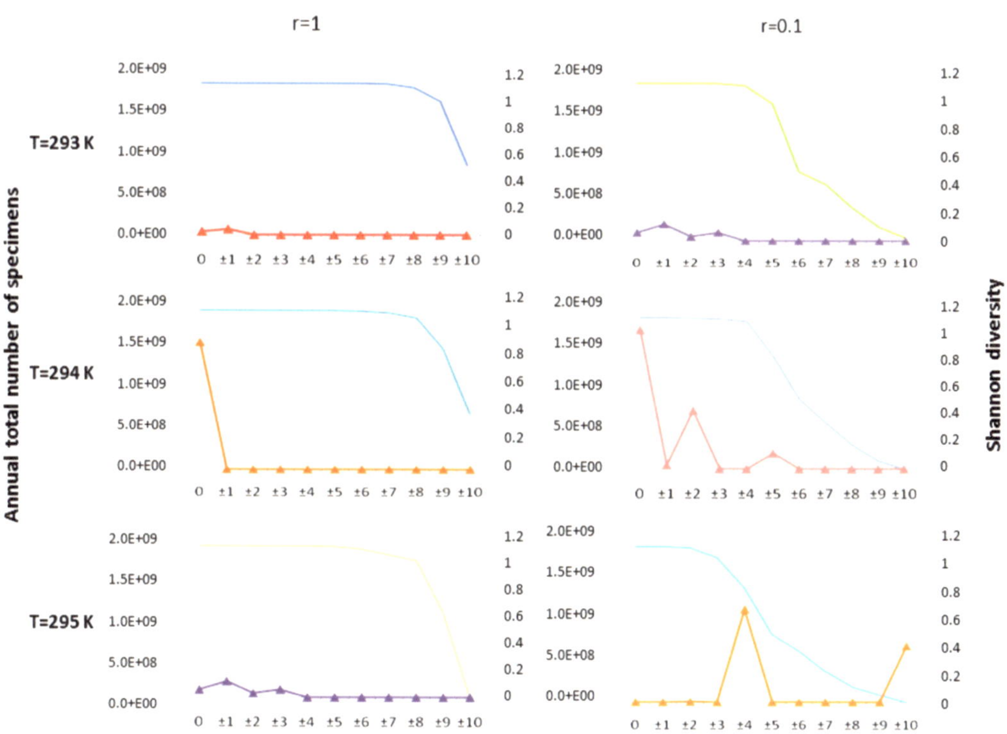

Fig. 9. The change in productivity and diversity value of the theoretical ecosystem in the function of daily random fluctuation in constant temperature environment with TEGMb model (The signed plots show the diversity values.)

At linear increasing temperature pattern the value of the use of resources does not decrease to zero value in case or larger gradients (0.005K/day and 0.01 K/day) and large random daily fluctuation (±7K). This decrease happens in case of smaller gradients. This is because the supergeneralists are less sensitive for the daily random fluctuation. This does not appear in case of slower ecosystems. (Figure 10)

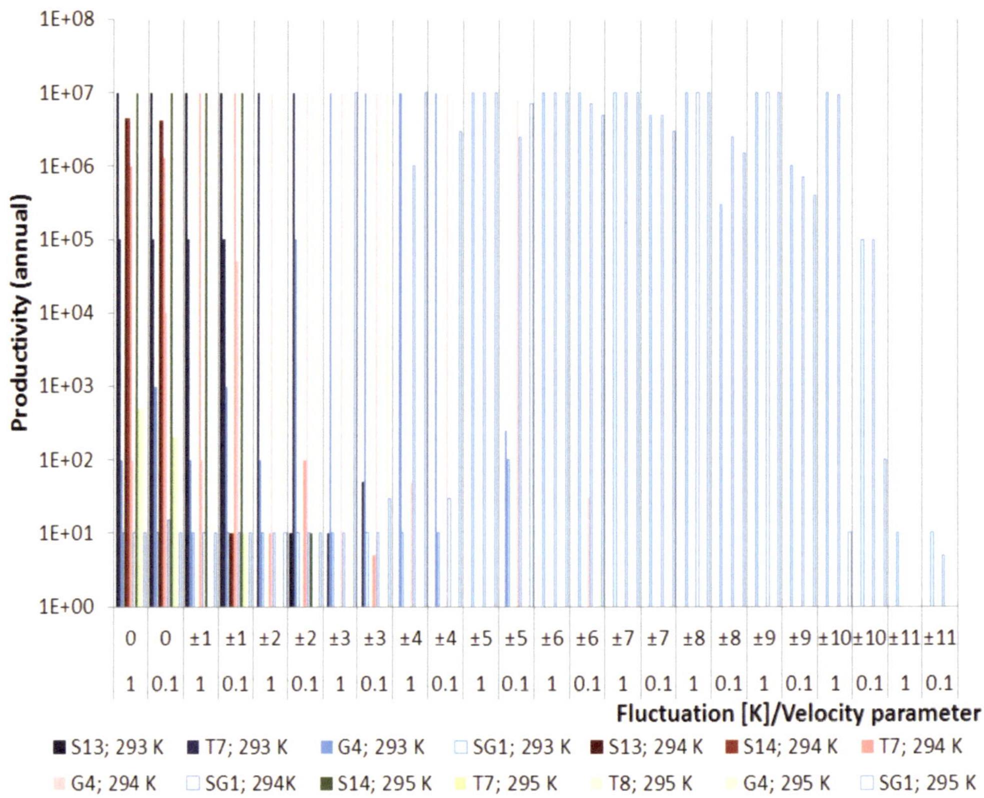

Fig. 10. Annual maximum number of specimens of the species on constant temperature with different velocity parameters (logarithmic representation) (Drégelyi-Kiss & Hufnagel, 2010)

5. Conclusion

The theoretical freshwater algae community reacts on the effect of temperature impulse in different ways depending on the magnitude of the impulse (1K-100K). If the ecosystem is affected by small or medium temperature impulses, then the composition of the evolving equilibrium state is differs from what could be expected. 1K impulse in temperature does not change the composition of the ecosystem in cases of 293K and 295K base-temperatures. So the specialists, as S13 and S14 are the strongest species during the competition. In 294K in case of 1K impulse the S14 specialist extruded the other species in the competition. In 294 K temperature (which is not an optimum value for the reproduction in either species) the given small or medium impulse results that the specialist wins the competition which has optimum closer to the temperature evolved during the impulse. Under small or medium temperature impulse the evolved new equilibrium differs significantly from the status before.

If the ecosystem being in equilibrium state at constant temperature is affected by 2K temperature step the productivity of specialists decreases fast, and the transient genre becomes the winner during the competition. In case of TEGMa the balance exists through 25 years but there is a slow exchange in species during the years under TEGMb simulations. This point out that the generalists, which are able to spread well, could exploit the changing habitat conditions (Dukes & Mooney, 1999). To analyse the results of the unit step function it is found that the composition of the ecosystem determines the diversity of the evolving new balance.

If the temperature changes linearly and slowly in time, then there is a competition between the specialists and the transient species during the days of simulation. The largest diversity values can be observed when the species has just exchanged with each other. Such a thing was observed by Sanford (1999), who examined the ecological system of starfish and shellfish with in-field and lab experiments. To analyse the results of the unit step function it is found that the composition of the ecosystem determines the diversity of the evolving new balance.

Increasing the daily fluctuation the community tries to adapt to its environment and many genre compete for the environmental elements with large number of specimens. In case of large noise the diversity is large and the annual total number of specimens is low, because a few genres could adapt to the environmental conditions. It is stated that the specialists reproducing in narrow temperature interval are dominant species in case of constant or slowly changing temperature pattern but these species disappear under small fluctuation in the temperature. As a result it is found, that species with narrow adaptation ability disappear, species with wide adaptation ability become dominant and the biodiversity decreases. The results of the simulations show that the way towards the equilibrium is different in cases of various disturbances. The composition of the ecosystem in equilibrium at a given time affects the evolved new equilibrium under disturbances.

The strategic model, so-called "TEGM" was adapted to field data (tactical model). The "tactical model" is a simulation model fitted to the observed temperature data set (Sipkay et al. 2009). The tactical models could be beneficial if the general functioning of ecosystems is in the focus.

6. Acknowledgement

This research was supported by the "Bolyai János" Research Fellowship (Hungarian Academy of Sciences) and the Research Assistant Fellowship (Corvinus University of Budapest). This work was supported by HAS-CUB "Adaptation to Climate Change" Research Group and the research project of the National Development Agency TÁMOP 4.2.1.B-09/1/KMR-2010-0005.

7. References

Christou, E.D. & Moraitou-Apostolopoulou, M. (1995). Metabolism and feeding of mesozooplankton at the eastern Mediterranean (Hellenic coastal waters), *Marine Ecology Progress Series* Vol. 126, pp. 39-48, ISSN 1616-1599

Connell, J.H. (1978). Diversity in tropical rain forests and coral reefs, *Science,* Vol. 199, pp. 1302 - 1310, ISSN 1095-9203

Dippner, J. W.; Kornilovs, G. & Sidrevics, L. (2000). Long-term variability of mesozooplankton in the Central Baltic Sea. *Journal of Marine Systems* Vol. 25, pp. 23-31, ISSN 0924-7963

Dobson, A.P.; Bradshaw, A.D. & Baker, A.J.M. (1997). Hopes for the future: Restoration ecology and conservation biology, *Science,* Vol. 277, pp. 515 - 522, ISSN 1095-9203

Drégelyi-Kiss, Á. & Hufnagel, L. (2009). Simulations of Theoretical Ecosystem Growth Model (TEGM) during various climate conditions. *Applied Ecology and Environmental Research*, Vol. 7, pp. 71-78, ISSN 1785-0037

Drégelyi-Kiss, Á. & Hufnagel, L. (2010). Effects of temperature-climate patterns on the production of some competitive species on grounds of modelling. *Environmental Modeling & Assessment*, Vol. 15, pp. 369-380, ISSN 1573-2967

Dukes, J. S. & Mooney, H.A. (1999). Does global change increase the success of biological invaders? *Trends in Ecology & Evolution* Vol. 14, pp. 135-139, ISSN 0169-5347

Easterling, D. R.; Meehl, G. A.; Parmesan, C.; Changnon, S. A.; Karl, T. R. & Mearns L. O. (2000). Climate Extremes: Observations, Modeling, and Impacts, *Science* Vol. 289, pp. 2068 - 2074, ISSN 1095-9203

Fischlin, A., Midgley, G.F., Price, J.T., Leemans, R., Gopal, B., Turley, C., Rounsevell, M.D.A., Dube, O.P., Tarazona, J. & Velichko, A.A. (2007). Ecosystems, their properties, goods, and services. In *Climate Change 2007: Impacts, Adaptation and Vulnerability*. Parry, M.L., Canziani, O.F., Palutikof, J.P., van der Linden P.J. & Hanson, C.E. (Eds.), pp. 211-272, Cambridge University Press, ISBN-13: 9780521705974, Cambridge

Fitter, A. H. & Fitter, R. S. R. (2002). Rapid Changes in Flowering Time in British Plants. *Science,* Vol. 296, pp. 1689 - 1691, ISSN 1095-9203

Gause, G.F. (1934). *The struggle for existence*, Williams & Wilkins, Baltimore, MD

Gray, John S. (1989). Effects on Environmental Stress on Species Rich Assemblages. *Biological Journal of the Linnean Society,* Vol. 37, pp. 19 - 32, ISSN 1095-8312

Grime, J. P. (1973). Competitive exclusion in herbaceous vegetation. *Nature,* Vol. 242, pp. 344-347, ISSN 0028-0836

Haffner, G D.; Harris, G.B. & Jarais, M.K. (1980). Physical variability and phytoplankton communities. III. Vertical structure in phytoplankton populations. *Archiv für Hydrobiologie*, Vol. 89, pp. 363 - 381, ISSN 0003-9136

Hardin, G. (1960). The competitive exclusion principle, *Science* Vol. 131, pp. 1292-1297, ISSN 1095-9203

Hufnagel, L.; Drégelyi-Kiss, Á., Drégelyi-Kiss, G. (2010). The effect of the reproductivity's velocity on the biodiversity of a theoretical ecosystem. *Applied Ecology and Environmental Research* Vol. 8, No. 2, pp. 119-131, ISSN 1785-0037

Hutchinson, G. E. (1957). Concluding remarks, *Cold Spring Harbour Symposia on Quantitative Biology* Vol. 22, pp. 415-427, ISSN 0091-7451

Iguchi, N. (2004). Spatial/temporal variations in zooplankton biomass and ecological characteristics of major species in the southern part of the Japan Sea: a review. *Progress in Oceanography,* Vol. 61, pp. 213-225, ISSN 0079-6611

Laska, G. (2001). The disturbance and vegetation dynamics: a review and an alternative framework. *Plant Ecology,* Vol. 15, pp. 77 - 99, ISSN 1433-8319

MacMahon, J.A. (1998): Empirical and theoretical ecology as a basis for restoration: an ecological success story. In: *Success, Limitations and Frontiers in Ecosystem Science,* M.L. Pace & P.M. Groffman, (eds.) pp. 220 - 246, Springer-Verlag, New York.

Magura, T.; Tóthmérész, B. & Elek, Z. (2004). Changes in Caribid Beetle Assemblages along an Urbanisation Gradient in the City of Debrecen, Hungary. *Landscape Ecology,* Vol. 19, pp.747-759, ISSN 0921-2973

Magura, T.; Tóthmérész, B.; & Hornung, E. (2006). Az urbanizáció hatása a talajfelszíni ízeltlábúakra, *Magyar Tudomány* Vol. 6, pp. 75-79, ISSN 0025-0325

Nechay, G. (2002): A biológiai sokféleség csökkenése In: *Vissza vagy hova (Útkeresés a fenntarthatóság felé Magyarországon),* Pálvölgyi, T, Nemes, Cs, Tamás, Zs (ed) pp. 36-46, Tertia, Budapest

Olrik, K. & Nauwerck, A. (1993). Stress and disturbance in the phytoplankton community of a shallow, hypertrophic lake. *Hydrobiologia,* Vol. 249, pp. 15 - 24, ISSN 0018-8158

Padisák, J. (1998). Sudden and gradual responses of phytoplankton to global climate change: case studies from two large, shallow lakes (Balaton, Hungary and the Neusiedlersee, Austria/Hungary) In: *Management of Lakes and Reservoirs during Global Change.* George, D. G, J. G, Jones, P. Puncochar, C. S. Reynolds, & D. W. Sutcliffe (eds.), pp- 111-125. Kluwer Academic Publishers, Dordrecht, Boston. London

Penuelas, J.; Prieto, P.; Beier, C.; Cesaraccio, C.; De Angelis, P.; de Dato, G.; Emmett, B.A.; Estiarte, M.; Gorissen, A.; Kovács-Láng, E.; Kröel-Dulay, Gy.; Garadnai, J.; Llorens, L.; Pellizzaro, G.; Riis-Nielsen, T.; Schmidt, I.K.; Sirca, C.; Sowerby, A.; Spano, D. & Tietema, A. (2007). Response of plant species richness and primary productivity in shrublands along a north-south gradient in Europe to seven years experimental warming and drought. Reductions in primary productivity in the heat and drought year of 2003, *Global Change Biology* Vol. 13, pp. 2563 - 2581, ISSN 1365-2486

Pickett, S. T. A. & Parker, V. T. (1997). Restoration as an ecosystem process: implications of the modern ecological paradigm. In: *Restoration Ecology and Sustainable Development.* Urbanska, K. M., N. R. Webb & P. J. Edwards. (eds.) pp. 17 - 23, Cambridge University Press, Cambridge.

Pokorádi L. (2008): *Rendszerek és folyamatok modellezése,* Campus Kiadó, Debrecen, ISBN 978-963-9822-06-1

Reynolds, C. S. (2006). *The ecology of Phytoplankton,* pp. 372 - 381, Cambrige University Press

Sanford, E. (1999). Regulation of Keystone Predation by Small Changes in Ocean Temperature, *Science* Vol. 283, pp. 2095 - 2097, ISSN 1095-9203

Scheffer, M.; Brovkin, V. & Cox, P. (2006): Positive feedback between global warming and atmospheric CO2 concentration inferred from past climate change, *Geophysical Research Letters,* Vol. 33, L10702, doi:10.1029/2005GL025044, ISSN 0094-8276

Sipkay Cs.; Kiss, K. T.; Vadadi-Fülöp, Cs. & Hufnagel, L. (2009): Trends in research on the possible effects of climate change concerning aquatic ecosystems with special emphasis on the modelling approach. *Applied Ecology and Environmental Research* Vol. 7, No.2, pp. 171-198. ISSN 1785-0037

Sipkay, Cs., Drégelyi-Kiss, Á., Horváth L., Garamvölgyi, Á., Kiss, K. T., Hufnagel, L. (2010). Community ecological effects of climate change. In *Climate change and variability*, Sciyo Books, www.sciyo.com, ISBN 978-953- 307-144-2

Sommer, U. (1995). An experimental test of the intermediate disturbance hypothesis using cultures of marine phytoplankton. *Limnology and Oceonography,* Vol. 40, pp. 1271–1277, ISSN 0024-3590

Vadadi-Fülöp, Cs.; Türei, D.; Sipkay, Cs.; Verasztó, Cs.; Drégelyi-Kiss, Á. & Hufnagel, L.(2009). Comparative assessment of climate change scenarios based on aquatic food web modelling. *Environmental Modeling and Assessment,* Vol. 14, No. 5, pp. 563-576, ISSN 1573-2967

Viner, B. & Kemp, L. (1983). The effect of vertical mixing on the phytoplankton of Lake Rotongaio (July 1979 -January 1981). *New Zealand Journal of Marine and Freshwater Research,* Vol. 17, pp. 407 – 422, ISSN 0028-8330

18

The Use and Misuse of Climatic Gradients for Evaluating Climate Impact on Dryland Ecosystems - an Example for the Solution of Conceptual Problems

Marcelo Sternberg[1], Claus Holzapfel[1,7], Katja Tielbörger[2], Pariente Sarah[3], Jaime Kigel[4], Hanoch Lavee[3], Aliza Fleischer[5], Florian Jeltsch[6] and Martin Köchy[6*]

1. Introduction

Current trends of emissions of greenhouse gases are expected to cause the global temperature to rise faster over the present and next century than during any previous period (Houghton et al., 1996, 2001; Zweirs, 2002). Climate models for the Middle East predict an increase in winter temperatures combined with changes in rainfall amounts and distribution (Ben-Gai et al., 1998; Black, 2009, Klafe and Bruins, 2009). These changes may alter ecosystem functioning, with direct effects on ecosystem, community and population processes such as plant litter decomposition, nutrient cycling, primary productivity, biodiversity, plant recruitment and survival (e.g., Aronson et al., 1993; Hobbie, 1996; Robinson et al., 1998; Sternberg et al., 1999; Chapin et al., 2000; Hughes, 2000; Sarah, 2004).

Considerable research has been directed at understanding the responses of terrestrial ecosystems to global environmental change. This topic is of great societal concern in the light of the potential impacts on the natural resources on which the human population depends on (Vitousek, 1994). Nevertheless, the challenge to predict ecosystem response to climate change is based on the multi-dimensional and multi-scale nature of the problem (Osmond et al., 2004). Complex ecological interactions make it difficult to extrapolate from individuals to communities and to predict the ecosystem response when only few levels of ecosystem organization are targeted. In addition, the lack of realistic climatic scenarios at relevant scales adds further complexity to the up-scaling process (Harvey, 2000).

Vast experimental research efforts have been invested in understanding the effects of global warming and CO_2 atmospheric enrichment on ecosystem functioning. These processes are considered key drivers of environmental change, particularly in northern latitudes.

*[1]*Department of Molecular Biology & Ecology of Plants, Faculty of Life Sciences, Tel Aviv University, Tel Aviv, Israel*
[2]*Department of Plant Ecology, Tübingen University, Tübingen, Germany*
[3]*Department of Geography, Bar-Ilan University, Ramat Gan, Israel*
[4]*Institute of Plant Sciences and Genetics in Agriculture, The Hebrew University of Jerusalem, Rehovot, Israel*
[5]*Department of Agricultural Economics and Management, The Hebrew University of Jerusalem, Rehovot, Israel*
[6]*Department for Plant Ecology and Nature Conservation, Potsdam University, Potsdam, Germany*
[7]*Department of Biological Sciences, Rutgers University, 101 Warren Street, Newark USA*

However, relatively little attention has been focused on assessing the responses of terrestrial ecosystems to potential changes in precipitation (Lavee et al., 1998; Svejar et al., 1999; Weltzin et al., 2003). This is probably because projected trends in precipitation changes differ widely between different regions of the world (Intergovernmental Panel on Climate Change, IPCC) therefore, downscaled climate scenarios are required. Precipitation changes are particularly important at mid-latitudes (e.g., around the Mediterranean Basin), where water availability is a key driver in ecosystem functioning, and where global circulation models agree that future precipitation will be lower than today (IPCC). In arid and semi-arid regions, anticipated changes in precipitation regimes may have an even greater impact on ecosystem dynamics than the separate or combined effects of rising temperatures and CO_2 levels. Therefore, studies focusing on the effects of changing patterns of rainfall on ecosystem functioning in these regions are much needed to improve our understanding of the impact of possible future climatic scenarios.

Current predictions of the effects of climate change on water-limited ecosystems are commonly based on empirical investigations of existing climatic gradients (Diaz and Cabido, 1997). However, such purely descriptive approaches alone do not provide sufficient information to enable accurate modelling of the effects of altered water availability caused by climate change. The greatest uncertainty stems from the assumption that climatic differences are the main single determinant of variation among communities along a gradient. There is little experimentally derived information that would allow mechanistic predictions of the impacts of these changes on natural plant and animal communities and ecosystem functioning (Walther et al., 2002; Dunne et al., 2004; Osmond et al., 2004).

Here we present an approach that was designed to overcome several of the major drawbacks of previous studies on the effects of climate change on natural ecosystems. The multidisciplinary project described here employs, in addition to observations along a natural aridity gradient, climate manipulations that are intended to close the gap between descriptive and experimental research approaches, and theoretical modelling. Our unique contribution is to base our study on integrative and complementary investigations of soil, overland flow, and vegetation and landscape processes, combined with consideration of the socio-economic impacts. We thus aim at a holistic approach to the assessment of the impact of climate changes on plant and human communities.

In the following we develop the rationale for combining different, complementary approaches in current climate change research. First, we review the advantages and drawbacks of commonly employed methodology. Based on this outline, we present an approach that takes these concerns into account. Finally, we present an ongoing research programme that is based on this rationale.

2. Methodology

2.1 Gradient approaches as space-for-time approaches

Natural climatic gradients, which include environmental factors such as altitude, topography, temperature and precipitation, provide a useful framework for studying the effects of climate change (Diaz and Cabido, 1997; Imeson and Lavee, 1998; Dunne et al., 2003). Comparisons of ecosystems and biotic communities along gradients provide powerful approaches to the investigation and understanding of the effects of climate variation on ecosystems (Le Houerou, 1990; Koch et al., 1995; Austin, 2002; Cocke et al.,

2005). Approaches based on aridity gradients have been frequently used to study Mediterranean ecosystems (e.g., Boyko, 1947; Holzapfel et al., 1992; Holzapfel et al., 2006; Imeson et al., 1998; Kutiel et al., 2000; de Bello et al., 2005).

The varying effects of climate change on vegetation along gradients can be investigated directly and indirectly. The direct approach involves monitoring of dynamic, long-term vegetation changes in permanent plots (Schmidt, 1988), which necessarily involves long study periods. However, in many cases it is necessary to derive conclusions about successional trajectories from short-term observations. This indirect approach typically involves extrapolation from spatially distinct sites that are expected to represent certain stages in a temporal succession, to temporal patterns; i.e., space-for-time substitution or 'chronsequences' (Pickett, 1989). For example, this approach has been used to investigate the effect of chronic additions from the atmosphere to natural ecosystems. By assuming that all forests were principally N limited before industrialization, current differences in forest health along gradients of N deposition have been attributed to eutrophication (e.g., Lovett and Rueth, 1999), while other causes often have been neglected (Binkley and Högberg, 1997). This indirect and static approach is necessary in several fields of scientific inquiry where direct observation of chronsequences are feasible (e.g., palaeobotany, archaeology and geology, among others).

One has to keep in mind that indirect approaches are based on deductions and not on evidence obtained by direct observations. These deductions are clearly dependent on the subjective selection of sites used as reference for changes in time. Likewise, extrapolations from patterns formed by climatic and ecological gradients to temporal changes - especially in the context of climate change - have often proved to be problematic, as one cannot establish causal relationships on the basis of correlative studies alone (Rastetter, 1996; Dunne et al., 2004).

"Space-for-Time" approaches that use existing environments as proxies for environments under future changed climate are not, in themselves, sufficient to predict changes in species interactions within communities. Such substitutions involve comparisons with communities that have come into balance with local climates over long periods, and the development of such balances is not to be expected in the context of the current rapid pace of climate change (Rastetter, 1996). Moreover, the Space-for-Time approach works under the assumption that except for the climate, all ecosystem components and environmental factors are equally important. Clearly, this is not the case, and the detection of a causal relationship between changes in climate and in ecosystems necessitates more complex approaches. The Space-for-Time approach also neglects the effects of the current fast rate of climate change on ecosystem and community functions that have evolved over long periods. Populations are not likely to vary and move in unison, in response to climate change, and important changes in community composition are to be anticipated (Parmesan, 1996; Walther et al., 2002).

An additional problem is that short-term studies cannot mimic long-term environmental changes. Even if conducted over periods of several years, such studies necessarily present only snap-shots of slowly developing actual changes in climate regimes. In order to overcome the above-mentioned shortcomings of a space-for time approach, such descriptive studies need to be complemented by experiments that enable causal inference, and by theoretical methods that enable the development of scenarios over longer time periods (Sutherland, 2006).

2.2 Experiment vs. gradient approaches: combining efforts

One logical solution for the above problems with simple Space-for-Time approaches is to use climate change experiments *in situ*. The **Experiment-for-Time** approach provides a realistic solution, as climate change will happen in a given place and will be imposed on the biotic communities and ecosystem functions present in that place. This approach mimics the short-term responses of biotic communities and environments and, thereby, directly provides the needed causal relationships between the changes in climate change and in the ecosystem.

Combinations of descriptive gradient studies and experimental approaches have been extremely rare to date (Dunne et al., 2004) and, to the best of our knowledge, have not yet been applied to arid and eastern Mediterranean ecosystems.

The central idea and conceptual scheme behind a combined space-for time and experimental design is illustrated in Fig. 1. Comparisons of given ecosystem and community parameters (e.g., soil properties, biomass production, flora and fauna composition, sediment transport, etc.) along the climatic gradient (e.g., the north-south vertical comparison in Fig. 1) reflect combined changes in the community and climatic conditions. Comparison of the same parameters between climate manipulations within the same site (horizontal comparison in Fig. 1) reflects the effects of changes in climate only. Furthermore, comparisons of control

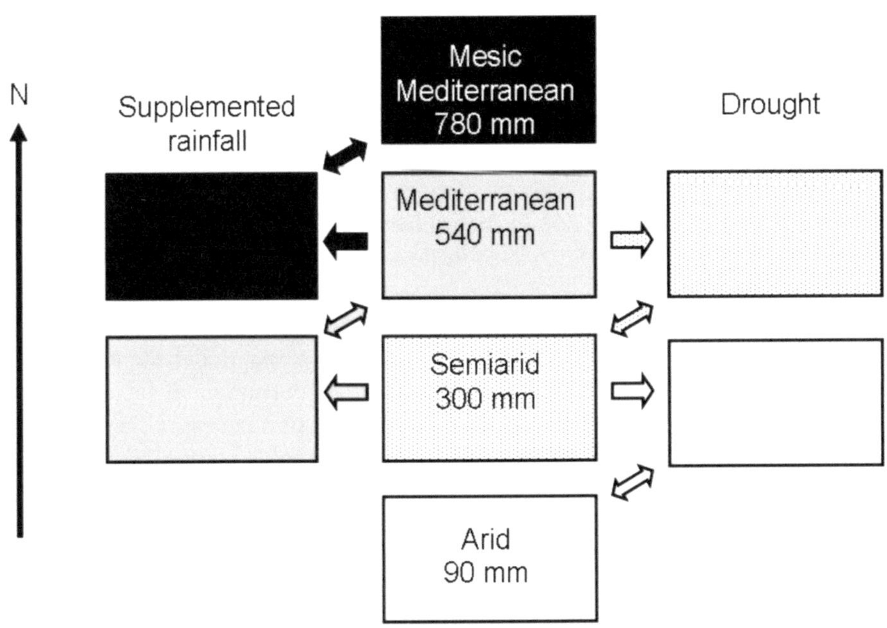

Fig. 1. Schematic illustration of comparisons of study sites along the gradient, with rainfall manipulations, representing the rationale of the experiment set up. Connecting arrows between squares illustrate meaningful comparisons (see text).

sites (i.e., natural precipitation) with manipulated climate treatments in adjacent sites along the gradient (diagonal comparison in Fig. 1) reflect mainly the effects of site differences, since their rainfall amounts tend to be similar. Thus, these three comparisons facilitate the separation of the effects of climate alone from the effects of site-specific community conditions. Furthermore, these comparisons highlight the interactions between climatic and other environmental effects. One prediction is that simple comparisons along an existing gradient will not necessarily enable direct forecasting of shifts in ecosystem conditions caused by climate change, since climatic and other environmental effects are likely to be confounded (Fukami and Wardle, 2005). Our research tests this prediction and evaluates the usefulness of this approach as compared with those used in previous studies of the effects of climate change.

The above outlined approach tests whether purely descriptive gradient approaches are sufficient to enable prediction of the effects of climate change. The present project is carried out within the framework of the research initiative, GLOWA–Jordan River (for details see: http://www.glowa-jordan-river.de). Besides contrasting direct and indirect approaches, as stated above, the project has two specific applied goals: (1) to understand the effects of global climate change on soil, run-off, populations, biotic communities and ecosystem properties and dynamics in ecosystems in Israel that range from Mediterranean to arid; and (2) to provide empirical data for modelling, economic analysis and prediction of ecosystem responses to climate change in an environmentally sensitive area.

2.3 Integration by modelling: extending the time-scale

The addition of experimental studies to the space-for-time approach should improve our understanding of relevant processes linked to short-term changes in precipitation patterns. However, this experimental approach is limited in space and time, and the limited number of years available for observation and experimental manipulation may not be sufficient for complete exploration of the long-term dynamics. This is because rainfall in more arid regions is highly variable and annual precipitation may be autocorrelated between consecutive years, which may lead to long-term fluctuations. In addition, technical and logistic constraints restrict the size of the area that can be manipulated in experiments to a scale that is necessarily smaller than scales relevant for land users and nature conservation. Furthermore, the type of climate-change scenario that can be mimicked by experiments is constrained by the practicability of active manipulation of the climate. Therefore, complex scenarios such as increased frequency of extreme events, as predicted for our study region (Kunstmann et al., unpubl.), cannot be directly addressed in the empirical study.

Modelling provides useful tools to close these gaps. Spatially explicit, stochastic simulation models have been found successful in using short-term and small-scale information to gain understanding of and to predict longer-term and larger-scale processes, as described by Jeltsch and Moloney (2003). Grid-based, modular simulation models of vegetation dynamics are especially able to link information on differing scales (Jeltsch et al., 1996, 1997; Jeltsch et al., 1999). However, in the type of climate change studies proposed in the present paper, phenomena on at least three different spatial scales have to be distinguished: (1) responses of individual plants, e.g., growth, seed production, mortality and, possibly, physiological adaptation mechanisms (e.g., Petru et al., 2006); (2) small-scale intra- and interspecific interactions between individuals of contrasting growth forms, among which interactions between herbaceous and woody vegetation, including competition and facilitation mechanisms, are of especial importance (Holzapfel et al., 2006); and (3) the effects of these interactions on vegetation pattern formation on the landscape level, with feedbacks to spatial processes such as runoff, soil moisture distribution and availability, fire, grazing, and other types of land use. At all these levels, the models need and use data obtained in the detailed field investigations and experiments, and thus also function as integrators of collected information.

It is, however, not feasible to collect all the necessary information on all scales in full detail. Therefore, we apply a cascade of models that differ in spatial and temporal resolution. With increasing scale, spatial resolution of data has to be aggregated to reduce the otherwise immense complexity of the model. Since, in contrast to previous studies, realistic climatic scenarios at biologically meaningful scales are produced within the larger framework of the GLOWA-Jordan River programme (see website), we will be able to utilize these scenarios for generating more realistic predictions. Field monitoring and experiments can be used for model testing and validation, and additional simulation experiments, such as increasing the frequency of extreme years, can be conducted that are not feasible in the experimental plots.

The combination of hierarchical modelling and experiments with the space for time approach provides a powerful strategy for gaining understanding of the impact of climatic changes by examining processes that link individuals, patches, and the landscape (up-scaling procedure). Only such a multi-scale approach is capable of predicting consequences of changed climatic conditions on a level that is relevant to land use and management.

2.4 Socio-economic evaluations: a need for a stronger link with nature

Linking the ecological and socio-economic approaches at the landscape level provides improved tools to analyze the impacts of global change on society and nature. Scientists are responding to the demands to estimate potential local impacts of climate change and to produce relevant information that can be used at regional and local scales, and can be related to public and natural needs (Cash and Moser, 2000). However, most of the papers that analyze this issue place little emphasis on the link between ecosystem processes and socio-economic impacts. This topic deserves to be investigated in more detail, in light of the multi-scale nature of the global climate change problem

The impact of climate change on agricultural activities and their economic consequences has received much attention from economists (Mendelsohn et al., 1994; Nordaus, 1994). The general approach is to assume changes in temperature and precipitation and to determine their effects on the income generated from agriculture. Fewer scientists have dealt with evaluation of the loss of welfare caused by changes in ecosystems (Layton and Brown, 2000). In both types of analysis changes in vegetation, soil and water processes are regarded as external agents; Layton and Brown (2000), for example, used arbitrary values of forest loss depicted in computer-enhanced photographs.

In the present study, we link the climate to changes in biomass and ecosystem processes via changes in the landscape, and hence to societal welfare. In order to link the changes in biomass to the welfare level of society a stated preference approach was used. In this approach a survey of the urban population was conducted, in which the biomass level was expressed in the form of landscape photos of the experimental stations (see Fleischer and Sternberg, 2006). Previous studies have shown that the recreational value of agricultural landscape can be much higher than the returns obtained from farming (Fleischer and Tsur, 2000). In our experimental stations, in which the land is mainly and extensively used for grazing, this difference is even greater, because emphasis was put mainly on the changes in utility that accrue to the individuals exposed to the landscape. Since these changes will occur only in the future we were able to use only a stated preference approach, within which we used a choice modelling approach. Data were collected by means of a face-to-face survey of a representative sample of the urban population in Israel, where more than 91% of the population are city dwellers. The landscape in the photos varied along the north-south environmental gradient, from a typical mesic Mediterranean, through Mediterranean, semiarid to arid. By using the Random Parameters Logit (RPL) model it was possible to determine the population's Willingness To Pay (WTP) in order to prevent changes to the landscape. The WTP was, in fact, the monetary value of the welfare loss to the population caused by the changing landscape. The novel feature of this research was in the use of photographs of the experimental stations along the north-south gradient; photographs that provided a simulation of the expected impact of the climate change on the landscape (for full details see Fleischer and Sternberg, 2006).

3. A case study along an aridity gradient

During 2001 four experimental sites were established in Israel, along a 245-km-long climatic gradient running from Galilee in the north to the Negev Desert in the south (Fig. 2). These sites represent respectively, Mesic Mediterranean (MM), Mediterranean (M), semiarid (S), and arid (A) climatic conditions (see Table 1). All the sites rest on the same calcareous (hard limestone) bedrock and are positioned on south-facing slopes, i.e., the drier aspect. The study sites were fenced to exclude the main domestic grazers, i.e., cattle, sheep and goats. The basic climate is Mediterranean, with mild and rainy winters (October-April) and prolonged rainless, hot summers. The plant-growing season is closely associated with the temporal distribution of rainfall. Germination of annuals, and growth of most perennials starts soon after the first rains, between October and December each year (Table 1).

3.1 Climatic manipulations: rationale and application

Experimental approaches to climate-change studies, particularly those that address the impacts of precipitation on ecosystems, typically involve the use of rainout shelters to exclude natural precipitation, and artificial irrigation to increase rainfall (Fay et al., 2000; Sarah and Rodeh, 2004). Rainout shelters provide control over the daily or seasonal timing and extent of dry and wet periods. Global climate change is predicted to alter rainfall patterns during the growing season, and this may lead to either reductions or increases in the total amount of precipitation in different cases, in addition to causing shifts in the temporal distribuion of the rainfall during the growing season. Such changes may affect numerous ecosystem processes, through the temporal and spatial redistribution of water, and ultimately may have impacts on rates of primary productivity and decomposition, as well as on biological diversity (Hulme, 2005).

Rainfall manipulations are applied in only two of the four sites. These sites are located in the two intermediate locations along the climatic gradient and they represent the transition from mesic to arid conditions: the Mediterranean (M) and the semi-desert (S) regions. The rationale of climatic manipulations at these sites is based on predicted potential climate-change scenarios in which the strongest changes occur in the transition zone between mesic Mediterranean and arid desert areas (Fig. 2). The mesic Mediterranean and the desert stations (MM and A) at the ends of the gradient, are kept under natural climatic conditions and serve as controls for the climate-manipulated areas.

Ecosystem type	Rainfall & CV (mm - %)	Temperature (°C) Min. Mean Max.	Elevation (a.s.l)	Soil type	Vegetation formation
Arid (N 30°52' E 34°46')	90 - 51	13.6 - 19.1 - 26.1	470 m	Desert Lithosol	Open vegetation dominated by small shrubs and semi-shrubs such as *Zygophyllum dumosum*, *Artemisia sieberi* and *Hammada scoparia* and sparsely growing desert annuals, geophytes and hemicryptophytes.
Semiarid (N 31°23' E 34°54')	300 - 37	13.2 – 18.4 – 24.8	590 m	Light Brown Rendzina	Dwarf-shrubs of *Sarcopoterium spinosum* and *Coridothymus capitatus* associated with herbaceous (chiefly annual) plant species
Mediterranean (N 31°42' E 35°3')	530 - 30	12.8 – 17.7 - 23.6	620 m	Terra Rossa	Dwarf-shrubland dominated by *Sarcopoterium spinosum* and a high diversity of herbaceous (mostly annual) plant species.
Mesic Mediterranean (N 33°0' E 35°14')	780 - 22	13.5 - 18.1 - 23.4	500 m	Montmorillonitic Terra Rossa	Closed oak maquis (*Quercus calliprinos*) and open garrigue formations dominated by shrubs (*e.g. Calicotome villosa, Sarcopoterium spinosum, Cistus* spp.) and associated herbaceous plants.

Table 1. Physical and biotic characteristics of the study sites along the aridity gradient. Temperature refers to annual means (mean minimum, mean and mean maximum) Rainfall coefficient of variance (CV) is presented as percent (Modified from Fleischer and Sternberg, 2006).

Two climate-change scenarios are currently being tested in the sites with winter, i.e., growing season, rainfall manipulations: increased rainfall and drought. These scenarios arebased on existing climate-change models for the region (Ben-Gai et al., 1998; Zangvil et al., 2003) and on scenarios generated within our own project (Kunstmann et al., unpubl.) and represent two extreme boundary conditions. In each site, plots of 10 × 25 m are subjected to simulations of either wetter or drier winters. There are three treatments per site: 1) control (natural conditions); 2) artificially augmented winter rainfall (30% more than average annual precipitation); and 3) winter drought (30% less than average annual precipitation), simulated by means of rainout shelters. The climatic manipulations are each applied to five plots.

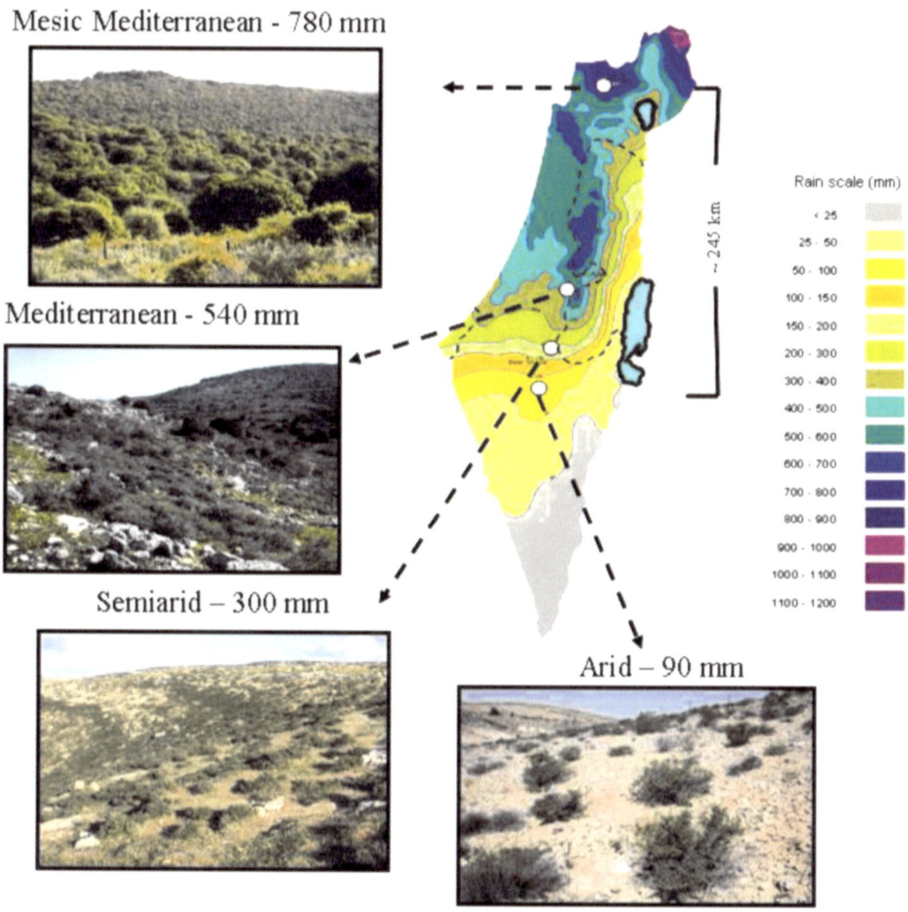

Fig. 2. Location of the experimental sites along the aridity gradient. Photographs: Claus Holzapfel

The rainfall treatments mimic the natural timing, frequency and intervals of rainfall events at the sites. Irrigation is applied at the end of each rainfall event by means of drizzle sprinklers. Drought treatment will be achieved by using plastic rainout shelters as described by Yahdjian and Sala (2002) to intercept 30% of the precipitation arriving at a site (Figure 3). These fixed rainout shelters utilize V-shaped bands of greenhouse plastic to intercept a given amount of rainfall. The bands are supported by a frame of galvanized aluminum (mean height = 2.5m) and cover a total surface of 25m x 10m. The roofs are angled and drain to gutters at the downslope edge of the roof. Gutters leading to collecting pipes that drain the water intercepted outsite the study plots. Sides of the rain shelters are open to allow for air movement and minimize temperature and humidity differences under and outside the shelter (Fay et al., 2000).

Summaries of the soil properties, overland flow and vegetation studies, and the measured parameters are presented in Table 2. The selection of the type of data collected is based on the need to couple between climate and vegetation responses and to cover a whole range of spatial, temporal and organizational scales.

Fig. 3. View of the study sites with the rainout shelter structures at the Mediterranean (a and b) and at the semiarid site (c and d). Photographs: Claus Holzapfel and Marcelo Sternberg.

4. Current directions

The results obtained from this innovative project provide useful tools for refining predictions of the responses of ecosystems ranging from Mediterranean to desert to global environmental change. The combination of methodologies that was achieved by up-scaling and down-scaling specifically targeted processes provides a more accurate means of coping with the great uncertainty inherent in studying global climate change. Moreover, the results of this project will lead to a better understanding of the mechanisms that determine changes in community structure, ecosystem functioning and soil erosion in these ecosystems. Another innovation is that the results we obtain will enable us to obtain realistic down-scaled climate scenarios for the region, to be integrated into our models (in collaboration with the climate modellers participating in the GLOWA - Jordan River Project - see web page). Finally, our results can be utilized directly for evaluating the socio-economic consequences of climate change for ecosystems. Although such an integrated approach has been advocated previously (Weltzin et al., 2003), to our knowledge, no similar study, which combines gradient and experimental approaches with modelling and socio-economic approaches, has addressed the effects of global climatic change in Mediterranean and desert ecosystems.

Parameter	Methology	Objective
Soil structure (aggregate stability and aggregate size distribution), organic matter content, soluble salts, electrical conductivity, pH, microbial enzymes (arylsulphatase activity) and nitrogen content in the soil (NO_2, NO_3 and NH_4).	Soil sampling three times a year – autumn, winter and spring – in different microhabitats: a) shrub understorey , b) open patches between shrubs , c) beneath rock fragments and d) tree understorey. Soil samples include two layers: 0-2 cm and 5-10 cm depth.	Understanding changes in soil properties linked to nutrient turnover and susceptibility to erosion.
Overland flow and sediment properties.	Small run-off plots set up in several microhabitats: a) shrub understorey; b) open patches between shrubs, and c) a combination of shrubs and open gaps. The area of each plot is about 0.15 m^2 and that of shrub-open gap matrix is about 0.3 m^2.	Redistribution of water-accepting patches (sink) and overland flow-contributing patches (source) as a strategy of water conservation.
Aboveground biomass, species richness, species diversity, species composition, plant density, phenology, soil seed banks, germination strategies, species interactions	Comparisons for all parameters are between microhabitats: a) shrub understorey; b) open patches between shrubs. Destructive and non-destructive biomass estimations using allometric estimations. Species richness, diversity and density per quadrat of seedlings is determined with repeated counts, taking into account separate germination events. Soil seed bank is collected from all stations every year, in autumn before the onset of rainfall and watering, for 3 consecutive years to deplete dormant seeds. Adaptation to the climate via germination strategies and adaptation to biotic conditions are studied for selected plant species and functional groups.	Understanding of mechanisms behind plant population and plant community dynamics. Baseline for theoretical predictions of species and community response to climate change.
Plant litter decomposition	Plant litter bag experiment including microhabitat comparisons (shrub understorey and open gaps between shrubs) along the climatic gradient. Reciprocal plant litter transplantation among study sites	Estimations of nutrient cycling and carbon turnover.

Table 2. Description of biotic and abiotic parameters measured in all study sites along the climatic gradient (includes the experimental climatic manipulations).

5. Acknowledgements

This research is funded in the framework of the GLOWA - Jordan River project by the German Federal Ministry of Education and Research (BMBF) in collaboration with the Israeli Ministry of Science and Technology (MOST).

6. References

Aronson, J., Floret, Ch., Le Floc'h. E., Ovalle, C., Pontaimer, R., 1993. Restoration and rehabilitation of degraded ecosystems II: Case studies in central Chile, southern Tunisia and northern Cameroon. Restoration Ecology 1, 8-11.

Austin, A.T., 2002. Differential effects of precipitation on production and decomposition along a rainfall gradient in Hawaii. Ecology 83, 328-338.

Ben-Gai, T., Bitan, A., Manes, A., Alpert, P., Rubin, S., 1998. Spatial and temporal changes in rainfall frequency distribution patterns in Israel. Theoretical and Applied Climatology 61, 177-190.

Binkley, D., Högberg, P., 1997. Does atmospheric deposition of nitrogen threaten Swedish forests? Forest Ecology and Management 92, 119-152.

Boyko, H., 1947. On the role of plants as quantitative climate indicators and the geo-ecological law of distribution. Journal of Ecology 35, 138-157.

Black, E. 2009 The impact of climate change on daily precipitation statistics in Jordan and Israel. Atmospheric Science Letters 10, 192-200.

Cash, D., Moser, S.C., 2000. Linking global and local scales: designing dynamic assessment and management processes. Global Environmental Change 10, 109-120.

Cocke, A.E., Fulé, P.Z., Crouse, J.E., 2005. Forest change on a steep mountain gradient after extended fire exclusion: San Francisco Peaks, Arizona, USA. Journal of Applied Ecology 42, 814–823.

Chapin, F.S., Zavaleta, E.S., Eviner, V.T., Naylor, R.L., Vitousek, P.M., Reynolds, H.L., Hooper, D.U., Lavorel, S., Sala, O.E., Hobbie, S.E., Mack, M.C., Diaz, S., 2000. Consequences of changing biodiversity. Nature 405, 234-242.

de Bello, F., Leps, J., Sebastia, M.T., 2005. Predictive value of plant traits to grazing along a climatic gradient in the Mediterranean. Journal of Applied Ecology 42, 824–833.

Diaz, S., Cabido, M., 1997. Plant functional types and ecosystem function in relation to global change. Journal of Vegetation Science 8, 463-474.

Dunne, J.A., Harte, J., Taylor, K.J., 2003. Subalpine meadow flowering phenology responses to climate change: Integrating experimental and gradient methods. Ecological Monographs 73, 69-86.

Dunne, J.A., Saleska, S.R., Fischer, M.L., Harte, J., 2004. Integrating experimental and gradient methods in ecological climate change research. Ecology 85, 904-916.

Fay, P.A., Carlisle, J.D., Knapp, A.K., Bair, J.M., Collins, S.L., 2000. Altering rainfall timing and quantity in a mesic grassland ecosystem: Design and performance of rainfall manipulation shelters. Ecosystems 3, 308-319.

Fleischer, A., Sternberg, M., 2006. The economic impact of global climate change on Mediterranean rangeland ecosystems: A Space-for-Time approach. Ecological Economics 59, 287-295.

Fleischer, A., Tsur, Y., 2000. Measuring the recreational value of agricultural landscape. European Review of Agricultural Economics 27, 385-398.

Fukami, T., Wardle, D.A., 2005. Long-term ecological dynamics: reciprocal insights from natural and anthropogenic gradients. Proceedings of the Royal Society B 272, 2105-2115.

Harvey, L.D.D., 2000. Upscaling in climate change research. Climatic Change 44, 225-263.
Hobbie, S.E., 1996. Temperature and plant species control over litter decomposition in Alaskan tundra. Ecological Monographs 66, 503-522.
Holzapfel, C., Schmidt, W., Shmida, A., 1992. Effects of human-caused disturbances on the flora along a Mediterranean-desert gradient. Flora 186, 261-270.
Holzapfel, C., Tielboerger, K., Parag, H., Kigel, J., Sternberg, M., 2006. Annual plant-shrub interactions along an aridity gradient in Israel. Basic and Applied Ecology 7, 268-279.
Houghton, J.T., Meira Filho, L.G., Bruce, J., Lee, H., Callander, B.A., Haites, E., Harris, E., Kattenberg, A., Maskell, K., 1996. Climate Change 1995. The Science of Climate Change. Cambridge University Press. Cambridge, UK.
Houghton, J.T., Ding, Y., Griggs, D.J., Noguer, M., van der Linden, P.J., Dai, X., Maskell, K., Johnson, C.A., 2001. Climate Change 2001: The Scientific Basis: Contribution of Working Group I to the Third Assessment Report of the Intergovernmental Panel on Climate Change. Cambridge University Press. Cambridge, UK.
Hughes, L., 2000. Biological consequences of global warming: is the signal already apparent? TREE 15, 56-61.
Hulme, P.E., 2005. Adapting to climate change: is there scope for ecological management in the face of a global threat? Journal of Applied Ecology 42, 784-794.
Imeson, A., Lavee, H., 1998. Soil erosion and climate change: the transect approach and the influence of scale. Geomorphology 23, 219-227.
Imeson, A., Lavee, H., Calvo-Cases, A., Cerda, A., 1998. The erosional response of calcareous soils along a climatological gradient in South-East Spain. Geomorphology 24, 3-16.
Jeltsch, F., Moloney, K.A., 2003. Spatially-explicit vegetation models: what have we learned? Progress in Botany. Volume 63. Esser, K., Lüttge, U., Beyschlag W., Hellwig, F. (Eds.), Springer, Berlin, New York, pp. 326-343.
Jeltsch, F., Milton, S.J., Dean, W.R.J., Rooyen, N., 1996. Tree spacing and coexistence in semiarid savannas. Journal of Ecology 84, 583-595.
Jeltsch, F., Milton, S.J., Dean, W.R.J., Rooyen, N., 1997. Analysing shrub encroachment in the southern Kalahari: A grid-based modelling approach. Journal of Applied Ecology 34, 1497-1508.
Jeltsch, F., Moloney, K., Milton, S.J., 1999. Detecting process from snapshot pattern: lessons from tree spacing in the southern Kalahari. Oikos 85, 451-466.
Kafle, H. K, Bruins, H. J. 2009. Climatic trends in Israel 1970–2002: warmer and increasing aridity inland. Climatic Change 96, 63-77.
Koch, G.W., Vitousek, P.M., Steffen, W.L., Walker, B.H., 1995. Terrestrial transects for global change research. Vegetatio 121, 53-65.
Kutiel, P., Kutiel, H., Lavee, H., 2000. Vegetation response to possible scenarios of rainfall variations along a Mediterranean–extreme arid climatic transect. Journal of Arid Environments 44, 277-290.
Lavee, H., Imeson, A., Sarah, P., 1998. The impact of climate change on geomorphology and desertification. Journal of Land Degradation and Development 9, 407-422.
Layton, D.F., Brown, G., 2000. Heterogeneous preferences regarding global climate change. The Review of Economics and Statistics 82, 616-624.
Le Houerou, H.N., 1990. Global change - vegetation, ecosystems, and land use in the southern Mediterranean basin by the mid-twenty-first century. Israel Journal of Botany 39, 481- 508
Lovett, G.M., Rueth, H., 1999. Soil nitrogen transformations in beech and maple stands along a nitrogen deposition gradient. Ecological Applications 9, 1330-1344.

Mendelsohn, R., Nordaus, W.D., Shaw, D., 1994. The impact of global warming on agriculture: a recardian analysis. American Economic Review 84, 753-771.
Nordaus, W.D., 1994. Managing the Global Commons: The Economics of Climate Change. MIT Press, Cambridge, MA.
Osmond, B., Ananyev, G., Berry, J., Langdon, C., Kolber, Z., Lin, G., Monson, R., Nichol, C., Rascher, U., Schurr, U., Smith, S., Yakir, D., 2004. Changing the way we think about global climate change research: scaling up in ecosystem experimental science. Global Change Biology 10, 393-407.
Parmesan, C., 1996. Climate and species' range. Nature 382, 765-766.
Petru, M., Tielboerger, K., Belkin, R., Sternberg, M., Jeltsch, F., 2006. Life history variation in an annual plant under two opposing selective forces along a steep climatic gradient. Ecography 29, 66-74.
Pickett, S.T.A., 1989. Space-for-time substitution as an alternative to long-term studies. In: G.E., Likens (Ed.), Long-Term Studies in Ecology: Approaches and Alternatives. Springer, New York, pp. 110-135.
Rastetter, E.B., 1996. Validating models of ecosystem response to global change. Bioscience 46, 190-198.
Robinson, C.H., Wookey, P.A., Lee, J.A., Callaghan, T.V., Press, M.C., 1998. Plant community responses to simulated environmental change at a high artic polar semi-desert. Ecology 79, 856–866.
Sarah, P., 2004. Nonlinearity of ecogeomorphic processes along Mediterranean-arid transect. Geomorphology 60, 303-317.
Sarah, P., Rodeh, Y., 2004. Soil structure variations under manipulations of water and vegetation. Journal of Arid Environments 58, 43-57.
Schmidt, W., 1988. An experimental study of old-field succession in relation to different environmental factors. Vegetatio 77, 103-114.
Sternberg, M., Brown, V.K., Masters, G.J., Clarke, I.P., 1999. Plant community dynamics in a calcareous grassland under climate change manipulations. Plant Ecology 143, 29-37.
Sutherland, W.J., 2006. Predicting the ecological consequences of environmental change: a review of the methods. Journal of Applied Ecology 43, 599-616.
Svejcar, T., Angell, R., Miller R., 1999. Fixed location rain shelters for studying precipitation effects on rangelands. Journal of Arid Environments 42, 187-193.
Vitousek, P.M., 1994. Beyond global warming: Ecology and global change. Ecology 75, 1861-1876.
Walther, G.R., Post, E., Convey, P., Menzel, A., Parmesan, C., Beebee, T.J.C, Fromentin, J.M., Hoegh-Guldberg, O., Bairlein, F., 2002. Ecological responses to recent climate change. Nature 416, 389-395.
Weltzin, J. F, Loik, M.E., Schwinning, S., Williams, D.G., Fay, P.A., Haddad, B.M., Harte, J., Huxman, T.E., Knapp, A.K., Lin, G.H, Pockman, W.T., Shaw, M.R., Small, E.E., Smith, M.D., Smith, S.D., Tissue, D.T., Zak, J.C., 2003. Assessing the response of terrestrial ecosystems to potential changes in precipitation. Bioscience 53, 941-952.
Yahdjian, L., and O. E. Sala. 2002. A rainout shelter design for intercepting different amounts of rainfall. Oecologia 133, 95-101.
Zangvil, A., Karas, S., Sasson, A., 2003. Connection between eastern Mediterranean seasonal mean 500 hPa height and sea-level pressure patterns and the spatial rainfall distribution over Israel. International Journal of Climatology 23, 1567-1576.
Zweirs, F.W., 2002. The 20-year forecast. Nature 416, 690-691.

Part 3

Changes in Alpine and Boreal Landscapes

19

Climate-Driven Change of the Stand Age Structure in the Polar Ural Mountains

Valeriy Mazepa, Stepan Shiyatov and Nadezhda Devi
Institute of Plant and Animal Ecology, Ural Branch Russian Academy of Sciences
Russia

1. Introduction

Most records across the Arctic show a widespread transition from cold conditions of the 19th century to warm conditions of the 20th century, with local warming of 1-3°C that average ~1.5°C across the Arctic domain (Jackson et al., 1997). The region has experienced higher warming rates over the last 30 years and the process has been accelerating at unprecedented rates over the last decade (Miller et al., 2010). The mean global temperatures of the last decade have been the warmest for the last 1000 years, and projections suggest a further increase in the average surface temperature around the world (Briffa et al., 1995; Esper & Schweingruber, 2004; IPCC, 2007). In the arctic and alpine regions, the estimated temperature anomalies for the past 100 years were twice as large as those averaged for the northern hemisphere (Kelly et al., 1982). Warming is expected to have large effects on global vegetation and plant distribution, particularly in ecosystems at high altitudes and high latitudes, where plant growth is mainly limited by temperature (Kittel et al., 2000; Becker & Bugmann, 2001; Shiyatov & Mazepa, 2007).

Today, spatiotemporal studies of forest-tundra and forest-grassland plant communities in high mountains are given considerable but deserved consideration because of the opportunity and necessity to evaluate their responses to what is widely considered to be anomalous 20th century warming (Kullman, 1990; Körner, 1999; Holtmeier, 2003). The montane plant communities growing at high latitudes are of particular interest because of the magnitude of climate change observed in instrumental records at these locations (Briffa & Jones, 1993). The pace and pattern of biotic response to climate variation on scales of decades to centuries is highly relevant to understanding the potential ecological consequences of anthropogenically induced climate change (Lloyd, 1997). Conducting such research in "natural" or largely undisturbed areas where human impact is minimal is highly desirable.

Patterns of plant population response to climate variation are most clearly observed at climatically determined ecotones, where the abiotic environment is the ultimate control over ecological processes (Lloyd & Graumlich, 1997). Following Körner (1999), the term "tree-line ecotone" is taken here to represent the transitional belt of mountain vegetation situated between the upper limit of single tree growth in the tundra and the upper limit of closed forests. This ecotone is wider than the subgoltsy belt, because it covers the lower part of the mountain tundra belt, where solitary woody plants are found.

In the Polar Urals, dead trees at high elevation are preserved *in situ* for more than a millennium. Numerous explorers (Sukachev, 1922; Gorodkov, 1926; Sochava, 1927; Andreev et al., 1935) observed a great number of dead trees and wood remains in various degrees of decomposition around the upper tree-line on the eastern slope of the Polar Ural Mountains. Such wood is especially abundant in the Sob River Basin. Dead trees located above the current tree-line ecotone provide evidence of the spatio-temporal dynamics of the forest-tundra communities in the recent past. The paleoecological record preserved in these dead trees is highly resolved, both spatially and temporally, providing a unique opportunity to reconstruct the precise history of actual changes in the structure of forest-tundra stands (Shiyatov, 1993, 2003; Mazepa, 2005).

Historic photographs as well as satellite remote sensing surveys documented that high-latitude ecosystems have changed considerably during the last century: in tundra regions of northern Alaska and central Russia, the abundance of shrubs has increased strongly (Myneni et al., 1997; Shvartsman et al., 1999; Silapaswan et al., 2001; Sturm et al., 2005), and tree-line as well as forest-tundra ecotones in North America, Scandinavia, Siberia, and Urals have been shifting north- and upwards (Kullman, 2002; Kharuk et al., 2002; Shiyatov, 2003; Moiseev & Shiyatov, 2003; Esper & Schweingruber, 2004; Shiyatov, 2009). Similarly, dendroecological studies show that trees at high latitudes and altitudes of the northern hemisphere have been growing better during the last decades (Paulsen et al., 2000). There are, however, also reports on decreasing tree growth in the drier regions of Interior Alaska (D'Arrigo et al., 2004) and Middle Siberia (Vaganov et al., 1999). In contrast to the growth of individual trees, little is known about the impact of climatic changes on the growth forms and growth strategies of trees. At the fringe of their distribution, boreal trees grow in a number of growth forms (creeping, single-, and multi-stem), and they are able to adapt their growth form to environmental changes (Gorchakovskiy & Shiyatov, 1985; Weisberg & Baker, 1995; Pereg & Payette, 1998; Goroshkevich & Kustova, 2002; Holtmeier, 2003; Mazepa & Devi, 2007; Devi et al., 2008). However, there is no quantitative understanding on the development and the timing of these growth forms, and to date only few attempts have been made to link the dynamics of growth forms to the changes in climatic conditions.

Forest expansions into former tundra could induce positive and negative feedbacks of ecosystems with climate (Chapin et al., 2000). While increasing tree biomass sequesters carbon from the atmosphere and thus would lead to a negative feedback, decreasing albedo through increasing tree cover would amplify atmospheric heating, particularly during the snow-covered season. For arctic Alaska, Chapin et al. (2005) estimated that the potential heating effect by expanding forests exceeds the climatic effects induced by a doubling of atmospheric CO_2. However, treeline advances and forest establishment lag behind anthropogenic climatic change, and rates of change differ strongly among sites, which add great uncertainties in predicting the feedbacks between terrestrial ecosystems and climate (Lloyd, 2005).

One such promising area is the Polar Ural Mountains. Important note has to be made with respect to the choice of the field site location. The eastern macroslope of the Polar Urals represents a pristine environment where tundra, forest, and shrub ecosystems have not been exposed to significant anthropogenic impact and show no signs of forest fires in more than 1000-year long tree-ring records (Shiyatov, 1965, 1986, 2003; Mazepa, 2005). These ecosystems experience the effect of natural disturbance factors; these are mainly related to climate variability/change. Furthermore, the routes of seasonal migration of reindeer herders go around this area because of the danger related to crossing of the Sob River and the railroad. Most backpackers, who travel to the Rai-Iz range, approach it from the western and northern slopes because of the proximity to railroad stations. From the eastern side, it is only possible to reach the foothills of the mountains using an all-terrain vehicle. Consequently, the proposed field monitoring area represents an ideally suited location for studies of climate impacts on pristine tundra, forest, and forest-tundra communities of the subarctic region.

This paper demonstrates how the age structure of forest-tundra stands within the former and current tree-line ecotone and the morphogenesis of Siberian larch in the Polar Urals have changed over the last millennium.

2. Materials and methods

2.1 Study area

The study area is located on the eastern slope of the Polar Ural Mountains, in the Sob River basin (66°46'- 66°55' N, 65°22'- 65°49' E), 50 km northwest of the Ob River and 30 km north of the Arctic Circle (Fig. 1). The bedrock basically consists of Paleozoic amphibolite and crystal granodiorite, which forms a series of peaks and depressions 1000-1200 m a.s.l. There are over 90 glaciers in the Polar Ural Mountains. Recent publications indicate a reduction in the volume of these mountain glaciers (Solomina, 1999), almost all of which are located in deep depressions and valleys on east- and north-facing slopes, on the leeward side of the mountains. They are classified as so-called hillside and kar glaciers and located below the snow line. This is possible because the snow is distributed by the wind and concentrated in depressions, which allows the glaciers to survive. Traces of glacial activity (i.e., lateral moraines and lakes of a glacial origin) are visible, and dating studies of the moraines have been reported by Yu. Martin (1970).

In the current study, meteorological data recorded at Salekhard meteorological stations (Fig. 1) have been used. Salekhard station is located 55 km southeast of the study area in the valley of the Ob River at an elevation of 35 m a.s.l. The record begins in 1883.

The mean annual air temperature for Salekhard station is about -6.7 °C; mean monthly minimum (January) and maximum (July) temperatures are about -24.4 and 13.8 °C, respectively. The mean frost-free period lasts about 64 days, and the growing season extends from mid-June to early August. Frosts are possible in all summer months, and short-duration positive temperatures can be experienced in all winter months. The mean annual precipitation is 500-600 mm, with 50% falling as snow.

The annual radiation balance is positive and about 42 J/cm^2. The study area is in a zone of continuous permafrost. The Polar Urals are influenced by a predominantly westerly air flow, with minimum wind speed observed in summer (average 5–6 m/s) and maximum in winter (average 9-10 m/s). Peak registered wind speeds can reach 40–50 m/s, a characteristic feature of west–east oriented valleys.

The study area is located within the tree-line ecotone and occupies deep valleys and lower slopes up to 200-350 m a.s.l. Within the tree-line ecotone (with occasional single trees found at 400-410 m a.s.l.) Siberian larch (*Larix sibirica* Ledeb.) stands of varying density predominate. These are found in association with Siberian spruce (*Picea obovata* Ledeb.) and mountain birch (*Betula tortuosa* Ledeb.). Patches of closed larch-spruce forests grow at lower elevations. There are also dense stands of tall shrubs (*Dushekia fruticosa* (Rupr.) Pouzar) and some species of Salix (*Salix lanata* L., *Salix philicifolia* L.).

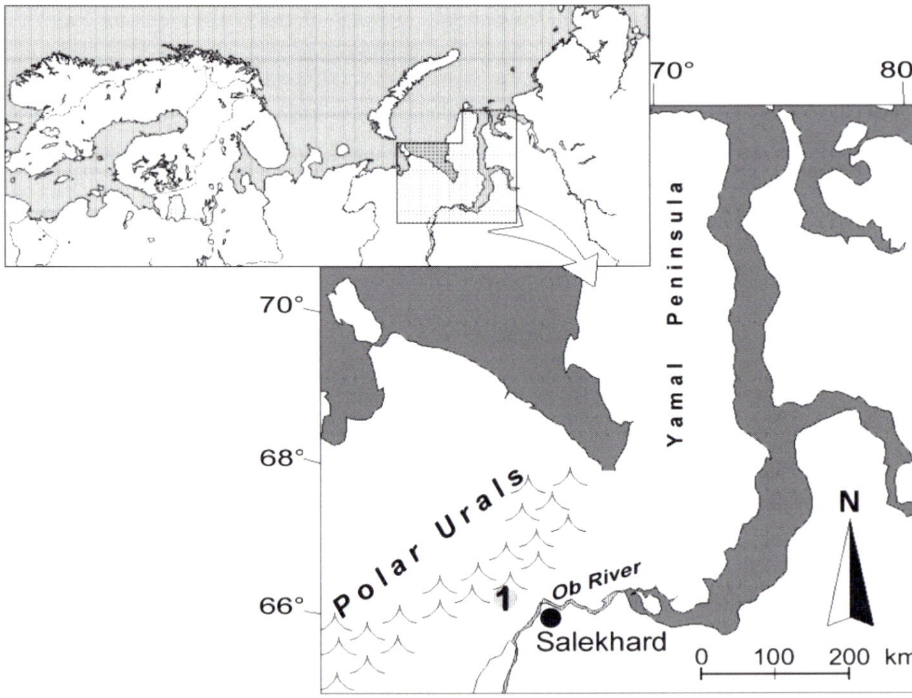

Fig. 1. Location of the study area (1) in the Polar Urals.

Larch growth and morphology is influenced by wind conditions and the effects of major snowdrifts (Fig. 2), which cause structural deviation from the arborescent, monopodial growth form and the development of stem anomalies (matgrowth, wedge, hedge, cornice, flag table tree, multi-stemmed growth forms).

Fig. 2. Different life growth forms of larch trees under wind and snowdrift conditions.

A great number of wood remnants on the ground, up to 60-80 m above the present tree-line ecotone, provide direct evidence of forest-tundra ecosystem changes. These remnants have been preserved for a long time, up to 1500 years, because of the low rate of wood decomposition in these severe climatic conditions (Fig. 3). These circumstances provide us with the possibility of extending the local tree-ring chronology back to AD 459, while at the same time accurately dating the life-spans of a large number of the living and dead trees.

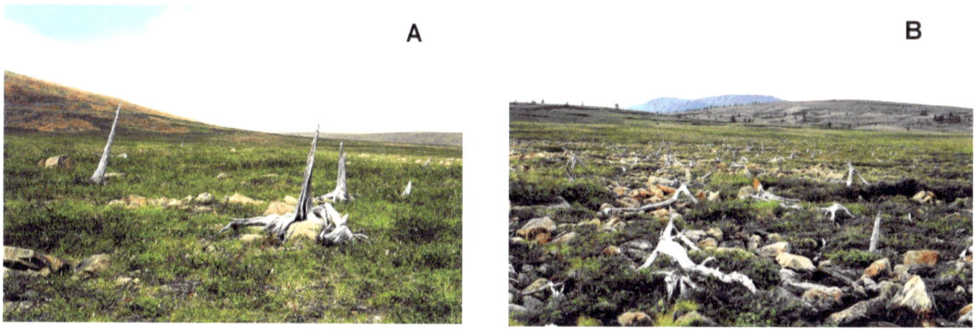

Fig. 3. Remnants of the larch trees which have died off during at the beginning (A) and the end of Little Ice Age (B).

2.2 Field sampling

As the experimental field framework we used altitudinal profiles/transects, laid out previously. Field data on forest stands were obtained by using 6 permanent, continuous altitudinal transects 300-1100 m long and 20-80 m wide. These transects were divided into squares of 20 m x 20 m length-wise and delineated by stone piles in grid corners. On one side of each transect, stone piles were numbered. The locations of all live trees, standing dead trees, and fallen dead trees as well as woody remnants were mapped. Transects typically begin at the upper tree-line of the past, where both live trees and tree remnants cannot be anymore found. The other part of all transects is located in the modern tree-line ecotone and begins in the area of sparse tree growth and open forests, extending further down slope to closed-canopy forest. Transect elevation range is approximately 220-250 m.

From each tree samples were collected for a dendrochronological estimation of their calendar life span and age. On each transect all dead trees and wood remnants were mapped and cross-sections were taken for tree-ring dating of life span. 2262 cross-sections in total from preserved remnants of dead trees were collected. The degree of decay (presence of sapwood, extent of center rot) was noted for all samples. Each specimen was examined prior to cutting to select the radius most likely to contain the maximum number of rings. Not all wood remnants on the altitudinal transect were suitable for dendrochronological analysis because of a high degree of decay, especially in the lower part of transects, where these remnants are found under moss cover. The total number of cross-sections obtained represented >90% of all wood remnants mapped. For approximately half of these remnants, samples were taken from ground-level stump remnants (virtually the upper root) because of the poor preservation of the stem wood.

In order to determine when multi-stemmed trees changed from creeping to upright growth, we cut disks of 23 tree clusters with 2-27 stems in cluster. Cross-sections were taken at the base of their horizontal creeping stems and again at the base of the vertical stem (Fig. 4).

Biomass of each growth form was estimated using selected 34 representative plants (21 single-stemmed, 10 multi-stemmed and 3 creeping). These trees were dug out including the full rooting system (except fine roots). Different compartments of biomass such as roots, stems, branches, needles, and cones were counted, measured, weighted, and sub-samples were taken to the laboratory to dry them at 106 ^{0}C.

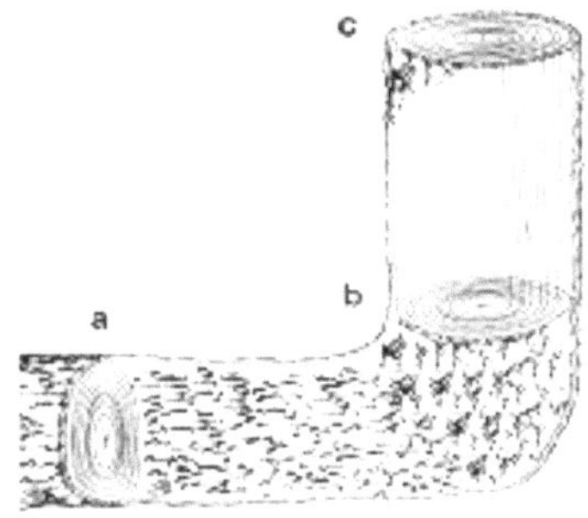

Fig. 4. Three places of wood sampling to determine linear growth rate and the dates when multi-stemmed trees have changed from creeping to upright growth.

2.3 Methods

The main source of information for the spatio-temporal reconstruction of stand structure in response to climate change was the observational record of the long-term monitoring program (about 50 years) in the studied area. The monitoring has been conducted using different methods, including *tree taxation, cartographic, photographic* and *dendrochronological* methods.

Forest taxation method includes measuring the main morphometric parameters of all standing trees and newly grown tree seedlings along pre-determined permanent altitudinal profiles/transects. The most reliable method of the tree cover change assessment in time and space is the identification of age structure of living and dead tree stands as well as tree remnants. These methodologies have been previously successfully tested in conditions of the Polar Urals (Shiyatov, Mazepa, 2007).

In order to evaluate the productivity of tree stands, the method of "representative plants" was used. Both aboveground and below-ground dry biomass was determined. Various compartments were represented: bark and sapwood of tree branches, stems, and roots as well as foliage and reproductive organs. The biomass of each plot within the transects was then estimated by calculating allometric function between tree diameter or sectional areas and the measured compartment biomasses, followed by multiplying these amounts with the mapped tree diameters (Fig. 5). Carbon storage in the biomass was calculated by multiplying biomass with 0.5 (Kobak, 1988).

We also made repeated *large-scale ecosystem mapping* in key areas of the field site to estimate horizontal and vertical shifts of the upper tree-line of various types of forest-tundra and forest communities using GIS technology. A new technique for estimating the shifts of the upper boundary of open and closed forest communities has been developed and tested using highland areas of the Polar Urals as a case study. The technique is based on estimation and GIS representation of boundaries at the beginning and end of the analysis period (Shiyatov et al., 2007).

The *photographic* technique is the method of taking repeated landscape photos from the same locations where historical photos had been taken. This provides valuable information on changes in structure and spatial locations of woody and shrubby vegetation over the past 80-100 years. High-mountainous areas with open views of landscape are ideally suited for using this technique: previous locations of photography can be easily determined using characteristic points on topographic maps for geolocation. The proposed site exhibits all necessary characteristics for using this method. Repeated photos had been taken here previously at each of such locations; the exact geographic coordinates were determined using accurate GPS-receivers.

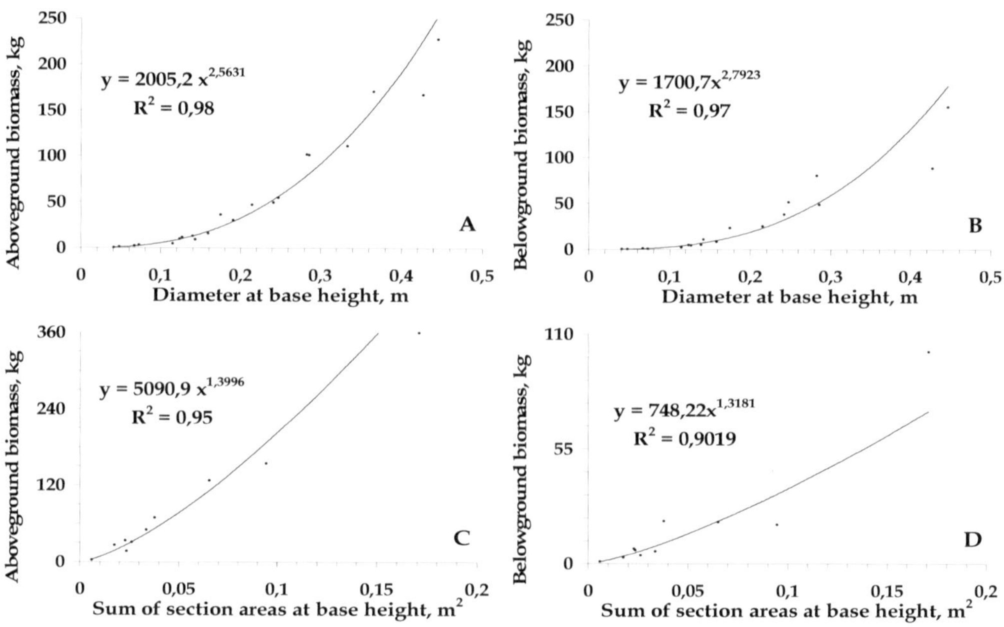

Fig. 5. Tree diameter (or sum of section areas), above- and belowground biomass relationships for single-stemmed (A and B) and multiple-stemmed trees (C and D).

2.3.1 Tree-ring methods

A particular attention was given to the *dendrochronological* method. Specifically, on the basis of tree-ring analysis, biological age and calendar life span of each individual tree can be estimated. The use of this technique in conditions of the Polar Urals, where remnants of dead trees remain intact over long periods of time (up to 1.5 thousand years), is especially promising. The method is also suitable because the correlation of the radial growth among trees over a huge area is very high from year to year.

We used this technique in the study area for a reconstruction of mean summer (June-July) air temperature of the past. A high reliability of the tree-ring reconstruction method in conditions of the Far North is achievable because of high correlations between tree-ring indexes and corresponding climatic factors. The theoretical basis and methods that was used for the reconstruction of climatic conditions of the past have been described in a monograph (Vaganov et al., 1996).

To determine the calendar year of each inner and outer ring, ring widths in each cross-section were measured and cross-dated against existing larch chronologies from the Polar Urals. Cross-dating enabled the assignment of a calendar year to each annual ring. This was accomplished by visually comparing dated and undated samples. A computer program, COFECHA (Holmes 1995), which statistically matches undated samples against samples for which exact dating has been established, was also used to confirm the dating.

Inner and outer ring dates for each cross-section established in this way do not represent the exact germination and mortality dates of the trees because of pre- and post-mortem decay of the wood. The innermost wood in many dead trees was also found to have decayed. For the altitudinal transect-1, laid out within the current tree-line ecotone, we made a detailed analysis for accuracy determination of exact germination and mortality dates of the trees (Fig. 6). It proved possible to cross-date 667 series from the 769 samples collected within the transect-1. The remaining 102 samples had too few rings (from 17 to 43).

A total of 85 cross-sections (average age 171 ± 78 years, ±l SE) had both inner and outer rings (bark rings). The life-span of these trees lasted until the 17th to 19th century (Fig. 6A and 6B), whereas 359 cross-sections (average age 147 ± 60 years) had inner rings only. Germination dates are bimodally distributed (Fig. 6C). Cross-sections from 16 trees (average age 138 ± 55 years) had outer rings (bark rings) only. Death dates are concentrated in the 19th century (Fig. 6D). The remaining 207 cross-sections (average age 118 ± 53 years) had neither inner nor outer rings.

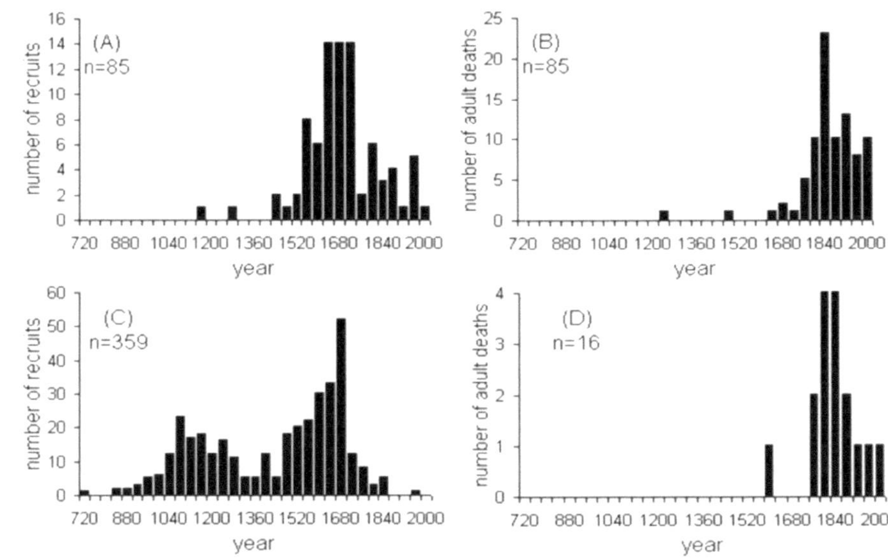

Fig. 6. The life-spans of sampled dead trees, which had both inner (A) and outer rings (B), inner rings only (C), and outer rings only (D).

For most cross-sections with no inner rings surviving, it was generally possible to measure the distance from the innermost measurable ring to the estimated pith position of the cross-section and estimate the number of missing rings using a circle template fit to the ring curvature. The estimate took into account the mean ring width for at least 10 years for the appropriate tree age and time period.

To estimate true establishment date, it is also necessary to account for the relationship between the true ages of living trees (established from hypocotyls) and the ages of the full cross-sections. This relationship was derived on the basis of measurements made on 21 selected trees growing near the transect-1. These trees were dug out of the ground to enable estimates of the biomass to be measured. The trees were ranked by basal diameter and height from 4 cm up to 45 cm and from 2 m up to 13 m, respectively. Sections were also cut from stems, at a height of 20-25 cm, 1.3 m, and every 1 m above the position of the hypocotyls. The dates of the innermost ring on each cross-section were determined. On average, the 9- to 14- year-old trees are 20-25 cm high and the 28- to 43-year-old trees are 1.3

m high. This relationship was used to estimate the establishment date of those trees surviving only as subfossils.

Because of postmortem decay and erosion of outer-surface wood, the outer ring date of these samples is in most cases not the true mortality date. In general, it was not possible to establish the "true" mortality date with great confidence; however, in some cases a closer estimate could be produced by using additional sections from the same remnants, but taken from different parts of the stumps.

3. Results

3.1 Stand age structure since AD 1500

The analysis of 2 altitudinal transects within the current tree-line ecotone has shown that the locations of the dead and living dated larch trees was not uniform (Fig. 7).

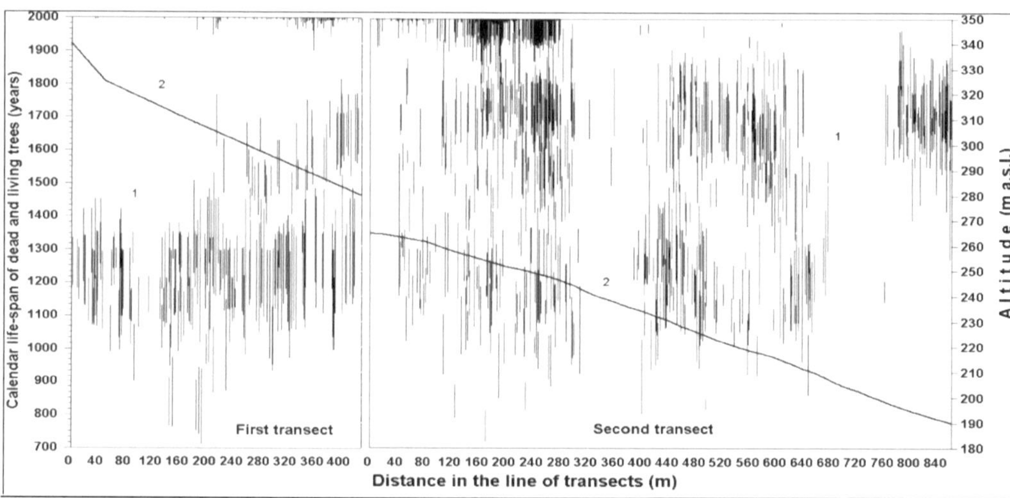

Fig. 7. Location of the dated larch remnants and living trees along transects and their life-span. Time of appearance of living trees in the bottom part transect-2 is not shown: (1) life span of trees; (2) altitude changing of transects.

The overwhelming majority of remnants occur within three parts of the second transect. The first forest island became established between 30-40 m and 300-320 m in distance, along transect. The second stand is situated from between 380-400 m and 640-660 m further down. The third stand occupies an area from 750-770 m up to the end of transect. Live trees grow in the same parts of transects. Such a spatial distribution of trees is caused by the particular micro relief in different parts of transects. The higher part of transect is almost at the uppermost point of the moraine. Here, there are no barriers from wind, and this site is similar to a wind-dominated ecological type of tree-line. At a distance of about 320 m from the beginning of transect-1, there is a slope that is formed by stone ledges. Here, during wintertime, increased snowdrifts up to 3-6 m in height accumulate. Such a significant depth of snow and the associated delay in the spring thaw that results from it are unfavorable factors for the survival of undergrowth. Absence of forest stands in these parts of transect is, therefore, caused by high snow depths in the wintertime. The same micro relief is found at a distance of about 680 m from the top of transect.

Wood macrofossils from the bottom part of the transect date mainly to the 16th to 19th centuries. On this site the more ancient remnants have not, in general, been preserved. The reason is that this part of transect is a closed forest stand with mature moss cover. Such conditions lead to increased decomposition of wood. The greatest number of undated samples was found in this part of transect, and their poor state of preservation meant that many samples could not be taken at all.

Using the calendar life-spans of the dead, changes in abundance through time were estimated by calculating the number of trees that were alive during each of the years from 1300 years ago to the present. Abundance estimates are subject to two specific errors. First, the sample is biased toward larger trees. Saplings and small adult trees are likely to decompose more rapidly than large trees, resulting in their systematic exclusion from the sample. Second, the record deteriorates back in time because of the decomposition of all wood. The sample therefore includes an increasing proportion of the original forest composition in more recent periods and would be expected, as a result of decomposition alone, to indicate a trend of increasing tree abundance with time. Although trees have been dated as far back as AD 720, the analysis has been truncated at AD 1000 to minimize the influence of the diminishing record (Fig. 8). An early peak of abundance is observed from the end of the 12th century to the second half of the 13th century. For this time interval the average stand density is estimated at 50-60 trees/ha. A correction was made in the observed tree abundance at the end of the 12th century and the second half of the 13th century.

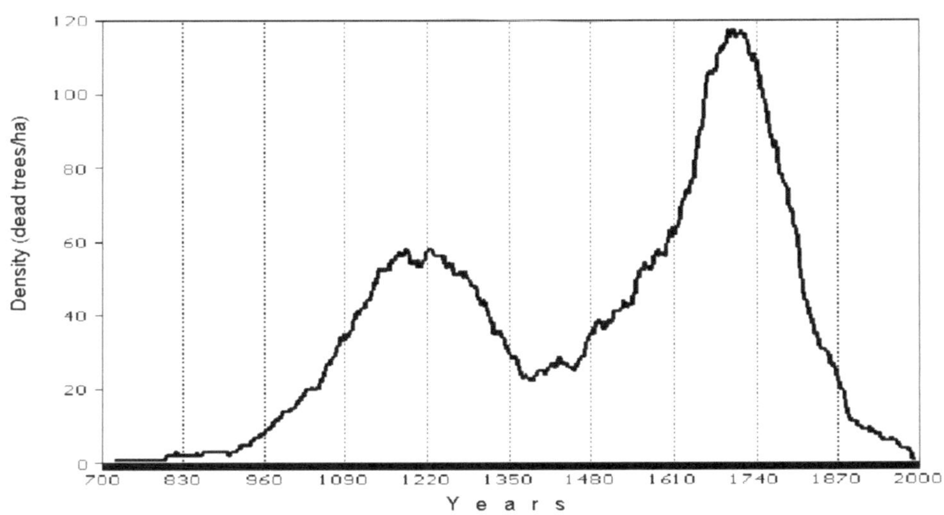

Fig. 8. Changes in stand density within the current tree-line ecotone.

This is based on the earlier observation that the greatest number of undated samples was found in the bottom part of transect, a result of poor preservation. This produces an "empty" area in the bottom right corner on (see Fig. 7). It is assumed, however, that trees would have been growing at these times in this part of transect, as they clearly grew in the upper part of transect. It has therefore been assumed that the life-span of most of the undated wood macrofossils from the lower part of transect can likely be dated from the 9th to 14th centuries. The estimate of tree abundance for the end of the 12th century and the second half of the 13th century is, therefore, increased up to 100-120 trees/ha.

Abundance began to decline after AD 1280-1300. Minimum tree abundance was reached during the interval 1360-1450. A second period of high tree abundance occurred between 1660 and 1740. The decline began again after 1730. Present-day living trees show that the second minimum of tree abundance was reached at the second half of the 19th century.
The proportional age structure of the trees in these transects has varied greatly during the last millennium (Fig.9). There is clear evidence that two distinct generations of trees appeared, between the 10th to 11th and 16th to 17th centuries. The percentage of total trees represented by these cohorts throughout their life-spans remained appreciable for 300-350 years. Some trees from the second generation are still alive today. The generation that appeared on the boundary between the 13th and 14th centuries was not so numerous. In 200-250 years the proportion of these trees was insignificant. The generation, which appeared in the 18th and 19th centuries, was even smaller. The number of these trees within the total age structure did not exceed 15-20%.

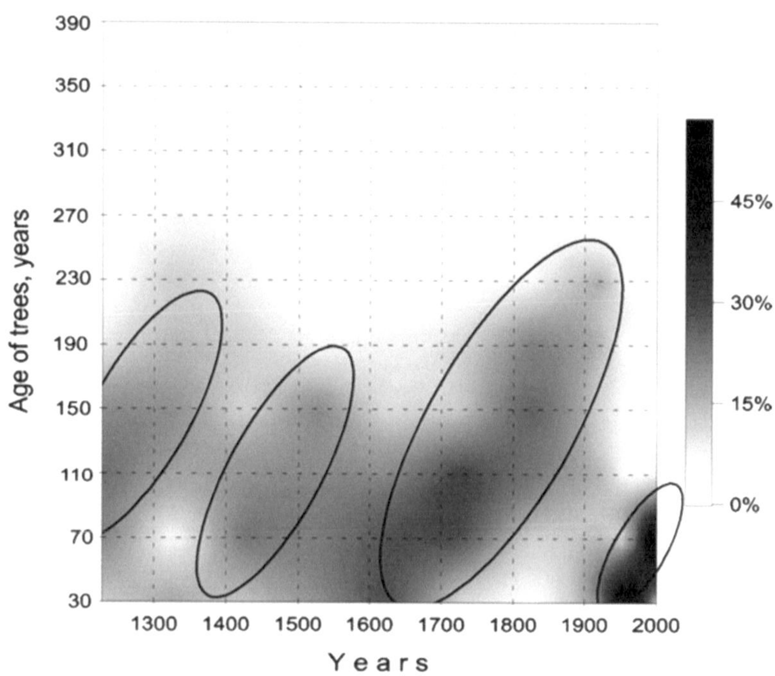

Fig. 9. Changing age structure of tree populations in the current tree-line ecotone. Distinct generations are marked by ellipses.

The analysis of 4 transects within the former tree-line ecotone, now being as a treeless site and 60-80 m above modern one, has shown that the locations of the dead, dated larch trees was more or less uniform because of microrelief (Fig. 10). Peak of abundance is observed from the end of the 11th century to the second half of the 14th century. The large generation in the 18th and 19th centuries did not appear.
On the base of life-span for dead trees (more than 1200 pieces) which grew above current tree-line ecotone the estimation of the most high-altitude position of tree-line for the last 1500 years was received (Fig. 11). The most high-altitude position of tree-line during the medieval climatic optimum was estimated at altitude 400-420 m a.s.l. This border occupied

the highest position during 13th and in the beginning of 14th centuries. After that the huge dying off of trees, decrease of sparse tree growth and light forests up to the beginning of 20th century has begun. Intensive decrease of this border and thinning of forest stands occurred in 15th and 16th centuries and especially in 19th century.

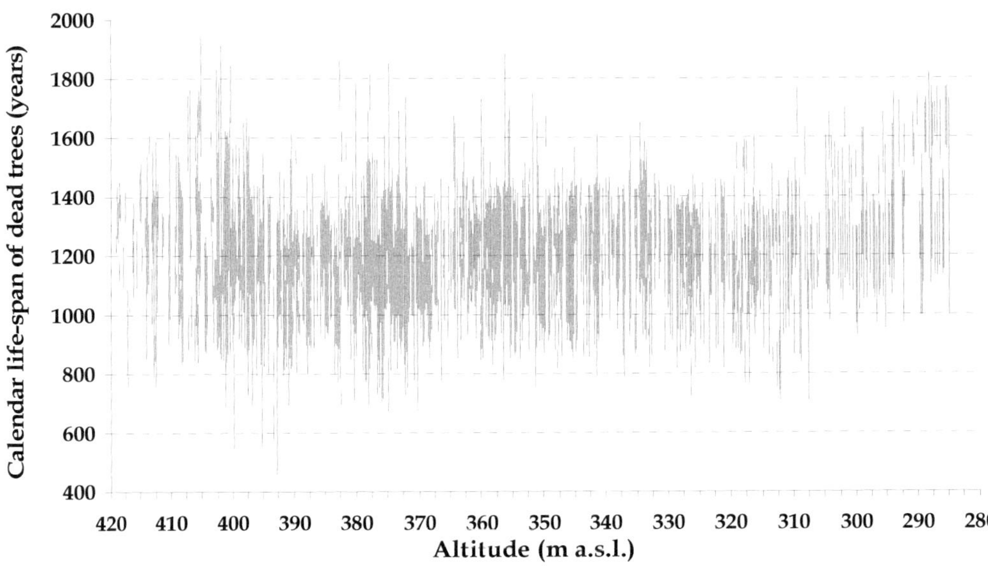

Fig. 10. Location of the dated larch remains along transects and their life-span. Transects laid out 80 m above current tree-line.

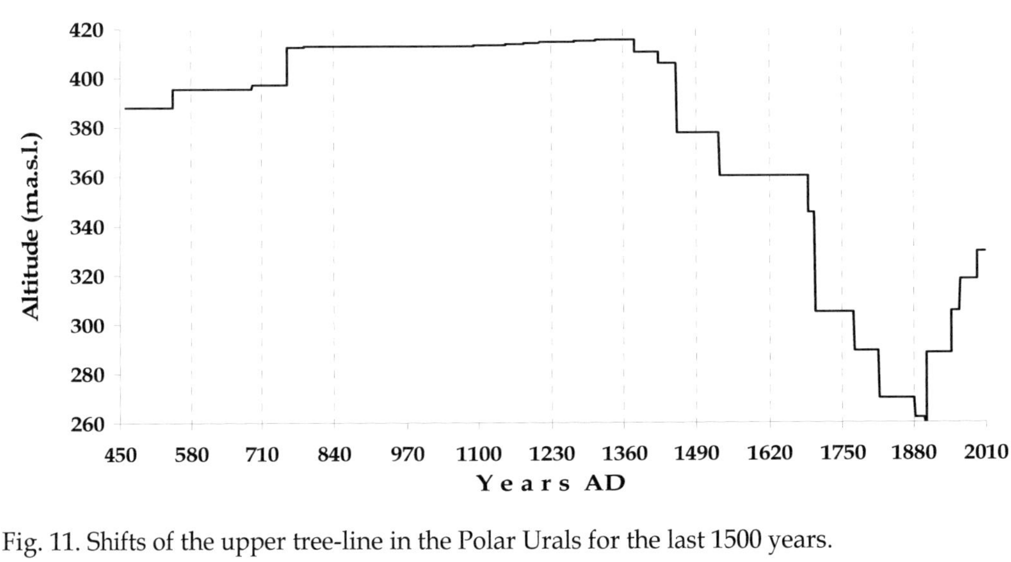

Fig. 11. Shifts of the upper tree-line in the Polar Urals for the last 1500 years.

The estimation of changes in forest stand biomass for the last millennium was received. The long trends of changes in dynamics of productivity are well expressed, which caused by century fluctuations of thermal conditions of summer months. It is clearly visible, that trends of biomass changes and tree-ring indexes are the same. It testifies to an one-orientation of these processes. High-altitude position of sparse tree growth, light forests and density of forest stands synchronously changed as well (Fig. 12).

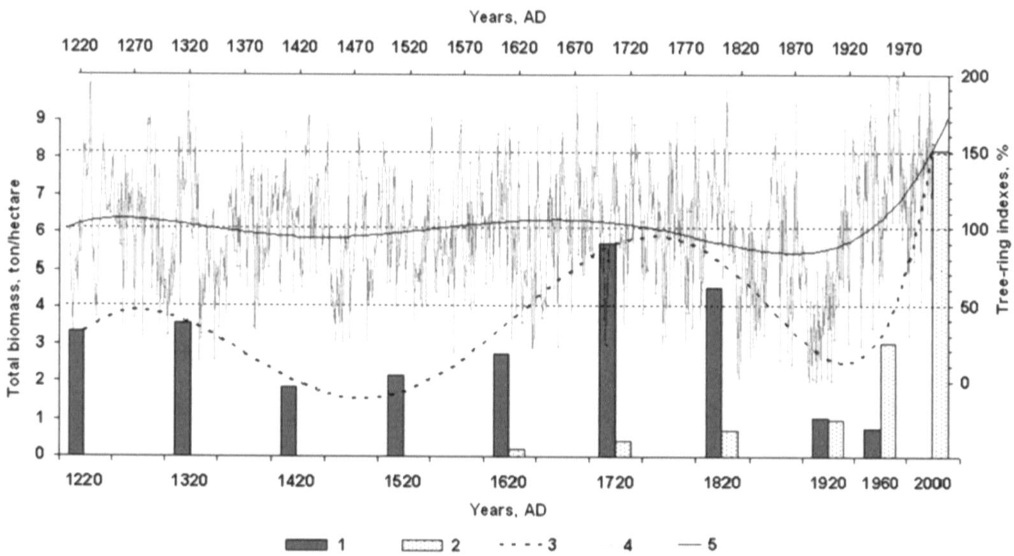

Fig. 12. Open forest's biomass changing in the last 800 years: (1) biomass of dead trees; (2) biomass of living trees; (3) trend of biomass changing; (4) tree-ring indexes; (5) trend of tree-ring indexes.

Previous studies have concluded that increases in tree-line elevation, and associated increases in tree abundance within the transient tree-line ecotone, are associated with extended warm periods (Tranquillini, 1979; Kullman, 1986).

The increase in tree abundance in the upper tree-line ecotone in the Polar Ural Mountains that has been dated to between AD 1150 and 1250 corresponds to a period of relatively warm June–July temperatures, as inferred from Siberian larch ring-width change adjacent to the study area. Early medieval times, from about 1100 to 1250, experienced warming above the long-term mean and relatively stable interannual temperature variability. Reconstructed June–July temperatures generally remained high until the late 1500s, with two major exceptions: the cool periods centered on 1300 and 1460. These times represent two of three periods with the most rapid decreases in temperature observed between adjacent 20-year periods (Graybill & Shiyatov, 1992). The episode of increasing tree abundance between 1660 and 1740 corresponds to a shorter period, which is dominated by positive temperature anomalies relative to the 1951-1970 mean. Reconstructed interannual temperature variations after 1500 are highest between 1750 and 1769, with a lesser peak occurring again in the late 19th and early 20th centuries.

The species structure of forest vegetation and climate in the Holocene was reconstructed on the basis of macroscopic plant remnants, botanical analysis of peat, radiocarbon dating (Koshkarova et al., 1999), and palynological, paleocarpological, and paleoentomological analyses of frozen deposits (Panova et al., 2003) in the Polar Ural peatland (Massif Rai-Iz).

The results showed that the upper forest limit repeatedly migrated upward for 220-400 m in the periods of warming and retreated during cold periods.
As the vertical gradient of summer air temperature in the Polar Urals is 0.7 °C/100 m, the temperature dependent upper boundary of the zone suitable for tree growth rose by approximately 100 m (Shiyatov, 2003). On the majority of slopes, however, trees failed to expand up to this potential elevation because of an insufficient seed supply to the tundra areas located in the upper part of the ecotone. As shown previously (Shiyatov, 1966), larch seeds in this region fall in June or July, when the snow cover disappears, and are carried by wind no farther than 40-60 m away from the fertile tree. Hence the spread of seeds up the slope is very slow. That is why many habitats suitable for tree growth have as yet remained vacant. This line of reasoning may also be used to explain more active natural afforestation within the tundra and transformation of sparse stands into closed forests in the lower part of the tree-line ecotone, where microclimatic and soil conditions are more favorable and seed supply is more abundant.

3.2 Twentieth-century stand dynamics
3.2.1 Growth form changes
The upward expansion of forests was accompanied by significant changes in tree growth forms (Fig. 13). Larches that had been growing in a creeping form since the 15th century started to grow upright at the beginning of the 20th century, with up to 20 stems per tree individual. More than 90% of the trees emerging after 1950 are single-stem ones.

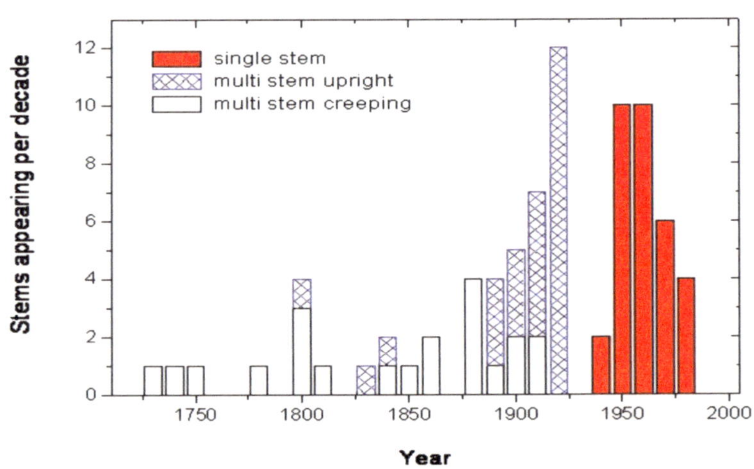

Fig. 13. Growth form changes of larch trees in exposed sites under snowpack by snow abrasion, low temperatures and wind.

The changes of multi-stemmed trees from creeping to vertical growth led to a 2- to 10-fold increase in radial growth of the creeping stems (Fig. 14). We mainly attribute these strong increases in ring width to the higher photosynthetically active needle area from several emerging vertical stems. More favorable climatic conditions might also have contributed to the growth enhancements, but the increases in ring widths of the horizontal stems of multi-stemmed trees were much greater and more abrupt than the climatic-driven increases in the ring widths of single-stemmed trees.

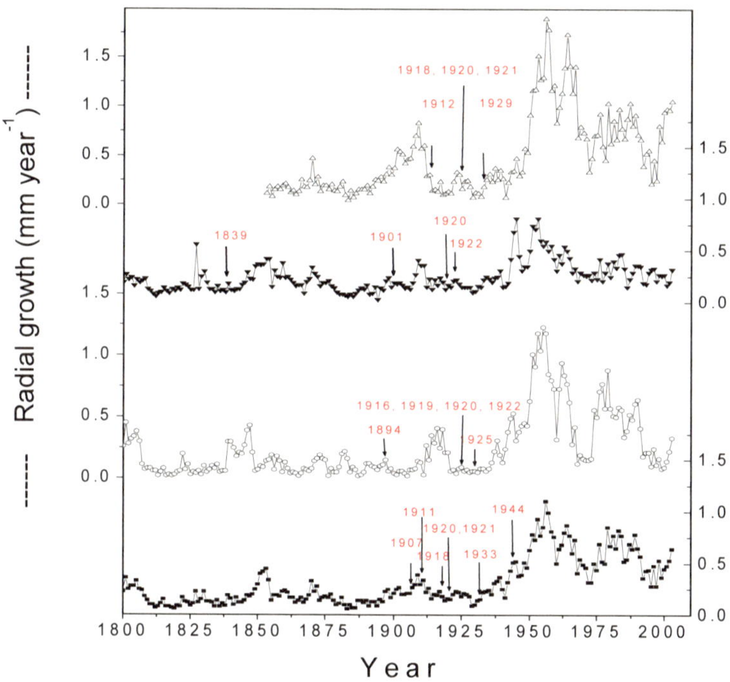

Fig. 14. Tree-ring widths of horizontal stems from multi-stem trees during the last two centuries. The arrows indicate the beginning of upright stem growth.

According to our tree-ring analysis, vertical stems of multi-stemmed larch trees emerged during a relatively short period, i.e. only a few decades (Figs 7 and 8). This contrasts with the behavior of 'mobile', 500-year-old tree islands in the Colorado Front Range, where windward edges are dying back and new stems are emerging over long periods on the leeward side (Marr, 1977; Benedict, 1984). Our age analysis of horizontal and vertical stems indicate that in the Polar Urals vertical growth of multi-stemmed larch is a much more recent phenomenon, and because climatic conditions have apparently been changing rapidly, multi-stemmed trees are very likely only a transitional growth form. Single-stemmed larches are dominating already now 50 m below current treeline. Therefore, when conditions are becoming slightly more favorable, larch seedlings can succeed as single stems, and they do not first grow in a creeping form before they start to grow upright. As multi-stems are a in transitional growth form, they are indicators for changing growth conditions that can easily be dated using dendrochronological methods.

3.2.2 Large-scale mapping

We concentrate our attention on the recent expansion of forest-tundra ecosystems, which began 80-90 years ago in connection with climate warming and moistening. Summer temperature (since 1921) at Salekhard weather station increased by 0.9 °C and winter temperature by 1.2 °C in comparison with the first period of observation (1883-1920). It means that summer temperature isotherm rose 120-130 m in altitude. Mean precipitation of summer months is increased from 146 to 178 mm and of winter months from 67 to 113 mm.

To estimate spatio-temporal changes of forest-tundra ecosystems, which occurred during the 20th century, special attention was given to description and large-scale mapping of forest-tundra ecosystems over the treeline ecotone. To present day the area of 5770 hectares was mapped. Two maps were developed which show the state of stands for the beginning and end of the 20th century (Fig. 15).

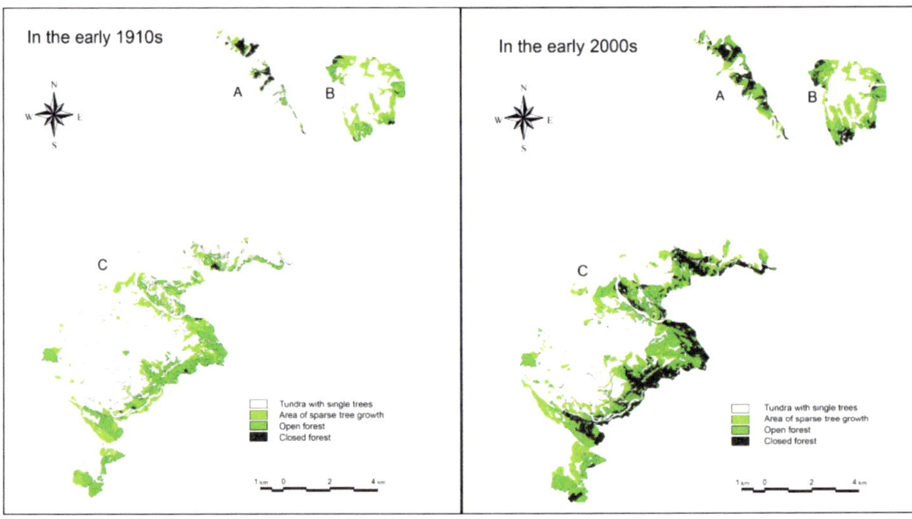

Fig. 15. The spatial distribution of different forest-tundra communities in the ecotone of the upper tree-line in the Polar Urals during 1910-s and 2000-s: (A and C) the area of the Rai-Iz massif and Mts. Chernaya and Malaya Chernaya; (B) the area of Mt. Slantcevaya.

During the last 90 years a significant afforestation of tree-line ecotone has occurred. The area under tundra with individual trees decreased from 4399 to 3306 ha or 25%, the area under sparse growth of trees increased from 662 to 775 ha or 15%, the area under open forests increased from 640 to 1066 ha or 40%. The most impressive changes were seen with closed forests. The area increased from 69 to 623 ha or 89% as a result of the transformation of sparse growth of trees and open forests into closed forests. The degree afforestation within the tree-line ecotone (including sparse growth of trees, open forests and closed forests) has increased from 1371 ha to 2464 ha or 45%.

Impressive changes have occurred in the structure and productivity of existing stands during the last 90 years. From the beginning of warming an intensive renewal larch and spruce occurred. Most of stands have become much denser and more productive (up to 2-5 times) and many tundra sites located within the treeline ecotone have been afforested. To date young generation of trees is formed, which come to upper wood canopy and occupies dominating position in the majority of stands. This generation is presented basically by single stem form of growing even on powerfully wind sites, while middle-aged generation is presented mainly multi-stem form of growing.

3.2.3 Repeat landscape photographs

To date, repeat photographs are made from 1200 points, the majority of which were used for quantitative and qualitative estimation of changes in the tree and shrub vegetation that have occurred over the past 30-50 years. During this period of time there was an intensive expansion of tree and shrub vegetation into mountain tundra. These changes occurred under the influence of the modern warming and moistening of climate (Fig. 16-23).

Fig. 16. The western extremity of the stand strip located at the bottom of a southern slope of Rai-Iz massif (66°50.853' N, 65°36.058' E, 295 m a.s.l.).

Fig. 17. General view of moraine on the left bank of Kerdomanshor River (66°50.326' N, 65°34.294' E, 226 m a.s.l.).

Fig. 18. The bottom of eastern slope of Malaya Chernaya Mountain (66°50.751' N, 65°32.770' E, 286 m a.s.l.).

Fig. 19. South-western slope of Yar-Keu Mountain (66°56.641' N, 65°43.668' E, 314 m a.s.l.).

Fig. 20. Spruce growth form changes from creeping to multi-stem growth (66°48.923'N, 65°34.310'E, 251 m a.s.l.).

Fig. 21. The top part of a lateral moraine (66°47.477' N, 65°30.778' E, 267 m a.s.l.).

Fig. 22. Southern slope of a hill of 429 m. Small river Orehyogan below (66°47.338' N, 65°23.601' E, 287 m a.s.l.).

Fig. 23. The uppermost light forest growing on a southern slope of height of 312.8 m (66°49.060' N, 65°32.262' E, 267 m a.s.l.).

4. Conclusion

The earlier analysis (Shiyatov 1993, Fig. 1) showed a similar conclusion based on 209 macro fossils and 16 living trees. A maximum tree abundance being reached in the middle of the 13th century. In the first half of the 15th century a deterioration of forest cover came to an end. In the 16th century and first half of the 17th century tree growing conditions were again more favourable, and in the second half of the 16th century another decline of the open larch forests ceased. In the second half of the 18th century all trees growing on the transect died. In the 19th century not a single living tree stood on the transect. The first larch seedlings reappeared at the beginning of the 20th century. These two independent analyses are, therefore, mutually supportive.

The upper boundary of open forests has markedly shifted both on slopes exposed in winter to strong westerly winds and in the areas where summer temperatures are the main limiting factors. This is evidence that the pattern of winds and the temperature conditions have become more favorable for the growth of trees over the past 90 years. The growing period begins earlier, and its duration has increased. This allows young larch shoots to complete the cycle of their growth and development and to prepare themselves for wintering under severe conditions. Before the recent climate warming, larch in windswept conditions with little snow was represented only by prostrate growth form. To date, prostrate plants have transformed into multi-stemmed trees up to 5-6 m high and young trees aged up to 50-60 years are mainly single-stemmed.

These data on the magnitude of altitudinal and horizontal shifts in the upper boundaries of open and closed forests in the Polar Urals confirm our previous findings concerning the expansion of tree vegetation to higher elevations due to climate warming and increasing humidity in the 20th century (Shiyatov et al., 2007). According to data from the Salekhard weather station the recent mean air summer temperature was 0.9 °C higher and winter was 1.2 °C higher than between 1883 and 1919; and the amount of precipitation also increased by 32 mm in summer and by 46 mm in winter. As the elevational lapse rate of summer air temperatures in the Polar Urals is 0.7 °C, the climate-dependent upper boundary of the zone suitable for tree growth could ascend approximately 100 m. This value is almost three times greater than the actual altitudinal shift of the upper boundary of open and closed forests. In other words, tree vegetation has not yet expanded to the climatic limit because of insufficient supply of larch seeds to tundra areas in the upper part of the treeline ecotone, although it is already close to this limit on some slopes.

Thus, the analysis of vertical and horizontal shifts in the upper boundaries of open and closed forests on the eastern macroslope of the Polar Urals provides evidence for widespread and active expansion of tree vegetation to higher elevations over the past 90 years, which is apparently explained by favorable changes in climatic conditions.

The improvement of climatic conditions in the first half of the 20th century resulted in the development of multi-stemmed life forms in Siberian larch through the formation of vertical stems from its prostrate and stem-shrub life forms. The active growth of multi-stemmed trees and consequent increase in their size during the 20th century resulted in increasing competition for light and nutrients between stems and crowns within the same clone. As a consequence, some stems dried out. Data on the time of tree emergence obtained in this study agree with the known fact that active forest regeneration takes place mainly in relatively warm climatic periods with higher humidity.

Historic photographs and tree age structure analyses indicate that remote areas of the forest-tundra ecotone in the Polar Urals have changed significantly during the 20th century. The distribution of Siberian larch has been shifting upwards by 20-60 m in altitude, and today young larch trees are growing in formerly treeless tundra. Since 1900, growth forms of larch have started to change from creeping growth that prevailed over centuries to upright growth as multi-stemmed trees. At the same time, single-stemmed trees have commenced to establish. At present, the single-stemmed larches are dominating the forest 50 m below the species line. The expanding close forest led to an increase of biomass by 40-75 t ha^{-1} and a carbon sequestration of approximately 20-40 gCm^{-2} yr^{-1} during the last century, which is, however, small compared with temperate forests. The forest expansion and change in growth forms coincided with a significant warming in summer and a doubling of winter precipitation, thus drastically changing snow conditions. In summary, our study shows that the ongoing climatic changes are altering the composition, structure, and growth forms of Siberian larch in undisturbed tree-line ecotones of the Polar Urals.

The absence of evidence of forest fires or other catastrophic phenomena in the study area and the similar absence of any evidence of significant human influence on the open larch forests strongly attest to the probability that forest dynamics in the region are the result of climatic changes alone. If the reconstruction of these tree dynamics describes the changing influence of climate in the Polar Urals Mountains during the period represented in this study, then the hypothesis that the expansion of forest is always associated with warm temperatures is generally supported by these results.

5. Acknowledgments

This research was financially supported by the Russian Foundation for Basic Research (Grants: 04-04-48687, 07-04-00847, 08-04-00208, 10-04-01484, 11-04-00623), in part by INTAS project 01-0052 and projects ГК 02.740.11.0279, 09-П-4-1004, 09-Т-4-1005. We thank Lyudmila Gorlanova and Natalia Ukhabina for laboratory assistance and students for field assistance.

6. References

Andreev, V.N.; Igoshina, K.N. & Leskov, A.I. (1935). Reindeer pastures and vegetation cover in the Polar Urals. *Soviet Deer Farming,* No.5, pp. 171-406. [In Russian].

Becker, A. & Bugmann, H. (2001). *Global Change and Mountain Regions: The Mountain Research Initiative.* IGBP Report 49. IGBP Secretariat, Stockholm, Sweden.

Benedict JB (1984) Rates of tree-island migration, Colorado Rocky Mountains, USA. *Ecology,* No. 65, pp. 820-823. ISSN: 0012-9658

Briffa, K.R. & Jones, P.D. (1993). Global surface air temperature variations during the twentieth century. Part 2. Implications for large-scale high-frequency palaeoclimatic studies. *The Holocene*, Vol.3, pp. 77-88, ISSN: 0959-6836

Briffa, K.R.; Jones, P.D.; Schweingruber, F.H.; Shiyatov, S.G. & Cook, E.R. (1995). Unusual twentieth-century summer warmth in the 1000-year temperature record from Siberia. *Nature,* Vol.376, pp. 156-159, ISSN: 0028-0836

Chapin, F.S. III.; McGuire, A.D.; Randerson, J. et al. (2000). Arctic and boreal ecosystems of western North America as components of the climate system. *Global Change Biology,* 6, 211–223. ISSN: 1365-2486

Chapin, F.S. III.; Sturm, M.; Serreze, M.S. et al. (2005). Role of landsurface changes in arctic summer warming. *Science*, Vol.310, pp. 657-660. ISSN: 0869-7078

D'Arrigo, R.D.; Kaufmann, R.K.; Davi, N.; Jacobi, G.C.; Laskowski, Ch.; Myreni, R.B. & Cherubini, P. (2004). Thresholds for warming-induced growth induced growth decline at elevational tree line in the Yukon Territory, Canada. *Global Biochemical Cycles*, 18, GB3021, doi: 10.1029/2004GB002249.

Devi, N.; Hagedorn, F.; Moiseev, P.A.; Bugmann, H.; Shiyatov, S.; Mazepa, V. & Rigling, A. (2008). Expanding forests and changing growth forms of Siberian larch at the Polar Urals treeline during the 20th century. *Global Change Biology*, Vol.14, pp. 1581-1591. ISSN: 1365-2486

Esper, J. & Schweingruber, F.H. (2004). Large-scale treeline changes recorded in Siberia. *Geophysical Research Letters,* 31, L06202, doi: 10.1029/2003GL019178.

Gorchakovskiy, P.L. & Shiyatov, S.G. (1985). *Phytoindication of the Environmental Conditions and Natural Processes in the Alpine Terrain.* Nauka, Moscow. [in Russian].

Gorodkov, B.N. (1926). The Polar Urals in the watershed of the Sob River. In *Proc. Bot. Mus. USSR Acad. Sci.* Vol.19, pp. 1-74. [In Russian].

Goroshkevich, S.N. & Kustova, E.A. (2002). Morphogenesis of the creeping life form of Siberian cedar at the upper limit of proliferation in West Sayan Mountains. *Russian Journal of Ecology*, No.4, pp. 243–249. ISSN: 1067-4136

Graybill, D.A., & Shiyatov, S.G. (1992). Dendroclimatic evidence from the northern Soviet Union. In Climate since A.D. 1500, R.S. Bradley & P.D. Jones, (Ed.), Routledge, London. pp. 393–414. ISBN: 0-415-07593-9.

Holmes, R.L. (1995). Dendrochronological Program Library [computer program]. Laboratory of Tree-Ring Research, the University of Arizona, Tucson, Ariz.

Holtmeier, F-K. (2003). *Mountain Timberlines. Ecology, Patchiness, and Dynamics.* Kluwer, Dordrecht. ISBN: 1-4020-1356-6

IPCC. (2007). *Climate Change 2007: The Physical Science Basis. Contribution of Working Group I to the Fourth Assessment Report of the Intergovernmental Panel on Climate Change.* In: Solomon, S., Qin, D., Manning, M., Chen, Z., Marquis, M., Averyt, K.B., Tignor, M., Miller, H.L. (Eds.). Cambridge University Press, Cambridge, United Kingdom and New York, 996 pp.

Jackson, S.T.; Overpeck, J.T.; Thompson, W.III.; Sharen, E.K. & Katherine, H.A. (1997). Mapped plant-macrofossil and pollen records of late quaternary vegetation change in Eastern North America. *Quaternary Science Reviews*, Vol.16, No.1, pp. 1-70. ISSN: 0277-3791

Kelly, P.M.; Jones, P.D.; Sear, C.B.; Cherry, B.S.G. & Tavakol R.K. (1982). Variations in surface air temperature: part 2. Arctic regions. 1881-1980. *Monthly Weather Review*, No.110, pp. 71–82. ISSN: 0027-0644

Kharuk, V.I.; Shiyatov, S.G.; Kashishke, E.; Fedotova, E.V.; Naurzbaev, M.M. (2002). Forest-tundra ecotone response to climate change. In: *Problems of ecological monitoring and ecosystem modeling.* Vol. XVIII, Sankt-Petersburg: Gidrometeoizdat, pp. 234-260. ISBN 5-286-00655-8

Kittel, T.G.F.; Steffen, W.L. & Chapin, F.S. (2000). Global and regional modelling of arctic-boreal vegetation distribution and its sensitivity to altered forces. *Global Change Biology*, No.6, pp. 1–18. ISSN: 1365-2486

Körner, Ch. (1999). *Alpine plant life.* Springer-Verlag, Berlin.
Kobak, K.I. (1988). *Biotic components of carbon circle.* Gidrometeoizdat, Leningrad. [In Russian].
Koshkarova, V.L., Karpenko, L.V., & Orlova, L.A. (1999). The Holocene dynamics of vegetation and the upper forest limit in the Polar Urals. Rus. J. Ecol., No.30, pp. 102–106. ISSN: 1067-4136
Kullman, L. (1986). Late Holocene reproductional patterns of Pinus sylvestris and Picea abies at the forest limit in central Sweden. Can. J. Bot. No.64, pp. 1682–1690. ISSN: 0008-4026.
Kullman, L. (1990). Dynamics of altitudinal tree-limits in Sweden: a review. *Nor. Geogr. Tidsskr.* B, No.44, pp. 103-116.
Kullman, L. (2002). Rapid recent range-margin rise of tree and shrub species in the Swedish Scandes. *Journal of Ecology,* No.90, pp. 68–77. ISSN: 1365-2745
Lloyd, A.H. (1997). Response of tree-line populations of foxtail pine *(Pinus balfouriana)* to climate variation over the last 1000 years. *Can. J. For. Res.* No.27, pp. 936-942. ISSN: 0045-5067
Lloyd, A.H. (2005). Ecological histories from Alaskan tree lines provide insight into future change. *Ecology,* No.86, pp. 1687–1695. ISSN: 0012-9658
Lloyd, A.H., & Graumlich, L.J. (1997). Holocene dynamics of treeline forests in the Sierra Neveda. *Ecology,* No.78, pp. 1199-1210. ISSN: 0012-9658
Marr, J.W. (1977). The development and movement of tree islands near the upper limit on tree growth in the Southern Rocky Mountains. *Ecology,* No.58, pp. 1159–1164. ISSN: 0012-9658
Martin, Yu.L. (1970). Lichenometric indication of time of rock outcrop's formation. *Russ. J. Ecol.* No.5, pp. 16-24. [In Russian].
Mazepa, V.S. (2005). Stand density in the last millennium at the upper timberline ecotone in the Polar Ural Mountains. *Canadian Journal of Forest Research,* Vol.35, pp. 2082–2091. ISSN: 0045-5067
Mazepa V.S. & Devi N.M. (2007). Development of multi-stemmed life forms of Siberian larch as an indicator of climate change in the timberline ecotone of the Polar Urals. *Russian Journal of Ecology,* Vol.38, No.6, pp. 440–443, ISSN: 1067-4136
Miller, G.H.; Alley, R.B.; Brigham-Grette, J.; Fitzpatrick, J.J.; Polyak, L.; Serreze, M.C. & White, J.W.C. (2010). Arctic amplification: can the past constrain the future? *Quaternary Science Reviews,* Vol.29, pp. 1779–1790. ISSN: 0277-3791
Moiseev, P.A. & Shiyatov, S.G. (2003). Vegetation dynamics at the treeline ecotone in the Ural Highlands, Russia. *Ecological Studies,* No.167, pp. 423–435. ISSN: 0070-8356
Myneni, R.B.; Keeling, C.D.; Tucker, C.J.; Ascrar, G. & Nemani, R.R. (1997). Increased plant growth in the northern high latitudes from 1981 to 1991. *Nature,* No.386, pp. 698–702. ISSN: 0028-0836
Panova, N.K., Jankovska, V., Korona, O.M., & Zinov'ev, E.V. (2003). The Holocene dynamics of vegetation and ecological conditions in the Polar Urals. Rus. J. Ecol. No.34, pp. 219–230. ISSN: 1067-4136.
Paulsen, J. Weber, U.M. & Körner, C. (2000). Tree growth near treeline: abrupt or gradual reduction with altitude? *Arctic, Antarctic, and Alpine Research,* No.32, pp. 14–20.

Pereg, D. & Payette, S. (1998). Development of black spruce growth forms at tree-line. *Plant Ecology*, No.138, pp. 137–147. ISSN: 0065-2113

Shiyatov, S.G. (1965). Age structure and generation of larch light forests in the treeline ecotone in the Ob river basin (Polar Urals). In: Proceedings of the Institute of Biology, Ural Division of the Academy of Sciences of the USSR: *Geography and Dynamics of Plant Community*. Sverdlovsk, No.42, pp. 81-96. [In Russian].

Shiyatov, S.G. (1966). Time of dispersion of Siberian larch seeds in the north-western part of its areal and role of that factor in mutual relation between forest and tundra. In Problems of physiology and geobotany. Publication of the Sverdlovsk Branch of the All-Union Botanical Society, No.4. pp. 109–113. [In Russian.]

Shiyatov, S.G. (1986). *Dendrochronology of treeline in the Urals.* Nauka Publishers, Moscow. (In Russian).

Shiyatov, S.G. (1993). The upper timberline dynamics during the last 1100 years in the Polar Urals Mountains. *Eur. Palaeoclim. Man*, No.4, pp. 195-203.

Shiyatov, S.G. (2003). Rates of change in the upper tree-line ecotone in the Polar Ural mountains. *Pages News*, No.11, pp. 8–10. ISSN: 1563-0803

Shiyatov, S.G. (2009). *Dynamics of Wood and Shrub Vegetation in the Polar Ural Mountains under the Modern Climate Change.* UB RAS. ISBN: 978-5-7691-2035-0, Ekaterinburg. [in Russian].

Shiyatov, S.G. & Mazepa, V.S. (2007). Climatogenic dynamics of forest–tundra vegetation at the Polar Urals. *Lesovedenie*, No.6, pp. 37–48. ISSN: 0024-1148 [in Russian].

Shiyatov, S.G.; Terent'ev, M.M., Fomin, V.V. & Zimmermann N.E. (2007). Altitudinal and Horizontal Shifts of the Upper Boundaries of Open and Closed Forests in the Polar Urals in the 20th Century. *Russian Journal of Ecology*, Vol.38, pp. 223–227. ISSN: 1067-4136

Shvartsman, Y.G.; Barzut, V.M.; Vidyakina, S.V. & Iglovsky, S.A. (1999). Climate variations and dynamic ecosystems of the Arkhangelsk region. Chemosphere. *Global Change Science*, No.1, pp. 417–428. ISSN: 1465-9972

Silapaswan, C.S., Verbyla, D.L. & McGuire, A.D. (2001). Land cover change on the Seward Peninsula: The use of remote sensing to evaluate the potential influences of climate warming on historical vegetation dynamics. *Canadian Journal of Remote Sensing*, No.27, pp. 542–554. ISSN: 0703-8992

Sochava, V.B. (1927). Botanical essay of the Polar Ural forests from the Nelkja River to the Kulga River. In: *Proc. Bot. Mus. USSR Acad. Sci.* No.21, pp. 1-71. [In Russian].

Solomina, O.N. (1999). *Mountain glaciation of northern Eurasia in the Holocene.* Scientific World, Moscow. [In Russian].

Sukachev, V.N. (1922). Climate and vegetation changes during the post-Tertiary time in the Siberian north. *Meteorol. Vestn.* No.1-4, pp. 25-43. [In Russian].

Sturm, M.; Schimel, J.; Michaelson, G. et al. (2005). Winter biological processes could help convert Arctic tundra to shrubland. *BioScience*, No.55, pp.17–26. ISSN: 0006-3568

Tranquillini, W. (1979). Physiological ecology of the alpine timberline. Springer-Verlag, Berlin.

Vaganov, E.A., Shiyatov, S.G. & Mazepa V.S. (1996). *Dendroclimatic Study in Ural-Siberian Subarctic.* Nauka, Siberian Publishing Firm RAS, ISBN: 5-02-031185-5, Novosibirsk. [In Russian].

Vaganov, E.A.; Hughes, M.K.; Kirdyanov, A.V.; Schweingruber, F.H. & Silkin, P.P. (1999). Influence of snowfall and melt timing on tree growth in subarctic Eurasia. *Nature,* 400, 149–151. ISSN: 0028-0836

Weisberg, P.J. & Baker, W.L. (1995). Spatial variation in tree seedling and krummholz growth in the forest - tundra ecotone of Rocky Mountain National Park, Colorado, U.S.A. *Arctic and Alpine Research,* No.27, pp. 116–129. ISSN: 1523-0430

20

Mountains Under Climate and Global Change Conditions – Research Results in the Alps

Oliver Bender, Axel Borsdorf, Andrea Fischer and Johann Stötter
Institute of Mountain Research: Man and Environment, Austrian Academy of Sciences
Austria

1. Introduction

Mountain regions are often marketed with statements suggesting "..., where time has stood still." Anyone who believes such statements is in for a disappointment as, of course, time does not stand still in the mountains, on the contrary: it is there that, behind the proffered Disneyland of mountain traditions, change often happens faster than in the lowlands. Global Change as climate change is hitting ecologically and economically sensitive mountain regions. Globalization reaches areas that adhere to a small-scale patchwork of social and economic structures, ill prepared for external impacts and global competition. It takes over traditional cultures and changes the way we interact, what we wear, what we eat, the music we listen to. Political agency is passed on to new integrative spatial units, to the global economy and political agreements. This shifts the responsibility for shaping our living environment to extranational bodies and reduces agency at national and regional level. In addition, coherent mountain regions are often spread across several states, which makes them potential or actual contended spaces.

Global Change research in mountain regions has to date focused on climate change (EEA, 2004; Kohler & Maselli, 2009) while the impact of globalization has not received anything like as much attention (Borsdorf et al., 2008; Jandl et al., 2009). In reality, both phenomena interact and can only be analysed and evaluated in an integrative manner.

This chapter aims to look at Global Change from different perspectives and in an integrative approach. Our case in point are the Alps and, within them, mainly the Austrian Alps where the authors concentrate their research. After analysing the status quo, we shall present ongoing political processes, programmes and research and define the issues. Climate change concerns all spheres of nature and different subareas of the anthroposphere. It is in the cryosphere that change becomes most clearly visible, especially on glaciers and in permafrost areas, but also in the valleys. These lower lying areas face increased hazards but also new economic opportunities for agriculture and tourism, which we will discuss before closing with an outlook on future challenges.

2. The Alps – a case in point of a mountain area and its political dimension

The Alps - as defined by the Alpine Convention - cover an area of roughly 190,000 km^2 and are populated by more than 14 million inhabitants plus another 120 million annual visitors (Bätzing, 2003). They provide resources for up to a third of the European population. Three large streams, the Rhine, the Rhone and the Po originate in the Alps, most of the water for the Danube comes from there, justifying their by-name of 'water tower of Europe'. This water is nourishment, process water and provider of clean and renewable energy as well as peak electricity for the European grid.

The Austrian share of the Alps makes up two thirds of the state territory but only 40% of Austrians live in the mountains. Permanent settlement is only possible up to ca. 2000 m a.s.l. This means that 75% of the Alpine area can only be settled seasonally or not at all. Human settlements thus reach the upper altitudinal limit and with it an ecologically highly sensitive border area. At the same time, the Alps represent a hotspot of biodiversity and an important gene pool as well as pools of cultural diversity and are a recreation area of global stature. Scientists may see them as open-air laboratories of Global Change. Here, especially, strategies for sustainable regional development are needed to minimize risks and to leverage opportunities. They can help maintain a unique area in its potency for inhabitants, visitors and for the extra-Alpine population that depends on its resources.

In November 1971, UNESCO established the 'Man and the Biosphere' research programme. It posed questions about the interaction between humans and their environment from a joint natural and social sciences point of view, questions that could be vital for our future. Importantly, the sixth project in this programme, 'Impact of human activities on mountain ecosystems' was dedicated to the mountains of the world. The first UN conference 'On the Human Environment' was held in Stockholm in June 1972. The recommendations from this meeting shaped the ensuing international mountain conferences. The same year saw the publication of the Club of Rome's famous report 'The Limits to Growth' (Meadows et al., 1972). Its much debated projections of exponential growth of populations, economy and technology started research into globalization.

Many international mountain conferences followed over the next years. In 1983, the mountain centre ICIMOD for all eight states of the Hindu Kush-Himalaya was established in Kathmandu; in 1986, the African Mountain Association formed in Ethiopia, followed in 1991 in Chile by the Andean Mountain Association. On this basis, the developing countries enthusiastically supported the proposal of a mountain chapter in the Agenda 21 at the third preparatory conference for Rio in the autumn of 1991. This chapter 13, 'Managing Fragile Ecosystems: Sustainable Mountain Development', was then passed unanimously at the so-called Earth Summit in Rio de Janeiro in June 1992. This amounts to a global recognition of the world's mountains, their resources and their problems (Messerli, 2010).

In 1991, the Alpine Convention was signed by eight Alpine states. It formed the first cross-border agreement based on the idea of sustainable regional development in a mountain area. A similar convention is currently being prepared for the Carpathians.

In 1997, the UN General Assembly for the evaluation of Agenda 21 was fundamental to rethinking the global significance of mountains. The delegates understood that mountains preserve a series of common goods, such as vast treasures of biological and cultural diversity, that they are water towers for an increasingly thirsty planet, sensitive indicators of climate and environmental change, vital recreation areas for an increasingly urbanized world population, sacred places in many cultures and religions and uniquely privileged regions for protected areas, but also sites of erosion, risks and disasters with damaging effects on adjacent lowlands. Based on this new understanding, the UN General Assembly decided on 10 November 1998 that the year 2002 should be the 'International Year of the Mountains', the same year as the World Summit on Sustainable Development in

Johannesburg. These developments greatly increased public awareness of the effects of climate change and globalization even for remote mountain regions.
In 11 March 2010, the UN General Assembly passed Resolution 64/205 on 'Sustainable Mountain Development'. Its 45 paragraphs include numerous references to common goods of the mountains, common concerns, as well as recommendations for governments and research. In 2012, ten years after the International Year of the Mountains, the conference Rio+20 (Stockholm+40) will be held again in Rio de Janeiro.

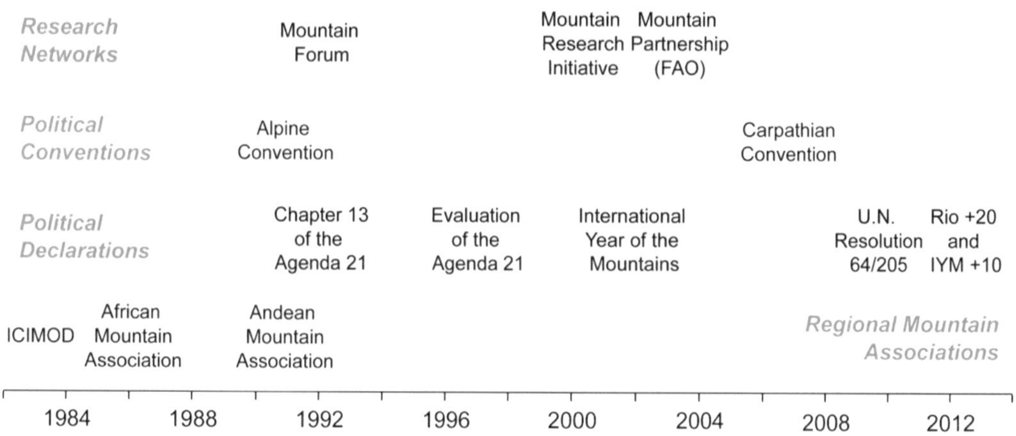

Fig. 1. The rise of mountain research within the framework of regional associations, political declarations and conventions and research networks (graphic design: K. Heinrich 2011)

This brief summary shows how mountain regions are attracting growing political attention. It also shows that the complex structures, processes and problems of mountain regions can only be tackled successfully on the basis of an inter- and transdisciplinary approach.
In order to strengthen mountain research, to profit from regional expertise and scientific results, to allow comparative analyses, to improve the lives of mountain people and to respond to the challenges of global change, mountain scientists have established mountain research, information and cooperation networks such as the Mountain Forum, the Mountain Partnership and the Mountain Research Initiative (Fig. 1). These in turn encouraged the formation of regional networks like Science for the Carpathians (S4C) and South Eastern European Mountain Research Network (SEEmore) (Borsdorf & Braun, 2008). While these networks are open for any initiative, the GLORIA network with over 90 research sites on summits in all major mountain ranges of the world focuses on monitoring climate change effects on mountain biodiversity (Grabherr et al., 2010).
In Innsbruck, the Research Focus Alpine Research, the Institute for Mountain Research: Man and Environment and, since 2010, the K1 alpS Centre for Climate Change Adaptation Technologies (in mountain regions) are collaborating within the "Innsbruck Mountain Competence" (see Borsdorf et al., 2010).

3. Climate change in the Alps

The Alps are among the most intensely studied mountain regions on earth. This is also true of fast changing weather and climate parameters such as air temperature, humidity, pressure, wind, precipitation and sunshine hours. While 'weather' denotes a snapshot of the

state of the atmosphere (timescale: minutes to days), 'climate' describes a median state of the atmosphere in a place on a timescale of at least three decades.
For the Alpine Space not only a large number of time series are available but also long time-series data, which are vital for analysing long-term trends and changes in frequency distributions. Instrumental records, i.e. recorded time series of meteorological and climatological parameters, go back more than 200 years. In the Alps, there is excellent additional information available on former climate states from so-called proxy data, such as speleothems, lake sediments and dendrochronology.

3.1 Instrumental period

For the Alps, the longest time series of the meteorological parameters air temperature and air pressure go back to the year 1760. Precipitation records started in 1800, those for cloudiness in the 1840s and sunshine duration records in the 1880s. Within the HISTALP project, records from 577 stations in the Greater Alpine Region (4–19° E, 43–49°N, 0–3500 m a.s.l.) were homogenized to account for changes in the location, method or instruments for measuring the parameters (Auer et al., 2007). The time series are available as station mode or gridded data and show an incremental increase in air temperature of +1.2°C for the 20th century. A first peak occurred at the end of the 1940s, the second increase started in the 1970s (1.3°C per 25 years). Calculated over several decades, the temperature trends are the same for all regions of the Alps. Nor are there significant differences between valley and summit stations. Overall, warming in the Greater Alpine Region has been twice as strong as the global mean. On a local/regional scale, however, short-term variabilities have been observed that are stronger than long-term trends (Böhm, 2008). Prominent examples of such short-term and localized deviations would be the extremely warm summer of 2003 or the high-precipitation winter of 1999 with its avalanche disasters. By themselves, such occurrences do not say anything about a change in climate, since they could, in principle, occur in any climate, if only very rarely.
Precipitation shows regionally differing long-term trends. While precipitation in the north-eastern part of the Alpine Arc is increasing by 10–15% per century, it is decreasing by the same percentage in the south-east. However, a short-term variability of ±50% must be superimposed on these trends and dominates them.
As regards possible future climate changes, we must assume a statistical change in the distribution of individual events (extremes and mean), which is important for several applications:

- agriculture: the number of frost days or the duration and frequency of droughts,
- flood calculation: extreme precipitation,
- calculation of avalanche protection measures, calculation of roof loads: snow depths,
- water management: runoff, i.e. precipitation, snowmelt, glacier runoff.

Currently it is not possible to predict how these and other relevant parameters will change with future climate change as the models return widely differing results. Böhm (2008) considers the following scenarios as well founded: a rise in temperature by 3°C (IPCC Scenario A1B) by the year 2100, more frequent positive temperature extremes, no change in storm occurrences in the Alps, a decrease in solid precipitation as well as a reduction in glacier area. Clear scientific pronouncements on other climate parameters are not feasible, especially on convective precipitation (thunderstorms, hail), which is notoriously difficult to capture in numerical models. This is also true of future precipitation events as the relevant models differ widely (Christensen et al., 2007).

3.2 Proxy data and historical records

Proxy parameters are very useful for understanding past climate changes better (Jansen et al., 2007) and they improve the accuracy of predictions for future climate scenarios. They provide information on the climate prior to the start of instrumental records. Tree rings, lake cores, moraines and speleothems show a statistical relation to means of climatological parameters for a given period. This connection stems from processes such as tree growth or dripstone formation, which are controlled by complicated chemical or biological mechanisms that are themselves statistically related to climate parameters.

Dendrochronological dating of *Pinus cembra* pieces, for instance, provides important climate information for the Alps, Alpine glacier dynamics and shifts in the tree line during the Holocene (Büntgen et al., 2005; Joerin et al., 2008; Nicolussi et al., 2005). Statements on climate change deduced from lake sediments go even further back in time. The layering of the sediments itself can yield information on the climate but also the activity of the biota living in those sediments (Ilyashuk et al., 2011; Schmidt et al., 2006).

A large number of written records on the climate of the past exist for the Alps, particularly on extreme events and climate parameters relating to tax-relevant aspects like harvests. Fliri (1998) and Jäger (2010) have put together an exemplary collection for Tyrol. These reports create an impressive insight into past climate change effects on human spheres of settlement and interest.

3.3 Climate change and cryosphere

Cryosphere means the frozen parts of the hydrosphere. In the Alps, these are the glaciers, seasonal snow cover as well as the ice in permafrost and ice caves.

Early studies of glaciers were triggered by catastrophic bursts of glacier lakes in the course of a cooling climate around the year 1600 (Nicolussi, 1990). From the middle of the 19th century, we find systematic investigations of Alpine glaciers (e.g. Schlagintweit & Schlagintweit, 1850). From around 1890, changes in the length of the glaciers were recorded annually (Patzelt, 1970); in the second half of the 20th century, scientists measured glacier mass balances in the Alps (Fischer, 2011), which were collected by the World Glacier Monitoring Service in Zurich. A global summary of changes in the cryosphere in general and the mountain glaciers in particular is included in the IPCC report (Lemke et al., 2007).

From their peak during the Little Ice Age around the middle of the 19th century, the glaciers of the Austrian Alps have lost circa 50% of their area (Gross, 1987). The changes in glacier area are documented by country in glacier inventories (Abermann et al., 2009; Lambrecht & Kuhn, 2007) largely based on remote sensing methods (Haeberli et al., 2007). After the last glacier-friendly period so far in the 1980s, which saw glacier advances, we can now observe an intensified retreat of Alpine glaciers. Related phenomena, such as the formation of supra- and peri-glacial lakes require new strategies in glacier monitoring (Paul et al., 2007) and trigger research on basic mechanisms (Huss et al., 2007).

By studying long glaciological time series we obtain base lines for interpreting past climate in high mountain areas where only few direct recordings of weather stations are available (Fischer, 2010; Vincent et al., 2004). Recent comparisons of glacier and climate data led to the development of models that allow an interpretation of glacier stands from the Holocene, documented by moraines, to extrapolate the climate of the time (e.g. Kerschner & Ivy-Ochs, 2008). Figures 2 and 3 illustrate the dramatic retreat of the Schlatenkees glacier, municipality of Matrei, East Tyrol.

Fig. 2. Schlatenkees 1890 (Photograph: V. Sella)

Fig. 3. Schlatenkees 2010 (Photograph: A. Fischer)

Alpine permafrost in general and permafrost in the Alps in particular, is a phenomenon that has only been known and scientifically investigated for a short time (Barsch, 1969). Concrete pronouncements on its extent are therefore still relatively imprecise (e.g. Harris et al., 2001; Krainer et al., 2007). What we know about the large-scale distribution of permafrost in the Alps stems from the Swiss Alps. Barsch (1978) puts the volume of permafrost in the Western Alps at 5-10 km^3. Haeberli (1975) suggests that permafrost can be assumed for areas above 2200 m a.s.l, above 2550 m a.s.l. permafrost is likely at the foot of slopes with a northern exposition and above 3000 m a.s.l. on steep south-facing slopes (Keller, 1987). Such statements are either based on morphological analyses of block glaciers as the most pronounced form of Alpine permafrost or, more recently, on modelling approaches using the energy balance in a simplified way as indicator for the existence of mountain permafrost (Hoelzle et al., 2001). For rock glaciers in Alto Adige, scientists were able to show clearly how active block glaciers have become inactive with loss of the ice from the middle of the 19th century (Stötter, 1994). Unlike the robust climate indicators of glaciers and permafrost, the seasonal snow cover is subject to large interannual fluctuations. If you take a series of studies on the duration and thickness of snow cover within the Alpine Space (Beniston, 1997) no homogenous picture emerges, unlike the results of other surveys in the northern hemisphere (e.g. Lemke et al., 2007). Given the regionally differentiated precipitation trends, confirmed by HISTALP data, this is hardly surprising. From a climatological perspective, it can be said that even in a tendentially warmer climate, winters with large amounts of snow will continue to occur, just as during a cooler climate there were some winters with little snow (Beniston, 2011; Jäger, 2010).

3.4 Climate change and natural hazards

Since climate and weather drive many Alpine natural hazards, it is hardly surprising to find scientists everywhere debating this issue of how natural hazards behave under conditions of climate change. In this context, we must distinguish between the medium- to long-term temporal dimension of the climate, which establishes the framework of a longer-term disposition, and the short-term dimension of weather conditions, which may trigger natural hazard events. If we look at it using a time-frame of decades, the impact of global warming for natural hazard processes takes priority, while a look at what triggers processes shows that changes in precipitation patterns are decisive.

In the course of rising temperatures, ice in permafrost melts. Together with shrinking glaciers, this means that the stabilizing effect of ice, both as ice cement in the substratum and as protective ice shield on the surface, is shifting to ever higher altitudes. At the same time, this process makes more and more unconsolidated sediment available for erosion. In its wake, tension fissures not only become more frequent but they also occur at higher altitudes (Stötter et al., 1996) and gravitational (rock) fall from over-steep walls also increases in frequency (Stötter, 2007). Add to this general development of disposition, which can be recorded reliably, a changed pattern of precipitation intensity as trigger, which cannot be statistically proven to date.

One interesting clue of a change in precipitation towards more intense precipitation events stems from the analysis of the flood events in the Alps during the last two decades and, on a larger spatial scale, in Central Europe. Since the 1990s, an increase in extreme events has been recorded in several river systems, with return periods of 100 years and more, which show a probability of recurrence in such short succession of $p < 0.001$ (Table 1).

River	Years of flood events			Return period (yrs.)
Rhine	1993	1995		3300
Danube	1999	2002		2700
Elbe	2002	2006		1000
Bregenzer Ache	1999	2002	2005	29000
Lech	1999	(2002)	2005	?

Table 1. Return period of cumulative large flood events on Central European rivers (Stötter et al., 2009)

Such mass occurrences of large events go hand in hand with equally large, and rising, damages. Here another aspect comes into play that not only explains but also drives the increase of risk. Since the middle of the 20th century, Tyrol, for example, has experienced a dramatic spatial expansion of settlement, commercial and transport infrastructure areas, driven by the shift from a predominantly agrarian society to a service society shaped by tourism. The number of exposed assets has risen in parallel with this development and can be read off key indicators such as increase in population, number of buildings, visitor beds and buildings used in tourism (Table 2). In sum, we can say that while changes in the risk of natural hazards show a clear correlation to climate change, they are possibly driven much more by socio-economic changes.

Indicator	Change factor
resident population	1.73
visitor beds	7.20
residential buildings	5.10
visitor accomm. buildings	6.12

Table 2. Change factor of key elements for damage potential in the period of 1951–2001 for Tyrol (Stötter, 2007)

4. Socio-economic change

4.1 Population and settlements

For centuries, population density in the Alps was roughly equivalent to the European average. With the beginning of industrialization, massive migration from the rural regions of the Alps set in (Bätzing, 2003). This so-called 'flight from the mountains' started in the French Alps and spread unevenly to other Alpine regions at different times and in varying intensity. At the end of the 20th century, significant depopulation still occurred in the Italian and, to a lesser degree, the east-Austrian part of the Alps. As it is mostly the young population who was and still is leaving the Alpine Space, such migration losses cannot be compensated for by a surplus of births and result in an overall decline in the number of inhabitants.

Current demographic change throughout Europe is characterized by lower fertility, increasing old-age overhang and mortality (second demographic transition). Apart from a few regions in the northern Alps (including Western Austria), current natural population growth in the Alpine Space is negative (cf. Tappeiner et al., 2008). Inter- and intraregional migration now make up the strongest component of population development, further deepening the spatial disparities between active and passive regions and between urban agglomerations and peripheral rural spaces (cf. Bätzing 2003; Tappeiner et al., 2008). Within the densely populated areas, this development is differentiated by sub- and post-suburbanizing processes (Fig. 4 and 5). Along the main valleys (e.g. the lower Inn valley in Tyrol) these processes have encouraged the formation of bands of settlement, commercial and transport infrastructure areas. Add to this urbanization processes along the Alpine fringe in the catchment of large extra-Alpine metropoles and in the inner-Alpine tourist centres (e.g. Chamonix, Davos, Kitzbühel) (Bätzing, 2003; cf. Tappeiner et al., 2008). Such developments result from changing preferences in establishing main and additional domiciles (amenity migration) and a greater reach in travel to work and to holiday homes (Perlik, 1999). In the main valleys and tourist centres on the Alpine fringe, settlement space is getting scarce, building land prices are rising and building coverage is becoming very compact (Bender, 2005). This is contrasted in peripheral areas with shrinking municipalities - all the way to complete desertion of villages in the Southern Alps - that find it ever harder to cover the costs of maintaining their infrastructure.

Fig. 4. and 5. Post-suburbanization in the municipality of Völs (along the Inn valley near Innsbruck, Tyrol) between 1940 and 2005 (Source: Orthoimages, Land Tirol, Scale 1:10,000)

As a result of climate change, the inhabitants of valleys and basins will increasingly suffer from unfavourable local climate conditions. These can be mitigated through suitable architectural and urban planning measures, such as high energy-efficiency of buildings, avoiding islands of warmth, keeping fresh air corridors open, landscaping and water retention. Such measures, however, will be hampered by the trend towards further densification of settlements, as communities run out of available building land and transport policy favours compact settlements. The required constructional measures represent not just an adaptation to climate change but are also aimed at lessening it (Hiess, 2010b; Kronberger et al., 2010; Wyss & Arlanch, 2009).

4.2 Agriculture and forestry

Ever since subsistence farming was abandoned, agriculture in the Alps has been dominated by grassland farming, including the special case of *Almwirtschaft,* a seasonal expansion of pastures into the subalpine altitudinal zone, which continues to this day. Arable farming, market gardening as well as fruit and wine growing (the latter in the Austrian Alps only in very small areas) only happens in the valleys and basins these days.

Given the difficult conditions of climate and terrain, mountain agriculture is becoming less and less competitive. In large parts of the Alps, this leads to the increasing abandonment of farms and marginal areas. In the Italian Alps in particular, we can observe a pronounced decline, while in the northern, German-speaking, Alpine countries and especially in Austria such processes are rare (cf. Borsdorf, 2005). Here, the agrarian cultural landscape has largely been retained, even if it has been adapted in many places to allow modern utilization (cf. Tyrolean Cultural Landscape Inventory; Land Tirol, n.d.). This is linked to the mountain farming subsidies, which started several decades ago, but also to the fact that Austrian mountain farms are mostly run as a part-time concern, i.e. in combination with other economic activities, mainly in tourism. Moreover, agricultural production, especially in the Alps, has been oriented to high quality products, viz. the high proportion of organic farming and regional origin certified products (Bender, 2010).

Climate change affects Alpine agriculture in positive and negative ways. Negative impacts include reduced yield or quality and declining yield security as a result of the likely increase in summer drought, plus a possible increase in extreme events such as heavy rain, storms and hail. Arable farming, fruit and wine growing can adapt to these problems by choosing less vulnerable varieties. While grassland in regions with up to 800 mm annual precipitation is likely to experience severe reductions in yield, higher areas with more than 800 mm precipitation can expect a rise in production potential (Kronberger et al., 2010), which benefits mountain grassland farming.

As agricultural subsidies strongly influence profitability and land-use distribution, a possible - and repeatedly discussed - reduction or reorientation of such subsidies would lead to significant changes. The effects of climate change are therefore going to be mainly noticed with crops that are less affected by EU market regulations. These include mainly wine and fruit growing, which also benefit from longer vegetation periods at higher altitudes, always presuming that irrigation in fruit growing can be guaranteed. For North and South Tyrol, farmers already assume a significant expansion of fruit growing areas at the expense of arable and grasslands (Fig. 6). It is impossible to predict whether total yields in Alpine agriculture will improve or decline as a result of climate change as there are too many parameters influencing results (warmth, dryness, pests, etc.).

Fig. 6. Designated fruit growing area near Glurns, Upper Vinschgau (South Tyrol). On the field in the foreground, irrigation equipment, indispensible for fruit growing in this area, has already been installed (Photograph: O. Bender 2011)

Despite the relatively stabile situation of agriculture in Austria as compared to that of southern Alpine countries, recent decades did see considerable losses of cultural landscapes, leading to an expansion of wooded areas, particularly in steep mid-slope areas of the montane zone. Figures from the cadastre show that in the 1990s alone the wooded area increased by 5.5% (Borsdorf & Bender, 2007). Mountain forests fulfil many functions for humans, particularly as protective forests and for timber production, but also as CO_2 sinks for climate protection. Forestry is increasingly making an effort to plant mixed forests suited to the individual location with varieties adapted to climate change. In the past, particularly near ore and salt processing sites in eastern Austria, fast growing fir trees were planted as monocultures. These are now being replaced by mixed forests. In many places, mountain forests are exposed not just to climate stress but also to additional factors such as game damage, immissions, pests, etc., which further increase the vulnerability of forest ecosystems to climate change. Overall, forestry in the Austrian Alps must be considered highly vulnerable to changes in climate (Kronberger et al., 2010).

At higher altitudes, more extensive mountain pasturing basically encourages a rise in the forest line, which has been held down artificially through grazing on land mostly cleared by burning. There are as yet no clear indications of tree stands rising solely as a result of changes in climate (cf. Nicolussi et al., 2005; Wieser et al., 2009; Fig. 7 and 8). The ecological conditions in the treeline ecotone are too complex and the response times of subalpine forest communities too long to draw direct conclusions (Borsdorf & Bender, 2007).

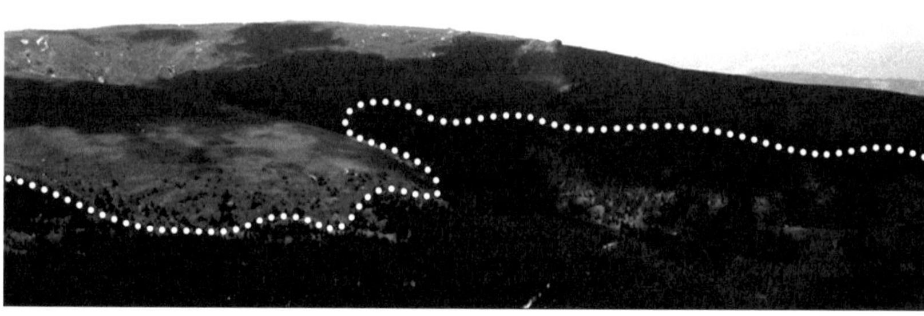

Fig. 7. and 8. Eastern flank of Koralpe in 1936 and 2004. The dotted line shows the current forest line (Source: Stützer, 2005)

4.3 Traffic and tourism

Within the European economic area, the Alps represent an obstacle to traffic. Since prehistoric times, this obstacle has had to be overcome by transit routes along the large cross valleys and across the passes. Since the end of the 18th century, these routes have been repeatedly expanded for roads and railways in response to increasing traffic flows. The large valleys have definitely benefitted from this transit traffic, but at the same time they are increasingly suffering from the accompanying environmental burdens (air pollution, noise, land grab and dissection of the landscape). The total traffic volume must be broken down into transit traffic, import traffic (including incoming tourism) and local traffic (including commuting in the agglomerations) (Permanent Secretariat of the Alpine Convention, 2007). Contrary to the often expressed view at local protest activities against transit traffic, regional indigenous traffic accounts for more than 70% of total road traffic (Gruber, 1994).

Traffic is today seen as a key driver of climate change. New transport projects that aim to shift traffic from road to rail (like the Brenner Base Tunnel in Tyrol) should also be motivated by the desire to protect the climate, even if, as the traffic breakdown shows, they can only solve a small part of the problem. In the Alps as well, the main focus should be on environment- and climate-friendly solutions to regional and urban traffic as well as on incoming tourism (cf. Hiess, 2010a).

The Alps are among the earliest tourist destinations and from the beginning the focus has been on the landscape. With the advent of mass tourism in the 1960s, tourism and the leisure industry have become a major economic factor in the rural areas of the Alps (12% of jobs, 16% of GDP), albeit with great differences between regions and municipalities (Bätzing, 2003). Within the last thirty years, a long-term trend towards skiing-based winter tourism has emerged for large parts of the Alps, with significantly higher added value than summer tourism. This development goes hand in hand with a knock-out competition between tourist destinations. Only municipalities that invest continuously in their tourism portfolio will be able to achieve growth in the future.

Climate change increasingly threatens winter tourism, at first in lower-lying areas, later on possibly in all tourist destinations. If we assume a temperature rise of 4°C, only 30% of winter sports places will be able to guarantee snow (Abegg et al., 2007). This development will, however, be subject to great regional differences. Clear losers in this scenario are the lower-lying skiing areas with pistes not exceeding 2500 m a.s.l. (Alpine fringe, large parts of the Eastern Alps), while higher destinations with pistes above 3000 m (glaciated regions of the Western Alps and places in the Central Alps) are in a naturally more favourable position to cope. About two decades ago, municipalities began to mitigate lack of snow by making artificial snow and have come to rely on snow canons more and more (Hahn, 2004; Fig. 9). Artificial snow makes the pistes mechanically more robust and better adapted to developments in skiing equipment and to the rising standards of the visitors. Undoubtedly this must be seen as an example of successful development and implementation of adaptation strategies to the challenges of climate change. According to Steiger (2011), only a significant increase in artificial snow making allows almost all of the 100+ skiing areas in Tyrol and South Tyrol to maintain skiing until the middle of the 21st century if we assume a moderate warming scenario (A1B). At the same time, this form of adaptation is linked to considerable environmental problems (Pröbstl, 2006) and, with rising temperatures, the demand for water and energy will increase further (Olefs et al., 2010). Moreover, it raises the question if rising costs for winter skiing operations can be set off by even higher prices for the tourism portfolio or whether they will have to be met by the public (Abegg 2011; Steiger & Mayer 2008).

Fig. 9. Snow-making with snow canons in Obergurgl, Tyrol (Photograph: A. Borsdorf 2005)

Summer and all year skiing areas on glaciers, established in the 1960s and 1970s, today operate almost exclusively from autumn to spring and no longer in summer as a result of glacier shrinkage (Fischer et al., 2011; Smiraglia et al., 2008). As regards tourist infrastructure (lift stations and pylons), glacier melt has been very effectively delayed by covering smaller areas with fleece to extend the maintenance intervals (Olefs & Fischer, 2008). For Alpine summer tourism, negative as well as positive effects can be deduced from regional climate scenarios. Negative effects include changes in the landscape, melting glaciers, thawing permafrost, drought and water scarcity as well as a potential increase in the risk of natural hazards (Abegg, 2011). Alpine walking and climbing routes will be ever more difficult to secure under these conditions. Overall, however, there is the chance of a revival of Alpine summer tourism (classic summer holiday) because the areas outside the Alps, where most tourists come from, and the Mediterranean competition destinations could well be much harder hit by global warming in the form of heat waves, water shortages, forest and bush fires, dying bathing lakes and rivers (cf. Bourdeau, 2008). If they want to take advantage of such a scenario, Alpine destinations must diversify away from a narrow orientation on winter sports and towards sustainable tourism. For such a transition to be handled successfully, it is vital to create the right awareness (Kronberger et al., 2010). This will take new communication strategies as well as comprehensive participatory processes involving all stakeholders as well as tourist operators and tourists themselves.

5. Conclusion

Mountain research started early with some Arab scientists and in Europe with Marco Polo and Christopher Columbus. The latter was fascinated by the mountains he observed in the New World and refers to them repeatedly in his work. Mathieu (2011) sees this work as a nodal point for mountain research that continued with the questionnaires of Felipe II which include six questions about mountains. The French geographer Philippe Buache was the first who drew a map that showed the mountains of the world as a cohesive system. Alexander von Humboldt provided another important impulse analysing the nature and hypsographic zones of the Andean Cordillera. With the rise of sciences in the 19th and 20th century, the Alps gradually became the most intensely investigated mountain range of the

world. However, the Alpine experience has to be seen in the light of comparison with other mountain ranges. Carl Troll based his research in the Andes, the Himalaya and the Alps.
This is why the example of the Alps covered in this chapter has to be seen in a global context. Climate change and globalization are world-wide phenomena and they meet fragile ecosystems and traditional societies and economies ill prepared for the impacts of these global processes. It cannot be the purpose of this chapter to make a detailed comparison, but it is quite clear that effects observed in the Alps may be even more drastic in mountain regions located in a socio-economic environment less well equipped to find and implement adaptations strategies.
As the examples from the Alps show, climate change and globalization meet complex man-environment systems in mountain regions. Glacier and permafrost retreat, water scarcity, soil erosion and land-use change, as well as loss of biodiversity are consequences of climate change on the regional scale of the mountain area. Climate change thus impacts on ecosystem services, not just for the societies in the mountains but also for the inhabitants of the adjacent lowlands. Globalization processes also exert an ever faster impact on ecosystems. Even more dramatic is their effect on the cultural landscape, on agriculture, population structure, mobility, as well as the urbanization and marginalization of peripheral mountain regions. Societies are affected in socio-economic, political and institutional terms and require decision making at regional and local level. Both subsystems interact in the sphere of land use and land management (Fig. 10).

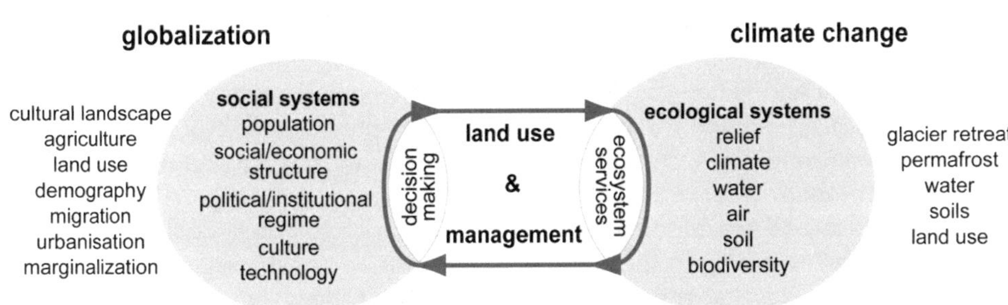

Fig. 10. Man-environment system in mountain regions (modified from: IGBP, 2005)

Since Global Change affects man-environment systems with greatly varying intensity and speed in different mountain regions, comprehensive long-term observation and monitoring programmes are necessary to capture it. One such programme is the global GLORIA initiative with its high-mountain ecology monitoring instruments (Grabherr et al., 2010; Pauli et al., 2007). It is not enough to study individual system elements of the natural- and the anthroposphere (single discipline approach) or the interaction between subsystems or even the entire man-environment system (interdisciplinary approach). When all is said and done, such analyses are always restricted to the production of system knowledge.
The challenges of climate change and globalization increasingly call for target knowledge to be provided to decision makers in politics and the economy. Everybody uses key terms such as sustainability, preserving biodiversity, safeguarding ecosystem services, disaster protection, establishing protected areas, ensuring economic capability, competitiveness and social coherence as things to strive for. These are the expression of a heightened awareness of social

and political responsibility, but as a rule they are not very specific and their implementation is not without conflict. Therefore it is vital to include stakeholders, not only in the creation of system knowledge but even more importantly in the assessment and transformation into target knowledge in truly transdisciplinary style. The mountain.TRIP project is making a valuable contribution to the dissemination of system and target knowledge to a variety of recipients (stakeholders and general populace) by facilitating the sharing of insights (Braun, 2010). Another important aspect is creating and disseminating management knowledge at different levels and across different regions. The DIAMONT project produced exemplary databases on steering instruments and best practices (Borsdorf et al., 2010).

Tourist advertising may talk about 'time standing still' but global climate change and globalization have triggered dramatic changes in the Alps that can be made visible using the indicators mentioned. In the cultural landscape, persistent structures used to be at work for a long time, but the process that has its roots in the beginnings of industrialization is forcing accelerating dynamics onto them, speeded up further in recent decades as a result of globalization.

Within the cultural space of the Alps, social and economic impact factors may dominate (cf. Slaymaker, 2001), but with each 'warm' year it becomes clearer that climate factors are gaining in significance. Climate change determines tourism just as much as the growth options for settlements and commercial areas, what is and is not a secure road and the routing of new roads and rail tracks. We may not yet perceive the full extent of this impact, yet the consequences are real and ever more significant. There is much to research!

6. References

Abegg, B. (2011). *Tourismus im Klimawandel.* CIPRA, Available from: http://www.cipra.org/pdfs/964_de/at_download/file

Abegg, B., Agrawala, S., Crick, F. & de Montfalcon, A. (2007). Climate change impacts and adaptation in winter tourism, In: *Climate change in the European Alps: Adapting winter tourism and natural hazards management*, Agrawala, S., pp. 25–60, OECD, ISBN 978-92-64-03168-5, Paris

Abermann, J., Lambrecht, A., Fischer, A. & Kuhn, M. (2009). Quantifying changes and trends in glacier area and volume in the Austrian Ötztal Alps (1969-1997-2006). *The Cryosphere,* Vol. 3, No. 2, (October 2009), pp. 205–215, ISSN 1994-0416

Auer, I., Böhm, R., Jurkovic, A., Lipa, W., Orlik, A., Potzmann, R., Schöner, W., Ungersböck, M., Matulla, C., Briffa, K., Jones, P.D., Efthymiadis, D., Brunetti, M., Nanni, T., Maugeri, M., Mercalli, L., Mestre, O., Moisselin, J.-M., Begert, M., Müller-Westermeier, G., Kveton, V., Bochnicek, O., Stastny, P., Lapin, M., Szalai, S., Szentimrey, T., Cegnar, T., Dolinar, M., Gajic-Capka, M., Zaninovic, K., Majstorovic, Z. & Nieplova, E. (2007). HISTALP - historical instrumental climatological surface time series of the Greater Alpine Region. *International Journal of Climatology,* Vol. 27, No. 1, (January 2007), pp. 17–46, ISSN 1097-0088

Barsch, D. (1969). Permafrost in der oberen subnivalen Stufe der Alpen. *Geographica Helvetica,* Vol. 24, No. 1, pp. 10–12, ISSN 0016-7312

Barsch, D. (1978). Active rock glaciers as indicators tor discontinuous alpine permafrost - an example from the Swiss Alps, *Proceedings of the Third International Conference on Permafrost,* Vol. 1, pp. 348–353, ISBN 0-660-01735-0, Edmonton, July 1978

Bätzing, W. (2003). *Die Alpen. Geschichte und Zukunft einer europäischen Kulturlandschaft* (2nd edition), Beck, ISBN 978-3-406-50185-0, München

Bender, O. (2005). Verdichtung oder Zersiedlung? Regionale Unterschiede in der Bautätigkeit und Siedlungsdynamik, In: *Das neue Bild Österreichs. Strukturen und Entwicklungen im Alpenraum und den Vorländern*, Borsdorf, A., pp. 122–123, ÖAW, ISBN 978-3-7001-3513-0, Wien

Bender, O. (2010). Entstehung, Entwicklung und Ende der alpinen Bergbauernkultur, In: *Über das Entstehen und die Endlichkeit physischer Prozesse, biologischer Arten und menschlicher Kulturen*, Heller, H., pp. 113–137, ISBN 978-3-643-50229-2, Wien, Berlin

Beniston, M. (1997). Variations of snow depth and duration in the Swiss Alps over the last 50 years: links to changes in large-scale climatic forcings. *Climatic Change*, Vol. 36, No. 3-4, (July 1997), pp. 281–300, ISSN 0165-0009

Beniston, M., Uhlmann, B., Goyette, S. & Lopez-Moreno, J.I. (2011). Will snow-abundant winters still exist in the Swiss Alps in an enhanced greenhouse climate? *International Journal of Climatology*, Vol. 31, doi: 10.1002/joc.2151, ISSN 0899-8418

Böhm, R. (2008). Harte und weiche Fakten zum Klimawandel. *Sachverständige*, Vol. 32, No. 4, pp. 170–177, ISSN 2075-3586

Borsdorf, A. (Ed.). (2005). *Das neue Bild Österreichs. Strukturen und Entwicklungen im Alpenraum und den Vorländern*, ÖAW, ISBN 978-3-7001-3513-0, Wien

Borsdorf, A. & Bender, O. (2007). Kulturlandschaftsverlust durch Verbuschung und Verwaldung im subalpinen und hochmontanen Höhenstockwerk: Die Folgen des klimatischen und sozioökonomischen Wandels, In: *Alpine Kulturlandschaft im Wandel. Hugo Penz zum 65. Geburtstag*, Innsbrucker Geographische Gesellschaft, pp. 29–50, ISBN 978-3-901182-86-0, Innsbruck

Borsdorf, A. & Braun V. (2008). The European and Global Dimension of Mountain Research – An Overview. *Revue de Géographie Alpine*, Vol. 96, No. 4, pp. 117–129, ISSN 1760-7426

Borsdorf, A., Stötter, J. & Veuillet, E. (Eds.). (2008). *Managing Alpine Future. Proceedings of the Innsbruck Conference October 15–17, 2007*, ÖAW, ISBN 978-3-7001-6571-2, Innsbruck

Borsdorf, A., Tappeiner, U. & Tasser E. (2010). Mapping the Alps. In: *Challenges for mountain regions. Tackling Complexity*, Borsdorf, A., Grabherr, G., Heinrich, K., Scott, B. & Stötter, J., pp. 186-191, Böhlau, Vienna

Bourdeau, P. (2008). The Alps in the age of new style tourism: between diversification and post-tourism? In: *Managing Alpine Future. Proceedings of the Innsbruck Conference October 15–17, 2007*, Borsdorf, A., Stötter, J. & Veulliet, E., pp. 81–86, ÖAW, ISBN 978-3-7001-6571-2, Innsbruck

Braun, F. (2010). Closing the gap between science and practice. mountain.TRIP - an EU project coordinated by IGF, In: *Challenges for mountain regions. Tackling Complexity*, Borsdorf, A., Grabherr, G., Heinrich, K., Scott, B. & Stötter, J., pp. 204–210, Böhlau, ISBN 978-3-205-78652-8, Vienna

Büntgen, U., Tegel, W., Nicolussi, K., McCormick, M., Frank, D., Trouet, V., Kaplan, J.O., Herzig, F., Heussner, K.-U., Wanner, H., Luterbacher, J. & Esper, J. (2011). 2500 Years of European Climate Variability and Human Susceptibility. *Science*, Vol. 331, No. 6017, (February 2011), pp. 578–582, ISSN 0036-8075

Christensen, J.H., Hewitson, B., Busuioc, A., Chen, A., Gao, X., Held, I., Jones, R., Kolli, R.K., Kwon, W.-T., Laprise, R., Magaña Rueda, V., Mearns, L., Menéndez, C.G., Räisänen, J., Rinke, A., Sarr A. & Whetton, P. (2007). Regional Climate Projections, In: *Climate Change 2007: The Physical Science Basis. Contribution of Working Group I to the Fourth Assessment Report of the Intergovernmental Panel on Climate Change,* Solomon, S., Qin, D., Manning, M., Chen, Z., Marquis, M., Averyt, K.B., Tignor M. & Miller H.L., pp. 847–940, Cambridge University Press, ISBN 978-0-521-88009-1, Cambridge, UK, New York, NY

EEA (Ed.). (2004). *Impacts of Europe's changing climate. An indicator-based assessment,* European Environment Agency, ISBN 978-92-9167-372-8, Copenhagen

Fischer, A. (2010). Glaciers and climate change: Interpretation of 50 years of direct mass balance of Hintereisferner. *Global and Planetary Change,* Vol. 71, No. 1-2, (March 2010), pp. 13–26, ISSN 0921-8181

Fischer, A. (2011). Long term glacier monitoring at the LTER test sites Hintereisferner, Kesselwandferner and Jamtalferner and other glaciers in Tyrol. *Plant Ecology and Diversity,* in press, ISSN 1755-0874

Fischer, A., Olefs, M. & Abermann, J. (2011). Glaciers, snow and ski tourism in Austria's changing climate. *Annals of Glaciology,* Vol. 52, No. 58, pp. 89–96, ISSN 0260-3055

Fliri, F. (1998). *Naturchronik von Tirol. Tirol, Oberpinzgau, Vorarlberg, Trentino.* Wagner, ISBN 3-7030-0313-8, Innsbruck

Grabherr, G., Pauli H. & Gottfried, M. (2010). A worldwide observation of effects on climate change on mountain ecosystems, In: *Challenges for mountain regions. Tackling Complexity,* Borsdorf, A., Grabherr, G., Heinrich, K., Scott, B. & Stötter, J., pp. 48–57, Böhlau, ISBN 978-3-205-78652-8, Vienna

Gross, G. (1987). Der Flächenverlust der Gletscher in Österreich 1850-1920-1969. *Zeitschrift für Gletscherkunde und Glazialgeologie,* Vol. 23, No. 2, pp. 131–141, ISSN 0044-2836

Gruber, R. (1994). Der Strassen- und Schienenverkehr im Alpenraum. Verkehrsaufkommen und Auswirkungen auf die Umwelt, In: *Verkehr in den Alpen: mehr als nur Transit,* Belletti, P., Nada, E. & Pastorelli, F., pp. 9–38, CIPRA Italia, ISBN 3-906521-28-1, Torino

Haeberli, W. (1975). Untersuchungen zur Verbreitung von Permafrost zwischen Flüelapass und Piz Grialetsch. *VAW-Mitteilungen,* Vol. 17, 182 pp., ISSN 0374-0056

Haeberli, W., Hoelzle, M., Paul, F. & Zemp, M. (2007). Integrated monitoring of mountain glaciers as key indicators of global climate change: the European Alps. *Annals of Glaciology,* Vol. 46, No. 1, (October 2007), pp. 150–160, ISSN 0260-3055

Hahn, F. (2004). *Künstliche Beschneiung im Alpenraum.* CIPRA, Available from: http://www.cipra.org/de/alpmedia/publikationen/2709/

Harris, C., Haeberli, W., Vonder Mühll, D. & King, L. (2001). Permafrost monitoring in the high mountains of Europe: the PACE Project in its global context. *Permafrost and Periglacial Processes,* Vol. 12, No. 1, (March 2001), pp. 3–11, ISSN 1099-1530

Hiess, H. (2010a). *Transport in climate change.* CIPRA, Available from: http://www.cipra.org/pdfs/826_en/at_download/file

Hiess, H. (2010b). *Spatial planning in climate change.* CIPRA Available from: http://www.cipra.org/pdfs/837_en/at_download/file

Hoelzle, M., Mittaz, C., Etzelmüller, B. & Haeberli, W. (2001). Surface energy fluxes and distribution models of permafrost in European mountain areas: an overview of current developments. *Permafrost and Periglacial Processes,* Vol. 12, No. 1, (March 2001), pp. 53–68, ISSN 1099-1530

Huss, M., Bauder, A., Werder, M., Funk, M. & Hock, R. (2007). Glacier-dammed lake outburst events of Gornersee, Switzerland. *Journal of Glaciology,* Vol. 53, No. 181, (March 2007), pp. 189–200, ISSN 1727-5652

IGBP - International Geosphere-Biosphere Programme - Secretariat (Ed.). (2005). Global Land Project: Science Plan and Implementation Strategy. *IGBP Report,* No. 53/*IHDP Report,* No. 19, 64 pp., ISSN 0284-8105

Ilyashuk, E.A., Koinig, K.A., Heiri, O., Ilyashuk, B.P. & Psenner, R. (2011). Holocene temperature variations at a high-altitude site in the Eastern Alps: a chironomid record from Schwarzsee ob Sölden, Austria. *Quaternary Science Reviews,* Vol. 30, No. 1-2, (January 2011), pp. 176–191, ISSN 0277-3791

Jäger, G. (2010). *Schwarzer Himmel - Kalte Erde - Weißer Tod. Eine kleine Agrar- und Klimageschichte von Tirol,* Wagner, ISBN 978-3-7030-0476-6, Innsbruck

Jandl, R., Borsdorf, A., van Miegroet, H., Lackner, R. & Psenner, R. (Eds.). (2009). *Global Change and Sustainable Development in Mountain Regions. COST Strategic Workshop,* Innsbruck University Press, ISBN 978-3-902571-97-7, Innsbruck

Jansen, E., Overpeck, J., Briffa, K.R., Duplessy, J.-C., Joos, F.V., Masson-Delmotte, D., Olago, B., Otto-Bliesner, W., Peltier, R., Rahmstorf, S., Ramesh, R., Raynaud, D., Rind, D., Solomina, O., Villalba, R. & Zhang, D. (2007). Palaeoclimate, In: *Climate Change 2007: The Physical Science Basis. Contribution of Working Group I to the Fourth Assessment Report of the Intergovernmental Panel on Climate Change,* Solomon, S., Qin, D., Manning, M., Chen, Z., Marquis, M., Averyt, K.B., Tignor, M. & Miller H.L., pp. 433–497, Cambridge University Press, ISBN 978-0-521-88009-1, Cambridge, UK, New York, NY

Joerin, U.E., Nicolussi, K., Fischer, A., Stocker, T.F. & Schlüchter, C. (2008). Holocene optimum events inferred from subglacial sediments at Tschierva Glacier, Eastern Swiss Alps. *Quaternary Science Reviews*, Vol. 27, No. 3-4, (February 2008), pp. 337–350, ISSN 0277-3791

Keller, F. (1987). Permafrost im Schweizerischen Nationalpark. *Jahresbericht der Naturforschenden Gesellschaft Graubünden*, Vol. 104, pp. 35–53, ISSN 0373-384X

Kerschner, H., Ivy-Ochs, S. (2008). Palaeoclimate from glaciers: Examples from the Eastern Alps during the Alpine Lateglacial and early Holocene. *Global and Planetary Change,* Vol. 60, No. 1-2, (January 2008), pp. 58–71, ISSN 0921-8181

Kohler, T. & Maselli, D. (Eds.). (2009). *Mountains and Climate Change - From Understanding to Action,* Centre for Development and Environment (CDE), ISBN 978-3-905835-16-8, Bern

Krainer, K., Mostler, W. & Spötl, C. (2007). Discharge from active rock glaciers, Austrian Alps: a stable isotope approach. *Austrian Journal of Earth Sciences,* Vol. 100, pp. 102–112, ISSN 0251-7493

Kronberger, B., Balas, M. & Prutsch, A. (Red.). (2010). *Auf dem Weg zu einer nationalen Anpassungsstrategie,* Policy Paper, Bundesministerium für Land- und Forstwirtschaft, Umwelt und Wasserwirtschaft (BMLFUW), 13.04.2011, Available from: http://umwelt.lebensministerium.at/filemanager/download/68173/

Lambrecht, A. & Kuhn, M. (2007). Glacier changes in the Austrian Alps during the last three decades, derived from the new Austrian glacier inventory. *Annals of Glaciology,* Vol. 46, No. 1, (October 2007), pp. 177–184, ISSN 0260-3055

Land Tirol (n.d.). Tiroler Kulturlandschaftsinventar, In: *TIRIS Kartendienste,* 27.04.2011, Available from: http://www.tirol.gv.at/themen/zahlen-und-fakten/statistik-tiris/tiris-kartendienste

Lemke, P., Ren, J., Alley, RB., Allison, I., Carrasco, J., Flato, G., Fujii, Y., Kaser, G., Mote, P., Thomas, RH. & Zhang, T. (2007). Observations: Changes in Snow, Ice and Frozen Ground, In: *Climate Change 2007: The Physical Science Basis. Contribution of Working Group I to the Fourth Assessment Report of the Intergovernmental Panel on Climate*

Change, Solomon, S., Qin, D., Manning, M., Chen, Z., Marquis, M., Averyt, K.B., Tignor, M. & Miller H.L., pp. 338–383, Cambridge University Press, ISBN 978-0-521-88009-1, Cambridge, UK, New York, NY

Mathieu, J. (2011). *Die dritte Dimension. Eine vergleichende Geschichte der Berge in der Neuzeit*, Schwabe, ISBN 978-3-7965-2711-1, Basel

Meadows, D.H., Meadows, D.L., Randers, J. & Behrens III, W.W. (1972). *The limits to growth: a report for The Club of Rome's project on the predicament of mankind*, Universe, ISBN 0-87663-165-0, New York, NY

Messerli, B. (2010). Mountains, their resources and risks - common goods or common concerns? In: *Challenges for mountain regions. Tackling Complexity*, Borsdorf, A., Grabherr, G., Heinrich, K., Scott, B. & Stötter, J., pp. 211–218, Böhlau, ISBN 978-3-205-78652-8, Vienna

Nicolussi, K. (1990). Bilddokumente zur Geschichte des Vernagtferners im 17. Jahrhundert. *Zeitschrift für Gletscherkunde und Glazialgeologie*, Vol. 26, No. 2, pp. 97–119, ISSN 0044-2836

Nicolussi, K., Kaufmann, M., Patzelt, G., van der Plicht, J. & Thurner, A. (2005). Holocene tree-line variability in the Kauner Valley, Central Eastern Alps, indicated by dendrochronological analysis of living trees and subfossil logs. *Vegetation History and Archaeobotany*, Vol. 14, No. 3, (August 2005), pp. 221–234, ISSN 0939-6314

Olefs, M. & Fischer, A. (2008). Comparative study of technical measures to reduce snow and ice ablation in Alpine glacier ski resorts. *Cold regions science and technology*, Vol. 52, No. 3, (May 2008), pp. 371–384, ISSN 0165-232X

Olefs, M., Fischer, A. & Lang, J. (2010). Boundary conditions for artificial snow production in the Austrian Alps. *Journal of Applied Meteorology and Climatology*, Vol. 49, No. 6, (June 2010), pp. 1096–1113, ISSN 1558-8424

Patzelt, G. (1970). Die Längenmessungen an den Gletschern der österreichischen Ostalpen 1890–1969. *Zeitschrift für Gletscherkunde und Glazialgeologie*, Vol. 6, No. 1-2, pp. 151–159, ISSN 0044-2836

Paul, F., Kääb, A. & Haeberli, W. (2007). Recent glacier changes in the Alps observed by satellite: Consequences for future monitoring strategies. *Global and Planetary Change*, Vol. 56, No. 1-2, (March 2007), pp. 111–122, ISSN 0921-8181

Pauli, H., Gottfried, M., Reiter, K., Klettner, C. & Grabherr, G. (2007). Signals of range expansions and contractions of vascular plants in the high Alps: observations (1994–2004) at the GLORIA master site Schrankogel, Tyrol, Austria. *Global Change Biology*, Vol. 13, No. 1, (January 2007), pp. 147–156, ISSN 1365-2486

Perlik, M. (1999). Urbanisationszonen in den Alpen. Ergebnis wachsender Pendeldistanzen. *Revue de Géographie Alpine*, Vol. 87, No. 2, pp. 147–165, ISSN 0035-1121

Permanent Secretariat of the Alpine Convention (Ed.). (2007). *Transport and Mobility in the Alps*, Alpine Convention, Available from: http://www.alpconv.org/documents/Permanent_Secretariat/web/RSAI/RSA_eng_20071128_low.pdf

Pröbstl, U. (2006). *Kunstschnee und Umwelt. Entwicklung und Auswirkungen der technischen Beschneiung*, Haupt, ISBN 978- 3-258-06936-0, Bern

Schlagintweit, H. & Schlagintweit, A. (1850). *Untersuchungen über die physicalische Geographie der Alpen in ihren Beziehungen zu den Phaenomenen der Gletscher, zur Geologie, Meteorologie und Pflanzengeographie*. Barth, Leipzig

Schmidt, R., Kamenik, C., Tessadri, R. & Koinig, K.A. (2006). Climatic changes from 12,000 to 4,000 years ago in the Austrian Central Alps tracked by sedimentological and biological proxies of a lake sediment core. *Journal of Paleolimnology*, Vol. 35, No. 3, (April 2006), pp. 491-505, ISSN 0921-2728

Slaymaker, O. (2001). Why so much concern about climate change and so little attention to land use change. *The Canadian Geographer*, Vol. 45, No. 1, (March 2001), pp. 71–78, ISSN 0008-3658

Smiraglia, C., Diolaiuti, G., Pelfini, M., Belò, M., Citterio, M., Carnielli, T. & D'Agata, C. (2008). Glacier changes and their impacts on mountain tourism: two case studies from the Italian Alps, In: *Darkening peaks: glacier retreat, science and society*, Orlove, B., Wiegandt, E. & Luckman, B.H., pp. 206–215, University of California Press, ISBN 978-0-520-25305-6, Berkeley, CA

Steiger, R. (2011). *Climate Change Impact on Skiing Tourism in Tyrol (Austria, Italy).* PhD Thesis, Universität Innsbruck

Steiger, R. & Mayer, M. (2008). Snowmaking and Climate Change: Future Options for Snow Production in Tyrolean Ski Resorts. *Mountain Research and Development*, Vol. 28, No. 3-4, (August/November 2008), pp. 292–298, ISSN 0276-4741

Stötter, J. (1994). *Veränderungen der Kryosphäre in Vergangenheit und Zukunft sowie Folgeerscheinungen – Untersuchungen in ausgewählten Hochgebirgsräumen im Vinschgau (Südtirol).* Habilitation Thesis, Ludwig-Maximilian-Universität München

Stötter, J. (2007). Zunahme des Schadenspotentials und Risikos in Tirol als Ausdruck der Kulturlandschaftsentwicklung seit den 1950er Jahren, In: *Alpine Kulturlandschaft im Wandel. Hugo Penz zum 65. Geburtstag*, Innsbrucker Geographische Gesellschaft, pp. 164–178, ISBN 978-3-901182-86-0, Innsbruck

Stötter, J., Maukisch, M., Simstich, J. & Belitz, K. (1996). Auswirkungen des zeitlich-räumlichen Wandels der Permafrostverteilung im Suldental (Ortlergebiet) auf das Gefährdungspotential durch Erosionsprozesse im Lockermaterial, *Tagungspublikation, Band 1, Internationales Symposium INTERPRAEVENT*, pp. 447–457, ISBN 3-901164-04-9, Garmisch-Partenkirchen, 1996

Stötter, J., Weck-Hannemann, H. & Veulliet, E. (2009). Global Change and Natural Hazards: New Challenges, New Strategies, In: *Sustainable Natural Hazard Management in Alpine Environments*, Veulliet, E., Stötter, J. & Weck-Hannemann, H., pp. 1–34, Springer, ISBN 978-3-642-03228-8, Heidelberg

Stützer, A. (2005): Bildsequenzen als Zeugen der Vegetationsdynamik in der subalpinen Höhenstufe der Koralpe (Kärnten/Österreich). *Wulfenia*, Vol. 12, pp. 127–138, ISSN 1561-882X

Tappeiner, U., Borsdorf, A. & Tasser, E. (Eds.). (2008). *Alpenatlas – Atlas des Alpes – Atlante delle Alpi – Atlas Alp – Mapping the Alps*, Spektrum, ISBN 978-3-8274-2004-6, Heidelberg

Vincent, C., Kappenberger, G., Valla, F., Bauder, A., Funk, M. & Le Meur, E. (2004). Ice ablation as evidence of climate change in the Alps over the 20th century. *Journal of Geophysical Research - Atmospheres*, Vol. 109, D10104, (May 2004), doi:10.1029/2003JD003857, ISSN 0148-0227

Wieser, G., Matyssek, R., Luzian, R., Zwerger, P., Pindur, P., Oberhuber, W. & Gruber, A. (2009). Effects of atmospheric and climate change at the timberline of the Central European Alps. *Annals of Forest Science*, Vol. 66, No. 4, (June 2009), pp. 402–412, ISSN 1286-4560

Wyss, A. & Arlanch, S. (2009). *Bauen und Sanieren im Klimawandel*, CIPRA, Available from: http://www.cipra.org/pdfs/782_de/at_download/file

21

Are Debris Floods and Debris Avalanches Responding Univocally to Recent Climatic Change – A Case Study in the French Alps

Jomelli, V.[1], Pavlova, I.[1], Utasse, M.[1], Chenet, M.[1], Grancher, D.[1], Brunstein, D.[1] and Leone, F.[2]
[1]*University Paris 1, Panthéon Sorbonne-CNRS*
[2]*University Montpellier 3*
France

1. Introduction

Debris flow is a dominant mass movement process in mountain areas all over the world and is a significant natural hazard. A classical distinction is made between a debris flood (DF) corresponding to a rapid, surging flow of water, heavily charged with debris in a steep channel, and a debris avalanche (DA) corresponding to a rapid or extremely rapid shallow flow of partially or fully saturated debris on a steep slope without confinement in an established channel (Hungr, 2005). In mountain areas like the Alps, the increase in human activity has resulted in increased risks of natural hazards such as debris flows. There is thus a growing demand for hazard zoning and debris flow protection. However, debris flows are caused by complex interactions between local topography, weather and sediment properties, making the understanding of debris flow activity very difficult. Because anticipated changes in climate may alter the dynamics of slope processes and the frequency or magnitude of extreme events, understanding the mechanisms that link climate and debris flow activity is the first step in any attempt at forecasting. Consequently, many studies have focused on the meteorological conditions that trigger debris flows in different environmental conditions in northern Europe (Innes, 1985; Rapp, 1995; Nyberg and Rapp, 1998) and in the Alps (Haeberli et al., 1990; Zimmerman & Haeberli, 1992; Rebetez et al., 1997). Triggering thresholds based on analyses of intense rainy events or long duration precipitation have been proposed for different spatial scales (Caine 1980; Guzzetti et al., 2008). An increase in temperatures and changes in the amount and frequency of rainfall have been observed in different mountain regions in the last few decades. Such changes in climate conditions could have an impact on the intensity and/or frequency of debris flows. However, only a few authors conducted detailed analyses of the impacts of climate change on DF activity to check the validity of this hypothesis. In British Columbia, Canada, Jakob & Lambert (2009) predicted an increase in the total number of debris flows by the end of the century due to increases in precipitation. From tree-ring series Stoffel & Beniston (2006) clearly show that the debris-flow frequency at Ritigraben (Swiss Alps) increased in the 1866–1895 period that followed the maximum extent of LIA glaciers and that events occurred most often in the early decades of the 20th century.

Surprisingly few papers have been devoted to debris avalanches and their relationships with meteorological or climatic conditions (Jomelli et al., 2004, 2009). In the French Alps, Pech & Jomelli (2001) underlined the active role of the apical cone of talus slopes in DA activity as the main factor responsible for asynchrony between sequences of heavy rainfall and DA triggering. Jomelli et al., (2007) observed a contrasted response of DA activity to recent climatic change depending on the geomorphic characteristics such as the lithology and the nature of stored debris. However, there are several geomorphic differences between DF and DA. For instance Blijenberg (1998) observed that the triggering of DF and DA is partly controlled by the relationship between slope angle and the intensity of precipitation in the triggering area. DF events occur in confined channels; hence slopes in the triggering zones may differ from those in the triggering zone of DA catchments. Moreover, sediment supply is *a priori* more limited in DA catchments than in DF catchments because most DF events transport sediment temporarily stored inside the channel (Veyrat & Menier, 2006). Consequently these differences in morphometrical and sedimentological characteristics may trigger a different response by DF and DA to climate change.

A natural hazard such as debris flows is defined as the result of a combination of hazards that correspond to the natural event and the vulnerability of elements exposed to this event, linked to human presence (Alexander, 2005). Thus, risk involves the exposure of populations and their infrastructures to a potentially damaging natural event. Concerning the risk of DF, most studies focused on understanding and reducing natural events, whereas vulnerability to DF is both a relatively new and innovative concept. Several studies have been conducted on the structural vulnerability of buildings (Hulsbergern & Carree, 1987; Alexander, 1988; Leone et al., 1996). For instance Fuchs et al., (2007) showed that vulnerability to DF in the Alps is highly dependent on the construction material used for structures exposed to this risk. However, few analyses have been made of the vulnerability of transport networks.

Existing studies on the vulnerability of communication networks focused on seismic risk, particularly following the Los Angeles (1994) and Kobe (1995) earthquakes (Hassani & Takada, 1995; Chang, 2000). However, new assessment methods have been developed for network disturbances, especially the risk of landslides, and these approaches can be transposed to DF hazard. For instance, Leone (1996, 2008) developed a method based on calculations of the rate of damage in order to establish a referential (damage matrix) for the economic assessment of the structural and functional vulnerability of transportation networks. This approach enabled physical damage to be distinguished from the resulting disruptions. Manche (2000) emphasized the latter aspect and developed a method for analyzing functional disturbances in mountainous areas, based on the concept of loss of accessibility. Here the aim was to highlight the fact that consequences may occur far from the affected area. Cleyze (2007) tackled these concepts in more detail, in particular the degree of damage to the physical infrastructures comprising the transportation network, but also the deterioration in the services provided by that network. However, existing studies focus on road networks, while few tackle the vulnerability of rail networks.

The aim of this paper is twofold: 1) to analyze the morphometrical characteristics of DF and DA catchments and to compare the response of these two processes to climatic change in recent decades (since the 1970s) in the French Alps; and 2) because the French alpine economy mainly depends on transportation networks, to compare the vulnerability of rail and road networks to recently documented debris flow activity.

2. Study area

DF and DA catchment areas are widespread throughout the French Alps. For this study, we selected 308 DF in the Arc and upper Isere river valleys close to the town of St Michel de Maurienne, and 111 DA in the massif des Ecrins around the Durance river close to the town of Briançon (45°00'S, 6°30'E) (Figs. 1-2). The Arc and upper Isere region consists mostly of schist and sandstone, while the Massif des Ecrins is a crystalline formation and mainly consists of granites and metamorphic rocks. Typical geomorphic DA characteristics have the following components: (1) a funnel-shaped debris source area consisting of broad steep granite and metamorphic walls (100-400 m in height), (2) a zone where debris levees are found on both sides of the flow track on scree slopes or transitional deposits (Jomelli & Francou, 2000; Pech & Jomelli 2001); and (3) a terminal part where the levees join to form a frontal lobe or a combination of several lobes (Van Steijn,1991; Van Steijn et al., 1988; Major, 1997). These DA deposits are located between 1 600 and 2 400 m asl which is close to the 0°C annual isotherm. DF catchments consist of steep non-vegetated slopes in the triggering zone, an incised channel with an intermittent water flow that can be eroded by debris flow events, and a gentle deposit zone situated between 450 and 1 900 m asl. In the period 1961–2000, the annual average temperature at Briançon and at Pralognan was respectively 6.1 °C and 5.5 °C, and annual average precipitation was 1 064 mm and 974 mm.

3. Data source

3.1 DF and DA surveys

DF data used in this study came from a survey conducted by the service *Restauration des Terrains de Montagne* (RTM) which was established by foresters in the 1900s and covers the entire French Alps. The organization has a departmental structure and for our study we used the database of Hautes-Alpes and Savoie departments. DF data were also collected from scientific and technical journals, monographs by local publishers, technical reports and unpublished documents from the archives of local authorities and state agencies stored at the RTM.

The inventory and cartography of DA was carried out by analyzing series of aerial photographs starting in 1970 and completed by field observations made every year between 1995 and 2010 (Jomelli et al., 2003). The scale of the aerial photographs ranged between 1: 15000 and 1: 30000. Well-defined debris avalanche deposits showing clear lobes and levees were dated by combining three approaches: analysis of aerial photographs, old documents from RTM, and from department archives. Some DAs were dated by dendrochronology: damage to *Larix deciduas* trees included broken tree trunks, impact scars, and stem tilting leading to the production of reaction wood. Samples were processed in accordance with standard methods of dendrogeomorphology (Stoffel et al., 2006).

Descriptors of each DF and DA event included an identification number for each catchment area, the year of the event, the elevation of the starting and runout zone, some morphometrical characteristics such as the difference (δh) in height (m) between the upper limit of the source area and the deposit (fig. 3). The volume of some recent DA deposits was estimated from the length, width, height of the deposit lobe measured in the field. In the case of DF deposits, we used rare documented events in the RTM data base.

3.2 Climate data

Sixteen meteorological stations at different locations (Fig. 1) and elevations were selected to characterize climatic conditions in the study area. Observed cumulative precipitation and minimum and maximum temperature data at a daily time scale were mostly available for 1971-2010 and in a few cases, for a longer period (table 1). Temperatures were analyzed to

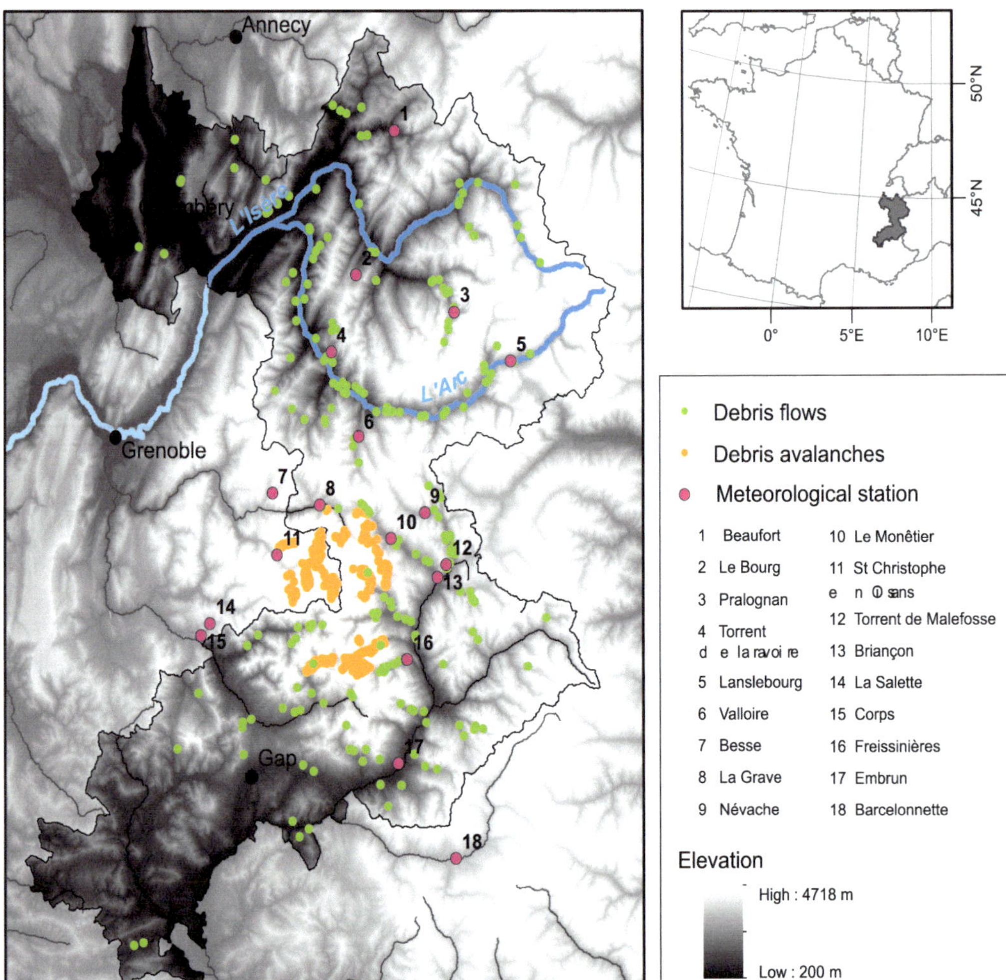

Fig. 1. Location map.

Fig. 2. Debris flood (left) and debris avalanche (right) catchments selected in this study that show different geomorphic and material properties in the triggering, transit and deposit zones.

document possible recent warming using t-tests. Precipitation was analyzed to identify changes in the intensity and frequency of precipitation. First we checked if daily precipitation above a threshold of, for example, 10 mm/d or 21 mm/d or 33 mm/d changed significantly in recent decades. To answer this question, a pertinent threshold had to be determined. For each station the choice of the threshold was determined by a mean residual plot. In a series of precipitation events we tried to identify the lowest threshold (for example 12 mm/d) above which precipitation events are considered as high intensity events i.e. extreme events. The lowest threshold was selected to analyze as many data as possible above it. A specific threshold was computed for each meteorological station. To characterize a possible trend in the intensity of precipitation for a given threshold, we computed the distribution of extreme events above the threshold and compared the parameters of the distribution estimated for the period 1970-1990 with those obtained for the period 1990-2010 for each station. The Generalized Pareto Distribution (GPD) i.e. the distribution of extreme

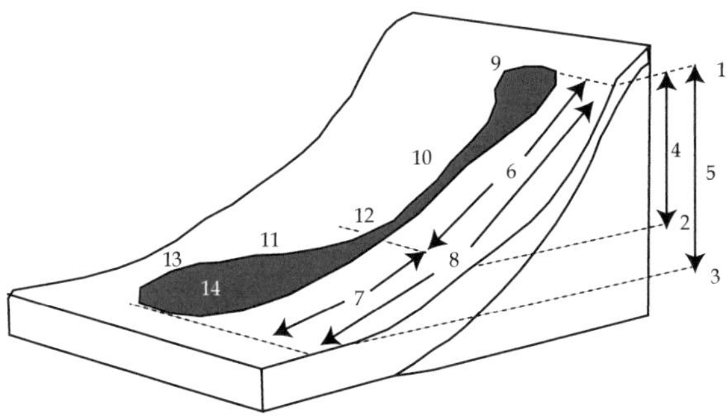

Fig. 3. Parameters measured in the DF and DA catchments (after Lorente et al., 2003 modified). 1: The elevation of the top of the debris flow. 2: the elevation at which debris flow deposition begins; 3: the elevation at the runout deposit ends; 4: difference in height (m) between the top and the transit zone; 5: difference in height (m) between the top and the base; 6: total length (m) of the debris flow between the upper part and the beginning of the deposit; 7: length (m) of the debris flow deposit from end of channel to the front; 8: the total length (m) of the landform; 9: average gradient at the top; 10: average gradient of the channel; 11: average gradient of the deposit; 12: average width of the channel; 13: average width of the deposit; 14: estimated volume of material mobilized by the debris flow.

events above a given threshold (Embrechts et al., 1997) was then fitted to daily precipitation data. The second step involved analyzing changes in the frequency of these extreme precipitation events since the 1970s. To compare the number of intense precipitations per year before and after 1990, we used a Poisson model because the number of intense rainfall events above a given threshold followed a Poisson distribution (Embrechts et al., 1997). This Poisson distribution can be defined as follows:

$$P(x) = \exp(-\lambda)\frac{\lambda^x}{x} \quad \text{where } x \in N \tag{1}$$

where λ is the parameter of Poisson law and x the number of events.

For each station, we estimated the λ parameter describing the distribution of the rainfall events between two distinct observation periods (1970-1990; 1990-2010). After testing, comparison of the λ parameters obtained for the two periods made it possible to identify significant variations in precipitation. These climate data were also used to link the triggering of debris flows with climate conditions. Different climatic parameters were calculated: mean monthly precipitation, number of rainy days per month, number of rainy days with daily cumulative rainfall greater than 10, 20 and 30 mm/day, monthly minimum and maximum temperatures.

Principal Components Analysis (PCA) was used to extract a common regional climate signal from the different meteorological stations. This analysis enabled us to reduce data dimensionality by performing covariance analysis between factors and to obtain non correlated factors, i.e. linear combinations of values. By using PCA separately for each temperature and precipitation characteristic, we generated new synthetic value of these parameters that combine the common components of the selected meteorological stations. The values of the principal coordinates were then used for further analyses in the logistic regression probability model described below.

4. Statistical method used to characterize factors that trigger DF and DA

To better understand the relationship between climate parameters and the occurrence of debris floods/avalanches, we used a logistic regression (LR) model (Aldrich & Nelson, 1984). LR analysis is often used to investigate the relationship between a set of explanatory variables such as meteorological factors and discrete responses such as event/non-event or presence/absence (Hosmer and Lemeshow, 2000). The logistic regression estimates probabilities of the occurrence of the event and non-event, depending on the explanatory variables. Our objective for the LR probability model at a yearly time scale was to find the best annual temperature and precipitation parameters that explain DF/DA triggering in the region. The first component of Pc values of the different climatic parameters were used as explanatory variables.
The dependent variable Yi is ordered and has values from 1 to k. The model based on cumulative probabilities is:

$$\text{Logit}(p_i) = f\left(\Pr\left(Y_i \leq i \middle| x\right)\right) = \alpha_i + \beta' x + e \quad (2)$$

where $f(x)$ is the logistic distribution function, i varies from 1 to k, the intercept α_i varies from $\alpha 1$ to αk-1, β' is the slope coefficient and e the error. The logistic distribution constrains the estimated probabilities to between 0 and 1. The cumulated probability pi of the occurrence i is calculated from the equation:

$$p_i = \frac{e^{\text{Logit}(p_i)}}{1 + e^{\text{Logit}(p_i)}} \quad (3)$$

To check the quality of the model, for each LR, several verification tests were computed and compared to select the most significant model results. First the probability of the adjusted model was tested against a test model. If this probability Pr > LR was less than the 0.05 significance threshold that was set, the contribution of the variable to the adjustment of the model was significant. Otherwise, it was removed from the model. Next, the estimated values, the corresponding standard deviation, Wald's Chi^2, the corresponding p-value and the confidence interval were displayed for the constant and for each variable of the model. The best model had the highest Chi^2 values. The table of standardized coefficients was used to compare the relative weights of the variables. The higher the absolute value of a coefficient, the greater the weight of the corresponding variable. When the confidence interval around standardized coefficients had a value of 0, the weight of a variable in the model was not significant. We also computed the percentage of well-classified observations for the different explanatory climatic parameters. If the final percentage score was higher than 50%, the model was considered to be significant. After all the tests, the best compilation of temperature and/or precipitation parameters was chosen based on the highest LR model coefficient values.

5. Estimation of transportation network vulnerability

To analyze the impacts of DF on the transportation network, we distinguished physical damage to roads (direct impacts) from functional disturbances i.e. the consequences of the physical damage (indirect impacts). The method consisted in compiling an inventory of disturbances to the transportation network documented by the RTM and the DDE (direction départementale de l'équipement) responsible for maintenance of the roadway network and SNCF in charge of the railway network. The intensity of these disturbances was then classified using the method proposed by Leone (1996) with modifications, and from a classification of the damage by the SNCF (société nationale des chemins de fer français) (ISRI index) (fig. 4). However, the difference between direct and indirect impacts was difficult to distinguish in the SNCF damage testimonies.

Dpmax	Type of damage to road
SD	No damage described (but functional perturbation)
D	Damage but no details
DP	Damaged bridge (road closed as precaution)
D1	Degradation of pavement
D2	Subsidence of the platform
D3	Obstruction of the roadway
D4	Destruction of the platform without the bridge breaking
D5	Destruction of the platform, bridge broken

PFmax	Functional disturbance due to damage to road
SP	No perturbation described (but physical damage)
P	Perturbation but no details
P1	Limitation of size of vehicle
PP	Traffic interrupted, duration not specified
P2	Temporary interruption of road traffic (some hours)
P3	Prolonged interruption of road traffic (some days)
P4	Long term or permanent interruption of road traffic (weeks, months)

ISRI	Type of incident on the railroad
G4	Simple detection of event
G3	No direct impact, requires repairs
G2.b	Potentially serious incident, requires works
G2.a	Major incident affecting regularity and safety of rail traffic, requires works
G1	Serious risk for rail traffic safety (1.a crash, derailment, 1.b crash, derailment avoided)

Fig. 4. Intensity of road and rail disturbances.

6. Results

6.1 Climatic analysis

In the French Alps, since the nineteen century the temperature has increased by 1 or 2°C (depending on the season and the location (Déqué, 2007). Table 1 shows that since the 1970s and independent of altitude, most meteorological stations recorded a significant increase in temperatures (Mann-Whitney test; 0.05 level) at a yearly time scale. This trend confirms other observations for the whole European Alps using homogenized instrumental time series (Böhm et al., 2001; Beniston et al., 1997). Analysis of seasonal data (results not shown) revealed that in spring, temperatures increased significantly (t-test) since the beginning of the 1970s at most stations. In summer, the mean value of this increase was around 0.9 °C. In autumn and spring, the increase in temperatures (0.6°C) was also significant (t-test) at most stations. In winter, there was a general significant increase (0.7 °C) over the whole period at all stations. Concerning changes in the number of freezing days since the 1970s, a significant decrease (t-test) in the number of freezing days (between 10-15%) was recorded at different stations.

Analysis of precipitation revealed trends in the intensity of summer precipitation and changes in their frequency above a certain threshold (Table 1). Changes in intensity/ frequency were analyzed over the whole period. At an annual time scale, only the stations at St Christophe en Oisans, Névache and Beaufort presented significant variations in intensity since 1970. The minimum threshold at which changes in the intensity were considered significant was 10 mm at Beaufort station, but 21 mm at Nevache. The frequency of rainfall events increased (Student's test; 0.05 level) at most stations except Corps, Briançon, Freissinière and Lanslebourg (Table 1). The minimum threshold at which changes in the frequency were considered significant was 10 mm at La Grave station, but 30 mm at St Christophe en Oisans. Analysis of seasonal data (results not shown) revealed that in spring the monthly precipitation mean increased significantly (t-test) since the beginning of the 1990s at most stations, whereas in summer, the mean did not reveal any significant change. In fall the increase in the monthly precipitation mean was significant (t-test) at most stations. In winter there was a significant increase in the whole period at all stations. Analysis of variations in the number of rainy days revealed a significant increase over the last decades at Lanslebourg, Pralognan, St Christophe en Oisans and Valloire.

6.2 Geomorphic characteristics

A comparison of the geomorphic characteristics of DA and DF catchments revealed significant differences. The mean value of the gradient where deposition started was 19.8°, with a wide range of gradients (15-27°) in granite DA catchments. This variance can be explained by the conditions under which DA occur. The angle of deposition can be strongly influenced by the presence of break angles in the length profile of the scree slope (Jomelli & Francou, 2000) or by forest patches, and by variations in water content. These values are close to those reported by Lorente et al (2003) in the Pyrenees. By comparison, DF deposition occurred on gentler slopes (< 20°) with the same lithology. The difference in angle between DF and DA can be explained by the fact that slope deposits on which DF and DA flow do not have the same sedimentological characteristics. Most DA deposits were recorded on scree slopes composed of a superficial coarse-grained openwork layer (Jomelli & Francou, 2000) that favors water infiltration, while DF deposits mainly occurred on slope deposits composed of a superficial massive layer, totally or partially filled with a sorted, fine-grained matrix.

The mean length of the DA deposits was 22.1 m with a minimum length of 5.8 m, and a maximum length of 55.6 m, which is much shorter than in DF catchments where deposits longer than 100 m were observed. Relatively large differences in the length of both types of debris flow deposits were observed due to the influence of local topography especially local differences in slope gradient. However, different lengths were also expected due to variations in water content. Another major difference was in the lowest altitude reached by the DF and DA deposits. DF ended at about 1 150 m (standard deviation = 426 m) while DA ended at about 1 820 m (standard deviation = 211 m).

Location	Eleva-tion		Precipita-tion					Tempera-ture	
		Period of observation	Frequency		Intensity		Period of observation	Yearly time scale	Summer time scale
			Significant trend	Threshold	Significant trend	Threshold		Significant Trend	Significa-nt Trend
St Christophe-en Oisans	1570	1963-2010d	Increase	30 mm	Yes	15mm	1961-2010 m	Increase	Increase
Embrun	849	1950-2010d	Increase	25mm	No		1950-2010d	Increase	Increase
Névache	1600	1961-2010d	Increase	30 mm	Yes	21mm	1978-2010d	Increase	Increas
Valloire	1460	1973-2010d	Increase	10mm	No		1983-2010d	Increase	Increase
Corps	1265	1947-2010d	No		No		No	Not analysed	Not analysed
La Grave	1780	1961-2010d	Increase	10mm	No		1961-2005d	Increase	Increase
La Salette	1770	1970-2010d	Increase	25mm	No		1970-2010d	Increase	Increase
Monétier les bains	1450	1950-2010d	Increase	30mm	-	Not available	1961-2010d	Increase	Increase
Barcelonnette	1155	1970-2010d	Increase	15mm	No		1970-2010d	Increase	Increase
Briançon	1324	1970-2010d	No		No		1970-2005d	Increase	Increase
Freissiniere	1320	1970-2002d	No		No		1970-2010d	Increase	Increase
Beaufort	1030	1970-2010d	Increase	25mm	Yes	10mm	1970-2010d	Increase	Increase
Lanslebourg	2000	1970-2010d	No		No		1970-2010d	Increase	Increase
Pralognan	1420	1970-2010d	Increase	15mm	No		1970-2010d	Increase	Increase
Bourg	865	1970-2010d	Increase	25mm	No		1970-2010d	Increase	Increase
Besse	1416	1950-2001d	Increase	20mm	No	Not available	1961-2010 m	Increase	Increase

Table 1. Analysis of significant variations (0.05 level) in precipitation and temperature since the 1960s for 16 stations in the French Alps (see figure 1); m–d = monthly and daily observation respectively

To estimate runout distance using morphometric parameters we used the simple formula of Vandre (1985), who found that runout distance is about 35–45% of the difference in height between the head of the source area and the point at which deposition starts (δh). The formula is:

$$\text{Runout} = \alpha \delta h \tag{4}$$

where α is an empirically derived fraction parameter expressing the ratio of runout to δh. In the case of DA, the α value was around 0.43 while for DF, it was around 0.605.

A very clear difference was observed in the volume of material mobilized by DF and DA (Brochot et al., 2002). The estimated volume of material mobilized based on 26 DA deposits averaged 130 m^3 and the maximum value was 472 m^3. Thus according to Innes (1983), these cases can be defined as "small scale" debris avalanches. These values are of the same order of magnitude as most debris avalanches cited in the literature (Blijenberg, 1998). Unfortunately, the volume of DF was rarely documented in the database. For 36 documented DF events, the volume averaged 35 000 m^3 with a maximum value of 500 000 m^3.

6.3 Changes in the frequency of DF and DA

The analysis of variations in the frequency of debris avalanches was based on 111 events that occurred since 1970. Debris avalanches are a very common geomorphic process with a mean recurrence interval of 3.7, meaning that more than one event occurred in one of the valleys every third year. However, recurrence intervals varied considerably between the catchments with a standard deviation of 21.9 for the whole region. We used a Student's t-test to check if changes in the frequency of debris avalanches were significant at a level of 0.125. Some catchments underwent a significant increase in the number of debris avalanches in the last 20 years compared to previous periods. However, in most cases the number of events per catchment was not high enough to reveal a significant change over the study period. We also tested if the number of events changed significantly over the last four decades at the regional scale. Results revealed no significant variation in the occurrence of DA. However this global result masks two different trends in DA activity depending on minimum elevation of the starting zone. At low altitudes (< 2 200 m), the number of DA decreased significantly since the 1990s, whereas at high altitudes (> 2 200 m) it has increased. The data were then classified according to the total length of the DA system. A significant decrease (0.01, Mann-Whitney test) in the number of debris avalanches was observed for the systems less than 600 m in length (Fig. 5). In contrast, there were no significant variations in the number of the debris avalanches longer than 600 m between the two periods (Fig. 5).

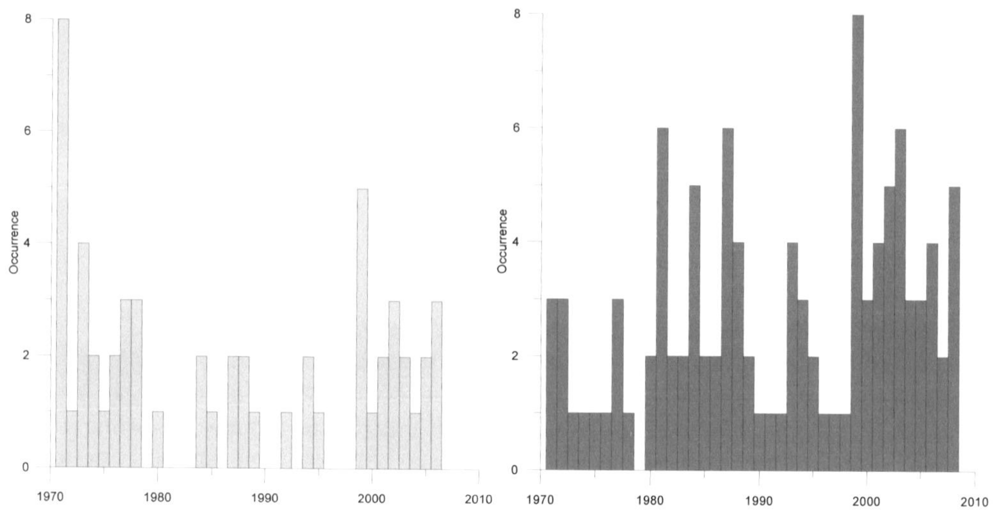

Fig. 5. Occurrence of debris avalanches over the last four decades according to the minimum elevation of the starting zone: left = below 2200 m asl; right = above 2200 m asl.

However, this trend should be interpreted with caution because the inventory of events cannot be considered as complete. Indeed, in a system with a high degree of activity, younger deposits can easily cover or sometimes erode older flows. The overall decadal frequency of debris avalanches is shown in figure 5. On average, two and three events per decade years were identified during the study period. A slight trend towards more debris avalanches was observed in the second 20-year period.

The same analysis was conducted on debris floods. Results revealed the same trend considering that most started at high elevations i.e. higher than 2 000 m asl. There was an increase in the number of debris flood since the end of the 1980s (fig. 6). About 5.8 events were triggered per year between 1970 and 1989 versus 13.3 between 1990 and 2010.

Catchment areas with high altitude triggering zones were more active. Even if there is probably a bias in the inventory (the data reported in the survey represent minimum frequencies) we believe that the higher debris flood activity may not only be the result of sample skewness because this trend was observed in most valleys for which we had a sufficient number of data and in valleys where the people responsible for the observation were the same for the two periods.

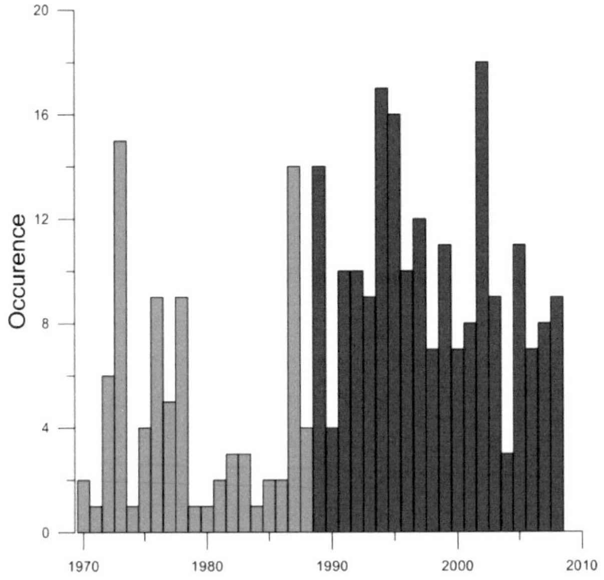

Fig. 6. Changes in debris flood activity since 1970.

6.4 Climate conditions responsible for DA and DF activity

We analyzed the climate conditions responsible for triggering DA. Binary logistic regression was performed yearly. We carried out the annual sum of the number of occurrence observed DA. The result was classified in two groups of equal size (obtained starting from a threshold of three annual releases) giving to the years with less than three occurrences the code 0, and to the years with at least three occurrences the code 1.

We tested all first components of the PAC of different parameters one by one and then combined them to identify most significant parameters in the temperature and precipitation series. After different combinations, the independent variables that gave the best fit were mean monthly temperature and number of daily rainfall events greater than 20 mm/day (*Nd20*) between May and October 1971-2010. The model is:

$$Logit(p_1) = 4.7 - (0.38 \times mean\ TX) - (0.5717 \times Nd20) + e \quad (5)$$

The percentage of correct predictions for the presence/absence of a DA event was higher than 75% (Table 2).

We conducted the same analysis for DF events. The number of occurrence was classified in two groups of equal size (obtained starting from a threshold of five annual releases) giving to the years with less than five occurrences the code 0, and to the years with at least five occurrences the code 1.

The independent variables that gave the best fit were mean minimum temperature (*Tn*) and the number rainy days (*Nrd*) between May and September 1971-2008. The model is:

$$Logit(p_1) = - 0.14 + (0{,}44 * Tn) + (0{,}39 * Nrd) + e \tag{6}$$

The percentage of correct predictions for the presence/absence of a DF event was higher than 65% (Table 2).

	Parameter	Value	Wald Chi square	Pr>Chi2	% correct 0	% correct 1	% correct_total
DA	NFR	0.38	2.36	<0.03	66.67	85.71	77.78
	Nd (20)	0.57	5.482	<0.007			
DF	Tn	0.504	5.958	<0.0001	74.81	54.69	69.79
	Nrd	0.476	5.598	<0.0001			

Table 2. Statistics for the two DA and DF models.

6.5 Impacts of DF and DA activity on the transportation network

The analysis of DF and DA events revealed clear differences in impacts on the transportation network. No death was reported for either DF or DA. However, the social impacts of DF were much greater than those of DA mainly due to the fact that the DF deposit zone was located at low altitudes i.e. at an altitude with permanent socio-economic activity. Damage caused by DA was mainly due to the deposition of rock debris on roads. However there were only a few cases and these affected secondary roads used by only a few cars. The indirect consequences of the damage was thus limited. By contrast, DF activity had both direct and indirect impacts on road and rail networks. To illustrate such impacts we selected two catchments that had undergone much DF activity over the last decades. The first DF catchment named "La Ravoire" is located in the Arc valley, south of Savoie department (figs. 1-7-8). The elevation ranges between 2 686 m in the upper part to 499 m at the confluence with the Arc river, with a average slope of 16-17°. The watershed covers 10,5 km² with a highly urbanized debris fan area of 0.2 km² at Pontamafrey village. La Ravoire drainage basin is composed of metamorphic and sedimentary rocks covered by superficial deposits.

The second DF catchment named "Le Malefosse" is located in the northeast of Hautes-Alpes department (figs.1-7). This catchment is rather small with a surface area of 1.8 km² and 0.3 km² for the debris fan. Altitudes range from 2 509 m in the upper part to 1 315 m at the confluence with the Durance river, with a high average slope of 21.8°. The geological conditions at La Malefosse catchment are similar to those at La Ravoire. The transportation networks affected by DF were as following for both catchments:

- the D1006 road in Savoie department linking France and Italy. The traffic is relatively high for this type of highway with 8515 vehicles per day (Conseil Général, 2010).
- the international high-speed train (TGV) linking Paris and Turin, and the regional line which links different towns in the valley, with a total of 71 trains per day (SNCF, 2010).
- the N94 road in Hautes-Alpes department; this is an international transit road linking the main towns in the French Alps (Briançon, Gap, Marseille, Grenoble) with Italy. In 2006, daily traffic was 9346 vehicles (DDE Hautes-Alpes).

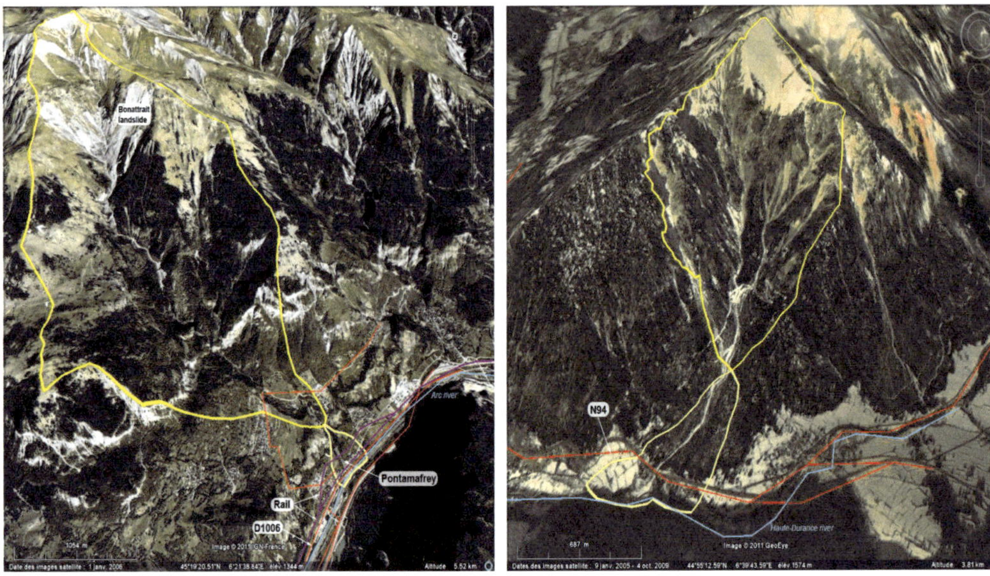

Fig. 7. Location map of La Ravoire catchment (left) and Malefosse catchment (right) showing impacted transportation networks.

Fig. 8. View of Pontamafrey village constructed on the debris cone of La Ravoire. La Ravoire channel crosses the village the international line of high-speed train (TGV) which links Paris to Turin and the road way D1006.

The survey of DF conducted by RTM-DDE and SNCF reported that 24 events were triggered since 1970 in the La Ravoire (fig. 9) catchment and that 14 had an impact on the transportation network (43% on the railway and 57% on roads). In the Malefosse catchment nine events were reported (fig. 9) of which 5 damaged roads. In both catchments, the majority of events that caused damage occurred between June and August. A slight trend towards more DF was observed in the more recent years in the Malefosse catchment while no trend was observed in La Ravoire. Of the events that caused damage in both catchments, 69% were triggered by consecutive rainy days (between 2 and 6 days of rain on average) recorded by the meteorological stations at Saint-Martin and Briançon. Six percent were triggered by a single rainy day with daily cumulated precipitation < 20 mm and 12% of events were triggered but no rainfall event was recorded at the weather stations. This can be explained by localized rainfall such as storms. Finally, no information concerning the weather was available for 13% of the events.

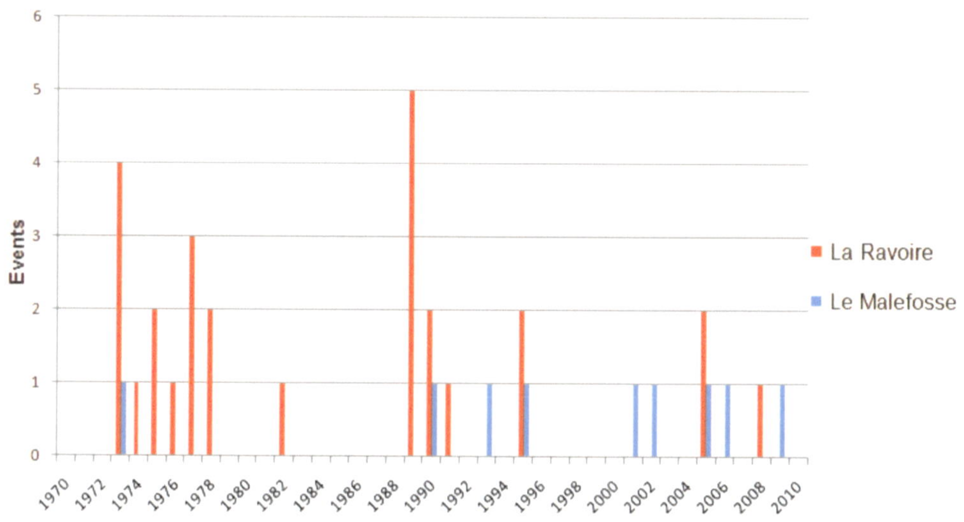

Fig. 9. Debris flow triggering at La Ravoire and Le Malefosse catchments since 1970.

Results of the analysis of damage caused by DF activity is shown in figure 10. In both catchments, the main physical damage was caused by rock debris transported by DF and deposited on roads in around 60% of documented cases in both catchments. The other damage to roads (21%) was reported but no details were given. The PFmax scale highlighted the lack of a precise description of the functional disturbance in about 60% of documented cases. This result can be explained by the fact that the evaluation was made just after the DF was triggered and it is often very difficult to predict the real consequences of impacts over a long period, and thus to quantify functional disturbances. However La Ravoire underwent a large proportion of serious perturbations with an interruption of road traffic of several hours. Finally, the ISRI scale revealed a large proportion of major incidents that affected the rail traffic and safety. Functional disturbances caused by this damage usually involved delays or temporary breaks in rail traffic. The consequences were operating losses and economic losses due to the immobilization of rolling stock. Moreover, in the case of a temporary interruption of rail traffic, the rail network owner (SNCF) must provide alternative means of transport (bus, taxi, transfer to other rail lines, etc.). The rail network thus appears to be the most vulnerable in terms of the intensity of disturbances caused by the La Ravoire torrent, although the number of impacts on the road network was higher.

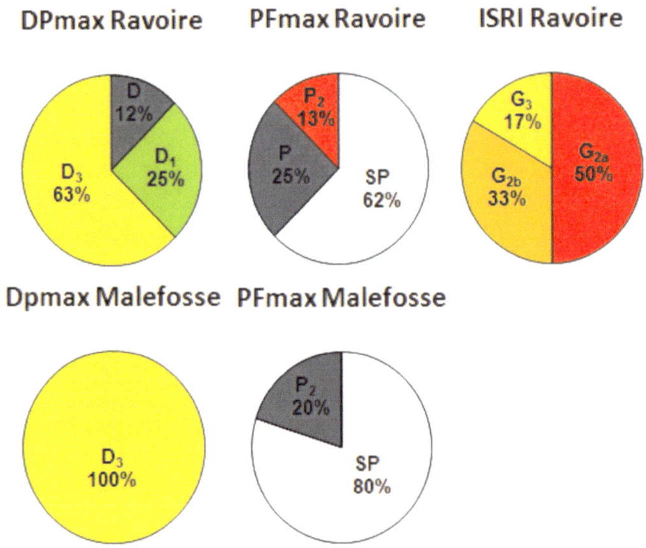

Fig. 10. Distribution of physical and functional damages (La Ravoire and Malefosse catchments) and rail incidents (La Ravoire catchment).

The sudden triggering of DF makes it difficult to manage and forecast the associated risk. Consequently, a large number of measures have been taken to reduce damage to infrastructures caused by DF in high risk areas. Pontamafrey village, which is located on the debris fan of La Ravoire, provides an illustration of possible network protection strategies (fig. 8). The village has a bypass system for the railroad (2.2 km long), with a drawbridge to allow trains to pass in the case of interruption of the normal railroad caused by a DF event. The road is also equipped with a drawbridge that is automatically activated by the torrent detection systems located upstream on the La Ravoire river. The total cost of these installations was 3,774,000 € (Cojean et al., 2002). However, the cost of damage caused by DF before the construction of the protective structures rose to 4,548,000 € ((Cojean et al., 2002). This example highlights the need for assessment and management systems to reduce network vulnerability.

7. Discussion

In this paper, we compared the geomorphic characteristics of debris floods and debris avalanches in the same alpine region and the relationships with climate conditions responsible for their triggering. The climatic trend observed in the French Alps was characterized by analyzing data on extreme summer rainfall events recorded daily at 16 stations located in Hautes-Alpes, Savoie and Alpes de Haute-Provence departments since 1970. According to the generalized Pareto law (GPD) our results showed that extreme summer rainfall events increased significantly in the region. In addition, there was a significant increase in annual and seasonal temperatures in the last 20 years combined with a significant reduction in the number of freezing days. The occurrence of DF and DA exhibited two different trends depending on elevation. At low altitudes (< 2 200 m) the number of DF and DA decreased significantly since the 1980s, whereas at high altitude (> 2 200 m), it increased. Over the same period there was a significant increase in DF activity.

However, despite this common trend in DF and DA activity, modeling revealed that the climatic parameters responsible for the triggering of DF and DA differed. The frequency of triggered DA depends mainly on summer rainy events greater than 20 mm/day and minimum temperatures while that of DF depends on mean summer temperatures and the number of rainy days. The role of intense or long duration precipitation in triggering debris flows has been known for a long time. The data we analyzed do not differ from this rule. However, it is difficult to interpret the fact that DA are mainly caused by extreme events while DF are mainly caused by long duration precipitation. Extreme rainy events were an explanatory factor in DF activity that was less significant than the number of rainy days. In addition, laboratory experiments and model results revealed that the morphometric characteristics of the catchment area (channel confinement for instance) play a significant role in the dynamics of debris flows and depositional structures. In particular, debris flow behavior evolves in response to changing pore pressures that depends on a combination of morphosedimentological characteristics and water infiltration (Major & Iverson 1999; Marr, et al., 2001). Consequently it is probable that the differences in slope inclination between DA and DF as well as the confinement/non-confinement of the channel may be an explanation (Hampton, 1975) but the scale at which measurements were made in this study did allow us to check this hypothesis. Indeed, we were unable to measure the local topography which has an influence on the behavior of debris flow initiation (Blijenberg, 1998). Differences between the granulometrical characteristics of DF and DA triggering zones which were impossible to quantify at the large spatial scale considered in this study as well as probable differences in the amount of precipitation between meteorological data collected at low elevations and data collected in the triggering zone may also be relevant. Indeed, summer storms responsible for extreme rainy events are often local and are consequently not always recorded by meteorological station pluviometers. It is also interesting to note that both DA and DF triggering were sensitive to temperature. This parameter is rarely considered as significant in the literature. Some cases studies in the Alps revealed a relationship between the rapid snow melt and/or glacier retreat and permafrost degradation induced by a temperature and the triggering of debris flows (Haeberli & Beniston, 1998; Wegmann et al., 1998; Imhof et al., 2000; Bardou & Delaloye, 2004). Permafrost degradation may be a relevant explanation for the high altitude DF and DA triggering selected in this study. As the mean altitude of the triggering zone of DA is about 2 200 m in this region close to the 0°C isotherm, the significant trend observed for high altitude DA activity (fig. 5) suggests a change in the mobilization of rock debris triggered by warmer temperatures. Such an increase in temperature at high altitudes has been demonstrated in recent decades by different authors working on homogenized series (Beniston et al., 1997; Diaz and Bradley, 1997; Böhm et al., 2001). It may also reflect the influence of temperature on the snow/rain limit with liquid precipitation occurring at high altitudes or possible changes in the duration of the snow cover, which may expose the high slopes to greater temperature variations. However one can wonder if this relationship is only due to the recent increase in temperature observed from the 1990s on. To test this hypothesis, we computed a logit analysis for the 1970-90 period. Results revealed that temperature was less important than rainfall but was nevertheless significant.

Another important aspect is the consequences of these changes in DF and DA activity for society. The analysis of DF and DA data revealed clear differences. No death was reported due to either DF or DA. However, the social impacts of DF were much greater than for DA mainly due to the fact that the DA deposit zone was located at low elevations, i.e. where there is permanent socio-economic activity.

8. Conclusion

The aim of this study was to analyze the morphometrical characteristics of DF and DA catchments and to compare the response of these two processes to climate change in recent decades (since the 1970s) in the French Alps. Two areas the Hautes-Alpes and Savoie regions, were selected. In the two areas, a total of 419 debris floods and debris avalanches occurred since the beginning of the 1970s. Significant geomorphic differences were observed between the two processes. The debris avalanches were "small scale" events according to the definition of Innes (1983). Located at high elevations, they occurred on steep slopes. By contrast, DF were much larger events triggered in larger catchment areas that spread to urbanized areas. A climate analysis was performed using data from 16 meteorological stations located in the area. This analysis revealed a significant increase in annual and seasonal temperatures in the 20 last years combined with a significant reduction in the number of freezing days. Changes in the frequency of summer rains were also observed. The response of debris flows and debris avalanches to this climate change was investigated. There was a significant increase in the occurrence of both processes at high elevations (> 2 200 m). In addition, logistic regression models were used to characterize the relationship between climate and the frequency of debris floods and debris avalanches. Results revealed that different climatic parameters are responsible for the triggering of DF and DA. We showed that the probably frequency of triggered DA depends mainly on summer rainy events greater than 20 mm/day while that of DF depends on the number of rainy days and mean seasonal temperature. Consequently, we observed a univocal response of DF and DA to recent climatic change because for both processes climate parameters responsible for their triggering changed. Finally an analysis of the socio-economic impacts of DF was performed. The results showed that such processes cause direct and indirect damages. By identifying the critical sections of road and rail networks and their strategic value in organizing the territory, this approach allowed us to characterize the types of impacts that are most frequently observed in these networks. Obstruction and degradation of roads with temporary interruption of traffic are the main types of damage. The cost of damage was also estimated. However, evaluating the vulnerability of the network has some limits, such as the lack of visibility and information available on the damage, particularly with respect to functional disturbances whose consequences extend beyond the area of the event.

9. Acknowledgment

Financial support was provided by the French ANR Scampei program n°08-Clim-010 and the EU/FP7 ACQWA Project and the EU/ERANET CIRCLE-MOUNTAIN ARNICA Project. Finally we would like to extend our thanks to RTM services for giving us debris flow survey databases.

10. References

Aldrich J.H. & Nelson F.D. (1984). Linear probability, logit, and probit models. *Series, Quantitative applications in the social sciences*, 45, Sage University Paper, Thousand Oaks. 95p.

Alexander, D.E. (1988). Developing a Landslide Damage Scale: the lessons of the Great Ancona Landslide of 1982. Ground Failure, n°4, 4.

Alexander, D.E. (2005). Vulnerability to Landslides. *Landslide Hazard and Risk, Whiley Chichester*, 23, 175-198.

Bardou, E. & Delaloye, R. (2004). Effects of ground freezing and snow avalanche deposits on debris flows in alpine environments. *Natural Hazards and Earth System Sciences*, 4, 519–530.

Beniston M., Diaz H. F., & Bradley R. S. (1997). Climatic change at high elevation sites: an overview. *Climatic Change*, 36, 233–25.

Blijenberg, H.M. (1998). *Rolling Stones? Triggering and frequency of hill slope debris flows in the Bachelard Valley, Southern French Alps*. PhD Thesis, Utrecht University.

Böhm, R. Auer, I. Brunetti, M. Maugeri, M. Nanni, R. & Schöner, W. (2001). Regional temperature variability in the European Alps: 1760-1998 from homogenized instrumental time series. *International Journal of Climatology*, 21, 1779-1801.

Brochot, S. Marchi, L. & Lang, M. (2002). L'estimation des volumes des laves torrentielles: méthodes disponibles et application au torrent du Poucet (Savoie). *Bulletin of Engineering Geology and the Environment* 61, 389–402.).

Caine, N. (1980). The rainfall intensity-duration control of shallow landslides and debris-flows. *Geografiska Annaler*, 62A, 23–27.

Conseil Général (2010). Bilan des Trafics et Circulation Hivernale de l'année 2009, département de la Savoie. Rapport Interne, 55.

Chang S.E. (2000). Disasters and transport systems: loss, recovery and competition at the Port of Kobe after the 1995 earthquake. *Journal of Transport Geography*, No. 8, 53–65.

Cleyze, J.F. (2007). La vulnérabilité structurelle des réseaux de transport dans un contexte de risque. PhD Thesis. Paris VII University, 540.

Cojean, R. Laugier, P. Le Mignon, N. Pollet, N. Thevenin, I. Fleurisson, J.A. (2002). Risques naturels et montagne : Mouvements de terrain et risques hydrologiques associés et induits. Evaluation des méthodes et moyens de protection et analyse des retour d'investissement. *Rapport final d'Amines-CGI*, 298.

Dequé, M. (2007). Frequency of precipitation and temperature extremes over France in an anthropogenic scenario: model results and statistical correction according to observed values. *Glob Planet Change*, 57, 16–26.

Diaz, H. F. & Bradley, R. S. (1997). Temperature Variations during the Last Century at High Elevation Sites. *Clim. Change*, 36, 253-279.

Embrechts P., Klüppelberg, C. & Mikosch, T. 1997. *Modelling Extremal Events for Insurance and Finance*. Springer.

Fuchs, S. Heiss, K. Hübl, J. (2007). Towards an empirical vulnerability function for use in debris flow risk assessment. *Natural Hazards and Earth System Sciences 7*, 495–506.

Guzzetti, F. Peruccacci, S. Rossi,M. & Stark, CP. (2008). The rainfall intensity-duration control of shallow landslides and debris flows: an update. *Landslides*, 5, 3–17.

Haeberli, W. Rickenmann, D. & Zimmerman, M. (1990). Investigation of 1987 debris flows in the Swiss Alps: general concept and geophysical soundings. *Hydrology in Mountainous regions. II Artificial reservoirs; Water and slopes*. IAHS publication, 194, 303-310, Proceedings of two Lausanne Symposia, August 1990.

Haeberli W. & Beniston M. (1998). Climate change and its impacts on glaciers and permafrost in the Alps. *Ambio*, 27, 258-265.

Hampton, M. (1975). Competence of fine-grained debris flows. *Journal of Sedimentary Research, 45, 834-844.*

Hassani N. & Takada S. (1995). Lifelines performance and damage during 17 Jan. 1995, South Hyogo Great Earthquake. *Proceedings of the first Iran–Japan Workshop on Recent Earthquakes in Iran and Japan*, Iran, 1809–1821.

Hulsbergern J.G. & Carree G.J. (1987). A method for estimating damage to buildings in subsiding areas. *Proccedings European Conference on Soil Mechanics and Foundation Engineering*, Dublin, Vol. 2, 699–702.

Hosmer D. & Lemeshow, S. (2000). *Applied Logistic Regression*, 2nd Edition Wiley.

Hungr, O. (2005). Classification and terminology, In: *Debris flow hazards and related phenomena*. Jakob, M., and Hungr, O. (eds), 10-23. Springer Berlin.

Imhof, M. Pierrenhumbert, G. Haeberli, W. & Kienholz, H. (2000). Permafrost investigation in the Schilthorn massif, Bernese Alps, Switzerland. *Permafrost and Periglacial Processes*, 11, 189–206.

Innes J.L. (1983). Debris flows. *Progress in Physical Geography*, 7, 469–501.

Innes J.L. (1985). Magnitude-frequency relations of debris flows in Northwest Europe. *Geografiska Annaler*, 67A, 23-32.

Jakob M. & Lambert S. (2009). Climate change effects on landslides along the southwest coast of British Columbia. *Geomorphology*, 107, 275-284.

Jomelli V. & Francou B. 2000. Comparing characteristics of rockfall talus and snow avalanche landforms in an alpine environment using a new methodological approach. *Geomorphology*, 35, 181-192.

Jomelli, V. Chochillon, C. Brunstein, D. & Pech P. (2003). Hillslope debris flows frequency since the beginning of the 20th Century in the Massif des Ecrins, In: *Debris flow hazards mitigation:* 127-137, Rickenmann & Chen (ed), Millpress, Rotterdam.

Jomelli, V. Pech, P. Chochillon, C. & Brunstein, D. (2004). Geomorphic variations of debris flows and recent climatic change in the French Alps. *Climatic Change*, 64, 77-102.

Jomelli, V. Brunstein, D. Grancher, D. & Pech, P. (2007). Is the response of hill slope debris flows to recent climate change univocal? A case study in the Massif des Ecrins (French Alps). *Climatic Change*, 85, 119-137.

Jomelli, V. Déqué, M. Brunstein, D. & Grancher, D. (2009). Impacts of future climate change (2070-2100) on debris flow occurrence: A case study in the Massif des Ecrins (French Alps). *Climatic Change* DOI 10.1007/s10584-009-9616-0.

Léone, F. (1996). *Concept de vulnérabilité appliqué à l'évaluation des risques générés par les phénomènes de mouvements de terrain*. PhD Thesis. Document du BRGM, n°250, ed B.RG.M. Orléans, 286.

Leone F., Asté J.P., Leroi E. (1996). Vulnerability assessment of elements exposed to mass movement: Working toward a better risk perception. Senneset K. (ed.): *Landslides*, Ed. Balkema, Rotterdam, 263-269.

Léone, F. (2008). Caractérisation des vulnérabilités aux catastrophes 'naturelles' : contribution à une évalutation géographique multirisque (Mouvements de terrain, séismes, tsunamis, éruptions volcaniques, cyclones). Vol.2- Mémoire d'Habilitation à Diriger des Recherches. Montpellier, 330.

Lorente, A. Beguer S. Bathurst, J.C. & Garcıa-Ruiz, J. M. (2003). Debris flow characteristics and relationships in the Central Spanish Pyrenees. *Natural Hazards and Earth System Sciences*, 3, 683–692.

Major, J. J. (1997). Depositional processes in large-scale debris flow experiments. *Journal of Geology*, 105, 345_68.

Major, J.J. & Iverson, R.M. (1999). Debris-flow deposition: Effects of pore-fluid pressure and friction concentrated at flow margins. *Geological Society of America Bulletin,* 111, 1424-1434.

Manche, Y. (2000). Analyse spatial et mise en place de systems d'information pour l'évaluation de la vulnérabilité des territoires de montagne face aux risqué naturels. PhD Thesis. Joseph Fourier - Grenoble I University, 175.

Marr, J.G. Harff, P.A. Shaumugam, G. & Parker, G. (2001). Experiments on subaqueous sandy gravity flows: The role of clay and water content in flow dynamics and depositional structures. *Geological Society of America Bulletin* ,113, 1377-1386.

Nyberg, R. & Rapp, A. (1998). Extreme erosional events and natural hazards in Scandinavian mountains. *Ambio,* 27, 292-299.

Pech, P. & Jomelli, V. (2001). Caractéristiques et rôle du cône apical dans le déclenchement des coulées de débris. *Géographie physique et Quaternaire*, 55, 47-61.

Rapp, A. (1995). Case Studies of geoprocesses and environmental change in mountains of northern Sweden. *Geografiska Annaler*, 77A, 189-196.

Rebetez, M. Lugon, R. & Baeriswyl, P.A. (1997). Climatic change and debris flows in high mountain regions: the case study of the Ritigraben Torrent (Swiss Alps). *Climatic Change,* 36, 371-389.

Stoffel, M. Bollschweiler, M. & Hassler, G. (2006). Differentiating past events on a cone influenced by debris-flow and snow avalanche activity—a dendrogeomorphological approach. *Earth Surf Process Landf* .,31,1424–1437.

SNCF (2010). Carnet journaliers année 2010 document interne, 32.

Stoffel, M. & Beniston, M. (2006). On the incidence of debris flows from the early Little Ice Age to a future greenhouse climate: A case study from the Swiss Alps. *Geophysical Research Letters*, Vol. 33, LXXXXX, doi:10.1029/ 2006GL026805

Vandre, B. C. (1985). Rudd Creek debris flow, In: *Delineation of landslide, flash flood and debris flow hazards in Utah,* edited by Bowles, D. S., Utah Water Res. Lab., Logan, Utah.

Van Steijn, H. (1991). Frequency of hill slope debris flows in part of the French Alps. *Turkish Bulletin of Geomorphology*, 19, 83-90.

Van Steijn, H., De Ruig, J. & Hoozemans, F. (1988). Morphological and mechanical aspects of debris flows in parts of the French Alps. *Zeitschrift für Geomorphologie*, 32, 143-161.

Veyrat-Charvillon,, S. & Menier, M. (2006).Stereophotogrammetry of archive data and topographic approaches to debris-flow torrent measurements: calculation of channel-sediment states and a partial sediment budget for Manival torrent (Isère, France). *Earth Surface Processes and Landforms,* 31, 201-219.

Wegmann, M. Gudmudsson, G. H. & Haeberli, W. (1998). Permafrost changes in rock walls and the retreat of Alpine glaciers: A thermal modelling approach. *Permafrost and Periglac. Process.*, 9, 23–33.

Zimmerman, M. & Haeberli, W. (1992). Climatic change and debris flow activity in high mountains areas. A case study in the Swiss Alps. *Catena,* suppl 22, 59-72.

22

Glaciers Shrinking in Nepal Himalaya

Samjwal R. Bajracharya, Sudan B. Maharjan
and Finu Shrestha
*International Centre for Integrated Mountain
Development (ICIMOD)
Nepal*

1. Introduction

Glaciers are repositories of information for climate change studies, as they are sensitive to global temperature and precipitation changes. Due to global warming the impact was directly influencing in the melting of the glaciers and enhancing in recent decades. The rapid melting of glaciers reduce the glacier area by which the glaciers are fragmented with increase in glacier number (Bajracharya et al., 2006a,b, 2007a,b, 2008, 2009a). The history of glacier study in Nepal is not old, it was just started by Fritz Műller in 1956; who visit Nepal as a participant in the Swiss Everest Expedition. During following years, the number of scientific expeditions has gradually increased. However, Nepal has no glaciers under long-term observation, though a number of fragmented and short studies have been made on the AX010 Glacier, Mera Glacier, Yala Glacier and Rikha Samba Glacier. The AX010 has the densest observations in terms of Glacier extent, mass balance, and ice flow (Fujita et al., 2001). The systematic investigation of glaciers in Nepal was first organized by Nagoya and Kyoto Universities of Japan. The Glaciological Expedition of Nepal (GEN), led by Higuchi (1976, 1977, 1978, 1980), carried out a series of field studies in coordination with Department of Hydrology and Meteorology (DHM) Nepal. The first detailed study of AX010 glacier was conducted in 1978/1979 (Ageta et al., 1980, 1992; Kadota et al., 1997), Yala Glacier was studied since the 1980s, and Rikha Samba Glacier has been surveyed intermittently since 1974 (Nakawo et al., 1976; Fugii et al., 1996; Fugita et al., 1997). The glaciers of Nepal was first mapped by ICIMOD in 2001 from the Indian Survey topographic maps published from 1963 to 1982. The maps were prepared from the aerial photographs of 1957 to 1959 with extensive field work (Mool et al.; 2001, 2005). The study revealed 3,252 glaciers with 5,323km^2 glacier area, which is almost 3.6% of the total land cover of Nepal. The second generation of glacier mapping of Nepal (Bajracharya et al., 2011 unpub.) was based on the satellite images of 2008/2009, which shows that the number of glacier has apparently increased but the total area has decreased drastically. The total number of glaciers in this survey shows 3808 glaciers with 4212km^2 glacier area and 346km^3 estimated ice reserves. The glacier area loss is about 20% in last 40 years. The glacier cover of Nepal reduces to 2.9% of total land cover in Nepal. The subsidence of glacier surface by 0.40m per year in Dudh Koshi basin is also reported since late 1960's due to the melting of the glaciers (Bolch, 2008). Bajracharya et al., 2008 has also reported the glacier retreat rate of 10 to 60m per year in Dudh Koshi basin. GEN, 2006 has studied many glaciers and reported the glacier retreat with different rate. The monitoring of glaciers in the high altitude of remote area with harsh climatic condition is thorny and cumbersome, hence; remote sensing approach is the best way of monitoring in the first round for the prioritization of detail mass balance study. The aim of the present study is to show the suitability of multi-temporal optical remote sensing data to map and monitor the glaciers and to illustrate the glacier retreat from different elevations. Moreover, the accuracy of multi-temporal Landsat data is addressed.

2. Study area

Nepal is one of the small, mountainous and landlocked South Asian countries extending between 26° 15′ to 30° 30′ N latitude and 80° 00′ to 88° 15′ E longitude (Fig. 1). The country's total area is 147,181km^2 and its length is 840km. The width of the country ranges from 90 to 230km (about 180km in average) from east to west.

Fig. 1. Location map of Nepal

It is bordered by massive countries, Tibet/China to the north and on the remaining three sides by India. The northern border runs along the crown of glaciated and snowy peaks with flat terrain to the southern border. The country has a great altitudinal variation ranging from 64m asl in the southeast to 8850m asl (Mt. Everest) in the north within the range of 150km, the greatest land-based relief in the world. The country is vulnerable to various hazards due to great elevation differences, fragile geological conditions, soft soil cover, steep river gradients and high intensity rainfall especially during the summer monsoon. An excessive monsoon rainfall usually triggers a variety of slope movements in Nepal often causing extensive damages to life and property.

3. Data and methods

3.1 Data

Landsat 7 ETM+

Glaciers show different spectral reflectance, which helps characterize them in satellite data to delineate glacier outline. The Landsat 7-ETM+ are freely downloadable and suitable for glacier mapping and monitoring. The Landsat images of 2008 and 2009 with least snow cover and no cloud cover were selected and acquired. The SLC-off was corrected with two images either SLC-off and SLC-on images or both cases the SLC-off images in which gaps is overlap each other from the Landsat Gapfill tools in the ENVI software.

Digital Elevation Model (DEM)

The glacier outlines when combined with a digital elevation model (DEM) to derive glacier parameters such as hypsometry, slope and elevation of crown and snout of the glaciers. The Shuttle Radar Topography Mission (SRTM) of 90m resolution data is used to derive the glacier parameters.

3.2 Methods

The methodology of semi-automated mapping of glaciers (Bajracharya et al., 2009 unpub.) is used to delineate the glacier boundary. The GLIMS ID is given to each one of the glacier polygons and the glacier attribute parameters were generated by combining the earth observation data and digital elevation model using GIS techniques.

The spectral uniqueness of glacier ice in the visible and near infrared (NIR) bands of the electromagnetic spectrum enables use of simple algorithms permitting semi-automatic mapping of glaciers (Frauenfelder et al., 2009; Paul, 2000), in contrast to a tedious fully manual approach. Among Clean ice type (CI-type) and Debris covered (DC-type) glaciers latter poses challenges in illuminating errors due to effect from surrounding materials. The threshold value of algorithm differs slightly in different scenes of the images. Bearing in mind the spatial scalability of features of interest, multi-band Landsat images enable extraction of information by standardized, uniform processing steps. The approach in mapping of CI-type and DC-type glaciers is different (Fig.2). The threshold value of NDSII for CI-type and mean slope for DC-type ideally maps all glacier pixels; further filtering using different variables such as NDVI (for vegetation), land & water mask (water bodies), mean hue, mean slope and mean altitude (glaciers), and then ultimately omitting polygons with areas less than 0.02 Km^2, eliminates most misclassified pixels and efficiently generates a product suitable for manual editing.

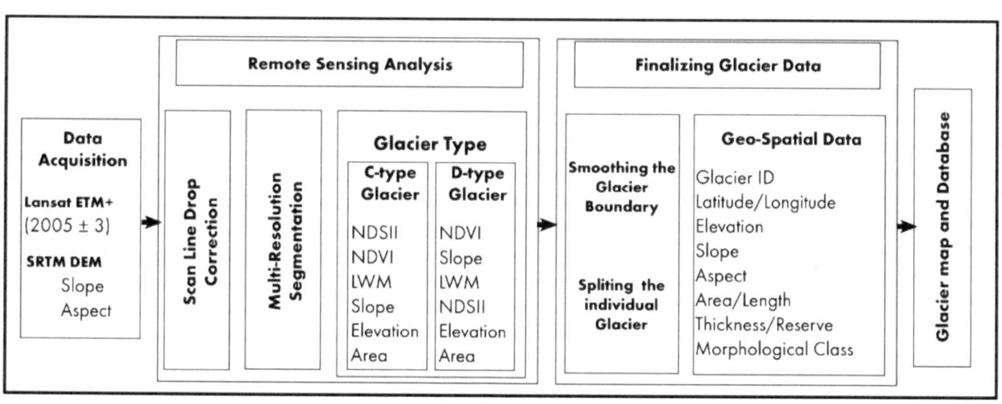

Fig. 2. Flow diagram of methodology for mapping of Clean Ice and Debris Cover glaciers using satellite images

4. Results

Altogether 3808 glaciers were mapped from Nepal. The total aerial extension of glaciers in Nepal is 4212km^2 with an estimated ice reserves of about 346km^3. The estimated ice reserves is higher in central and eastern Nepal than in western Nepal, but the numbers of glaciers are higher in central and western Nepal than eastern Nepal (Fig. 3).

The rapid melting of glaciers fragmented the glaciers with the result of increase in number of glaciers and decrease in glacier area. The overall glacier area of Nepal is reduced from 3.6% to 2.9% of total land of Nepal from 1970 to 2008 (Bajracharya et al., 2010). The largest glacier in the ICIMOD inventory of 2001 was Ktr_gr 193 in Kanchenjunga of Tamor sub-basin, which had a total glacier area of 94km^2 but due to the shrinkage of glacier it has broke down in to two glaciers and the larger part is having the area of around 77km^2, that means the Ktr_gr 193 is no longer the largest glacier in Nepal. The new glacier data shows that the Ngojumba Glacier of Dudh Koshi sub-basin is the largest glacier in Nepal. However, the area of Ngojumba Glacier is also reduced from 82.6km^2 to 80.7 km^2.

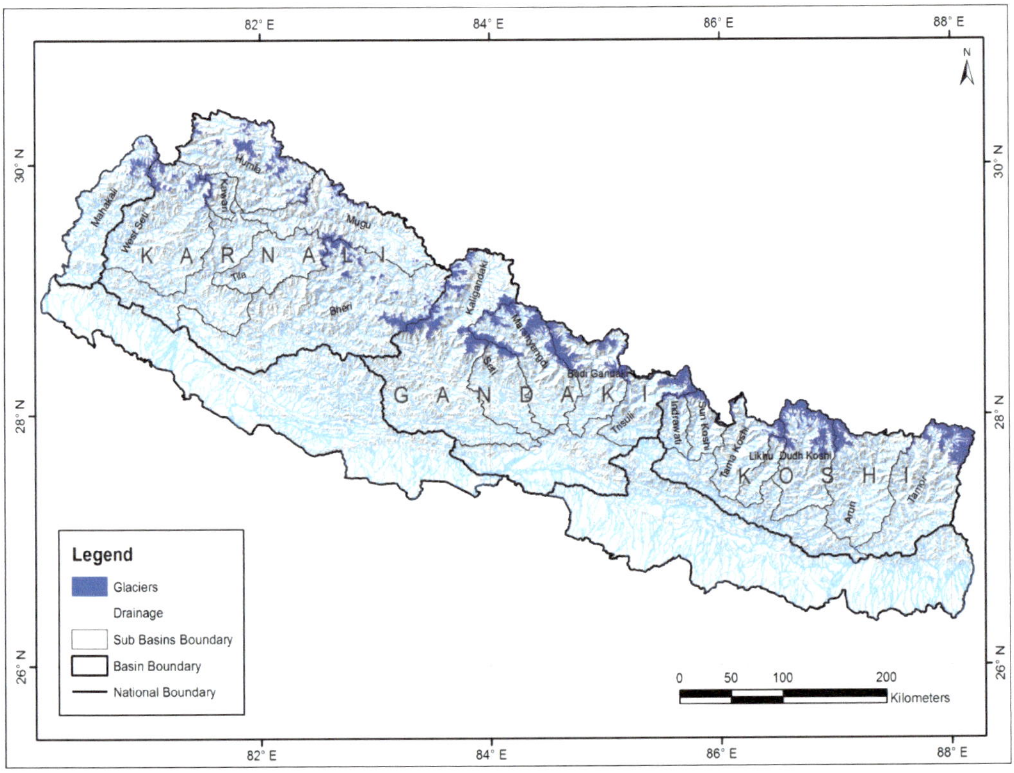

Fig. 3. Glaciers of Nepal

4.1 Change in glacier number, area and estimated ice reserves

The glacier study in 2001 revealed that some 3,252 glaciers covering a total area of 5,323km^2 and including 481km^3 of estimated ice reserves occur in the Nepal Himalaya (Table 1). This was the first information on glaciers of Nepal, which served the important baseline information on glaciers. As the time and source materials were different, the glacier information was heterogeneous and is not matching in the present context due to rapid melting of glaciers as an impact of global warming. The present glacier inventory is based on the single source of Landsat satellite images of 2008 and 2009. The overall glaciers mapped in Nepal are 3808 in number and 4212 km² of glacier area with 346km³ estimated ice reserves.

SN	Sub-basins	Basins	Glacier Inventory 2001			Glacier Inventory 2010		
			Data source: Topomap of 1963 to 1982 and Images of 1999			Data source: Landsat images of 2008/2009		
			No	Area (km²)	Ice reserve (km³)	No	Area (km²)	Ice reserve (km³)
1	Tamor	Koshi	261	474.15	56.64	261	407.33	45.66
2	Arun		91	216.07	23.47	108	163.81	16.70
3	Dudh		278	482.20	51.01	287	415.10	41.62
4	Tama		80	109.69	10.37	85	93.75	9.03
5	Sun Koshi		23	74.56	7.19	39	57.66	4.81
6	Indrawati		18	22.98	1.44	37	17.91	0.79
7	Likhu		28	30.19	1.94	26	24.30	1.53
8	Trishuli	Gandaki	74	246.65	27.47	163	2 31.66	22.82
9	Budhi		180	442.14	40.40	242	387.40	33.67
10	Marsyangdi		311	614.31	54.99	385	545.38	44.01
11	Seti		61	164.48	16.88	45	72.10	8.35
12	Kali		399	562.67	51.65	502	562.93	43.06
13	Bheri	Karnali	452	583.40	47.77	400	397.33	27.91
14	Mugu		254	220.39	12.52	205	127.78	6.40
15	Tila		58	54.69	3.75	66	39.21	2.05
16	Kawari		39	53.33	3.29	48	33.37	1.78
17	Humla		424	534.53	36.006	474	361.71	20.90
18	West Seti		134	294.13	24.48	268	160.85	8.47
19	Mahakali		87	143.34	10.06	167	112.23	7.13
	Nepal		**3252**	**5323.9**	**481.32**	**3808**	**4212.03**	**346.7**

Table 1. Summary of change in glaciers in 2001 and 2010 Inventory

The inventory of 2010 shows that the number of glaciers in Nepal has increased by about 17% compared to 2001 inventory (Fig. 4). The number of glaciers increased largely due to actual shrinking and fragmentation of the glaciers as an impact of global warming, although the overall glacier area and ice reserve has been decreased by 21% and 28% respectively (Fig. 5 and 6). The glacier area of Nepal is decreasing 30km² per year since 1970s. If this trend continues, the glaciers of Nepal will disappear in 140 Years or by 2150AD. As mentioned previously, it cannot be excluded that some of the apparent change between surveys is an artifact of the differing data bases, methodologies, and analysts. Further work is needed to sort through this completely and arrive at final conclusions about actual changes in the glacier numbers, areas, and volumes.

4.2 Change in hypsography

Glacier Area - Altitude of 100m bin is calculated based on the SRTM DEM of 90m resolution in both glacier inventories. During the conversion of glacier polygon from raster to vector, there was a loss of 2% of glacier area which is rectified in the final glacier database.

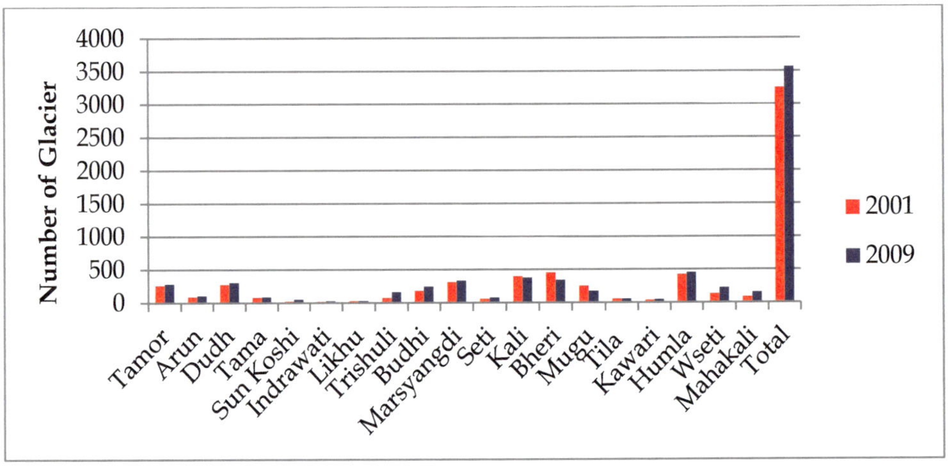

Fig. 4. Comparison of number of glaciers mapped in 2001 and 2010 inventories of Nepal.

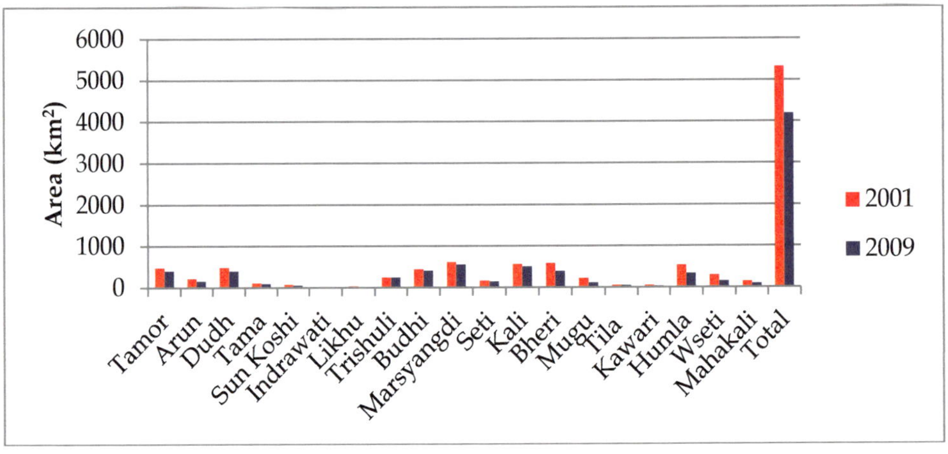

Fig. 5. Comparison of glacierized area mapped in 2001 and 2010 inventories of Nepal.

4.2.1 Glacier hypsography of 2001 inventory

The glaciers were mapped from the elevation 2425 to 8485m asl in Nepal (Fig. 7a). The lowest and highest elevations of glaciers are found from the Gandaki basin and Koshi basin respectively. The lowest elevations of glaciers were mapped from 3536m asl, 3637 and 3940m asl in Karnali, Mahakali and Koshi basins respectively (Fig. 7). The glaciers are densely populated at the elevations of 5000m asl to 6000m asl with the total glacier area of about 3700km^2. The elevations from 5200 to 5500m asl contain about 400km^2 in each 100m bin whereas the elevations below 4500 m asl and above 6300m asl contain less than 100km^2 in each 100m bin.

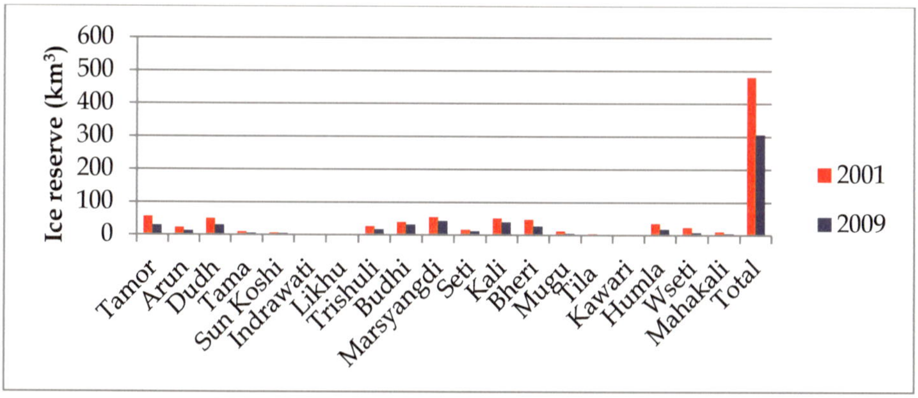

Fig. 6. Comparison of estimated ice reserves in 2001 and 2010 inventories of Nepal

4.2.2 Glacier hypsography of 2010 inventory

The glacier polygons were delineated from the elevations ranging from above 3200m asl to 8485m asl with lowest in Gandaki basin and highest in Koshi basin as in 2001 inventory. A comparison of the glacier area-altitude distribution for Nepal shows a maximum glacierized area is 320km^2 at 5500 to 5600 elevation, however more than 100km^2 are found from 4800 to 6300m asl in each 100m bin (Fig. 7). The highest glacier area in Koshi basin is 100km^2 at elevation 5400-5500m asl, 140km^2 at 5800-5900m asl for Gandak basin, 110km^2 at 5300-5400m asl for Karnali basin and about 10km^2 at 5000-5100m asl for Mahakali basin.

4.2.3 Change in glacier hypsography

By comparing the glaciers hypsographs of 2001 and 2010 glacier inventory stick out the disappearance of glaciers from the elevations below 3200m asl. It is also noticed that the glacier area has been decreased only from below the elevation 5800m asl of Nepal (Fig. 7). There are no changes in glacier area of Nepal above 5800m asl (Fig. 7a). However, the threshold value of unchanged glacier area is 5700m asl for Koshi basin, 5900m asl for Gandaki basin and 5600m asl for Karnali basin (Fig. 7). In contrary to this the glacier area is increased in Mahakali River basin above 5500m asl but the glacier area has shrunk comparatively very high at the elevations of 5300 to 4000m asl. The glacier area is reduced maximum from the elevation 5000 to 5500m asl in Nepal. The glacier area loss percentage is higher in Koshi and Mahakali basins, moderate in Karnali basin and low in Gandaki basin. However the cumulative glacier area loss shows constant in Koshi, Gandaki, Karnali and throughout Nepal except in Mahakali basin (Fig. 8). The glacier retreat rate is higher in Karnali and Mahakali River basins below 5000m asl compared to other basins of Nepal. The low retreat rate in lower elevations in other basins might be due to the presence of debris cover.

4.3 Decadal glacier area change in Langtang valley of Trishuli basin

Based on the developed methodology, an attempt of a decadal change of glaciers in Langtang sub-basin of Trishuli River was made. For this, base maps of glacier polygons were established from the recent satellite image. The glacier polygons derived from Landsat 7 ETM+ image of 2009 were used as a base map of glacier to that year and by overlaying the glacier polygons of 2009 on the older satellite images of 1977, 1988 and 2000, the polygons were edited and modified to that years separately. In this way the glaciers of 1977, 1988, 2000 and 2009 were generated and compared for the change graphically and spatially.

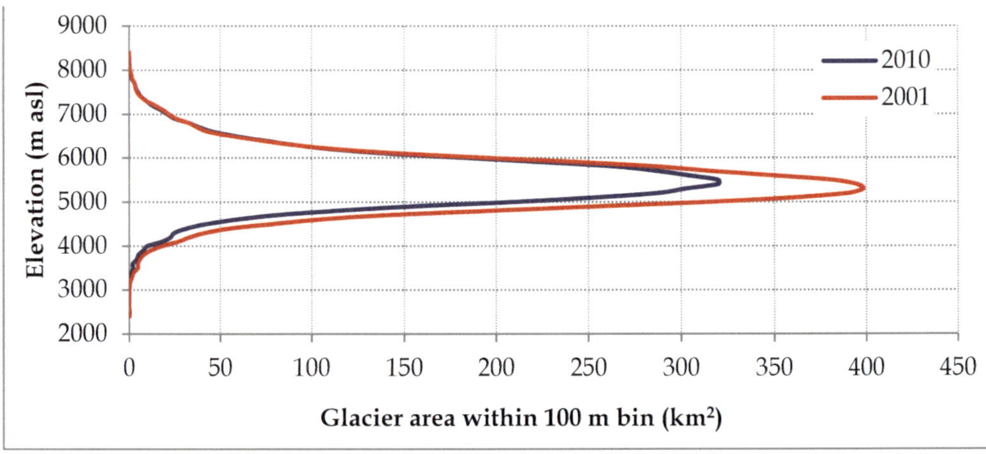

a. Nepal

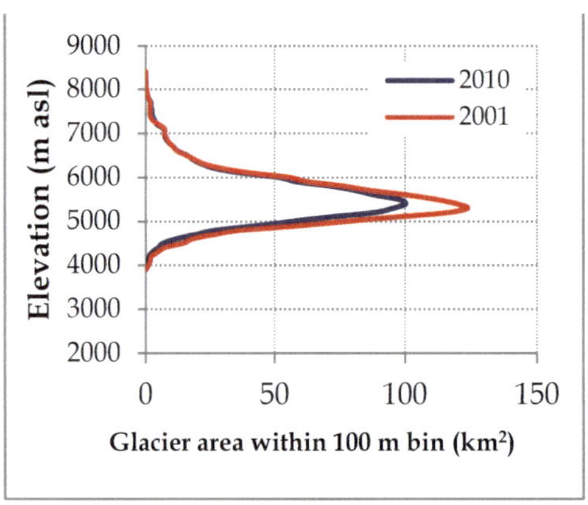

b. Koshi

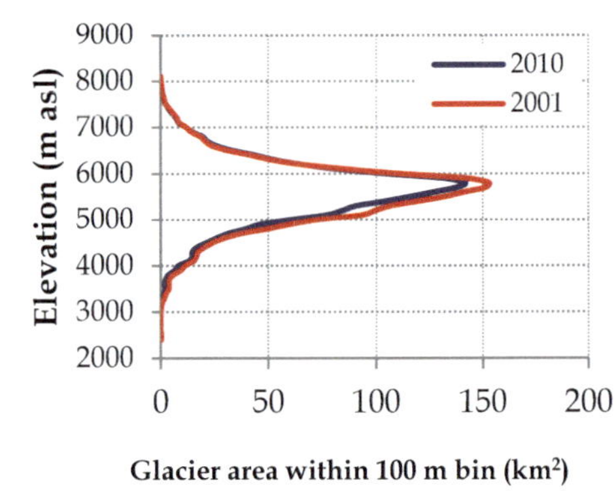

c. Gandaki

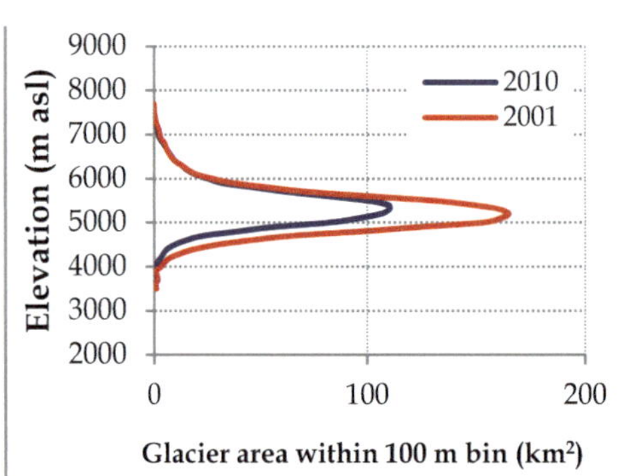

d. Karnali

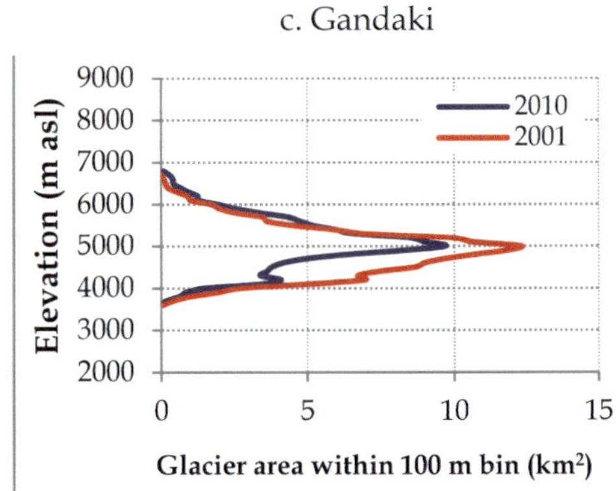

e. Mahakali

Fig. 7. Hypsograph of glaciers of 2001 and 2010 inventory.

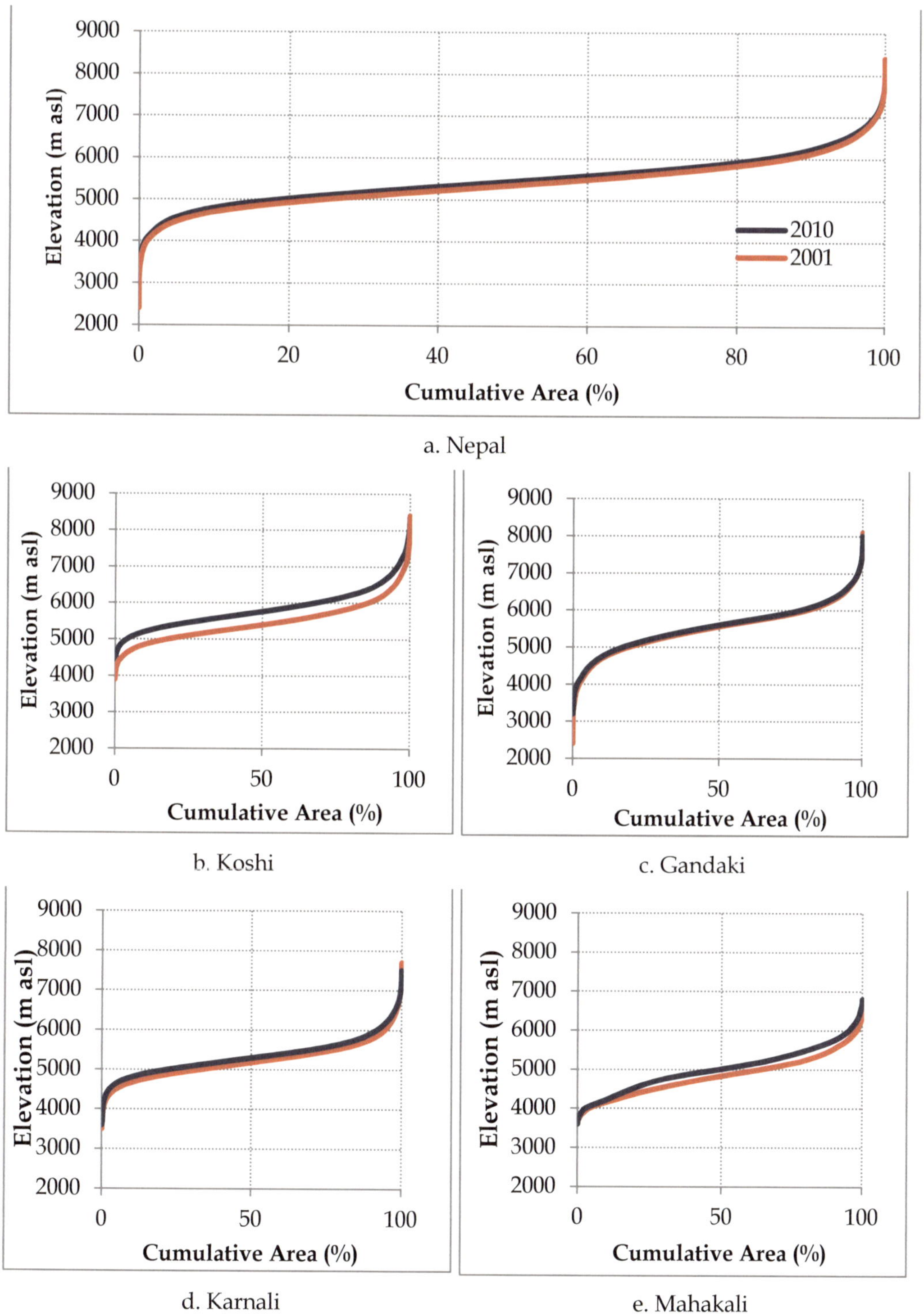

Fig. 8. Comparative cumulative area - altitude of glaciers in 2001 and 2010 inventory.

The Langtang sub-basin is a small northeast-southwest elongated basin, tributary of Trishuli River north of Kathmandu and bordered with China to the north. The basin encompasses an area of 554km^2. The basin contains 192km^2 of glacier area in 1977 and changes to 171km^2 in 1988, 152km^2 in 2000 and 142km^2 in 2009. In 32 years from 1977 to 2009 the glacier area is reduced by 26%. The decrease in glacier area is recorded below the elevation 5800m asl. Due to shrinkage of glaciers the snout elevations of Lirung and Kimjung Glaciers shifted vertically 100m upward and the snout of Yala Glacier by 100 to 150m (Fig. 9).

5. Discussion

The data sources and methods of the two inventories differ, until further study is undertaken of the older survey, conclusions regarding temporal changes are considered tentative. Questions remain about whether the increase in numbers of glaciers is due mainly to (1) actual fragmentation of glaciers, (2) reduced snow cover in the more recent survey, or (3) a more detailed mapping in the more recent survey. With either the second or third explanations, it would be likely that the first survey overestimated the glacier area and underestimated the total number of glaciers, whereas under the first explanation the changes documented between the two surveys are real. A great deal of evidence has accumulated to show that the long valley glaciers of Nepal have, for the most part, retreated drastically in the past few decades, and since these glaciers contribute so strongly to the total areas occupied by glaciers, the decreased glacier area between the two periods of the two inventories is very likely to be correct. However, a focused scrutiny of the earlier survey, and a cross comparison of the two surveys in sample areas, is needed to gain full confidence in the documented changes indicated here, especially with regard to the number of glaciers. In any case, the change in number of glaciers occurs mainly at the smaller glacier sizes, which are much less significant than the big glaciers with respect to their contributions to total glacier area, estimated ice reserves, and river flow.

Consideration below of just one example, that of what was once Nepal's largest glacier, Ktr_gr 193, which fragmented into two glaciers subsequent to the first survey, shows that this phenomenon of fragmentation is not restricted to small glaciers. Nevertheless, a detailed quantitative comparison of the two inventories is needed and recommended for the near future. A good approach could be a blind re-inventory, using the same methodology and same analysts of a small sample of glaciers using the old data used in the first inventory, and then comparison with the first inventory results.

The elevations of the glaciers in Nepal range from 8485 to 3273m asl. The glacier snout elevation ranges from basin to basin, the lowest elevation of glacier extension is about 3300m asl in Gandaki basin in central Nepal, about 3600m asl in Karnali basin of western Nepal and about 4000m asl in Koshi basin of eastern Nepal.

Space based monitoring of glaciers is a valuable aid to climate change scientist, policy makers and civil engineers assigned to water resources and climate change researchers, though these methods have their limitations as well (Fujita et al. 2009; Kargel et al. 2011). Accurate knowledge of overall glaciers will enhance each nation's abilities to deal with the water resources. A digital GIS repository of relevant data then can inform policy makers to make sensible decisions on responsible development and land management.

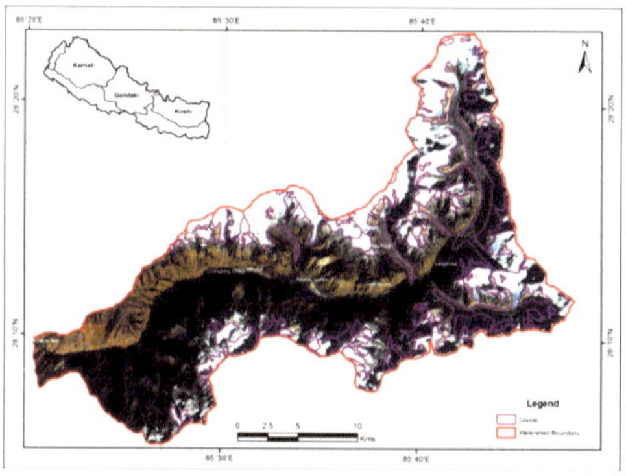

Fig. 9a. Glaciers in 1977

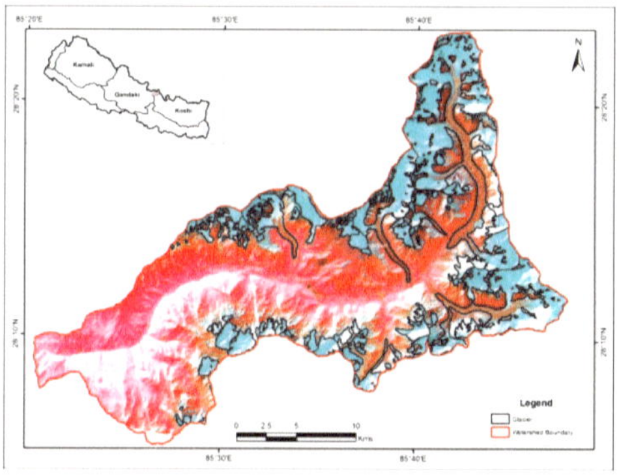

Fig. 9b. Glaciers in 1988

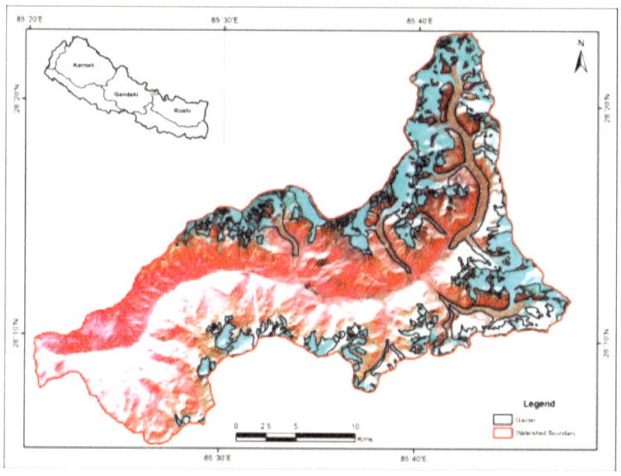

Fig. 9c. Glaciers in 2000

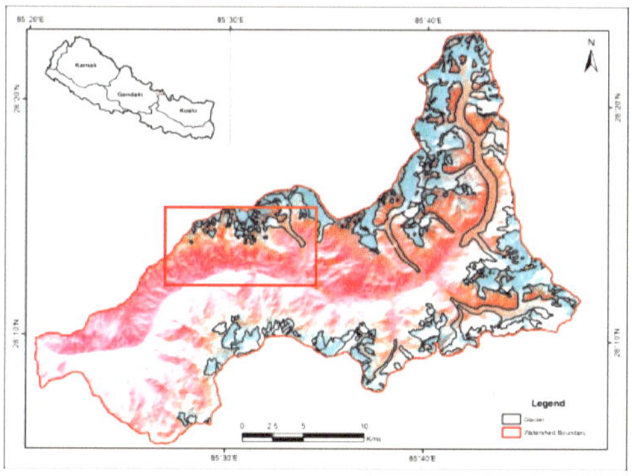

Fig. 9d. Glaciers in 2009

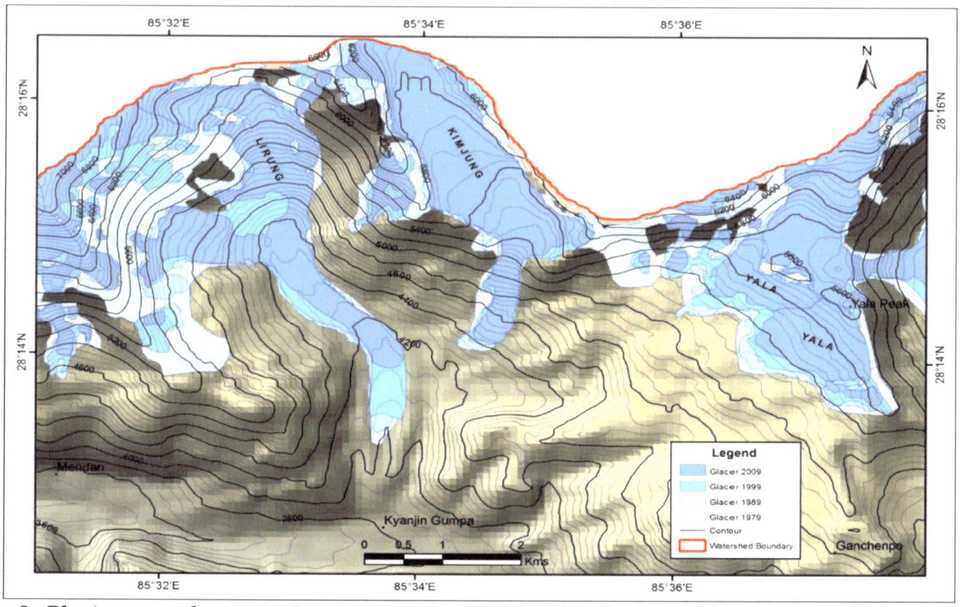

Fig. 9. Glacier area change in Lirung, Kimjung and Yala Glaciers of Langtang Valley.

6. Conclusions

The Himalayan region is one of the most dynamic, fragile and complex mountain ranges in the world due to tectonic activity and a rich diversity of climates, hydrology and ecology. The high Himalayan region is the fresh water tower of south Asia. The melt of snow and ice from these towers feeding the ten largest river systems in Asia, on which over 1.3 billion peoples are dependent. Glaciers and snowmelt runoff from the mountains is of course just a part of the water supplies of these rivers. However, in a chronically water-short and in places an over-populated region, these alpine sources of water are very important, especially for populations in the proximal river valleys. In the context of global warming the glaciers are melting rapidly causing the shrinking, subsidence and retreating of glaciers with the result of expansion and formation of glacial lakes to the stage of potential glacial lake outburst floods. The glacier retreat phenomena has been taking place rapidly in recent decades, with the common and widespread fear of too much water (GLOFs) and too little water (glacier retreat). If it continues for long period definitely there will be negative balance of ice with the deficiency of fresh water resources and increase in the frequency of GLOF disasters in near future. In some places, increased melting of glaciers might provide a temporary surge in water supplies lasting some decades, and in other places, retreat and disappearance of glaciers can cause a disappearance of certain types of mountain hazards and disasters. In general, however, rapidly changing glaciers are both caused by instabilities in climate and further cause new instabilities in the natural system. The instabilities can impact people severely, especially those residing within or near the mountains. For the quick delivery of glacier data a semi-automatic method was developed using the advanced remote sensing and geographic information systems tools and techniques. The primary glacier data upon analysis was performed consist of the Landsat satellite images of 2008 and 2009. This study will help to understand the glaciers dynamics in the Himalaya. Some imperative findings of this study are listed below:

- Due to shrinkage of glaciers, the number of glaciers has increased due to fragmentation of glaciers. There has been about a 17% increase in the numbers of glaciers compared to the inventory of 2001.
- The area and estimated ice reserves of glacier have declined 21% and 28%, respectively, compared to the inventory of 2001.
- The glaciers below 3200m asl has been disappeared from Gandaki basin.
- The glacier area is shrinking and retreating only below the 5800m asl in Nepal.
- The glaciers are advancing in Mahakali basin above 5800m asl but the retreat rate is higher in Mahakali and Karnali below the elevation 5000m asl.
- The glacier retreat rate is high at the elevations 5000 to 5500m asl. The low retreat rate below 5000m asl may be due to the presence of debris cover glaciers.

7. Acknowledgement

We are grateful to Mr. Basanta Shrestha, Division Head, MENRIS for his persistent guidance, inspiration and support. We wish to express our sincere gratitude to Mr. Pradeep Mool, Remote Sensing Specialist and coordinator of Cryosphere from IWHM for his indispensable support during the preparation of this manuscript.

8. References

Ageta, Y. & Kadota, T. (1992). 'Predictions of changes of glacier mass balance in the Nepal Himalaya and Tibetan Plateau: a case study of air temperature increase for three glaciers.' Ann. Glaciol; 16, 89–94

Ageta, Y.; Ohata, T.; Tanaka, Y.; Ikegami, K. & Higuchi, K. (1980). 'Mass balance of glacier AX010 in Shorong Himal, east Nepal during the summer monsoon season.' Seppyo, 41, 34–41

Bajracharya, SR.; Maharjan, SB. & Shrestha F. (2011). The status of glaciers in the Nepal Himalayas. ICIMOD. unpub. 19

Bajracharya, SR.; Maharjan, SB.; Shrestha F.; Khattak, GA.; Wanqin, G.; Shiyin,L.; Junfeng, W. & Xiaojun, Y. (2010). The status of glaciers in the Hindu Kush-Himalayas. ICIMOD. unpub. 543

Bajracharya, SR. & Mool, PK. (2009a). 'Glaciers, glacial lakes and glacial lake outburst floods in the Everest region. Nepal.' Annals of Glaciology, 50 (53) London, UK. 81 - 86

Bajracharya, SR.; Gurung, DR.; Uddin, K.; Mool, PR. & Shrestha, BR. (2009b). 'Semi-automatic glacier mapping and monitoring usiing remote sensing data and techniques: Hands-on training manual' ICIMOD (unpub.), 178

Bajracharya, SR.; Mool, PK. & Shrestha BR. (2008). 'Global climate change and melting of Himalayan glaciers.' In Ranade, PS; ed. Melting glaciers and rising sea levels: impacts and implications, Hyderabad, India, Icfai University Press, 28–46

Bajracharya, SR.; Mool, PK. & Shrestha, BR. (2007a). Impact of climate change on Himalayan glaciers and glacial lakes: case studies on GLOF and associated hazards in Nepal and Bhutan. ICIMOD, Nepal. 119

Bajracharya, SR. & Mool, PK. (2007b). 'Melting glaciers in the Himalaya.' Mountain Forum Secretarial C/O ICIMOD, Bulletin, volume VII Issue 2, July 2007, 5 - 6

Bajracharya, SR.; Mool, PK. & Shrestha, BR. (2006a). 'The Impact of Global Warming on the Glaciers of the Himalaya.' In Proceedings of the International Symposium on Geodisasters, Infrastructure Management and Protection of World Heritage Sites, 25-26 Nov 2006, pp. 231-242. Kathmandu: Nepal Engineering College, National Society for Earthquake Technology Nepal, and Ehime University Japan

Bajracharya, SR. & Mool, PK. (2006b). Impact of Global Climate Change from 1970s to 2000s on the Glaciers and Glacial Lakes in Tamor Basin, Eastern Nepal. Unpublished report for ICIMOD, Kathmandu

Bolch, T.; Buchroithner, MF.; Peters, J.; Baessler, M. & Bajracharya, S. (2008). 'Identification of glacier motion and potentially dangerous glacial lakes in the Mt. Everest region/Nepal using spaceborne imagery.' Nat. Hazards Earth Syst. Sci; 8, 1329-1340

Frauenfelder, R. & Kääb, A. (2009). 'Glacier mapping from multi-temporal optical remote sensing data within the Brahmaputra river basin'. 33rd Int. Symp. on Remote Sensing of Environment. Italy

Fujita, K.; Sakai, A.; Nuimura, T.; Yamaguchi, S. & Sharma, RR. (2009). 'Recent changes in Imja Glacial Lake and its damming moraine in the Nepal Himalaya revealed by in situ surveys and multi-temporal ASTER imagery.' IOP. Environ. Res. Lett. 4, 7

Fujita, K.; Kadota, T.; Rana, B.; Kayastha, RB. & Ageta, Y. (2001). 'Shrinkage of Glacier AX010 in Shorong region, Nepal Himalayas in the 1990s'. Bull. Glaciol. Res; 18, 51-54

Fujita, K.; Nakawo, M.; Fujii, Y. & Paudyal, P. (1997). 'Changes in glaciers in Hidden Valley, Mukut Himal, Nepal Himalayas, from 1974 to 1994'. Jr. Glaciology, 43(145), 583-588

Fugii, Y.; Fugita, K. & Paudyal, P. (1996). Glaciological Research in Hidden Valley, Mukut Himal in 1994. Bulletin of Glacier Research, 14, 7-11

GEN.; CREH.; NU.; DHM (2006).; Data Report 4 (2001-2004). & GEN. (2001-2002). Imja Glacier Lake in Khumbu, East Nepal. Kathmandu: DHAS and DHM

Higuchi K, eds. (1980). Glaciers and climates of Nepal Himalayas-Report of the Glaciological Expedition of Nepal-Pt. 4: Seppyo, v. 41, special iss; 111 p.HMG/ADB/FINNIDA (MOPE) 1998. Land-use pattern of Nepal

Higuchi K, eds. (1978). Glaciers and climates of Nepal Himalayas-Report of the Glaciological Expedition of Nepal-Pt. 3: Seppyo, v. 40, special iss; 84

Higuchi K, eds. (1977). Glaciers and climates of Nepal Himalayas-Report of the Glaciological Expedition of Nepal-Pt. 2: Seppyo, v. 39, special iss; 67

Higuchi K, eds. (1976). Glaciers and climates of Nepal Himalayas-Report of the Glaciological Expedition to Nepal: Seppyo, v. 38, special iss; 130

Kadota, T.; Fujita, K.; Seko K.; Kayastha, RB. & Ageta, Y. (1997). 'Monitoring and Prediction of Shrinkage of a Small Glacier in the Nepal Himalayas'. In Annals of Glaciology, 24: 90-94

Kargel, J.; Furfaro, R.; Kaser, G.; Leonard, G.; Fink, W.; Huggel, C.; Kääb, A.; Raup, B.; Reynolds, J.; Wolfe, D. & Zapata, M. (2011). ASTER Imaging and Analysis of Glacier Hazards, Chapter 15 in Land Remote Sensing and Global Environmental Change: NASA's Earth Observing System and the Science of Terra and Aqua, B. Ramachandran, Christopher O. Justice, and M.J. Abrams (Eds.), pp. 325-373, Springer, New York

Mool, PK.; Bajracharya, SR. & Shrestha, BR. (2005). Inventory of Glaciers and Glacial lakes and the Identification of Potential Glacial Lake Outburst Floods (GLOFs) Affected by Global Warming in the Mountains of India, Pakistan and China/TAR (APN 2004-03-CMY-Campbell). APN newsletter, October 2005, vol.11, issue 4, 6-7

Mool, PK.; Bajracharya, SR. & Joshi, SP. (2001). Inventory of glaciers, glacial lakes, and glacial lake outburst flood monitoring and early warning system in the Hindu Kush-Himalayan Region, Nepal. ICIMOD and UNEP/RRC-AP, 364.

Nakawo, M.; Fugii, Y. & Shrestha, ML. (1976). Flow of Glaciers in Hidden Valley, Mukut Himal. Seppyo, Journal of Japanese Society of Snow and Ice, 38, Special Issue, 39-43

Paul, F. (2000). Evaluation of different methods for glacier mapping using Landsat TM EARSeL eProceedings No. 1, 239-245

23

Subglacial and Proglacial Ecosystem Responses to Climate Change

Jacob C. Yde[1], Teresa G. Bárcena[2] and Kai W. Finster[3]
[1]*Faculty of Engineering and Science, Sogn and Fjordane University College,*
[2]*Department of Forest and Landscape Ecology, University of Copenhagen,*
[3]*Department of Biological Sciences, University of Aarhus,*
[1]*Norway*
[2,3]*Denmark*

1. Introduction

Life beneath glaciers requires efficient adaptive strategies in order to thrive in these extreme habitats that are characterized by nutrient and energy limitation, low temperatures and reduced water availability. Together with the deep subsurface and the deep oceanic basins, subglacial (beneath glaciers) environments are among the last biosphere-exploration frontiers on Earth. At present we have only a vague idea about the types of organisms that live beneath ice sheets and glaciers as well as their characteristics (Willerslev *et al.*, 1999, 2007). As a consequence we do not know what will happen to them as climate amelioration causes glaciers worldwide to recede and disappear. What we observe today in almost all glacierised regions of the World is that previous subglacial environments change into proglacial (in front of glaciers) environments, where the environmental conditions are very different. In this chapter we will provide an overview of our current knowledge on the subglacial and proglacial ecosystems, and highlight potential consequences of climate change on these habitats.

The multidisciplinary study of subglacial and proglacial ecosystems, is only in its infancy and the literature is rapidly expanding. The reported results are not only relevant for our understanding of biological diversity and adaptability in general, but they also provide the foundation for hypotheses on how we imagine life survived during the periods where the Earth was totally ice-covered (the so-called Snowball Earth glaciations) and how life may survive on Earth in case of a global nuclear disaster (the so-called Nuclear Winter). In fact, the results may seed far-reaching theories on whether subglacial environments may host past or present extra-terrestrial (astrobiological) habitats on Mars, Jupiter's moon Europa, Saturn's moon Titan, and even on planets and moons beyond our solar system (Des Marais *et al.*, 2008). In addition, cold-adapted organisms may also provide a valuable source for enzymes that may have biotechnological applications (Reddy *et al.*, 2009).

A prerequisite for life as we know it is liquid water required for biochemical processes, electrochemical gradients and transfer of chemical substances. The occurrence of water in subglacial environments is largely dependent on the subglacial thermal regime. Beneath cold-based glaciers, i.e. glaciers where the sole is frozen to the substratum, liquid water is a limited resource and may be primarily restricted to a few subglacial channels. However, it has been argued that liquid water is present at subfreezing temperatures in the micron-diameter vein network between ice crystals (Price, 2000) and as a nanometer-thick water film interface between ice and entrained mineral grains (Price, 2007). These two potential habitats may both contain ion-rich solutions that allow immobilized microbial life to extract energy from redox reactions (Price, 2007). Rohde and Price (2007) have proposed the existence of a third habitat within ice crystals, where metabolism is achieved by diffusion of small molecules through the ice crystal structure. It is likely that microorganisms in these habitats go into a state of 'survival metabolism' or 'dormancy', sufficient to repair DNA damage as it occurs but not sufficient to grow (Price, 2009). In addition, embedded viable endospores, i.e. bacterial spores, and bacterial endospore formers have been observed in ice cores (Christner *et al.*, 2003; Sheridan *et al.*, 2003; Miteva *et al.*, 2004; Miteva and Brenchley, 2005; Yung *et al.*, 2007). These endospores may return to a metabolic state when the conditions for water and nutrient access improve, but the long-term accumulation of damages DNA may not be repairable. Hence, it is likely that the most successful strategy for surviving in sub-zero environments on glaciation-interglaciation scales (~ 10^5 years) is to enter a survival metabolism state rather than convert to endospores (Price, 2009). If subfreezing environments, containing endospores and metabolizing microbes, undergo environmental changes in the form of thawing and accessibility to essential substrates, the surviving microbes may experience rapid growth and found a new microbial consortium adapted to unfrozen subglacial conditions (Humlum *et al.*, 2005).

Beneath polythermal glaciers (i.e. glaciers where the peripheral areas are frozen to the substratum while the central part of the glacier bed is at the pressure-melting point of ice) and temperate glaciers (i.e. glaciers where the entire bed is unfrozen), water flow is routed along various pathways from large fast-flow channels to thin slow-flow water films or as groundwater flow within subglacial aquifers (e.g., Benn and Evans, 2010). Subglacial drainage systems are very dynamic, and spatial and temporal changes occur frequently. Microorganisms that live beneath these types of glaciers must therefore be able to adapt to environmental changes in order to survive. The most common causes of death for microorganisms besides desiccation due to freezing may be cold shock, exhaustion of nutrients, and α-particle radiation from decay of U and Th-containing minerals in subglacial rocks and sediments (Price, 2009).

The origins of subglacial and proglacial microbial communities are closely related to the history of glaciers and, thus, to regional and global climate changes (Figure 1A). When glaciers grow during a cold climatic period, they increase in volume and advance across their proglacial forelands towards a state where their mass balances are in equilibrium with the prevailing climate conditions (Figure 1B). Opposite, when the climate changes towards a warmer period glaciers recede and previous subglacial environments become proglacial environment (Figure 1C). In this context, it must be emphasized that not all glacier fluctuations are directly linked to short-term (10^1 - 10^4 years) climate change, although climate is the dominant control on long-term (~ 10^5 years) scales (Yde and Paasche, 2010). Nevertheless, proglacial forelands may harbor sediment, soil, forest, lake, or marine ecosystems that, when overridden by glacier advance, provide founding microbial communities for subglacial ecosystems (Figure 1B). Additional microorganisms may be added by melt-out of microorganisms that have been buried in snow on the glacier surface and transported in ice to the glacier bed, or by subglacial deposition of microorganisms that have been entrained in meltwater on the glacier surface and transported to the glacier bed via crevasses or glacier wells (moulins) (Hodson *et al.*, 2008). Since the end of the Little Ice Age, the global climate has ameliorated and glaciers around the world have receded or even disappeared (IPCC, 2007). As a consequence, large areas have been deglaciated and subglacial ecosystems are now exposed to proglacial conditions. If global warming accelerates or continues at the present rate for centuries, it is easy to imagine that many subglacial communities, which have survived in niches beneath glaciers during several interglaciation periods, may disappear.

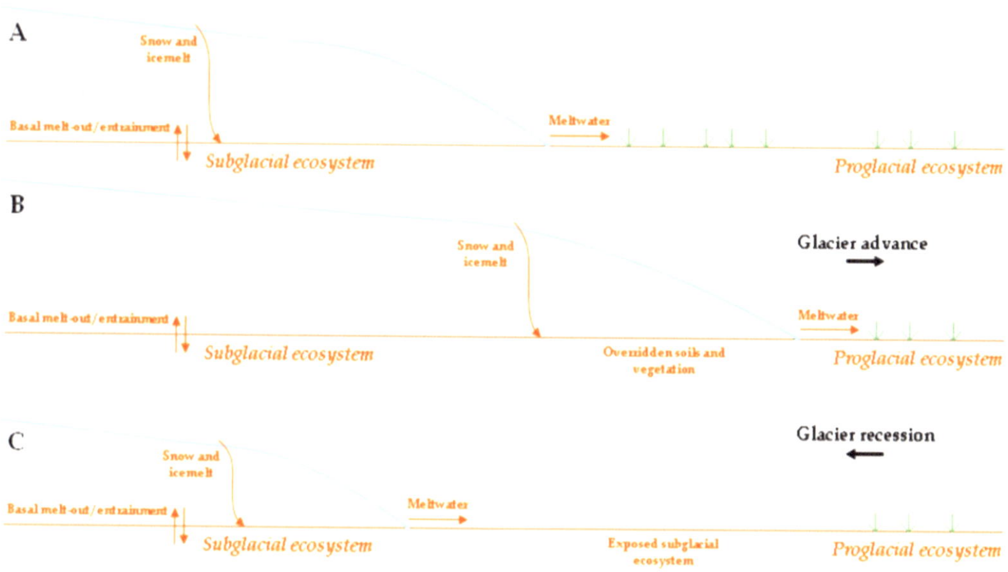

Fig. 1. Potential pathways for the flow of microorganisms in subglacial and adjacent ecosystems. (A) The subglacial ecosystem receives microorganisms from the glacier surface via snow- and icemelt and from basal melt-out, and exports microorganisms through outflow meltwater and basal entrainment; (B) during glacier advance, microorganisms in overridden soils and vegetation are incorporated into the subglacial ecosystem; whereas (C) during glacier recession, subglacial microorganisms become exposed to atmospheric conditions and the system changes into a proglacial ecosystem.

2. Microbial influence on greenhouse gas cycles

Microorganisms play a major role in the cycling of macronutrients, such as carbon and nitrogen, through a variety of aerobic and anaerobic processes. These processes include respiration, methane oxidation and production, nitrification and denitrification, sulfur oxidation and reduction, and are relevant in climatic terms since they directly affect greenhouse gas (GHG) budgets on a global scale. Therefore, focus on carbon dioxide (CO_2), methane (CH_4) and nitrous oxide (N_2O) has increased considerably in recent years due to the warming potential of these gases and their importance for the present and future climate.

Much attention has been given to the anthropogenic sources of major GHGs and how human emissions affect their budgets, but the microbially-mediated pathways governing

CO_2, CH_4 and N_2O fluxes in various environments have also attracted increasing research efforts. At present, the physiology of many of the involved microorganisms and their responses to changing environmental conditions are not yet fully understood, especially when it comes to extreme habitats as it is the case for subglacial and proglacial ecosystems.

Due to the harsh physical environmental conditions commonly found in subglacial and proglacial environments, biological enzymatic processes slow down due to the low temperatures and limited nutrient supply, forcing microorganisms to develop adaptive strategies that enable them to harvest available substrates. Therefore microorganisms found in these habitats are often characterized as oligotrophs, i.e. organisms that thrive in low nutrient level environments.

In subglacial and proglacial environments, GHGs are produced and assimilated by microorganisms in a number of ways (Figure 2). Generation of CO_2 occurs during autotrophic and heterotrophic respiration and microbial oxidation of organic carbon. Anoxic conditions may occur both in subglacial (Wadham *et al.*, 2008) and proglacial (Wadham *et al.*, 2007) environments, favoring fermentative formation of CH_4 from substrates such as hydrogen gas and CO_2 by methanogenic microorganisms (Boyd *et al.*, 2010, in press). However, uptake of CH_4 by methanotrophs also seems to be a common process on proglacial forefields (Bárcena *et al.*, 2011). Not much is known about generation and assimilation of N_2O in subglacial and proglacial environments. Studies of Midtre Lovénbreen, Svalbard, have provided indirect evidence of subglacial nitrification, denitrification and microbial assimilation of ammonia (Wynn *et al.*, 2006, 2007; Hodson *et al.*, 2010), indicating active microbial influence on nitrogen cycling.

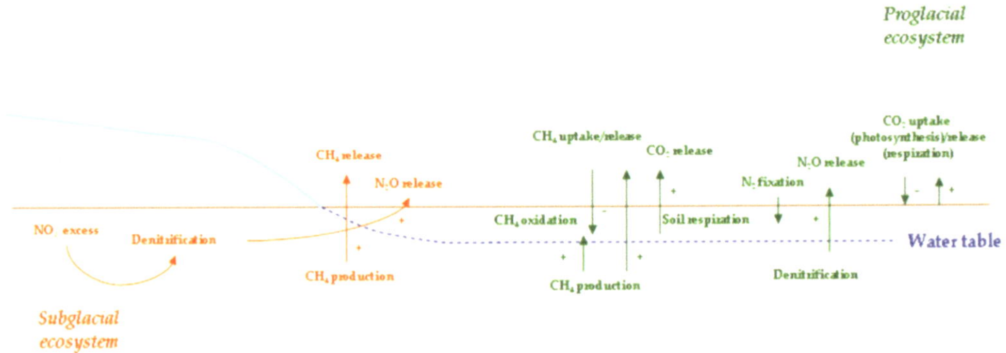

Fig. 2. Conceptual overview of potential processes to greenhouse gas fluxes in subglacial and proglacial ecosystems.

3. Subglacial ecosystems

The study on subglacial ecosystems has started within the last decade and is rapidly growing. At present, most investigations have focused on the basic characteristics, such as who is there (biodiversity) and how do they obtain energy and nutrients (metabolism). So far, non-photosynthetic *Bacteria* from a variety of phylogenetic groups have been described from subglacial environments (Christner *et al.*, 2008; Hodson *et al.*, 2008), and as psychrophilic *Archaea* (Finster, 2008) and viruses (Anesio *et al.*, 2007) are known to exist in other cold environments there are high expectations of their presence in subglacial microbial communities. In addition, studies of glacial meltwater and basal ice have shown that basidiomycetous yeasts inhabit subglacial ecosystems (Buzzini *et al.*, 2005; Butinar *et al.*, 2007, 2009; de Garcia *et al.*, 2007; Turchetti *et al.*, 2008), and Kornobis *et al.* (2010) suggest that endemic groundwater arthropods on Iceland have survived in subglacial refugia during glaciations. Thus, the consortium of microorganisms beneath glaciers is not only restricted to prokaryotes but also includes eukaryotes. In the following, we will focus on the biodiversity and metabolic processes of *Bacteria* as this domain is likely to dominate in subglacial ecosystems, but first we will examine the physical conditions and ice-bed interface processes that define the subglacial environment.

3.1 Physical environmental conditions

The subglacial zone consists of three components: (1) basal ice, (2) subglacial meltwater, and (3) the substratum. Basal ice is defined as ice formed by processes occurring at the lower part of glaciers and ice sheets (for comprehensive reviews of basal ice, see Knight (1997) and Hubbard *et al.* (2009)). Basal ice is generally composed of a series of ice layers with distinctive characteristics (basal ice facies) formed by different processes at the ice-bed interface, which are able to incorporate significant amounts of debris into the basal ice. The most common processes are *regelation* where ice is formed by pressure-melting on the up-glacier side and refreezing on the lee side of minor bedrock obstacles; *ice infiltration* where low porewater pressure in unconsolidated sediments allows ice to pressure-melt and refreeze into subglacial debris; and *net adfreezing* where the freezing front migrates downwards into water-saturated subglacial debris as a consequence of heat deficit at the ice bed. All these processes result in incorporation of sediments, solutes and microorganisms into the basal ice. After formation, basal ice may undergo metamorphic processes within the ice layer, which cause recrystallisation of ice and redistribution of incorporated particles. Basal ice is also disturbed by deformation and thrusting due to ice-flow extension and compression stress-formation and stress-release.

The configuration of the subglacial hydrological system controls atmospheric oxygen concentrations and, hence, redox potentials (Eh) (Tranter *et al.*, 2005). The main channelized system rapidly routs snow and ice meltwater from the glacier surface through the glacier with a residence time of hours to a few days. Microorganisms inhabiting the channelized drainage system and its hyporheic zone (the channel marginal zone) are likely to have access to oxygen and other atmospheric gasses, and a steady supply of flushed nutrients and organic material from the surface during the ablation season. Allochthonous (derived from adjacent environments) microorganisms originating from supraglacial ecosystems, e.g. cryoconite hole habitats, are also exposed to potential in-wash into the subglacial hydrological system where they must compete with autochthonous (indigenous) microorganisms, providing an important coupling between the two ecosystems (Sharp *et al.*, 1999; Hodson *et al.*, 2008).

The hyporheic zone defines a transition between the channelized drainage system and the distributed drainage system, which is characterized by water-saturated, pressurized conditions in a tortuous slow-flowing network. Water flow and nutrient exchange between the channelized and distributed systems passes the hyporheic zone, making it a particular favorable habitat for subglacial ecosystems (Tranter *et al.*, 2005). In the distributed drainage system oxygen is released by geothermal or frictional melting of bubble-containing glacier ice and consumed by microbially mediated chemical weathering reactions (Tranter *et al.*, 2002). However, the distributed drainage system is almost isolated from atmospheric gasses and is generally considered to be anoxic (e.g., Sharp *et al.*, 1999).

The temperature at the ice bed is constantly close to zero degrees Celsius. There may be slightly higher temperatures in infiltrating water from tributary streams that enter glaciers along the lateral ice-margins, but beneath most parts of warm-based glaciers the temperature is at the pressure-melting point (0 – -2°C). It is impossible for light to penetrate the ice thickness all the way to the subglacial zone, prohibiting photosynthesis to take place.

3.2 Subglacial microbiology

Subglacial microbial communities have now been described from a number of glaciers worldwide. The main focus has been to present total cell count numbers and phylogenetic analyses of 16S ribosomal DNA sequences of isolates. A short overview is provided below and presented in Table 1.

3.2.1 Greenland

The bottom 13 m of basal ice from the deep Greenland ice core (GISP2) has attracted significant attention as this provides a rare opportunity to study subglacial ice sheet microbiology. Sheridan *et al.* (2003) found 6.1 x 10^7 and 9.1 x 10^7 cells ml^{-1} in two duplicate samples, while Tung *et al.* (2006) counted about 10^{11} cells g^{-1} of sediment. Isolated anaerobic and aerobic enrichment culture studies show a diverse community represented by several major phylogenetic groups: α-, β- and γ-*Proteobacteria*, *Thermus-Deinococcus*, *Bacteroides*, *Eubacterium*, *Clostridium*, *Fusobacterium* and *Actinobacteria* (Sheridan *et al.*, 2003; Miteva *et al.*, 2004; Miteva and Brenchley, 2005). The CH_4 concentration in basal ice at GRIP and GISP2 ice cores is much higher (6,000 and 12,000 ppm, respectively) than atmospheric concentrations (1.8 ppm) (Tison *et al.*, 1998; Price and Sowers, 2004), and Tung *et al.* (2005, 2006) have detected methanogens that are likely to be responsible for producing excess CH_4 concentrations. Similarly, Fe-reducers are inferred to be responsible for excess CO_2 concentrations (Souchez *et al.*, 1995) in the GRIP ice core (Tung *et al.*, 2005, 2006). At the Greenland Ice Sheet margin, Yde *et al.* (2010) found a total cell count 2.3 x 10^8 cells cm^{-3} in basal ice and sequenced a clone library dominated by α-, β- and δ-*Proteobacteria*, *Bacteroidetes* and *Firmicutes*, but also including *Actinobacteria*, *Acidobacteria*, *Gemmatimonadetes*, *Chloroflexi*, *Caldiserica*, and spore-formers. The presence of Fe-reducers in the form of *Rhodoferax* and *Geobacter* sequences indicates a high potential for Fe reduction, but lithotrophic β-*Proteobacteria* Fe-oxidizers within the genera *Gallionella* and *Ferritrophicum* were also identified.

3.2.2 Iceland

Microbial communities have been described in two of the three subglacial volcanic lakes beneath the Vatnajökull Ice Cap. Gaidos *et al.* (2004) retrieved lake water and tephra sediments from the Grímsvötn caldera lake with cell counts of 2 x 10^4 cells ml^{-1} and 4 x 10^7 cells g^{-1}, respectively. They primarily identified isolates affiliated with β-, γ- and ε-*Proteobacteria*. At western Skaftá lake, Gaidos *et al.* (2009) examined anoxic bottom lake water in detail and discovered a wide variety of bacterial phylotypes including α-, β-, γ-, δ- and ε-*Proteobacteria*, *Thermus*-Deinococcus, *Bacteroidetes*, *Clostridium*, *Bacillus*, *Actinobacteria*, *Caldilinea*, *Erysipelotrichi*, *Gemmatimonadetes*, *Verrucomicrobia* and *Chlamydiae*. In the anoxic bottom waters of this subglacial lake it seems that acetogenic bacteria thieve instead of methanogens. The cell counts were 4.7-5.7 x 10^5 cells ml^{-1}.

3.2.3 Svalbard

Subglacial sediments from below Werenskioldbreen and Torellbreen have been analyzed to characterize their microbial community structures (Kaštovská *et al.*, 2007). Viable cyanobacteria and microalgae were found in addition to bacterial cells. Bacterial cell counts showed 2.1×10^8 cells g^{-1} of sediment.

3.2.4 Alaska

The subglacial ecosystem beneath Bench Glacier has shown to include α-, β-, γ-, δ- and ε-*Proteobacteria*, *Bacteroidetes*, *Holophaga/Acidobacteria*, and *Spirochaeta* (Skidmore *et al.*, 2005). The clone library consisted almost exclusively of *Proteobacteria* (97% of clones) with β-*Proteobacteria* being the dominant group (68% of clones). Cell counts in snowmelt, ice-marginal and subglacial bulk meltwater and subglacial borehole water ranged between 6.6×10^4 and 3.7×10^5 cells ml^{-1}.

3.2.5 Canada

John Evans Glacier on Ellesmere Island, Nunavut, has the currently best-examined subglacial ecosystem. Skidmore *et al.* (2000) conducted aerobic and anaerobic incubations to show the presence of heterotrophs, NO_3-reducers, SO_4-reducers and methanogens. The microbial community comprised α-, β- and γ-*Proteobacteria*, *Bacteroidetes*, *Holophaga/Acidobacteria*, *Planctomycetales*, *Actinobacteria* and *Verrucomicrobia* (Skidmore *et al.*, 2005), although a later study using different gene primers and sequence techniques only found β-*Proteobacteria*, *Bacteroidetes* and *Actinobacteria* (Cheng and Foght, 2007). The subglacial community was different from supraglacial and proglacial communities and, thus, appeared to be adapted to subglacial conditions (autochthonous) rather than allochthonous (Bhatia *et al.*, 2006).

3.2.6 The European Alps

Basal ice and sediment at Glacier de Tsanfleuron and subglacial meltwater at Haut Glacier d'Arolla, Switzerland, were analyzed for microbiological activity by Sharp *et al.* (1999). Total cell counts ranged between 9.3×10^5 and 5.9×10^7 cells ml^{-1} at Glacier de Tsanfleuron and 5.3×10^4 and 1.8×10^6 cells ml^{-1} at Haut Glacier d'Arolla. Incubation experiments with basal ice from Glacier de Tsanfleuron revealed a considerable production of SO_4^{2-}, leading Sharp *et al.* (1999) to infer that microbial-mediated sulfide oxidation enhances the oxidation rate in subglacial environments.

3.2.7 New Zealand

Unfrozen subglacial sediments and basal ice were collected at Fox Glacier and Franz Josef Glacier by Foght *et al.* (2004). The total cell counts in the sediment samples were 2.3 and 7.4×10^6 cells g^{-1} sediment at Fox Glacier and Franz Josef Glacier, respectively. Phylogenetic analyses detected α- and β-*Proteobacteria* and *Actinobacteria* at Fox Glacier, and α- and β-*Proteobacteria*, *Thermus-Deinococcus*, *Bacteroidetes*, *Actinobacteria* and *Firmicutes* at Franz Josef Glacier.

3.2.8 Antarctica

In Antarctica there have been microbiological investigations in different subglacial environments at the West and East Antarctic Ice Sheets and in basal ice at a local glacier. At the West Antarctic Ice sheet, unfrozen subglacial sediment from beneath Kamb Ice Stream (formerly Ice Stream C) had a cell abundance of 2-4 x 10^5 cells g^{-1} wet sediment (corrected for 15 months storage growth from cell counts of 1.5-2.7 x 10^7 cells g^{-1}) and a clone library represented by α- and β-*Proteobacteria* and *Actinobacteria* (Lanoil *et al.*, 2009).

At Taylor Glacier, an outlet glacier from the East Antarctic Ice Sheet, the saline Fe-rich anoxic subglacial discharge called Blood Falls has been investigated for microbial composition. Cell counts ranged between 0.3 x 10^4 and 7.6 x 10^5 cells ml^{-1} (Mikuchi *et al.*, 2004, 2009; Mikuchi and Priscu, 2007). The clone library consists of α-, β-, γ-, and δ-*Proteobacteria, Bacteroidetes* and *Chloroflexi* (Mikuchi and Priscu, 2007). At Victoria Upper Glacier, a local glacier in the McMurdo Dry Valleys, Klassen (2009) counted 2.7 x 10^3 cells ml^{-1} in the lowest part of the basal ice and identified α- and β-*Proteobacteria, Thermus-Deinococcus, Bacteroidetes, Arthrobacter, Flavobacterium, Frigoribacterium, Janthinobacterium, Kocuria* and *Microbacterium*. Detailed studies of heterotrophic, aerobic *Hymenobacter*-like strains (belonging to *Bacteroidetes*) have revealed non-vertical evolution, i.e. not by diversification by descent, within this genus (Klassen and Foght, 2011).

The potential for lacustrine ecosystems in subglacial lakes beneath the East Antarctic Ice Sheet has received much attention (e.g. Price, 2000; Siegert *et al.*, 2001, 2003). At the largest of the subglacial lakes, Lake Vostok, an ice core has been drilled into the accretion ice (refrozen lake water) facies above the lake. This ice contained 2-3 x 10^2 cells ml^{-1} (Karl *et al.*, 1999) and a low biodiversity comprising α- and β-*Proteobacteria* and *Actinobacteria,* which likely derived from lake water (Priscu *et al.*, 1998). At present, there has not been any sampling of Antarctic subglacial lakes in order to avoid contamination of these pristine ecosystems.

	α-Proteobacteria	*β-Proteobacteria*	*γ-Proteobacteria*	*δ-Proteobacteria*	*ε-Proteobacteria*	*Thermus-Deinococcus*	*Bacteroidetes*	*Firmicutes*	*Actinobacteria*	Others
GIS interior	✓	✓	✓			✓	✓	✓	✓	✓
GIS margin	✓	✓		✓			✓	✓	✓	✓
GL, Iceland		✓	✓		✓					
SL, Iceland	✓	✓	✓	✓	✓	✓	✓		✓	✓
BG, Alaska	✓	✓	✓	✓	✓		✓			✓
JEG, Canada	✓	✓		✓			✓		✓	✓
FG, New Zealand	✓	✓							✓	
FJG, New Zealand	✓	✓				✓	✓	✓	✓	
KIS, WAIS	✓	✓							✓	
VUG, Antarctica	✓	✓				✓	✓			✓
TG, EAIS	✓	✓	✓	✓			✓			✓
LV, EAIS	✓	✓							✓	

Table 1. Identified *Bacteria* phyla in subglacial environments; Greenland Ice Sheet (GIS), Grímsvötn caldera lake (GL), Skaftá subglacial lake (SL), Bench Glacier (BG), John Evans Glacier (JEG), Fox Glacier (FG), Franz Josef Glacier (FJG), Kamb Ice Stream (KIS) belonging to the West Antarctic Ice Sheet, Victoria Upper Glacier (VUG), Taylor Glacier (TG) and Lake Vostok accretion ice (LV) belonging to the East Antarctic Ice Sheet. See references in the text.

3.2.9 Microbial diversity

From this summary it seems clear that lithotrophic β-*Proteobacteria* dominates the microbial diversity in many subglacial ecosystems, and α-*Proteobacteria*, *Bacteroidetes* and *Actinobacteria* are relatively common (Table 1). It can also be concluded that subglacial ecosystems may host a wide variety of microbes, including aerobic and microaerophilic heterotrophs; obligate and facultative aerobes and anaerobes; Fe-reducers; denitrifiers; sulfate-reducers and sulfur-oxidizers; homoacetogens; methanogens; strict psychrophiles and psychrotolerants; and spore-formers and non-spore-formers. However, it is noteworthy that while some phyla and genera are abundant in some subglacial environments, they may be rare or absent in others (Skidmore *et al.*, 2005). Also, the total cell counts indicate that subglacial sediments contain a higher number of microbial cells than debris-poor basal ice. The reasons for this may be due to a higher area of debris surfaces where microbes can metabolize, a higher nutrient transport in liquid water flow and the presence of an initial preglacial microbial population that has become glacier-covered.

3.3 Subglacial biogeochemical processes

Throughout the last decade it has become clear that microorganisms have a strong impact on mineral weathering rates and biogeochemical cycling in subglacial environments (Sharp *et al.*, 1999). Microbial mediation may increase the rate of redox reactions such as sulfide oxidation. Where the hydrological drainage system is distributed and the water residence time is long, atmospheric CO_2 and O_2 in meltwater may become exhausted (Tranter *et al.*, 2005). Hence, anoxia is likely to occur in many subglacial environments, and microbially-produced CO_2 may be a significant driver for subglacial carbonation reactions (Wadham *et al.*, 2010).

3.3.1 Organic matter

Organic matter (OM) is a potential metabolic substrate for heterotrophic microorganisms. OM is incorporated into the subglacial substratum when glaciers advance across their proglacial forelands (Barker *et al.*, 2009). The amount of subglacial OM is heterogeneously distributed and relict soils (e.g. Humlum *et al.*, 2005) and vegetation (e.g. Knudsen *et al.*, 2008) may be preserved *in situ* beneath glaciers. For instance, during the formation of Pleistocene continental ice sheets such as the Laurentide Ice Sheet and the Scandinavian Ice Sheet, boreal forests, tundra and organic-rich lake sediments were overridden and their organic content became available for subglacial oxidation reactions (Wadham *et al.*, 2008). The organic carbon (OC) content can be used as a proxy for the amount of OM. In basal ice at the margin of the Greenland Ice Sheet, Yde *et al.* (2010) found that OC comprised 0.15 – 0.21% by mass of the total debris content. This is in accordance with previous findings of 0.27% OC by mass in suspended sediment emanating from Kuannersuit Glacier on Disko Island, West Greenland (Yde *et al.*, 2005), and indicates that OC is potentially available for heterotrophic metabolism after centuries of glacier cover (Yde *et al.*, 2010). Even after about 100.000 years of glacier cover, OC of 0.002% has been detected beneath the Greenland Ice Sheet (Tung *et al.*, 2006). This subglacial reservoir of organic carbon may be metabolized by methanogens, producing subglacial methane that may be releasing to the atmospheric (Wadham *et al.*, 2008). We will discuss this in more detail later, but the overall impression is that organic matter, and particularly organic carbon, may be a valuable nutrient resource for subglacial microorganisms, influence biogeochemical weathering rates and be subglacially converted to greenhouse gasses.

3.3.2 Biogeochemical weathering

Interpretations of the chemical composition of subglacial meltwater often reveal circumstantial evidences of microbially-mediated chemical weathering reactions (Tranter *et al.*, 2002; Hodson *et al.*, 2004, 2005; Skidmore *et al.*, 2005, 2010; Wynn *et al.*, 2006, 2007; Wadham *et al.*, 2010). The exact effect of microbial activity on the total solute flux from glaciers have not yet been quantified because it requires very detailed knowledge on biodiversity and metabolic rates, biotic and abiotic geochemical weathering rates, subglacial thermodynamic processes, hydrology, and the temporal and spatial fluctuations of these variables over the sampling period. However, as more and more information becomes available we are approaching a state where realistic conceptual models may simulate solute fluxes from glacierized catchments based on scenarios with entirely abiotic and assumed biotic geochemical reactions.

3.4 Subglacial ecosystem responses to climate change

Subglacial ecosystems may be sensitive to climate-driven environmental changes such as (1) changes in the basal thermal regime; (2) changes in the hydrological regime; or (3) changes in glacier extent. In the High Arctic, small glaciers may become thinner as a consequence of negative net mass balances during a warming climate. This may change their basal thermal regimes from polythermal or temperate to cold-based because the isolation effects of a thinner glacier cover are unable to keep the basal temperature at the pressure melting point. This will terminate the distributed drainage regime and thereby reduce the accessibility of water-transported nutrient supply to the subglacial communities. Changes in liquid and solid precipitation, the magnitude and frequency of precipitation events, and contribution from ice melt may also affect the hydrological regime and supply of nutrients to subglacial habitats. Unfortunately, we know little about how subglacial ecosystems will respond to these changes due to the inaccessibility of the glacier bed. It is, however, easier to study how atmospherically-exposed subglacial communities change following glacier recession. This will be discussed in the following section.

4. Proglacial ecosystems

Proglacial landscapes consist of a variety of landforms such as glaciofluvial outwash plains, moraine systems, till and dead-ice deposits, bedrock exposures, proglacial lakes and rivers. These landforms host different forms of life that either origin from the subglacial communities or later settlement of pioneering species. Hence, the proglacial landscape allows us to study how the microbial consortium changes from being dominated by subglacial structures to gradually adapt to proglacial environmental conditions. It also allows us to examine how colonizers and soil development processes affect biogeochemical processes and GHG fluxes over time.

The transition from subglacial to proglacial ecosystem is closely linked to climate warming via the dynamic response of glaciers. When the ecosystem has become proglacial its adaption rate is primarily controlled by the time since exposure (e.g. Bárcena *et al.*, 2010). Similar non-glacial processes are well known in the proglacial area after the withdrawal of glaciers and are generally in geological literature referred to as paraglacial processes (Church and Ryder, 1972). An interesting challenge is to assess and quantify the climatic feedback due to net microbiologically-driven changes in CO_2, CH_4 and N_2O fluxes. To date, we have limited knowledge on how proglacial ecosystems respond to climate change, but research on proglacial ecosystems is rapidly growing and we start to get an impression of microbial diversity and biogeochemical processes from a number of proglacial sites worldwide.

In the following, we focus on proglacial ecosystems, where bacterial diversity and biogeochemical processes have been examined along deglaciation gradients. However, it must be emphasized that changes in proglacial microbial communities cannot be isolated from effects of pioneering plants (Tscherko *et al.*, 2003; Miniaci *et al.*, 2007) and soil invertebrates (Doblas-Miranda *et al.*, 2008), but at present there is little knowledge on the coupling between invasive eukaryotes and microbial processes (Bernasconi *et al.*, 2008).

4.1 Glacier forefields

The term *glacier forefields* generally refers to newly-formed landscapes in front of glaciers that recently have been exposed after glacier retreat (Matthews, 1992). Glacier forefields often provide excellent opportunities to study temporal and spatial variations along a chronological deglaciation transect, referred to as a chronosequence. The glacial recession history since the termination of the Little Ice Age can be reconstructed from old maps and expedition accounts, aerial photos and satellite imagery. This allows a quantitative assessment of the ecological variations of primary successional communities between sampling sites along transects as a consequence of time since deglaciation, without needing long-term observations from the same sites.

4.1.1 Microbial community structures

It is now generally believed that the initial stage after deglaciation is dominated by a development in heterotrophic communities, which decompose allochthonous organic carbon deposits previously overridden by glacier advance (Tscherko *et al.*, 2003; Bardgett *et al.*, 2007). This is followed by increased fixation of carbon by autotrophs that build-up an organic carbon pool, which sustains the heterotrophic community when the ancient organic carbon pool is exhausted (Bardgett *et al.*, 2007). The low availability of carbon and nitrogen is likely to be limiting factors in the early deglaciation stage (Yoshitake *et al.*, 2007), but as the proglacial ecosystem ages the nitrogen pool increases, probably due to fixation of nitrogen by a growing community of nitrogen-fixing bacteria such as *Cyanobacteria* (Nemergut *et al.*, 2007). Also, a molecular study by Kandeler *et al.* (2006) has shown changes in gene composition and abundance of the denitrifying community along a chronosequence. This could indicate a coupling between nitrogen fixers and denitrifiers, however this needs to be confirmed by further studies.

Little is known about the sulfur cycle in glacier forefields. Schmalenberger and Noll (2010) found a surprisingly high diversity of desulfonating *Bacteria* in young proglacial ecosystems at Damma Gletscher, Switzerland, indicating that atmospheric deposition or glacier-derived particles may inoculate glacier forefields. Also, the community structure of desulfonating genetic phylotypes varied along the chronosequence, suggesting that the desulfonating communities change with age based on their affiliation with plant host rhizospheres (Schmalenberger and Noll, 2010).

The potential role of glacier forefields as either CH_4 sinks or sources is very interesting in context to proglacial ecosystem responses to climate change. A study by Bárcena *et al.* (2010) has shown CH_4 production in young deglaciated sediments, in accordance with the methanogenesis observed in subglacial ecosystems (Wadham *et al.*, 2008). However, the microbial community changes from net methanogenic to net methanotrophic as a function

of time since deglaciation (Bárcena *et al.*, 2010). The highest CH_4 consumption rate of 0.76 μg_{CH4} h^{-1} m^{-2} was measured at the top of the Little Ice Age moraine (Bárcena *et al.*, 2010), where incubation experiments at 10°C and 22°C showed atmospheric CH_4 rates of 1.2 and 2.1 nmol CH_4 day^{-1} g^{-1} sediment, respectively (Bárcena *et al.*, 2011). Similar to the denitrifying and desulfonating community studies, the high-affinity methanotrophic diversity also changes along the chronosequence, showing the highest diversity at the oldest sites.
Total cell counts in proglacial ecosystems have been conducted at a number of sites. At Damma Gletscher chronosequence the total cell count increased from 8.2×10^7 to 1.5×10^9 cells g^{-1} sediment (youngest to oldest), whereas at near Rotfirn Gletscher, Switzerland, the total cell count increased from 1.1×10^8 to 1.9×10^9 cells g^{-1} sediment (Sigler and Zeyer, 2002). The magnitude of about 1×10^8 cells g^{-1} sediment is confirmed by total cell counts from six Swiss glacier forefields, including Damma Gletscher (Lazzaro *et al.*, 2009). At five glacier forefields near Ny-Ålesund, Svalbard, the mean total cell count was 6.5×10^7 cells g^{-1} (Kaštovská *et al.*, 2005), and at Pindari Glacier in the Himalayas it ranged between 2.2 and 8.7×10^8 cells g^{-1} sediment (Shivaji *et al.*, 2011). Based on these studies, the magnitude range of proglacial communities is within 10^7-10^9 cells g^{-1} sediment.

4.1.2 Soil development

Several studies on glacier forefield ecology have mainly focused on processes involving the establishment and development of soils after a glacier's recession (Cooper, 1923; Field, 1947; Crocker and Mayor, 1955; Evans, 1999; Hodkinson *et al.*, 2003; Egli *et al.*, 2006; He and Tang, 2007; Breen and Lévesque, 2008; Moreau *et al.*, 2008; Schmidt *et al.*, 2008; Strauss *et al.*, 2009), mostly based on approaches regarding plant colonization patterns and changes in soil properties along a chronosequence.
During the last decade, interest in microbial processes taking place in proglacial areas has increased due to the essential role of microbiology in nutrient cycling (Sigler *et al.*, 2002; Bekku *et al.*, 2004; Kaštovská *et al.*, 2005; Kandeler *et al.*, 2006; Bardgett *et al.*, 2007; Hämmerli *et al.*, 2007; Lazzaro *et al.*, 2009; Schütte *et al.*, 2010). The study of soil development is critical in glacier forefield ecology research. Pedological information may elucidate important aspects of a chronosequence such as age of glacial and glaciofluvial deposits and define successional patterns that determine the maturity of different zones in a forefield. In the early stage after deglaciation, soils are merely disorganized accumulations of moraine debris whose significant properties show no regular variation with depth. With the deposition of the stranded glacial till in a stable position, climate and vegetation begin to modify the soils, and to impart them other characteristics (Crocker and Mayor, 1955). Nevertheless, it is essential to address the role of microorganisms in soils from early after deglaciation, since inputs of nutrients and organic matter during ecosystem development are assumed to be dominated by microbial carbon and nitrogen fixation (Schmidt, 2008).
Soil formation on glacier forelands has both autogenic (endogenous) and allogenic (exogenous) components. The autogenic component refers to biotic driven variations; while allogenic refers to changes driven by external environmental factors. Soils in these types of environments are characteristic for being developed on relatively coarse-textured till material (e.g. Bárcena *et al.*, 2011) and the imprint from parent material is highly significant, especially in early stages after glacial recession.

4.1.3 Proglacial ecosystem responses to climate change

Proglacial ecosystems are likely to be affected by climate-driven environmental changes in sediment temperature and moisture. The transition from subglacial to proglacial ecosystem is of particular interest since it may involve significant changes in net GHG sinks and sources. The study by Bárcena *et al.* (2010) indicates a rapid change from a subglacial methanogenic community to a methanotrophic community in newly-exposed deglaciated areas. This change is unlikely to have any effect on global climate as the CH_4 consumption rates are low compared to other CH_4 fluxes, but it has a significant ecological impact on these cold ecosystems. Future studies should also include investigations of CO_2, CH_4 and N_2O flux rates along chronosequences.

5. Conclusion

Both subglacial and proglacial ecosystems are likely to be affected by climate-forcing environmental changes. In subglacial environments microorganisms are sensitive to changes in the basal thermal regime, in the hydrological regime, and in glacier extent. The transition from subglacial to proglacial ecosystem due to glacier recession is of particular interest because studies indicate a climatic feedback response in the form of rapid changes from subglacial methanogenic communities to methanotrophic communities in deglaciated areas. In proglacial environments the rate of soil development is linked to climate change, and proglacial chronosequence studies have shown to provide essential assessments of how microbial communities change as a function of time since deglaciated.

At present we know very little about these harsh glacial ecosystems. Therefore, initial research efforts have attempted to provide qualitative and quantitative data on microbial biodiversity and abundance, microbial-mediated geochemical weathering processes, and primary succession rates. A review of current literature on subglacial biodiversity shows that lithotrophic β-*Proteobacteria* dominates the microbial diversity in many subglacial ecosystems, and α-*Proteobacteria, Bacteroidetes* and *Actinobacteria* are relatively common. However, it is clear that subglacial ecosystems host a wide variety of microbes, including aerobic and microaerophilic heterotrophs; obligate and facultative aerobes and anaerobes; Fe-reducers; denitrifiers; sulfate-reducers and sulfur-oxidizers; homoacetogens; methanogens; strict psychrophiles and psychrotolerants; and spore-formers and non-spore-formers.

This research topic is under rapid development, and future research will give more detailed information on the linkages between biodiversity, environmental conditions and biogeochemical processes. Especially, more research on ecological changes during transition from subglacial to proglacial ecosystems is needed in order to understand how the climatic feedback mechanisms of these ecosystems. So far, focus has been on CH_4 fluxes, but future studies should also include investigations of CO_2 and N_2O flux rates.

6. References

Anesio, A.M., Mindl. B., Laybourn-Parry, J., Hodson, A.J. & Sattler, B. (2007). Viral dynamics in cryoconite holes on a high Arctic glacier (Svalbard). *Journal of Geophysical Research*, Vol. 112, G04S31, doi:10.1029/2006JG000350.

Bárcena, T.G., Yde, J.C. & Finster, K.W. (2010). Methane flux and high-affinity methanotrophic diversity along the chronosequence of a receding glacier in Greenland. *Annals of Glaciology*, Vol. 51, No. 56, pp. 23-31.

Bárcena, T.G., Finster, K.W. & Yde, J.C. (2011). Spatial patterns of soil development, methane oxidation and methanotrophic diversity along a receding glacier forefield,

Southeast Greenland. *Arctic, Antarctic, and Alpine Research.* Vol. 43, No. 2, pp. 178-188

Barker, J.D., Sharp, M.J. & Turner, R.J. (2009). Using synchronous fluorescence spectroscopy and principal components analysis to monitor dissolved organic matter dynamics in a glacier system. *Hydrological Processes*, Vol. 23, pp. 1487-1500.

Bardgett, R.D., Richter, A., Bol, R., Garnett, M.H., Bäumler, R., Xu, X., Lopez-Capel, E., Manning, D.A.C., Hobbs, P.J., Hartley, I.R. & Wanek, W. (2007). Heterotrophic microbial communities use ancient carbon following glacial retreat. *Biology Letters*, Vol. 3, pp. 487-490.

Bekku, Y.S., Nakatsubo, T., Kume, A. & Koizumi, H. (2004). Soil microbial biomass, respiration rate, and temperature dependence on a successional glacier foreland in Ny-Ålesund, Svalbard. *Arctic, Antarctic, and Alpine Research*, Vol. 36, No. 4, pp. 395-399.

Benn, D.I. & Evans, D.J.A. (2010). *Glaciers and Glaciation.* Second edition, Hodder Education, ISBN 978-0-340-90579-1, Abingdon, United Kingdom.

Bernasconi, S.M. & BIGLINK project members (2008). Weathering, soil formation and initial ecosystem evolution on a glacier forefield : a case study from the Damma Glacier, Switzerland. *Mineralogical Magazine*, Vol. 72, No. 1, pp. 19-22.

Bhatia, M., Sharp, M. & Foght, J. (2006). Distinct bacterial communities exist beneath a High Arctic polythermal glacier. *Applied and Environmental Microbiology*, Vol. 72, No. 9, pp. 5838-5845.

Boyd, E.S., Skidmore, M., Mitchell, A.C., Bakermans, C. & Peters, J.W. (2010). Methanogenesis in subglacial sediments. *Environmental Microbiology Reports*, Vol. 2, pp. 685-692.

Boyd, E.S., Lange, R.K., Mitchell, A.C., Havig, J.R., Hamilton, T.L., Lafrenière, M.J., Shock, E.L., Peters, J.W. & Skidmore, M. (in press). Diversity, abundance, and potential activity of nitrifying and nitrate-reducing microbial assemblages in a subglacial ecosystem. *Applied and Environmental Microbiology.*

Breen, K. & Lévesque, E. (2008). The influence of biological soil crusts on soil characteristics along a High Arctic glacier foreland, Nunavut, Canada. *Arctic, Antarctic, and Alpine Research*, Vol. 40, pp. 287-297.

Butinar, L., Spencer-Martins, I. & Gunde-Cimerman, N. (2007). Yeasts in high Arctic glaciers: the discovery of a new habitat for eukaryotic microorganisms. *Antonie van Leeuwenhoek*, Vol. 91, pp. 277-289.

Butinar, L., Sonjak, S. & Gunde-Cimerman, N. (2009). Fungi in High Arctic glaciers. In: *New Permafrost and Glacier Research*, M.I. Krugger & H.P. Stern (Eds.), pp. 237-264, Nova Science Publishers, ISBN 978-1-60692-616-1, New York, NY, USA.

Buzzini, P., Turchetti, B., Diolaiuti, G., D'Agata, C., Martini, A. & Smiraglia, C. (2005). Culturable yeasts in meltwaters draining from two glaciers in the Italian Alps. *Annals of Glaciology*, Vol. 40, pp. 119-122.

Cheng, S.M. & Foght, J.M. (2007). Cultivation-independent and –dependent characterization of Bacteria resident beneath John Evans Glacier. *FEMS Microbiology Ecology*, Vol. 59, pp. 318-330.

Christner, B.C., Skidmore, M.L., Priscu, J.C., Tranter, M. & Foreman, C.M. (2008). Bacteria in subglacial environments. In: *Psychrophiles : from biodiversity to biotechnology*, R.

Margesin, F. Schinner, J.-C. Marx & C. Gerday (Eds.), pp. 51-71, Springer, Berlin Heidelberg, Germany.

Church, M. & Ryder, J.M. (1972). Paraglacial sedimentation: a consideration of fluvial processes conditioned by glaciations. *Geological Society of America Bulletin*, Vol. 83, pp. 3059-3071.

Cooper, W.S. (1923). The recent ecological history of Glacier Bay, Alaska. *Ecology*, Vol. 4, pp. 93-128.

Crocker, R.L. & Mayor, J. (1955). Soil development in relation to vegetation and surface age at Glacier Bay, Alaska. *Journal of Ecology*, Vol. 43, pp. 427-448.

De Garcia, V., Brizzio, S., Libkind, D., Buzzini, P. & van Broock, M. (2007). Biodiversity of cold-adapted yeasts from glacial meltwater rivers in Patagonia, Argentina. *FEMS Microbiology Ecology*, Vol. 59, pp. 331-341.

Des Marais, D.J., Nuth III, J.A., Allamandola, L.J., Boss, A.P., Farmer, J.D., Hoehler, T.M., Jakosky, B.M., Meadows, V.S., Pohorille, A., Runnegar, B. & Spormann, A.M. (2008). The NASA Astrobiology Roadmap. *Astrobiology*, Vol. 8, pp. 715-730.

Doblas-Miranda, E., Wardle, D.A., Peltzer, D.A. & Yeates, G.W. (2008). Changes in the community structure and civersity of soil invertebrates across the Franz Josef Glacier chronosequence. *Soil Biology and Biochemistry*, Vol. 40, pp. 1069-1081.

Duc, L., Noll, M., Meier, B.E., Bürgmann, H. & Zeyer, J. (2009). High diversity of diazotrophs in the forefield of a receding Alpine glacier. *Microbial Ecology*, Vol. 57, pp. 179-190.

Egli, M., Wernli, M., Kneisel, C. & Haeberli, W. (2006). Melting glaciers and soil development in the proglacial area Morteratsch (Swiss Alps): I. Soil type chronosequence. *Arctic, Antarctic, and Alpine Research*, Vol. 38, pp. 499-509.

Evans, D.J.A. (1999). A soil chronosequence from Neoglacial moraines in western Norway. *Geografiska Annaler*, Vol. 81A, pp. 47-62.

Field, W.O. Jr. (1947). Glacier recession in Muir Inlet, Glacier Bay, Alaska. *Geographical Reviews*, Vol. 37, pp. 369-399.

Finster, K. (2008). Anaerobic Bacteria and Archaea in cold ecosystems. In: *Psychrophiles : from biodiversity to biotechnology*, R. Margesin, F. Schinner, J.-C. Marx & C. Gerday (Eds.), pp. 103-119, Springer, Berlin Heidelberg, Germany.

Foght, J., Aislabie, J., Turner, S., Brown, C.E., Ryburn, J., Saul, D.J. & Lawson, W. (2004). Culturable Bacteria in subglacial sediments and ice from two southern hemisphere glaciers. *Microbial Ecology*, Vol. 47, pp. 329-340.

Gaidos, E., Lanoil, B., Thorsteinsson, T., Graham, A., Skidmore, M., Han, S.-K., Rust, T. & Popp, B. (2004). A viable microbial community in a subglacial volcanic crater lake, Iceland. *Astrobiology*, Vol. 4, No. 3, pp. 327-344.

Gaidos, E., Marteinsson, V., Thorsteinsson, T., Jóhannesson, T., Rúnarsson, Á.R., Stefansson, A., Glazer, B., Lanoil, B., Skidmore, M., Han, S., Miller, M., Rusch, A. & Foo, W. (2009). An oligarchic microbial assemblage in the anoxic bottom waters of a volcanic subglacial lake. *The ISME Journal*, Vol. 3, pp. 486-497.

Hämmerli, A., Waldhuber, S., Miniaci, C. Zeyer, J. & Bunge, M. (2007). Local expansion and selection of soil bacteria in a glacier forefield. *European Journal of Soil Science*, Vol. 58, pp. 1437-1445.

He, L. & Tang, Y. (2007). Soil development along primary succession sequences on moraines of Hailuogou Glacier, Gongga Mountain, Sichuan, China. *Catena*, Vol. 72, 259-269.

Hodkinson, I., Coulson, S.J. & Webb, N.R. (2003). Community assembly along proglacial chronosequences in the high Arctic: vegetation and soil development in north-west Svalbard. *Journal of Ecology*, Vol. 91, pp. 655-663.

Hodson, A.J., Mumford, P. & Lister, D. (2004). Suspended sediment and phosphorus in proglacial rivers: bioavailability and potential impacts upon the P status of ice-marginal receiving waters. *Hydrological Processes*, Vol. 18, pp. 2409-2422.

Hodson, A.J., Mumford, P.N., Kohler, J. & Wynn, P.M. (2005). The High Arctic glacial ecosystem: new insights from nutrient budgets. *Biogeochemistry*, Vol. 72, pp. 233-256.

Hodson, A., Anesio, A.M., Tranter, M., Fountain, A., Osborn, M., Priscu, J., Laybourn-Parry, J. & Sattler, B. (2008). Glacial ecosystems. *Ecological Monographs*, Vol. 78, pp. 41-67.

Hodson, A., Roberts, T.J., Engvall, A.-C., Holmén, K. & Mumford, P. (2010). Glacier ecosystem response to episodic nitrogen enrichment in Svalbard, European High Arctic. *Biogeochemistry*, Vol. 98, pp. 171-184.

Hubbard, B.P., Cook, S.J. & Coulson, H. (2009). Basal ice facies: a review and unifying approach. *Quaternary Science Research*, Vol. 28, pp. 1956-1969.

Humlum, O., Elberling, B., Hormes, A., Fjordheim, K., Hansen, O.H. & Heinemeier, J. (2005). Late-Holocene glacier growth in Svalbard, documented by subglacial relict vegetation and living soil microbes. *The Holocene*, Vol. 15, pp. 396-407.

IPCC (2007). *IPCC Fourth Assessment Report: Climate Change 2007*. Cambridge University Press, Cambridge, United Kingdom and New York, NY, USA.

Kandeler, E., Deiglmayr, K., Tscherko, D., Bru, D. & Philippot, L. (2006). Abundance of *narG*, *nirS*, *nirK*, and *nosZ* genes of denitrifying Bacteria during primary successions of a glacier foreland. *Applied and Environmental Microbiology*, Vol. 72, No. 9, pp. 5957-5962.

Karl, D.M., Bird, D.F., Björkman, K., Houlihan, T., Shackelford, R. & Tupas L. (1999). Microorganisms in the accreted ice of Lake Vostok, Antarctica. *Science*, Vol. 286, pp. 2144-2147.

Kaštovská, K., Elster, J., Stibal, M., & Šantrůčkova, H. (2005). Microbial assemblages in soil microbial succession after glacial retreat in Svalbard (High Arctic). *Microbial Ecology*, Vol. 50, pp. 396-407.

Kaštovská, K., Stibal, M., Šabacká, M., Černá, B., Šantrůčkova, H. & Elster, J. (2007). Microbial community structure and ecology of subglacial sediments in two polythermal Svalbard glaciers characterized by epifluorescence microscopy and PLFA. *Polar Biology*, Vol. 30, pp. 277-287.

Klassen, J.L. (2009). *Carotenoid diversity in novel Hymenobacter strains isolated from Victoria Upper Glacier, Antarctica, and implications for the evolution of microbial carotenoid biosynthesis*. PhD dissertation, Department of Biological Sciences, University of Alberta, Edmonton, Canada.

Klassen, J.L. & Foght, J.M. (2011). Characterization of *Hymenobacter* isolates from Victoria Upper Glacier, Antarctica reveals five new species and substantial non-vertical evolution within this genus. *Extremophiles*, Vol. 15, pp. 45-57.

Kornobis, E., Pálsson, S., Kristjánsson, B.K. & Svavarsson, J. (2010). Molecular evidence of the survival of subterranean amphipods (Arthropoda) during Ice Age underneath glaciers in Iceland. *Molecular Ecology*, Vol. 19, pp. 2516-2530.

Knight, P.G. (1997). The basal ice layer of glaciers and ice sheets. *Quaternary Science Research*, Vol. 16, pp. 975-993.

Knudsen, N.T., Nørnberg, P. Yde, J.C., Hasholt, B. & Heinemeier, J. (2008). Recent marginal changes of the Mittivakkat Glacier, Southeast Greenland and the discovery of remains of reindeer (*Rangifer tarandus*), polar bear (*Ursus maritimus*) and peaty material. *Danish Journal of Geography*, Vol. 108, pp. 141-146.

Lanoil, B., Skidmore, M., Priscu, J.C., Han, S., Foo, W., Vogel, S.W., Tulaczyk, S. & Engelhardt, H. (2009). Bacteria beneath the West Antarctic Ice Sheet. *Environmental Microbiology*, Vol. 11, No. 3, pp. 609-615.

Lazzaro, A., Abegg, C. & Zeyer, J. (2009). Bacterial community structure of glacier forefields on siliceous and calcareous bedrock. *European Journal of Soil Science*, Vol. 60, pp. 860-870.

Mader, H.M., Pettitt, M.E., Wadham, J.L., Wolff, E.W. & Parkes, R.J. (2006). Subsurface ice as a microbial habitat. *Geology*, Vol. 34, No. 3, pp. 169-172.

Matthews, J.A. (1992). *The ecology of recently-deglaciated terrain: a geoecological approach to glacier forelands and primary succession*. Cambridge University Press, Cambridge, United Kingdom.

Mikucki, J.A., Foreman, C.M., Sattler, B., Lyons, W.B. & Priscu, J.C. (2004). Geomicrobiology of Blood Falls: An iron-rich saline discharge at the terminus of the Taylor Glacier, Antarctica. *Aquatic Geochemistry*, Vol. 10, pp. 199-220.

Mikucki, J.A. & Priscu, J.C. (2007). Bacterial diversity associated with Blood Falls, a subglacial outflow from the Taylor Glacier, Antarctica. *Applied and Environmental Microbiology*, Vol. 73, No. 12, pp. 4029-4039.

Mikucki, J.A., Pearson, A., Johnston, D.T., Turchyn, A.V., Farquhar, J., Schrag, D.P., Anbar, A.D., Priscu, J.C. & Lee, P.A. (2009). A contemporary microbially maintained subglacial ferrous 'ocean'. *Science*, Vol. 324, pp. 397-400.

Miniaci, C., Bunge, M., Duc, L., Edwards, I., Bürgmann, H. & Zeyer, J. (2007). Effects of pioneering plants on microbial structures and functions in a glacier forefield. *Biology and Fertility of Soils*, Vol. 44, pp. 289-297.

Miteva, V.I., Sheridan, P.P. & Brenchley, J.E. (2004). Phylogenetic and physiological diversity of microorganisms isolated from a deep Greenland glacier ice core. *Applied and Environmental Microbiology*, Vol. 70, pp. 202-213.

Miteva, V.I. & Brenchley, J.E. (2005). Detection and isolation of ultrasmall microorganisms from a 120 000 year old Greenland glacier ice core. *Applied Environmental Microbiology*, Vol. 71, pp. 7806-7818.

Moreau, M., Mercier, D., Laffly, D. & Roussel, E. (2008). Impacts of recent paraglacial dynamics on plant colonization: a case study of Midtre Lovénbreen foreland, Spitsbergen (79°N). *Geomorphology*, Vol. 95, pp. 48-60.

Nemergut, D.R., Anderson, S.P., Cleveland, C.C., Martin, A.P., Miller, A.E., Seimon, A. & Schmidt, S.K. (2007). Microbial community succession in an unvegetated, recently deglaciated soil. *Microbial Ecology*, Vol. 53, pp. 110-122.

Price, P.B. (2000). A habitat for psychrophiles in deep Antarctic ice. *Proceedings of the National Academy of Sciences*, Vol. 97, pp. 1247-1251.

Price, P.B. & Sowers, T. (2004). Temperature dependence of metabolic rates for microbial growth, maintenance, and survival. *Proceedings of the National Academy of Science*, Vol. 101, pp. 4631-4636.

Price, P.B. (2007). Microbial life in glacial ice and implications for a cold origin of life. *FEMS Microbiology Ecology*, Vol. 59, pp. 217-231.

Price, P.B. (2009). Microbial genesis, life and death in glacial ice. *Canadian Journal of Microbiology*, Vol. 55, pp. 1-11.

Priscu, J.C., Fritsen, C.H., Adams, E.E., Giovannoni, S.J., Paerl, H.W., McKay, C.P., Doran, P.T., Gordon, D.A., Lanoil, B.D. & Pinckney, J.L. (1998). Perennial Antarctic lake ice: An oasis for life in a polar desert. *Science*, Vol. 280, pp. 2095-2098.

Reddy, P.V.V., Rao, S.S.S.N., Pratibha, M.S., Sailaja, B., Kavya, B., Manorama, R.R., Singh, S.M., Srinivas, T.N.R. & Shivaji, S. (2009). Bacterial diversity and bioprospecting for cold-active enzymes from culturable bacteria associated with sediment from a melt water stream of Midtre Lovénbreen glacier, an Arctic glacier. *Research in Microbiology*, Vol. 160, pp. 538-546.

Rohde, R.A. & Price, P.B. (2007). Diffusion-controlled metabolism for long-term survival of single isolated microorganisms trapped within ice crystals. *Proceedings of the National Academy of Science*, Vol. 104, pp. 16592-16597.

Schmalenberger, A. & Noll, M. (2010). Shifts in desulfonating bacterial communities along a soil chronosequence in the forefield of a receding glacier. *FEMS Microbiology Ecology*, Vol. 71, pp. 208-217.

Schmidt, S.K., Reed, S.C., Nemergut, D.R., Grandy, S.A., Cleveland, C.C., Weintraub, M.N., Hill, A.W., Costello, E.K., Meyer, A.F. Neff, J.C. & Martin, A.M. (2008). The earliest stages of ecosystem succession in high-elevation (5000 meters above sea level), recently deglaciated soils. *Proceedings of the Royal Society B: Biological Sciences*, Vol. 275, pp. 2793-2802.

Schütte, U.M.E., Abdo, Z., Foster, J., Ravel, J., Bunge, J., Solheim, B. & Forney, L.J. (2010). Bacterial diversity in a glacier foreland of the high Arctic. *Molecular Ecology*, Vol. 19, pp. 54-66.

Sharp, M., Parkes, J., Cragg, B., Fairchild, I.J., Lamb, H. & Tranter, M. (1999). Widespread bacterial populations at glacier beds and their relationship to rock weathering and carbon cycling. *Geology*, Vol. 27, No. 2, pp. 107-110.

Sheridan, P.P., Miteva, V.I. & Brenchley, J.E. (2003). Phylogenetic analysis of anaerobic psychrophilic enrichment cultures obtained from a Greenland glacier ice core. *Applied and Environmental Microbiology*, Vol. 69, pp. 2153-2160.

Shivaji, S., Pratibha, M.S., Sailaja, B., Kishore, K.H., Singh, A.K., Begum, Z., Anarasi, U., Prabagaram, S.R., Reddy, G.S.N. & Srinivas, T.N.R. (2011). Bacterial diversity of soil in the vicinity of Pindari glacier, Himalayan mountain ranges, India, using culturable bacteria and soil 16S rRNA gene clones. *Extremophiles*, Vol. 15, pp. 1-22.

Siegert, M.J., Ellis-Evans, J.C., Tranter, M., Mayer, C., Petit, J.-R., Salamatin, A. & Priscu, J.C. (2001). Physical, chemical and biological processes in Lake Vostok and other Antarctic subglacial lakes. *Nature*, Vol. 414, pp. 603-609.

Siegert, M.J., Tranter, M., Ellis-Evans, J.C., Priscu, J.C. & Lyons, W.B. (2003). The hydrochemistry of Lake Vostok and the potential for life in Antarctic subglacial lakes. *Hydrological Processes*, Vol. 17, pp. 795-814.

Sigler, W.V., Crivii, S. & Zeyer, J. (2002). Bacterial succession in glacial forefield soils characterized by community structure, activity and opportunistic growth dynamics. *Microbial Ecology*, Vol. 44, pp. 306-316.

Sigler, W.V. & Zeyer, J. (2002). Microbial diversity and activity along the forefields of two receding glaciers. *Microbial Ecology*, Vol. 43, pp. 397-407.

Skidmore, M., Foght, J.M. & Sharp, M.J. (2000). Microbial life beneath a High Arctic glacier. *Applied and Environmental Microbiology*, Vol. 66, No. 8, pp. 3214-3220.

Skidmore, M., Anderson, S.P., Sharp, M., Foght, J. & Lanoil, B.D. (2005). Comparison of microbial community compositions of two subglacial environments reveals a possible role for microbes in chemical weathering processes. *Applied and Environmental Microbiology*, Vol. 71, pp. 6986-6997.

Skidmore, M., Tranter, M., Tulaczyk, S. & Lanoil, B. (2010). Hydrochemistry of ice stream beds - evaporitic or microbial effects? *Hydrological Processes*, Vol. 24, pp. 517-523.

Souchez, R., Lemmens, M. & Chappellaz, J. (1995). Flow-induced mixing in the GRIP basal ice deduced from the CO_2 and CH_4 records. *Geophysical Research Letters*, Vol. 22, pp. 41-44.

Strauss, S.L., Ruhland, C.T. & Day, T.A. (2009). Trends in soil characteristics along a recently deglaciated foreland on Anvers Island, Antarctic Peninsula. *Polar Biology*, Vol. 32, No. 12, pp. 1779-1788.

Tison, J.-L., Souchez, R., Wolff, E.W., Moore, J.C., Legrand, M.R. & de Angelis, M. (1998). Is a periglacial biota responsible for enhanced dielectric response in basal ice from the Greenland ice core project ice core? *Journal of Geophysical Research*, Vol. 103, pp. 18885-18894.

Tranter, M., Sharp, M.J., Lamb, H.R., Brown, G.H., Hubbard, B.P. & Willis, I.C. (2002). Geochemical weathering at the bed of Haut Glacier d'Arolla, Switzerland - a new model. *Hydrological Processes*, Vol. 16, pp. 959-993.

Tranter, M., Skidmore, M. & Wadham, J. (2005). Hydrological controls on microbial communities in subglacial environments. *Hydrological Processes*, Vol. 19, pp. 995-998.

Tscherko, D., Rustemeier, J., Richter, A., Wanek, W. & Kandeler, E. (2003). Functional diversity of the soil microflora in primary succession across two glacier forelands in the Central Alps. *European Journal of Soil Science*, Vol. 54, pp. 685-696.

Tung, H.C., Bramall, N.E. & Price, B.P. (2005). Microbial origin of excess of methane in glacial ice and implications for life on Mars. *Proceedings of the National Academy of Science*, Vol. 102, pp. 18292-18296.

Tung, H.C., Price, P.B., Bramall, N.E. & Vrdoljak, G. (2006). Microorganisms metabolizing on clay grains in 3-km-deep Greenland basal ice. *Astrobiology*, Vol. 6, No. 1, pp. 69-86.

Turchetti, B., Buzzini, P., Goretti, M., Branda, E., Diolaiuti, G., D'Agata, C., Smiraglia, C. & Vaughan-Martini, A. (2008). Psychrophilic yeasts in glacial environments of Alpine glaciers. *FEMS Microbiology Ecology*, Vol. 63, pp. 73-83.

Wadham, J.L., Cooper, R.J., Tranter, M. & Bottrell, S. (2007). Evidence for widespread anoxia in the proglacial zone of an Arctic glacier. *Chemical Geology*, Vol. 243, pp. 1-15.

Wadham, J., Tranter, M., Tulaczyk, S. & Sharp, M. (2008). Subglacial methanogenesis: A potential climatic amplifier? *Global Biogeochemical Cycles*, Vol. 22, GB2021, doi:10.1029/2007GB002951.

Wadham, J.L., Tranter, M., Skidmore, M., Hodson, A.J., Priscu, J., Lyons, W.B., Sharp, M., Wynn, P. & Jackson, M. (2010). Biogeochemical weathering under ice: Size matters. *Global Biogeochemical Cycles*, Vol. 24, GB3025, doi:10.1029/2009GB003688.

Willerslev, E., Hansen, A.J., Christensen, B., Steffensen, J.P. & Arctander, P. (1999). Diversity of Holocene life forms in fossil glacier ice. *Proceedings of the National Academy of Science*, Vol. 96, pp. 8017-8021.

Willerslev, E., Cappellini, E., Boomsma, W., Nielsen, R., Hebsgaard, M.B., Brand, T.B., Hofreiter, M., Bunce, M., Poinar, H.N., Dahl-Jensen, D., Johnsen, S., Steffensen, J.P., Bennike, O., Schwenninger, J.-L., Nathan, R., Armitage, S., de Hoog, C.-J., Alfimov, V., Christl, M., Beer, J., Muscheler, R., Barker, J., Sharp, M., Penkman, K.E.H., Haile, J., Taberlet, P., Gilbert, M.T.P., Casoli, A., Campani, E. & Collins, M.J. (2007). Ancient biomolecules from deep ice cores reveal a forested Southern Greenland. *Science*, Vol. 317, pp. 111-114.

Wynn, P.M., Hodson, A. & Heaton, T. (2006). Chemical and isotopic switching within the subglacial environment of a High Arctic glacier. *Biogeochemistry*, Vol. 78, pp. 173-193.

Wynn, P.M., Hodson, A.J., Heaton, T.H.E. & Chenery, S.R. (2007). Nitrate production beneath a High Arctic glacier, Svalbard. *Chemical Geology*, Vol. 244, pp. 88-102.

Yde, J.C., Knudsen, N.T. & Nielsen, O.B. (2005). Glacier hydrochemistry, solute provenance, and chemical denudation at a surge-type glacier, Disko Island, West Greenland. *Journal of Hydrology*, Vol. 300, pp. 172-187.

Yde, J.C., Finster, K.W., Raiswell, R., Steffensen, J.P., Heinemeier, J., Olsen, J., Gunnlaugsson, H.P. & Nielsen, O.B. (2010). Basal ice microbiology at the margin of the Greenland ice sheet. *Annals of Glaciology*, Vol. 51, No. 56, pp. 71-79.

Yde, J.C. & Paasche, Ø. (2010). Reconstructing climate change: Not all glaciers suitable. *EOS*, Vol. 91, pp. 189-190.

Yoshitake, S., Uchida, M., Koizumi, H. & Nakatsubo, T. (2007). Carbon and nitrogen limitation of soil microbial respiration in a High Arctic successional glacier foreland near Ny-Ålesund, Svalbard. *Polar Research*, Vol. 26, pp. 22-30.

Yung, P.T., Shafaat, H.S., Connon, S.A. & Ponce, A. (2007). Quantification of viable endospores from a Greenland ice core. *FEMS Microbiology Ecology*, Vol. 59, pp. 300-306.

24

Why Do We Expect Glacier Melting to Increase Under Global Warming?

Roger J. Braithwaite
University of Manchester
UK

1. Introduction

Media stories about global warming almost always mention "melting glaciers" and their effects upon global sealevel. The reader might therefore ask why the title of this chapter includes a question mark. It may seem blindingly obvious that global warming will cause a rise in global sealevel with a substantial contribution from melting glaciers, but the reasons are less obvious despite the copious literature. For example, the IPCC assessment reports 1991, 1996, 2001 and 2007 all include many references to published papers on glaciers (IPCC, 2011). These are admirable summaries of *who* has said *what*, or *who* has done *what*, but they do not explain *why*. My purpose for the present chapter is to provide a clear narrative on *why* we expect glacier melting to increase with any change in temperature, whether due to global warming or to natural fluctuations. By its very nature, the *why* of increased glacier melting must also answer the question of *how much extra melting?*

A simple and direct relation between glacier melt and air temperature has not always been as self-evident as it may appear today. For example, Hoinkes (1955) wrote: *In recent years many authors, on the basis of careful studies, have come to the conclusion that summer temperature is to be regarded as the most important factor influencing the behaviour of glaciers* and he quotes four references to support this statement, to which I could add many more. Hoinkes (1955) then goes on to say: *This result is not in contradiction to the results of the measurements which are given here (according to which radiation is the main source of energy for the ablation of the alpine glaciers) so long as it is not combined with the idea that the greater heat exchange from air to ice during a hot summer is sufficient to account for the greater ablation.* Braithwaite (1981) discusses the fallacy behind this statement from Hoinkes (1955) that seems to conflate large average values of radiative energy with large variations in melt energy. The present chapter demonstrates empirically that higher melt is associated with higher air temperature. I do this in three stages: (1) correlating daily melt with daily air temperature for some Arctic and/or Greenland locations, (2) linking the results to the wider literature on the degree-method, and (3) showing that recent changes in glacier mass balance in the Alps are consistent with higher air temperatures in and around the Alps. I refer the reader to Kuhn (1979), Braithwaite (1980 and 1981), Ambach (1988), Braithwaite and Olesen (1990a and 1990b) and Braithwaite (1995) for the theoretical interpretation of the melt-temperature relation in terms of the energy balance at the glacier surface.

2. Short-term variations in glacier melt

I start the narrative by considering studies of short-term variations in glacier melt that were made in Arctic Canada (1960-63), South Greenland (1979-83), West Greenland (1980-86) and North Greenland (1993-94). See Fig. 1 and Table 1 for locations and periods. The high arctic bias of the measurements should be obvious with four of the six sites being in the region of year-round sea ice cover leading to a relatively dry continental climate (Braithwaite, 2005).

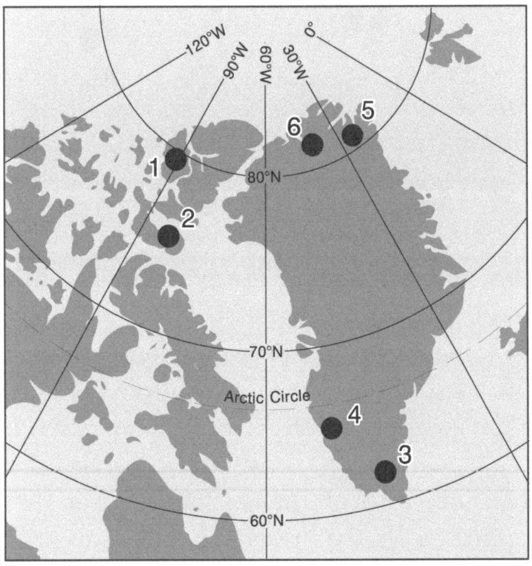

Fig. 1. Locations of the six sites in the present study. (1) is White Glacier, Axel Heiberg Island, (2) is Sverdrup Glacier, Devon Island, (3) is Nordbogletscher, South Greenland, (4) is Qamanârssûp sermia, West Greenland, (5) is Kronprins Christians Land, North Greenland, and (6) is Hans Tavsen Ice Cap, North Greenland.

The data consist of daily, or nearly daily, measurements made on stakes drilled into the ice near to a climate station, yielding a continuous record of air temperature measurements made 1.5-2 m above the surface, located on or very near the glacier. As a technical point, I should say that the stake measurements strictly refer to ice *ablation*, i.e. material lost by the glacier surface, rather than *melt* in the strict sense. However, the difference between ablation and melt is small in mass-balance terms under the conditions considered, i.e. with relatively small sublimation, condensation or re-freezing, and is negligible compared with stake measurement errors. You can therefore treat ablation and melt as almost synonymous for the present chapter but I will use the term ablation when referring to measurements made with stakes drilled into the ice.

Braithwaite (1981) analysed the data from Arctic Canada (see Section 2.1) and concluded that there was a useful relation between ice ablation and air temperature. When Braithwaite joined the staff of the Geological Survey of Greenland (GGU) in 1979, he persuaded colleagues Poul Clement and Ole B. Olesen to add daily ablation measurements to routine programmes at Nordbogletscher and Qamanârssûp sermia, see Sections 2.2 and 2.3 below.

In 1993 and 1994 Braithwaite was able to make relatively short visits to North Greenland (see Section 2.4), which is normally difficult to access, and gave high priority to daily measurements of ablation as part of energy balance studies (Braithwaite et al, 1998a and 1998b).

Site	Latitude (° N)	Longitude (° W)	Altitude (m a.s.l.)	Periods
1). White Glacier, Axel Heiberg Island	79	91	200	July-August 1960 June-August 1961 July 1962
2). Sverdrup Glacier, Devon Island	76	83	300	July-August 1963
3). Nordbogletscher, South Greenland	61	45	880	June-August 1979-1983
4). Qamanârssûp sermia, West Greenland	64	49	790	June-August 1980-1986
5). Kronprins Christians Land, North Greenland	80	24	380	July 1993
6). Hans Tavsen Ice Cap, North Greenland	83	36	540	July-August 1994

Table 1. Locations of the sites used for this study.

For this chapter, we represent the relation between daily ablation a_t and daily mean temperature T_t by the simple linear equation below:

$$a_t = a + \beta . T_t + e_t \quad (1)$$

The subscript t denotes the day of record and e_t represents the error in the equation. If we have a series of measurements for a_t and T_t covering N days, we can evaluate the intercept a and slope β parameters using the well-known least-square algorithm of linear regression, available in many computer data packages. According to this, a and β parameter values are chosen to minimise the variance of the error term e_t. The square root of the error variance, i.e. standard deviation of the error term, is often called the root mean square error, or RMSE. If the RMSE is relatively small compared with the fluctuating values of ablation we can say that air temperature is a good predictor of ablation. Alternatively, the correlation coefficient associated with the regression equation (1) should be relatively high.

We can regard a series of parallel ablation and temperature data for N days as a statistical sample. We cannot expect to find the same a and β values for different samples although we might hope that they will be similar to each other. For the study of the ablation-temperature relationship, the hypotheses are (1) the correlation coefficient should be relatively high (or the RMSE should be relatively low) for the sample in question, and (2) the a parameters from different samples should be similar if not identical, and same should hold for the β parameters. The proposed ablation-temperature relation would be useful if both (1) and (2) were true such that we could accurately calculate glacier ablation for situations where it is not measured.

2.1 Arctic Canada

I took the data from the work of Fritz Müller (1926-80) and his colleagues, working on White Glacier, (Axel Heiberg Island, NWT, Canada) and Sverdrup Glacier (Devon Island, NWT, Canada). Müller and Keeler (1969) discuss the accurate measurement of daily, or nearly daily, ablation in connection with studies of energy balance at the glacier surface. Their ablation measurements involved attempts to measure accurately both the lowering of the glacier surface by mass loss and the loss of material from within the surface layer, i.e. changes in effective density of the surface material. The formation of a low-density "weathering crust" and its decay a few hours, or even a few days, later depends upon the weather conditions. For example, selective absorption of global radiation around grain boundaries or dirt particles, e.g. on relatively sunny days, leads to formation of the weathering crust such that more ice ablates than is indicated by the measurement of surface lowering. The weathering crust disappears when stormy/overcast weather follows sunny weather: measurement of surface lowering then overestimate the amount of ablation. Figs 1 to 3 in Müller and Keeler (1969) nicely illustrate these processes.

For the regression analyses underlying Table 2, Müller and Keeler (1969) give the daily ablation for White Glacier 1961 and 1962, and for Sverdup Glacier 1963, and I extracted daily ablation for White Glacier 1960 from Andrews (1964). I found the corresponding air temperature data in Andrews (1964), Müller and Roskin-Sharlin (1967), Haven et al (1965) and Keeler (1964). Fritz Müller and colleagues set an admirable standard in documenting their work by providing extensive tables of data for possible use by later researchers.

Location	Year	Month	α (mm d^{-1})	β (mm d^{-1} K^{-1})	Days	ρ	RMSE (mm d^{-1})
White Gl.	1960	Periods	4	5.67	16	0.67	±13
White Gl.	1961	June	7	3.74	14	0.37	±7
White Gl.	1961	July	6	4.64	31	0.75	±8
White Gl.	1961	Aug,	-13	8.51	18	0.77	±16
White Gl.	1962	Periods	-3	7.89	11	0.84	±11
Sverdrup Gl.	1963	July	10	2.18	23	0.20	±15
Sverdrup Gl.	1963	Aug.	-14	10.36	10	0.89	±9
		Combined	1	6.13	123	0.74	±13

Table 2. Regression equation linking daily ablation and temperature for two sites in Arctic Canada.

Correlation coefficients between daily ablation and daily mean temperature (Table 2) are generally reasonably high, i.e. greater than the 0.71 that corresponds to "explanation" of 50% of the ablation variance. However, depending on the way in which the samples are sub-divided, low correlations ρ can also occur, e.g. for 14 days in June 1961 for White Glacier and for 23 days in July 1963 at Sverdrup Glacier. In these cases, the low correlation coefficients coincide with relatively large values of intercept α and low values of slope β. This phenomenon is a property of regression lines that tend to the horizontal as correlation coefficients tend to zero. The low values of correlation coefficients for these periods could reflect excessive measurement errors in the ablation data or a real lack of temperature-dependence in the energy balance for these periods. I will explore this issue in future work.

I am reluctant to calculate confidence intervals for the α, β and ρ parameters for the small samples in Table 2 as the background theory assumes that the e_t should be purely random. This is probably not the case for various reasons, e.g. serial correlation of measurement errors and persistence of certain weather types over many days. The solution is to pool all the data into a single regression equation. This produces a larger sample of 123 days, which should be more reliable from the statistical point of view, but suppresses any real differences between the individual series. The bottom line in Table 2 shows the parameters for this combined sample that represent a rough average of the individual samples. The underlying pattern is a small positive intercept (1 mm d^{-1}) and a slope of about 6 mm d^{-1} K^{-1}.

2.2 Nordbogletscher, South Greenland

The ablation data from Arctic Canada (Section 2.1) involve measurements of surface lowering together with attempts to measure density changes within the glacier surface. Müller and Keeler (1969) fully discuss the latter but the measurements seem very tedious to make and are probably not very reliable. We therefore decided to concentrate solely on measurements of surface lowering at Nordbogletscher and Qamanârssûp sermia, and to treat variations in surface density as an unavoidable error.

Year	Month	α (mm d^{-1})	β (mm d^{-1} K^{-1})	Days	ρ	RMSE (mm d^{-1})
1979	July	15	4.53	31	0.48	±21
	Aug.	10	4.64	30	0.60	±13
1980	June	4	7.58	30	0.81	±14
	July	13	4.74	31	0.57	±9
	Aug.	15	2.13	31	0.23	±14
1981	June	10	6.38	29	0.78	±14
	July	1	6.86	31	0.75	±16
	Aug.	1	5.02	31	0.83	±8
1982	June	14	5.54	30	0.82	±9
	July	6	6.38	31	0.70	±10
	Aug.	6	6.36	31	0.83	±10
1983	June	4	6.38	30	0.80	±12
	July	8	5.07	28	0.86	±9
	Aug.	8	5.22	31	0.85	±10
	Combined	7	5.68	425	0.74	±13

Table 3. Regression equation linking daily ablation and temperature at Nordbogletscher, South Greenland.

The field team made the ablation measurements at Nordbogletscher on one stake close to the edge of the ice and we take the temperature data from the nearby base camp. Air temperature above the melting point generally decreases as one proceeds onto a glacier (Braithwaite, 1980: Braithwaite et al., 2002) but the present measurement site is close enough to the ice edge not to show such "cooling effect". Ablation was measured every day in early evening, while temperature data refer to a 24-hour day in local time. This difference in timing introduces an extra small error into the melt-temperature correlation.

Once again, as in Table 2, correlation coefficients between daily ablation and daily mean temperature for Nordbogletscher (Table 3) are generally reasonably high, e.g. 0.70 to 0.86, but there are also periods with low correlations 0.23, 0.48, 0.57 and 0.60. The latter values are associated with relatively high intercept (10-15 mm d^{-1}) and low slope values (2.13 to 4.74 mm d^{-1} K^{-1}) as in Table 2. However, a high intercept (14 mm d^{-1}) also occur in June 1982 with a high correlation (0.82) when the slope is not especially low (5.54 mm d^{-1} K^{-1}).

The combined sample for all data, covering a total of 425 days, shows a slightly higher positive intercept (7 mm d^{-1}) and similar slope (about 6 mm d^{-1} K^{-1}) compared to the Arctic Canada results in Table 2. It is interesting that the respective combined samples for Arctic

Canada and for Nordbogletscher show identical root-mean square errors (±13 mm d^{-1}) despite not measuring variations in surface density at the latter site.

2.3 Qamanârssûp sermia, West Greenland

The ablation measurements at Qamanârssûp sermia were made on three stakes, within a few metres of each other and close to the edge of the ice while temperature data are taken from the nearby base camp. The results from Qamanârssûp sermia (Table 4) show generally similar patterns to the other sites (Tables 2 and 3). There are some high correlations (0.70 to 0.95) but also some low correlations (0.43 to 0.60). Both intercept α and slope β seem more variable than in the previous cases although the combined sample shows a small positive intercept (3 mm d^{-1}) with only a slightly higher slope of about 8 mm d^{-1} K^{-1}.

Year	Month	α (mm d^{-1})	β (mm d^{-1} K^{-1})	Days	ρ	RMSE (mm d^{-1})
1980	June	2	7.35	10	0.95	±9
	July	-6	7.70	28	0.49	±19
	Aug.	-20	10.34	23	0.79	±12
1981	June	4	7.24	26	0.85	±13
	July	-19	10.99	28	0.72	±23
	Aug.	4	6.09	31	0.76	±14
1982	June	6	6.13	17	0.74	±13
	July	0	8.88	27	0.81	±13
	Aug.	2	8.90	29	0.87	±16
1983	June	9	7.85	25	0.82	±18
	July	5	8.54	25	0.70	±24
	Aug.	7	6.87	29	0.70	±17
1984	June	9	4.76	21	0.58	±16
	July	-10	9.60	23	0.75	±20
	Aug.	-5	9.32	26	0.77	±17
1985	June	2	9.76	22	0.80	±20
	July	21	4.96	23	0.43	±16
	Aug.	9	6.61	25	0.76	±15
1986	June	10	4.74	27	0.60	±20
	July	-7	9.12	31	0.78	±18
	Aug.	-8	9.32	28	0.80	±19
	Combined	3	7.68	524	0.78	±18

Table 4. Regression equation linking daily ablation and temperature at Qamanârssûp sermia, West Greenland.

We had expected that measuring ablation at three stakes and averaging the results would give us slightly more accurate data than at Nordbogletscher where we used only one stake. It is, therefore, rather disappointing that the root-mean square error at Qamanârssûp sermia is actually slightly higher than at Nordbogletscher, i.e. ±18 compared with ±13 mm d^{-1}.

2.4 North Greenland

The ablation measurements in Kron Prins Christians Land (KPCL) and Hans Tavsen Ice Cap (HTIC) were made on 10 stakes, and air temperature data were taken from a station a few

metres from the stakes. The march of modern technology was marked by the fact that the temperature data were recorded on a digital data logger while earlier studies used data from thermographs checked by manual readings of mercury-in-glass thermometers.
The results from North Greenland show very low negative intercept and high slope in one case and small positive intercept with relatively low slope in the other case. When the two samples are pooled, the overall pattern is for a small positive intercept (3 mm d^{-1}) and a slope of about 7 mm d^{-1} K^{-1}.

Location	Year	Month	α (mm d^{-1})	β (mm d^{-1} K^{-1})	Days	ρ	RMSE (mm d^{-1})
KPCL	1993	July	-14	13.27	20	0.76	±6
HTIC	1994	July-Aug.	3	5.20	35	0.88	±6
		Combined	3	6.95	55	0.78	±10

Table 5. Regression equation linking daily ablation and mean temperature for two sites in North Greenland.

We had expected that measuring ablation at ten stakes and averaging the results would give us more accurate data than at Nordbogletscher and Qamanârssûp sermia, and the root-mean square error is indeed slightly lower than in the previous cases, i.e. only ±10 mm d^{-1} for the combined sample. More worrying is that the 10-stake measurements show that ablation varies by about ±10% of mean ablation within a few metres, probably due to micro-scale variations in albedo (Konzelmann and Braithwaite, 1995: Braithwaite et al., 1998b). This implies a ±10% error in any quantity calculated from stake measurements, including degree-day factors (see below).

2.5 Ablation-temperature correlation

The daily measurements of ablation in the four cases, covering a total of 1,127 days, show reasonably high correlations (0.74, 0.74, 0.78 and 0.78) such that air temperature variations explain slightly more than 50% of daily ablation variations. Errors in the measurements probably account for quite a sizeable percentage of the unexplained variance. The intercepts in the regression lines are slightly positive (1 to 7 mm d^{-1}) and the slopes are about 6 to 8 mm d^{-1} K^{-1} in round figures.
For readers who prefer graphs to numbers, Fig. 2 shows plots of ablation versus temperature for the four cases. Overall, the ablation-temperature relations are remarkable consistent despite differences in geographical setting from 61 to 83 °N. The occurrence of some negative ablation values in Figs 2b and 2c are clear signs of measurement errors as ablation should never be less than zero. The 95% confidence intervals around the regression lines give an impression of possible sampling errors in the regression lines. In the following section, I try to generalize this overall relation and compare it to results from other areas. However, I do intend in future to look at the energy balance variations between the different periods in Tables 2 to 5 to see if I can explain the apparent anomalies of low or high slope in the ablation-temperature regression equations.
The daily ablation measurements were very laborious to make as they involved manual measurements by human operators every day. This meant that fieldworkers had to live for many weeks on, or near, the glaciers. This probably explains why nobody has attempted to replicate the measurements elsewhere, e.g. in Svalbard. If we are realistic, we have to agree that glaciologists may never again spend such long continuous periods in the field although

new technology may allow unattended ablation measurements over periods of many months (Bøggild et al, 2004: Hulth, 2011). One purpose in writing this chapter is to motivate others to make similar ablation measurements in key areas.

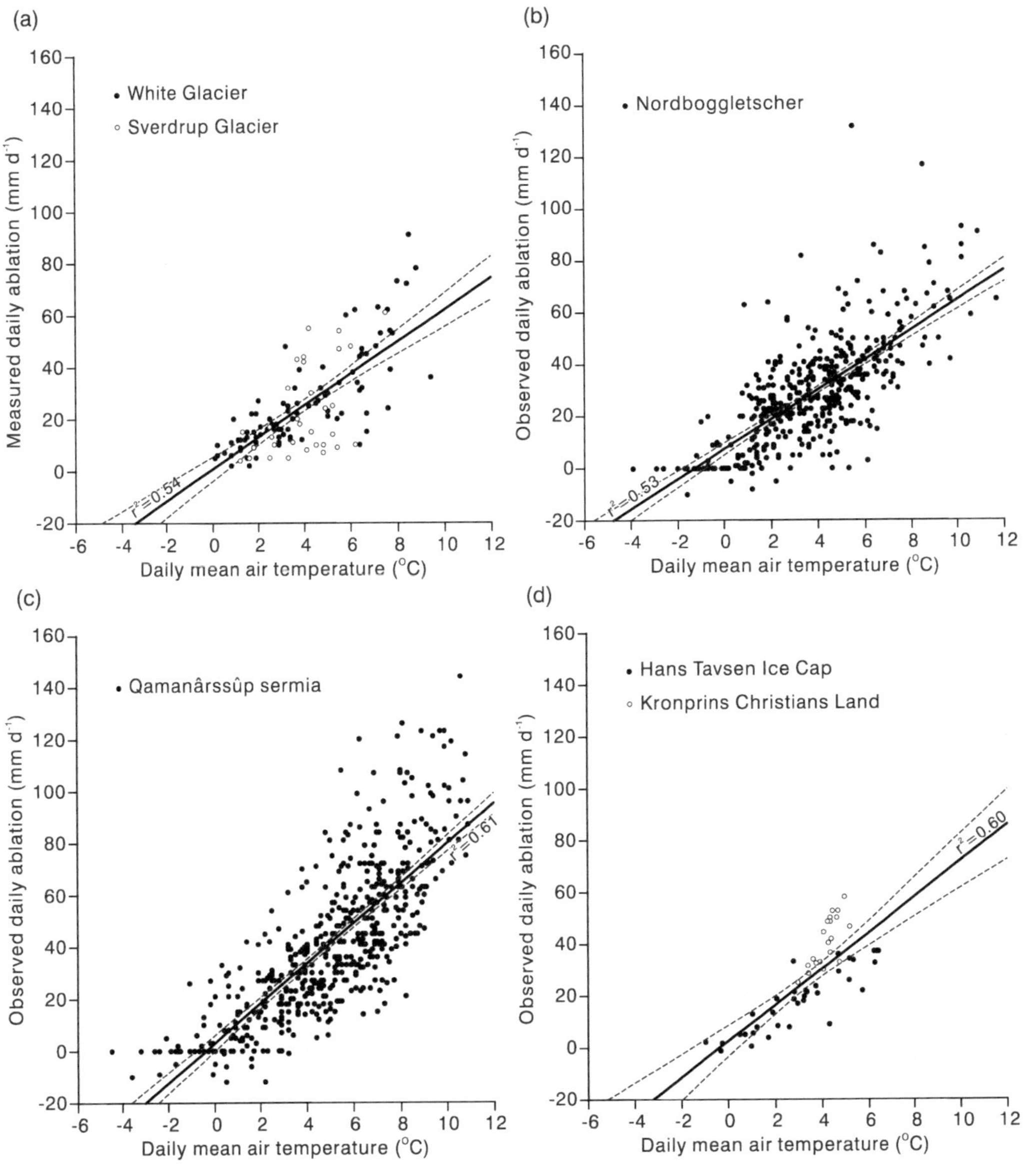

Fig. 2. Daily ablation versus daily mean temperature for: (a) Arctic Canada, (b) South Greenland, (c) West Greenland and (d) North Greenland.

3. Ablation and degree-day total

The intercept α in equation (1) represents the ablation that occurs when daily mean temperature equals 0 °C. Our statistical analysis shows that α is generally positive but quite small, i.e. there is relatively little ablation when daily mean air temperature is at 0 °C or below. Melt may occur at low temperature if energy from global radiation is high enough to supply melt energy as well as maintaining a heat flux from the glacier surface to the overlying atmosphere (Ohmura, 1981: Kuhn, 1987). However, another explanation lies in the choice of daily mean temperature as our independent variable. For example, air temperature may be above the melting point for part of the day with substantial ablation even if the daily mean temperature is at or below the freezing point (Arnold and MacKay, 1964). We can overcome this problem by only considering air temperatures that are above the melting point. This leads to the degree-day approach, whereby melt during any period is assumed proportional to the sum of positive temperatures during the same period. The approach is well established in hydrology (De Walle and Rango, 2008).

3.1 Degree-day model

With modern data loggers, the sum of positive temperatures during any period can be achieved simply by adding successive hourly values and dividing the total by 24. Data at Nordbogletscher and Qamanârssûp sermia were collected by thermographs, which are now obsolete, supplemented by readings of maximum and minimum temperatures by mercury-in-glass thermometers. A reasonable estimate of the positive degree-day total for each day at these stations can be achieved as the sum of positive values of daily mean temperature (counted twice), daily maximum and daily minimum temperatures, and dividing the resulting daily sum by 4. Calculated in this way, the positive degree-day total is identical to daily mean temperature for high temperatures and zero for low temperatures. There is an intermediate region where daily degree-day total is already positive while daily mean temperature is negative. The extent of this intermediate region is determined by the daily temperature range.

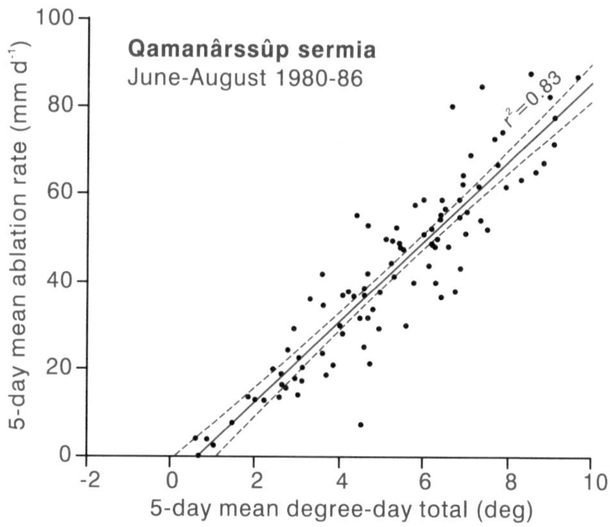

Fig. 3. 5-day averages of ablation and daily degree-day total at Qamanârssûp sermia, West Greenland.

There is a relatively good correlation between ablation and daily degree-day total at Qamanârssûp sermia, as calculated above, but the correlation is not much better than that between ablation and daily mean temperature (Fig. 2c). The apparently negative values of ablation in Fig. 2c are clear evidence of measurement errors as ablation cannot be less than

zero if accurately measured. There are similar values for Nordbogletscher (not shown). Braithwaite et al (1998b) avoided similar situations for North Greenland data by an objective method of screening gross errors, e.g. due to simple misreading of the measurement scale by several centimetres. However, for technical reasons, I cannot apply the same method retrospectively to the Nordbogletscher and Qamanârssûp sermia data.
As an alternative, I applied 5-day averaging to both daily ablation and daily degree-day total to see if this improves the correlation, which is obviously the case (Fig. 3). A particular type of measurement errors, involving a gross misreading on one visit to the stake, will affect ablation data for that day and the following day, and the averaging should eliminates this error if the two data points happen to fall within the same 5-day "window" for averaging. For this, or other reasons, there are no spurious negative values of ablation in Fig. 3. Krenke and Khodakov (1966) commented on a similar improvement in ablation measurements made over a few days. The results from Nordbogletscher are similar but I omit the graphs in the interests of conciseness.

3.2 Degree-day factor

Scatter plots like Fig. 3 demonstrate the validity of the degree-day approach for situations where daily ablation readings are available. This is not often the case. More commonly, workers visit the glacier at intervals of weeks or months and measure ablation totals for this time interval and then compare them with the degree-day total for the same period. The ratio of these longer-term totals is the degree-day factor. Estimates of degree-day factor are available for a number of locations (Table 6) and show widespread variations. We cannot regard the conditions underlying the listed values as being uniform and, no doubt, some of the variations in Table 6 will be due to methodological differences as well as real differences in glacier-climate conditions. However, the result that degree-day factors for snow are generally lower than for ice seems plausible, and Braithwaite (1995) explains this in energy balance terms. Ambach (1988) and Braithwaite (1995) also show that degree-day factors for melting ice may be quite large for low temperatures but not for high temperatures.
Many workers cite Ohmura (2001) for a physical explanation of the melt-temperature relation but Ohmura (2001) overlooks the temperature sensitivity of the different energy balance terms that is relevant rather than their absolute magnitudes. Ohmura (2001) is correct in saying that incoming longwave radiation is generally greater than the sensible heat flux to glacier surface, but the temperature sensitivity of sensible heat flux is generally greater than the temperature sensitivity of longwave radiation (Kuhn, 1979: Braithwaite, 1981: Ambach, 1988: Braithwaite, 1995) and hence accounts for a greater share of the degree-day factor.
I regard the results in Table 6 as "work in progress" and I hope that more data will become available representing a wider range of conditions. In particular, we can now recognize the effects of debris-cover and sublimation on ice ablation (Zhang et al. 2006) so future tables may not be so simply divided into results for "snow" and for "ice". Until we get more data, I regard the present results as a safe basis for three hypotheses that we can test in future studies:

1. We can reliably calculate snow and ice ablation from degree-day totals,
2. Degree-day factors for snow ablation are generally about 3 to 5 mm d^{-1}, and
3. Degree-day factors for ice ablation are generally about 6 to 8 mm d^{-1} K^{-1}.

In the following section, I discuss the modelling of glacier mass balance using the proposed relationship between ablation and positive degree-day total.

Location	Type	Mean	St. devn.	Sample size
De Quervain (1979)	Snow	4.2	±1.0	28 melt seasons
Braithwaite & Olesen (1988)	Snow	2.5	-	1 glacier
Laumann and Reeh (1993)	Snow	3.5 to 4.5	-	3 glaciers
Jóhannesson et al. (1993)	Snow	2.8 to 5.7	-	3 glaciers
Vincent and Vallon (1997)	Snow	3.8	-	1 glacier
Hock (1999)	Snow	4.4	-	1 glacier
Hock (2003)	Snow	5.1	±2.2	18 sites
Anderson et al (2006)	Snow	4.6	-	1 glacier
Radic (2008)	Snow	4.8	±1.5	44 glaciers
Braithwaite (2008)	Snow	4.1	±1.5	66 glaciers
Shea et al (2009)	Snow	3.0	±0.4	9 glaciers
Schytt (1964)	Ice	13.8	-	1 glacier
Orheim (1970)	Ice	6.1 & 6.5	-	2 Seasons at 1 site
Braithwaite (1977)	Ice	5.4	± 2.3	4 glaciers
Braithwaite (1981)	Ice	5.5 to 7.8	-	2 glaciers
Ambach (1988)	Ice	18.6		1 site
Braithwaite & Olesen (1988)	Ice	7.2	-	1 glacier
Laumann and Reeh (1993)	Ice	5.5 to 6.0	-	3 glaciers
Jóhannesson et al. (1993)	Ice	6.4 to 7.7	-	
Vincent and Vallon (1997)	Ice	6.2	-	1 glacier
Hock (1999)	Ice	6.3	-	1 glacier
Hock (2003)	Ice	8.9	±3.7	32 sites
Anderson et al (2006)	Ice	7.2	-	1 glacier
Zhang et al (2006)	Ice	6.5	±3.7	15 glaciers
Radic (2008)	Ice	7.3	±2.6	44 glaciers
Shea et al (2009)	Ice	4.6	±0.6	9 glaciers

Table 6. Mean and standard deviation of degree-day factors from different studies. Units are mm d^{-1} K^{-1}.

4. Modelling glacier mass balance

According to above results, we can calculate the snow or ice ablation for any period and location if we know the degree-day total, which we calculate as the sum of positive temperatures at the same location and period. If measured data are not available, we can estimate them by extrapolation of temperatures from some location where they are measured, e.g. from a weather station at lower altitude in the same region as the glacier. However, Braithwaite (1984) notes that it is tedious to find and store daily temperatures, or sub-daily temperatures, if one only needs to add them up to form a single total, and he suggests a simple method for estimating monthly degree-day totals from monthly mean and standard deviation of temperature. To do this, we assume that temperatures within the month are normally distributed about the mean temperature with a standard deviation s, and the degree-day total is given by a numerical integration of temperatures above 0 °C multiplied by their probabilities. The area under the Normal probability curve that lies above 0 °C is the duration of positive temperatures in days.

The paper by Braithwaite (1984), combined with the values of degree-day factor proposed by Braithwaite and Olesen (1989), was quite literally "seminal" in that many workers later developed glacier mass balance models where ablation is calculated from the degree-day approach. See papers by Reeh (1991), Huybrechts et al. (1991), Lauman and Reeh (1993), Jóhannesson (1993), Van der Veen (1999, pages 355-363), Braithwaite and Zhang (1999a and 2000), Braithwaite et al. (2002), De Woul and Hock (2005), Anderson et al (2006 & 2010), and Radic and Hock (2007 & 2011) for examples. Tarasov and Peltier (1997) describe the degree-day approach as *the standard methodology for parameterization of ablation over both glaciers and the Greenland ice sheet*. This might be a slight exaggeration but Carlov and Greve (2005) regard the method as sufficiently useful to merit them developing a more efficient algorithm for the many repeated calculations needed for very long-term simulations of the Greenland ice sheet. In some of these models, ablation is calculated directly from monthly mean temperature while other workers follow a variant of the method where monthly mean temperature is estimated from annual mean and annual amplitude of temperature. The latter approach, developed by Reeh (1991), is especially interesting as it links the degree-day approach to several empirical studies where workers plot annual ablation, equal to annual accumulation at the ELA, as a nonlinear function of summer mean temperature as first suggested by Ahlmann (1924) and extended by Ohmura et al (1992) and Braithwaite (2008).

4.1 Calculating ablation and snow accumulation

Figs 4a and 4b illustrate the performance of the Braithwaite (1984) model. Fig. 4a shows monthly snow and ice ablation as functions of monthly mean temperature for suitable values of degree-day factor. Fig. 4b actually shows the calculated probability of temperatures under 0 °C in the month but we can interpret this as the ratio of snow accumulation to total precipitation if precipitation rate is constant throughout the month. The choice of standard deviation s affects the precise shape of the curves (Braithwaite, 1984). Recent results from Greenland (Fausto et al, 2009) suggest that standard deviation might generally be somewhat lower than assumed in Fig. 4. I intend to revisit this issue myself in the near future, and to extend the Braithwaite (1984) model to explicitly include variations in daily temperature range, which is relatively small at high latitudes but may be quite large at lower latitudes.

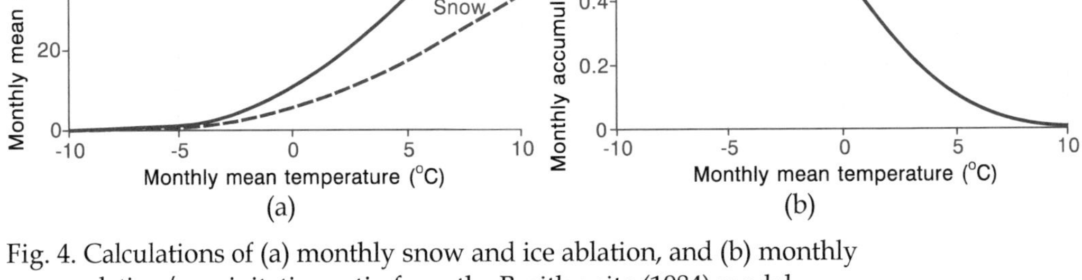

Fig. 4. Calculations of (a) monthly snow and ice ablation, and (b) monthly accumulation/precipitation ratio from the Braithwaite (1984) model.

Figs. 4a and 4b show that the range -10 to +10 °C in monthly mean temperature is the critical range for glacier-climate conditions. If monthly mean temperature is less than -10 °C in the warmest month, there will be no melting, which seems to be the situation over most of the Antarctic. As temperature rises there is a slightly exponential rise in melting as more and more days experience temperatures of over 0 °C and there is melting on every day in the month with monthly mean greater than 10 °C. Fig. 4b shows that there no days with below-freezing temperatures in months with mean temperature greater than 10 °C. This may explain the rough coincidence of the climatic tree line with the +10 °C July isotherm in the Northern Hemisphere, i.e. there is at least one frost-free month at the tree line.

4.2 Tuning the model

Fig. 5 sketches the operation of the glacier mass-balance model of Braithwaite and Zhang (1999a and 2000) and Braithwaite et al (2002). Monthly values of air temperature and precipitation are extrapolated to the glacier from a nearby weather station or gridded climatology (top left of the diagram). Degree-day factors for ice and snow are specified (very top right of figure). The model calculates a temporary value of annual balance (top right of figure) by summing monthly ablation and accumulation according to Figs 4a and 4b. The model then compares the computed mass balance with observed mass balance and adjusts the precipitation in successive small steps until the computed and observed mass balance agree closely. This represents the "tuning" of the model to fit observed mass balance (bottom right of figure) and we can now use the model for experiments to explore the sensitivity of glacier mass balance to changes in different factors (bottom left of figure).

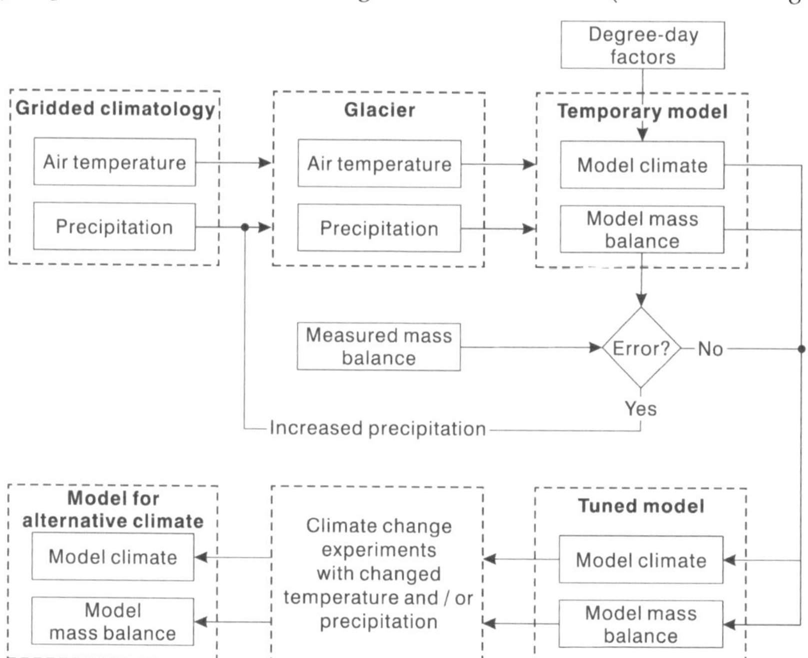

Fig. 5. Flow diagram illustrating the principles of a glacier mass-balance model based on the degree-day approach. From Braithwaite et al (2002).

We can tune the mass-balance model in Fig. 5 to fit an observed profile of mass-balance as a function of altitude. However, such data are probably only available for about 100 glaciers, i.e.

where workers have both made and published the necessary measurements. This limits our ability to apply the model but we can greatly extend it by noting that all we really need is a known value of mass-balance somewhere on the glacier. We do know that mass balance must be zero at the equilibrium line altitude (ELA) of the glacier and we can apply the model in Fig. 5 to the estimated ELA of the glacier. We can estimate this with an accuracy of about ±100 m for many thousands of glaciers using the "median" elevation parameter in the World Glacier Inventory (Braithwaite and Raper, 2009). This allowed Braithwaite and Raper (2007) to tune the mass-balance model for seven glacier regions with good coverage in the World Glacier Inventory, and then to extrapolate results to all mountain glaciers in the world to estimate 21st Century sea-level rise from melting glaciers (Raper and Braithwaite, 2006).

4.3 Temperature sensitivity of glacier mass balance

Common experiments on mass-balance models include changing air temperature by +1 °K, either throughout the whole year or just for the summer (June to August) , or changing annual precipitation by +10% while holding temperature constant. A number of sensitivity studies have shown that precipitation must increase by 20-40% to offset the effects of a +1 K temperature increase.

Fig. 6 shows the sensitivity of mass balance to a +1 K temperature change throughout the whole year for the seven glacier regions studied by Braithwaite and Raper (2007), who fitted the mass balance model (Fig. 5) to the estimated average ELA for each half-degree latitude/longitude grid cell in the region. The circles denote the average values for the N grid cells in each region, and the error bars denote standard deviations around these averages. As Braithwaite and Raper (2007) could not be completely certain of the correct values of degree-factors to use, they made calculations for low, medium and high estimates of the degree-factor for ice (6, 7 and 8 mm d^{-1} K^{-1}) in the hope that the true values will be somewhere within the range of results.

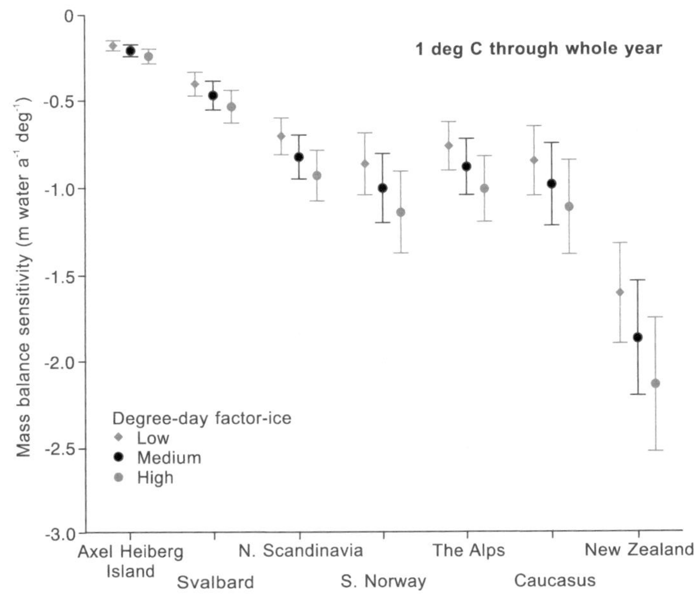

Fig. 6. Temperature sensitivity for glacier mass balance in seven regions calculated with a degree-day based mass-balance model. From Braithwaite and Raper (2007).

Braithwaite and Raper (2007) tuned the model with climate data from the gridded climatology of New et al (1997). This climatology is centred on averages for the 30-year period 1961-1990 so we can interpret Fig. 6 as a prediction of how mass balance will change in each region if the temperature increases by 1 K compared with the 1961-1990 averages. Despite uncertainties about degree-day factor, it is clear that mass-balance sensitivity is highly variable between regions, varying by almost an order of magnitude between Axel Heiberg Island, in the high arctic, and New Zealand. We can interpret this in terms of contrast between extreme continental and extreme maritime conditions as suggested by a number of workers. Svalbard is apparently somewhat less continental than Axel Heiberg Island while Northern Scandinavia, Southern Norway, the Alps and the Caucasus are relatively similar to each other on the continental/maritime scale.

For the purposes of the present Chapter, I take the results shown in Fig. 6 as a definite prediction that the mass balances of glaciers in the Alps will decrease by somewhat less than 1 m a^{-1} for each +1 K temperature change from the 1961-1990 average. I focus here on the Alps because it is the mountain region with best coverage of mass balance and climate data and I test this prediction with data from the Alps in the following sections.

You can regard the model in Fig. 5 as a method for "upscaling" daily measurements of ablation and air temperature, e.g. as described in Figs 2 to 5, to variations in annual balance. So, according to Fig. 6, degree-day factors of 6 to 8 mm d^{-1} K^{-1}, combined with a +1 K temperature increase, should be equivalent to a mass balance change of up to 1 m w.e. a^{-1} for Alpine glaciers.

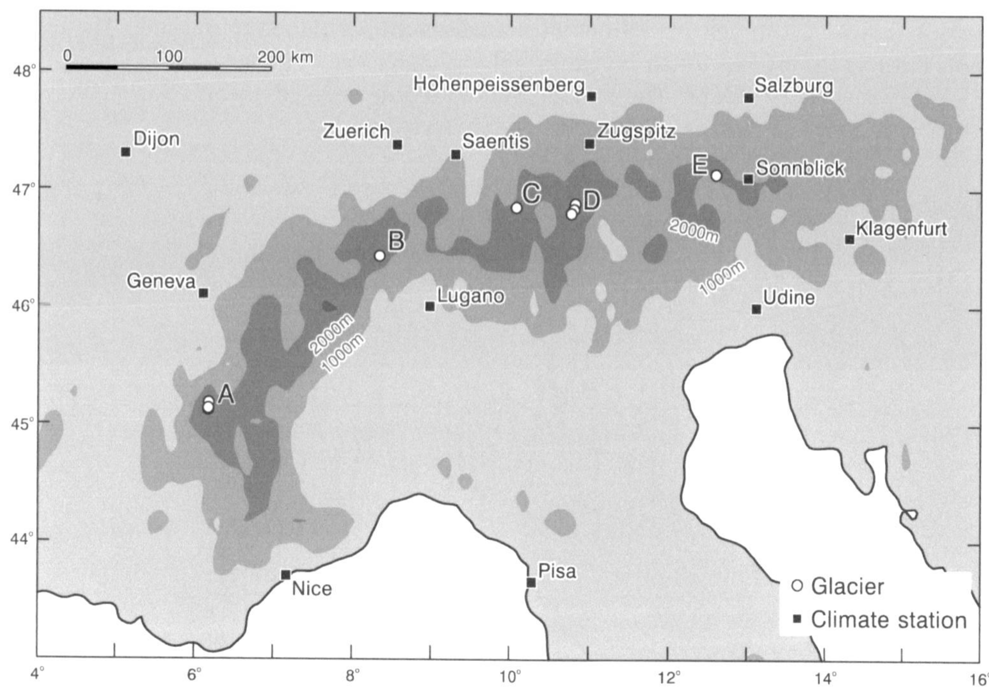

Fig. 7. Locations of glaciers and weather stations in the Alps with long records. (A) denotes Glacier de Sarennes and Glacier de Saint Sorlin; (B) denotes Griesgletscher; (C) denotes Silvrettagletscher; (D) denotes Hintereisferner, Kesselwandferner, Vernagferner; (E) denotes Sonnblikkees.

5. The Alps

The Alps occupy a key location in the middle of a continent with high levels of culture and education. It is no surprise therefore that the Alps, including its glaciers, have been a focus of scientific study for several centuries. However, actual measurement of glacier mass-balance by stakes and snow pits, i.e. the so-called "direct" glaciological method developed by H. W. Ahlmann (1889-1974), was relatively late coming to the Alps. For example, the longest continuous series is from Glacier de Sarennes that started in 1948 and continues today. Similar studies started at Limmern/Plattalvagletscher in 1948 but stopped in 1988. During the 1950's and 1960's mass-balance studies started on a number of glaciers and continue today.
The World Glacier Monitoring Service (WGMS, 2011) collects mass-balance data from the individual field workers and re-distributes them to potential users. For the present study, we need glaciers with full records throughout the whole 30-year period 1961-1990 and continuing up to nearly the present day, i.e. the balance year 2008/9 which is currently the latest year in the WGMS dataset. This is so we can compare recent mass-balance values with those in the climatic base period (1961-1990). I can identify mass-balance data for 37 Alpine glaciers although most of these have short records of only a few years. There are six glaciers with a full record for all years 1961-2009, and there are a further two glaciers that we can use because they have nearly full records, e.g. Griesgletscher for 1962-2009 and Vernagtferner for 1965-2009. Fig. 7 shows the location of these eight glaciers used in this study.

5.1 Mass-balance variations

Figure 8 shows year-to-year variations in mass balance for the eight glaciers. The circles denote the average balance for each year for the eight glaciers, or for six or seven glaciers for the very first years of record. The bars denote the corresponding standard deviations for each year, and give a measure of glacier-to-glacier variation.

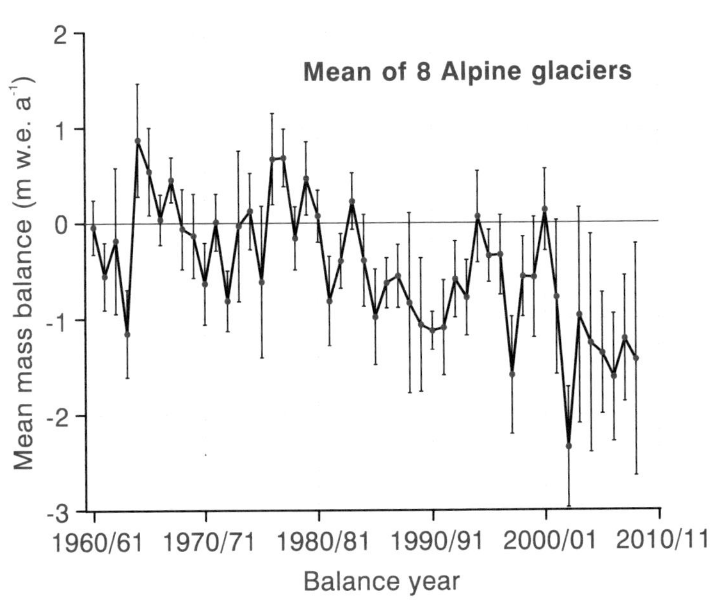

Fig. 8. Mass balance variations 1961-2009 for eight glaciers in the Alps with long records.

However, the year-to-year variations in balance are larger than these glacier-to-glacier variations, suggesting that the eight glaciers have very similar (but not identical) year-to-year variations. For example, principal component analysis (PCA) shows that a single principal component explains 80% of the total variance and a second component explains a further 10%. These two components are heavily loaded on mass balance for eastern and western Alpine glaciers respectively. The general agreement in year-to-year variations for the different glaciers agrees with Reynaud (1980) who suggested homogeneous mass-balance variations across the whole Alps.

Mass balance fluctuated around a constant base from 1961 to about 1985 and then becomes more negative in later years (Fig. 8). Positive balances occurred in just under half the years up to 1984 and then become very rare with positive balances only in 1995 and 2001. The highly negative balance in 2002/2003 is noteworthy as most of Europe suffered from a heat wave in the summer of 2003 (Schar et al, 2004). If we interpret the balance variations in Fig, 8 as mainly due to variations in ablation, it is obvious that ablation has increased markedly in recent years, especially after 2001. Does this increase in ablation agree quantitatively with an increase in air temperature according to our earlier hypothesis?

5.2 Temperature variations

The NASA/GISS (2011) website is a convenient source of temperature data from established weather stations. In and around the Alps, I could identify 13 weather stations (Fig. 7) covering long periods of record including the period of interest, 1961-2009. Temperature data are already available up to March 2011 (as of writing on 25 May 2011) but we only have glacier mass-balance data up to 2008/09 at present. Seven of the stations are noteworthy in that they have temperature records extending back well into the 19th Century. Three stations are located at 2500 m a.s.l. or above (Saentis, Zugspitz and Sonnblick) and may therefore reflect conditions near to the glacier ELA, i.e. roughly 2500-3000 m a.s.l. for the Alps. The 13 weather stations are situated at different locations and altitudes and obviously do not have the same temperatures. However, if we express temperatures at each station as deviations (or anomalies) from their 1961-1990 averages, the year-to-year variations in anomalies are remarkably similar. The three high-altitude stations also show very similar time variations to those at the ten lower-lying stations. These results show a pattern of very similar temperature variations over the whole region of the Alps in agreement with Bohm et al (2001).

Period	Summer	Mean	S.D.	Cases	%
1961-1990	June-August	0.00	±0.74	389	99.7 %
1991-2009	June-August	1.35	±1.12	239	96.8 %
Difference		1.35			
1961-1990	June-September	0.00	±0.75	388	99.5 %
1991-2009	June-September	1.02	±0.99	239	96.8 %
Difference		1.02			
1961-1990	May-September	0.00	±0.73	389	99.7 %
1991-2009	May-September	1.11	±0.89	239	96.8 %
Difference		1.11			

Table 7. Mean and standard deviation (S.D.) of summer mean temperatures (°C) at 13 weather stations for different lengths of summer.

Table 7 shows the mean and standard deviation of temperature anomalies for the 13 weather stations for the 30-year period 1961-1990 and for the 19 years 1991-2009. The mean temperature anomaly for 1961-1990 is obviously zero by definition of "anomaly". The column marked "%" denotes the percentage of expected data that are actually present. The 1961-1990 record is almost complete for the expected 30 × 13 = 390 records except for one missing month at one station. There are slightly more missing data for the later period. The standard climate definition of "summer" is June-August but Alpine glaciers generally have a longer melt season (Braithwaite et al, 2002) and I therefore calculated the anomalies for different lengths of summer to see if this is a critical issue. For any sensible choice of summer length, the period 1991-2009 is clearly on average 1.02 to 1.11 °C warmer than the base period 1961-1990. A three-month summer (June to August) is too short to cover the full melting period on a typical Alpine glacier which is more like 120-150 days, i.e. 4 to 5 months. I therefore adopt the average of temperature changes for 4- and 5-month summers, i.e. +1.07 K, for the analysis in the next section.

5.3 Temperature sensitivity of Alpine glacier mass balance

Table 8 shows the mean and standard deviation of mass balance for the eight glaciers and for the two periods 1961-1990 and 1991-2009. For all eight glaciers, the mean mass balance for the second period is much more negative than for the first period.

Glacier	1961-1990 Mean	1961-1990 S.D.	1961-1990 N	1991-2009 Mean	1991-2009 S.D.	1991-2009 N	Diff.	$\Delta b/\Delta T$
	m a^{-1}	m a^{-1}		m a^{-1}	m a^{-1}		m a^{-1}	m a^{-1}K^{-1}
Saint Sorlin	-0.20	±0.83	30	-1.31	±1.04	19	-1.11	-1.04
Sarennes	-0.57	±0.90	30	-1.64	±1.29	19	-1.08	-1.01
Griesgletscher	-0.35	±0.76	29	-1.13	±0.68	19	-0.78	-0.73
Silvrettagletscher	+0.03	±0.71	30	-0.60	±0.66	19	-0.63	-0.59
Hintereisferner	-0.33	±0.56	30	-0.98	±0.44	19	-0.65	-0.61
Kesselwandferner	+0.05	±0.40	30	-0.34	±0.49	19	-0.40	-0.37
Vernagtferner	-0.09	±0.41	26	-0.70	±0.47	19	-0.62	-0.58
Sonnblickkees	-0.08	±0.75	30	-0.77	±0.87	19	-0.69	-0.65
Mean	-0.19			-0.93			-0.74	-0.70
S.D.	±0.21			±0.42			±0.24	±0.23

Table 8. Mean and standard deviation (S.D.) of mass balance for eight glaciers for two different periods. Diff. is the average between means for the two periods and $\Delta b/\Delta T$ is the estimated mass-balance sensitivity assuming a temperature change of +1.07 K.

The standard deviation of mass balance in the period 1961-1990 varies greatly from relatively high values in the western Alps (Saint Sorlin and Sarennes), medium values in the central Alps, low values in the eastern Alps and a medium value in the far-eastern Alps. This pattern reflects the different maritime/continental character of the different glaciers as lower standard deviations of mass balance occur in more continental environments (Braithwaite and Zhang, 1999b)

We can divide the differences in mass balance for the two periods by the estimated increase in mean temperature between the two periods, estimated to be +1.07 K, to obtain estimates of the mass balance sensitivity. These range from relatively large values, i.e. around -1.0 m a^{-1} K^{-1}, for the two western glaciers, to a relatively small value for Kesselwandferner, i.e. about -0.4 m a^{-1} K^{-1}. The overall average mass balance sensitivity for the eight glaciers is about -0.7 ± 0.23 m a^{-1} K^{-1}. This is in good agreement with the range of values predicted by the model for the Alps (Fig. 6) for low-medium degree-day factor (6-7 mm d^{-1} K^{-1}) but somewhat smaller than predicted for high degree-day factor (8 mm d^{-1} K^{-1}).

The model predictions in Fig. 6 are averages for all half-degree latitude/longitude grid cells in the Alps, i.e. covering the whole range of Alpine climates, while the eight glaciers are only a fairly small sample and may biased to more continental glaciers. However, there is also a methodological reason why the degree-day model should somewhat overestimate mass balance changes. This is because we calibrate the model with data for present-day glaciers and then change the temperature by +1 K while keeping the present-day area distribution of the glacier. In the modelling literature, we call the resulting mass-balance sensitivity the "static sensitivity". However, as a result of the increased melting, glacier areas in the second period are already smaller than in the first, thus reducing the negative mass balance somewhat. Given enough time, without any further temperature increase, the glaciers will arrive at a new equilibrium with zero mass balance. The time scale for this to occur for Alpine glaciers is of the order of 10^2 years (Raper et al, 2000: Raper and Braithwaite, 2009).

In the above discussion, I have attributed the whole difference in mass balance for the two periods to the effects of changing temperature. However, precipitation changes may be responsible for a part of the observed mass balance change, although sensitivity studies with the degree-day model (Braithwaite et al. 2002: Braithwaite and Raper, 2007) suggest this can only be a relatively small part. If we wanted to verify this empirically, we would have to use precipitation data at much higher spatial resolution than the temperature data used here as correlation distances for precipitation are much shorter than those for temperature.

6. Conclusion

The increasingly negative mass balance for Alpine glaciers and the recent rise in temperature in and around the Alps should be no surprise to the reader as both have been reported by other workers. It is, however, noteworthy that the temperature-sensitivity of the mass-balance change is in good agreement with the prediction in Fig. 6. This implies that the range of degree-day factors used in the model (6-7 mm d^{-1} K^{-1}) is also valid for the Alps. These parameters are inferred from daily ablation-temperature correlations in the high arctic and in Greenland (Tables 2 to 5) and from secondary data, covering a wider range of geographical conditions (Table 6). In other words, we can explain a recent trend to increasingly negative mass balance in the Alps in terms of a theory based on measurements made at other places, and even a few decades ago.

If we accept the validity of the degree-day model, which the results seem to demonstrate, we can be confident that high rates of melting on Alpine glaciers will continue in the future as long as temperatures continue to rise as predicted by the theory of global warming. This is not a conclusion that pleases me as it means that Alpine glaciers will largely disappear in the coming century. On the other hand, if temperatures do not rise any further, for any reason, the mass balances of Alpine glaciers will tend towards zero as the glaciers tend to a new equilibrium.

In strict logic, the above conclusion only applies to Alpine glaciers, and I should test the predictions for other areas, e.g. as shown in Fig. 6. Northern Scandinavia and southern Norway have relatively good coverage of mass balance records so I might be able to use the same approach as here. Other areas, with more restricted data, may need more flexibility. For example, instead of using a 30-year reference period (1961-1990) it might be possible to use a shorter reference period for which there may be more mass-balance records. Aside from glaciers with mass-balance measurements from stakes and snow pits, we could also apply the present approach to glaciers where longer-term mass changes are available from geodetic methods, including increasingly high precision survey by satellites.

7. Acknowledgment

The School of Environment and Development (SED), University of Manchester supported preparation of the present chapter through an Honorary Senior Research Fellowship and Sarah Raper, Manchester Metropolitan University (MMU), commented on a first draft. Norma and Peter Braithwaite first stimulated my interest in nature and the great outdoors. I have talked about glacier melting with many people for over 40 years. I can especially mention Walther Ambach, Ludwig Braun, David Collins, Mark Dyurgerov, Wilfried Haeberli, Edward Hannah, Regine Hock, Philip Hughes, Kang Ersi, Thomas Konzelmann, Michael Kuhn, Herbert Lang, Martin Laternser, Liu Shiyin, Christoph Marty, Mark Meier, Fritz Müller, Hans Müller, Felix Ng, Hans Oerlemans, Atsumu Ohmura, Simon Ommanney, Gunnar Østrem, Sarah Raper, Niels Reeh, Hans Röthlisberger, Shi Yafeng, Konrad Steffan, Wilfred Theakstone, Peter Worsley, Xie Zichu, Gordon Young, Zhang Yong and Zhang Yu. I also thank former Danish colleagues Jan Andsbjerg, Ole Bendixen, Carl Bøggild, Poul Clement, Martin Ghisler, Tvis N. Knudsen, Ole B. Olesen, Henrik Højmark Thomsen and Anker Weidick.

8. References

Ahlmann, H.W. (1924). Le niveau de glaciation comme function de l'accumulation d'humidité sous forme solide. *Geografiska Annaler* 6, 223-272, ISSN 0435-3676

Ambach, W. (1988) Interpretation of the positive-degree-days factor by heat balance characteristics - West Greenland. *Nordic Hydrology* 19, 4, 217-224, ISSN 0029-1277

Anderson, B., Lawson, W., Owens, I. & Goodsell, B. (2006). Past and future mass balance of 'Ka Roimata o Hine Hukatere' Franz Josef Glacier, New Zealand. *Journal of Glaciology* 52, 179, 597-607, ISSN 0022-1430

Anderson, B., Mackintosh, A., Stumm, D., George, L., Kerr, T., Winter-Billington, A. & Fitzsimons, S. (2010). Climate sensitivity of a high-precipitation glacier in New Zealand. *Journal of Glaciology* 56, 195, 114-128, ISSN 0022-1430

Andrews, R. H. (1964). Meteorology No. 1. Meteorology & heat balance of the ablation area, White Glacier, Canadian Arctic Archipelago - summer 1960. *Axel Heiberg Island Research Reports, McGill University, Montreal, Jacobsen-McGill Arctic Research Expedition 1959-1962.*

Arnold, K. C. & MacKay, D. K. (1964). Different methods of calculating mean daily temperatures, their effects on degree-day totals in the high Arctic and their significance to glaciology. *Geographical Bulletin* 21, 123-129.

Bohm, R., Auer, I., Brunetti, M., Mauger, M., Nanni, T. & Schoner, W. (2001). Regional temperature variability in the European Alps: 1760-1998 from homogenized instrumental time series. *International Journal of Climatology* 21, 14, 1779-1801, ISSN 0899-8418

Braithwaite, R. J. (1977). Air temperature and glacier ablation - a parametric approach. *Ph.D. Thesis*, McGill University, Montreal, Canada.

Braithwaite, R. J. (1980). Glacier energy balance and air temperature: comments on a paper by Dr M. Kuhn. *Journal of Glaciology* 25, 93, 501-503, ISSN 0022-1430

Braithwaite, R. J. (1981). On glacier energy balance, ablation, and air temperature. *Journal of Glaciology* 27, 97, 381-391, ISSN 0022-1430

Braithwaite, R. J. (1984). Calculation of degree-days for glacier-climate research. *Zeitschrift für Gletscherkunde und Glazialgeologie* 20, 1-8, ISSN 0044-2836

Braithwaite, R. J. (1995). Positive degree-day factors for ablation on the Greenland ice sheet studied by energy-balance modelling. *Journal Glaciology* 41, 137, 153-160, ISSN 0022-1430.

Braithwaite, R. J. 2005. Mass-balance characteristics of arctic glaciers. *Annals of Glaciology* 42, 225-229, ISSN 0260-3055.

Braithwaite, R. J. (2008). Temperature and precipitation climate at the equilibrium-line altitude of glaciers expressed by the degree-day factor for melting snow. *Journal of Glaciology* 54, 186, 437-444, ISSN 0022-1430

Braithwaite, R. J. & Olesen, O. B. (1988). Winter accumulation reduces summer ablation on Nordbogletscher, South Greenland. *Zeitschrift für Gletscherkunde und Glazialgeologie* 24, 1, 21-30, ISSN 0044-2836

Braithwaite, R. J. & Olsen, O. B. (1989). Calculation of glacier ablation from air temperature, West Greenland. In: Oerlemans, J. (ed.) *Glacier Fluctuations and Climatic Change*. Kluwer. Dordrecht. 219-233, ISBN 0-7923-0110-2

Braithwaite, R. J. & Olesen, O. B. (1990a). Response of the energy balance on the margin of the Greenland ice sheet to temperature changes. *Journal of Glaciology* 36, 123, 217-221, ISSN 0022-1430

Braithwaite, R.J. & Olesen, O. B. (1990b). Increased ablation at the margin of the Greenland ice sheet under a greenhouse-effect climate. *Annals of Glaciology* 14: 20-22, ISSN 0260-3055

Braithwaite, R. J., Konzelmann, T., Marty, C. & Olesen, O. B. (1998a). Reconnaissance study of glacier energy balance in North Greenland 1993-94. *Journal of Glaciology* 44, 147, 239-247, ISSN 0022-1430

Braithwaite, R. J., Konzelmann, T., Marty, C. & Olesen, O. B. (1998b). Errors in daily ablation measurements in northern Greenland, 1993-94, and their implications for glacier-climate studies. *Journal of Glaciology* 44, 148, 583-588, ISSN 0022-1430

Braithwaite, R. J. & Zhang, Y. (1999a). Modelling changes in glacier mass balance that might occur as a result of climate changes. *Geografiska Annaler* 81A, 4, 489-496, ISSN 0435-3676

Braithwaite, R. J. & Zhang, Y. (1999b). Relationship between interannual variability of glacier mass balance and climate. *Journal of Glaciology* 45, 151, 456-462, ISSN 0022-1430

Braithwaite, R. J. & Zhang, Y. (2000). Sensitivity of mass balance of five Swiss glaciers to temperature changes assessed by tuning a degree-day model. *Journal of Glaciology* 46, 152, 7-14, ISSN 0022-1430

Braithwaite, R. J., Zhang, Y. & Raper, S. C. B. (2002). Temperature sensitivity of the mass balance of mountain glaciers and ice caps as a climatological characteristic. *Zeitschrift für Gletscherkunde und Glazialgeologie* 38, 1, 35-61, ISSN 0044-2836

Braithwaite, R. J. & Raper, S. C. B. (2007). Glaciological conditions in seven contrasting regions estimated with the degree-day model. *Annals of Glaciology* 46, 297-302, ISSN 0260-3055

Braithwaite, R. J. & Raper, S. C. B. (2009). Estimating equilibrium-line altitude (ELA) from glacier inventory data. *Annals of Glaciology* 50, 53, 127-132, ISSN 0260-3055

Bøggild, C. E., Olesen, O. B., Ahlstrøm, A. P. & Jørgensen, P. (2004). Automatic glacier measurements using pressure transducers. *Journal of Glaciology* 50, 169,303-305, ISSN 0022-1430

Carlov, R. & Greve, R. (2005). A semi-analytical solution for the positive degree-day model with stochastic temperature variations. *Journal of Glaciology* 51, 172, 173-175, ISSN 0022-1430

De Quervain, M. (1979). Schneedeckenablation und Gradtag in Versuchsfeld Weissfluhjoch. *Mitteilungen VAW/ETH Zürich* 41, 215-232, ISSN 0374-0056

De Walle, D. R. & Rango, A. (2008). *Principles of snow hydrology*. Cambridge, ISBN 978-0-521-82362-3, Cambridge University Press.

De Woul, M. & Hock, R. (2005). Static mass-balance sensitivity of Arctic glaciers and ice caps using a degree-day approach. *Annals of Glaciology* 42, 217-224, ISSN 0260-3055

Fausto, R. S., Ahlstrøm, A. P., van As, D., Bøggild, C. E. & Johnsen, S. J. (2009). A new present-day temperature parameterization for Greenland. *Journal of Glaciology* 55, 189, 95-105, ISSN 0022-1430

Havens, J. M., Müller, F. & Wilmot, G. C. (1965). Meteorology No. 4. Comparative meteorological survey and short-term heat balance study of the White Glacier, Canadian Arctic Archipelago - summer 1962. *Axel Heiberg Island Research Reports, McGill University, Montreal, Jacobsen-McGill Arctic Research Expedition 1959-1962.*

Hock, R. (1999). A distributed temperature index ice and snow melt model including potential direct solar radiation. *Journal of Glaciology* 45, 149, 101-111, ISSN 0022-1430

Hock, R. (2003). Temperature index melt modelling in mountain areas. *Journal of Hydrology* 282, 1-4, 104-115, ISSN 0022-1694

Hoinkes, H. C. (1955). Measurements of ablation and heat balance on Alpine glaciers. *Journal of Glaciology* 2, 17, 497-501, ISSN 0022-1430

Hulth, J. (2011). Using a draw-wire sensor to continuously monitor glacier melt. *Journal of Glaciology* 57, 201, 922-924, ISSN 0022-1430

Huybrechts, P. Letréguilly, A. & Reeh, N. (1991). The Greenland ice sheet and Greenhouse warming. *Palaeogeography, Palaeoclimatology and Palaeoecology* 89, 4, 399-412, ISSN 0921-8181

IPCC (2011) *Intergovernmental Panel on Climate Change,* Accessed 27 April 2011, <http://www.ipcc.ch>

Jóhannesson, T., Sigursson, O., Laumann, T. & Kennett, M. (1993). Degree-day mass balance modelling with applications to glaciers in Iceland, Norway and Greenland. *Journal of Glaciology* 41, 138, 345-358, ISSN 0022-1430

Keeler, C. M. (1964). Relationship between climate, ablation and run-off on the Sverdrup Glacier, 1963, Devon Island, N.W.T. Montreal, Arctic Institute of North America. *Arctic Institute of North America Research Paper* 27

Konzelmann, T. & Braithwaite, R. J. (1995). Variations of ablation, albedo and energy balance at the margin of the Greenland ice sheet, Kronprins Christian Land, eastern north Greenland. *Journal of Glaciology* 41, 137, 174-182, ISSN 0022-1430

Krenke, A. N. & Khodakov, V. G. (1966). On the correlation between glacier melting and air temperature (in Russian with English summary). *Materialy Glyatsiologicheskikh Issledovanii* 12, 153-164.

Kuhn, M. (1979). On the computation of heat-transfer coefficients from energy-balance gradients on a glacier. *Journal of Glaciology* 22, 87, 263-272, ISSN 0022-1430

Kuhn, M. (1987). Micro-meteorological conditions for snow melt. *Journal of Glaciology* 33, 113, 24-26, ISSN 0022-1430

Laumann, T. & Reeh, N. (1993). Sensitivity to climate change of the mass balance of glaciers in southern Norway. *Journal of Glaciology* 39, 133, 656-665, ISSN 0022-1430

Müller, F. & Keeler, C. M. (1969). Errors in short-term ablation measurements on melting ice surfaces. *Journal of Glaciology* 8, 52, 91-105, ISSN 0022-1430

Müller, F. & Roskin-Sharlin, N. (1967). Meteorology No. 3. A high arctic climate study on Axel Heiberg Island, Canadian Arctic Archipelago - Summer 1961. *Axel Heiberg Island Research Reports, McGill University, Montreal, Jacobsen-McGill Arctic Research Expedition 1959-1962.*

NASA/GISS (2011). *GISS Surface Temperature Analysis (GISTEMP), Accessed 27 April,* <http://data.giss.nasa.gov/gistemp/>

New, M., Hulme, M. & Jones, P. (1999). Representing twentieth century space-time climatic variability. I. Development of a 1961-1990 mean monthly climatology. *Journal of Climate* 12, 829-856, ISSN 0894-8755

Oerlemans, J., Anderson, B., Hubbard, A., Huybrechts, P., Jóhannesson, T., Knap, W. H., Schmeits, M., Stroeven, A. P., van de Wal, R.S.W., Wallinga, J. & Zuo, Z. (1998). Modelling the response of glaciers to climate warming. *Climate Dynamics* 14, 267-274, ISSN 0930-7575

Ohmura, A. (1981). Climate and energy balance on Arctic tundra, Axel Heiberg Island, Canadian Arctic Archipelago, spring and summer 1969, 1970 and 1972. *Zürcher Geographische Schriften* 3. 445 pp.

Ohmura, A. (2001). Physical basis for the temperature-based melt-index method. *Journal of Applied Meteorology* 40, 4, 753-761, ISSN 0894-8763.

Ohmura, A., Kasser, P. & Funk, M. (1992). Climate at the equilibrium line of glaciers. *Journal of Glaciology* 38, 130, 397-411, ISSN 0022-1430

Orheim, O. (1970). Glaciological investigations of Store Supphellebre, west-Norway. *Norsk Polarinstitutt Skrifter* 151.

Radic, V. (2008). Modeling future sea level rise from melting glaciers. *Ph.D. Thesis,* University of Alaska, Fairbanks, USA.

Radic, V., Hock, R. & Oerlemans, J. (2007). Volume-area scaling vs flowline modelling in glacier volume projections. *Annals of Glaciology* 46, 234-240, ISSN 0260-3055

Radic, V. & Hock, R. (2011). Regionally differentiated contribution of mountain glaciers and ice caps to future sea-level rise. *Nature Geoscience* 4, 2, 91-94, ISSN 1752-0894

Raper, S. C. B. & Braithwaite, R. J. (2006). Low sea level rise projections from mountain glaciers and icecaps under global warming. *Nature* 439, 311-313, ISSN 0028-0836

Raper, S. C. B. & Braithwaite, R. J. (2009). Glacier volume response time and its links to climate and topography based on a conceptual model of glacier hypsometry. *The Cryosphere* 3, 183-194, ISSN 1994-0416

Raper, S. C. B., Brown, O. & Braithwaite, R. J. (2000). A geometric glacier model for sea-level change calculations. *Journal of Glaciology* 46, 154, 357-368, ISSN 0022-1430

Reeh, N. (1991). Parameterization of melt rate and surface temperature on the Greenland ice sheet. *Polarforschung* 59, 3, 113-128.

Reynaud, L. (1980). Can the linear balance model be extended to the whole Alps? *International Association of Hydrological Sciences Publication* 126 (Riederalp Workshop 1978 - World Glacier Inventory), 273-247.

Schar, C., Vidale, P. L., Luthi, D., Frei, C., Haeberli, C., Liniger, M. A. & Appenzeller, C. (2004). The role of increasing temperature variability in European summer heatwaves. *Nature* 427, 332-336, ISSN 0028-0836

Schytt, V. (1964). Scientific results of the Swedish Glaciological Expedition to Nordaustlandet, Spitsbergen, 1957 and 1958. *Geografiska Annaler* 46, 3, 243-281, ISSN 0435-3676

Shea, J. M., Moore, D. R. & Stahl, K. (2009). Derivation of melt factors from glacier-mass balance records in western Canada. *Journal of Glaciology* 55, 189, 147-162, ISSN 0022-1430

Tarasov, L. & Peltier, W. R. (1997). Terminating the 100 kyrs ice age cycle. *Journal of Geophysical Research* 102, D18, 21,665-21,693, IDSN XY462

van der Veen, C. J. (1999). *Fundamental of glacier dynamics*. Balkema, ISBN 90 5410 471 6, Rotterdam.

Vincent, C. & Vallon, M. (1997). Meteorological controls on glacier mass balance: empirical relations suggested by measurements on Glacier de Sarennes, France. *Journal of Glaciology* 43, 143, 131-137, ISSN 0022-1430

WGMS (2011). World Glacier Monitoring Service (WGMS), Accessed 27 April 2011, <http://www.wgms.ch>

Zhang, Y, Liu, S., Xie, C.W. & Ding, Y. (2006). Observed degree-day factors and their spatial variations on glaciers in western China. *Annals of Glaciology* 43, 301-306, ISSN 0260-3055.

25

Estimation of the Sea Level Rise by 2100 Resulting from Changes in the Surface Mass Balance of the Greenland Ice Sheet

Xavier Fettweis, Alexandre Belleflamme, Michel Erpicum, Bruno Franco and Samuel Nicolay
University of Liège
Belgium

1. Introduction

The Surface Mass Balance (SMB) can be seen, in first approximation, as the water mass gained by the winter snowfall accumulation minus the mass lost by the meltwater run-off in summer. The mass gain from rainfall as well as the mass loss from erosion from the net water fluxes (the sum of the evaporation, sublimation, deposition and condensation) and from the wind (blowing snow) are negligible in the SMB equation of the Greenland Ice Sheet (GrIS) compared to the snowfall and the melt (Box et al., 2004). The ice sheet mass balance takes also into account the mass loss from iceberg calving. Consequences of a warmer climate on the Greenland ice sheet SMB will be a thickening inland, due to increased solid precipitation, and a thinning at the Greenland ice sheet periphery, due to an increasing surface melt. A climatic warming increases the snow and ice melting in summer but it enhances also evaporation above the ocean. This leads to higher moisture transport inland and, consequently, higher precipitation. The response of the iceberg calving to the climate change could be an acceleration of the glacier flow (Nick et al., 2009; Zwally et al., 2002) but these projections are very uncertain (Sundal et al., 2011) and a lot of developments are still needed in the glaciology models for improving our knowledge and modelling of the Greenland ice sheet dynamics. That is why we will focus our study only on the SMB of the Greenland ice sheet.

The IPCC (Intergovernmental Panel on Climate Change) projects, in response to global warming induced by human activities, that the run-off increase will exceed the precipitation increase and therefore that the currently observed surface melting of the Greenland ice sheet (Fettweis et al., 2011b; Tedesco et al., 2011; Van den Broeke et al., 2009) will continue and intensify during the next decades (IPCC, 2007). An increasing freshwater flux from the Greenland ice sheet melting could perturb the thermohaline circulation (by reducing the density contrast driving this last one) in the North Atlantic including the drift which tempers the European climate. In addition, an enduring Greenland ice sheet melting, combined with the thermal expansion of the oceans and the melt of continental glaciers, will raise the sea level with well-known consequences for countries such as the Netherlands, Bangladesh,... The contribution of the Greenland ice sheet SMB decrease to the sea level rise is currently evaluated to be 5-10 cm by 2100 (Gregory and Huybrechts, 2006; Fettweis et al.,

2008; Franco et al., 2011; IPCC, 2007; Mernild et al., 2010). However, despite their importance for the global climate, large uncertainties remain in these estimations because most of the current studies are based mainly on outputs from atmosphere–ocean general circulation models (called hereafter global models) not well suited to the polar regions and with a typical horizontal resolution of 300 km while the width of the ablation zone is lower than 100km.

In response to these uncertainties, the regional modelling is the ideal tool to provide some more precise answers to understand and quantify how the Greenland ice sheet SMB will respond to climate change. In particular, the regional climate model MAR (for *Modèle Atmosphérique Régional*) fully coupled with a snow model and extensively validated to simulate the SMB of the Greenland ice sheet (Fettweis et al., 2005; Fettweis, 2007; Fettweis et al., 2011b; Lefebre et al., 2003; 2005), is designated to study the Greenland climate at high resolution (25 km), with a physic developed for and well adapted to polar regions. Until now, the only future projections at high-resolution of the Greenland ice sheet SMB have been carried out with models not taking into account the atmosphere-snow feedbacks occurring above the melt area as the well known surface albedo positive feedback (Mernild et al., 2008; 2010). Therefore, we propose here to carry out future projections of Greenland climate with the MAR model coupled with a snow model and forced by two scenarios of greenhouse gas emissions made for the next IPCC assessment report (AR5). This work fits in the ICE2SEA project (http://www.ice2sea.eu) of the 7th Framework Program (FP7) which aims to improve the projections of the continental ice melting contribution to sea level rise.

After a brief description in Sect. 2 of the regional model we used, Sect. 3 evaluates, by comparison with the reanalysis and other IPCC AR5 global models (see Table 1), the global model CanESM2 which is used for forcing the MAR future projections. Sect. 4 validates the CanESM2-forced MAR simulation over the current climate (1970–1999) in respect to the ERA-40 forced MAR simulation. Future SMB projections of CanESM2-forced MAR simulations are analysed in Sect. 5. Finally, future projections of sea level rise coming from the Greenland ice sheet SMB decrease are given in Sect. 6.

2. Data

The model used here is the regional climate model MAR coupled to the 1-D Surface Vegetation Atmosphere Transfer scheme SISVAT (Soil Ice Snow Vegetation Atmosphere Transfer) (Gallée and Schayes, 1994). The snow-ice part of SISVAT, based on the CEN (*Centre d'Etudes de la Neige*) snow model called CROCUS (Brun et al., 1992), is a one-dimensional multi-layered energy balance model that determines the exchanges between the sea ice, the ice sheet surface, the snow-covered tundra, and the atmosphere. It consists of a thermodynamic module, a water balance module taking into account the meltwater refreezing, a turbulence module, a snow metamorphism module, a snow/ice discretization module, and an integrated surface albedo module (Gallée et al., 2001). The blowing snow model, currently under development, is not yet used here. SISVAT does not contain an ice dynamics module. Therefore, a fixed ice sheet mask (i.e. a fixed land surface classification of the pixels representing the Greenland ice sheet) and topography are assumed for simulating both current and future climates. This explains why we only present here the SMB and not the ice sheet mass balance of the Greenland ice sheet.

The MAR version used here is the one from Fettweis et al. (2011b) calibrated to best compare with the satellite derived melt extent over 1979-2009. Compared to the set-up of Fettweis et al. (2005; 2011b), a new tundra/ice mask is used here based on the Greenland land surface classification mask from Jason Box (http://bprc.osu.edu/wiki/Jason_Box_Datasets) and the filtering of the Bamber et al. (2001) based topography is reduced by a factor two. The integration domain (shown in green in Fig. 1) as well as the spatial resolution (25 km) are the same than in Fettweis et al. (2005).

For studying the current climate, the ERA-40 reanalysis (1958–1999) and after that the ERA-INTERIM reanalysis (2000–2010) from the European Centre for Medium Range Weather Forecasts (ECMWF) are used to initialize the meteorological fields at the beginning of the MAR simulation in September 1957 and to force every 6 hours the MAR lateral boundaries with temperature, specific humidity and wind components during the simulation. The Sea Surface Temperature and the sea-ice cover are also prescribed by the reanalysis. No corrections are applied to the MAR outputs.

Model ID	Institutes, Country
BCC-CSM1-1	Beijing Climate Center, China
CanESM2	Canadian Centre for Climate Modelling and Analysis, Canada
CNRM-CM5	Centre National de Recherches Météorologiques et Centre Europeen de Recherche et Formation Avancées en Calcul Scientifique, France
GISS-E2	NASA Goddard Institute for Space Studies, USA
HadGEM2-ES	Met Office Hadley Centre, UK
INMCM4	Institute for Numerical Mathematics, Russia
IPSL-CM5A-LR	Institut Pierre Simon Laplace, France

Table 1. List of the seven IPCC AR5 global models available in June 2011 and used in this paper. These outputs are provided by the World Climate Research Programme's (WCRP's) Coupled Model Intercomparison Project phase 5 (CMIP5) multi-model dataset (http://cmip-pcmdi.llnl.gov/cmip5).

For computing future projections, we use the 6 hourly outputs of CanESM2 (see Table 2). The daily sea surface temperature and sea ice cover from CanESM2 are used to force the ocean surface conditions in SISVAT. The CanESM2-forced MAR simulation starts in September 1964 and lasts till December 2005 with the outputs from the Historical experiment (representing the current climate). The MAR future projections (2006-2100) use the CanESM2 outputs from two new scenarios (called RCP for Representative Concentration Pathways) of greenhouse gases concentration that will be used in the next IPCC report:

- **RCP 4.5**: mid-range scenario corresponding to a radiative forcing of +4.5 w/m^2 at stabilization in 2100. This scenario corresponds to an increase of the atmospheric greenhouse gas concentration during the 21st century to a level of 850 CO_2 equivalent p.p.m. by 2100.
- **RCP 8.5**: pessimistic scenario corresponding to a radiative forcing of > +8.5 w/m^2 in 2100. This scenario corresponds to an increase of the atmospheric greenhouse gas concentration during the 21st century to a level of > 1370 CO_2 equivalent p.p.m. by 2100.

We refer to Moss et al. (2010) for more details about the RCP scenarios.

3. Comparison of CanESM2 with other IPCC AR5 models

The aim of this section is to gauge the ability of the IPCC AR5 global models to reproduce the present-day climate conditions over Greenland. A good representation of the current climate is a necessary, but not sufficient condition for the global model ability to simulate

MAR Forcing	Periode
ERA-40 reanalysis	1957-1999
ERA-INTERIM reanalysis	2000-2010
Historical experiment from CanESM2	1964-2005
RCP 4.5 experiment from CanESM2	2006-2100
RCP 8.5 experiment from CanESM2	2064-2100

Table 2. Summary of the different forcings used in the MAR simulations.

future climate change. Indeed, a model that fails to reproduce the current climate generates projections that lack in reliability and validity.
We will show that CanESM2 used for forcing MAR is one of the most suitable global models currently available from the CMIP5 data base for simulating the current climate (1970–1999) over Greenland in respect to the ERA-40 reanalysis. In addition, its future projections of temperatures are in the range of the other IPCC AR5 global models.

3.1 Evaluation over current climate

The global models listed in Table 1 from the CMIP5 data base are evaluated with the aim to use them as forcing of a regional model. The general atmospheric circulation in the regional model is fully induced by the global model-based boundaries forcing while the surface conditions (except the sea surface temperature and sea ice cover which are used as forcing) simulated by the global model do not impact the results of the regional model. This means that the regional model is not able to correct biases in the general circulation coming from the global models. In addition, Fettweis et al. (2011a) showed that the general circulation at 500hPa (gauged by the geopotentiel height) correlates significantly with the surface melt anomalies. Therefore, only the average state of the free atmosphere (here at 500hPa) is evaluated here.
As validation, the global model outputs at 500hPa coming from the Historical experiment (representing the current climate) are compared to the ERA-40 reanalysis over 1970–1999 (considered here as the reference period). Franco et al. (2011) showed that at 500hPa, both ERA-40 and NCEP-NCAR (National Centers for Environmental Prediction–National Center for Atmospheric Research) reanalyses compare very well and so a comparison with only the ERA-40 reanalysis is enough here. The fields which are evaluated are:

- **the mean temperature over the melt season** (from May to September and called summer hereafter). A global model-based temperature bias is propagated through the regional model boundaries inducing a similar temperature bias in the regional model which impacts the amount of simulated surface melt. Only CanESM2 satisfactorily simulates the mean summer temperature although it is too cold in the South and too warm in the North (see Fig. 1). HadGEM2-ES and INMCM4 are 2°C too warm at the north of the GrIS and the other global models are rather several degrees too cold in summer. The observed doubling of the amount of surface melt since 30 years corresponds to a warming of 2-3 °C (Fettweis, 2007; Fettweis et al., 2011a;b). Thus, it suggests that a bias in summer of several degrees at the regional model boundaries is not acceptable.
- **the mean temperature over the accumulation season** (from October to April and called winter hereafter). The global model-based temperature at the regional model boundaries impacts the maximum moisture content of the air masses and so it impacts the amount of precipitation simulated by the regional model. Except INMCM4, the global models are too cold in winter. The highest biases are observed for most of the models in the South-East of Greenland.

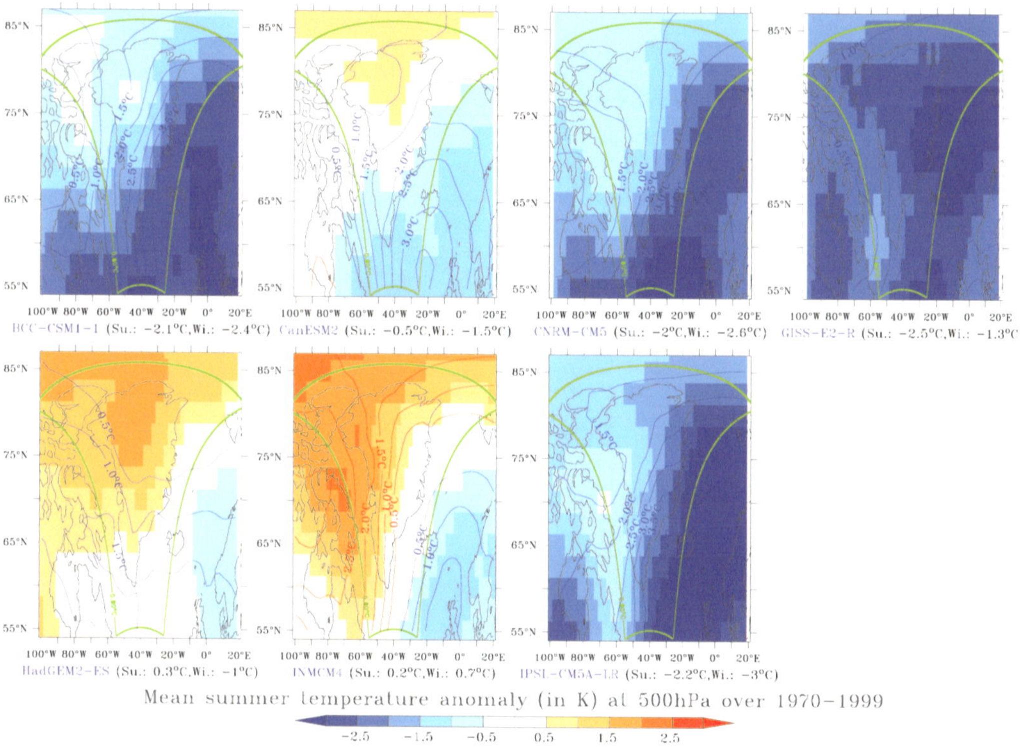

Fig. 1. Difference between the summer (May-Sep) mean 500 hPa temperature simulated by the global models for the Historical experiment and from the ERA-40 reanalysis over 1970–1999 (reference period). The solid lines in red (positive bias) and in blue (negative bias) plot the winter (Oct-Apr) mean 500 hPa temperature difference with regard to the ERA-40 reanalysis over 1970–1999. Units are degrees. The mean bias for winter (Wi.) and summer (Su.) is given in brackets. Biases above 1°C in Summer and 2°C in winter are statistically significant i.e. two times higher than the ERA-40 temperature standard deviation. Finally, the boundaries of the MAR integration domain are shown in green.

- **the annual mean wind speed**. Most of humidity comes into the regional model domain from the southern boundary situated in the storm tracks associated to the Icelandic Low. The wind speed at the regional model boundaries impacts the moisture advection into the integration domain and so the precipitation simulated by the regional model. The wind speed simulated by HadGEM2-ES compares the best with the reanalysis (see Fig. 2). CanESM2 overestimates the wind speed in the south of the domain while the other global models rather underestimate it.
- **the annual geopotential height**. This last one reflects the main general circulation pattern i.e. an eastward general circulation from the North American continent, adopting a north-west direction over Baffin Bay before reaching the western coast of Greenland, and generating a north-eastward circulation over central Greenland. In southern Greenland, the regional circulation is more influenced by meridional fluxes. Biases at the regional model boundaries in the direction of the main flows impact the precipitation pattern. Only GISS-E2-R does not clearly show an atmospheric circulation pattern which is consistent with the reanalysis. In addition, along the East Greenland coast, BCC-CSM1-1,

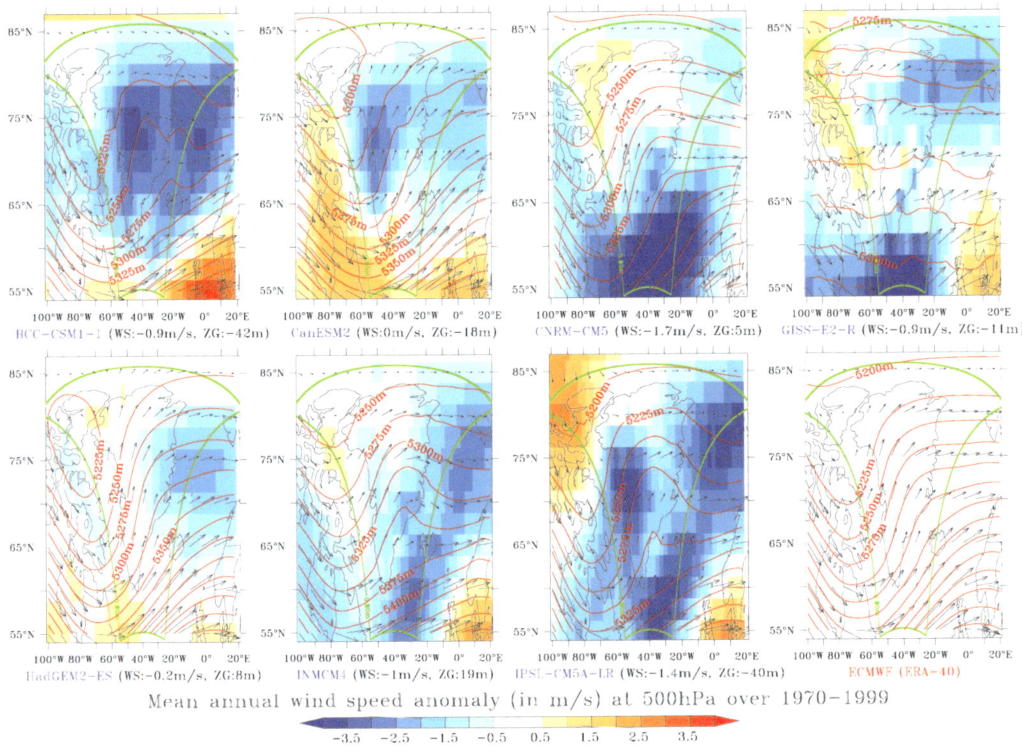

Fig. 2. Difference between the annual mean 500 hPa wind speed (in m/s) simulated by the global models for the Historical experiment and from the ERA-40 reanalysis over 1970–1999. The solid lines in red and arrows in black plot respectively the yearly mean 500 hPa geopotential height in meters and the wind vectors. Finally, as last plot, the mean 500 hPa geopotential height from the ERA-40 reanalysis is given as reference as well as the wind vectors. The mean bias of wind speed (WS) in m/s and geopotential height (ZG) in m are given in brackets.

INMCM4 and IPSL-CM5A-LR simulate a north-west component of the general circulation in disagreement with the reanalysis.

HadGEM2-ES (resp. CanESM2) is the top-performing model in simulating the present-day wind speed and circulation (resp. the summer temperature) at 500hPa. INMCM4 is too warm and the other global models are too cold and underestimate the wind speed. Knowing the importance of the global model summer temperature biases in the amount of the melt simulated by the regional model, CanESM2 has been selected as the most suitable candidate for forcing the MAR model. However, simulations of MAR forced by HadGEM2-ES should be performed in future because the general circulation and the winter temperatures simulated by HadGEM2-ES are better than the one simulated by CanESM2. However, HadGEM2-ES is too warm in summer. The main biases of CanESM2 are a wind speed overestimation (resp. underestimation) in the south (resp. north-west) of Greenland and a cold bias in winter.

3.2 Future projections

In Fig. 3, we see first that the run of CanESM2 used for forcing MAR captures well the temperature variability coming for the NCEP-NCAR reanalysis with a sharp increase of

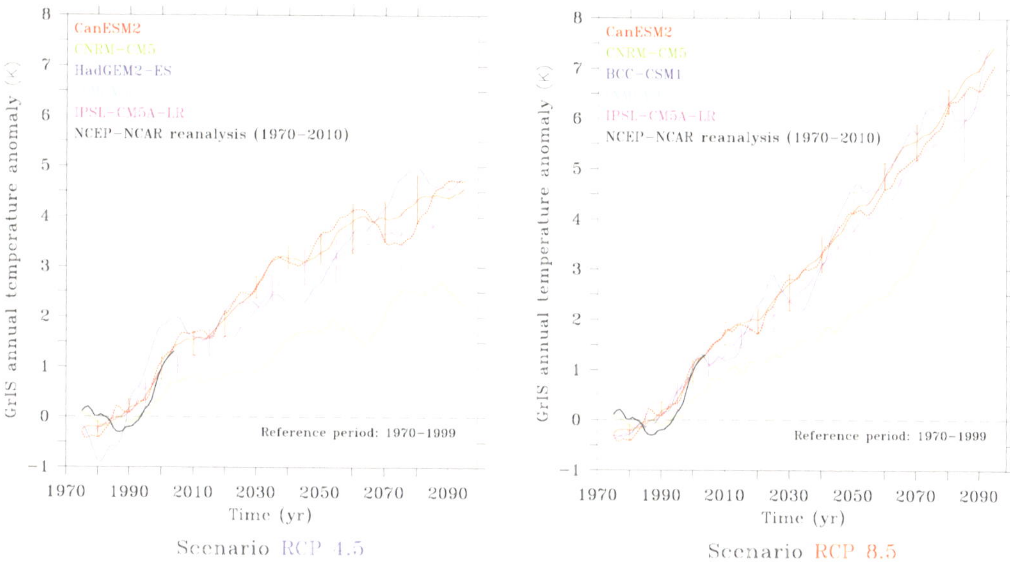

Fig. 3. Time series of the current and future projections of the Greenland ice sheet annual temperature anomalies (in respect to the 1970–1999 period) simulated by the global models for the RCP 4.5 (left) and RCP 8.5 (right) scenario and by the NCEP-NCAR reanalysis (representing the observed temperature variability of the last 40 years). A 10-year running mean is applied here and the Greenland ice sheet temperature variability is taken over an area covering Greenland: 60° W≤longitude≤20° W and 60° N≤latitude≤88° N. When several runs are available for a model (CanESM2 and IPSL-CM5A-LR), the average over all runs is shown and the error bar gives the range around this average. Finally, the run (called *r1i1p1*) of CanESM2 used for forcing MAR is shown in dotted red.

the temperature in the end of 1990's in respect to the 1970–1999 period. HadGEM2-ES simulates this temperature increase too early in the 1990's while INMCM4 and CNRM-CM5 underestimate it. About the future projections, we see that the CanESM2 future projections compare well with those ones from BCC-CSM1-1, HadGEM2-ES and IPSL-CM5A-LR while INMCM4 and CNRM-CM5 seem to be less sensitive over Greenland to the greenhouse gas concentration increase. At the end of the century, most of the global models project an annual temperature increase in the Greenland area of 4-5°C (resp. 7-8°C) for the RCP 4.5 (resp. RCP 8.5) scenario.

4. SMB evaluation of the CanESM2 forced MAR over current climate

As for the global models, it is important to evaluate, over the present-day climate, the regional model forced by the most suitable global model selected in the previous section. Before discussing future projections, we have first to check if MAR forced by the Historical experiment from CanESM2 is able to simulate the current surface conditions over the Greenland ice sheet. That is why, the CanESM2 forced MAR simulation is compared here over 1970–1999 with the ERA-40 forced MAR simulation considered as the reference simulation for the currently observed SMB over the Greenland ice sheet (Fettweis et al., 2011b).

4.1 Average annual rates of the SMB components

At the scale of the whole Greenland ice sheet, MAR forced by CanESM2 over 1970–1999 underestimates a few the SMB in respect to the ERA-40 forced run due to an underestimation of the winter accumulation (Oct-Apr) and an overestimation of the run-off (See Table 3). However, these biases are not significant and are included in the natural variability. Finally, the solid and liquid precipitation in Summer (May-Sep) as well as the net water fluxes compare well.

MAR forced by	SMB	Winter (WSF) Snowfall	Summer (SSF) Snowfall	Rainfall (RF)	Run-off (RU)	Meltwater (ME)	Net water fluxes (SU)
ERA-40	460±106	387±46	288±34	27±5	237±61	419±84	6±1
CanESM2	392±102	335±31	2891±42	36±8	263±78	421±107	5±2

Table 3. Average and standard deviation of the annual surface mass balance components simulated by MAR forced by ERA-40 and by CanESM2 over 1970–1999. Units are Gt/yr. The surface mass balance (SMB) equation is here SMB = WSF + SSF + RF - RU - SU. The run-off is the part of not refrozen water from both surface melt and rainfall reaching the ocean. Finally, the blowing snow sublimation is not taken into account in our simulation.

4.2 Temporal variability and trend of the SMB components

In average over 1970–1999, the CanESM2 forced MAR is significantly too cold in winter in respect to the ERA-40 forced run (Fig. 4) while the annual cycle of the mean Greenland ice sheet temperature is well reproduced. In summer, MAR simulations using both forcings compare well. This is in full agreement with the CanESM2 temperature biases versus ERA-40 shown in Fig 1.

The mean seasonal cycle of the melt area simulated by MAR forced by ERA-40 and CanESM2 compare very well knowing that the ERA-40 forced MAR simulation compare well with the microwave satellite derived one according to Fettweis et al. (2011b). However, although this is not significant, there is a delay of about 5 days between the CanESM2 and the ERA-40 forced simulations. The annual cycle of the bare ice area (i.e. the ablation zone where most of the melt takes place due to the lower albedo of bare ice compared to melting snow (Tedesco et al., 2011)) is also shown in Fig. 4. There is a significant overestimation through the whole melting season of the bare ice area if MAR is forced by CanESM2. This bias, due to an underestimation of the winter accumulation listed in Table 3, is explained in the next section.

Despite of the SMB components biases quoted in Section 4.1, the CanESM2 forced run reproduces very well the ERA-40 forced SMB components trends i.e. a decreasing of the SMB since the beginning of the 2000's associated to a quasi constant snowfall variability over 1970-2005 and an increasing trend of the run-off in agreement with the summer temperature changes (see Fig. 5).

4.3 Spatial variability

The CanESM2 forced SMB simulated by MAR is significantly too low in the north-west of the Greenland ice sheet and too high in the south and along the north-eastern coast (see Fig. 6). The negative anomalies in the north-west are a conjunction of negative anomalies in snowfall and positive anomalies in run-off induced by biases in summer temperature and winter accumulation. In addition to impact directly the gained mass, a too low winter accumulation decreases the snowpack height above bare ice in the ablation zone which induces premature

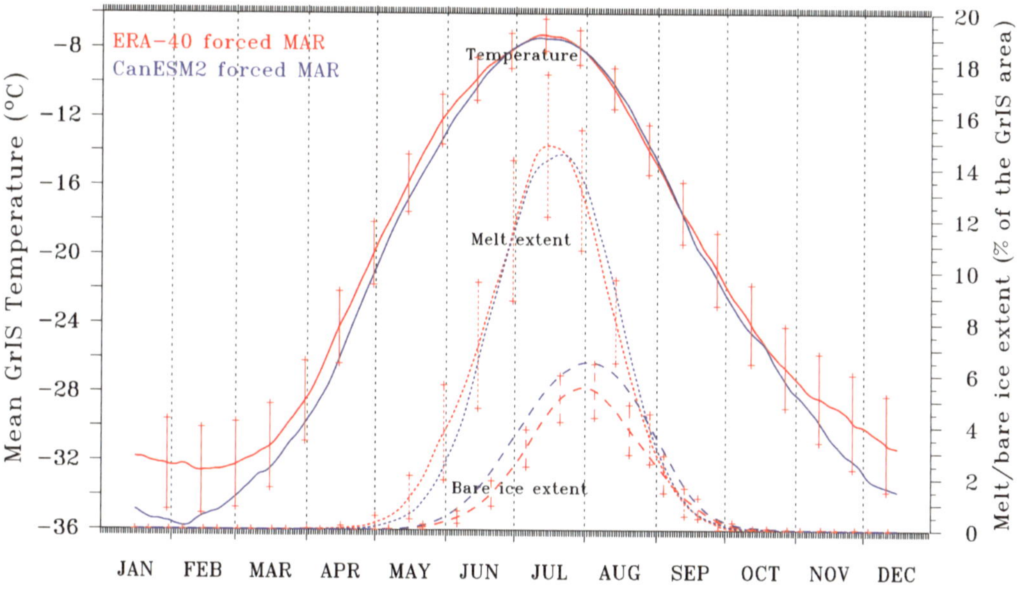

Fig. 4. Average over 1970–1999 of the mean Greenland ice sheet temperature annual cycle (left Y-axis) as well as the melt and bare ice extent simulated by MAR forced by ERA-40 (in red) and by CanESM2 (in blue) (right Y-axis). The melt extent is defined as the area where the daily meltwater production is higher than 8 mmWE according to Fettweis et al. (2011b) and the bare ice extent is the area where the surface snow density is higher than 900 kg/m^3. Finally, a 30 days running mean is applied here and the error bars plot the standard deviation over 1970–1999 around the ERA-40 forced mean.

bare ice exposure in summer, reducing the surface albedo and then enhancing the surface melt. This snowfall negative anomaly is due to an underestimation by CanESM2 of the south-westerly flow, common in this area (see Fig. 2), impacting the amount of moisture which is advected by MAR on this area. In the south of the ice sheet, in contrary, the westerly flow is overestimated by CanESM2 enhancing the precipitation in MAR. It should be remembered that biases at 500hPa in the large scale forcing are not corrected by the regional model even in the interior of its integration domain.

In addition to the winter accumulation induced biases, the run-off anomalies reflect the summer temperature biases i.e. MAR forced by CanESM2 is too cold in summer along the eastern and southern coasts and too warm along the western and northern coasts. Similar patterns of temperature biases are present in CanESM2 itself as shown in Fig. 2. In addition, the warm bias in the north-west of the ablation zone is enhanced by the surface albedo feedback as a consequence of the overestimation of the bare ice exposure. However, it should be noted that these temperature biases are statistically not significant.

As conclusion, the negative SMB anomaly in the north-west of the ice sheet is compensated by the overestimation of the SMB in the south of the ice sheet if MAR is forced by CanESM2 in respect to the ERA-40 forced simulation. This explains why the SMB simulated by MAR forced by CanESM2 and by ERA-40 compare very well at the scale of the whole Greenland ice sheet.

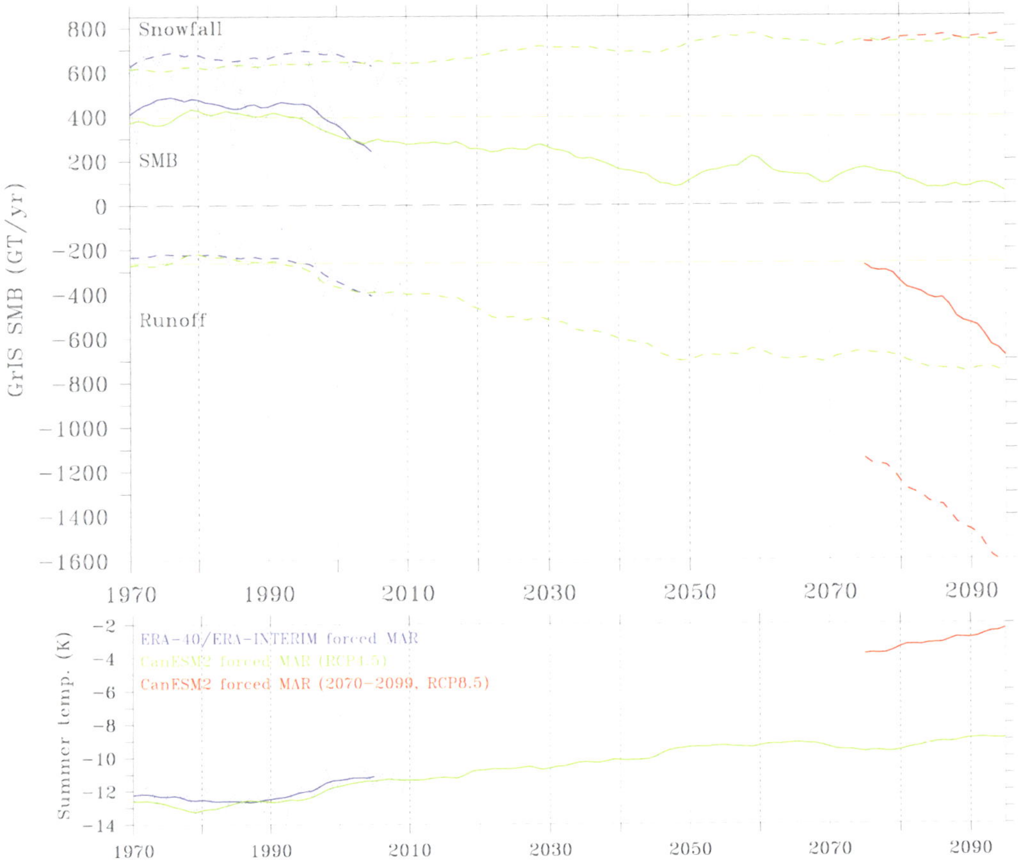

Fig. 5. Time series of the annual Greenland ice sheet snowfall , run-off SMB (top in GT/yr) and summer temperature (below in °C) simulated by MAR forced by ERA-40 over 1970–1999 (in blue), by ERA-INTERIM over 2000-2010 (in blue) and by the CanESM2 based Historical experiment over 1970-2005 (in green), by the RCP4.5 experiment over 2006-2099 (in green) and by the RCP8.5 experiment over 2070-2099 (in red). A 10 year running mean is applied here. Finally, the annual values of the ERA-40/ERA-INTERIM forced simulation (in dotted red) as well as the 1970–1999 average of the CanESM2 forced simulation (in dashed blue) are also shown

5. SMB future projection from CanESM2 forced MAR

5.1 Trends of the SMB components

In respect to the 1970–1999 period simulated by MAR forced by CanESM2, MAR projects for 2070–2099 a decreasing of the SMB of $\sim +75\%$ (resp. $\sim +215\%$) if it is forced by the CanESM2 RCP 4.5 (resp. RCP 8.5) experiment (see Table 4 and Fig. 5). For both scenarios, MAR projects a precipitation increase ($\sim +25\%$ for RCP 4.5 and $\sim +40\%$ for RCP 8.5) that is not large enough to compensate the run-off increase ($\sim +175\%$ for RCP 4.5 and $\sim +420\%$ for RCP 8.5) according to IPCC (2007). The snowfall increase should occur mainly in winter. In summer, the precipitation increased concerns only the rainfall because a part of the solid precipitation should fall in liquid phase due to the rising temperature. This one enhances the melt, knowing that rainfall moistens the snowpack which decreases the surface albedo.

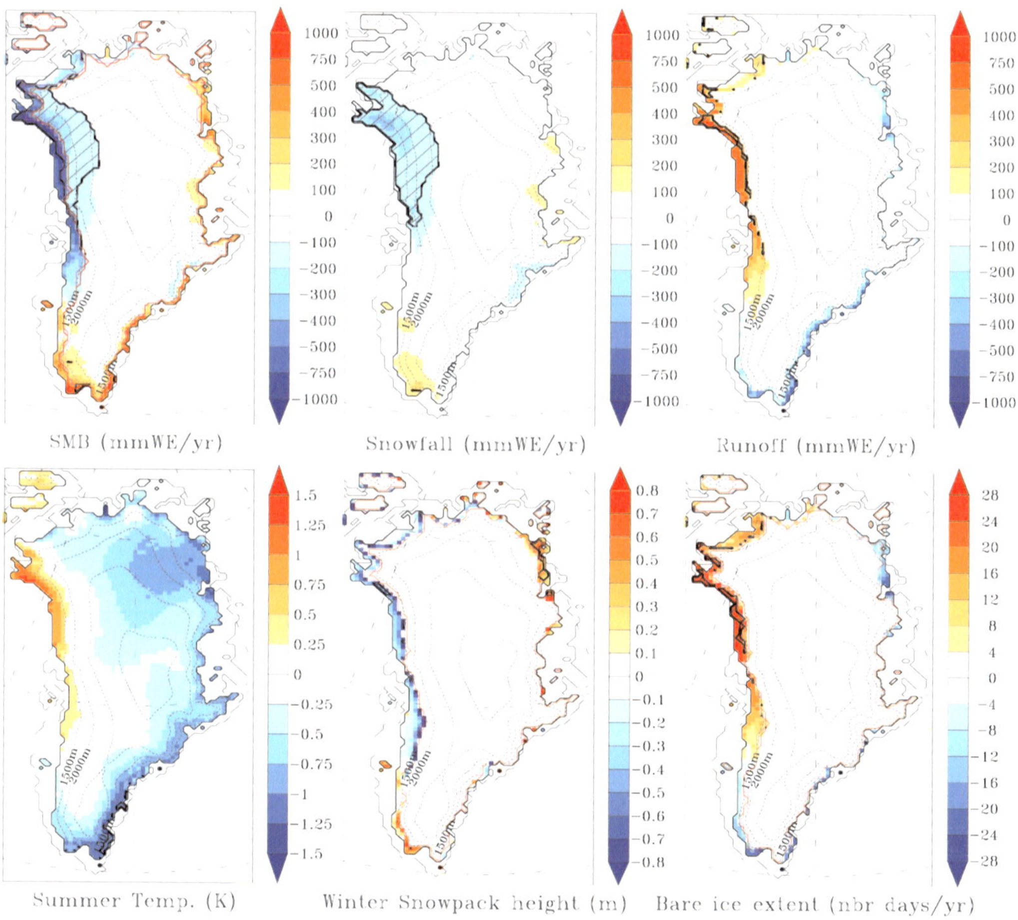

Fig. 6. Difference between the mean 1970–1999 SMB (in mm Water Equivalent/yr), snowfall (in mmWE/yr), run-off (in mmWE/yr), summer near surface temperature (in °C), winter snowpack height (in meter) above bare ice the first of May (.i.e. at the beginning of the summer) and bare ice extent (in number of days where bare ice appears at the surface per year) simulated by MAR forced by CanESM2 and forced by ERA-40. Significant anomalies (those at least twice the magnitude of the 1970–1999 standard deviation) are hatched and bounded by bold black lines. Finally, the mean equilibrium line altitude (i.e. the altitude where the SMB is null) from MAR forced by ERA-40 and forced by CanESM2 is shown respectively in mauve and in red. This altitude line delimits the ablation zone (where the SMB is negative in average) to the accumulation zone (where the SMB is positive).

Scenario	SMB	WSF	SSF	RF	RU	ME	SU	Temperature
RCP 4.5	−294±128	+86±35	+ 20±36	+ 53±10	+458±97	+558±123	−5±2	+3.9±0.8°C
RCP 8.5	−839±245	+277±58	−148±26	+120±35	+1101±271	+1285±302	−14±5	+9.1±0.9°C

Table 4. Anomaly of the annual mean 2070–2099 SMB components simulated by MAR forced by the CanESM2 based RCP 4.5 experiment and RCP 8.5 experiment in respect to the Historical experiment over 1970–1999 (in Gt/yr). The acronyms are described in Table 3. Finally, the annual mean temperature anomalies are also given in °C

The ERA-40/ERA-INTERIM forced time series (Fig 5) shows notably the record low SMB rate simulated by MAR in 2010 and confirmed by the observations (Mernild et al., 2011; Tedesco et al., 2011). The very low SMB rates simulated these last years (2007, 2010) with ERA-INTERIM as forcing should be common at the end of this century with the RCP 4.5 experiment. These observed rates were a combination of lower snowfall and higher run-off than normal. At the end of this century, the run-off and temperature should be higher than those observed the last year but the heavier snowfall in winter should compensate a part of the increase of melt in summer. For RCP 8.5 experiment, MAR simulates never observed negative SMB rates since 50 years (Fettweis et al., 2011a). The snowfall changes are similar to the ones projected for RCP 4.5 but the run-off increase is twice larger than the one projected for RCP 4.5. Finally, the SMB decrease is quasi interrupted in 2050 for RCP 4.5 while the SMB is still decreasing in 2100 with RCP 8.5.
In average, over 2070–2099, the main melt season should still occur between May and September for both scenarios but the melt extent should reach Mid-July 33% of the Greenland ice sheet area for RCP 4.5 (resp. 50% for RCP 8.5) compared to 15% for the current climate (Fig. 4), while the bare ice extent should reach respectively 16% and 22% of the ice sheet area compared to 6% for the present-day climate.

5.2 Spatial changes

In respect to the current climate simulated by MAR forced by CanESM2, MAR projects that for 2070–2099 (RCP 4.5 scenario), SMB should significantly decrease along the ice sheet margin in response to a significant run-off increase resulting from rising summer temperature and that it should slightly increase in the interior of the ice sheet due to heavier snowfall (Fig. 7). The warming is enhanced in the north of the ice sheet because of the decrease of sea ice concentration in the closest parts of the Arctic Ocean according to Franco et al. (2011). In the south of the ice sheet, while the total precipitation is increasing, a part of snowfall becomes rainfall, which enhances the melt. Finally, the bare ice should be exposed in the ablation zone (delimited by the equilibrium line altitude) one month longer by comparison with the current climate. The patterns of changes over 2070–2099 from the RCP 8.5 scenario (not shown here) are similar than those ones from the RCP 4.5 scenario but in more extreme.

6. Future sea level rise projection from SMB changes

According to Fettweis et al. (2008), projected SMB anomalies can be estimated from mean June-July-August (JJA) temperature (T) and annual snowfall (SF) anomalies computed by CanESM2 on the Greenland area by this multiple regression:

$$\Delta SMB \simeq a\frac{T_{RCPxx} - \overline{T_{\text{Historical}}}}{stdev(T_{\text{Historical}})} + b\frac{SF_{RCPxx} - \overline{SF_{\text{Historical}}}}{stdev(SF_{\text{Historical}})} \quad (1)$$

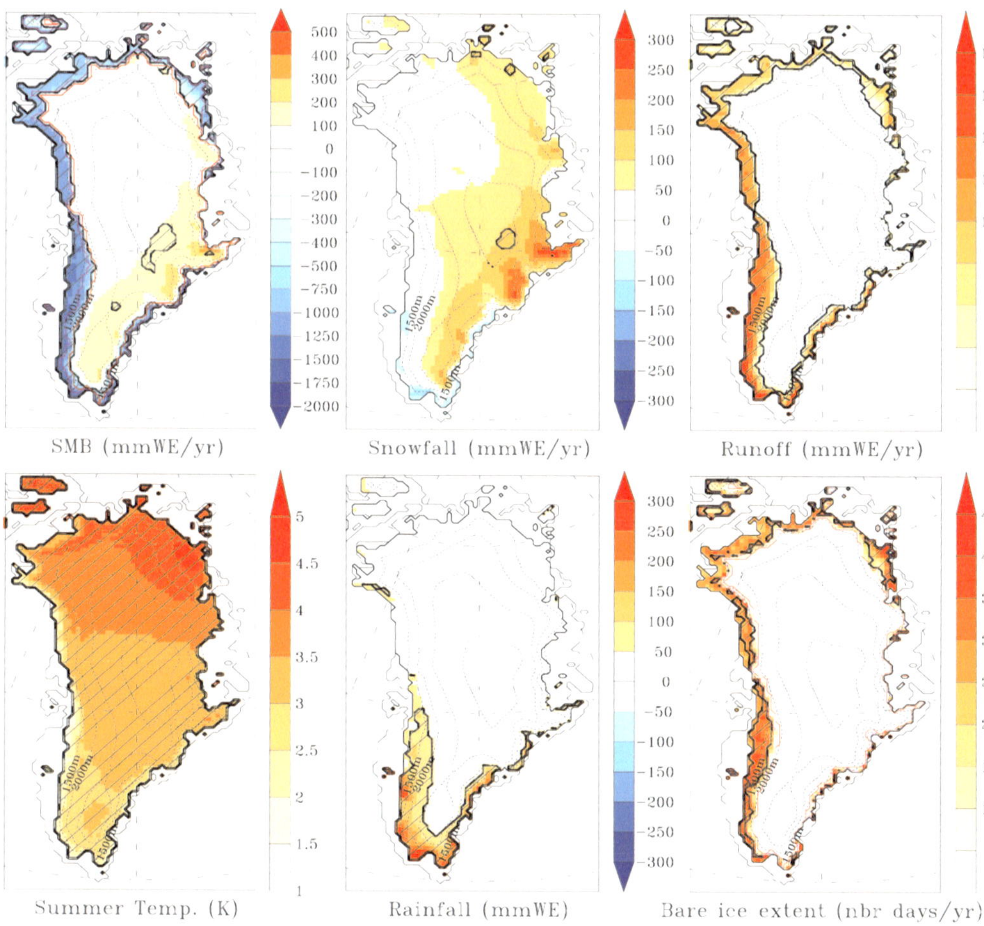

Fig. 7. Difference between the mean 2070–2099 (RCP 4.5 experiment) and the mean 1970-1999 (Historical experiment) SMB (in mm Water Equivalent/yr), snowfall (in mmWE/yr), run-off (in mmWE/yr), summer near surface temperature (in °C), rainfall (in mmWE/yr) and bare ice extent (in number of days where bare ice appears at the surface per year) simulated by MAR forced by CanESM2. Significant anomalies (those at least twice the magnitude of the 1970–1999 standard deviation) are hatched and bounded by bold black lines. Finally, the mean equilibrium line altitude from MAR forced by CanESM2 over 1970–1999 (Historical scenario) and forced by CanESM2 over 2070–2099 (RCP 4.5 scenario) are shown in mauve and in red respectively.

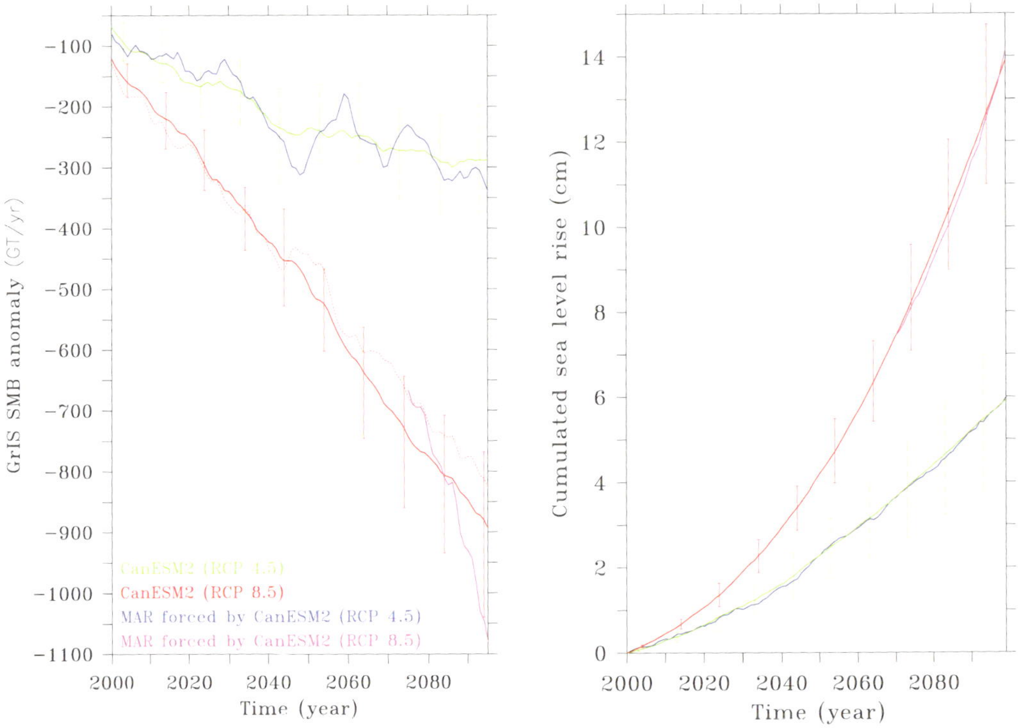

Fig. 8. Left) Future projection of anomalies of the Greenland ice sheet SMB simulated by MAR and retrieved from CanESM2 temperature and precipitation anomalies for the RCP 4.5 and RCP 8.5 scenarios in respect to the 1970–1999 period. For the CanESM2 based time series, the average over all the five available runs is plotted in solid line and the error bar gives the range around this average. The run (called *r1i1p1*) of CanESM2 used for forcing MAR is plotted in dotted line. Finally, a 10-year running mean is applied here and units are GT/yr. Right) The corresponding cumulated sea level rise from SMB changes is shown in cm. The computations use an ocean area of 361 million km^2.

where a and b are two constant parameters, T_{RCPxx} (resp. SF_{RCPxx}) is the JJA temperature (resp. the annual snowfall) from the CanESM2 outputs for the RCP 4.5 or the RCP 8.5 scenario, $\overline{T_{\text{Historical}}}$ is the mean JJA temperature over 1970–1999 from the CanESM2 outputs for the Historical experiment and $stdev(T_{\text{Historical}})$ is the standard deviation around this average. The values of the parameters a and b can be found for a fixed value of the ratio $k = a/b$ by imposing that the standard deviation of the SMB estimated by the CanESM2 temperature and precipitation time series from the Historical experiment (1970–1999) is 106 $km^3\,yr^{-1}$ as simulated by MAR forced by ERA-40 over this period. Further details and caveats about the multiple regression model and the parameters it uses are given in Fettweis et al. (2008). The values of $k = a/b$ used here (-1.25 for RCP 4.5 and -5.0 for RCP 8.5) are chosen in order to have the best comparison between the projected sea level rise time series simulated by MAR and the one retrieved from the CanESM2 outputs as shown in Fig. 8. Different values of k are needed for both RCP 4.5 and RCP 8.5 scenarios i.e. the weight of the temperature variability against the precipitation variability in the SMB variability is not the same. For RCP 4.5, the

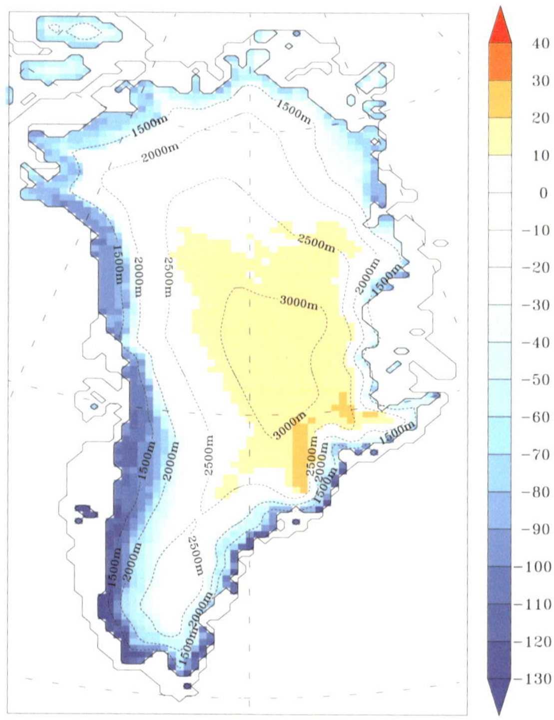

Fig. 9. Cumulated surface height anomaly (in meters) from 2000 to 2100 simulated by MAR for the RCP 4.5 scenario. The anomalies are computed in respect to the 1970–1999 period.

10-year running means of MAR and CanESM2 SMB projections compare well and the value of k used is in agreement with Fettweis et al. (2008). It is not the case for RCP 8.5. This means that the response of SMB to extreme rising temperature as projected by RCP 8.5 seems to be not a linear function of the temperature anomalies as assumed in Eq. 1 but rather a quadratic or exponential function due to the positives feedbacks (e.g. the albedo) which accelerate the surface melt. However, future projections of SMB simulated by MAR forced by RCP 8.5 over 2006-2069 are needed to confirm this hypothesis.

Although the large uncertainties in Eq. 1 resulting from the choice of a fixed value for $k = a/b$ and the hypothesis of linearity in the temperature dependence, the 10-year running means of MAR and CanESM2 SMB projections give the same sea level rise at the end of this century (see Fig. 8). The use of Eq. 1 with the four other runs from CanESM2 (in addition to the run used to force MAR) allows to estimate uncertainties coming from the forcing in our future projections. In 2100, we project a sea level rise coming from changes in Greenland ice sheet SMB estimated to be $\sim +6.5 \pm 1.5$ cm and $\sim +14 \pm 2$ cm for the CanESM2 based RCP 4.5 and RCP 8.5 experiment respectively. The projections for the RCP 4.5 scenario (+4°C in 2070–2099 compared to 1970–1999) are in full agreement with the projections made for the SRES A1B scenario by Franco et al. (2011) while the projections for the RCP 8.5 scenario are out of range from previous studies (IPCC, 2007) because the warming projected by RCP 8.5 scenario (+9°C) is larger than the most pessimistic scenario used in the IPCC AR4.

In addition to the uncertainties linked to the models/scenarios, it should be noted that these projections do not take into account changes in ice dynamics and surface topography as described in Gregory and Huybrechts (2006). Only changes in SMB are considered here and the topography of the Greenland ice sheet is fixed during the simulation although successive annual negative Greenland ice sheet mass rates induce a decrease of the surface height in the ablation zone as shown in Fig. 9 and so, a warming and an acceleration of the melt. For the RCP 4.5 scenario, the cumulated surface height anomalies could be higher than 100 m in 2100 inducing a warming of about 1°C for these pixels in dry atmospheric conditions.

7. Conclusion

Future projections of the Greenland ice sheet SMB were carried out over 2006–2100 for the scenarios RCP 4.5 and RCP 8.5 with the regional climate model MAR forced by the global model CanESM2. The CanESM2 model has been chosen because it is one of the global models from the CMIP5 data base which simulates the best the current climate at 500hPa over Greenland in respect to the European reanalysis. The differences between MAR forced by the ERA-40 reanalysis and forced by CanESM2 over 1970–1999 are not statistically significant except in the north-west of the ice sheet. However, at the scale of the whole Greenland ice sheet, MAR forced by ERA-40 and CanESM2 over current climate compare very well.

As future projections, MAR simulates a significant decrease of the SMB along the ice sheet margin due to increasing melt and a low SMB increase in the interior of the ice sheet due to heavier snowfall. At the scale of the ice sheet, the increase of precipitation does not compensate the increase of melt and MAR simulates a mean surface mass loss of about ~ -300 (resp. ~ -800) GT/yr over 2070–2099 for the RCP 4.5 (resp. RCP 8.5) scenario in respect to current climate (1970–1999). These surface mass losses correspond to cumulated sea level rises of about $\sim +6.5 \pm 1.5$ and $\sim +14 \pm 2$ cm in 2100 respectively. The error ranges in these projections estimate the uncertainties coming from CanESM2 used to force these regional projections. The uncertainties coming from MAR are difficult to evaluate in view of lack of in situ data of SMB at the scale of the whole Greenland ice sheet. Knowing that the future projections of CanESM2 are in the range of other global models of the CMIP5 multi-model data base and that the SMB simulated by MAR compare well with rates coming from other regional models (Fettweis, 2007), this shows that the biggest uncertainties for sea level rise come rather from the used scenario than from the models themselves. However, future projections of MAR forced by another global model (e.g. HadGEM2-ES) from the CMIP5 data base will allow to evaluate better the uncertainties coming from the forcing model.

It is important to note that these projections of sea level rise from Greenland ice sheet mass loss do not take into account changes in ice dynamics (influencing the mass changes from iceberg calving) and in surface topography which could amplify the deglaciation of Greenland due to the positive elevation feedback. At the end of this century, according to the RCP 4.5 scenario, cumulated anomalies of surface height could reach 150m in some areas along the ice sheet margin, emphasizing the necessity of taking into account changes in topography. That is why it will be very interesting to couple MAR with an ice sheet model in addition to the coupling with a snow model to evaluate the feedbacks due to elevation and ice mask changes to the SMB. Moreover, this double coupling will allow to evaluate changes in total Greenland ice sheet mass balance and so to evaluate the sea level rise coming from all changes of mass of the ice sheet.

8. Acknowledgment

Xavier Fettweis is a postdoctoral researcher of the Belgian National Fund for Scientific Research. For their roles in producing, coordinating, and making available the CMIP5 model output, we acknowledge the climate modelling groups (listed in Table 1 of this paper), the World Climate Research Programme's (WCRP) Working Group on Coupled Modelling (WGCM), and the Global Organization for Earth System Science Portals (GO-ESSP). Finally, this work was partly supported by funding from the ice2sea programme from the European Union 7th Framework Programme, grant number 226375 (ice2sea contribution number 046).

9. References

Bamber, J.L., R.L. Layberry, S.P. Gogenini (2001): A new ice thickness and bed data set for the Greenland ice sheet 1: Measurement, data reduction, and errors, Journal of Geophysical Research, 106 (D24): 33773-33780.

Brun, E., David, P., Sudul, M., and Brunot, G.: A numerical model to simulate snowcover stratigraphy for operational avalanche forecasting (1992), J. Glaciol., 38, 13–22.

Box, J. E., Bromwich, D. H., and Bai, L.-S. (2004): Greenland ice sheet surface mass balance for 1991–2000: application of Polar MM5 mesoscale model and in-situ data, J. Geophys. Res., 109(D16), D16105, doi:10.1029/2003JD004451.

Gallée, H. and Schayes, G. (1994): Development of a three-dimensional meso-γ primitive equations model, Mon. Wea. Rev., 122, 671–685.

Gallée, H., Guyomarc'h, G. and Brun, E. (2001): Impact of the snow drift on the Antarctic ice sheet surface mass balance: possible sensitivity to snow-surface properties, Boundary-Layer Meteorol., 99, 1–19.

Gregory, J. and Huybrechts, P. (2006): Ice-sheet contributions to future sea-level change, Philos. T. R. Soc. A, 364, 1709–1731.

Fettweis, X., Gallée, H., Lefebre, L., and van Ypersele, J.-P. (2005): Greenland surface mass balance simulated by a regional climate model and comparison with satellite derived data in 1990–1991, Clim. Dyn., 24, 623–640, doi:10.1007/s00382-005-0010-y.

Fettweis, X. (2007): Reconstruction of the 1979–2006 Greenland ice sheet surface mass balance using the regional climate model MAR, The Cryosphere, 1, 21-40, doi:10.5194/tc-1-21-2007.

Fettweis X., Hanna E., Gallé H., Huybrechts P., Erpicum M. (2008): Estimation of the Greenland ice sheet surface mass balance for the 20th and 21st centuries, The Cryosphere, 2, 117-129.

Fettweis X., Mabille G., Erpicum M., Nicolay, S., and van den Broeke M. (2011a): The 1958–2009 Greenland ice sheet surface melt and the mid-tropospheric atmospheric circulation, Clim. Dynam., Vol. 36 (1-2), 139-159, doi:10.1007/s00382-010-0772-8.

Fettweis, X., Tedesco, M., van den Broeke, M., and Ettema, J. (2011b): Melting trends over the Greenland ice sheet (1958–2009) from spaceborne microwave data and regional climate models, The Cryosphere, 5, 359-375, doi:10.5194/tc-5-359-2011.

Franco B., Fettweis X., Erpicum, M. and Nicolay S.(2011): Present and future climates of the Greenland ice sheet according to the IPCC AR4 models, Clim. Dynam., Vol. 36 (9), 1897-1918, doi:10.1007/s00382-010-0779-1.

IPCC, 2007: Climate Change 2007: The Physical Science Basis. Contribution of Working Group I to the Fourth Assessment Report of the Intergovernmental Panel on Climate Change [Solomon, S., D. Qin, M. Manning, Z. Chen, M. Marquis, K.B. Averyt, M.Tignor and

H.L. Miller (eds.)]. Cambridge University Press, Cambridge, United Kingdom and New York, NY, USA.

Lefebre, F., Gallée, H., van Ypersele, J., and Greuell, W. (2003): Modeling of snow and ice melt at ETH-camp (west Greenland): a study of surface albedo, J. Geophys. Res., 108(D8), 4231, doi:10.1029/2001JD001160.

Lefebre, F., Fettweis, X., Gallée, H., van Ypersele, J., Marbaix, P., Greuell, W., and Calanca, P. (2005): Evaluation of a high-resolution regional climate simulation over Greenland, Clim. Dyn., 25, 99–116, doi:10.1007/s00382-005-0005-8.

Mernild, S.H., Liston, G.E., and Hasholt, B. (2008): East Greenland freshwater run-off to the Greenland-Iceland-Norwegian Seas 1999-2004 and 2071-2100, Hydrological Processes, 22(23), 4571–4586.

Mernild, Sebastian H., Glen E. Liston, Christopher A. Hiemstra, Jens H. Christensen, (2010): Greenland Ice Sheet Surface Mass-Balance Modeling in a 131-Yr Perspective, 1950–2080. J. Hydrometeor, 11, 3–25, doi: 10.1175/2009JHM1140.1.

Mernild, S. H., Knudsen, N. T., Lipscomb, W. H., Yde, J. C., Malmros, J. K., Hasholt, B., and Jakobsen, B. H. (2011): Increasing mass loss from Greenland's Mittivakkat Gletscher, The Cryosphere, 5, 341-348, doi:10.5194/tc-5-341-2011.

Moss, R.H., Edmonds, J.A., Hibbard, K., Manning, M., Rose, S.K., van Vuuren, D.P., Carter, T.R., Emori, S., Kainuma, M., Kram, T., Meehl, G., Mitchell, J., Nakicenovic, N., Riahi, K., Smith, S., Stouffer, R.J., Thomson, A., Weyant, J. and Wilbanks, T. (2010): The next generation of scenarios for climate change research and assessment. Nature, doi:10.1038/nature08823.

Nick, F. M., Vieli, A., Howat, I. M. and Joughin, I (2009). Large-scale changes in Greenland outlet glacier dynamics triggered at the terminus. Nature Geoscience 2, 110-114, doi:10.1038/ngeo394.

Sundal AV, Shepherd A, Nienow P, Hanna E, Palmer S, Huybrechts P. (2011) Melt-induced speed-up of Greenland ice sheet offset by efficient subglacial drainage, Nature, 469(7331), 521–524, doi:10.1038/nature09740.

Tedesco, M., Fettweis, X., van den Broeke, M., van de Wal, R., Smeets, P., van de Berg, W. J., Serreze, M., and Box, J. (2011). The role of albedo and accumulation in the 2010 melting record in Greenland. Environmental Research Letters, 6(1).

Van den Broeke, M. R., Bamber, J., Ettema, J., Rignot, E., Schrama, E., van de Berg, W. J., van Meijgaard, E., Velicogna, I., and Wouters, B. (2009): Partitioning recent Greenland mass loss, Science, 326, 984–986.

Zwally, J. H., Abdalati, W., Herring, T., Larson, K., Saba, J., and Steffen, K. (2002): Surface Melt-Induced Acceleration of Greenland Ice-Sheet Flow, Science, 297, 218–222.